T0235219

MARKOV CHAINS AND STOCHASTIC STABILITY

Second Edition

Meyn and Tweedie is back!

The bible on Markov chains in general state spaces has been brought up to date to reflect developments in the field since 1996 – many of them sparked by publication of the first edition.

The pursuit of more efficient simulation algorithms for complex Markovian models, or algorithms for computation of optimal policies for controlled Markov models, has opened new directions for research on Markov chains. As a result, new applications have emerged across a wide range of topics including optimization, statistics, and economics. New commentary and an epilogue by Sean Meyn summarize recent developments, and references have been fully updated.

This second edition reflects the same discipline and style that marked out the original and helped it to become a classic: proofs are rigorous and concise, the range of applications is broad and knowledgeable, and key ideas are accessible to practitioners with limited mathematical background.

"This second edition remains true to the remarkable standards of scholarship established by the first edition . . . a very welcome addition to the literature."

Peter W. Glynn
Prologue to the Second Edition

MARKOV CHAINS AND STOCHASTIC STABILITY

Second Edition

SEAN MEYN AND RICHARD L. TWEEDIE

CAMBRIDGE
UNIVERSITY PRESS

CAMBRIDGE
UNIVERSITY PRESS

University Printing House, Cambridge CB2 8BS, United Kingdom

Cambridge University Press is part of the University of Cambridge.

It furthers the University's mission by disseminating knowledge in the pursuit of
education, learning and research at the highest international levels of excellence.

www.cambridge.org
Information on this title: www.cambridge.org/9780521731829

© S. Meyn and R. L. Tweedie 2009

This publication is in copyright. Subject to statutory exception
and to the provisions of relevant collective licensing agreements,
no reproduction of any part may take place without the written
permission of Cambridge University Press.

First edition published in 1993 by Springer-Verlag
Second edition published 2009
3rd printing 2013

A catalogue record for this publication is available from the British Library

ISBN 978-0-521-73182-9 Paperback

Cambridge University Press has no responsibility for the persistence or accuracy of
URLs for external or third-party internet websites referred to in this publication,
and does not guarantee that any content on such websites is, or will remain, accurate
or appropriate.

Contents

Asterisks (*) mark sections from the first edition that have been revised or augmented in the second edition.

List of figures

Prologue to the second edition

Markov Chains and Stochastic Stability is one of those rare instances of a young book that has become a classic. In understanding why the community has come to regard the book as a classic, it should be noted that all the key ingredients are present. Firstly, the material that is covered is both interesting mathematically and central to a number of important applications domains. Secondly, the core mathematical content is nontrivial and had been in constant evolution over the years and decades prior to the first edition's publication; key papers were scattered across the literature and had been published in widely diverse journals. So, there was an obvious need for a thoughtful and well-organized book on the topic. Thirdly, and most important, the topic attracted two authors who were research experts in the area and endowed with remarkable skill in communicating complex ideas to specialists and applications-focused users alike, and who also exhibited superb taste in deciding which key ideas and approaches to emphasize.

When the first edition of the book was published in 1993, Markov chains already had a long tradition as mathematical models for stochastically evolving dynamical systems arising in the physical sciences, economics, and engineering, largely centered on discrete state space formulations. A great deal of theory had been developed related to Markov chain theory, both in discrete state space and general state space. However, the general state space theory had grown to include multiple (and somewhat divergent) mathematical strands, having much to do with the fact that there are several natural (but different) ways that one can choose to generalize the fundamental countable state concept of irreducibility to general state space. Roughly speaking, one strand took advantage of topological ideas, compactness methods, and required Feller continuity of the transition kernel. The second major strand, starting with the pioneering work of Harris in the 1950s, subsequently amplified by Orey, and later simplified through the beautiful contributions of Nummelin, Athreya, and Ney in the 1970s, can be viewed as an effort to understand general state space Markov chains through the prism of regeneration. Thus, Meyn and Tweedie had to make some key decisions regarding the general state space tools that they would emphasize in the book. The span of time that has elapsed since this book's publication makes clear that they chose well.

While offering an excellent and accessible discussion of methods based on topological machinery, the book focuses largely on the more widely applicable and more easily used concept of regeneration in general state space. In addition, the book recognizes the central role that Foster–Lyapunov functions play in verifying recurrence and bounding the moments and expectations that arise naturally in development of the theory of

Markov chains. In choosing to emphasize these ideas, the authors were able to offer the community, and especially practitioners, a convenient and easily applied roadmap through a set of concepts and ideas that had previously been accessible only to specialists. Sparked by the publication of the first edition of this book, there has subsequently been an explosion in the number of papers involving applications of general state space Markov chains.

As it turns out, the period that has elapsed since publication of the first edition also fortuitously coincided with the rapid development of several key applications areas in which the tools developed in the book have played a fundamental role. Perhaps the most important such application is that of Markov chain Monte Carlo (MCMC) algorithms. In the MCMC setting, the basic problem at hand is the construction of an efficient algorithm capable of sampling from a given target distribution, which is known up to a normalization constant that is not numerically or analytically computable. The idea is to produce a Markov chain having a unique stationary distribution that coincides with the target distribution. Constructing such a Markov chain is typically easy, so one has many potential choices. Since the algorithm is usually initialized with an initial distribution that is atypical of equilibrium behavior, one then wishes to find a chain that converges to its steady state rapidly. The tools discussed in this book play a central role in answering such questions. General state space Markov chain ideas also have been used to great effect in other rapidly developing algorithmic contexts such as machine learning and in the analysis of the many randomized algorithms having a time evolution described by a stochastic recursive sequence. Finally, many of the performance engineering applications that have been explored over the past fifteen years leverage off this body of theory, particularly those results that have involved trying to make rigorous the connection between stability of deterministic fluid models and stability of the associated stochastic queueing analogue. Given the ubiquitous nature of stochastic systems or algorithms described through stochastic recursive sequences, it seems likely that many more applications of the theory described in this book will arise in the years ahead. So, the marketplace of potential consumers of this book is likely to be a healthy one for many years to come.

Even the appendices are testimony to the hard work and exacting standards the authors brought to this project. Through additional (and very useful) discussion, these appendices provide readers with an opportunity to see the power of the concepts of stability and recurrence being exercised in the setting of models that are both mathematically interesting and of importance in their own right. In fact, some readers will find that the appendices are a good way to quickly remind themselves of the methods that exist to establish a particular desired property of a Markov chain model.

This second edition remains true to the remarkable standards of scholarship established by the first edition. As noted above, a number of applications domains that are consumers of this theory have developed rapidly since the publication of the first edition. As one would expect with any mathematically vibrant area, there have also been important theoretical developments over that span of time, ranging from the exploration of these ideas in studying large deviations for additive functionals of Markov chains to the generalization of these concepts to the setting of continuous time Markov processes. This new edition does a splendid job of making clear the most important such developments and pointing the reader in the direction of the key references to be studied in each area. With the background offered by this book, the reader who wishes to explore these recent theoretical developments is well positioned both to read the literature and to creatively apply these ideas to the problem at hand. All the elements that made the first edition of *Markov Chains and Stochastic Stability* a classic are here in the second edition, and it will no doubt be a very welcome addition to the literature.

Peter W. Glynn
Palo Alto

Preface to the second edition

A new edition of Meyn & Tweedie – *what for?*

The majority of topics covered in this book are well established. Ancient topics such as the Doeblin decomposition and even more modern concepts such as f-regularity are mature and not likely to see much improvement. Why then is there a need for a new edition?

Publication of this book in the Cambridge Mathematical Library is a way to honor my friend and colleague Richard Tweedie. The memorial article [103] contains a survey of his contributions to applied probability and statistics and an announcement of the initiation of the *Tweedie New Researcher Award Fund*.[1] Royalties from the book will go to Catherine Tweedie and help to support the memorial fund.

Richard would be very pleased to know that our book will be placed on the shelves next to classics in the mathematical literature such as Hardy, Littlewood, and Pólya's *Inequalities* and Zygmund's *Trigonometric Series*, as well as more modern classics such as Katznelson's *An Introduction to Harmonic Analysis* and Rogers and Williams' *Diffusions, Markov Processes and Martingales*.

Other reasons for this new edition are less personal.

Motivation for topics in the book has grown along with growth in computer power since the book was last printed in March of 1996. The need for more efficient simulation algorithms for complex Markovian models, or algorithms for computation of optimal policies for controlled Markov models, has opened new directions for research on Markov chains [29, 113, 10, 245, 27, 267]. It has been exciting to see new applications to diverse topics including optimization, statistics, and economics.

Significant advances in the theory took place in the decade that the book was out of print. Several chapters end with new commentary containing explanations regarding changes to the text, or new references. The final chapter of this new edition contains a partial roadmap of new directions of research on Markov models since 1996. The new Chapter 20 is divided into three sections:

Section 20.1: *Geometric ergodicity and spectral theory*   Topics in Chapters 15 and 16 have seen tremendous growth over the past decade. The operator-theoretic framework of Chapter 16 was obviously valuable at the time this chapter was written. We could not have known then how many new directions for research this framework

---

[1]The Tweedie New Researcher Award Fund is now managed by the Institute of Mathematical Statistics <www.imstat.org/awards/tweedie.html>.

would support. Ideally I would rewrite Chapters 15 and 16 to provide a more cohesive treatment of geometric ergodicity, and explain how these ideas lead to foundations for multiplicative ergodic theory, Lyapunov exponents, and the theory of large deviations. This will have to wait for a third edition or a new book devoted to these topics. In its place, I have provided in Section 20.1 a brief survey of these directions of research.

Section 20.2: *Simulation and MCMC*   Richard Tweedie and I became interested in these topics soon after the first edition went to print. Section 20.2 describes applications of general state space Markov chain techniques to the construction and analysis of simulation algorithms, such as the control variate method [10], and algorithms found in reinforcement learning [29, 379].

Section 20.3: *Continuous time models*   The final section explains how theory in continuous time can be generated from discrete time counterparts developed in this book. In particular, all of the ergodic theorems in Part III have precise analogues in continuous time.

The significance of Poisson's equation was not properly highlighted in the first edition. This is rectified in a detailed commentary at the close of Chapter 17, which includes a menu of applications, and new results on existence and uniqueness of solutions to Poisson's equation, contained in Theorems 17.7.1 and 17.7.2, respectively.

The multi-step drift criterion for stability described in Section 19.1 has been improved, and this technique has found many applications. The resulting "fluid model" approach to stability of stochastic networks is one theme of the new monograph [267]. Extensions of the techniques in Section 19.1 have found application to the theory of stochastic approximation [40, 39], and to Markov chain Monte Carlo (MCMC) [100].

It is surprising how few errors have been uncovered since the first edition went to print. Section 2.2.3 on the gumleaf attractor contained errors in the description of the figures. There were other minor errors in the analysis of the forward recurrence time chains in Section 10.3.1, and the coupling bound in Theorem 16.2.4. The term *limiting variance* is now replaced by the more familiar *asymptotic variance* in Chapter 17, and starting in Chapter 9 the term *norm-like* is replaced with the more familiar *coercive*.

Words of thanks

Continued support from the National Science Foundation is gratefully acknowledged. Over the past decade, support from *Control, Networks and Computational Intelligence* has funded much of the theory and applications surveyed in Chapter 20 under grants ECS 940372, ECS 9972957, ECS 0217836, and ECS 0523620. The NSF grant DMI 0085165 supported research with Shane Henderson that is surveyed in Section 20.2.1.

It is a pleasure to convey my thanks to my wonderful editor Diana Gillooly. It was her idea to place the book in the Cambridge Mathematical Library series. In addition to her work "behind the scenes" at Cambridge University Press, Diana dissected the manuscript searching for typos or inconsistencies in notation. She provided valuable advice on structure, and patiently answered all of my questions.

Jeffrey Rosenthal has maintained the website for the online version of the first edition at probability.ca/MT. It is reassuring to know that this resource will remain in place "till death do us part."

In the preface to the first edition, we expressed our thanks to Peter Glynn for his correspondence and inspiration. I am very grateful that our correspondence has continued over the past 15 years. Much of the material contained in the surveys in the new Chapter 20 can be regarded as part of "transcripts" from our many discussions since the book was first put into print.

I am very grateful to Ioannis Kontoyiannis for collaborations over the past decade. Ioannis provided comments on the new edition, including the discovery of an error in Theorem 16.2.4. Many have sent comments over the years. In particular, Vivek Borkar, Jan van Casteren, Peter Haas, Lars Hansen, Galin Jones, Aziz Khanchi, Tze Lai, Zhan-Qian Lu, Abdelkader Mokkadem, Eric Moulines, Gareth Roberts, Li-Ming Wu, and three graduates from the University of Oslo – Tore W. Larsen, Arvid Raknerud, and Øivind Skare – all pointed out errors that have been corrected in the new edition, or suggested recent references that are now included in the updated bibliography.

Sean Meyn
Urbana-Champaign

Preface to the first edition (1993)

Books are individual and idiosyncratic. In trying to understand what makes a good book, there is a limited amount that one can learn from other books; but at least one can read their prefaces, in hope of help.

Our own research shows that authors use prefaces for many different reasons.

Prefaces can be explanations of the role and the contents of the book, as in Chung [71] or Revuz [326] or Nummelin [303]; this can be combined with what is almost an apology for bothering the reader, as in Billingsley [37] or Çinlar [59]; prefaces can describe the mathematics, as in Orey [309], or the importance of the applications, as in Tong [388] or Asmussen [9], or the way in which the book works as a text, as in Brockwell and Davis [51] or Revuz [326]; they can be the only available outlet for thanking those who made the task of writing possible, as in almost all of the above (although we particularly like the familial gratitude of Resnick [325] and the dedication of Simmons [355]); they can combine all these roles, and many more.

This preface is no different. Let us begin with those we hope will use the book.

Who wants this stuff anyway?

This book is about Markov chains on general state spaces: sequences Φ_n evolving randomly in time which remember their past trajectory only through its most recent value. We develop their theoretical structure and we describe their application.

The theory of general state space chains has matured over the past twenty years in ways which make it very much more accessible, very much more complete, and (we at least think) rather beautiful to learn and use. We have tried to convey all of this, and to convey it at a level that is no more difficult than the corresponding countable space theory.

The easiest reader for us to envisage is the long-suffering graduate student, who is expected, in many disciplines, to take a course on countable space Markov chains.

Such a graduate student should be able to read almost all of the general space theory in this book without any mathematical background deeper than that needed for studying chains on countable spaces, provided only that the fear of seeing an integral rather than a summation sign can be overcome. Very little measure theory or analysis is required: virtually no more in most places than must be used to define transition probabilities. The remarkable Nummelin–Athreya–Ney regeneration technique, together with coupling methods, allows simple renewal approaches to almost all of the hard results.

Courses on countable space Markov chains abound, not only in statistics and mathematics departments, but in engineering schools, operations research groups and even business schools. This book can serve as the text in most of these environments for a one-semester course on more general space applied Markov chain theory, provided that some of the deeper limit results are omitted and (in the interests of a fourteen-week semester) the class is directed only to a subset of the examples, concentrating as best suits their discipline on time series analysis, control and systems models or operations research models.

The prerequisite texts for such a course are certainly at no deeper level than Chung [72], Breiman [48], or Billingsley [37] for measure theory and stochastic processes, and Simmons [355] or Rudin [345] for topology and analysis.

Be warned: we have not provided numerous illustrative unworked examples for the student to cut teeth on. But we have developed a rather large number of thoroughly worked examples, ensuring applications are well understood; and the literature is littered with variations for teaching purposes, many of which we reference explicitly.

This regular interplay between theory and detailed consideration of application to specific models is one thread that guides the development of this book, as it guides the rapidly growing usage of Markov models on general spaces by many practitioners.

The second group of readers we envisage consists of exactly those practitioners, in several disparate areas, for all of whom we have tried to provide a set of research and development tools: for engineers in control theory, through a discussion of linear and nonlinear state space systems; for statisticians and probabilists in the related areas of time series analysis; for researchers in systems analysis, through networking models for which these techniques are becoming increasingly fruitful; and for applied probabilists, interested in queueing and storage models and related analyses.

We have tried from the beginning to convey the applied value of the theory rather than let it develop in a vacuum. The practitioner will find detailed examples of transition probabilities for real models. These models are classified systematically into the various structural classes as we define them. The impact of the theory on the models is developed in detail, not just to give examples of that theory but because the models themselves are important and there are relatively few places outside the research journals where their analysis is collected.

Of course, there is only so much that a general theory of Markov chains can provide to all of these areas. The contribution is in general qualitative, not quantitative. And in our experience, the critical qualitative aspects are those of stability of the models. Classification of a model as stable in some sense is the first fundamental operation underlying other, more model-specific, analyses. It is, we think, astonishing how powerful and accurate such a classification can become when using only the apparently blunt instruments of a general Markovian theory: we hope the strength of the results described here is equally visible to the reader as to the authors, for this is why we have chosen stability analysis as the cord binding together the theory and the applications of Markov chains.

We have adopted two novel approaches in writing this book. The reader will find key theorems announced at the beginning of all but the discursive chapters; if these are understood then the more detailed theory in the body of the chapter will be better motivated, and applications made more straightforward. And at the end of the book we have constructed, at the risk of repetition, "mud maps" showing the crucial equivalences between forms of stability, and we give a glossary of the models we evaluate. We trust both of these innovations will help to make the material accessible to the full range of readers we have considered.

What's it all about?

We deal here with Markov chains. Despite the initial attempts by Doob and Chung [99, 71] to reserve this term for systems evolving on countable spaces with both discrete and continuous time parameters, usage seems to have decreed (see for example Revuz [326]) that Markov chains move in discrete time, on whatever space they wish; and such are the systems we describe here.

Typically, our systems evolve on quite general spaces. Many models of practical systems are like this; or at least, they evolve on $\mathbb{R}^k$ or some subset thereof, and thus are not amenable to countable space analysis, such as is found in Chung [71], or Çinlar [59], and which is all that is found in most of the many other texts on the theory and application of Markov chains.

We undertook this project for two main reasons. Firstly, we felt there was a lack of accessible descriptions of such systems with any strong applied flavor; and secondly, in our view the theory is now at a point where it can be used properly in its own right, rather than practitioners needing to adopt countable space approximations, either because they found the general space theory to be inadequate or the mathematical requirements on them to be excessive.

The theoretical side of the book has some famous progenitors. The foundations of a theory of general state space Markov chains are described in the remarkable book of Doob [99], and although the theory is much more refined now, this is still the best source of much basic material; the next generation of results is elegantly developed in the little treatise of Orey [309]; the most current treatments are contained in the densely packed goldmine of material of Nummelin [303], to whom we owe much, and in the deep but rather different and perhaps more mathematical treatise by Revuz [326], which goes in directions different from those we pursue.

None of these treatments pretend to have particularly strong leanings towards applications. To be sure, some recent books, such as that on applied probability models by Asmussen [9] or that on nonlinear systems by Tong [388], come at the problem from the other end. They provide quite substantial discussions of those specific aspects of general Markov chain theory they require, but purely as tools for the applications they have to hand.

Our aim has been to merge these approaches, and to do so in a way which will be accessible to theoreticians and to practitioners both.

So what else is new?

In the preface to the second edition [71] of his classic treatise on countable space Markov chains, Chung, writing in 1966, asserted that the general space context still had had "little impact" on the the study of countable space chains, and that this "state of mutual detachment" should not be suffered to continue. Admittedly, he was writing of continuous time processes, but the remark is equally apt for discrete time models of the period. We hope that it will be apparent in this book that the general space theory has not only caught up with its countable counterpart in the areas we describe, but has indeed added considerably to the ways in which the simpler systems are approached.

There are several themes in this book which instance both the maturity and the novelty of the general space model, and which we feel deserve mention, even in the restricted level of technicality available in a preface. These are, specifically,

(i) the use of the *splitting technique*, which provides an approach to general state space chains through regeneration methods;

(ii) the use of "Foster–Lyapunov" *drift criteria*, both in improving the theory and in enabling the classification of individual chains;

(iii) the delineation of appropriate *continuity conditions* to link the general theory with the properties of chains on, in particular, Euclidean space; and

(iv) the development of *control model* approaches, enabling analysis of models from their deterministic counterparts.

These are not distinct themes: they interweave to a surprising extent in the mathematics and its implementation.

The key factor is undoubtedly the existence and consequences of the Nummelin splitting technique of Chapter 5, whereby it is shown that if a chain $\{\Phi_n\}$ on a quite general space satisfies the simple "φ-irreducibility" condition (which requires that for some measure φ, there is at least positive probability from *any* initial point x that one of the Φ_n lies in any set of positive φ-measure; see Chapter 4), then one can induce an artificial "regeneration time" in the chain, allowing all of the mechanisms of discrete time renewal theory to be brought to bear.

Part I is largely devoted to developing this theme and related concepts, and their practical implementation.

The splitting method enables essentially all of the results known for countable space to be replicated for general spaces. Although that by itself is a major achievement, it also has the side benefit that it forces concentration on the aspects of the theory that depend, not on a countable space which gives regeneration at every step, but on a single regeneration point. Part II develops the use of the splitting method, amongst other approaches, in providing a full analogue of the positive recurrence/null recurrence/transience trichotomy central in the exposition of countable space chains, together with consequences of this trichotomy.

In developing such structures, the theory of general space chains has merely caught up with its denumerable progenitor. Somewhat surprisingly, in considering asymptotic results for positive recurrent chains, as we do in Part III, the concentration on a single regenerative state leads to stronger ergodic theorems (in terms of total variation convergence), better rates of convergence results, and a more uniform set of equivalent conditions for the strong stability regime known as positive recurrence than is typically realised for countable space chains.

The outcomes of this splitting technique approach are possibly best exemplified in the case of so-called "geometrically ergodic" chains.

Let τ_C be the hitting time on any set C: that is, the first time that the chain Φ_n returns to C; and let $P^n(x, A) = \mathsf{P}(\Phi_n \in A \mid \Phi_0 = x)$ denote the probability that the chain is in a set A at time n given it starts at time zero in state x, or the "n-step transition probabilities", of the chain. One of the goals of Part II and Part III is to link conditions under which the chain returns quickly to "small" sets C (such as finite or compact sets), measured in terms of moments of τ_C, with conditions under which the probabilities $P^n(x, A)$ converge to limiting distributions.

Here is a taste of what can be achieved. We will eventually show, in Chapter 15, the following elegant result:

The following conditions are all equivalent for a φ-irreducible "aperiodic" (see Chapter 5) chain:

(A) *For some one "small" set C, the return time distributions have geometric tails; that is, for some $r > 1$*

$$\sup_{x \in C} \mathsf{E}_x[r^{\tau_C}] < \infty.$$

(B) *For some one "small" set C, the transition probabilities converge geometrically quickly; that is, for some $M < \infty, P^\infty(C) > 0$ and $\rho_C < 1$*

$$\sup_{x \in C} |P^n(x, C) - P^\infty(C)| \le M \rho_C^n.$$

(C) *For some one "small" set C, there is "geometric drift" towards C; that is, for some function $V \ge 1$ and some $\beta > 0$*

$$\int P(x, dy)V(y) \le (1 - \beta)V(x) + \mathbb{I}_C(x).$$

Each of these implies that there is a limiting probability measure π, a constant $R < \infty$ and some uniform rate $\rho < 1$ such that

$$\sup_{|f| \le V} \left| \int P^n(x, dy)f(y) - \int \pi(dy)f(y) \right| \le RV(x)\rho^n$$

where the function V is as in (C).

This set of equivalences also displays a second theme of this book: not only do we stress the relatively well-known equivalence of hitting time properties and limiting results, as between (A) and (B), but we also develop the equivalence of these with the one-step "Foster–Lyapunov" drift conditions as in (C), which we systematically derive for various types of stability.

As well as their mathematical elegance, these results have great pragmatic value. The condition (C) can be checked directly from P for specific models, giving a powerful applied tool to be used in classifying specific models. Although such drift conditions have been exploited in many continuous space applications areas for over a decade, much of the formulation in this book is new.

The "small" sets in these equivalences are vague: this is of course only the preface! It would be nice if they were compact sets, for example; and the continuity conditions we develop, starting in Chapter 6, ensure this, and much beside.

There is a further mathematical unity, and novelty, to much of our presentation, especially in the application of results to linear and nonlinear systems on $\mathbb{R}^k$. We formulate many of our concepts first for deterministic analogues of the stochastic systems, and we show how the insight from such deterministic modeling flows into appropriate criteria for stochastic modeling. These ideas are taken from control theory, and forms of control of the deterministic system and stability of its stochastic generalization run in tandem. The duality between the deterministic and stochastic conditions is indeed almost exact, provided one is dealing with φ-irreducible Markov models; and the continuity conditions above interact with these ideas in ensuring that the "stochasticization" of the deterministic models gives such φ-irreducible chains.

Breiman [48] notes that he once wrote a preface so long that he never finished his book. It is tempting to keep on, and rewrite here all the high points of the book.

We will resist such temptation. For other highlights we refer the reader instead to the introductions to each chapter: in them we have displayed the main results in the chapter, to whet the appetite and to guide the different classes of user. Do not be fooled: there are many other results besides the highlights inside. We hope you will find them as elegant and as useful as we do.

Who do we owe?

Like most authors we owe our debts, professional and personal. A preface is a good place to acknowledge them.

The alphabetically and chronologically younger author began studying Markov chains at McGill University in Montréal. John Taylor introduced him to the beauty of probability. The excellent teaching of Michael Kaplan provided a first contact with Markov chains and a unique perspective on the structure of stochastic models.

He is especially happy to have the chance to thank Peter Caines for planting him in one of the most fantastic cities in North America, and for the friendship and academic environment that he subsequently provided.

In applying these results, very considerable input and insight has been provided by Lei Guo of Academia Sinica in Beijing and Doug Down of the University of Illinois. Some of the material on control theory and on queues in particular owes much to their collaboration in the original derivations.

He is now especially fortunate to work in close proximity to P.R. Kumar, who has been a consistent inspiration, particularly through his work on queueing networks and adaptive control. Others who have helped him, by corresponding on current research, by sharing enlightenment about a new application, or by developing new theoretical ideas, include Venkat Anantharam, A. Ganesh, Peter Glynn, Wolfgang Kliemann, Laurent Praly, John Sadowsky, Karl Sigman, and Victor Solo.

The alphabetically later and older author has a correspondingly longer list of influences who have led to his abiding interest in this subject. Five stand out: Chip Heathcote and Eugene Seneta at the Australian National University, who first taught the enjoyment of Markov chains; David Kendall at Cambridge, whose own fundamental work exemplifies the power, the beauty and the need to seek the underlying simplicity of such processes; Joe Gani, whose unflagging enthusiasm and support for the interaction of real theory and real problems has been an example for many years; and probably most significantly for the developments in this book, David Vere-Jones, who has shown an uncanny knack for asking exactly the right questions at times when just enough was known to be able to develop answers to them.

It was also a pleasure and a piece of good fortune for him to work with the Finnish school of Esa Nummelin, Pekka Tuominen and Elja Arjas just as the splitting technique was uncovered, and a large amount of the material in this book can actually be traced to the month surrounding the First Tuusula Summer School in 1976. Applying the methods over the years with David Pollard, Paul Feigin, Sid Resnick and Peter Brockwell has also been both illuminating and enjoyable; whilst the ongoing stimulation and encouragement to look at new areas given by Wojtek Szpankowski, Floske Spieksma, Chris Adam and Kerrie Mengersen has been invaluable in maintaining enthusiasm and energy in finishing this book.

By sheer coincidence both of us have held Postdoctoral Fellowships at the Australian National University, albeit at somewhat different times. Both of us started much of our own work in this field under that system, and we gratefully acknowledge those most useful positions, even now that they are long past.

More recently, the support of our institutions has been invaluable. Bond University facilitated our embryonic work together, whilst the Coordinated Sciences Laboratory of the University of Illinois and the Department of Statistics at Colorado State University have been enjoyable environments in which to do the actual writing.

Support from the National Science Foundation is gratefully acknowledged: grants ECS 8910088 and DMS 9205687 enabled us to meet regularly, helped to fund our students in related research, and partially supported the completion of the book.

Writing a book from multiple locations involves multiple meetings at every available opportunity. We appreciated the support of Peter Caines in Montréal, Bozenna and Tyrone Duncan at the University of Kansas, Will Gersch in Hawaii, Götz Kersting and Heinrich Hering in Germany, for assisting in our meeting regularly and helping with far-flung facilities.

Peter Brockwell, Kung-Sik Chan, Richard Davis, Doug Down, Kerrie Mengersen, Rayadurgam Ravikanth, and Pekka Tuominen, and most significantly Vladimir Kalashnikov and Floske Spieksma, read fragments or reams of manuscript as we produced them, and we gratefully acknowledge their advice, comments, corrections and encouragement. It is traditional, and in this case as accurate as usual, to say that any remaining infelicities are there despite their best efforts.

Rayadurgam Ravikanth produced the sample path graphs for us; Bob MacFarlane drew the remaining illustrations; and Francie Bridges produced much of the bibliography and some of the text. The vast bulk of the material we have done ourselves: our debt to Donald Knuth and the developers of LaTeX is clear and immense, as is our debt to Deepa Ramaswamy, Molly Shor, Rich Sutton and all those others who have kept software, email and remote telematic facilities running smoothly.

Lastly, we are grateful to Brad Dickinson and Eduardo Sontag, and to Zvi Ruder and Nicholas Pinfield and the Engineering and Control Series staff at Springer, for their patience, encouragement and help.

And finally ...

And finally, like all authors whether they say so in the preface or not, we have received support beyond the call of duty from our families. Writing a book of this magnitude has taken much time that should have been spent with them, and they have been unfailingly supportive of the enterprise, and remarkably patient and tolerant in the face of our quite unreasonable exclusion of other interests.

They have lived with family holidays where we scribbled proto-books in restaurants and tripped over deer whilst discussing Doeblin decompositions; they have endured sundry absences and visitations, with no idea of which was worse; they have seen come and go a series of deadlines with all of the structure of a renewal process.

They are delighted that we are finished, although we feel they have not yet adjusted to the fact that a similar development of the continuous time theory clearly needs to be written next.

So to Belinda, Sydney and Sophie; to Catherine and Marianne: with thanks for the patience, support and understanding, this book is dedicated to you.

Part I

COMMUNICATION
and
REGENERATION

Chapter 1

Heuristics

This book is about Markovian models, and particularly about the structure and stability of such models. We develop a theoretical basis by studying Markov chains in very general contexts; and we develop, as systematically as we can, the applications of this theory to applied models in systems engineering, in operations research, and in time series.

A Markov chain is, for us, a collection of random variables $\Phi = \{\Phi_n : n \in T\}$, where T is a countable time set. It is customary to write T as $\mathbb{Z}_+ := \{0, 1, \ldots\}$, and we will do this henceforth.

Heuristically, the critical aspect of a Markov model, as opposed to any other set of random variables, is that it is forgetful of all but its most immediate past. The precise meaning of this requirement for the evolution of a Markov model in time, that the future of the process is independent of the past given only its present value, and the construction of such a model in a rigorous way, is taken up in Chapter 3. Until then it is enough to indicate that for a process Φ, evolving on a space X and governed by an overall probability law P, to be a time-homogeneous Markov chain, there must be a set of "transition probabilities" $\{P^n(x, A), x \in \mathsf{X}, A \subset \mathsf{X}\}$ for appropriate sets A such that for times n, m in $\mathbb{Z}_+$

$$\mathsf{P}(\Phi_{n+m} \in A \mid \Phi_j, j \leq m; \Phi_m = x) = P^n(x, A); \tag{1.1}$$

that is, $P^n(x, A)$ denotes the probability that a chain at x will be in the set A after n steps, or transitions. The independence of P^n on the values of $\Phi_j, j \leq m$, is the Markov property, and the independence of P^n and m is the time-homogeneity property.

We now show that systems which are amenable to modeling by discrete time Markov chains with this structure occur frequently, especially if we take the state space of the process to be rather general, since then we can allow auxiliary information on the past to be incorporated to ensure the Markov property is appropriate.

1.1   A range of Markovian environments

The following examples illustrate this breadth of application of Markov models, and a little of the reason why stability is a central requirement for such models.

(a) The cruise control system on a modern motor vehicle monitors, at each time point k, a vector $\{X_k\}$ of inputs: speed, fuel flow, and the like (see Kuo [230]). It calculates a control value U_k which adjusts the throttle, causing a change in the values of the environmental variables X_{k+1} which in turn causes U_{k+1} to change again. The multidimensional process $\Phi_k = \{X_k, U_k\}$ is often a Markov chain (see Section 2.3.2), with new values overriding those of the past, and with the next value governed by the present value. All of this is subject to measurement error, and the process can never be other than stochastic: stability for this chain consists in ensuring that the environmental variables do not deviate too far, within the limits imposed by randomness, from the pre-set goals of the control algorithm.

(b) A queue at an airport evolves through the random arrival of customers and the service times they bring. The numbers in the queue, and the time the customer has to wait, are critical parameters for customer satisfaction, for waiting room design, for counter staffing (see Asmussen [9]). Under appropriate conditions (see Section 2.4.2), variables observed at arrival times (either the queue numbers, or a combination of such numbers and aspects of the remaining or currently uncompleted service times) can be represented as a Markov chain, and the question of stability is central to ensuring that the queue remains at a viable level. Techniques arising from the analysis of such models have led to the now familiar single-line multi-server counters actually used in airports, banks and similar facilities, rather than the previous multi-line systems.

(c) The exchange rate X_n between two currencies can be and is represented as a function of its past several values $X_{n-1}, \ldots, X_{n-k}$, modified by the volatility of the market which is incorporated as a disturbance term W_n (see Krugman and Miller [222] for models of such fluctuations). The autoregressive model

$$X_n = \sum_{j=1}^{k} \alpha_j X_{n-j} + W_n$$

central in time series analysis (see Section 2.1) captures the essential concept of such a system. By considering the whole k-length vector $\Phi_n = (X_n, \ldots, X_{n-k+1})$, Markovian methods can be brought to the analysis of such time-series models. Stability here involves relatively small fluctuations around a norm; and as we will see, if we do not have such stability, then typically we will have instability of the grossest kind, with the exchange rate heading to infinity.

(d) Storage models are fundamental in engineering, insurance and business. In engineering one considers a dam, with input of random amounts at random times, and a steady withdrawal of water for irrigation or power usage. This model has a Markovian representation (see Section 2.4.3 and Section 2.4.4). In insurance, there is a steady inflow of premiums, and random outputs of claims at random times. This model is also a storage process, but with the input and output reversed when compared to the engineering version, and also has a Markovian representation (see Asmussen [9]). In business, the inventory of a firm will act in a manner between these two models, with regular but sometimes also large irregular withdrawals, and irregular ordering or replacements, usually triggered by levels of stock reaching threshold values (for an early but still relevant overview see Prabhu [322]). This also has, given appropriate assumptions, a Markovian representation. For all of these, stability is essentially the requirement that the chain stays in "reasonable values": the stock does not overfill the warehouse, the dam does not overflow, the claims do not swamp the premiums.

(e) The growth of populations is modeled by Markov chains, of many varieties. Small homogeneous populations are branching processes (see Athreya and Ney [12]); more coarse analysis of large populations by time series models allows, as in (c), a Markovian representation (see Brockwell and Davis [51]); even the detailed and intricate cycle of the Canadian lynx seem to fit a Markovian model [287], [388]. Of these, only the third is stable in the sense of this book: the others either die out (which is, trivially, stability but a rather uninteresting form); or, as with human populations, expand (at least within the model) forever.

(f) Markov chains are currently enjoying wide popularity through their use as a tool in simulation: Gibbs sampling, and its extension to Markov chain Monte Carlo methods of simulation, which utilise the fact that many distributions can be constructed as invariant or limiting distributions (in the sense of (1.16) below), has had great impact on a number of areas (see, as just one example, [312]). In particular, the calculation of posterior Bayesian distributions has been revolutionized through this route [359, 381, 385], and the behavior of prior and posterior distributions on very general spaces such as spaces of likelihood measures themselves can be approached in this way (see [112]): there is no doubt that at this degree of generality, techniques such as we develop in this book are critical.

(g) There are Markov models in all areas of human endeavor. The degree of word usage by famous authors admits a Markovian representation (see, amongst others, Gani and Saunders [136]). Did Shakespeare have an unlimited vocabulary? This can be phrased as a question of stability: if he wrote forever, would the size of the vocabulary used grow in an unlimited way? The record levels in sport are Markovian (see Resnick [325]). The spread of surnames may be modeled as Markovian (see [78]). The employment structure in a firm has a Markovian representation (see Bartholomew and Forbes [18]). This range of examples does not imply all human experience is Markovian: it does indicate that if enough variables are incorporated in the definition of "immediate past", a forgetfulness of all but that past is a reasonable approximation, and one which we can handle.

(h) Perhaps even more importantly, at the current level of technological development, telecommunications and computer networks have inherent Markovian representations (see Kelly [199] for a very wide range of applications, both actual and potential, and Gray [144] for applications to coding and information theory). They may be composed of sundry connected queueing processes, with jobs completed at nodes, and messages routed between them; to summarize the past one may need a state space which is the product of many subspaces, including countable subspaces, representing numbers in queues and buffers, uncountable subspaces, representing unfinished service times or routing times, or numerous trivial 0-1 subspaces representing available slots or wait-states or busy servers. But by a suitable choice of state space, and (as always) a choice of appropriate assumptions, the methods we give in this book become tools to analyze the stability of the system.

Simple spaces do not describe these systems in general. Integer or real-valued models are sufficient only to analyze the simplest models in almost all of these contexts.

The methods and descriptions in this book are for chains which take their values in a virtually arbitrary space X. We do not restrict ourselves to countable spaces, nor even to Euclidean space $\mathbb{R}^n$, although we do give specific formulations of much of our theory in both these special cases, to aid both understanding and application.

One of the key factors that allows this generality is that, for the models we consider, there is no great loss of power in going from a simple to a quite general space. The reader interested in any of the areas of application above should therefore find that the structural and stability results for general Markov chains are potentially tools of great value, no matter what the situation, no matter how simple or complex the model considered.

1.2   Basic models in practice

1.2.1   The Markovian assumption

The simplest Markov models occur when the variables Φ_n, $n \in \mathbb{Z}_+$, are independent. However, a collection of random variables which is independent certainly fails to capture the essence of Markov models, which are designed to represent systems which *do* have a past, even though they depend on that past only through knowledge of the most recent information on their trajectory.

As we have seen in Section 1.1, the seemingly simple Markovian assumption allows a surprisingly wide variety of phenomena to be represented as Markov chains. It is this which accounts for the central place that Markov models hold in the stochastic process literature. For once some limited independence of the past is allowed, then there is the possibility of reformulating many models so the dependence is as simple as in (1.1).

There are two standard paradigms for allowing us to construct Markovian representations, even if the initial phenomenon appears to be non-Markovian.

In the first, the dependence of some model of interest $Y = \{Y_n\}$ on its past values may be non-Markovian but still be based only on a finite "memory". This means that the system depends on the past only through the previous $k + 1$ values, in the probabilistic sense that

$$\mathsf{P}(Y_{n+m} \in A \mid Y_j, j \leq n) = \mathsf{P}(Y_{n+m} \in A \mid Y_j, j = n, n-1, \ldots, n-k). \qquad (1.2)$$

Merely by reformulating the model through defining the vectors

$$\Phi_n = \{Y_n, \ldots, Y_{n-k}\}$$

and setting $\boldsymbol{\Phi} = \{\Phi_n, n \geq 0\}$ (taking obvious care in defining $\{\Phi_0, \ldots, \Phi_{k-1}\}$), we can define from Y a Markov chain $\boldsymbol{\Phi}$. The motion in the first coordinate of $\boldsymbol{\Phi}$ reflects that of Y, and in the other coordinates is trivial to identify, since Y_n becomes $Y_{(n+1)-1}$, and so forth; and hence Y can be analyzed by Markov chain methods.

Such *state space* representations, despite their somewhat artificial nature in some cases, are an increasingly important tool in deterministic and stochastic systems theory, and in linear and nonlinear time series analysis.

As the second paradigm for constructing a Markov model representing a non-Markovian system, we look for so-called *embedded regeneration points*. These are times at which the system forgets its past in a probabilistic sense: the system viewed at such time points is Markovian even if the overall process is not.

Consider as one such model a storage system, or dam, which fills and empties. This is rarely Markovian: for instance, knowledge of the time since the last input, or the size of previous inputs still being drawn down, will give information on the current level of the dam or even the time to the next input. But at that very special sequence of times when the dam is empty and an input actually occurs, the process may well "forget the past", or "regenerate": appropriate conditions for this are that the times between inputs and the size of each input are independent. For then one cannot forecast the time to the next input when at an input time, and the current emptiness of the dam means that there is no information about past input levels available at such times. The dam content, viewed at these special times, can then be analyzed as a Markov chain.

"Regenerative models" for which such "embedded Markov chains" occur are common in operations research, and in particular in the analysis of queueing and network models.

State space models and regeneration time representations have become increasingly important in the literature of time series, signal processing, control theory, and operations research, and not least because of the possibility they provide for analysis through the tools of Markov chain theory. In the remainder of this opening chapter, we will introduce a number of these models in their simplest form, in order to provide a concrete basis for further development.

1.2.2   State space and deterministic control models

One theme throughout this book will be the analysis of stochastic models through consideration of the underlying deterministic motion of specific (non-random) realizations of the input driving the model.

Such an approach draws on both control theory, for the deterministic analysis; and Markov chain theory, for the translation to the stochastic analogue of the deterministic chain.

We introduce both of these ideas heuristically in this section.

Deterministic control models

In the theory of deterministic systems and control systems we find the simplest possible Markov chains: ones such that the next position of the chain is determined completely as a function of the previous position.

Consider the deterministic linear system on $\mathbb{R}^n$, whose "state trajectory" $\boldsymbol{x} = \{x_k, \ k \in \mathbb{Z}_+\}$ is defined inductively as

$$x_{k+1} = Fx_k \tag{1.3}$$

where F is an $n \times n$ matrix.

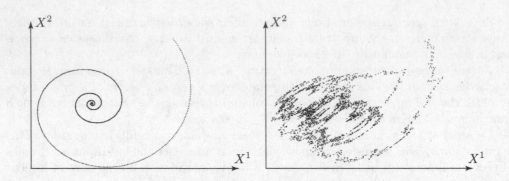

Figure 1.1: At left is a sample path generated by the deterministic linear model on $\mathbb{R}^2$. At right is a sample path from the linear state space model on $\mathbb{R}^2$ with Gaussian noise.

Clearly, this is a multidimensional Markovian model: even if we know all of the values of $\{x_k, k \leq m\}$ then we will still predict x_{m+1} in the same way, with the same (exact) accuracy, based solely on (1.3) which uses only knowledge of x_m.

At left in Figure 1.1 we show a sample path corresponding to the choice of F as $F = I + \Delta A$ with I equal to a 2×2 identity matrix, $A = \begin{pmatrix} -0.2, & 1 \\ -1, & -0.2 \end{pmatrix}$ and $\Delta = 0.02$. It is instructive to realize that two very different types of behavior can follow from related choices of the matrix F. The trajectory spirals in, and is intuitively "stable"; but if we read the model in the other direction, the trajectory spirals out, and this is exactly the result of using F^{-1} in (1.3).

Thus, although this model is one without any built-in randomness or stochastic behavior, questions of stability of the model are still basic: the first choice of F gives a stable model, the second choice of F^{-1} gives an unstable model.

A straightforward generalization of the linear system of (1.3) is the *linear control model*. From the outward version of the trajectory in Figure 1.1, it is clearly possible for the process determined by F to be out of control in an intuitively obvious sense. In practice, one might observe the value of the process, and influence it either by adding on a modifying "control value" either independently of the current position of the process or directly based on the current value. Now the state trajectory $\boldsymbol{x} = \{x_k\}$ on $\mathbb{R}^n$ is defined inductively not only as a function of its past, but also of such a (deterministic) control sequence $\boldsymbol{u} = \{u_k\}$ taking values in, say, $\mathbb{R}^p$.

Formally, we can describe the linear control model by the postulates (LCM1) and (LCM2) below.

If the control value u_{k+1} depends at most on the sequence $\dot{x}_j, j \leq k$ through x_k, then it is clear that the LCM(F,G) model is itself Markovian.

However, the interest in the linear control model in our context comes from the fact that it is helpful in studying an associated Markov chain called the *linear state space model*. This is simply (1.4) with a certain random choice for the sequence $\{u_k\}$, with u_{k+1} independent of $x_j, j \leq k$, and we describe this next.

Deterministic linear control model

Suppose $x = \{x_k\}$ is a process on $\mathbb{R}^n$ and $u = \{u_n\}$ is a process on $\mathbb{R}^p$, for which x_0 is arbitrary and for $k \geq 1$

(LCM1) there exists an $n \times n$ matrix F and an $n \times p$ matrix G such that for each $k \in \mathbb{Z}_+$,

$$x_{k+1} = Fx_k + Gu_{k+1}; \tag{1.4}$$

(LCM2) the sequence $\{u_k\}$ on $\mathbb{R}^p$ is chosen deterministically.

Then x is called the *linear control model driven by* F, G, or the LCM(F,G) model.

The linear state space model

In developing a stochastic version of a control system, an obvious generalization is to assume that the next position of the chain is determined as a function of the previous position, but in some way which still allows for uncertainty in its new position, such as by a random choice of the "control" at each step. Formally, we can describe such a model by

---

Linear state space model

Suppose $X = \{X_k\}$ is a stochastic process for which

(LSS1) there exists an $n \times n$ matrix F and an $n \times p$ matrix G such that for each $k \in \mathbb{Z}_+$, the random variables X_k and W_k take values in $\mathbb{R}^n$ and $\mathbb{R}^p$, respectively, and satisfy inductively for $k \in \mathbb{Z}_+$,

$$X_{k+1} = FX_k + GW_{k+1}$$

where X_0 is arbitrary;

(LSS2) the random variables $\{W_k\}$ are independent and identically distributed (i.i.d), and are independent of X_0, with common distribution $\Gamma(A) = \mathsf{P}(W_j \in A)$ having finite mean and variance.

Then X is called the linear state space model driven by F, G, or the LSS(F,G) model, with *associated* control model LCM(F,G).

Such linear models with random "noise" or "innovation" are related to both the simple deterministic model (1.3) and also the linear control model (1.4).

There are obviously two components to the evolution of a state space model. The matrix F controls the motion in one way, but its action is modulated by the regular input of random fluctuations which involve both the underlying variable with distribution Γ, and its adjustment through G. At left in Figure 1.1 we show a sample path corresponding to the same matrix F, $G = \binom{2.5}{2.5}$, and with Γ taken as a bivariate Normal, or Gaussian, distribution $N(0,1)$. This indicates that the addition of the noise variables W can lead to types of behavior very different to that of the deterministic model, even with the same choice of the function F.

Such models describe the movements of airplanes, of industrial and engineering equipment, and even (somewhat idealistically) of economies and financial systems [3, 57]. Stability in these contexts is then understood in terms of return to level flight, or small and (in practical terms) insignificant deviations from set engineering standards, or minor inflation or exchange-rate variation. Because of the random nature of the noise we cannot expect totally unvarying systems; what we seek to preclude are explosive or wildly fluctuating operations.

We will see that, in wide generality, if the linear control model LCM(F,G) is stable in a deterministic way, and if we have a "reasonable" distribution Γ for our random control sequences, then the linear state space LSS(F,G) model is also stable in a stochastic sense.

In Chapter 2 we will describe models which build substantially on these simple structures, and which illustrate the development of Markovian structures for linear and nonlinear state space model theory.

We now leave state space models, and turn to the simplest examples of another class of models, which may be thought of collectively as models with a regenerative structure.

1.2.3 The gamblers ruin and the random walk

Unrestricted random walk

At the roots of traditional probability theory lies the problem of the gambler's ruin.

One has a gaming house in which one plays successive games; at each time point, there is a playing of a game, and an amount won or lost: and the successive totals of the amounts won or lost represent the fluctuations in the fortune of the gambler.

It is common, and realistic, to assume that as long as the gambler plays the same game each time, then the winnings W_k at each time k are i.i.d.

Now write the total winnings (or losings) at time k as Φ_k. By this construction,

$$\Phi_{k+1} = \Phi_k + W_{k+1}. \tag{1.5}$$

It is obvious that $\boldsymbol{\Phi} = \{\Phi_k : k \in \mathbb{Z}_+\}$ is a Markov chain, taking values in the real line $\mathbb{R} = (-\infty, \infty)$; the independence of the $\{W_k\}$ guarantees the Markovian nature of the chain $\boldsymbol{\Phi}$.

In this context, stability (as far as the gambling house is concerned) requires that $\boldsymbol{\Phi}$ eventually reaches $(-\infty, 0]$; a greater degree of stability is achieved from the same perspective if the time to reach $(-\infty, 0]$ has finite mean. Inevitably, of course, this stability is also the gambler's ruin.

Such a chain, defined by taking successive sums of i.i.d. random variables, provides a model for very many different systems, and is known as random walk.

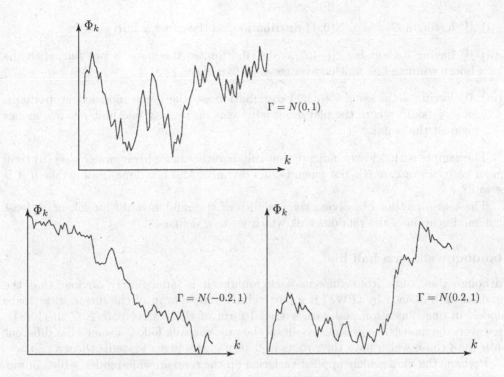

Figure 1.2: Random walk sample paths from three different models. The increment distributions is $\Gamma = N(0,1)$ for the path shown at top. The increment distribution is $\Gamma = N(-0.2,1)$ for the path shown on the lower left, and $\Gamma = N(+0.2,1)$ for the path shown on the lower right.

Random walk

Suppose that $\boldsymbol{\Phi} = \{\Phi_k; k \in \mathbb{Z}_+\}$ is a collection of random variables defined by choosing an arbitrary distribution for Φ_0 and setting for $k \in \mathbb{Z}_+$

(RW1)
$$\Phi_{k+1} = \Phi_k + W_{k+1}$$

where the W_k are i.i.d. random variables taking values in $\mathbb{R}$ with

$$\Gamma(-\infty, y] = \mathsf{P}(W_n \leq y). \qquad (1.6)$$

Then $\boldsymbol{\Phi}$ is called *random walk* on $\mathbb{R}$.

In Figure 1.2 we give sets of three sample paths of random walks with different distributions for Γ: all start at the same value but we choose for the winnings on each game

(i) W having a Gaussian $N(0, 1)$ distribution, so the game is fair;

(ii) W having a Gaussian $N(-0.2, 1)$ distribution, so the game is not fair, with the house winning one unit on average each five plays;

(iii) W having a Gaussian $N(0.2, 1)$ distribution, so the game modeled is, perhaps, one of "skill" where the player actually wins on average one unit per five games against the house.

The sample paths clearly indicate that ruin is rather more likely under case (ii) than under case (iii) or case (i): but when is ruin certain? And how long does it take if it is certain?

These are questions involving the stability of the random walk model, or at least that modification of the random walk which we now define.

Random walk on a half line

Although they come from different backgrounds, it is immediately obvious that the random walk defined by (RW1) is a particularly simple form of the linear state space model, in one dimension and with a trivial form of the matrix pair F, G in (LSS1). However, the models traditionally built on the random walk follow a somewhat different path than those which have their roots in deterministic linear systems theory.

Perhaps the most widely applied variation on the random walk model, which immediately moves away from a linear structure, is the random walk on a half line.

Random walk on a half line

Suppose $\Phi = \{\Phi_k; k \in \mathbb{Z}_+\}$ is defined by choosing an arbitrary distribution for Φ_0 and taking

(RWHL1)
$$\Phi_{k+1} = [\Phi_k + W_{k+1}]^+ \tag{1.7}$$

where $[\Phi_k + W_{k+1}]^+ := \max(0, \Phi_k + W_{k+1})$ and again the W_k are i.i.d. random variables taking values in $\mathbb{R}$ with $\Gamma(-\infty, y] = \mathsf{P}(W \leq y)$.

Then Φ is called *random walk on a half line*.

This chain follows the paths of a random walk, but is held at zero when the underlying random walk becomes non-positive, leaving zero again only when the next positive value occurs in the sequence $\{W_k\}$.

In Figure 1.3 we again give sets of sample paths of random walks on the half line $[0, \infty)$, corresponding to those of the unrestricted random walk in the previous section. The difference in the proportion of paths which hit, or return to, the state $\{0\}$ is again clear.

We shall see in Chapter 2 that random walk on a half line is both a model for storage systems and a model for queueing systems. For all such applications there are similar

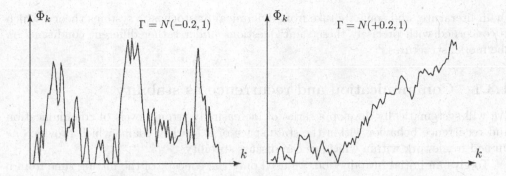

Figure 1.3: Random walk paths reflected at zero. The increment distribution is $\Gamma = N(-0.2, 1)$ for the plot shown on the left, and $\Gamma = N(+0.2, 1)$ for the plot shown on the right.

concerns and concepts of the structure and the stability of the models: we need to know whether a dam overflows, whether a queue ever empties, whether a computer network jams. In the next section we give a first heuristic description of the ways in which such stability questions might be formalized.

1.3 Stochastic stability for Markov models

What is "stability"?

It is a word with many meanings in many contexts. We have chosen to use it partly because of its very diffuseness and lack of technical meaning: in the stochastic process sense it is not well defined, it is not constraining, and it will, we hope, serve to cover a range of similar but far from identical "stable" behaviors of the models we consider, most of which have (relatively) tightly defined technical meanings.

Stability is certainly a basic concept. In setting up models for real phenomena evolving in time, one ideally hopes to gain a detailed quantitative description of the evolution of the process based on the underlying assumptions incorporated in the model. Logically prior to such detailed analyses are those questions of the structure and stability of the model which require qualitative rather than quantitative answers, but which are equally fundamental to an understanding of the behavior of the model. This is clear even from the behavior of the sample paths of the models considered in the section above: as parameters change, sample paths vary from reasonably "stable" (in an intuitive sense) behavior, to quite "unstable" behavior, with processes taking larger or more widely fluctuating values as time progresses.

Investigation of specific models will, of course, often require quite specific tools: but the stability and the general structure of a model can in surprisingly wide-ranging circumstances be established from the concepts developed purely from the Markovian nature of the model.

We discuss in this section, again somewhat heuristically (or at least with minimal technicality: some "quotation-marked" terms will be properly defined later), various general stability concepts for Markov chains. Some of these are traditional in the Markov chain literature, and some we take from dynamical or stochastic systems theory, which is concerned with precisely these same questions under rather different conditions on the model structures.

1.3.1   Communication and recurrence as stability

We will systematically develop a series of increasingly strong levels of communication and recurrence behavior within the state space of a Markov chain, which provide one unified framework within which we can discuss stability.

To give an initial introduction, we need only the concept of the hitting time from a point to a set: let

$$\tau_A := \inf(n \geq 1 : \Phi_n \in A)$$

denote the first time a chain reaches the set A. This will be infinite for those paths where the set A is never reached.

In one sense the least restrictive form of stability we might require is that the chain does not in reality consist of two chains: that is, that the collection of sets which we can reach from different starting points is not different. This leads us to first define and study

(I) φ-*irreducibility* for a general space chain, which we approach by requiring that the space supports a measure φ with the property that for every starting point $x \in \mathsf{X}$

$$\varphi(A) > 0 \Rightarrow \mathsf{P}_x(\tau_A < \infty) > 0$$

where P_x denotes the probability of events conditional on the chain beginning with $\Phi_0 = x$.

This condition ensures that all "reasonable sized" sets, as measured by φ, can be reached from every possible starting point.

For a countable space chain φ-irreducibility is just the concept of irreducibility commonly used [59, 71], with φ taken as counting measure.

For a state space model φ-irreducibility is related to the idea that we are able to "steer" the system to every other state in $\mathbb{R}^n$. The linear control LCM(F,G) model is called *controllable* if for any initial states x_0 and any other $x^\star \in \mathsf{X}$, there exists $m \in \mathbb{Z}_+$ and a sequence of control variables $(u_1^\star, \ldots, u_m^\star) \in \mathbb{R}^p$ such that $x_m = x^\star$ when $(u_1, \ldots, u_m) = (u_1^\star, \ldots, u_m^\star)$. If this does not hold then for some starting points we are in one part of the space forever; from others we are in another part of the space. Controllability, and analogously irreducibility, preclude this.

Thus under irreducibility we do not have systems so unstable in their starting position that, given a small change of initial position, they might change so dramatically that they have no possibility of reaching the same set of states.

A study of the wide-ranging consequences of such an assumption of irreducibility will occupy much of Part I of this book: the definition above will be shown to produce remarkable solidity of behavior.

The next level of stability is a requirement, not only that there should be a possibility of reaching like states from unlike starting points, but that reaching such sets of states should be guaranteed eventually. This leads us to define and study concepts of

(II) *recurrence*, for which we might ask as a first step that there is a measure φ guaranteeing that for every starting point $x \in \mathsf{X}$

$$\varphi(A) > 0 \Rightarrow \mathsf{P}_x(\tau_A < \infty) = 1, \tag{1.8}$$

and then, as a further strengthening, that for every starting point $x \in \mathsf{X}$

$$\varphi(A) > 0 \Rightarrow \mathsf{E}_x[\tau_A] < \infty. \tag{1.9}$$

These conditions ensure that reasonable sized sets are reached with probability one, as in (1.8), or even in a finite mean time as in (1.9). Part II of this book is devoted to the study of such ideas, and to showing that for irreducible chains, even on a general state space, there are solidarity results which show that either such uniform (in x) stability properties hold, or the chain is unstable in a well-defined way: there is no middle ground, no "partially stable" behavior available.

For deterministic models, the recurrence concepts in (II) are obviously the same. For stochastic models they are definitely different. For "suitable" chains on spaces with appropriate topologies (the T-chains introduced in Chapter 6), the first will turn out to be entirely equivalent to requiring that "evanescence", defined by

$$\{\boldsymbol{\Phi} \to \infty\} = \bigcap_{n=0}^{\infty} \{\boldsymbol{\Phi} \in O_n \text{ infinitely often}\}^c \tag{1.10}$$

for a countable collection of open pre-compact sets $\{O_n\}$, has zero probability for all starting points; the second is similarly equivalent, for the same "suitable" chains, to requiring that for any $\varepsilon > 0$ and any x there is a compact set C such that

$$\liminf_{k \to \infty} P^k(x, C) \geq 1 - \varepsilon \tag{1.11}$$

which is *tightness* [36] of the transition probabilities of the chain.

All these conditions have the heuristic interpretation that the chain returns to the "center" of the space in a recurring way: when (1.9) holds then this recurrence is faster than if we only have (1.8), but in both cases the chain does not just drift off (or evanesce) away from the center of the state space.

In such circumstances we might hope to find, further, a long-term version of stability in terms of the convergence of the distributions of the chain as time goes by. This is the third level of stability we consider. We define and study

(III) the limiting, or *ergodic*, behavior of the chain: and it emerges that in the stronger recurrent situation described by (1.9) there is an "invariant regime" described by a measure π such that if the chain starts in this regime (that is, if Φ_0 has distribution π) then it remains in the regime, and moreover if the chain starts in some other regime then it converges in a strong probabilistic sense with π as a limiting distribution.

In Part III we largely confine ourselves to such ergodic chains, and find both theoretical and pragmatic results ensuring that a given chain is at this level of stability. For whilst the construction of solidarity results, as in Parts I and II, provides a vital underpinning to the use of Markov chain theory, it is the consequences of that stability, in the form of powerful ergodic results, that makes the concepts of very much more than academic interest.

Let us provide motivation for such endeavors by describing, with a little more formality, just how solid the solidarity results are, and how strong the consequent ergodic theorems are. We will show, in Chapter 13, the following:

Theorem 1.3.1. *The following four conditions are equivalent:*

(i) *The chain admits a unique probability measure π satisfying the invariant equations*

$$\pi(A) = \int \pi(dx)P(x, A), \qquad A \in \mathcal{B}(\mathsf{X}); \tag{1.12}$$

(ii) *There exists some "small" set $C \in \mathcal{B}(\mathsf{X})$ and $M_C < \infty$ such that*

$$\sup_{x \in C} \mathsf{E}_x[\tau_C] \leq M_C; \tag{1.13}$$

(iii) *There exists some "small" set C, some $b < \infty$ and some non-negative "test function" V, finite φ-almost everywhere, satisfying*

$$\int P(x, dy)V(y) \leq V(x) - 1 + b\mathbb{I}_C(x), \qquad x \in \mathsf{X}; \tag{1.14}$$

(iv) *There exists some "small" set $C \in \mathcal{B}(\mathsf{X})$ and some $P^\infty(C) > 0$ such that as $n \to \infty$*

$$\liminf_{n \to \infty} \sup_{x \in C} |P^n(x, C) - P^\infty(C)| = 0 \tag{1.15}$$

Any of these conditions implies, for "aperiodic" chains,

$$\sup_{A \in \mathcal{B}(\mathsf{X})} |P^n(x, A) - \pi(A)| \to 0, \qquad n \to \infty, \tag{1.16}$$

for every $x \in \mathsf{X}$ for which $V(x) < \infty$, where V is any function satisfying (1.14).

Thus "local recurrence" in terms of return times, as in (1.13) or "local convergence" as in (1.15) guarantees the uniform limits in (1.16); both are equivalent to the mere existence of the invariant probability measure π; and moreover we have in (1.14) an exact test based only on properties of P for checking stability of this type.

Each of (i)–(iv) is a type of stability: the beauty of this result lies in the fact that they are completely equivalent. Moreover, for this irreducible form of Markovian system, it is further possible in the "stable" situation of this theorem to develop asymptotic results, which ensure convergence not only of the distributions of the chain, but also of very general (and not necessarily bounded) functions of the chain (Chapter 14); to develop global rates of convergence to these limiting values (Chapter 15 and Chapter 16); and to link these to Laws of Large Numbers or Central Limit Theorems (Chapter 17).

Together with these consequents of stability, we also provide a systematic approach for establishing stability in specific models in order to utilize these concepts. The extension of the so-called "Foster–Lyapunov" criteria as in (1.14) to all aspects of stability, and application of these criteria in complex models, is a key feature of our approach to stochastic stability.

These concepts are largely classical in the theory of countable state space Markov chains. The extensions we give to general spaces, as described above, are neither so well known nor, in some cases, previously known at all.

The heuristic discussion of this section will take considerable formal justification, but the end-product will be a rigorous approach to the stability and structure of Markov chains.

1.3.2   A dynamical system approach to stability

Just as there are a number of ways to come to specific models such as the random walk, there are other ways to approach stability, and the recurrence approach based on ideas from countable space stochastic models is merely one. Another such is through deterministic dynamical systems.

We now consider some traditional definitions of stability for a deterministic system, such as that described by the linear model (1.3) or the linear control model LCM(F,G).

One route is through the concepts of a *(semi) dynamical system*: this is a triple $(T, \mathcal{X}, d)$ where $(\mathcal{X}, d)$ is a metric space, and $T: \mathcal{X} \to \mathcal{X}$ is, typically, assumed to be continuous. A basic concern in dynamical systems is the structure of the *orbit* $\{T^k x : k \in \mathbb{Z}_+\}$, where $x \in \mathcal{X}$ is an *initial condition* so that $T^0 x := x$, and we define inductively $T^{k+1} x := T^k(Tx)$ for $k \geq 1$.

There are several possible dynamical systems associated with a given Markov chain. The dynamical system which arises most naturally if X has sufficient structure is based directly on the transition probability operators P^k. If μ is an initial distribution for the chain (that is, if Φ_0 has distribution μ), one might look at the trajectory of distributions $\{\mu P^k : k \geq 0\}$, and consider this as a dynamical system $(P, \mathcal{M}, d)$ with $\mathcal{M}$ the space of Borel probability measures on a topological state space X, d a suitable metric on $\mathcal{M}$, and with the operator P defined as in (1.1) acting as $P: \mathcal{M} \to \mathcal{M}$ through the relation

$$\mu P(\cdot) = \int_X \mu(dx) P(x, \cdot), \qquad \mu \in \mathcal{M}.$$

In this sense the Markov transition function P can be viewed as a deterministic map from $\mathcal{M}$ to itself, and P will induce such a dynamical system if it is suitably continuous. This interpretation can be achieved if the chain is on a suitably behaved space and has the *Feller* property that $Pf(x) := \int P(x, dy) f(y)$ is continuous for every bounded continuous f, and then d becomes a weak convergence metric (see Chapter 6).

As in the stronger recurrence ideas in (II) and (III) in Section 1.3.1, in discussing the stability of Φ, we are usually interested in the behavior of the terms P^k, $k \geq 0$, when k becomes large. Our hope is that this sequence will be bounded in some sense, or converge to some fixed probability $\pi \in \mathcal{M}$, as indeed it does in (1.16).

Four traditional formulations of stability for a dynamical system, which give a framework for such questions, are

(i) *Lagrange stability*: for each $x \in \mathcal{X}$, the orbit starting at x is a precompact subset of $\mathcal{X}$. For the system $(P, \mathcal{M}, d)$ with d the weak convergence metric, this is exactly tightness of the distributions of the chain, as defined in (1.11);

(ii) *Stability in the sense of Lyapunov*: for each initial condition $x \in \mathcal{X}$,

$$\lim_{y \to x} \sup_{k \geq 0} d(T^k y, T^k x) = 0,$$

where d denotes the metric on $\mathcal{X}$. This is again the requirement that the long-term behavior of the system is not overly sensitive to a change in the initial conditions;

(iii) *Asymptotic stability*: there exists some fixed point x^* so that $T^k x^* = x^*$ for all k, with trajectories $\{x_k\}$ starting near x^* staying near and converging to x^* as $k \to \infty$. For the system $(P, \mathcal{M}, d)$ the existence of a fixed point is exactly equivalent to the existence of a solution to the invariant equations (1.12);

(iv) *Global asymptotic stability*: the system is stable in the sense of Lyapunov and for some fixed $x^* \in \mathcal{X}$ and every initial condition $x \in \mathcal{X}$,

$$\lim_{k \to \infty} d(T^k x, x^*) = 0. \tag{1.17}$$

This is comparable to the result of Theorem 1.3.1 for the dynamical system $(P, \mathcal{M}, d)$.

Lagrange stability requires that any limiting measure arising from the sequence $\{\mu P^k\}$ will be a probability measure, rather as in (1.16).

Stability in the sense of Lyapunov is most closely related to irreducibility, although rather than placing a global requirement on every initial condition in the state space, stability in the sense of Lyapunov only requires that two initial conditions which are sufficiently close will then have comparable long term behavior. Stability in the sense of Lyapunov says nothing about the actual boundedness of the orbit $\{T^k x\}$, since it is simply continuity of the maps $\{T^k\}$, uniformly in $k \geq 0$. An example of a system on $\mathbb{R}$ which is stable in the sense of Lyapunov is the simple recursion $x_{k+1} = x_k + 1$, $k \geq 0$. Although distinct trajectories stay close together if their initial conditions are similarly close, we would not consider this system stable in most other senses of the word.

The connections between the probabilistic recurrence approach and the dynamical systems approach become very strong in the case where the chain is both Feller and φ-irreducible, and when the irreducibility measure φ is related to the topology by the requirement that the support of φ contains an open set.

In this case, by combining the results of Chapter 6 and Chapter 18, we get for suitable spaces

Theorem 1.3.2. *For a φ-irreducible "aperiodic" Feller chain with* $\operatorname{supp} \varphi$ *containing an open set, the dynamical system $(P, \mathcal{M}, d)$ is globally asymptotically stable if and only if the distributions $\{P^k(x, \cdot)\}$ are tight as in (1.11); and then the uniform ergodic limit (1.16) holds.*

This result follows, not from dynamical systems theory, but by showing that such a chain satisfies the conditions of Theorem 1.3.1; these Feller chains are an especially useful subset of the "suitable" chains for which tightness is equivalent to the properties described in Theorem 1.3.1, and then, of course, (1.16) gives a result rather stronger than (1.17).

Embedding a Markov chain in a dynamical system through its transition probabilities does not bring much direct benefit, since results on dynamical systems in this level of generality are relatively weak. The approach does, however, give insights into ways of thinking of Markov chain stability, and a second heuristic to guide the types of results we should seek.

1.4 Commentary

This book does not address models where the time set is continuous (when Φ is usually called a Markov *process*), despite the sometimes close relationship between discrete and continuous time models: see Chung [71] or Anderson [4] for the classical countable space approach.

On general spaces in continuous time, there are a totally different set of questions that are often seen as central: these are exemplified in Sharpe [352], although the interested reader should also see Meyn and Tweedie [279, 280, 278] for recent results which are much closer in spirit to, and rely heavily on, the countable time approach followed in this book.

There has also been considerable recent work over the past two decades on the subject of more generally indexed Markov models (such as Markov *random fields*, where T is multidimensional), and these are also not in this book. In our development Markov chains always evolve through time as a scalar, discrete quantity.

The question of what to call a Markovian model, and whether to concentrate on the denumerability of the space or the time parameter in using the word "chain", seems to have been resolved in the direction we take here. Doob [99] and Chung [71] reserve the term chain for systems evolving on countable spaces with both discrete and continuous time parameters, but usage seems to be that it is the time set that gives the "chaining". Revuz [326], in his Notes, gives excellent reasons for this.

The examples we begin with here are rather elementary, but equally they are completely basic, and represent the twin strands of application we will develop: the first, from deterministic to stochastic models via a "stochasticization" within the same functional framework has analogies with the approach of Stroock and Varadhan in their analysis of diffusion processes (see [378, 377, 168]), whilst the second, from basic independent random variables to sums and other functionals traces its roots back too far to be discussed here. Both these models are close to identical at this simple level. We give more diverse examples in Chapter 2.

We will typically use X and X_n to denote state space models, or their values at time n, in accordance with rather long established conventions. We will then typically use lower case letters to denote the values of related deterministic models. Regenerative models such as random walk are, on the other hand, typically denoted by the symbols Φ and Φ_n, which we also use for generic chains.

The three concepts described in (I)–(III) may seem to give a rather limited number of possible versions of "stability". Indeed, in the various generalizations of deterministic dynamical systems theory to stochastic models which have been developed in the past three decades (see for example Kushner [232] or Khas'minskii [206]) there have been many other forms of stability considered. All of them are, however, qualitatively similar, and fall broadly within the regimes we describe, even though they differ in detail.

It will become apparent in the course of our development of the theory of irreducible chains that in fact, under fairly mild conditions, the number of different types of behavior is indeed limited to precisely those sketched above in (I)–(III). Our aim is to unify many of the partial approaches to stability and structural analysis, to indicate how they are in many cases equivalent, and to develop both criteria for stability to hold for individual models, and limit theorems indicating the value of achieving such stability.

With this rather optimistic statement, we move forward to consider some of the specific models whose structure we will elucidate as examples of our general results.

Chapter 2

Markov models

The results presented in this book have been written in the desire that practitioners will use them. We have tried therefore to illustrate the use of the theory in a systematic and accessible way, and so this book concentrates not only on the theory of general space Markov chains, but on the application of that theory in considerable detail.

We will apply the results which we develop across a range of specific applications: typically, after developing a theoretical construct, we apply it to models of increasing complexity in the areas of systems and control theory, both linear and nonlinear, both scalar and vector valued; traditional "applied probability" or operations research models, such as random walks, storage and queueing models, and other regenerative schemes; and models which are in both domains, such as classical and recent time series models.

These are not given merely as "examples" of the theory: in many cases, the application is difficult and deep of itself, whilst applications across such a diversity of areas have often driven the definition of general properties and the links between them. Our goal has been to develop the analysis of applications on a step-by-step basis as the theory becomes richer throughout the book.

To motivate the general concepts, then, and to introduce the various areas of application, we leave until Chapter 3 the normal and necessary foundations of the subject, and first introduce a cross-section of the models for which we shall be developing those foundations.

These models are still described in a somewhat heuristic way. The full mathematical description of their dynamics must await the development in the next chapter of the concepts of transition probabilities, and the reader may on occasion benefit by moving to some of those descriptions in parallel with the outlines here.

It is also worth observing immediately that the descriptive definitions here are from time to time supplemented by other assumptions in order to achieve specific results: these assumptions, and those in this chapter and the last, are collected for ease of reference in Appendix C.

As the definitions are developed, it will be apparent immediately that very many of these models have a random additive component, such as the i.i.d. sequence $\{W_n\}$ in both the linear state space model and the random walk model. Such a component goes by various names, such as error, noise, innovation, disturbance or increment sequence, across the various model areas we consider. We shall use the nomenclature relevant to the context of each model.

We will save considerable repetitive definition if we adopt a global convention immediately to cover these sequences.

Error, noise, disturbance, innovation, and increments

Suppose $W = \{W_n\}$ is labeled as an error, noise, innovation, disturbance or increment sequence. Then this has the interpretation that the random variables $\{W_n\}$ are independent and identically distributed, with distribution identical to that of a generic variable denoted W.

We will systematically denote the probability law of such a variable W by Γ.

It will also be apparent that many models are defined inductively from their own past in combination with such innovation sequences. In order to commence the induction, initial values are needed. We adopt a second convention immediately to avoid repetition in defining our models.

Initialization

Unless specifically defined otherwise, the initial state $\{\Phi_0\}$ of a Markov model will be taken as independent of the error, noise, innovation, disturbance or increments process, and will have an arbitrary distribution.

2.1   Markov models in time series

The theory of time series has been developed to model a set of observations developing in time: in this sense, the fundamental starting point for time series and for more general Markov models is virtually identical. However, whilst the Markov theory immediately assumes a short-term dependence structure on the variables at each time point, time series theory concentrates rather on the parametric form of dependence between the variables.

The time series literature has historically concentrated on linear models (that is, those for which past disturbances and observations are combined to form the present observation through some linear transformation) although recently there has been greater emphasis on nonlinear models. We first survey a number of general classes of linear models and turn to some recent nonlinear time series models in Section 2.2.

It is traditional to denote time series models as a sequence $X = \{X_n : n \in \mathbb{Z}_+\}$, and we shall follow this tradition.

2.1.1 Simple linear models

The first class of models we discuss has direct links with deterministic linear models, state space models and the random walk models we have already introduced in Chapter 1.

We begin with the simplest possible "time series" model, the scalar autoregression of order one, or AR(1) model on $\mathbb{R}^1$.

Simple linear model

The process $\boldsymbol{X} = \{X_n, n \in \mathbb{Z}_+\}$ is called the *simple linear model,* or *AR(1) model* if

(SLM1)   for each $n \in \mathbb{Z}_+$, X_n and W_n are random variables on $\mathbb{R}$, satisfying
$$X_{n+1} = \alpha X_n + W_{n+1},$$
for some $\alpha \in \mathbb{R}$;

(SLM2) $\boldsymbol{W} = \{W_n\}$ is an error sequence with distribution Γ on $\mathbb{R}$.

The simple linear model is trivially Markovian: the independence of X_{n+1} from $X_{n-1}, X_{n-2}, \ldots$ given $X_n = x$ follows from the construction rule (SLM1), since the value of W_n does not depend on any of $\{X_{n-1}, X_{n-2} \ldots\}$ from (SLM2).

The simple linear model can be viewed in one sense as an extension of the random walk model, where now we take some proportion or multiple of the previous value, not necessarily equal to the previous value, and again add a new random amount (the "noise" or "error") onto this scaled random value. Equally, it can be viewed as the simplest special case of the linear state space model LSS(F,G), in the scalar case with $F = \alpha$ and $G = 1$.

In Figure 2.1 we give sets of sample paths of linear models with different values of the parameter α.

The choice of this parameter critically determines the behavior of the chain. If $|\alpha| < 1$ then the sample paths remain bounded in ways which we describe in detail in later chapters, and the process $\boldsymbol{X}$ is inherently "stable": in fact, ergodic in the sense of Section 1.3.1 (III) and Theorem 1.3.1, for reasonable distributions Γ. But if $|\alpha| > 1$ then $\boldsymbol{X}$ is unstable, in a well-defined way: in fact, evanescent with probability one, in the sense of Section 1.3.1 (II), if the noise distribution Γ is again reasonable.

2.1.2 Linear autoregressions and ARMA models

In the development of time series theory, simple linear models are usually analyzed as a subset of the class of autoregressive models, which depend in a linear manner on their past history for a fixed number $k \geq 1$ of steps in the past.

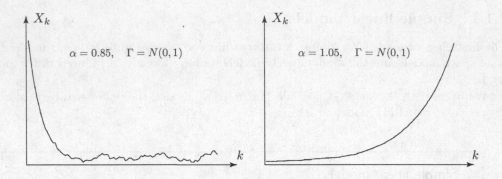

Figure 2.1: Shown on the left is a sample path from the linear model with $\alpha = 0.85$, and shown on the right is a sample path obtained with $\alpha = 1.05$. The increment distribution is $N(0,1)$ in each case.

Autoregressive model

A process $\boldsymbol{Y} = \{Y_n\}$ is called a (scalar) *autoregression of order k*, or AR(k) model, if it satisfies, for each set of initial values $(Y_0, \ldots, Y_{-k+1})$,

(AR1)   for each $n \in \mathbb{Z}_+$, Y_n and W_n are random variables on $\mathbb{R}$ satisfying inductively for $n \geq 1$

$$Y_n = \alpha_1 Y_{n-1} + \alpha_2 Y_{n-2} + \ldots + \alpha_k Y_{n-k} + W_n,$$

for some $\alpha_1, \ldots, \alpha_k \in \mathbb{R}$;

(AR2) $\boldsymbol{W}$ is an error sequence on $\mathbb{R}$.

The collection $\boldsymbol{Y} = \{Y_n\}$ is generally not Markovian if $k > 1$, since information on the past (or at least the past in terms of the variables $Y_{n-1}, Y_{n-2}, \ldots, Y_{n-k}$) provides information on the current value Y_n of the process. But by the device mentioned in Section 1.2.1, of constructing the multivariate sequence

$$X_n = (Y_n, \ldots, Y_{n-k+1})^\top$$

and setting $\boldsymbol{X} = \{X_n, n \geq 0\}$, we define $\boldsymbol{X}$ as a Markov chain whose first component has exactly the sample paths of the autoregressive process. Note that the general convention that X_0 has an arbitrary distribution implies that the first k variables $(Y_0, \ldots, Y_{-k+1})$ are also considered arbitrary.

The autoregressive model can then be viewed as a specific version of the vector-valued linear state space model LSS(F,G). For by (AR1),

$$X_n = \begin{bmatrix} \alpha_1 & \cdots & \cdots & \alpha_k \\ 1 & & & 0 \\ & \ddots & & \vdots \\ 0 & & 1 & 0 \end{bmatrix} X_{n-1} + \begin{bmatrix} 1 \\ 0 \\ \vdots \\ 0 \end{bmatrix} W_n. \tag{2.1}$$

The same technique for producing a Markov model can be used for any linear model which admits a finite-dimensional description. In particular, we take the following general model:

Autoregressive moving-average model

The process $Y = \{Y_n\}$ is called an *autoregressive moving-average process of order* (k, ℓ), or ARMA(k, ℓ) model, if it satisfies, for each set of initial values $(Y_0, \ldots, Y_{-k+1}, W_0, \ldots, W_{-\ell+1})$,

(ARMA1)   for each $n \in \mathbb{Z}_+$, Y_n and W_n are random variables on $\mathbb{R}$, satisfying, inductively for $n \geq 1$,

$$Y_n = \alpha_1 Y_{n-1} + \alpha_2 Y_{n-2} + \cdots + \alpha_k Y_{n-k}$$

$$+ W_n + \beta_1 W_{n-1} + \beta_2 W_{n-2} + \cdots + \beta_\ell W_{n-\ell},$$

for some $\alpha_1, \ldots, \alpha_k, \beta_1, \ldots, \beta_\ell \in \mathbb{R}$;

(ARMA2) W is an error sequence on $\mathbb{R}$.

In this case more care must be taken to obtain a suitable Markovian description of the process. One approach is to take

$$X_n = (Y_n, \ldots, Y_{n-k+1}, W_n, \ldots, W_{n-\ell+1})^\top.$$

Although the resulting state process X is Markovian, the dimension of this realization may be overly large for effective analysis. A realization of lower dimension may be obtained by defining the stochastic process Z inductively by

$$Z_n = \alpha_1 Z_{n-1} + \alpha_2 Z_{n-2} + \cdots + \alpha_k Z_{n-k} + W_n. \tag{2.2}$$

When the initial conditions are defined appropriately, it is a matter of simple algebra and an inductive argument to show that

$$Y_n = Z_n + \beta_1 Z_{n-1} + \beta_2 Z_{n-2} + \cdots + \beta_\ell Z_{n-\ell},$$

Hence the probabilistic structure of the ARMA(k, ℓ) process is completely determined by the Markov chain $\{(Z_n, \ldots, Z_{n-k+1})^\top : n \in \mathbb{Z}_+\}$ which takes values in $\mathbb{R}^k$.

The behavior of the general ARMA(k, ℓ) model can thus be placed in the Markovian context, and we will develop the stability theory of this, and more complex versions of this model, in the sequel.

2.2   Nonlinear state space models*

In discrete time, a general (semi) dynamical system on $\mathbb{R}$ is defined, as in Section 1.3.2, through a recursion of the form

$$x_{n+1} = F(x_n), \qquad n \in \mathbb{Z}_+, \tag{2.3}$$

for some continuous function $F \colon \mathbb{R} \to \mathbb{R}$. Hence the simple linear model defined in (SLM1) may be interpreted as a linear dynamical system perturbed by the "noise" sequence W.

The theory of time series is in this sense closely related to the general theory of dynamical systems: it has developed essentially as that subset of stochastic dynamical systems theory for which the relationships between the variables are linear, and even with the nonlinear models from the time series literature which we consider below, there is still a large emphasis on linear substructures.

The theory of dynamical systems, in contrast to time series theory, has grown from a deterministic base, considering initially the type of linear relationship in (1.3) with which we started our examples in Section 1.2, but progressing to models allowing a very general (but still deterministic) relationship between the variables in the present and in the past, as in (2.3). It is in the more recent development that "noise" variables, allowing the system to be random in some part of its evolution, have been introduced.

Nonlinear state space models are stochastic versions of dynamical systems where a Markovian realization of the model is both feasible and explicit: thus they satisfy a generalization of (2.3) such as

$$X_{n+1} = F(X_n, W_{n+1}), \qquad k \in \mathbb{Z}_+, \tag{2.4}$$

where W is a noise sequence and the function $F \colon \mathbb{R}^n \times \mathbb{R}^p \to \mathbb{R}^n$ is smooth (C^∞): that is, all derivatives of F exist and are continuous.

2.2.1   Scalar nonlinear models

We begin with the simpler version of (2.4) in which the random variables are scalar.

Scalar nonlinear state space model

The chain $X = \{X_n\}$ is called a *scalar nonlinear state space model on $\mathbb{R}$ driven by F*, or SNSS(F) model, if it satisfies

(SNSS1)   for each $n \geq 0$, X_n and W_n are random variables on $\mathbb{R}$, satisfying, inductively for $n \geq 1$,

$$X_n = F(X_{n-1}, W_n),$$

for some smooth (C^∞) function $F \colon \mathbb{R} \times \mathbb{R} \to \mathbb{R}$;

(SNSS2)   the sequence W is a disturbance sequence on $\mathbb{R}$, whose marginal distribution Γ possesses a density γ_w supported on an open set O_w.

The independence of X_{n+1} from $X_{n-1}, X_{n-2}, \ldots$ given $X_n = x$ follows from the rules (SNSS1) and (SNSS2), and ensures as previously that X is a Markov chain.

As with the linear control model (LCM1) associated with the linear state space model (LSS1), we will analyze nonlinear state space models through the associated deterministic "control models". Define the sequence of maps $\{F_k \colon \mathbb{R} \times \mathbb{R}^k \to \mathbb{R} : k \geq 0\}$ inductively by setting $F_0(x) = x, F_1(x_0, u_1) = F(x_0, u_1)$ and for $k > 1$

$$F_k(x_0, u_1, \ldots, u_k) = F(F_{k-1}(x_0, u_1, \ldots, u_{k-1}), u_k). \tag{2.5}$$

We call the deterministic system with trajectories

$$x_k = F_k(x_0, u_1, \ldots, u_k), \qquad k \in \mathbb{Z}_+, \tag{2.6}$$

the *associated control model* CM(F) for the SNSS(F) model, provided the deterministic control sequence $\{u_1, \ldots, u_k, k \in \mathbb{Z}_+\}$ lies in the set O_w, which we call the *control set* for the scalar nonlinear state space model.

To make these definitions more concrete we define two particular classes of scalar nonlinear models with specific structure which we shall use as examples on a number of occasions.

The first of these is the *bilinear model*, so called because it is linear in each of its input variables, namely the immediate past of the process and a noise component, whenever the other is fixed: but their joint action is multiplicative as well as additive.

Simple bilinear model

The chain $X = \{X_n\}$ is called the *simple bilinear model* if it satisfies

(SBL1)   for each $n \geq 0$, X_n and W_n are random variables on $\mathbb{R}$, satisfying for $n \geq 1$,
$$X_n = \theta X_{n-1} + b X_{n-1} W_n + W_n$$

where θ and b are scalars, and the sequence W is an error sequence on $\mathbb{R}$.

The bilinear process is thus a SNSS(F) model with F given by

$$F(x, w) = \theta x + bxw + w, \tag{2.7}$$

where the control set $O_w \subseteq \mathbb{R}$ depends upon the specific distribution of W.

In Figure 2.2 we give a sample path of a scalar nonlinear model with

$$F(x, w) = (0.707 + w)x + w$$

and with $\Gamma = N(0, \frac{1}{2})$. This is the simple bilinear model with $\theta = 0.707$ and $b = 1$. One can see from this simulation that the behavior of this model is quite different from that of any linear model.

The second specific nonlinear model we shall analyze is the scalar first-order *SETAR model*. This is piecewise linear in contiguous regions of $\mathbb{R}$, and thus while it may serve as an approximation to a completely nonlinear process, we shall see that much of its analysis is still tractable because of the linearity of its component parts.

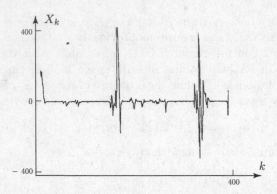

Figure 2.2: Simple bilinear model path with $F(x, w) = (0.707 + w)x + w$

SETAR model

The chain $\boldsymbol{X} = \{X_n\}$ is called a *scalar self-exciting threshold autoregression (SETAR) model* if it satisfies

(SETAR1)   for each $1 \leq j \leq M$, X_n and $W_n(j)$ are random variables on $\mathbb{R}$, satisfying, inductively for $n \geq 1$,

$$X_n = \phi(j) + \theta(j)X_{n-1} + W_n(j), \qquad r_{j-1} < X_{n-1} \leq r_j,$$

where $-\infty = r_0 < r_1 < \cdots < r_M = \infty$ and $\{W_n(j)\}$ forms an i.i.d. zero-mean error sequence for each j, independent of $\{W_n(i)\}$ for $i \neq j$.

Because of lack of continuity, the SETAR models do not fall into the class of nonlinear state space models, although they can often be analyzed using essentially the same methods. The SETAR model will prove to be a useful example on which to test the various stability criteria we develop, and the overall outcome of that analysis is gathered together in Section B.2.

2.2.2   Multidimensional nonlinear models

Many nonlinear processes cannot be modeled by a scalar Markovian model such as the SNSS(F) model. The more general multidimensional model is defined quite analogously.

Nonlinear state space model

Suppose $\boldsymbol{X} = \{X_k\}$, where

(NSS1)   for each $k \geq 0$, X_k and W_k are random variables on $\mathbb{R}^n$, $\mathbb{R}^p$ respectively, satisfying inductively for $k \geq 1$,

$$X_k = F(X_{k-1}, W_k),$$

for some smooth (C^∞) function $F \colon \mathsf{X} \times O_w \to \mathsf{X}$, where X is an open subset of $\mathbb{R}^n$ and O_w is an open subset of $\mathbb{R}^p$;

(NSS2)   the random variables $\{W_k\}$ are a disturbance sequence on $\mathbb{R}^p$, whose marginal distribution Γ possesses a density γ_w which is supported on an open set O_w.

Then $\boldsymbol{X}$ is called a *nonlinear state space model driven by F, or NSS(F) model, with control set* O_w.

The general nonlinear state space model can often be analyzed by the same methods that are used for the scalar SNSS(F) model, under appropriate conditions on the disturbance process $\boldsymbol{W}$ and the function F.

It is a central observation of such analysis that the structure of the NSS(F) model (and of course its scalar counterpart) is governed under suitable conditions by an associated deterministic control model, defined analogously to the linear control model and the linear state space model.

Control model CM(F)

(CM1)   The deterministic system

$$x_k = F_k(x_0, u_1, \ldots, u_k), \qquad k \in \mathbb{Z}_+, \tag{2.8}$$

where the sequence of maps $\{F_k : \mathsf{X} \times O_w^k \to \mathsf{X} : k \geq 0\}$ is defined by (2.5), is called the *associated control system* for the NSS(F) model and is denoted CM(F) provided the deterministic control sequence $\{u_1, \ldots, u_k, k \in \mathbb{Z}_+\}$ lies in the control set $O_w \subseteq \mathbb{R}^p$.

The general ARMA model may be generalized to obtain a class of nonlinear models, all of which may be "Markovianized", as in the linear case.

Nonlinear autoregressive moving-average model

The process $\boldsymbol{Y} = \{Y_n\}$ is called a *nonlinear autoregressive moving-average process of order* (k, ℓ) if the values $Y_0, \ldots, Y_{k-1}$ are arbitrary and

(NARMA1)   for each $n \geq 0$, Y_n and W_n are random variables on $\mathbb{R}$, satisfying, inductively for $n \geq k$,

$$Y_n = G(Y_{n-1}, Y_{n-2}, \ldots, Y_{n-k}, W_n, W_{n-1}, W_{n-2}, \ldots, W_{n-\ell})$$

where the function $G \colon \mathbb{R}^{k+\ell+1} \to \mathbb{R}$ is smooth (C^∞);

(NARMA2)   the sequence $\boldsymbol{W}$ is an error sequence on $\mathbb{R}$.

As in the linear case, we may define

$$X_n = (Y_n, \ldots, Y_{n-k+1}, W_n, \ldots, W_{n-\ell+1})^\top$$

to obtain a Markovian realization of the process $\boldsymbol{Y}$. The process $\boldsymbol{X}$ is Markovian, with state space $\mathsf{X} = \mathbb{R}^{k+\ell}$, and has the general form of an NSS(F) model, with

$$X_n = F(X_{n-1}, W_n), \qquad n \in \mathbb{Z}_+. \tag{2.9}$$

2.2.3   The gumleaf attractor

The gumleaf attractor is an example of a nonlinear model such as those which frequently occur in the analysis of control algorithms for nonlinear systems, some of which are briefly described below in Section 2.3. In an investigation of the pathologies which can reveal themselves in adaptive control, a specific control methodology which is described in Section 2.3.2, Mareels and Bitmead [247] found that the closed loop system dynamics in an adaptive control application can be described by the simple recursion

$$v_n = \frac{1}{v_{n-1}} - \frac{1}{v_{n-2}}, \qquad n \in \mathbb{Z}_+. \tag{2.10}$$

Here v_n is a "closed loop system gain" which is a simple function of the output of the system which is to be controlled. Figure 2.3 (a) shows a plot of $\boldsymbol{v}$ over 40,000 time steps. The sample path behavior is similar to that observed for the simple bilinear model in Figure 2.2. It is extremely bursty, but appears to be stationary.

We can obtain an NSS(F) model with $x_n = \binom{v_n}{v_{n-1}}$ and $F\binom{x^a}{x^b} = \binom{1/x^a - 1/x^b}{x^a}$. However, in view of the extremely large values observed in simulations, we perform a one-to-one transformation as follows. Define for $z \in \mathbb{R}^2$,

$$[z] = (1 + \|z\|)^{-1} \binom{z_1}{z_2},$$

so that the components of $[z]$ lie within the open unit disk in $\mathbb{R}^2$ for any $z \in \mathbb{R}^2$. Following this transformation we obtain the nonlinear state space model

$$x_n = \binom{x_n^a}{x_n^b} = F\binom{x_{n-1}^a}{x_{n-1}^b} = \left[\binom{1/x_{n-1}^a - 1/x_{n-1}^b}{x_{n-1}^a} \right]. \tag{2.11}$$

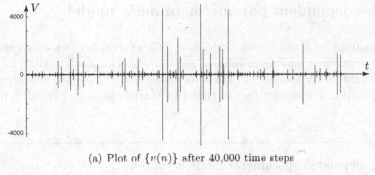

(a) Plot of $\{v(n)\}$ after 40,000 time steps

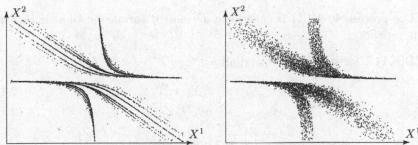

(b) Shown on the left is the gumleaf attractor, and on the right is the gumleaf attractor perturbed by noise.

Figure 2.3: The gumleaf attractor

A typical sample path of this model is given on the left hand side of Figure 2.3 (b). In this figure 40,000 consecutive sample points of $\{x_n\}$ have been indicated by points to illustrate the qualitative behavior of the model. Because of its similarity to some Australian flora, the authors call the resulting plot the *gumleaf attractor*. Ydstie in [410] also finds that such chaotic behavior can easily occur in adaptive systems.

One way that noise can enter the model (2.11) is to perturb (2.10) by noise. The resulting two-dimensional recursion becomes

$$X_n = \begin{pmatrix} X_n^a \\ X_n^b \end{pmatrix} = \left[\begin{pmatrix} 1/X_{n-1}^a - 1/X_{n-1}^b \\ X_{n-1}^a \end{pmatrix} + \begin{pmatrix} W_n \\ 0 \end{pmatrix} \right], \qquad (2.12)$$

where W is i.i.d.. The special case where for each n the disturbance W_n is uniformly distributed on $[-\frac{1}{2}, \frac{1}{2}]$ is illustrated on the right in Figure 2.3 (b). As in the previous figure, we have plotted 40,000 values of the sequence X which takes values in $\mathbb{R}^2$. Note that the qualitative behavior of the process remains similar to the noise-free model, although some of the detailed behavior is "smeared out" by the noise.

The analysis of general models of this type is a regular feature in what follows, and in Chapter 7 we give a detailed analysis of the path structure that might be expected under suitable assumptions on the noise and the associated deterministic model.

2.2.4   The dependent parameter bilinear model

As a simple example of a multidimensional nonlinear state space model, we will consider the following dependent parameter bilinear model, which is closely related to the simple bilinear model introduced above. To allow for dependence in the parameter process, we construct a two-dimensional process so that the Markov assumption will remain valid.

The dependent parameter bilinear model

The process $\boldsymbol{\Phi} = \binom{\boldsymbol{\theta}}{\boldsymbol{Y}}$ is called the *dependent parameter bilinear model* if it satisfies

(DBL1)   for some $|\alpha| < 1$ and all $k \in \mathbb{Z}_+$,

$$
\begin{aligned}
Y_{k+1} &= \theta_k Y_k + W_{k+1}, & (2.13) \\
\theta_{k+1} &= \alpha \theta_k + Z_{k+1}; & (2.14)
\end{aligned}
$$

(DBL2)   the joint process $(\boldsymbol{Z}, \boldsymbol{W})^\top$ is a disturbance sequence on $\mathbb{R}^2$, $\boldsymbol{Z}$ and $\boldsymbol{W}$ are mutually independent, and the distributions Γ_w and Γ_z of W, Z respectively possess densities which are lower semicontinuous – recall that a function h from X to $\mathbb{R}$ is lower semicontinuous if

$$
\liminf_{y \to x} h(y) \geq h(x), \qquad x \in \mathsf{X}.
$$

It is assumed that W has a finite second moment, and that $\mathsf{E}[\log(1+|Z|)] < \infty$.

This is described by a two-dimensional $\mathrm{NSS}(F)$ model, where the function F is of the form

$$
F\left(\binom{Y}{\theta}, \binom{Z}{W} \right) = \begin{pmatrix} \alpha\theta + Z \\ \theta Y + W \end{pmatrix}. \tag{2.15}
$$

As usual, the control set $O_w \subseteq \mathbb{R}^2$ depends upon the specific distribution of W and Z.

A plot of the joint process $\binom{Y}{\theta}$ is given in Figure 2.4. In this simulation we have $\alpha = 0.933$, $W_k \sim N(0, 0.14)$ and $Z_k \sim N(0, 0.01)$.

The dark line is a plot of the parameter process $\boldsymbol{\theta}$, and the lighter, more explosive path is the resulting output $\boldsymbol{Y}$. One feature of this model is that the output oscillates rapidly when θ_k takes on large negative values, which occurs in this simulation for time values between 80 and 100.

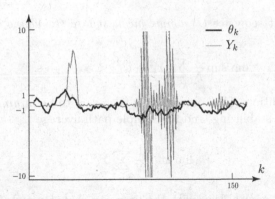

Figure 2.4: Dependent parameter bilinear model paths with $\alpha = 0.933$, $W_k \sim N(0, 0.14)$ and $Z_k \sim N(0, 0.01)$

2.3 Models in control and systems theory

2.3.1 Choosing controls

In Section 2.2, we defined deterministic control systems, such as (2.5), associated with Markovian state space models. We now begin with a general control system, which might model the dynamics of an aircraft, a cruise control in an automobile, or a controlled chemical reaction, and seek ways to choose a control to make the system attain a desired level of performance.

Such control laws typically involve feedback; that is, the input at a given time is chosen based upon present output measurements, or other features of the system which are available at the time that the control is computed. Once such a control law has been selected, the dynamics of the controlled system can be complex. Fortunately, with most control laws, there is a representation (the "closed loop" system equations) which gives rise to a Markovian state process $\boldsymbol{\Phi}$ describing the variables of interest in the system. This additional structure can greatly simplify the analysis of control systems.

We can extend the AR models of time series to an ARX (*autoregressive with exogenous variables*) system model defined for $k \geq 1$ by

$$Y_k + \alpha_1(k)Y_{k-1} + \cdots + \alpha_{n_1}(k)Y_{k-n_1} = \beta_1(k)U_{k-1} + \cdots + \beta_{n_2}(k)U_{k-n_2} + W_k \quad (2.16)$$

where we assume for this discussion that the output process $\boldsymbol{Y}$, the input process (or exogenous variable sequence) $\boldsymbol{U}$, and the disturbance process $\boldsymbol{W}$ are all scalar valued, and initial conditions are assigned at $k = 0$.

Let us also assume that we have random coefficients $\alpha_j(k), \beta_j(k)$ rather than fixed coefficients at each time point k. In such a case we may have to estimate the coefficients in order to choose the exogenous input $\boldsymbol{U}$.

The objective in the design of the control sequence $\boldsymbol{U}$ is specific to the particular application. However, it is often possible to set up the problem so that the goal becomes a problem of regulation: that is, to make the output as small as possible. Given the stochastic nature of systems, this is typically expressed using the concepts of sample mean square stabilizing sequences and minimum variance control laws.

We call the input sequence U *sample mean square stabilizing* if the input-output process satisfies

$$\limsup_{N\to\infty} \frac{1}{N} \sum_{k=1}^{N} [Y_k^2 + U_k^2] < \infty \qquad \text{a.s.}$$

for every initial condition. The control law is then said to be *minimum variance* if it is sample mean square stabilizing, and the sample path average

$$\limsup_{N\to\infty} \frac{1}{N} \sum_{k=1}^{N} Y_k^2 \qquad (2.17)$$

is minimized over all control laws with the property that, for each k, the input U_k is a function of $Y_k, \ldots, Y_0$, and the initial conditions.

Such controls are often called "causal", and for causal controls there is some possibility of a Markovian representation. We now specialize this general framework to a situation where a Markovian analysis through state space representation is possible.

2.3.2 Adaptive control

In *adaptive control*, the parameters $\{\alpha_i(k), \beta_i(k)\}$ are not known *a priori*, but are partially observed through the input-output process. Typically, a parameter estimation algorithm, such as recursive least squares, is used to estimate the parameters on-line in implementations. The control law at time k is computed based upon these estimates and past output measurements.

As an example, consider the system model given in equation (2.16) with all of the parameters taken to be independent of k, and let

$$\theta = (-\alpha_1, \ldots, -\alpha_{n_1}, \beta_1, \ldots, \beta_{n_2})$$

denote the time invariant parameter vector. Suppose for the moment that the parameter θ is known. If we set

$$\phi_{k-1}^{\top} := (Y_{k-1}, \ldots, Y_{k-n_1}, U_{k-1}, \ldots, U_{k-n_2}),$$

and if we define for each k the control U_k as the solution to

$$\phi_k^{\top} \theta = 0, \qquad (2.18)$$

then this will result in $Y_k = W_k$ for all k. This control law obviously minimizes the performance criterion (2.17) and hence is a minimum variance control law if it is sample mean square stabilizing.

It is also possible to obtain a minimum variance control law, even when θ is not available directly for the computation of the control U_k. One such algorithm (developed in [142]) has a recursive form given by first estimating the parameters through the following stochastic gradient algorithm:

$$\begin{aligned}
\hat{\theta}_k &= \hat{\theta}_{k-1} + r_{k-1}^{-1} \phi_{k-1} Y_k, \\
r_k &= r_{k-1} + \|\phi_k\|^2;
\end{aligned} \qquad (2.19)$$

the new control U_k is then defined as the solution to the equation

$$\phi_k^\top \hat{\theta}_k = 0.$$

With $X_k \in \mathsf{X} := \mathbb{R}_+ \times \mathbb{R}^{2(n_1+n_2)}$ defined as

$$X_k := \begin{pmatrix} r_k^{-1} \\ \phi_k \\ \hat{\theta}_k \end{pmatrix}$$

we see that $\boldsymbol{X}$ is of the form $X_{k+1} = F(X_k, W_{k+1})$, where $F \colon \mathsf{X} \times \mathbb{R} \to \mathsf{X}$ is a rational function, and hence $\boldsymbol{X}$ is a Markov chain.

To illustrate the results in stochastic adaptive control obtainable from the theory of Markov chains, we will consider here and in subsequent chapters the following ARX(1) random parameter, or state space, model.

Simple adaptive control model

The *simple adaptive control model* is a triple $\boldsymbol{Y}, \boldsymbol{U}, \boldsymbol{\theta}$ where

(SAC1)   the output sequence $\boldsymbol{Y}$ and parameter sequence $\boldsymbol{\theta}$ are defined inductively for any input sequence $\boldsymbol{U}$ by

$$
\begin{aligned}
Y_{k+1} &= \theta_k Y_k + U_k + W_{k+1}, & (2.20) \\
\theta_{k+1} &= \alpha \theta_k + Z_{k+1}, \qquad k \geq 1, & (2.21)
\end{aligned}
$$

where α is a scalar with $|\alpha| < 1$;

(SAC2)   the bivariate disturbance process $\begin{pmatrix} \boldsymbol{Z} \\ \boldsymbol{W} \end{pmatrix}$ is Gaussian and satisfies

$$
\begin{aligned}
\mathsf{E}[\begin{pmatrix} Z_n \\ W_n \end{pmatrix}] &= \begin{pmatrix} 0 \\ 0 \end{pmatrix}, \\
\mathsf{E}[\begin{pmatrix} Z_n \\ W_n \end{pmatrix}(Z_k, W_k)] &= \begin{pmatrix} \sigma_z^2 & 0 \\ 0 & \sigma_w^2 \end{pmatrix}\delta_{n-k}, \qquad n \geq 1;
\end{aligned}
$$

(SAC3)   the input process satisfies $U_k \in \mathcal{Y}_k$, $k \in \mathbb{Z}_+$, where $\mathcal{Y}_k = \sigma\{Y_0, \ldots, Y_k\}$. That is, the input U_k at time k is a function of past and present output values.

The time varying parameter process $\boldsymbol{\theta}$ here is not observed directly but is partially observed through the input and output processes $\boldsymbol{U}$ and $\boldsymbol{Y}$.

The ultimate goal with such a model is to find a mean square stabilizing, minimum variance control law. If the parameter sequence $\boldsymbol{\theta}$ were completely observed then this goal could be easily achieved by setting $U_k = -\theta_k Y_k$ for each $k \in \mathbb{Z}_+$, as in (2.18).

Since $\boldsymbol{\theta}$ is only partially observed, we instead obtain recursive estimates of the parameter process and choose a control law based upon these estimates. To do this we note that by viewing $\boldsymbol{\theta}$ as a state process, as defined in [57], then because of the assumptions made on $(\boldsymbol{W}, \boldsymbol{Z})$, the conditional expectation

$$\hat{\theta}_k := \mathsf{E}[\theta_k \mid \mathcal{Y}_k]$$

is computable using the Kalman filter (see [253, 240]) provided the initial distribution of (U_0, Y_0, θ_0) for (2.20), (2.21) is Gaussian.

In this scalar case, the Kalman filter estimates are obtained recursively by the pair of equations

$$\hat{\theta}_{k+1} = \alpha\hat{\theta}_k + \alpha\frac{\Sigma_k(Y_{k+1} - \hat{\theta}_k Y_k - U_k)Y_k}{\Sigma_k Y_k^2 + \sigma_w^2},$$

$$\Sigma_{k+1} = \sigma_z^2 + \frac{\alpha^2\sigma_w^2\Sigma_k}{\Sigma_k Y_k^2 + \sigma_w^2}.$$

When $\alpha = 1$, $\sigma_w = 1$ and $\sigma_z = 0$, so that $\theta_k = \theta_0$ for all k, these equations define the recursive least squares estimates of θ_0, similar to the gradient algorithm described in (2.19).

Defining the parameter estimation error at time n by $\tilde{\theta}_n := \theta_n - \hat{\theta}_n$, we have that $\tilde{\theta}_k = \theta_k - \mathsf{E}[\theta_k \mid \mathcal{Y}_k]$, and $\Sigma_k = \mathsf{E}[\tilde{\theta}_k^2 \mid \mathcal{Y}_k]$ whenever $\tilde{\theta}_0$ is distributed $N(0, \Sigma_0)$ and Y_0 and Σ_0 are constant (see [270] for more details).

We use the resulting parameter estimates $\{\hat{\theta}_k : k \geq 0\}$ to compute the "certainty equivalence" adaptive minimum variance control $U_k = -\hat{\theta}_k Y_k$, $k \in \mathbb{Z}_+$. With this choice of control law, we can define the closed loop system equations.

Closed loop system equations

The *closed loop system equations* are

$$\tilde{\theta}_{k+1} = \alpha\tilde{\theta}_k - \alpha\Sigma_k Y_{k+1}Y_k(\Sigma_k Y_k^2 + \sigma_w^2)^{-1} + Z_{k+1}, \qquad (2.22)$$

$$Y_{k+1} = \tilde{\theta}_k Y_k + W_{k+1}, \qquad (2.23)$$

$$\Sigma_{k+1} = \sigma_z^2 + \alpha^2\sigma_w^2\Sigma_k(\Sigma_k Y_k^2 + \sigma_w^2)^{-1}, \qquad k \geq 1, \qquad (2.24)$$

where the triple $\Sigma_0, \tilde{\theta}_0, Y_0$ is given as an initial condition.

The closed loop system gives rise to a nonlinear state space model of the form (NSS1). It follows then that the triple

$$\Phi_k := (\Sigma_k, \tilde{\theta}_k, Y_k)^\top, \qquad k \in \mathbb{Z}_+, \qquad (2.25)$$

is a Markov chain with state space $\mathsf{X} = [\sigma_z^2, \frac{\sigma_z^2}{1-\alpha^2}] \times \mathbb{R}^2$. Although the state space is not open, as required in (NSS1), when necessary we can restrict the chain to the interior of X to apply the general results which will be developed for the nonlinear state space model.

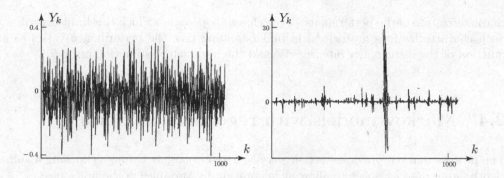

Figure 2.5: Output Y of the SAC model. The sample path shown on the left was obtained using $\sigma_z = 0.2$, and the one shown on the right used $\sigma_z = 1.1$. In each case $\alpha = 0.99$ and $\sigma_w = 0.1$

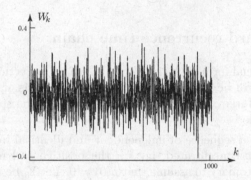

Figure 2.6: Disturbance W for the SAC model: $N(0, 0.01)$ Gaussian white noise

In Figure 2.5 we have illustrated two typical sample paths of the output process Y, identical but for the different values of σ_z chosen. The disturbance process W in both instances is i.i.d. $N(0, 0.01)$; that is, $\sigma_w = 0.1$. A typical sample path of W is given in Figure 2.6.

In both simulations we take $\alpha = 0.99$. In the "stable" case shown on the left we have $\sigma_z = 0.2$. In this case the output Y is barely distinguishable from the noise W. In the second simulation, where $\sigma_z = 1.1$, we see that the output exhibits occasional large bursts due to the more unpredictable behavior of the parameter process.

As we develop the general theory of Markov processes we will return to this example to obtain fairly detailed properties of the closed loop system described by (2.22)-(2.24).

In Chapter 16 we characterize the mean square performance (2.17): when the parameter σ_z^2 which defines the parameter variation is strictly less than unity, the limit supremum is in fact a limit in this example, and this limit is independent of the initial conditions of the system.

This limit, which is the expectation of Y_0 with respect to an invariant measure, cannot be calculated exactly due to the complexity of the closed loop system equations.

Using invariance, however, we may obtain explicit bounds on the limit, and give a characterization of the performance of the closed loop system which this limit describes. Such characterizations are helpful in understanding how the performance varies as a function of the disturbance intensity W and the parameter estimation error $\widetilde{\theta}$.

2.4 Markov models with regeneration times

The processes in the previous section were Markovian largely through choosing a sufficiently large product space to allow augmentation by variables in the finite past.

The chains we now consider are typically Markovian using the second paradigm in Section 1.2.1, namely by choosing specific *regeneration times* at which the past is forgotten. For more details of such models see Feller [114, 115] or Asmussen [9].

2.4.1 The forward recurrence time chain

A chain which is a special form of the random walk chain in Section 1.2.3 is the *renewal process*. Such chains will be fundamental in our later analysis of the structure of even the most general of Markov chains, and here we describe the specific case where the state space is countable.

Let $\{Y_1, Y_2, \ldots\}$ be a sequence of independent and identical random variables, with distribution function p concentrated, not on the positive and negative integers, but rather on $\mathbb{Z}_+$. It is customary to assume that $p(0) = 0$. Let Y_0 be a further independent random variable, with the distribution of Y_0 being a, also concentrated on $\mathbb{Z}_+$. The random variables

$$Z_n := \sum_{i=0}^{n} Y_i$$

form an increasing sequence taking values in $\mathbb{Z}_+$, and are called a *delayed renewal process*, with a being the delay in the first variable: if $a = p$ then the sequence $\{Z_n\}$ is merely referred to as a renewal process.

As with the two-sided random walk, Z_n is a Markov chain: not a particularly interesting one in some respects, since it is evanescent in the sense of Section 1.3.1 (II), but with associated structure which we will use frequently, especially in Part III.

With this notation we have $\mathsf{P}(Z_0 = n) = a(n)$ and by considering the value of Z_0 and the independence of Y_0 and Y_1, we find

$$\mathsf{P}(Z_1 = n) = \sum_{j=0}^{n} a(j)p(n - j).$$

To describe the n-step dynamics of the process $\{Z_n\}$ we need convolution notation.

Convolutions

We write $a * b$ for the *convolution* of two sequences a and b given by

$$a * b\,(n) := \sum_{j=0}^{n} b(j)a(n-j) = \sum_{j=0}^{n} a(j)b(n-j)$$

and a^{k*} for the kth convolution of a with itself.

By decomposing successively over the values of the first n variables $Z_0, \ldots, Z_{n-1}$ and using the independence of the increments Y_i we have that

$$\mathsf{P}(Z_k = n) = a * p^{k*}\,(n).$$

Two chains with appropriate regeneration associated with the renewal process are the *forward recurrence time chain*, sometimes called the residual lifetime process, and the *backward recurrence time chain*, sometimes called the age process.

Forward and backward recurrence time chains

If $\{Z_n\}$ is a discrete time renewal process, then the *forward recurrence time chain* $\boldsymbol{V}^+ = V^+(n), n \in \mathbb{Z}_+$, is given by

(RT1) $V^+(n) := \inf(Z_m - n : Z_m > n), \qquad n \geq 0,$

and the *backward recurrence time chain* $\boldsymbol{V}^- = V^-(n), n \in \mathbb{Z}_+$, is given by

(RT2) $V^-(n) := \inf(n - Z_m : Z_m \leq n), \qquad n \geq 0.$

The dynamics of motion for $\boldsymbol{V}^+$ and $\boldsymbol{V}^-$ are particularly simple.

If $V^+(n) = k$ for $k > 1$ then, in a purely deterministic fashion, one time unit later the forward recurrence time to the next renewal has come down to $k - 1$. If $V^+(n) = 1$ then a renewal occurs at $n + 1$: therefore the time to the next renewal has the distribution p of an arbitrary Y_j, and this is the distribution also of $V^+(n+1)$. For the backward chain, the motion is reversed: the chain increases by one, or ages, with the conditional probability of a renewal failing to take place, and drops to zero with the conditional probability that a renewal occurs. We define the laws of these chains formally in Section 3.3.1.

The regeneration property at each renewal epoch ensures that both $\boldsymbol{V}^+$ and $\boldsymbol{V}^-$ are Markov chains; and, unlike the renewal process itself, these chains are stable under straightforward conditions, as we shall see.

Renewal theory is traditionally of great importance in countable space Markov chain theory: the same is true in general spaces, as will become especially apparent in Part

III. We only use those aspects which we require in what follows, but for a much fuller treatment of renewal and regeneration see Kingman [208] or Lindvall [239].

2.4.2   The GI/G/1, GI/M/1 and M/G/1 queues

The theory of queueing systems provides an explicit and widely used example of the random walk models introduced in Section 1.2.3, and we will develop the application of Markov chain and process theory to such models, and related storage and dam models, as another of the central examples of this book.

These models indicate for the first time the need, in many physical processes, to take care in choosing the timepoints at which the process is analyzed: at some "regeneration" time-points, the process may be "Markovian", whilst at others there may be a memory of the past influencing the future.

In the modeling of queues, to use a Markov chain approach we can make certain distributional assumptions (and specifically assumptions that some variables are exponential) to generate regeneration times at which the Markovian forgetfulness property holds. We develop such models in some detail, as they are fundamental examples of the use of regeneration in utilizing the Markovian assumption.

Let us first consider a general queueing model to illustrate why such assumptions may be needed.

Queueing model assumptions

Suppose the following assumptions hold.

(Q1)   Customers arrive into a service operation at timepoints $T_0 = 0$, $T_0 + T_1$, $T_0 + T_1 + T_2, \ldots$ where the interarrival times T_i, $i \geq 1$, are independent and identically distributed random variables, distributed as a random variable T with $G(-\infty, t] = \mathsf{P}(T \leq t)$.

(Q2)   The nth customer brings a job requiring service S_n where the service times are independent of each other and of the interarrival times, and are distributed as a variable S with distribution $H(-\infty, t] = \mathsf{P}(S \leq t)$.

(Q3)   There is one server and customers are served in order of arrival.

Then the system is called a *GI/G/1 queue*.

The notation and many of the techniques here were introduced by Kendall [200, 201]: GI for general independent input, G for general service time distributions, and 1 for a single server system. There are many ways of analyzing this system: see Asmussen [9] or Cohen [76] for comprehensive treatments.

Let $N(t)$ be the number of customers in the queue at time t, including the customers being served. This is clearly a process in continuous time. A typical sample path for $\{N(t), t \geq 0\}$, under the assumption that the first customer arrives at $t = 0$, is shown

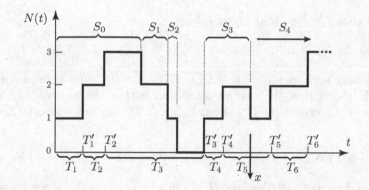

Figure 2.7: Typical sample path of the single server queue

in Figure 2.7, where we denote by T_i', the arrival times

$$T_i' = T_1 + \cdots + T_i, \quad i \geq 1, \tag{2.26}$$

and by S_i' the sums of service times

$$S_i' = S_0 + \cdots + S_i, \quad i \geq 0. \tag{2.27}$$

Note that, in the sample path illustrated, because the queue empties at S_2', due to $T_3' > S_2'$, the point $x = T_3' + S_3$ is not S_3', and the point $T_4' + S_4$ is not S_4', and so on.

Although the process $\{N(t)\}$ occurs in continuous time, one key to its analysis through Markov chain theory is the use of *embedded Markov chains*.

Consider the random variable $N_n = N(T_n'-)$, which counts customers immediately before each arrival. By convention we will set $N_0 = 0$ unless otherwise indicated. We will show that under appropriate circumstances for $k \geq -j$

$$\mathsf{P}(N_{n+1} = j + k \mid N_n = j, N_{n-1}, N_{n-2}, \ldots, N_0) = p_k, \tag{2.28}$$

regardless of the values of $\{N_{n-1}, \ldots, N_0\}$. This will establish the Markovian nature of the process, and indeed will indicate that it is a random walk on $\mathbb{Z}_+$.

Since we consider $N(t)$ immediately before every arrival time, N_{n+1} can only increase from N_n by one unit at most; hence, equation (2.28) holds trivially for $k > 1$.

For N_{n+1} to increase by one unit we need there to be no departures in the time period $T_{n+1}' - T_n'$, and obviously this happens if the job in progress at T_n' is still in progress at T_{n+1}'.

It is here that some assumption on the service times will be crucial. For it is easy to show, as we now sketch, that for a general GI/G/1 queue the probability of the remaining service of the job in progress taking any specific length of time depends, typically, on when the job began. In general, the past history $\{N_{n-1}, \ldots, N_0\}$ will provide information on when the customer began service, and this in turn provides information on how long the customer will continue to be served.

To see this, consider, for example, a trajectory such as that up to $(T_1'-)$ on Figure 2.7, where $\{N_n = 1, N_{n-1} = 0, \ldots\}$. This tells us that the current job began exactly at the arrival time T'_{n-2}, so that (as at (T'_2-))

$$P(N_{n+1} = 2 \mid N_n = 1, N_{n-1} = 0) = P(S_{n-2} > T_{n+1} + T_n \mid S_{n-2} > T_n). \qquad (2.29)$$

However, a history such as $\{N_n = 1, N_{n-1} = 1, N_{n-2} = 0\}$, such as occurs up to (T'_5-) on Figure 2.7, shows that the current job began within the interval (T'_n, T'_{n-1}), and so for some $z < T_n$ (given by $T'_5 - x$ on Figure 2.7), the behavior at (T'_6-) has the probability

$$P(N_{n+1} = 2 \mid N_n = 1, N_{n-1} = 1, N_{n-2} = 0) = P(S_n > T_{n+1} + z \mid S_n > z).$$

It is clear that for most distributions H of the service times S_i, if we know $T_{n+1} = t$ and $T_n = t' > z$

$$P(S_n > t + z \mid S_n > z) \neq P(S_n > t + t' \mid S_n > t'); \qquad (2.30)$$

so $N = \{N_n\}$ is not a Markov chain, since from equation (2.29) and equation (2.30) the different information in the events $\{N_n = 1, N_{n-1} = 0\}$ and $\{N_n = 1, N_{n-1} = 1, N_{n-2} = 0\}$ (which only differ in the past rather than the present position) leads to different probabilities of transition.

There is one case where this does not happen. If both sides of (2.30) are identical so that the time until completion of service is quite independent of the time already taken, then the extra information from the past is of no value.

This leads us to define a specific class of models for which N is Markovian.

GI/M/1 assumption

(Q4)   If the distribution of service times is exponential with

$$H(-\infty, t] = 1 - e^{-\mu t}, \quad t \geq 0,$$

then the queue is called a GI/M/1 queue.

Here the M stands for Markovian, as opposed to the previous "general" assumption.

If we can now make assumption (Q4) that we have a GI/M/1 queue, then the well-known "loss of memory" property of the exponential shows that, for any t, z,

$$P(S_n > t + z \mid S_n > z) = e^{-\mu(t+z)}/e^{-\mu z} = e^{-\mu t}.$$

In this way, the independence and identical distribution structure of the service times show that, no matter which previous customer was being served, and when their service started, there will be some z such that

$$P(N_{n+1} = j + 1 \mid N_n = j, N_{n-1}, \ldots) = P(S > T + z \mid S > z)$$

$$= \int_0^\infty e^{-\mu t} G(dt)$$

independent of the value of z in any given realization, as claimed in equation (2.28).

This same reasoning can be used to show that, if we know $N_n = j$, then for $0 < i \leq j$, we will find $N_{n+1} = i$ provided $j - i + 1$ customers *leave* in the interarrival time (T'_n, T'_{n+1}). This corresponds to $(j - i + 1)$ jobs being completed in this period, and the $(j - i + 1)^{\text{th}}$ job continuing past the end of the period. The probability of this happening, using the forgetfulness of the exponential, is *independent* of the amount of time the service is in place at time T'_n has already consumed, and thus N is Markovian.

A similar construction holds for the chain $N^* = \{N_n^*\}$ defined by taking the number in the queue immediately after the nth service time is completed. This will be a Markov chain provided the number of arrivals in each service time is independent of the times of the arrivals prior to the beginning of that service time. As above, we have such a property if the inter-arrival time distribution is exponential, leading us to distinguish the class of M/G/1 queues, where again the M stands for a Markovian inter-arrival assumption.

M/G/1 assumption

(Q5)   If the distribution of inter-arrival times is exponential with

$$G(-\infty, t] = 1 - e^{-\lambda t}, \quad t \geq 0,$$

then the queue is called an M/G/1 queue.

The actual probabilities governing the motion of these queueing models will be developed in Chapter 3.

2.4.3   The Moran dam

The theory of storage systems provides another of the central examples of this book, and is closely related to the queueing models above.

The storage process example is one where, although the time of events happening (that is, inputs occurring) is random, between those times there is a deterministic motion which leads to a Markovian representation at the input times which always form regeneration points.

A simple model for storage (the "Moran dam" [288, 9]) has the following elements. We assume there is a sequence of *input times* $T_0 = 0$, $T_0 + T_1$, $T_0 + T_1 + T_2, \ldots$, at which there is input into a storage system, and that the *inter-arrival times* T_i, $i \geq 1$, are independent and identically distributed random variables, distributed as a random variable T with $G(-\infty, t] = \mathsf{P}(T \leq t)$.

At the nth input time, the amount of input S_n has a distribution $H(-\infty, t] = \mathsf{P}(S_n \leq t)$; the input amounts are independent of each other and of the interarrival times. Between inputs, there is steady withdrawal from the storage system, at a rate r: so that in a time period $[x, x + t]$, the stored contents drop by an amount rt since there is no input.

When a path of the contents process reaches zero, the process continues to take the value zero until it is replenished by a positive input.

This model is a simplified version of the way in which a dam works; it is also a model for an inventory, or for any other similar storage system.

The basic storage process operates in continuous time: to render it Markovian we analyze it at specific time points when it (probabilistically) regenerates, as follows.

Simple storage models

(SSM1)   For each $n \geq 0$ let S_n and T_n be independent random variables on $\mathbb{R}$ with distributions H and G as above.

(SSM2)   Define the random variables

$$\Phi_{n+1} = [\Phi_n + S_n - J_n]^+,$$

where the variables $\{J_n\}$ are independent and identically distributed, with

$$\mathsf{P}(J_n \leq x) = G(-\infty, x/r] \tag{2.31}$$

for some $r > 0$.

Then the chain $\mathbf{\Phi} = \{\Phi_n\}$ represents the contents of a storage system at the times $\{T_n-\}$ immediately before each input, and is called the *simple storage model*.

The independence of S_{n+1} from $S_{n-1}, S_{n-2}, \ldots$ and the construction rules (SSM1) and (SSM2) ensure as before that $\{\Phi_n\}$ is a Markov chain: in fact, it is a specific example of the random walk on a half line defined by (RWHL1), in the special case where

$$W_n = S_n - J_n, \quad n \in \mathbb{Z}_+.$$

It is an important observation here that, in general, the process sampled at other time points (say, at regular time points) is *not* a Markov system, since it is crucial in calculating the probabilities of the future trajectory to know how much earlier than the chosen time point the last input point occurred: by choosing to examine the chain embedded at precisely those pre-input times, we lose the memory of the past. This was discussed in more detail in Section 2.4.2.

We define the mean input by $\alpha = \int_0^\infty x\, H(dx)$ and the mean output between inputs by $\beta = \int_0^\infty rx\, G(dx)$. In Figure 2.8 we give two sample paths of storage models with different values of the parameter ratio α/β. The behavior of the sample paths is quite different for different values of this ratio, which will turn out to be the crucial quantity in assessing the stability of these models.

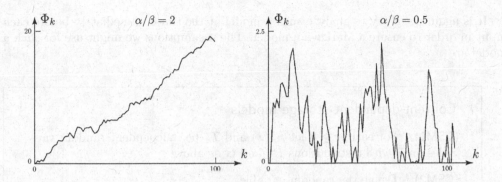

Figure 2.8: Storage system paths. The plot shown on the left uses $\alpha/\beta = 2$, and on the right $\alpha/\beta = 0.5$. In each case $r = 1$.

2.4.4 Content-dependent release rules

As with time series models or state space systems, the linearity in the Moran storage model is clearly a first approximation to a more sophisticated system.

There are two directions in which this can be taken without losing the Markovian nature of the model.

Again assume there is a sequence of input time points $T_0 = 0$, $T_0 + T_1$, $T_0 + T_1 + T_2, \ldots$, and that the inter-arrival times T_i, $i \geq 1$, are independent and identically distributed random variables, with distribution G.

Then one might assume that, if the contents at the nth input time are given by $\Phi_n = x$, the amount of input $S_n(x)$ has a distribution given by $H_x(-\infty, t] = \mathsf{P}(S_n(x) \leq t)$ dependent on x; the input amounts remain independent of each other and of the interarrival times.

Alternatively, one might assume that between inputs, there is withdrawal from the storage system, at a rate $r(x)$ which also depends on the level x at the moment of withdrawal. This assumption leads to the conclusion that, if there are no inputs, the deterministic time to reach the empty state from a level x is

$$R(x) = \int_0^x [r(y)]^{-1} dy. \tag{2.32}$$

Usually we assume $R(x)$ to be finite for all x. Since R is strictly increasing the inverse function $R^{-1}(t)$ is well defined for all t, and it follows that the drop in level in a time period t with no input is given by

$$J_x(t) = x - q(x, t)$$

where

$$q(x, t) = R^{-1}(R(x) - t).$$

This enables us to use the same type of random walk calculation as for the Moran dam.

As before, when a path of this storage process reaches zero, the process continues to take the value zero until it is replenished by a positive input.

It is again necessary to analyze such a model at the times immediately before each input in order to ensure a Markovian model. The assumptions we might use for such a model are

---

Content-dependent storage models

(CSM1) For each $n \geq 0$ let $S_n(x)$ and T_n be independent random variables on $\mathbb{R}$ with distributions H_x and G as above.

(CSM2) Define the random variables

$$\Phi_{n+1} = [\Phi_n - J_n + S_n(\Phi_n - J_n)]^+,$$

where the variables $\{J_n\}$ are independently distributed, with

$$\mathsf{P}(J_n \leq y \mid \Phi_n = x) = \int G(dt)\mathsf{P}(J_x(t) \leq y). \qquad (2.33)$$

Then the chain $\boldsymbol{\Phi} = \{\Phi_n\}$ represents the contents of the storage system at the times $\{T_n-\}$ immediately before each input, and is called the *content-dependent storage model*.

Such models are studied in [157, 53]. In considering the connections between queueing and storage models, it is then immediately useful to realize that this is also a model of the waiting times in a model where the service time varies with the level of demand, as studied in [56].

2.5   Commentary*

We have skimmed the Markovian models in the areas in which we are interested, trying to tread the thin line between accessibility and triviality. The research literature abounds with variations on the models we present here, and many of them would benefit by a more thorough approach along Markovian lines.

For many more models with time series applications, the reader should see Brockwell and Davis [51], especially Chapter 12; Granger and Anderson for bilinear models [143]; and for nonlinear models see Tong [388], who considers models similar to those we have introduced from a Markovian viewpoint, and in particular discusses the bilinear and SETAR models. Linear and bilinear models are also developed by Duflo in [102], with a view towards stability similar to ours. For a development of general linear systems theory the reader is referred to Caines [57] for a control perspective, or Aoki [5] for a view towards time series analysis.

Bilinear models have received a great deal of attention in recent years in both time series and systems theory. The dependent parameter bilinear model defined by (2.14, 2.13) is called a doubly stochastic autoregressive process of order 1, or DSAR(1), in

Tjøstheim [386]. Realization theory for related models is developed in Guégan [146] and Mittnik [285], and the papers Pourahmadi [321], Brandt [44], Meyn and Guo [275], and Karlsen [195] provide various stability conditions for bilinear models.

The idea of analyzing the nonlinear state space model by examining an associated control model goes back to Stroock and Varadhan [378] and Kunita [227, 228] in continuous time. In control and systems models, linear state space models have always played a central role, while nonlinear models have taken a much more significant role over the past decade: see Kumar and Varaiya [225], Duflo [102], and Caines [57] for a development of both linear adaptive control models, and (nonlinear) controlled Markov chains.

The embedded regeneration time approach has been enormously significant since its introduction by Kendall in [200, 201]. There are many more sophisticated variations than those we shall analyze available in the literature. A good recent reference is Asmussen [9], whilst Cohen [76] is encyclopedic.

The interested reader will find that, although we restrict ourselves to these relatively less complicated models in illustrating the value of Markov chain modeling, virtually all of our general techniques apply across more complex systems. As one example, note that the stability of models which are state dependent, such as the content-dependent storage model of Section 2.4.4, has only recently received attention [56], but using the methods developed in later chapters it is possible to characterize it in considerable detail [277, 279, 280].

The storage models described here can also be thought of, virtually by renaming the terms, as models for state-dependent inventories, insurance models, and models of the residual service in a GI/G/1 queue. To see the last of these, consider the amount of service brought by each customer as the input to the "store" of work to be processed, and note that the server works through this store of work at a constant rate.

The residual service can be, however, a somewhat minor quantity in a queueing model, and in Section 3.5.4 below we develop a more complex model which is a better representation of the dynamics of the GI/G/1 queue.

Added in second printing: In the last two years there has been a virtual explosion in the use of general state space Markov chains in simulation methods, and especially in Markov chain Monte Carlo methods which include Metropolis–Hastings and Gibbs sampling techniques, which were touched on in Chapter 1.1(f). Any future edition will need to add these to the collection of models here and examine them in more detail: the interested reader might look at the recent results [63, 290, 360, 333, 328, 256, 335], which all provide examples of the type of chains studied in this book.

Commentary for the second edition: More recent examples of analysis of Metropolis–Hastings and Gibbs sampling techniques based on methods in this book can be found in [330, 331, 79, 184, 125, 181]. The interested reader can find in Section 20.2 a summary of simulation techniques based on the theory contained in this book.

Chapter 3

Transition probabilities

As with all stochastic processes, there are two directions from which to approach the formal definition of a Markov chain.

The first is via the process itself, by constructing (perhaps by heuristic arguments at first, as in the descriptions in Chapter 2) the sample path behavior and the dynamics of movement in time through the state space on which the chain lives. In some of our examples, such as models for queueing processes or models for controlled stochastic systems, this is the approach taken. From this structural definition of a Markov chain, we can then proceed to define the probability laws governing the evolution of the chain.

The second approach is via those very probability laws. We define them to have the structure appropriate to a Markov chain, and then we must show that there is indeed a process, properly defined, which is described by the probability laws initially constructed. In effect, this is what we have done with the forward recurrence time chain in Section 2.4.1.

From a practitioner's viewpoint there may be little difference between the approaches. In many books on stochastic processes, such as Çinlar [59] or Karlin and Taylor [194], the two approaches are used, as they usually can be, almost interchangeably; and advanced monographs such as Nummelin [303] also often assume some of the foundational aspects touched on here to be well understood.

Since one of our goals in this book is to provide a guide to modern general space Markov chain theory and methods for practitioners, we give in this chapter only a sketch of the full mathematical construction which provides the underpinning of Markov chain theory.

However, we also have as another, and perhaps somewhat contradictory, goal the provision of a thorough and rigorous exposition of results on general spaces, and for these it is necessary to develop both notation and concepts with some care, even if some of the more technical results are omitted.

Our approach has therefore been to develop the technical detail in so far as it is relevant to specific Markov models, and where necessary, especially in techniques which are rather more measure theoretic or general stochastic process theoretic in nature, to refer the reader to the classic texts of Doob [99], and Chung [71], or the more recent exposition of Markov chain theory by Revuz [326] for the foundations we need. Whilst such an approach renders this chapter slightly less than self-contained, it is our hope that the gaps in these foundations will be either accepted or easily filled by such external sources.

Our main goals in this chapter are thus

(i) to demonstrate that the dynamics of a Markov chain $\{\Phi_n\}$ can be completely defined by its one step "transition probabilities"

$$P(x, A) = \mathsf{P}(\Phi_n \in A \mid \Phi_{n-1} = x),$$

which are well defined for appropriate initial points x and sets A;

(ii) to develop the functional forms of these transition probabilities for many of the specific models in Chapter 2, based in some cases on heuristic analysis of the chain and in other cases on development of the probability laws; and

(iii) to develop some formal concepts of hitting times on sets, and the "strong Markov property" for these and related stopping times, which will enable us to address issues of stability and structure in subsequent chapters.

We shall start first with the formal concept of a Markov chain as a stochastic process, and move then to the development of the transition laws governing the motion of the chain; and complete the cycle by showing that if one starts from a set of possible transition laws then it is possible to move from these to a chain which is well defined and governed by these laws.

3.1   Defining a Markovian process

A *Markov chain* $\Phi = \{\Phi_0, \Phi_1, \ldots\}$ is a particular type of *stochastic process* taking, at times $n \in \mathbb{Z}_+$, values Φ_n in a *state space* X.

We need to know and use a little of the language of stochastic processes. A discrete time stochastic process Φ on a state space is, for our purposes, a collection $\Phi = (\Phi_0, \Phi_1, \ldots)$ of random variables, with each Φ_i taking values in X; these random variables are assumed measurable individually with respect to some given σ-field $\mathcal{B}(\mathsf{X})$, and we shall in general denote elements of X by letters $x, y, z, \ldots$ and elements of $\mathcal{B}(\mathsf{X})$ by A, B, C.

When thinking of the process as an entity, we regard values of the whole chain Φ itself (called *sample paths* or *realizations*) as lying in the *sequence* or *path* space formed by a countable product $\Omega = \mathsf{X}^\infty = \prod_{i=0}^\infty \mathsf{X}_i$, where each X_i is a copy of X equipped with a copy of $\mathcal{B}(\mathsf{X})$. For Φ to be defined as a random variable in its own right, Ω will be equipped with a σ-field $\mathcal{F}$, and for each state $x \in \mathsf{X}$, thought of as an initial condition in the sample path, there will be a probability measure P_x such that the probability of the event $\{\Phi \in A\}$ is well defined for any set $A \in \mathcal{F}$; the initial condition requires, of course, that $\mathsf{P}_x(\Phi_0 = x) = 1$.

The triple $\{\Omega, \mathcal{F}, \mathsf{P}_x\}$ thus defines a stochastic process since $\Omega = \{\omega_0, \omega_1, \ldots : \omega_i \in \mathsf{X}\}$ has the product structure to enable the projections ω_n at time n to be well defined realizations of the random variables Φ_n.

Many of the models we consider (such as random walk or state space models) have stochastic motion based on a separately defined sequence of underlying variables, namely a noise or disturbance or innovation sequence W. We will slightly abuse notation by using $\mathsf{P}(W \in A)$ to denote the probability of the event $\{W \in A\}$ without specifically defining the space on which W exists, or the initial condition of the chain: this could be part of the space on which the chain Φ is defined or it could be separate. No confusion should result from this usage.

Prior to discussing specific details of the probability laws governing the motion of a chain Φ, we need first to be a little more explicit about the structure of the state space X on which it takes its values. We consider, systematically, three types of state spaces in this book:

State space definitions

(i) The state space X is called *countable* if X is discrete, with a finite or countable number of elements, and with $\mathcal{B}(\mathsf{X})$ the σ-field of all subsets of X.

(ii) The state space X is called *general* if it is equipped with a countably generated σ-field $\mathcal{B}(\mathsf{X})$.

(iii) The state space X is called *topological* if it is equipped with a locally compact, separable, metrizable topology with $\mathcal{B}(\mathsf{X})$ as the Borel σ-field.

It may on the face of it seem odd to introduce quite general spaces before rather than after topological (or more structured) spaces.

This is however quite deliberate, since (perhaps surprisingly) we rarely find the extra structure actually increasing the ease of approach. From our point of view, we introduce topological spaces largely because specific applied models evolve on such spaces, and for such spaces we will give specific interpretations of our general results, rather than extending specific topological results to more general contexts.

For example, after framing general properties of sets, we identify these general properties as holding for compact or open sets if the chain is on a topological space; or after framing general properties of Φ, we develop the consequences of these when Φ is suitably continuous with respect to the topology considered.

The first formal introduction of such topological concepts is given in Chapter 6, and is exemplified by an analysis of linear and nonlinear state space models in Chapter 7. Prior to this we concentrate on countable and general spaces: for purposes of exposition, our approach will often involve the description of behavior on a countable space, followed by the development of analogous behavior on a general space, and completed by specialization of results, where suitable, to more structured topological spaces in due course.

For some readers, countable space models will be familiar: nonetheless, by developing the results first in this context, and then the analogues for the less familiar general space processes on a systematic basis we intend to make the general context more accessible. By then specializing where appropriate to topological spaces, we trust the results will be found more applicable for, say, those models which evolve on multidimensional Euclidean space $\mathbb{R}^k$, or one of its subsets.

There is one caveat to be made in giving this description. One of the major observations for Markov chains is that in many cases, the full force of a countable space is not needed: we merely require one "accessible atom" in the space, such as we might have with the state $\{0\}$ in the storage models in Section 2.4.1. To avoid repetition we will often assume, especially later in the book, not the full countable space structure but just the existence of one such point: the results then carry over with only notational changes to the countable case.

In formalizing the concept of a Markov chain we pursue this pattern now, first developing the countable space foundations and then moving on to the slightly more complex basis for general space chains.

3.2 Foundations on a countable space

3.2.1 The initial distribution and the transition matrix

A discrete time Markov chain $\mathbf{\Phi}$ on a countable state space is a collection $\mathbf{\Phi} = \{\Phi_0, \Phi_1, \ldots\}$ of random variables, with each Φ_i taking values in the countable set X. In this countable state space setting, $\mathcal{B}(\mathsf{X})$ will denote the set of all subsets of X.

We assume that for any *initial distribution* μ for the chain, there exists a probability measure which denotes the law of $\mathbf{\Phi}$ on $(\Omega, \mathcal{F})$, where $\mathcal{F}$ is the product σ-field on the sample space $\Omega := \mathsf{X}^\infty$. However, since we have to work with several initial conditions simultaneously, we need to build up a probability space for each initial distribution.

For a given initial probability distribution μ on $\mathcal{B}(\mathsf{X})$, we construct the probability distribution P_μ on $\mathcal{F}$ so that $\mathsf{P}_\mu(\Phi_0 = x_0) = \mu(x_0)$ and for any $A \in \mathcal{F}$,

$$\mathsf{P}_\mu(\mathbf{\Phi} \in A \mid \Phi_0 = x_0) = \mathsf{P}_{x_0}(\mathbf{\Phi} \in A) \tag{3.1}$$

where P_{x_0} is the probability distribution on $\mathcal{F}$ which is obtained when the initial distribution is the point mass δ_{x_0} at x_0.

The defining characteristic of a *Markov chain* is that its future trajectories depend on its present and its past only through the current value.

To commence to formalize this, we first consider only the laws governing a trajectory of fixed length $n \geq 1$. The random variables $\{\Phi_0 \ldots \Phi_n\}$, thought of as a sequence, take values in the space $\mathsf{X}^{n+1} = \mathsf{X}_0 \times \cdots \times \mathsf{X}_n$, the $(n+1)$-fold product of copies X_i of the countable space X, equipped with the product σ-field $\mathcal{B}(\mathsf{X}^{n+1})$ which consists again of all subsets of X^{n+1}.

The conditional probability

$$\mathsf{P}_{x_0}^n(\Phi_1 = x_1, \ldots, \Phi_n = x_n) := \mathsf{P}_{x_0}(\Phi_1 = x_1, \ldots, \Phi_n = x_n), \tag{3.2}$$

defined for any sequence $\{x_0, \ldots, x_n\} \in \mathsf{X}^{n+1}$ and $x_0 \in \mathsf{X}$, and the initial probability distribution μ on $\mathcal{B}(\mathsf{X})$ completely determine the distributions of $\{\Phi_0, \ldots, \Phi_n\}$.

Countable space Markov chain

The process $\Phi = (\Phi_0, \Phi_1, \ldots)$, taking values in the path space $(\Omega, \mathcal{F}, \mathsf{P})$, is a *Markov chain* if for every n, and any sequence of states $\{x_0, x_1, \ldots, x_n\}$,

$$\mathsf{P}_\mu(\Phi_0 = x_0, \Phi_1 = x_1, \Phi_2 = x_2, \ldots, \Phi_n = x_n)$$

$$= \mu(x_0)\mathsf{P}_{x_0}(\Phi_1 = x_1)\mathsf{P}_{x_1}(\Phi_1 = x_2)\cdots\mathsf{P}_{x_{n-1}}(\Phi_1 = x_n). \tag{3.3}$$

The probability μ is called the *initial distribution* of the chain.

The process Φ is a *time-homogeneous* Markov chain if the probabilities $\mathsf{P}_{x_j}(\Phi_1 = x_{j+1})$ depend only on the values of x_j, x_{j+1} and are independent of the timepoints j.

By extending this in the obvious way from events in X^n to events in X^∞ we have that the initial distribution, followed by the probabilities of transitions from one step to the next, completely define the probabilistic motion of the chain.

If Φ is a time-homogeneous Markov chain, we write

$$P(x, y) := \mathsf{P}_x(\Phi_1 = y);$$

then the definition (3.3) can be written

$$\mathsf{P}_\mu(\Phi_0 = x_0, \Phi_1 = x_1, \ldots, \Phi_n = x_n)$$

$$= \mu(x_0)P(x_0, x_1)P(x_1, x_2)\cdots P(x_{n-1}, x_n), \tag{3.4}$$

or equivalently, in terms of the conditional probabilities of the process Φ,

$$\mathsf{P}_\mu(\Phi_{n+1} = x_{n+1} \mid \Phi_n = x_n, \ldots, \Phi_0 = x_0) = P(x_n, x_{n+1}). \tag{3.5}$$

Equation (3.5) incorporates both the "loss of memory" of Markov chains and the "time homogeneity" embodied in our definitions. It is possible to mimic this definition, asking that the $\mathsf{P}_{x_j}(\Phi_1 = x_{j+1})$ depend on the time j at which the transition takes place; but the theory for such inhomogeneous chains is neither so ripe nor so clean as for the chains we study, and we restrict ourselves solely to the time-homogeneous case in this book.

For a given model we will almost always define the probability P_{x_0} for a fixed x_0 by defining the one-step transition probabilities for the process, and building the overall distribution using (3.4).

This is done using a *Markov transition matrix*.

Transition probability matrix

The matrix $P = \{P(x,y), x, y \in \mathsf{X}\}$ is called a *Markov transition matrix* if

$$P(x,y) \geq 0, \quad \sum_{z \in \mathsf{X}} P(x,z) = 1, \qquad x, y \in \mathsf{X}. \qquad (3.6)$$

We define the usual matrix iterates $P^n = \{P^n(x,y), x, y \in \mathsf{X}\}$ by setting $P^0 = I$, the identity matrix, and then taking inductively

$$P^n(x,z) = \sum_{y \in \mathsf{X}} P(x,y) P^{n-1}(y,z). \qquad (3.7)$$

In the next section we show how to take an initial distribution μ and a transition matrix P and construct a distribution P_μ so that the conditional distributions of the process may be computed as in (3.1), and so that for any x, y,

$$\mathsf{P}_\mu(\Phi_n = y \mid \Phi_0 = x) = P^n(x,y) \qquad (3.8)$$

For this reason, P^n is called the *n-step transition matrix*. For $A \subseteq \mathsf{X}$, we also put

$$P^n(x,A) := \sum_{y \in A} P^n(x,y).$$

3.2.2   Developing Φ from the transition matrix

To define a Markov chain from a transition function we first consider only the laws governing a trajectory of fixed length $n \geq 1$. The random variables $\{\Phi_0, \ldots, \Phi_n\}$, thought of as a sequence, take values in the space $\mathsf{X}^{n+1} = \mathsf{X}_0 \times \cdots \times \mathsf{X}_n$, equipped with the σ-field $\mathcal{B}(\mathsf{X}^{n+1})$ which consists of all subsets of X^{n+1}.

Define the distributions P_x of Φ inductively by setting, for each fixed $x \in \mathsf{X}$

$$\begin{aligned}
\mathsf{P}_x(\Phi_0 = x) &= 1, \\
\mathsf{P}_x(\Phi_1 = y) &= P(x,y), \\
\mathsf{P}_x(\Phi_2 = z, \Phi_1 = y) &= P(x,y)P(y,z),
\end{aligned}$$

and so on. It is then straightforward, but a little lengthy, to check that for each fixed x, this gives a consistent set of definitions of probabilities P_x^n on $(\mathsf{X}^n, \mathcal{B}(\mathsf{X}^n))$, and these distributions can be built up to an overall probability measure P_x for each x on the set $\Omega = \prod_{i=0}^\infty \mathsf{X}_i$ with σ-field $\mathcal{F} = \bigvee_{i=0}^\infty \mathcal{B}(\mathsf{X}_i)$, defined in the usual way. Once we prescribe an initial measure μ governing the random variable Φ_0, we can define the overall measure by

$$\mathsf{P}_\mu(\Phi \in A) := \sum_{x \in \mathsf{X}} \mu(x) \mathsf{P}_x(\Phi \in A)$$

to govern the overall evolution of Φ. The formula (3.1) and the interpretation of the transition function given in (3.8) follow immediately from this construction.

A careful construction is in Chung [71], Chapter I.2. This leads to

Theorem 3.2.1. *If* X *is countable, and*

$$\mu \doteq \{\mu(x), x \in \mathsf{X}\}, \qquad\qquad P = \{P(x, y), x, y \in \mathsf{X}\}$$

are an initial measure on X *and a Markov transition matrix satisfying (3.6) then there exists a Markov chain* $\mathbf{\Phi}$ *on* $(\Omega, \mathcal{F})$ *with probability law* P_μ *satisfying*

$$\mathsf{P}_\mu(\Phi_{n+1} = y \mid \Phi_n = x, \ldots, \Phi_0 = x_0) = P(x, y).$$

□

3.3 Specific transition matrices

In practice models are often built up by constructing sample paths heuristically, often for quite complicated processes, such as the queues in Section 2.4.2 and their many ramifications in the literature, and then calculating a consistent set of transition probabilities. Theorem 3.2.1 then guarantees that one indeed has an underlying stochastic process for which these probabilities make sense.

To make this more concrete, let us consider a number of the models with Markovian structure introduced in Chapter 2, and illustrate how their transition probabilities may be constructed on a countable space from physical or other assumptions.

3.3.1 The forward and backward recurrence time chains

Recall that the forward recurrence time chain $\boldsymbol{V}^+$ is given by

$$V^+(n) := \inf(Z_m - n : Z_m > n), \qquad n \geq 0$$

where Z_n is a renewal sequence as introduced in Section 2.4.1.

The transition matrix for $\boldsymbol{V}^+$ is particularly simple. If $V^+(n) = k$ for some $k > 0$, then after one time unit $V^+(n + 1) = k - 1$. If $V^+(n) = 1$, then a renewal occurs at $n+1$ and $V^+(n+1)$ has the distribution p of an arbitrary term in the renewal sequence. This gives the sub-diagonal structure

$$P = \begin{bmatrix} p(1) & p(2) & p(3) & p(4) & \cdots \\ 1 & 0 & 0 & & \cdots \\ 0 & \ddots & \ddots & & \\ & 0 & 1 & 0 & \\ \vdots & \vdots & & \ddots & \ddots \end{bmatrix}$$

The backward recurrence time chain $\boldsymbol{V}^-$ has a similarly simple structure. For any $n \in \mathbb{Z}_+$, let us write

$$\overline{p}(n) = \sum_{j \geq n+1} p(j). \qquad\qquad (3.9)$$

Write $M = \sup(m \geq 1 : p(m) > 0)$; if $M < \infty$ then for this chain the state space $\mathsf{X} = \{0, 1, \ldots, M-1\}$; otherwise $\mathsf{X} = \mathbb{Z}_+$. In either case, for $x \in \mathsf{X}$ we have (with Y as a generic increment variable in the renewal process)

$$
\begin{aligned}
P(x, x+1) &= \mathsf{P}(Y > x+1 \mid Y > x) = \overline{p}(x+1)/\overline{p}(x), \\
P(x, 0) &= \mathsf{P}(Y = x+1 \mid Y > x) = p(x+1)/\overline{p}(x),
\end{aligned}
\tag{3.10}
$$

and zero otherwise. This gives a superdiagonal matrix of the form

$$
P = \begin{bmatrix}
b(1) & 1-b(1) & 0 & 0 & \cdots \\
b(2) & 0 & 1-b(2) & 0 & \cdots \\
b(3) & 0 & \ddots & 1-b(3) & \\
\vdots & \vdots & & & \ddots & \ddots
\end{bmatrix}
$$

where we have written $b(j) = p(j+1)/\overline{p}(j)$.

These particular chains are a rich source of simple examples of stable and unstable behaviors, depending on the behavior of p; and they are also chains which will be found to be fundamental in analyzing the asymptotic behavior of an arbitrary chain.

3.3.2 Random walk models

Random walk on the integers

Let us define the random walk $\Phi = \{\Phi_n; n \in \mathbb{Z}_+\}$ by setting, as in (RW1), $\Phi_n = \Phi_{n-1} + W_n$ where now the increment variables W_n are i.i.d. random variables taking only integer values in $\mathbb{Z} = \{\ldots, -1, 0, 1, \ldots\}$. As usual, write $\Gamma(y) = \mathsf{P}(W = y)$.

Then for $x, y \in \mathbb{Z}$, the state space of the random walk,

$$
\begin{aligned}
P(x, y) &= \mathsf{P}(\Phi_1 = y \mid \Phi_0 = x) \\
&= \mathsf{P}(\Phi_0 + W_1 = y \mid \Phi_0 = x) \\
&= \mathsf{P}(W_1 = y - x) \\
&= \Gamma(y - x).
\end{aligned}
\tag{3.11}
$$

The random walk is distinguished by this translation invariant nature of the transition probabilities: the probability that the chain moves from x to y in one step depends only on the difference $x - y$ between the values.

Random walks on a half line

It is equally easy to construct the transition probability matrix for the random walk on the half line $\mathbb{Z}_+$, defined in (RWHL1).

Suppose again that $\{W_i\}$ takes values in $\mathbb{Z}$, and recall from (RWHL1) that the random walk on a half line obeys

$$
\Phi_n = [\Phi_{n-1} + W_n]^+.
\tag{3.12}
$$

Then for $y \in \mathbb{Z}_+$, the state space of the random walk on a half line, we have as in (3.11) that for $y > 0$

$$
P(x, y) = \Gamma(y - x);
\tag{3.13}
$$

whilst for $y = 0$,

$$
\begin{aligned}
P(x,0) &= \mathsf{P}(\Phi_0 + W_1 \leq 0 \mid \Phi_0 = x) \\
&= \mathsf{P}(W_1 \leq -x) \\
&= \Gamma(-\infty, -x].
\end{aligned}
$$

The simple storage model

The storage model given by (SSM1)–(SSM2) is a concrete example of the structure in (3.13) and (3.14), provided the release rate is $r = 1$, the inter-input times take values $n \in \mathbb{Z}_+$ with distribution G, and the input values are also integer valued with distribution H.

The random walk on a half line describes the behavior of this storage model, and its transition matrix P therefore defines its one-step behavior. We can calculate the values of the increment distribution function Γ in a different way, in terms of the basic parameters G and H of the models, by breaking up the possibilities of the input time and the input size: we have

$$
\begin{aligned}
\Gamma(x) &= \mathsf{P}(S_n - J_n = x) \\
&= \sum_{i=0}^{\infty} H(i) G(x + i).
\end{aligned}
$$

We have rather forced the storage model into our countable space context by assuming that the variables concerned are integer valued. We will rectify this in later sections.

3.3.3   Embedded queueing models

The GI/M/1 queue

The next context in which we illustrate the construction of the transition matrix is in the modeling of queues through their embedded chains.

Consider the random variable $N_n = N(T_n'-)$, which counts customers immediately before each arrival in a queueing system satisfying (Q1)–(Q3).

We will first construct the matrix $P = (P(x,y))$ corresponding to the number of customers $\mathbf{N} = \{N_n\}$ for the GI/M/1 queue; that is, the queue satisfying (Q4).

Proposition 3.3.1. *For the GI/M/1 queue, the sequence $\mathbf{N} = \{N_n, n \geq 0\}$ can be constructed as a Markov chain with state space $\mathbb{Z}_+$ and transition matrix*

$$
P = \begin{bmatrix}
q_0 & p_0 & & & \\
q_1 & p_1 & p_0 & & \text{\Large 0} \\
q_2 & p_2 & p_1 & p_0 & \\
\vdots & \vdots & \vdots & \ddots & \ddots
\end{bmatrix}
$$

where $q_j = \sum_{i=j+1}^{\infty} p_i$, and

$$
p_0 = \mathsf{P}(S > T) = \int_0^{\infty} e^{-\mu t}\, G(dt), \tag{3.14}
$$

$$
\begin{aligned}
p_j &= \mathsf{P}\{S_j' > T > S_{j-1}'\} \\
&= \int_0^{\infty} \{e^{-\mu t}(\mu t)^j / j!\}\, G(dt), \qquad j \geq 1.
\end{aligned} \tag{3.15}
$$

Hence **N** *is a random walk on a half line.*

PROOF   In Section 2.4.2 we established the Markovian nature of the increases at $T_n'-$, in (2.28), under the assumption of exponential service times.

Since we consider $N(t)$ immediately before every arrival time, N_{n+1} can only increase from N_n by one unit at most; hence for $k > 1$ it is trivial that

$$P(N_{n+1} = j + k \mid N_n = j, N_{n-1}, N_{n-2}, \ldots, N_0) = 0. \qquad (3.16)$$

The independence and identical distribution structure of the service times show as in Section 2.4.2 that, no matter which previous customer was being served, and when their service started,

$$P(N_{n+1} = j + 1 \mid N_n = j, N_{n-1}, N_{n-2}, \ldots, N_0) = \int_0^\infty e^{-\mu t} \, G(dt) = p_0 \qquad (3.17)$$

as shown in equation (2.31). This establishes the upper triangular structure of P.

If $N_n = j$, then for $0 < i \le j$, we have $N_{n+1} = i$ provided exactly $(j - i + 1)$ jobs are completed in an inter-arrival period. It is an elementary property of sums of exponential random variables (see, for example, Çinlar [59], Chapter 4) that for any t, the number of services completed in a time $[0, t]$ is Poisson with parameter μt, so that

$$P(S_0 + \cdots + S_{j+1} > t > S_0 + \cdots + S_j) = e^{-\mu t} (\mu t)^j / j! \qquad (3.18)$$

from which we derive (3.15).

It remains to show that $P(j, 0) = q_j = \sum_{i=j+1}^\infty p_i$; but this follows analogously with equation (3.15), since the queue empties if more than $(j+1)$ customers complete service between arrivals.

Finally, to assert that $N = \{N_n\}$ can actually be constructed in its entirety as a Markov chain on $\mathbb{Z}_+$, we appeal to the general results of Theorem 3.2.1 above to build **N** from the probabilistic building blocks $P = (P(i, j))$, and any initial distribution μ.   □

The M/G/1 queue

Next consider the random variables N_n^*, which count customers immediately after each service time ends in a queueing system satisfying (Q1)–(Q3).

We showed in Section 2.4.2 that this is Markovian when the inter-arrival times are exponential: that is, for an M/G/1 model satisfying (Q5).

Proposition 3.3.2. *For the M/G/1 queue, the sequence* $\mathbf{N}^* = \{N_n^*, n \ge 0\}$ *can be constructed as a Markov chain with state space* $\mathbb{Z}_+$ *and transition matrix*

$$P = \begin{bmatrix} q_0 & q_1 & q_2 & q_3 & q_4 & \cdots \\ q_0 & q_1 & q_2 & q_3 & q_4 & \cdots \\ & q_0 & q_1 & q_2 & q_3 & \cdots \\ & & q_0 & q_1 & q_2 & \cdots \\ \vdots & \vdots & \vdots & \ddots & \ddots & \end{bmatrix}$$

where for each $j \geq 0$

$$q_j = \int_0^\infty \{e^{-\lambda t}(\lambda t)^j/j!\} \, H(dt) \qquad j \geq 1. \tag{3.19}$$

Hence $\mathbf{N}^*$ is similar to a random walk on a half line, but with a different modification of the transitions away from zero.

PROOF   Exactly as in (3.18), the expressions q_k represent the probabilities of k arrivals occurring in one service time with distribution H, when the inter-arrival times are independent exponential variables of rate λ.   □

3.3.4   Linear models on the rationals

The discussion of the queueing models above not only gives more explicit examples of the abstract random walk models, but also indicates how the Markov assumption may or may not be satisfied, depending on how the process is constructed: we need the exponential distributions for the basic building blocks, or we do not have probabilities of transition independent of the past.

In contrast, for the simple scalar linear AR(1) models satisfying (SLM1) and (SLM2), the Markovian nature of the process is immediate. The use of a countable space here is in the main inappropriate, but some versions of this model do provide a good source of examples and counterexamples which motivate the various topological conditions we introduce in Chapter 6. Recall then that for an AR(1) model X_n and W_n are random variables on $\mathbb{R}$, satisfying

$$\mathsf{X}_n = \alpha \mathsf{X}_{n-1} + W_n,$$

for some $\alpha \in \mathbb{R}$, with the "noise" variables $\{W_n\}$ independent and identically distributed. To use the countable structure of Section 3.2 we might assume, as with the storage model in Section 3.3.2 above, that α is integer valued, and the noise variables are also integer valued.

Or, if we need to assume a countable structure on X we might, for example, find a better fit to reality by supposing that the constant α takes a rational value; and that the generic noise variable W also has a distribution on the rationals $\mathbb{Q}$, with $\mathsf{P}(W = q) = \Gamma(q)$, $q \in \mathbb{Q}$. We then have, in a very straightforward manner

Proposition 3.3.3. *Provided $x_0 \in \mathbb{Q}$, the sequence $\mathbf{X} = \{X_n, n \geq 0\}$ can be constructed as a time homogeneous Markov chain on the countable space $\mathbb{Q}$, with transition probability matrix*

$$\begin{aligned} P(r,q) &= \mathsf{P}(X_{n+1} = q \mid X_n = r) \\ &= \Gamma(q - \alpha r), \qquad r, q \in \mathbb{Q}. \end{aligned}$$

PROOF   We have established that $\mathbf{X}$ is Markov. Clearly, from (SLM1), when $X_0 \in \mathbb{Q}$, the value of X_1 is in $\mathbb{Q}$ also; and $P(r, q)$ merely describes the fact that the chain moves from r to αr in a deterministic way before adding the noise with distribution W.

Again, once we have $P = \{P(r, q), r, q \in \mathbb{Q}\}$, we are guaranteed the existence of the Markov chain X, using the results of Theorem 3.2.1 with P as transition probability matrix. $\qquad\square$

This autoregression highlights immediately the shortcomings of the countable state space structure. Although $\mathbb{Q}$ is countable, so that in a formal sense we can construct a linear model satisfying (SLM1) and (SLM2) on $\mathbb{Q}$ in such a way that we can use countable space Markov chain theory, it is clearly more natural to take, say, α as real and the variable W as real valued also, so that X_n is real valued for any initial $x_0 \in \mathbb{R}$.

To model such processes, and the more complex autoregressions and nonlinear models which generalize them in Chapter 2, and which are clearly Markovian but continuous valued in conception, we need a theory for continuous-valued Markov chains. We turn to this now.

3.4 Foundations for general state space chains

3.4.1 Developing Φ from transition probabilities

The countable space approach guides the development of the theory we shall present in this book for a much broader class of Markov chains, on quite general state spaces: it is one of the more remarkable features of this seemingly sweeping generalization that the great majority of the countable state space results carry over virtually unchanged, without assuming any detailed structure on the space.

We let X be a general set, and $\mathcal{B}(\mathsf{X})$ denote a countably generated σ-field on X: when X is topological, then $\mathcal{B}(\mathsf{X})$ will be taken as the Borel σ-field, but otherwise it may be arbitrary.

In this case we again start from the one-step transition probabilities and construct Φ much as in Theorem 3.2.1.

Transition probability kernels

If $P = \{P(x, A), x \in \mathsf{X}, A \in \mathcal{B}(\mathsf{X})\}$ is such that

(i) for each $A \in \mathcal{B}(\mathsf{X})$, $P(\,\cdot\,, A)$ is a non-negative measurable function on X and

(ii) for each $x \in \mathsf{X}$, $P(x, \,\cdot\,)$ is a probability measure on $\mathcal{B}(\mathsf{X})$,

then we call P a *transition probability kernel* or *Markov transition function*.

On occasion, as in Chapter 6, we may require that a collection $T = \{T(x, A), x \in \mathsf{X}, A \in \mathcal{B}(\mathsf{X})\}$ satisfies (i) and (ii), with the exception that $T(x, \mathsf{X}) \leq 1$ for each x: such a collection is called a *substochastic* transition kernel. In the other direction, there will be times when we need to consider completely non-probabilistic mappings $K \colon \mathsf{X} \times \mathcal{B}(\mathsf{X}) \to$

$\mathbb{R}_+$ with $K(x, \cdot)$ a measure on $\mathcal{B}(\mathsf{X})$ for each x, and $K(\cdot, B)$ a measurable function on X for each $B \in \mathcal{B}(\mathsf{X})$. Such a map is called a *kernel* on $(\mathsf{X}, \mathcal{B}(\mathsf{X}))$.

We now imitate the development on a countable space to see that from the transition probability kernel P we can define a stochastic process with the appropriate Markovian properties, for which P will serve as a description of the one-step transition laws.

We first define a finite sequence $\Phi = \{\Phi_0, \Phi_1, \dots, \Phi_n\}$ of random variables on the product space $\mathsf{X}^{n+1} = \prod_{i=0}^n \mathsf{X}_i$, equipped with the product σ-field $\bigvee_{i=0}^n \mathcal{B}(\mathsf{X}_i)$, by an inductive procedure.

For any measurable sets $A_i \subseteq \mathsf{X}_i$, we develop the set functions $\mathsf{P}_x^n(\cdot)$ on X^{n+1} by setting, for a fixed starting point $x \in \mathsf{X}$ and for the "cylinder sets" $A_1 \times \cdots \times A_n$

$$\mathsf{P}_x^1(A_1) \;=\; P(x, A_1),$$

$$\mathsf{P}_x^2(A_1 \times A_2) \;=\; \int_{A_1} P(x, dy_1) P(y_1, A_2),$$

$$\vdots$$

$$\mathsf{P}_x^n(A_1 \times \cdots \times A_n) \;=\; \int_{A_1} P(x, dy_1) \int_{A_2} P(y_1, dy_2) \cdots P(y_{n-1}, A_n).$$

These are all well defined by the measurability of the integrands $P(\cdot, \cdot)$ in the first variable, and the fact that the kernels are measures in the second variable.

If we now extend P_x^n to all of $\bigvee_0^n \mathcal{B}(\mathsf{X}_i)$ in the usual way [37] and repeat this procedure for increasing n, we find

Theorem 3.4.1. *For any initial measure μ on $\mathcal{B}(\mathsf{X})$, and any transition probability kernel $P = \{P(x, A), x \in \mathsf{X}, A \in \mathcal{B}(\mathsf{X})\}$, there exists a stochastic process $\Phi = \{\Phi_0, \Phi_1, \dots\}$ on $\Omega = \prod_{i=0}^{\infty} \mathsf{X}_i$, measurable with respect to $\mathcal{F} = \bigvee_{i=0}^{\infty} \mathcal{B}(\mathsf{X}_i)$, and a probability measure P_μ on $\mathcal{F}$ such that $\mathsf{P}_\mu(B)$ is the probability of the event $\{\Phi \in B\}$ for $B \in \mathcal{F}$; and for measurable $A_i \subseteq \mathsf{X}_i, i = 0, \dots, n$, and any n*

$$\mathsf{P}_\mu(\Phi_0 \in A_0, \Phi_1 \in A_1, \dots, \Phi_n \in A_n) \tag{3.20}$$

$$= \int_{y_0 \in A_0} \cdots \int_{y_{n-1} \in A_{n-1}} \mu(dy_0) P(y_0, dy_1) \cdots P(y_{n-1}, A_n).$$

PROOF   Because of the consistency of definition of the set functions P_x^n, there is an overall measure P_x for which the P_x^n are finite-dimensional distributions, which leads to the result: the details are relatively standard measure-theoretic constructions, and are given in the general case by Revuz [326], Theorem 2.8 and Proposition 2.11; whilst if the space has a suitable topology, as in (MC1), then the existence of Φ is a straightforward consequence of Kolmogorov's Consistency Theorem for construction of probabilities on topological spaces.   □

The details of this construction are omitted here, since it suffices for our purposes to have indicated why transition probabilities generate processes, and to have spelled out that the key equation (3.20) is a reasonable representation of the behavior of the process in terms of the kernel P.

We can now formally define

Markov chains on general spaces

The stochastic process $\mathbf{\Phi}$ is called a time-homogeneous Markov chain with transition probability kernel $P(x, A)$ and initial distribution μ if the finite dimensional distributions of $\mathbf{\Phi}$ satisfy (3.20) for each n.

3.4.2   The n-step transition probability kernel

As on countable spaces the *n-step transition probability kernel* is defined iteratively. We set $P^0(x, A) = \delta_x(A)$, the Dirac measure defined by

$$\delta_x(A) = \begin{cases} 1 & x \in A \\ 0 & x \notin A, \end{cases} \tag{3.21}$$

and, for $n \geq 1$, we define inductively

$$P^n(x, A) = \int_{\mathsf{X}} P(x, dy) P^{n-1}(y, A), \quad x \in \mathsf{X}, A \in \mathcal{B}(\mathsf{X}). \tag{3.22}$$

We write P^n for the n-step transition probability kernel $\{P^n(x, A), x \in \mathsf{X}, A \in \mathcal{B}(\mathsf{X})\}$: note that P^n is defined analogously to the n-step transition probability matrix for the countable space case.

As a first application of the construction equations (3.20) and (3.22), we have the celebrated Chapman–Kolmogorov equations. These underlie, in one form or another, virtually all of the solidarity structures we develop.

Theorem 3.4.2. *For any m with $0 \leq m \leq n$,*

$$P^n(x, A) = \int_{\mathsf{X}} P^m(x, dy) P^{n-m}(y, A), \quad x \in \mathsf{X}, \; A \in \mathcal{B}(\mathsf{X}). \tag{3.23}$$

PROOF   In (3.20), choose $\mu = \delta_x$ and integrate over sets $A_i = \mathsf{X}$ for $i = 1, \ldots, n-1$; and use the definition of P^m and P^{n-m} for the first m and the last $n - m$ integrands.   □

We interpret (3.23) as saying that, as $\mathbf{\Phi}$ moves from x into A in n steps, at any intermediate time m it must take (obviously) some value $y \in \mathsf{X}$; and that, being a Markov chain, it forgets the past at that time m and moves the succeeding $(n - m)$ steps with the law appropriate to starting afresh at y. We can write equation (3.23) alternatively as

$$\mathsf{P}_x(\Phi_n \in A) = \int_{\mathsf{X}} \mathsf{P}_x(\Phi_m \in dy) \mathsf{P}_y(\Phi_{n-m} \in A). \tag{3.24}$$

Exactly as the one-step transition probability kernel describes a chain $\mathbf{\Phi}$, the m-step kernel (viewed in isolation) satisfies the definition of a transition kernel, and thus defines a Markov chain $\mathbf{\Phi}^m = \{\Phi_n^m\}$ with transition probabilities

$$\mathsf{P}_x(\Phi_n^m \in A) = P^{mn}(x, A). \tag{3.25}$$

This, and several other transition functions obtained from P, will be used widely in the sequel.

Skeletons and resolvents

The chain $\mathbf{\Phi}^m$ with transition law (3.25) is called the *m-skeleton* of the chain $\mathbf{\Phi}$.

The *resolvent* K_{a_ε} is defined for $0 < \varepsilon < 1$ by

$$K_{a_\varepsilon}(x, A) := (1 - \varepsilon) \sum_{i=0}^{\infty} \varepsilon^i P^i(x, A), \qquad x \in \mathsf{X}, \ A \in \mathcal{B}(\mathsf{X}). \qquad (3.26)$$

The Markov chain with transition function K_{a_ε} is called the K_{a_ε}-*chain*.

This nomenclature is taken from the continuous time literature, but we will see that in discrete time the m-skeletons and resolvents of the chain also provide a useful tool for analysis.

There is one substantial difference in moving to the general case from the countable case, which flows from the fact that the kernel P^n can no longer be viewed as symmetric in its two arguments.

In the general case the kernel P^n operates on quite different entities from the left and the right. As an operator P^n acts on both bounded measurable functions f on X and on σ-finite measures μ on $\mathcal{B}(\mathsf{X})$ via

$$P^n f(x) = \int_{\mathsf{X}} P^n(x, dy) f(y), \qquad \mu P^n(A) = \int_{\mathsf{X}} \mu(dx) P^n(x, A),$$

and we shall use the notation $P^n f, \mu P^n$ to denote these operations. We shall also write

$$P^n(x, f) := \int P^n(x, dy) f(y) := \delta_x P^n f$$

if it is notationally convenient. In general, the functional notation is more compact: for example, we can rewrite the Chapman–Kolmogorov equations as

$$P^{m+n} = P^m P^n, \qquad m, n \in \mathbb{Z}_+.$$

On many occasions, though, where we feel that the argument is more transparent when written in full form we shall revert to the more detailed presentation.

The form of the Markov chain definitions we have given to date concern only the probabilities of events involving $\mathbf{\Phi}$. We now define the expectation operation E_μ corresponding to P_μ.

For cylinder sets we define E_μ by

$$\mathsf{E}_\mu[\mathbb{I}_{A_0 \times \cdots \times A_n}(\Phi_0, \ldots, \Phi_n)] := \mathsf{P}_\mu(\{\Phi_0, \ldots, \Phi_n\} \in A_0 \times \cdots \times A_n),$$

where $\mathbb{I}_B$ denotes the indicator function of a set B. We may extend the definition to that of $\mathsf{E}_\mu[h(\Phi_0, \Phi_1, \ldots)]$ for any measurable bounded real-valued function h on Ω by requiring that the expectation be linear.

By linearity of the expectation, we can also extend the Markovian relationship (3.20) to express the Markov property in the following equivalent form. We omit the details, which are routine.

Proposition 3.4.3. *If Φ is a Markov chain on $(\Omega, \mathcal{F})$, with initial measure μ, and $h \colon \Omega \to \mathbb{R}$ is bounded and measurable, then*

$$\mathsf{E}_\mu[h(\Phi_{n+1}, \Phi_{n+2}, \ldots) \mid \Phi_0, \ldots, \Phi_n; \ \Phi_n = x] = \mathsf{E}_x[h(\Phi_1, \Phi_2, \ldots)]. \qquad (3.27)$$

□

The formulation of the Markov concept itself is made much simpler if we develop more systematic notation for the information encompassed in the past of the process, and if we introduce the "shift operator" on the space Ω.

For a given initial distribution, define the σ-field

$$\mathcal{F}_n^\Phi := \sigma(\Phi_0, \ldots, \Phi_n) \subseteq \mathcal{B}(\mathsf{X}^{n+1})$$

which is the smallest σ-field for which the random variable $\{\Phi_0, \ldots, \Phi_n\}$ is measurable. In many cases, $\mathcal{F}_n^\Phi$ will coincide with $\mathcal{B}(\mathsf{X}^n)$, although this depends in particular on the initial measure μ chosen for a particular chain.

The *shift operator* θ is defined to be the mapping on Ω defined by

$$\theta(\{x_0, x_1, \ldots, x_n, \ldots\}) = \{x_1, x_2, \ldots, x_{n+1}, \ldots\}.$$

We write θ^k for the kth iterate of the mapping θ, defined inductively by

$$\theta^1 = \theta, \qquad \theta^{k+1} = \theta \circ \theta^k, \quad k \geq 1.$$

The shifts θ^k define operators on random variables H on $(\Omega, \mathcal{F}, \mathsf{P}_\mu)$ by

$$(\theta^k H)(w) = H \circ \theta^k(\omega).$$

It is obvious that $\Phi_n \circ \theta^k(\omega) = \Phi_{n+k}$. Hence if the random variable H is of the form $H = h(\Phi_0, \Phi_1, \ldots)$ for a measurable function h on the sequence space Ω then

$$\theta^k H = h(\Phi_k, \Phi_{k+1}, \ldots)$$

Since the expectation $\mathsf{E}_x[H]$ is a measurable function on X, it follows that $\mathsf{E}_{\Phi_n}[H]$ is a random variable on $(\Omega, \mathcal{F}, \mathsf{P}_\mu)$ for any initial distribution. With this notation the equation

$$\mathsf{E}_\mu[\theta^n H \mid \mathcal{F}_n^\Phi] = \mathsf{E}_{\Phi_n}[H] \qquad \text{a.s. } [\mathsf{P}_\mu] \qquad (3.28)$$

valid for any bounded measurable h and fixed $n \in \mathbb{Z}_+$, describes the *time-homogeneous Markov property* in a succinct way.

It is not always the case that $\mathcal{F}_n^\Phi$ is complete: that is, contains every set of P_μ-measure zero. We adopt the following convention as in [326]. For any initial measure μ we say that an event A occurs P_μ-a.s. to indicate that A^c is a set contained in an element of $\mathcal{F}_n^\Phi$ which is of P_μ-measure zero.

If A occurs P_x-a.s. for all $x \in \mathsf{X}$ then we write that A occurs P_*-a.s.

3.4.3   Occupation, hitting and stopping times

The distributions of the chain $\boldsymbol{\Phi}$ at time n are the basic building blocks of its existence, but the analysis of its behavior concerns also the distributions at certain random times in its evolution, and we need to introduce these now.

Occupation times, return times and hitting times

(i) For any set $A \in \mathcal{B}(\mathsf{X})$, the *occupation time* η_A is the number of visits by $\boldsymbol{\Phi}$ to A after time zero, and is given by

$$\eta_A := \sum_{n=1}^{\infty} \mathbb{I}\{\Phi_n \in A\}.$$

(ii) For any set $A \in \mathcal{B}(\mathsf{X})$, the variables

$$\begin{aligned} \tau_A &:= \min\{n \geq 1 : \Phi_n \in A\}, \\ \sigma_A &:= \min\{n \geq 0 : \Phi_n \in A\} \end{aligned}$$

are called the *first return* and *first hitting* times on A, respectively.

For every $A \in \mathcal{B}(\mathsf{X})$, η_A^i, τ_A and σ_A are obviously measurable functions from Ω to $\mathbb{Z}_+ \cup \{\infty\}$.

Unless we need to distinguish between different returns to a set, then we call τ_A and σ_A the return and hitting times on A respectively. If we do wish to distinguish different return times, we write $\tau_A(k)$ for the random time of the kth visit to A: these are defined inductively for any A by

$$\begin{aligned} \tau_A(1) &:= \tau_A, \\ \tau_A(k) &:= \min\{n > \tau_A(k-1) : \Phi_n \in A\}. \end{aligned} \tag{3.29}$$

Analysis of $\boldsymbol{\Phi}$ involves the kernel U defined as

$$\begin{aligned} U(x, A) &:= \sum_{n=1}^{\infty} P^n(x, A) \\ &= \mathsf{E}_x[\eta_A] \end{aligned} \tag{3.30}$$

which maps $\mathsf{X} \times \mathcal{B}(\mathsf{X})$ to $\mathbb{R} \cup \{\infty\}$, and the return time probabilities

$$\begin{aligned} L(x, A) &:= \mathsf{P}_x(\tau_A < \infty) \\ &= \mathsf{P}_x(\boldsymbol{\Phi} \text{ ever enters } A). \end{aligned} \tag{3.31}$$

In order to analyze numbers of visits to sets, we often need to consider the behavior after the first visit τ_A to a set A (which is a random time), rather than behavior after fixed times. One of the most crucial aspects of Markov chain theory is that the "forgetfulness" properties in equation (3.20) or equation (3.27) hold, not just for fixed times n, but for the chain interrupted at certain random times, called *stopping times*, and we now introduce these ideas.

Stopping times

A function $\zeta \colon \Omega \to \mathbb{Z}_+ \cup \{\infty\}$ is a *stopping time* for Φ if for any initial distribution μ the event $\{\zeta = n\} \in \mathcal{F}_n^\Phi$ for all $n \in \mathbb{Z}_+$.

The first return and the hitting times on sets provide simple examples of stopping times.

Proposition 3.4.4. *For any set $A \in \mathcal{B}(\mathsf{X})$, the variables τ_A and σ_A are stopping times for Φ.*

PROOF   Since we have

$$\{\tau_A = n\} \;\; = \;\; \cap_{m=1}^{u-1}\{\Phi_m \in A^c\} \cap \{\Phi_n \in A\} \in \mathcal{F}_n^\Phi, \quad n \geq 1,$$
$$\{\sigma_A = n\} \;\; = \;\; \cap_{m=0}^{n-1}\{\Phi_m \in A^c\} \cap \{\Phi_n \in A\} \in \mathcal{F}_n^\Phi, \quad n \geq 0,$$

it follows from the definitions that τ_A and σ_A are stopping times.   □

We can construct the full distributions of these stopping times from the basic building blocks governing the motion of Φ, namely the elements of the transition probability kernel, using the Markov property for each fixed $n \in \mathbb{Z}_+$. This gives

Proposition 3.4.5.   (i) *For all $x \in \mathsf{X}$, $A \in \mathcal{B}(\mathsf{X})$*

$$\mathsf{P}_x(\tau_A = 1) = P(x, A),$$

and inductively for $n > 1$

$$\mathsf{P}_x(\tau_A = n) \;\; = \;\; \int_{A^c} P(x, dy)\mathsf{P}_y(\tau_A = n - 1)$$
$$= \;\; \int_{A^c} P(x, dy_1) \int_{A^c} P(y_1, dy_2) \cdots$$
$$\int_{A^c} P(y_{n-2}, dy_{n-1})P(y_{n-1}, A).$$

(ii) *For all $x \in \mathsf{X}$, $A \in \mathcal{B}(\mathsf{X})$*

$$\mathsf{P}_x(\sigma_A = 0) = \mathbb{I}_A(x),$$

and for $n \geq 1$, $x \in A^c$

$$\mathsf{P}_x(\sigma_A = n) = \mathsf{P}_x(\tau_A = n).$$

□

If we use the kernel I_B defined as $I_B(x, A) := \mathbb{I}_{A \cap B}(x)$, we have, in more compact functional notation,

$$\mathsf{P}_x(\tau_A = k) = [(PI_{A^c})^{k-1}P](x, A).$$

From this we obtain the formula

$$L(x, A) := \sum_{k=1}^{\infty} [(PI_{A^c})^{k-1}P](x, A)$$

for the return time probability to a set A starting from the state x.

The simple Markov property (3.28) holds for any bounded measurable h and fixed $n \in \mathbb{Z}_+$. We now extend (3.28) to stopping times.

If ζ is an arbitrary stopping time, then the fact that our time set is $\mathbb{Z}_+$ enables us to define the random variable Φ_ζ by setting $\Phi_\zeta = \Phi_n$ on the event $\{\zeta = n\}$. For a stopping time ζ the property which tells us that the future evolution of Φ after the stopping time depends only on the value of Φ_ζ, rather than on any other past values, is called the strong Markov property.

To describe this formally, we need to define the σ-field $\mathcal{F}_\zeta^\Phi := \{A \in \mathcal{F} : \{\zeta = n\} \cap A \in \mathcal{F}_n^\Phi, n \in \mathbb{Z}_+\}$, which describes events which happen "up to time ζ".

For a stopping time ζ and a random variable $H = h(\Phi_0, \Phi_1, \ldots)$ the shift θ^ζ is defined as

$$\theta^\zeta H = h(\Phi_\zeta, \Phi_{\zeta+1}, \ldots),$$

on the set $\{\zeta < \infty\}$. The required extension of (3.28) is then

---

Strong Markov property

We say Φ has the *strong Markov property* if for any initial distribution μ, any real-valued bounded measurable function h on Ω, and any stopping time ζ,

$$\mathsf{E}_\mu[\theta^\zeta H \mid \mathcal{F}_\zeta^\Phi] = \mathsf{E}_{\Phi_\zeta}[H] \quad \text{a.s. } [\mathsf{P}_\mu], \tag{3.32}$$

on the set $\{\zeta < \infty\}$.

Proposition 3.4.6. *For a Markov chain Φ with discrete time parameter, the strong Markov property always holds.*

PROOF   This result is a simple consequence of decomposing the expectations on both sides of (3.32) over the set where $\{\zeta = n\}$, and using the ordinary Markov property, in the form of equation (3.28), at each of these fixed times n.   □

We are not always interested only in the times of visits to particular sets. Often the quantities of interest involve conditioning on such visits being in the future.

Taboo probabilities

We define the n-step *taboo probabilities* as

$$_A P^n(x, B) := \mathsf{P}_x(\Phi_n \in B, \tau_A \geq n), \qquad x \in \mathsf{X}, \ A, B \in \mathcal{B}(\mathsf{X}).$$

The quantity $_A P^n(x, B)$ denotes the probability of a transition to B in n steps of the chain, "avoiding" the set A. As in Proposition 3.4.5 these satisfy the iterative relation

$$_A P^1(x, B) = P(x, B)$$

and for $n > 1$

$$_A P^n(x, B) = \int_{A^c} P(x, dy) \, _A P^{n-1}(y, B), \qquad x \in \mathsf{X}, \ A, B \in \mathcal{B}(\mathsf{X}), \tag{3.33}$$

or, in operator notation, $_A P^n(x, B) = [(PI_{A^c})^{n-1} P](x, B)$.

We will also use extensively the notation

$$U_A(x, B) := \sum_{n-1}^{\infty} {_A P^n(x, B)}, \qquad x \in \mathsf{X}, \ A, B \in \mathcal{B}(\mathsf{X}); \tag{3.34}$$

note that this extends the definition of L in (3.31) since

$$U_A(x, A) = L(x, A), \qquad x \in \mathsf{X}.$$

3.5   Building transition kernels for specific models

3.5.1   Random walk on a half line

Let Φ be a random walk on a half line, where now we do not restrict the increment distribution to be integer valued. Thus $\{W_i\}$ is a sequence of i.i.d. random variables taking values in $\mathbb{R} = (-\infty, \infty)$, with distribution function $\Gamma(A) = \mathsf{P}(W \in A), A \in \mathcal{B}(\mathbb{R})$.

For any $A \subseteq (0, \infty)$, we have by the arguments we have used before

$$
\begin{aligned}
P(x, A) &= \mathsf{P}(\Phi_0 + W_1 \in A \mid \Phi_0 = x) \\
&= \mathsf{P}(W_1 \in A - x) \\
&= \Gamma(A - x),
\end{aligned}
\tag{3.35}
$$

whilst

$$
\begin{aligned}
P(x, \{0\}) &= \mathsf{P}(\Phi_0 + W_1 \leq 0 \mid \Phi_0 = x) \\
&= \mathsf{P}(W_1 \leq -x) \\
&= \Gamma(-\infty, -x].
\end{aligned}
\tag{3.36}
$$

These models are often much more appropriate in applications than random walks restricted to integer values.

3.5.2   Storage and queueing models

Consider the Moran dam model given by (SSM1)–(SSM2), in the general case where $r > 0$, the inter-input times have distribution G; and the input values have distribution H.

The model of a random walk on a half line with transition probability kernel P given by (3.36) defines the one-step behavior of the storage model. As for the integer-valued case, we calculate the distribution function Γ explicitly by breaking up the possibilities of the input time and the input size, to get a similar convolution form for Γ in terms of G and H:

$$
\begin{aligned}
\Gamma(A) &= \mathsf{P}(S_n - J_n \in A) \\
&= \int_0^\infty G(A/r + y/r)\, H(dy),
\end{aligned} \tag{3.37}
$$

where as usual the set $A/r := \{y : ry \in A\}$.

The model (3.37) is of a storage system, and we have phrased the terms accordingly. The same transition law applies to the many other models of this form: inventories, insurance models, and models of the residual service in a GI/G/1 queue, which were mentioned in Section 2.5.

In Section 3.5.4 below we will develop the transition probability structure for a more complex system which can also be used to model the dynamics of the GI/G/1 queue.

3.5.3   Renewal processes and related chains

We now consider a real-valued renewal process: this extends the countable space version of Section 2.4.1 and is closely related to the residual service time mentioned above.

Let $\{Y_1, Y_2, \ldots\}$ be a sequence of independent and identical random variables, now with distribution function Γ concentrated, not on the whole real line nor on $\mathbb{Z}_+$, but rather on $\mathbb{R}_+$. Let Y_0 be a further independent random variable, with the distribution of Y_0 being Γ_0, also concentrated on $\mathbb{R}_+$. The random variables

$$
Z_n := \sum_{i=0}^n Y_i
$$

are again called a *delayed renewal process*, with Γ_0 being the distribution of the delay described by the first variable. If $\Gamma_0 = \Gamma$ then the sequence $\{Z_n\}$ is again referred to as a renewal process.

As with the integer-valued case, write $\Gamma_0 * \Gamma$ for the convolution of Γ_0 and Γ given by

$$
\Gamma_0 * \Gamma\,(dt) := \int_0^t \Gamma(dt - s)\, \Gamma_0(ds) = \int_0^t \Gamma_0(dt - s)\, \Gamma(ds) \tag{3.38}
$$

and Γ^{n*} for the nth convolution of Γ with itself. By decomposing successively over the values of the first n variables $Z_0, \ldots, Z_{n-1}$ we have that

$$
\mathsf{P}(Z_n \in dt) = \Gamma_0 * \Gamma^{n*}\,(dt)
$$

and so the *renewal measure* given by $U(-\infty, t] = \sum_0^\infty \Gamma^{n*}(-\infty, t]$ has the interpretation

$$U[0, t] = \mathsf{E}_0[\text{number of renewals in } [0, t]]$$

and

$$\Gamma_0 * U[0, t] = \mathsf{E}_{\Gamma_0}[\text{number of renewals in } [0, t]],$$

where E_0 refers to the expectation when the first renewal is at 0, and E_{Γ_0} refers to the expectation when the first renewal has distribution Γ_0.

It is clear that Z_n is a Markov chain: its transition probabilities are given by

$$P(x, A) = \mathsf{P}(Z_n \in A \mid Z_{n-1} = x) = \Gamma(A - x)$$

and so Z_n is a random walk. It is not a very stable one, however, as it moves inexorably to infinity with each new step.

The forward and backward recurrence time chains, in contrast to the renewal process itself, exhibit a much greater degree of stability: they grow, then they diminish, then they grow again.

Forward and backward recurrence time chains

If $\{Z_n\}$ is a renewal process with no delay, then we call the process

(RT3) $V^+(t) := \inf(Z_n - t : Z_n > t, \ n \geq 1), \ t \geq 0,$

the *forward recurrence time process*; and for any $\delta > 0$, the discrete time chain $\boldsymbol{V}_\delta^+ = \{V_\delta^+(n) = V^+(n\delta), \ n \in \mathbb{Z}_+\}$ is called the *forward recurrence time δ-skeleton*.

We call the process

(RT4) $V^-(t) := \inf(t - Z_n : Z_n \leq t, \ n \geq 1), \ t \geq 0,$

the *backward recurrence time process*; and for any $\delta > 0$, the discrete time chain $\boldsymbol{V}_\delta^- = \{V_\delta^-(n) = V^-(n\delta), \ n \in \mathbb{Z}_+\}$ is called the *backward recurrence time δ-skeleton*.

No matter what the structure of the renewal sequence (and in particular, even if Γ is not exponential), the forward and backward recurrence time δ-skeletons $\boldsymbol{V}_\delta^+$ and $\boldsymbol{V}_\delta^-$ are Markovian.

To see this for the forward chain, note that if $x > \delta$, then the transition probabilities P^δ of $\boldsymbol{V}_\delta^+$ are merely

$$P^\delta(x, \{x - \delta\}) = 1$$

whilst if $x \leq \delta$ we have, by decomposing over the time and the index of the last renewal in the period after the current forward recurrence time finishes, and using the independence of the variables Y_i

$$
\begin{aligned}
P^\delta(x, A) &= \int_0^{\delta-x} \sum_{n=0}^{\infty} \Gamma^{n*}(dt)\Gamma(A - [\delta - x] - t) \\
&= \int_0^{\delta-x} U(dt)\Gamma(A - [\delta - x] - t). \quad\quad (3.39)
\end{aligned}
$$

For the backward chain we have similarly that for all x

$$
\mathsf{P}(V^-(n\delta) = x + \delta \mid V^-((n-1)\delta) = x) = \Gamma(x + \delta, \infty)/\Gamma(x, \infty),
$$

whilst for $dv \subset [0, \delta]$

$$
\mathsf{P}(V^-(n\delta) \in dv \mid V^-((n-1)\delta) = x) = \int_x^{x+\delta} \Gamma(du)U(dv - (u - x) - \delta)\frac{\Gamma(v, \infty)}{[\Gamma(x, \infty)]^{-1}}.
$$

3.5.4   Ladder chains and the GI/G/1 queue

The GI/G/1 queue satisfies the conditions (Q1)–(Q3). Although the residual service time process of the GI/G/1 queue can be analyzed using the model (3.37), the more detailed structure involving actual numbers in the queue in the case of general (i.e. non-exponential) service and input times requires a more complex state space for a Markovian analysis.

We saw in Section 3.3.3 that when the service time distribution H is exponential, we can define a Markov chain by

$$
N_n = \{ \text{ number of customers at } T_n' -, n = 1, 2, \ldots \},
$$

whilst we have a similarly embedded chain after the service times if the inter-arrival time is exponential. However, the numbers in the queue, even at the arrival or departure times, are not Markovian without such exponential assumptions.

The key step in the general case is to augment $\{N_n\}$ so that we do get a Markov model. This augmentation involves combining the information on the numbers in the queue with the information in the residual service time

To do this we introduce a bivariate "ladder chain" on a "ladder" space $\mathbb{Z}_+ \times \mathbb{R}$, with a countable number of rungs indexed by the first variable and with each rung constituting a copy of the real line.

This construction is in fact more general than that for the GI/G/1 queue alone, and we shall use the ladder chain model for illustrative purposes on a number of occasions.

Define the Markov chain $\mathbf{\Phi} = \{\Phi_n\}$ on $\mathbb{Z}_+ \times \mathbb{R}$ with motion defined by the transition probabilities $P(i, x; j \times A)$, $i, j \in \mathbb{Z}_+$, $x \in \mathbb{R}$, $A \in \mathcal{B}(\mathbb{R})$ given by

$$
\begin{aligned}
P(i, x; j \times A) &= 0, \quad j > i + 1, \\
P(i, x; j \times A) &= \Lambda_{i-j+1}(x, A), \quad j = 1, \ldots, i + 1, \quad\quad (3.40) \\
P(i, x; 0 \times A) &= \Lambda_i^*(x, A).
\end{aligned}
$$

where each of the Λ_i, Λ_i^* is a substochastic transition probability kernel on $\mathbb{R}$ in its own right.

The translation invariant and "skip-free to the right" nature of the movement of this chain, incorporated in (3.41), indicates that it is a generalization of those random walks which occur in the GI/M/1 queue, as delineated in Proposition 3.3.1. We have

$$
P = \begin{bmatrix} \Lambda_0^* & \Lambda_0 & & & \\ \Lambda_1^* & \Lambda_1 & \Lambda_0 & & \huge 0 \\ \Lambda_2^* & \Lambda_2 & \Lambda_1 & \Lambda_0 & \\ \vdots & \vdots & \vdots & \ddots & \ddots \end{bmatrix}
$$

where now the Λ_i, Λ_i^* are substochastic transition probability kernels rather than mere scalars.

To use this construction in the GI/G/1 context we write

$$\Phi_n = (N_n, R_n), \quad n \geq 1,$$

where as before N_n is the number of customers at $T_n' -$ and

$$R_n = \{\text{total residual service time in the system at } T_n'+\} :$$

then $\boldsymbol{\Phi} = \{\Phi_n; n \in \mathbb{Z}_+\}$ can be realised as a Markov chain with the structure (3.41), as we now demonstrate by constructing the transition kernel P explicitly.

As in (Q1)–(Q3) let H denote the distribution function of service times, and G denote the distribution function of inter-arrival times; and let $Z_1, Z_2, Z_3, \ldots$ denote an undelayed renewal process with $Z_n - Z_{n-1} = S_n$ having the service distribution function H, as in (2.27). This differs from the process of completion points of services in that the latter may have longer intervals when there is no customer present, after completion of a busy cycle.

Let R_t denote the forward recurrence time in the renewal process $\{Z_k\}$ at time t in this process, i.e., $R_t = Z_{N(t)+1} - t$, where $N(t) = \sup\{n : Z_n \leq t\}$ as in (RT3). If $R_0 = x$ then $Z_1 = x$. Now write

$$P_n^t(x, y) = \mathsf{P}(Z_n \leq t < Z_{n+1}, R_t \leq y \mid R_0 = x) \qquad (3.41)$$

for the probability that, in this renewal process n "service times" are completed in $[0, t]$ and that the residual time of current service at t is in $[0, y]$, given $R_0 = x$.

With these definitions it is easy to verify that the chain $\boldsymbol{\Phi}$ has the form (3.41) with the specific choice of the substochastic transition kernels Λ_i, Λ_i^* given by

$$\Lambda_n(x, [0, y]) = \int_0^\infty P_n^t(x, y) \, G(dt) \qquad (3.42)$$

and

$$\Lambda_n^*(x, [0, y]) = \left[\sum_{n+1}^\infty \Lambda_j(x, [0, \infty)) \right] H[0, y]. \qquad (3.43)$$

3.5.5 State space models

The simple nonlinear state space model is a very general model and, consequently, its transition function has an unstructured form until we make more explicit assumptions in particular cases. The general functional form which we construct here for the scalar $SNSS(F)$ model of Section 2.2.1 will be used extensively, as will the techniques which are used in constructing its form.

For any bounded and measurable function $h\colon \mathsf{X} \to \mathbb{R}$ we have from (SNSS1),

$$h(X_{n+1}) = h(F(X_n, W_{n+1})).$$

Since $\{W_n\}$ is assumed i.i.d. in (SNSS2) we see that

$$\begin{aligned} Ph\,(x) &= \mathsf{E}[h(X_{n+1}) \mid X_n = x] \\ &= \mathsf{E}[h(F(x, W))] \end{aligned}$$

where W is a generic noise variable. Since Γ denotes the distribution of W, this becomes

$$Ph\,(x) = \int_{-\infty}^{\infty} h(F(x, w))\, \Gamma(dw)$$

and by specializing to the case where $h = \mathbb{I}_A$, we see that for any measurable set A and any $x \in \mathsf{X}$,

$$P(x, A) = \int_{-\infty}^{\infty} \mathbb{I}\{F(x, w) \in A\}\, \Gamma(dw).$$

To construct the k-step transition probability, recall from (2.5) that the transition maps for the $SNSS(F)$ model are defined by setting $F_0(x) = x$, $F_1(x_0, w_1) = F(x_0, w_1)$, and for $k \geq 1$,

$$F_{k+1}(x_0, w_1, \ldots, w_{k+1}) = F(F_k(x_0, w_1, \ldots, w_k), w_{k+1})$$

where x_0 and w_i are arbitrary real numbers. By induction we may show that for any initial condition $X_0 = x_0$ and any $k \in \mathbb{Z}_+$,

$$X_k = F_k(x_0, W_1, \ldots, W_k),$$

which immediately implies that the k-step transition function may be expressed as

$$\begin{aligned} P^k(x, A) &= \mathsf{P}(F_k(x, W_1, \ldots, W_k) \in A) \\ &= \int \cdots \int \mathbb{I}\{F_k(x, w_1, \ldots, w_k) \in A\}\, \Gamma(dw_1) \cdots \Gamma(dw_k) \end{aligned} \qquad (3.44)$$

3.6   Commentary

The development of foundations in this chapter is standard. The existence of the excellent accounts in Chung [71] and Revuz [326] renders it far less necessary for us to fill in specific details.

The one real assumption in the general case is that the σ-field $\mathcal{B}(\mathsf{X})$ is countably generated. For many purposes, even this condition can be relaxed, using the device of "admissible σ-fields" discussed in Orey [309], Chapter 1. We shall not require, for the models we develop, the greater generality of non-countably generated σ-fields, and leave this expansion of the concepts to the reader if necessary.

The Chapman–Kolmogorov equations, simple though they are, hold the key to much of the analysis of Markov chains. The general formulation of these dates to Kolmogorov [215]: David Kendall comments [204] that the physicist Chapman was not aware of his role in this terminology, which appears to be due to work on the thermal diffusion of grains in a non-uniform fluid.

The Chapman–Kolmogorov equations indicate that the set P^n is a semi-group of operators just as the corresponding matrices are, and in the general case this observation enables an approach to the theory of Markov chains through the mathematical structures of semi-groups of operators. This has proved a very fruitful method, especially for continuous time models. However, we do not pursue that route directly in this book, nor do we pursue the possibilities of the matrix structure in the countable case.

This is largely because, as general non-negative operators, the P^n often do not act on useful spaces for our purposes. The one real case where the P^n operate successfully on a normed space occurs in Chapter 16, and even there the space only emerges after a probabilistic argument is completed, rather than providing a starting point for analysis.

Foguel [122, 124] has a thorough exposition of the operator-theoretic approach to chains in discrete time, based on their operation on L^1 spaces. Vere-Jones [405, 406] has a number of results based on the action of a matrix P as a non-negative operator on sequence spaces suitably structured, but even in this countable case results are limited. Nummelin [303] couches many of his results in a general non-negative operator context, as does Tweedie [394, 395], but the methods are probabilistic rather than using traditional operator theory.

The topological spaces we introduce here will not be considered in more detail until Chapter 6. Very many of the properties we derive will actually need less structure than we have imposed in our definition of "topological" spaces: often (see for example Tuominen and Tweedie [391]) all that is required is a countably generated topology with the T_1 separability property. The assumptions we make seem unrestrictive in practice, however, and avoid occasional technicalities of proof.

Hitting times and their properties are of prime importance in all that follows. On a countable space Chung [71] has a detailed account of taboo probabilities, and much of our usage follows his lead and that of Nummelin [303], although our notation differs in minor ways from the latter. In particular our τ_A is, regrettably, Nummelin's S_A and our σ_A is Nummelin's T_A; our usage of τ_A agrees, however, with that of Chung [71] and Asmussen [9], and we hope is the more standard.

The availability of the strong Markov property is vital for much of what follows. Kac is reported as saying [50] that he was fortunate, for in his day all processes had the strong Markov property: we are equally fortunate that, with a countable time set, all chains still have the strong Markov property.

The various transition matrices that we construct are well known. The reader who is not familiar with such concepts should read, say, Çinlar [59], Karlin and Taylor [194] or Asmussen [9] for these and many other not dissimilar constructions in the queueing and storage area. For further information on linear stochastic systems the reader is referred to Caines [57]. The control and systems areas have concentrated more intensively on controlled Markov chains which have an auxiliary input which is chosen to control the state process $\mathbf{\Phi}$. Once a control is applied in this way, the "closed loop system" is frequently described by a Markov chain as defined in this chapter. Kumar and Varaiya [225] is a good introduction, and the article by Arapostathis et al. [6] gives an excellent and up-to-date survey of the controlled Markov chain literature.

Chapter 4

Irreducibility

This chapter is devoted to the fundamental concept of irreducibility: the idea that all parts of the space can be reached by a Markov chain, no matter what the starting point. Although the initial results are relatively simple, the impact of an appropriate irreducibility structure will have wide-ranging consequences, and it is therefore of critical importance that such structures be well understood.

The results summarized in Theorem 4.0.1 are the highlights of this chapter from a theoretical point of view. An equally important aspect of the chapter is, however, to show through the analysis of a number of models just what techniques are available in practice to ensure the initial condition of Theorem 4.0.1 ("φ-irreducibility") holds, and we believe that these will repay equally careful consideration.

Theorem 4.0.1. *If there exists an "irreducibility" measure φ on $\mathcal{B}(\mathsf{X})$ such that for every state x*

$$\varphi(A) > 0 \Rightarrow L(x, A) > 0 \tag{4.1}$$

then there exists an essentially unique "maximal" irreducibility measure ψ on $\mathcal{B}(\mathsf{X})$ such that

(i) *for every state x we have $L(x, A) > 0$ whenever $\psi(A) > 0$, and also*

(ii) *if $\psi(A) = 0$, then $\psi(\bar{A}) = 0$, where*

$$\bar{A} := \{y : L(y, A) > 0\};$$

(iii) *if $\psi(A^c) = 0$, then $A = A_0 \cup N$ where the set N is also ψ-null, and the set A_0 is absorbing in the sense that*

$$P(x, A_0) \equiv 1, \qquad x \in A_0.$$

PROOF   The existence of a measure ψ satisfying the irreducibility conditions (i) and (ii) is shown in Proposition 4.2.2, and that (iii) holds is in Proposition 4.2.3.   □

The term "maximal" is justified since we will see that φ is absolutely continuous with respect to ψ, written $\psi \succ \varphi$, for every φ satisfying (4.1); here the relation of absolute continuity of φ with respect to ψ means that $\psi(A) = 0$ implies $\varphi(A) = 0$.

Verifying (4.1) is often relatively painless. State space models on $\mathbb{R}^k$ for which the noise or disturbance distribution has a density with respect to Lebesgue measure will typically have such a property, with φ taken as Lebesgue measure restricted to an open set (see Section 4.4, or in more detail, Chapter 7); chains with a regeneration point α reached from everywhere will satisfy (4.1) with the trivial choice of $\varphi = \delta_\alpha$ (see Section 4.3).

The extra benefit of defining much more accurately the sets which are avoided by "most" points, as in Theorem 4.0.1 (ii), or of knowing that one can omit ψ-null sets and restrict oneself to an absorbing set of "good" points as in Theorem 4.0.1 (iii), is then of surprising value, and we use these properties again and again. These are however far from the most significant consequences of the seemingly innocuous assumption (4.1): far more will flow in Chapter 5, and thereafter.

The most basic structural results for Markov chains, which lead to this formalization of the concept of irreducibility, involve the analysis of communicating states and sets. If one can tell which sets can be reached with positive probability from particular starting points $x \in \mathsf{X}$, then one can begin to have an idea of how the chain behaves in the longer term, and then give a more detailed description of that longer term behavior.

Our approach therefore commences with a description of communication between sets and states which precedes the development of irreducibility.

4.1   Communication and irreducibility: Countable spaces

When X is general, it is not always easy to describe the specific points or even sets which can be reached from different starting points $x \in \mathsf{X}$. To guide our development, therefore, we will first consider the simpler and more easily understood situation when the space X is countable; and to fix some of these ideas we will initially analyze briefly the communication behavior of the random walk on a half line defined by (RWHL1), in the case where the increment variable takes on integer values.

4.1.1   Communication: random walk on a half line

Recall that the random walk on a half line Φ is constructed from a sequence of i.i.d. random variables $\{W_i\}$ taking values in $\mathbb{Z} = (\ldots, -2, -1, 0, 1, 2, \ldots)$, by setting

$$\Phi_n = [\Phi_{n-1} + W_n]^+. \tag{4.2}$$

We know from Section 3.3.2 that this construction gives, for $y \in \mathbb{Z}_+$,

$$
\begin{aligned}
P(x, y) &= \mathsf{P}(W_1 = y - x), \\
P(x, 0) &= \mathsf{P}(W_1 \le -x).
\end{aligned}
\tag{4.3}
$$

In this example, we might single out the set $\{0\}$ and ask: can the chain ever reach the state $\{0\}$?

It is transparent from the definition of $P(x, 0)$ that $\{0\}$ can be reached with positive probability, and in one step, provided the distribution Γ of the increment $\{W_n\}$ has an infinite negative tail. But suppose we have, not such a long tail, but only $P(W_n < 0) > 0$, with, say,

$$\Gamma(w) = \delta > 0 \tag{4.4}$$

for some $w < 0$. Then we have for any x that after $n \geq |x/w|$ steps,

$$P_x(\Phi_n = 0) \geq P(W_1 = w, W_2 = w, \ldots, W_n = w) = \delta^n > 0$$

so that $\{0\}$ is always reached with positive probability.

On the other hand, if $P(W_n < 0) = 0$ then it is equally clear that $\{0\}$ cannot be reached with positive probability from any starting point other than 0. Hence $L(x, 0) > 0$ for all states x or for none, depending on whether (4.4) holds or not.

But we might also focus on points other than $\{0\}$, and it is then possible that a number of different sorts of behavior may occur, depending on the distribution of W. If we have $P(W = y) > 0$ for all $y \in \mathbb{Z}$ then from any state there is positive probability of Φ reaching any other state at the next step. But suppose we have the distribution of the increments $\{W_n\}$ concentrated on the even integers, with

$$P(W = 2y) > 0, \qquad P(W = 2y + 1) = 0, \qquad y \in \mathbb{Z},$$

and consider any odd valued state, say w. In this case w cannot be reached from any even valued state, even though from w itself it is possible to reach every state with positive probability, via transitions of the chain through $\{0\}$.

Thus for this rather trivial example, we already see X breaking into two subsets with substantially different behavior: writing $\mathbb{Z}_+^0 = \{2y, y \in \mathbb{Z}_+\}$ and $\mathbb{Z}_+^1 = \{2y + 1, y \in \mathbb{Z}_+\}$ for the set of non-negative even and odd integers respectively, we have

$$\mathbb{Z}_+ = \mathbb{Z}_+^0 \cup \mathbb{Z}_+^1,$$

and from $y \in \mathbb{Z}_+^1$, every state may be reached, whilst for $y \in \mathbb{Z}_+^0$, only states in $\mathbb{Z}_+^0$ may be reached with positive probability.

Why are these questions of importance?

As we have already seen, the random walk on a half line above is one with many applications: recall that the transition matrices of $N = \{N_n\}$ and $N^* = \{N_n^*\}$, the chains introduced in Section 2.4.2 to describe the number of customers in GI/M/1 and M/G/1 queues, have exactly the structure described by (4.3).

The question of reaching $\{0\}$ is then clearly one of considerable interest, since it represents exactly the question of whether the queue will empty with positive probability. Equally, the fact that when $\{W_n\}$ is concentrated on the even integers (representing some degenerate form of batch arrival process) we will always have an even number of customers has design implications for number of servers (do we always want to have two?), waiting rooms and the like.

But our efforts should and will go into finding conditions to preclude such oddities, and we turn to these in the next section, where we develop the concepts of communication and irreducibility in the countable space context.

4.1.2 Communicating classes and irreducibility

The idea of a Markov chain Φ reaching sets or points is much simplified when X is countable and the behavior of the chain is governed by a transition probability matrix

$P = P(x, y)$, $x, y \in \mathsf{X}$. There are then a number of essentially equivalent ways of defining the operation of communication between states.

The simplest is to say that state x *leads to state* y, which we write as $x \to y$, if $L(x, y) > 0$, and that two distinct states x and y in X *communicate*, written $x \leftrightarrow y$, when $L(x, y) > 0$ and $L(y, x) > 0$. By convention we also define $x \to x$.

The relation $x \leftrightarrow y$ is often defined equivalently by requiring that there exists $n(x, y) \geq 0$ and $m(y, x) \geq 0$ such that $P^n(x, y) > 0$ and $P^m(y, x) > 0$; that is, $\sum_{n=0}^{\infty} P^n(x, y) > 0$ and $\sum_{n=0}^{\infty} P^n(y, x) > 0$.

Proposition 4.1.1. *The relation "$\leftrightarrow$" is an equivalence relation, and so the equivalence classes $C(x) = \{y : x \leftrightarrow y\}$ cover X, with $x \in C(x)$.*

PROOF   By convention $x \leftrightarrow x$ for all x. By the symmetry of the definition, $x \leftrightarrow y$ if and only if $y \leftrightarrow x$.

Moreover, from the Chapman–Kolmogorov relationships (3.23) we have that if $x \leftrightarrow y$ and $y \leftrightarrow z$ then $x \leftrightarrow z$. For suppose that $x \to y$ and $y \to z$, and choose $n(x, y)$ and $m(y, z)$ such that $P^n(x, y) > 0$ and $P^m(y, z) > 0$. Then we have from (3.23)

$$P^{n+m}(x, z) \geq P^n(x, y) P^m(y, z) > 0$$

so that $x \to z$: the reverse direction is identical.   □

Chains for which all states communicate form the basis for future analysis.

Irreducible spaces and absorbing sets

If $C(x) = \mathsf{X}$ for some x, then we say that X (or the chain $\{X_n\}$) is *irreducible*.

We say $C(x)$ is *absorbing* if $P(y, C(x)) = 1$ for all $y \in C(x)$.

When states do not all communicate, then although each state in $C(x)$ communicates with every other state in $C(x)$, it is possible that there are states $y \in [C(x)]^c$ such that $x \to y$. This happens, of course, if and only if $C(x)$ is not absorbing.

Suppose that X is not irreducible for Φ. If we reorder the states according to the equivalence classes defined by the communication operation, and if we further order the classes with absorbing classes coming first, then we have a decomposition of P such as that depicted in Figure 4.1.

Here, for example, the blocks $C(1)$, $C(2)$ and $C(3)$ correspond to absorbing classes, and block D contains those states which are not contained in an absorbing class. In the extreme case, a state in D may communicate only with itself, although it must lead to some other state from which it does not return. We can write this decomposition as

$$\mathsf{X} = \left(\sum_{x \in I} C(x) \right) \cup D \tag{4.5}$$

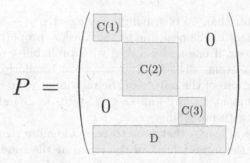

$$P = \begin{pmatrix} \boxed{C(1)} & & & 0 \\ & \boxed{C(2)} & & \\ 0 & & \boxed{C(3)} & \\ & \multicolumn{2}{c}{D} & \end{pmatrix}$$

Figure 4.1: Block decomposition of P into communicating classes

where the sum is of disjoint sets.

This structure allows chains to be analyzed, at least partially, through their constituent irreducible classes. We have

Proposition 4.1.2. *Suppose that $C := C(x)$ is an absorbing communicating class for some $x \in \mathsf{X}$. Let P_C denote the matrix P restricted to the states in C. Then there exists an irreducible Markov chain Φ_C whose state space is restricted to C and whose transition matrix is given by P_C.*

PROOF   We merely need to note that the elements of P_C are positive, and

$$\sum_{y \in C} P(x, y) \equiv 1, \qquad x \in C,$$

because C is absorbing: the existence of Φ_C then follows from Theorem 3.2.1, and irreducibility of Φ_C is an obvious consequence of the communicating class structure of C.   □

Thus for non-irreducible chains, we can analyze at least the absorbing subsets in the decomposition (4.5) as separate chains.

The virtue of the block decomposition described above lies largely in this assurance that any chain on a countable space can be studied assuming irreducibility. The "irreducible absorbing" pieces $C(x)$ can then be put together to deduce most of the properties of a reducible chain.

Only the behavior of the remaining states in D must be studied separately, and in analyzing stability D may often be ignored. For let J denote the indices of the states for which the communicating classes are not absorbing. If the chain starts in $D = \bigcup_{y \in J} C(y)$, then one of two things happens: either it reaches one of the absorbing sets $C(x), x \in \mathsf{X} \backslash J$, in which case it gets absorbed: or, as the only other alternative, the chain leaves every finite subset of D and "heads to infinity".

To see why this might hold, observe that, for any fixed $y \in J$, there is some state $z \in C(y)$ with $P(z, [C(y)]^c) = \delta > 0$ (since $C(y)$ is not an absorbing class), and $P^m(y, z) = \beta > 0$ for some $m > 0$ (since $C(y)$ is a communicating class). Suppose that in fact the chain returns a number of times to y: then, on each of these returns, one has a probability greater than $\beta\delta$ of leaving $C(y)$ exactly $m+1$ steps later, and this probability is independent of the past due to the Markov property.

Now, as is well known, if one tosses a coin with probability of a head given by $\beta\delta$ infinitely often, then one eventually actually gets a head: similarly, one eventually leaves the class $C(y)$, and because of the nature of the relation $x \leftrightarrow y$, one never returns.

Repeating this argument for any finite set of states in D indicates that the chain leaves such a finite set with probability one.

There are a number of things that need to be made more rigorous in order for this argument to be valid: the forgetfulness of the chain at the random time of returning to y, giving the independence of the trials, is a form of the strong Markov property in Proposition 3.4.6, and the so-called "geometric trials argument" must be formalized, as we will do in Proposition 8.3.1 (iii).

Basically, however, this heuristic sketch is sound, and shows the directions in which we need to go: we find absorbing irreducible sets, and then restrict our attention to them, with the knowledge that the remainder of the states lead to clearly understood and (at least from a stability perspective) somewhat irrelevant behavior.

4.1.3   Irreducible models on a countable space

Some specific models will illustrate the concepts of irreducibility. It is valuable to notice that, although in principle irreducibility involves P^n for all n, in practice we usually find conditions only on P itself that ensure the chain is irreducible.

The forward recurrence time model

Let p be the increment distribution of a renewal process on $\mathbb{Z}_+$, and write

$$r = \sup(n : p(n) > 0). \tag{4.6}$$

Then from the definition of the forward recurrence time chain it is immediate that the set $A = \{1, 2, \ldots, r\}$ is absorbing, and the forward recurrence time chain restricted to A is irreducible: for if $x, y \in A$, with $x > y$ then $P^{x-y}(x, y) = 1$ whilst

$$P^{y+r-x}(y, x) > P^{y-1}(y, 1)p(r)P^{r-x}(r, x) = p(r) > 0. \tag{4.7}$$

Queueing models

Consider the number of customers $\mathbf{N}$ in the GI/M/1 queue. As shown in Proposition 3.3.1, we have $P(x, x+1) = p_0 > 0$, and so the structure of $\mathbf{N}$ ensures that by iteration, for any $x > 0$

$$P^x(0, x) > P(0, 1)P(1, 2) \cdots P(x-1, x) = [p_0]^x > 0.$$

But we also have $P(x, 0) > 0$ for any $x \geq 0$: hence we conclude that for any pair $x, y \in \mathsf{X}$, we have

$$P^{y+1}(x, y) > P(x, 0)P^y(0, y) > 0.$$

Thus the chain $\mathbf{N}$ is irreducible no matter what the distribution of the inter-arrival times.

A similar approach shows that the embedded chain $\mathbf{N}^*$ of the M/G/1 queue is always irreducible.

Unrestricted random walk

Let d be the greatest common divisor of $\{n : \Gamma(n) > 0\}$. If we have a random walk on $\mathbb{Z}$ with increment distribution Γ, each of the sets $D_r = \{md + r, m \in \mathbb{Z}\}$ for each $r = 0, 1, \ldots, d - 1$ is absorbing, so that the chain is not irreducible.

However, provided $\Gamma(-\infty, 0) > 0$ and $\Gamma(0, \infty) > 0$ the chain is irreducible when restricted to any one D_r. To see this we can use Lemma D.7.4: since $\Gamma(md) > 0$ for all $m > m_0$ we only need to move m_0 steps to the left and then we can reach all states in D_r above our starting point in one more step. Hence this chain admits a finite number of irreducible absorbing classes.

For a different type of behavior, let us suppose we have an increment distribution on the integers, $\mathsf{P}(W_n = x) > 0$, $x \in \mathbb{Z}$, so that $d = 1$; but assume the chain itself is defined on the whole set of rationals $\mathbb{Q}$.

If we start at a value $q \in \mathbb{Q}$ then Φ "lives" on the set $C(q) = \{n + q, n \in \mathbb{Z}\}$, which is both absorbing and irreducible: that is, we have $P(q, C(q)) = 1$, $q \in \mathbb{Q}$, and for any $r \in C(q)$, $P(r, q) > 0$ also.

Thus this chain admits a countably infinite number of absorbing irreducible classes, in contrast to the behavior of the chain on the integers.

4.2 ψ-Irreducibility

4.2.1 The concept of φ-irreducibility

We now wish to develop similar concepts of irreducibility on a general space X. The obvious problem with extending the ideas of Section 4.1.2 is that we cannot define an analogue of "$\leftrightarrow$", since, although we can look at $L(x, A)$ to decide whether a set A is reached from a point x with positive probability, we cannot say in general that we return to single states x.

This is particularly the case for models such as the linear models for which the n-step transition laws typically have densities; and even for some of the models such as storage models where there is a distinguished reachable point, there are usually no other states to which the chain returns with positive probability.

This means that we cannot develop a decomposition such as (4.5) based on a countable equivalence class structure: and indeed the question of existence of a so-called "Doeblin decomposition"

$$\mathsf{X} = \left(\sum_{x \in I} C(x) \right) \cup D, \tag{4.8}$$

with the sets $C(x)$ being a countable collection of absorbing sets in $\mathcal{B}(\mathsf{X})$ and the "remainder" D being a set which is in some sense ephemeral, is a non-trivial one. We shall not discuss such reducible decompositions in this book although, remarkably, under a variety of reasonable conditions such a countable decomposition does hold for chains on quite general state spaces.

Rather than developing this type of decomposition structure, it is much more fruitful to concentrate on irreducibility analogues. The one which forms the basis for much modern general state space analysis is φ-irreducibility.

φ-Irreducibility for general space chains

We call $\Phi = \{\Phi_n\}$ *φ-irreducible* if there exists a measure φ on $\mathcal{B}(\mathsf{X})$ such that, whenever $\varphi(A) > 0$, we have $L(x, A) > 0$ for all $x \in \mathsf{X}$.

There are a number of alternative formulations of φ-irreducibility. Define the transition kernel

$$K_{a_{\frac{1}{2}}}(x, A) := \sum_{n=0}^{\infty} P^n(x, A) 2^{-(n+1)}, \qquad x \in \mathsf{X}, \ A \in \mathcal{B}(\mathsf{X}); \qquad (4.9)$$

this is a special case of the resolvent of Φ introduced in Section 3.4.2, and which we consider in Section 5.5.1 in more detail. The kernel $K_{a_{\frac{1}{2}}}$ defines for each x a probability measure equivalent to $I(x, A) + U(x, A) = \sum_{n=0}^{\infty} P^n(x, A)$, which may be infinite for many sets A.

Proposition 4.2.1. *The following are equivalent formulations of φ-irreducibility:*

(i) *for all $x \in \mathsf{X}$, whenever $\varphi(A) > 0$, $U(x, A) > 0$;*

(ii) *for all $x \in \mathsf{X}$, whenever $\varphi(A) > 0$, there exists some $n > 0$, possibly depending on both A and x, such that $P^n(x, A) > 0$;*

(iii) *for all $x \in \mathsf{X}$, whenever $\varphi(A) > 0$ then $K_{a_{\frac{1}{2}}}(x, A) > 0$.*

PROOF   The only point that needs to be proved is that if $L(x, A) > 0$ for all $x \in A^c$ then, since $L(x, A) = P(x, A) + \int_{A^c} P(x, dy) L(y, A)$, we have $L(x, A) > 0$ for all $x \in \mathsf{X}$: thus the inclusion of the zero-time term in $K_{a_{\frac{1}{2}}}$ does not affect the irreducibility. $\square$

We will use these different expressions of φ-irreducibility at different times without further comment.

4.2.2   Maximal irreducibility measures

Although seemingly relatively weak, the assumption of φ-irreducibility precludes several obvious forms of "reducible" behavior. The definition guarantees that "big" sets (as measured by φ) are always reached by the chain with some positive probability, no matter what the starting point: consequently, the chain cannot break up into separate "reduced" pieces.

For many purposes, however, we need to know the reverse implication: that "negligible" sets B, in the sense that $\varphi(B) = 0$, are avoided with probability one from most starting points. This is by no means the case in general: any non-trivial restriction of an irreducibility measure is obviously still an irreducibility measure, and such restrictions can be chosen to give zero weight to virtually any selected part of the space.

For example, on a countable space if we only know that $x \to x^*$ for every x and some specific state $x^* \in \mathsf{X}$, then the chain is δ_{x^*}-irreducible.

This is clearly rather weaker than normal irreducibility on countable spaces, which demands two-way communication. Thus we now look to measures which are extensions, not restrictions, of irreducibility measures, and show that the φ-irreducibility condition extends in such a way that, if we do have an irreducible chain in the sense of Section 4.1, then the natural irreducibility measure (namely counting measure) is generated as a "maximal" irreducibility measure.

The maximal irreducibility measure will be seen to define the range of the chain much more completely than some of the other more arbitrary (or pragmatic) irreducibility measures one may construct initially.

Proposition 4.2.2. *If Φ is φ-irreducible for some measure φ, then there exists a probability measure ψ on $\mathcal{B}(\mathsf{X})$ such that*

(i) *Φ is ψ-irreducible;*

(ii) *for any other measure φ', the chain Φ is φ'-irreducible if and only if $\psi \succ \varphi'$;*

(iii) *if $\psi(A) = 0$, then $\psi\{y : L(y, A) > 0\} = 0$;*

(iv) *the probability measure ψ is equivalent to*

$$\psi'(A) := \int_{\mathsf{X}} \varphi'(dy) K_{a_{\frac{1}{2}}}(y, A),$$

for any finite irreducibility measure φ'.

PROOF   Since any probability measure which is equivalent to the irreducibility measure φ is also an irreducibility measure, we can assume without loss of generality that $\varphi(\mathsf{X}) = 1$. Consider the measure ψ constructed as

$$\psi(A) := \int_{\mathsf{X}} \varphi(dy) K_{\frac{1}{2}}(y, A). \tag{4.10}$$

It is obvious that ψ is also a probability measure on $\mathcal{B}(\mathsf{X})$. To prove that ψ has all the required properties, we use the sets

$$\bar{A}(k) = \left\{ y : \sum_{n=1}^{k} P^n(y, A) > k^{-1} \right\}.$$

The stated properties now involve repeated use of the Chapman–Kolmogorov equations. To see (i), observe that when $\psi(A) > 0$, then from (4.10), there exists some k such that $\varphi(\bar{A}(k)) > 0$, since $\bar{A}(k) \uparrow \left\{ y : \sum_{n \geq 1} P^n(y, A) > 0 \right\} = \mathsf{X}$. For any fixed x, by φ-irreducibility there is thus some m such that $P^m(x, \bar{A}(k)) > 0$. Then we have

$$\sum_{n=1}^{k} P^{m+n}(x, A) = \int_{\mathsf{X}} P^m(x, dy)\left(\sum_{n=1}^{k} P^n(y, A)\right) \geq k^{-1} P^m(x, \bar{A}(k)) > 0,$$

which establishes ψ-irreducibility.

Next let φ' be such that $\boldsymbol{\Phi}$ is φ'-irreducible. If $\varphi'(A) > 0$, we have $\sum_n P^n(y, A) > 0$ for all y, and by its definition $\psi(A) > 0$, whence $\psi \succ \varphi'$. Conversely, suppose that the chain is ψ-irreducible and that $\psi \succ \varphi'$. If $\varphi'\{A\} > 0$ then $\psi\{A\} > 0$ also, and by ψ-irreducibility it follows that $K_{a_{\frac{1}{2}}}(x, A) > 0$ for any $x \in \mathsf{X}$. Hence $\boldsymbol{\Phi}$ is φ'-irreducible, as required in (ii).

Result (iv) follows from the construction (4.10) and the fact that any two maximal irreducibility measures are equivalent, which is a consequence of (ii).

Finally, we have that

$$\int_{\mathsf{X}} \psi(dy) P^m(y, A) 2^{-m} = \int_{\mathsf{X}} \varphi(dy) \sum_n P^{m+n}(y, A) 2^{-(n+m+1)} \leq \psi(A)$$

from which the property (iii) follows immediately. $\square$

Although there are other approaches to irreducibility, we will generally restrict ourselves, in the general space case, to the concept of φ-irreducibility; or rather, we will seek conditions under which it holds. We will consistently use ψ to denote an arbitrary maximal irreducibility measure for $\boldsymbol{\Phi}$.

ψ-Irreducibility notation

(i) The Markov chain is called ψ-*irreducible* if it is φ-irreducible for some φ and the measure ψ is a *maximal irreducibility* measure satisfying the conditions of Proposition 4.2.2.

(ii) We write
$$\mathcal{B}^+(\mathsf{X}) := \{A \in \mathcal{B}(\mathsf{X}) : \psi(A) > 0\}$$
for the sets of positive ψ-measure; the equivalence of maximal irreducibility measures means that $\mathcal{B}^+(\mathsf{X})$ is uniquely defined.

(iii) We call a set $A \in \mathcal{B}(\mathsf{X})$ *full* if $\psi(A^c) = 0$.

(iv) We call a set $A \in \mathcal{B}(\mathsf{X})$ *absorbing* if $P(x, A) = 1$ for $x \in A$.

The following result indicates the links between absorbing and full sets. This result seems somewhat academic, but we will see that it is often the key to showing that very many properties hold for ψ-almost all states.

Proposition 4.2.3. *Suppose that* $\boldsymbol{\Phi}$ *is* ψ-*irreducible. Then*

(i) *every absorbing set is full,*

(ii) *every full set contains a non-empty, absorbing set.*

PROOF   If A is absorbing, then were $\psi(A^c) > 0$, it would contradict the definition of ψ as an irreducibility measure: hence A is full.

Suppose now that A is full, and set

$$B = \{y \in \mathsf{X} : \sum_{n=0}^{\infty} P^n(y, A^c) = 0\}.$$

We have the inclusion $B \subseteq A$ since $P^0(y, A^c) = 1$ for $y \in A^c$. Since $\psi(A^c) = 0$, from Proposition 4.2.2 (iii) we know $\psi(B) > 0$, so in particular B is non-empty. By the Chapman–Kolmogorov relationship, if $P(y, B^c) > 0$ for some $y \in B$, then we would have

$$\sum_{n=0}^{\infty} P^{n+1}(y, A^c) \geq \int_{B^c} P(y, dz) \Big\{ \sum_{n=0}^{\infty} P^n(z, A^c) \Big\}$$

which is positive: but this is impossible, and thus B is the required absorbing set.   □

If a set C is absorbing and if there is a measure ψ for which

$$\psi(B) > 0 \Rightarrow L(x, B) > 0, \qquad x \in C,$$

then we will call C an absorbing ψ-irreducible set.

Absorbing sets on a general space have exactly the properties of those on a countable space given in Proposition 4.1.2.

Proposition 4.2.4. *Suppose that A is an absorbing set. Let P_A denote the kernel P restricted to the states in A. Then there exists a Markov chain Φ_A whose state space is A and whose transition matrix is given by P_A. Moreover, if Φ is ψ-irreducible then Φ_A is ψ-irreducible.*

PROOF   The existence of Φ_A is guaranteed by Theorem 3.4.1 since $P_A(x, A) \equiv 1, x \in A$. If Φ is ψ-irreducible then A is full and the result is immediate by Proposition 4.2.3.   □

The effect of these two propositions is to guarantee the effective analysis of restrictions of chains to full sets, and we shall see that this is indeed a fruitful avenue of approach.

4.2.3   Uniform accessibility of sets

Although the relation $x \leftrightarrow y$ is not a generally useful one when X is uncountable, since $P^n(x, y) = 0$ in many cases, we now introduce the concepts of "accessibility" and, more usefully, "uniform accessibility" which strengthens the notion of communication on which ψ-irreducibility is based.

We will use uniform accessibility for chains on general and topological state spaces to develop solidarity results which are almost as strong as those based on the equivalence relation $x \leftrightarrow y$ for countable spaces.

Accessibility

We say that a set $B \in \mathcal{B}(X)$ is *accessible* from another set $A \in \mathcal{B}(X)$ if $L(x, B) > 0$ for every $x \in A$;

We say that a set $B \in \mathcal{B}(X)$ is *uniformly accessible* from another set $A \in \mathcal{B}(X)$ if there exists a $\delta > 0$ such that

$$\inf_{x \in A} L(x, B) \geq \delta; \tag{4.11}$$

and when (4.11) holds we write $A \rightsquigarrow B$.

The critical aspect of the relation "$A \rightsquigarrow B$" is that it holds uniformly for $x \in A$. In general the relation "$\rightsquigarrow$" is non-reflexive although clearly there may be sets A, B such that A is uniformly accessible from B and B is uniformly accessible from A.

Importantly, though, the relationship is transitive. In proving this we use the notation

$$U_A(x, B) = \sum_{n=1}^{\infty} {_A}P^n(x, B), \qquad x \in X, \ A, B \in \mathcal{B}(X);$$

introduced in (3.34).

Lemma 4.2.5. *If $A \rightsquigarrow B$ and $B \rightsquigarrow C$, then $A \rightsquigarrow C$.*

PROOF   Since the probability of ever reaching C is greater than the probability of ever reaching C after the first visit to B, we have

$$\inf_{x \in A} U_C(x, C) \geq \inf_{x \in A} \int_B U_B(x, dy) U_C(y, C) \geq \inf_{x \in A} U_B(y, B) \inf_{x \in B} U_C(y, C) > 0$$

as required.   □

We shall use the following notation to describe the communication structure of the chain.

Communicating sets

The set $\bar{A} := \{x \in X : L(x, A) > 0\}$ is the set of points from which A is accessible.

The set $\bar{A}(m) := \{x \in X : \sum_{n=1}^{m} P^n(x, A) \geq m^{-1}\}$.

The set $A^0 := \{x \in X : L(x, A) = 0\} = [\bar{A}]^c$ is the set of points from which A is not accessible.

Lemma 4.2.6. *The set $\bar{A} = \cup_m \bar{A}(m)$, and for each m we have $\bar{A}(m) \rightsquigarrow A$.*

PROOF   The first statement is obvious, whilst the second follows by noting that for all $x \in \bar{A}(m)$ we have

$$L(x, A) \geq \mathsf{P}_x(\tau_A \leq m) \geq m^{-2}.$$

$\square$

It follows that if the chain is ψ-irreducible, then we can find a countable cover of X with sets from which any other given set A in $\mathcal{B}^+(\mathsf{X})$ is uniformly accessible, since $\bar{A} = \mathsf{X}$ in this case.

4.3 ψ-Irreducibility for random walk models

One of the main virtues of ψ-irreducibility is that it is even easier to check than the standard definition of irreducibility introduced for countable chains. We first illustrate this using a number of models related to random walk.

4.3.1   Random walk on a half line

Let $\mathbf{\Phi}$ be a random walk on the half line $[0, \infty)$, with transition law as in Section 3.5. The communication structure of this chain is made particularly easy because of the "atom" at $\{0\}$.

Proposition 4.3.1. *The random walk on a half line $\mathbf{\Phi} = \{\Phi_n\}$ with increment variable W is φ-irreducible, with $\varphi(0, \infty) = 0$, $\varphi(\{0\}) = 1$, if and only if*

$$\mathsf{P}(W < 0) = \Gamma(-\infty, 0) > 0; \tag{4.12}$$

and in this case if C is compact then $C \rightsquigarrow \{0\}$.

PROOF   The necessity of (4.12) is trivial. Conversely, suppose for some δ, $\varepsilon > 0$, $\Gamma(-\infty, -\varepsilon) > \delta$. Then for any n, if $x/\varepsilon < n$,

$$P^n(x, \{0\}) \geq \delta^n > 0.$$

If $C = [0, c]$ for some c, then this implies for all $x \in C$ that

$$\mathsf{P}_x(\tau_0 \leq c/\varepsilon) \geq \delta^{1 + c/\varepsilon}$$

so that $C \rightsquigarrow \{0\}$ as in Lemma 4.2.6. $\square$

It is often as simple as this to establish φ-irreducibility: it is not a difficult condition to confirm, or rather, it is often easy to set up "grossly sufficient" conditions such as (4.12) for φ-irreducibility.

Such a construction guarantees φ-irreducibility, but it does not tell us very much about the motion of the chain. There are clearly many sets other than $\{0\}$ which the chain will reach from any starting point. To describe them in this model we can easily construct the maximal irreducibility measure. By considering the motion of the chain after it reaches $\{0\}$ we see that $\mathbf{\Phi}$ is also ψ-irreducible, where

$$\psi(A) = \sum_n P^n(0, A) 2^{-n};$$

we have that ψ is maximal from Proposition 4.2.2.

4.3.2   Storage models

If we apply the result of Proposition 4.3.1 to the simple storage model defined by (SSM1) and (SSM2), we will establish ψ-irreducibility provided we have

$$\mathsf{P}(S_n - J_n < 0) > 0.$$

Provided there is some probability that no input takes place over a period long enough to ensure that the effect of the increment S_n is eroded, we will achieve δ_0-irreducibility in one step. This amounts to saying that we can "turn off" the input for a period longer than s whenever the last input amount was s, or that we need a positive probability of the input remaining turned off for longer than s/r. One sufficient condition for this is obviously that the distribution H have infinite tails.

Such a construction may fail without the type of conditions imposed here. If, for example, the input times are deterministic, occurring at every integer time point, and if the input amounts are always greater than unity, then we will not have an irreducible system: in fact we will have, in the terms of Chapter 9 below, an evanescent system which always avoids compact sets below the initial state.

An underlying structure as pathological as this seems intuitively implausible, of course, and is in any case easily analyzed. But in the case of content-dependent release rules, it is not so obvious that the chain is always φ-irreducible. If we assume $R(x) = \int_0^x [r(y)]^{-1} dy < \infty$ as in (2.32), then again if we can "turn off" the input process for longer than $R(x)$ we will hit $\{0\}$; so if we have

$$\mathsf{P}(T_i > R(x)) > 0$$

for all x we have a δ_0-irreducible model. But if we allow $R(x) = \infty$ as we may wish to do for some release rules where $r(x) \to 0$ slowly as $x \to 0$, which is not unrealistic, then even if the inter-input times T_i have infinite tails, this simple construction will fail. The empty state will never be reached, and some other approach is needed if we are to establish φ-irreducibility.

In such a situation, we will still get μ^{Leb}-irreducibility, where μ^{Leb} is Lebesgue measure, if the inter-input times T_i have a density with respect to μ^{Leb}: this can be determined by modifying the "turning off" construction above. Exact conditions for φ-irreducibility in the completely general case appear to be unknown to date.

4.3.3   Unrestricted random walk

The random walk on a half line, and the various applications of it in storage and queueing, have a single state reached from all initial points, which forms a natural candidate to generate an irreducibility measure. The unrestricted random walk requires more analysis, and is an example where the irreducibility measure is not formed by a simple regenerative structure.

For unrestricted random walk $\mathbf{\Phi}$ given by

$$\Phi_{k+1} = \Phi_k + W_{k+1},$$

and satisfying the assumption (RW1), let us suppose the increment distribution Γ of $\{W_n\}$ has an absolutely continuous part with respect to Lebesgue measure μ^{Leb} on $\mathbb{R}$, with a density γ which is positive and bounded from zero at the origin; that is, for some $\beta > 0, \delta > 0$,

$$P(W_n \in A) \geq \int_A \gamma(x)\, dx,$$

and

$$\gamma(x) \geq \delta > 0, \qquad |x| < \beta.$$

Set $C = \{x : |x| \leq \beta/2\}$: if $B \subseteq C$, and $x \in C$ then

$$\begin{aligned} P(x, B) &= P(W_1 \in B - x) \\ &\geq \int_{B-x} \gamma(y)\, dy \\ &\geq \delta \mu^{\text{Leb}}(B). \end{aligned}$$

But now, exactly as in the previous example, from any x we can reach C in at most $n = 2|x|/\beta$ steps with positive probability, so that μ^{Leb} restricted to C forms an irreducibility measure for the unrestricted random walk.

Such behavior might not hold without a density. Suppose we take Γ concentrated on the rationals $\mathbb{Q}$, with $\Gamma(r) > 0$, $r \in \mathbb{Q}$. After starting at a value $r \in \mathbb{Q}$ the chain $\boldsymbol{\Phi}$ "lives" on the set $\{r + q, q \in \mathbb{Q}\} = \mathbb{Q}$ so that $\mathbb{Q}$ is absorbing. But for any $x \in \mathbb{R}$ the set $\{x + q, q \in \mathbb{Q}\} = x + \mathbb{Q}$ is also absorbing, and thus we can produce, for this random walk on $\mathbb{R}$, an uncountably infinite number of absorbing irreducible sets.

It is precisely this type of behavior we seek to exclude for chains on a general space, by introducing the concepts of ψ-irreducibility above.

4.4 ψ-Irreducible linear models

4.4.1 Scalar models

Let us consider the scalar autoregressive AR(k) model

$$Y_n = \alpha_1 Y_{n-1} + \alpha_2 Y_{n-2} + \cdots + \alpha_k Y_{n-k} + W_n,$$

where $\alpha_1, \ldots, \alpha_k \in \mathbb{R}$, as defined in (AR1). If we assume the Markovian representation in (2.1), then we can determine conditions for ψ-irreducibility very much as for random walk.

In practice the condition most likely to be adopted is that the innovation process W has a distribution Γ with an everywhere positive density. If the innovation process is Gaussian, for example, then clearly this condition is satisfied. We will see below, in the more general Proposition 4.4.3, that the chain is then μ^{Leb}-irreducible regardless of the values of $\alpha_1, \ldots, \alpha_k$.

It is however not always sufficient for φ-irreducibility to have a density only positive in a neighborhood of zero. For suppose that W is uniform on $[-1, 1]$, and that $k = 1$ so we have a first order autoregression. If $|\alpha_1| \leq 1$ the chain will be $\mu^{\text{Leb}}_{[-1,1]}$-irreducible under such a density condition: the argument is the same as for the random walk. But if $|\alpha_1| > 1$, then once we have an initial state larger than $(|\alpha_1| - 1)^{-1}$, the chain will monotonically "explode" towards infinity and will not be irreducible.

This same argument applies to the general model (2.1) if the zeros of the polynomial $A(z) = 1 - \alpha_1 z^1 - \cdots - \alpha_k z^k$ lie outside of the closed unit disk in the complex plane $\mathbb{C}$. In this case $Y_n \to 0$ as $n \to \infty$ when W_n is set equal to zero, and from this observation it follows that it is possible for the chain to reach $[-1, 1]$ at some time in the future from every initial condition. If some root of $A(z)$ lies within the open unit disk in $\mathbb{C}$ then again "explosion" will occur and the chain will not be irreducible.

Our argument here is rather like that in the dam model, where we considered deterministic behavior with the input "turned off". We need to be able to drive the chain deterministically towards a center of the space, and then to be able to ensure that the random mechanism ensures that the behavior of the chain from initial conditions in that center are comparable.

We formalize this for multidimensional linear models in the rest of this section.

4.4.2   Communication for linear control models

Recall that the linear control model LCM(F,G) defined in (LCM1) by $x_{k+1} = Fx_k + Gu_{k+1}$ is called *controllable* if for each pair of states $x_0, x^\star \in \mathsf{X}$, there exists $m \in \mathbb{Z}_+$ and a sequence of control variables $(u_1^\star, \ldots, u_m^\star) \in \mathbb{R}^p$ such that $x_m = x^\star$ when $(u_1, \ldots, u_m) = (u_1^\star, \ldots, u_m^\star)$, and the initial condition is equal to x_0.

This is obviously a concept of communication between states for the deterministic model: we can choose the inputs u_k in such a way that all states can be reached from any starting point. We first analyze this concept for the deterministic control model then move on to the associated linear state space model LSS(F,G), where we see that controllability of LCM(F,G) translates into ψ-irreducibility of LSS(F,G) under appropriate conditions on the noise sequence.

For the LCM(F,G) model it is possible to decide explicitly using a finite procedure when such control can be exerted. We use the following rank condition for the pair of matrices (F, G):

Controllability for the linear control model

Suppose that the matrices F and G have dimensions $n \times n$ and $n \times p$, respectively.

(LCM3)   The matrix

$$C_n := [F^{n-1}G \mid \cdots \mid FG \mid G] \qquad (4.13)$$

is called the *controllability matrix*, and the pair of matrices (F, G) is called *controllable* if the controllability matrix C_n has rank n.

It is a consequence of the Cayley Hamilton Theorem, which states that any power F^k is equal to a linear combination of $\{I, F, \ldots, F^{n-1}\}$, where n is equal to the dimension of F (see [57] for details), that (F, G) is controllable if and only if

$$[F^{k-1}G \mid \cdots \mid FG \mid G]$$

has rank n for some $k \in \mathbb{Z}_+$.

Proposition 4.4.1. *The linear control model LCM(F,G) is controllable if the pair (F, G) satisfy the rank condition (LCM3).*

PROOF   When this rank condition holds it is straightforward that in the LCM(F,G) model any state can be reached from any initial condition in k steps using some control sequence $(u_1, \ldots, u_k)$, for we have by

$$x_k = F^k x_0 + [F^{k-1}G \mid \cdots \mid FG \mid G] \begin{pmatrix} u_1 \\ \vdots \\ u_k \end{pmatrix} \tag{4.14}$$

and the rank condition implies that the range space of the matrix $[F^{k-1}G \mid \cdots \mid FG \mid G]$ is equal to $\mathbb{R}^n$. □

This gives us as an immediate application

Proposition 4.4.2. *The autoregressive AR(k) model may be described by a linear control model (LCM1), which can always be constructed so that it is controllable.*

PROOF   For the linear control model associated with the autoregressive model described by (2.1), the state process x is defined inductively by

$$x_n = \begin{bmatrix} \alpha_1 & \cdots & \cdots & \alpha_k \\ 1 & & & 0 \\ & \ddots & & \vdots \\ 0 & & 1 & 0 \end{bmatrix} x_{n-1} + \begin{bmatrix} 1 \\ 0 \\ \vdots \\ 0 \end{bmatrix} u_n,$$

and we can compute the controllability matrix C_n of (LCM3) explicitly:

$$C_n = [F^{n-1}G \mid \cdots \mid FG \mid G] = \begin{bmatrix} \eta_{k-1} & \cdots & \eta_2 & \eta_1 & 1 \\ \vdots & & & 1 & 0 \\ \eta_2 & & & & \vdots \\ \eta_1 & 1 & & & \vdots \\ 1 & 0 & \cdots & \cdots & 0 \end{bmatrix}$$

where we define $\eta_0 = 1$, $\eta_i = 0$ for $i < 0$, and for $j \geq 2$,

$$\eta_j = \sum_{i=1}^{k} \alpha_i \eta_{j-i}.$$

The triangular structure of the controllability matrix now implies that the linear control system associated with the AR(k) model is controllable. □

4.4.3 Gaussian linear models

For the LSS(F,G) model

$$X_{k+1} = FX_k + GW_{k+1}$$

described by (LSS1) and (LSS2) to be ψ-irreducible, we now show that it is sufficient that the associated LCM(F,G) model be controllable and the noise sequence $\boldsymbol{W}$ have a distribution that in effect allows a full cross-section of the possible controls to be chosen. We return to the general form of this in Section 6.3.2 but address a specific case of importance immediately. The Gaussian linear state space model is described by (LSS1) and (LSS2) with the additional hypothesis

---

Disturbance for the Gaussian state space model

(LSS3)   The noise variable W has a Gaussian distribution on $\mathbb{R}^p$ with zero mean and unit variance: that is, $W \sim N(0, I)$, where I is the $p \times p$ identity matrix.

If the dimension p of the noise were the same as the dimension n of the space, and if the matrix G were full rank, then the argument for scalar models in Section 4.4 would immediately imply that the chain is μ^{Leb}-irreducible. In more general situations we use controllability to ensure that the chain is μ^{Leb}-irreducible.

Proposition 4.4.3. *Suppose that the LSS(F,G) model is Gaussian and the associated control model is controllable.*

Then the LSS(F,G) model is φ-irreducible for any non-trivial measure φ which possesses a density on $\mathbb{R}^n$, Lebesgue measure is a maximal irreducibility measure, and for any compact set A and any set B with positive Lebesgue measure we have $A \rightsquigarrow B$.

PROOF   If we can prove that the distribution $P^k(x, \cdot)$ is absolutely continuous with respect to Lebesgue measure, and has a density which is everywhere positive on $\mathbb{R}^n$, it will follow that for any φ which is non-trivial and also possesses a density, $P^k(x, \cdot) \succ \varphi$ for all $x \in \mathbb{R}^n$: for any such φ the chain is then φ-irreducible. This argument also shows that Lebesgue measure is a maximal irreducibility measure for the chain.

Under condition (LSS3), for each deterministic initial condition $x_0 \in \mathsf{X} = \mathbb{R}^n$, the distribution of X_k is also Gaussian for each $k \in \mathbb{Z}_+$ by linearity, and so we need only to prove that $P^k(x, \cdot)$ is not concentrated on some lower dimensional subspace of $\mathbb{R}^n$. This will happen if and only if the variance of the distribution $P^k(x, \cdot)$ is of full rank for each x.

We can compute the mean and variance of X_k to obtain conditions under which this occurs. Using (4.14) and (LSS3), for each initial condition $x_0 \in \mathsf{X}$ the conditional mean of X_k is easily computed as

$$\mu_k(x_0) := \mathsf{E}_{x_0}[X_k] = F^k x_0 \tag{4.15}$$

and the conditional variance of X_k is given independently of x_0 by

$$\Sigma_k := \mathsf{E}_{x_0}[(X_k - \mu_k(x_0))(X_k - \mu_k(x_0))^\top] = \sum_{i=0}^{k-1} F^i GG^\top F^{i\top}. \qquad (4.16)$$

Using (4.16), the variance of X_k has full rank n for some k if and only if the *controllability grammian*, defined as

$$\sum_{i=0}^{\infty} F^i GG^\top F^{i\top}, \qquad (4.17)$$

has rank n. From the Cayley Hamilton Theorem again, the conditional variance of X_k has rank n for some k if and only if the pair (F, G) is controllable and, if this is the case, then one can take $k = n$.

Under (LSS1)–(LSS3), it thus follows that the k-step transition function possesses a smooth density; we have $P^k(x, dy) = p_k(x, y)dy$ where

$$p_k(x, y) = (2\pi|\Sigma_k|)^{-k/2} \exp\left\{-\tfrac{1}{2}(y - F^k x)^\top \Sigma_k^{-1}(y - F^k x)\right\} \qquad (4.18)$$

and $|\Sigma_k|$ denotes the determinant of the matrix Σ_k. Hence $P^k(x, \cdot)$ has a density which is everywhere positive, as required, and this implies finally that for any compact set A and any set B with positive Lebesgue measure we have $A \rightsquigarrow B$. □

Assuming, as we do in the result above, that W has a density which is everywhere positive is clearly something of a sledge hammer approach to obtaining ψ-irreducibility, even though it may be widely satisfied. We will introduce more delicate methods in Chapter 7 which will allow us to relax the conditions of Proposition 4.4.3.

Even if (F, G) is not controllable then we can obtain an irreducible process, by appropriate restriction of the space on which the chain evolves, under the Gaussian assumption. To define this formally, we let $\mathsf{X}_0 \subset \mathsf{X}$ denote the *range space* of the controllability matrix:

$$\begin{aligned} \mathsf{X}_0 &= \mathcal{R}([F^{n-1}G \mid \cdots \mid FG \mid G]) \\ &= \left\{\sum_{i=0}^{n-1} F^i Gw_i : w_i \in \mathbb{R}^p\right\}, \end{aligned}$$

which is also the range space of the controllability grammian. If $x_0 \in \mathsf{X}_0$ then so is $Fx_0 + Gw_1$ for any $w_1 \in \mathbb{R}^p$. This shows that the set X_0 is absorbing, and hence the LSS(F,G) model may be restricted to X_0.

The restricted process is then described by a linear state space model, similar to (LSS1), but evolving on the space X_0 whose dimension is strictly less than n. The matrices (F_0, G_0) which define the dynamics of the restricted process are a controllable pair, so that by Proposition 4.4.3, the restricted process is μ^{Leb}-irreducible.

4.5 Commentary

The communicating class concept was introduced in the initial development of countable chains by Kolmogorov [216] and used systematically by Feller [114] and Chung [71] in developing solidarity properties of states in such a class.

The use of ψ-irreducibility as a basic tool for general chains was essentially developed by Doeblin [93, 95], and followed up by many authors, including Doob [99], Harris [155], Chung [70], Orey [308]. Much of their analysis is considered in greater detail in later chapters. The maximal irreducibility measure was introduced by Tweedie [394], and the result on full sets is given in the form we use by Nummelin [303]. Although relatively simple they have wide-ranging implications.

Other notions of irreducibility exist for general state space Markov chains. One can, for example, require that the transition probabilities

$$K_{\frac{1}{2}}(x,\cdot) = \sum_{n=0}^{\infty} P^n(x,\cdot)2^{-(n+1)}$$

all have the same null sets. In this case the maximal measure ψ will be equivalent to $K_{\frac{1}{2}}(x,\cdot)$ for every x. This was used by Nelson [291] and Šidák [353] to derive solidarity properties for general state space chains similar to those we will consider in Part II. This condition, though, is hard to check, since one needs to know the structure of $P^n(x,\cdot)$ in some detail; and it appears too restrictive for the minor gains it leads to.

In the other direction, one might weaken φ-irreducibility by requiring only that, whenever $\varphi(A) > 0$, we have $\sum_n P^n(x,A) > 0$ only for φ-almost all $x \in \mathsf{X}$. Whilst this expands the class of "irreducible" models, it does not appear to be noticeably more useful in practice, and has the drawback that many results are much harder to prove as one tracks the uncountably many null sets which may appear. Revuz [326] Chapter 3 has a discussion of some of the results of using this weakened form.

The existence of a block decomposition of the form

$$\mathsf{X} = \left(\sum_{x \in I} C(x) \right) \cup D$$

such as that for countable chains, where the sum is of disjoint irreducible sets and D is in some sense ephemeral, has been widely studied. A recent overview is in Meyn and Tweedie [281], and the original ideas go back, as so often, to Doeblin [95], after whom such decompositions are named. Orey [309], Chapter 9, gives a very accessible account of the measure-theoretic approach to the Doeblin decomposition.

Application of results for ψ-irreducible chains has become more widespread recently, but the actual usage has suffered a little because of the somewhat inadequate available discussion in the literature of practical methods of verifying ψ-irreducibility. Typically the assumptions are far too restrictive, as is the case in assuming that innovation processes have everywhere positive densities or that accessible regenerative atoms exist (see for example Laslett et al. [237] for simple operations research models, or Tong [388] in time series analysis).

The detailed analysis of the linear model begun here illustrates one of the recurring themes of this book: the derivation of stability properties for stochastic models by consideration of the properties of analogous controlled deterministic systems. The methods described here have surprisingly complete generalizations to nonlinear models. We will come back to this in Chapter 7 when we characterize irreducibility for the NSS(F) model using ideas from nonlinear control theory.

Irreducibility, whilst it is a cornerstone of the theory and practice to come, is nonetheless rather a mundane aspect of the behavior of a Markov chain. We now explore some far more interesting consequences of the conditions developed in this chapter.

Chapter 5

Pseudo-atoms

Much Markov chain theory on a general state space can be developed in complete analogy with the countable state situation when X contains an *atom* for the chain Φ.

Atoms

A set $\alpha \in \mathcal{B}(\mathsf{X})$ is called an *atom* for Φ if there exists a measure ν on $\mathcal{B}(\mathsf{X})$ such that

$$P(x, A) = \nu(A), \quad x \in \alpha.$$

If Φ is ψ-irreducible and $\psi(\alpha) > 0$ then α is called an *accessible atom*.

A single point α is always an atom. Clearly, when X is countable and the chain is irreducible then every point is an accessible atom.

On a general state space, accessible atoms are less frequent. For the random walk on a half line as in (RWHL1), the set $\{0\}$ is an accessible atom when $\Gamma(-\infty, 0) > 0$: as we have seen in Proposition 4.3.1, this chain has $\psi(\{0\}) > 0$. But for the random walk on $\mathbb{R}$ when Γ has a density, accessible atoms do not exist.

It is not too strong to say that the single result which makes general state space Markov chain theory as powerful as countable space theory is that there exists an "artificial atom" for φ-irreducible chains, even in cases such as the random walk with absolutely continuous increments. The highlight of this chapter is the development of this result, and some of its immediate consequences.

Atoms are found for "strongly aperiodic" chains by constructing a "split chain" $\check{\Phi}$ evolving on a split state space $\check{\mathsf{X}} = \mathsf{X}_0 \cup \mathsf{X}_1$, where X_0 and X_1 are copies of the state space X, in such a way that

(i) the chain Φ is the marginal chain of $\check{\Phi}$, in the sense that $\mathsf{P}(\Phi_k \in A) = \mathsf{P}(\check{\Phi}_k \in A_0 \cup A_1)$ for appropriate initial distributions, and

(ii) the "bottom level" X_1 is an accessible atom for $\check{\Phi}$.

The existence of a splitting of the state space in such a way that the bottom level is an atom is proved in the next section. The proof requires the existence of so-called "small sets" C, which have the property that there exists an $m > 0$, and a minorizing measure ν on $\mathcal{B}(\mathsf{X})$ such that for any $x \in C$,

$$P^m(x, B) \geq \nu(B). \tag{5.1}$$

In Section 5.2, we show that, provided the chain is ψ-irreducible

$$\mathsf{X} = \bigcup_1^\infty C_i$$

where each C_i is small: thus we have that the splitting is always possible for such chains.

Another non-trivial consequence of the introduction of small sets is that on a general space we have a finite cyclic decomposition for ψ-irreducible chains: there is a cycle of sets $D_i, i = 0, 1, \ldots, d-1$ such that

$$\mathsf{X} = N \cup \bigcup_0^{d-1} D_i$$

where $\psi(N) = 0$ and $P(x, D_i) \equiv 1$ for $x \in D_{i-1} \pmod{d}$. A more general and more tractable class of sets called petite sets are introduced in Section 5.5: these are used extensively in the sequel, and in Theorem 5.5.7 we show that every petite set is small if the chain is aperiodic.

5.1   Splitting φ-irreducible chains

Before we get to these results let us first consider some simpler consequences of the existence of atoms.

As an elementary first step, it is clear from the proof of the existence of a maximal irreducibility measure in Proposition 4.2.2 that we have an easy construction of ψ when X contains an atom.

Proposition 5.1.1. *Suppose there is an atom α in X such that $\sum_n P^n(x, \alpha) > 0$ for all $x \in \mathsf{X}$. Then α is an accessible atom and Φ is ν-irreducible with $\nu = P(\alpha, \cdot)$.*

PROOF   We have, by the Chapman–Kolmogorov equations, that for any $n \geq 1$

$$
\begin{aligned}
P^{n+1}(x, A) &\geq \int_\alpha P^n(x, dy) P(y, A) \\
&= P^n(x, \alpha) \nu(A)
\end{aligned}
$$

which gives the result by summing over n.   □

The uniform communication relation "$\rightsquigarrow A$" introduced in Section 4.2.3 is also simplified if we have an atom in the space: it is no more than the requirement that there is a set of paths to A of positive probability, and the uniformity is automatic.

Proposition 5.1.2. *If $L(x, A) > 0$ for some state $x \in \alpha$, where α is an atom, then $\alpha \rightsquigarrow A$.* $\square$

In many cases the "atoms" in a state space will be real atoms: that is, single points which are reached with positive probability.

Consider the level in a dam in any of the storage models analyzed in Section 4.3.2. It follows from Proposition 4.3.1 that the single point $\{0\}$ forms an accessible atom satisfying the hypotheses of Proposition 5.1.1, even when the input and output processes are continuous.

However, our reason for featuring atoms is not because some models have singletons which can be reached with probability one: it is because even in the completely general ψ-irreducible case, by suitably extending the probabilistic structure of the chain, we are able to artificially construct sets which have an atomic structure and this allows much of the critical analysis to follow the form of the countable chain theory.

This unexpected result is perhaps the major innovation in the analysis of general Markov chains in the last two decades. It was discovered in slightly different forms, independently and virtually simultaneously, by Nummelin [301] and by Athreya and Ney [13].

Although the two methods are almost identical in a formal sense, in what follows we will concentrate on the Nummelin splitting, touching only briefly on the Athreya–Ney random renewal time method as it fits less well into the techniques of the rest of this book.

5.1.1   Minorization and splitting

To construct the artificial atom or regeneration point involves a probabilistic "splitting" of the state space in such a way that atoms for a "split chain" become natural objects.

In order to carry out this construction we need to consider sets satisfying the following

---

Minorization condition

For some $\delta > 0$, some $C \in \mathcal{B}(\mathsf{X})$ and some probability measure ν with $\nu(C^c) = 0$ and $\nu(C) = 1$

$$P(x, A) \geq \delta \mathbb{I}_C(x)\nu(A), \qquad A \in \mathcal{B}(\mathsf{X}), \; x \in \mathsf{X}. \tag{5.2}$$

The form (5.2) ensures that the chain has probabilities uniformly bounded below by multiples of ν for every $x \in C$. The crucial question is, of course, whether any chains ever satisfy the minorization condition. This is answered in the positive in Theorem 5.2.2 below: for φ-irreducible chains "small sets" for which the minorization condition holds exist, at least for the m-skeleton. The existence of such small sets is a deep and difficult result: by indicating first how the minorization condition provides the promised atomic structure to a split chain, we motivate rather more strongly the development of Theorem 5.2.2.

In order to construct a split chain, we split both the space and all measures that are defined on $\mathcal{B}(\mathsf{X})$.

We first split the space X itself by writing $\check{\mathsf{X}} = \mathsf{X} \times \{0, 1\}$, where $\mathsf{X}_0 := \mathsf{X} \times \{0\}$ and $\mathsf{X}_1 := \mathsf{X} \times \{1\}$ are thought of as copies of X equipped with copies $\mathcal{B}(\mathsf{X}_0)$, $\mathcal{B}(\mathsf{X}_1)$ of the σ-field $\mathcal{B}(\mathsf{X})$

We let $\mathcal{B}(\check{\mathsf{X}})$ be the σ-field of subsets of $\check{\mathsf{X}}$ generated by $\mathcal{B}(\mathsf{X}_0)$, $\mathcal{B}(\mathsf{X}_1)$: that is, $\mathcal{B}(\check{\mathsf{X}})$ is the smallest σ-field containing sets of the form $A_0 := A \times \{0\}$, $A_1 := A \times \{1\}$, $A \in \mathcal{B}(\mathsf{X})$.

We will write $x_i, i = 0, 1$ for elements of $\check{\mathsf{X}}$, with x_0 denoting members of the upper level X_0 and x_1 denoting members of the lower level X_1. In order to describe more easily the calculations associated with moving between the original and the split chain, we will also sometimes call X_0 the *copy* of X, and we will say that $A \in \mathcal{B}(\mathsf{X})$ is a *copy* of the corresponding set $A_0 \subseteq \mathsf{X}_0$.

If λ is any measure on $\mathcal{B}(\mathsf{X})$, then the next step in the construction is to *split* the measure λ into two measures on each of X_0 and X_1 by defining the measure λ^* on $\mathcal{B}(\check{\mathsf{X}})$ through

$$\left. \begin{array}{rcl} \lambda^*(A_0) & = & \lambda(A \cap C)[1 - \delta] + \lambda(A \cap C^c) \\ \lambda^*(A_1) & = & \lambda(A \cap C)\delta \end{array} \right\} \tag{5.3}$$

where δ and C are the constant and the set in (5.2). Note that in this sense the splitting is dependent on the choice of the set C, and although in general the set chosen is not relevant, we will on occasion need to make explicit the set in (5.2) when we use the split chain.

It is critical to note that λ is the marginal measure induced by λ^*, in the sense that for any A in $\mathcal{B}(\mathsf{X})$ we have

$$\lambda^*(A_0 \cup A_1) = \lambda(A). \tag{5.4}$$

In the case when $A \subseteq C^c$, we have $\lambda^*(A_0) = \lambda(A)$; only subsets of C are really effectively split by this construction.

Now the third, and most subtle, step in the construction is to split the chain Φ to form a chain $\check{\Phi}$ which lives on $(\check{\mathsf{X}}, \mathcal{B}(\check{\mathsf{X}}))$. Define the split kernel $\check{P}(x_i, A)$ for $x_i \in \check{\mathsf{X}}$ and $A \in \mathcal{B}(\check{\mathsf{X}})$ by

$$\check{P}(x_0, \cdot) = P(x, \cdot)^*, \qquad\qquad x_0 \in \mathsf{X}_0 \backslash C_0; \tag{5.5}$$

$$\check{P}(x_0, \cdot) = [1 - \delta]^{-1}[P(x, \cdot)^* - \delta\nu^*(\cdot)], \qquad\qquad x_0 \in C_0; \tag{5.6}$$

$$\check{P}(x_1, \cdot) = \nu^*(\cdot), \qquad\qquad x_1 \in \mathsf{X}_1. \tag{5.7}$$

where C, δ and ν are the set, the constant and the measure in the minorization condition.

Outside C the chain $\{\check{\Phi}_n\}$ behaves just like $\{\Phi_n\}$, moving on the "top" half X_0 of the split space. Each time it arrives in C, it is "split"; with probability $1 - \delta$ it remains in C_0, with probability δ it drops to C_1. We can think of this splitting of the chain as tossing a δ-weighted coin to decide which level to choose on each arrival in the set C where the split takes place.

When the chain remains on the top level its next step has the modified law (5.6). That (5.6) is always non-negative follows from (5.2). This is the sole use of the minorization condition, although without it this chain cannot be defined.

Note here the whole point of the construction: the bottom level X_1 is an atom, with $\varphi^*(X_1) = \delta\varphi(C) > 0$ whenever the chain Φ is φ-irreducible. By (5.3) we have $\check{P}^n(x_i, X_1 \setminus C_1) = 0$ for all $n \geq 1$ and all $x_i \in \check{X}$, so that the atom $C_1 \subseteq X_1$ is the only part of the bottom level which is reached with positive probability. We will use the notation

$$\check{\alpha} := C_1 \tag{5.8}$$

when we wish to emphasize the fact that all transitions out of C_1 are identical, so that C_1 is an atom in $\check{X}$.

5.1.2   Connecting the split and original chains

The splitting construction is valuable because of the various properties that $\check{\Phi}$ inherits from, or passes on to, Φ. We give the first of these in the next result.

Theorem 5.1.3. *The following correspondences hold for the split and original chains:*

(i) *The chain Φ is the marginal chain of $\{\check{\Phi}_n\}$: that is, for any initial distribution λ on $\mathcal{B}(X)$ and any $A \in \mathcal{B}(X)$,*

$$\int_X \lambda(dx) P^k(x, A) = \int_{\check{X}} \lambda^*(dy_i) \check{P}^k(y_i, A_0 \cup A_1). \tag{5.9}$$

(ii) *The chain Φ is φ-irreducible if $\check{\Phi}$ is φ^*-irreducible; and if Φ is φ-irreducible with $\varphi(C) > 0$ then $\check{\Phi}$ is ν^*-irreducible, and $\check{\alpha}$ is an accessible atom for the split chain.*

PROOF   **(i)**   From the linearity of the splitting operation we only need to check the equivalence in the special case of $\lambda = \delta_x$, and $k = 1$. This follows by direct computation. We analyze two cases separately.

Suppose first that $x \in C^c$. Then, by (5.5) and (5.4),

$$\int_{\check{X}} \delta_x^*(dy_i) \check{P}(y_i, A_0 \cup A_1) = \check{P}(x_0, A_0 \cup A_1) = P(x, A).$$

On the other hand suppose $x \in C$. Then, from (5.6), (5.7) and (5.4) again,

$$\int_{\check{X}} \delta_x^*(dy_i) \check{P}(y_i, A_0 \cup A_1)$$

$$= (1-\delta)\check{P}(x_0, A_0 \cup A_1) + \delta\check{P}(x_1, A_0 \cup A_1)$$

$$= (1-\delta)\Big[[1-\delta]^{-1}[P^*(x, A_0 \cup A_1) - \delta\nu^*(A_0 \cup A_1)]\Big] + \delta\nu^*(A_0 \cup A_1)$$

$$= P(x, A).$$

(ii)   If the split chain is φ^*-irreducible it is straightforward that the original chain is φ-irreducible from (i). The converse follows from the fact that $\check{\alpha}$ is an accessible atom if $\varphi(C) > 0$, which is easy to check, and Proposition 5.1.1.   □

The following identity will prove crucial in later development. For any measure μ on $\mathcal{B}(\mathsf{X})$ we have

$$\int_{\check{\mathsf{X}}} \mu^*(dx_i)\check{P}(x_i,\,\cdot\,) = \left(\int_{\mathsf{X}} \mu(dx)P(x,\,\cdot\,)\right)^* \tag{5.10}$$

or, using operator notation, $\mu^*\check{P} = (\mu P)^*$. This follows from the definition of the $*$ operation and the transition function $\check{P}$, and is in effect a restatement of Theorem 5.1.3 (i).

Since it is only the marginal chain $\boldsymbol{\Phi}$ which is really of interest, we will usually consider only sets of the form $\check{A} = A_0 \cup A_1$, where $A \in \mathcal{B}(\mathsf{X})$, and we will largely restrict ourselves to functions on $\check{\mathsf{X}}$ of the form $\check{f}(x_i) = f(x_i)$, where f is some function on X; that is, $\check{f}$ is identical on the two copies of X. By (5.9) we have for any k, any initial distribution λ, and any function $\check{f}$ identical on X_0 and X_1

$$\mathsf{E}_\lambda[f(\Phi_k)] = \check{\mathsf{E}}_{\lambda^*}[\check{f}(\check{\Phi}_k)].$$

To emphasize this identity we will henceforth denote $\check{f}$ by f, and $\check{A}$ by A in these special instances. The context should make clear whether A is a subset of X or $\check{\mathsf{X}}$, and whether the domain of f is X or $\check{\mathsf{X}}$.

The minorization condition ensures that the construction in (5.6) gives a probability law on $\check{\mathsf{X}}$. A similar construction can also be carried out under the seemingly more general minorization requirement that there exists a function $h(x)$ with $\int h(x)\varphi(dx) > 0$, and a measure $\nu(\cdot)$ on $\mathcal{B}(\mathsf{X})$ such that

$$P(x,A) \geq h(x)\nu(A), \quad x \in \mathsf{X}, A \in \mathcal{B}(\mathsf{X}). \tag{5.11}$$

The details are, however, slightly less easy than for the approach we give above although there are some other advantages to the approach through (5.11): the interested reader should consult Nummelin [303] for more details.

The construction of a split chain is of some value in the next several chapters, although much of the analysis will be done directly using the small sets of the next section. The Nummelin splitting technique will, however, be central in our approach to the asymptotic results of Part III.

5.1.3   A random renewal time approach

There is a second construction of a "pseudo-atom" which is formally very similar to that above. This approach, due to Athreya and Ney [13], concentrates, however, not on a "physical" splitting of the space but on a random renewal time.

If we take the existence of the minorization (5.2) as an assumption, and if we also assume

$$L(x,C) \equiv 1, \quad x \in \mathsf{X} \tag{5.12}$$

we can then construct an almost surely finite random time $\tau \geq 1$ on an enlarged probability space such that $\mathsf{P}_x(\tau < \infty) = 1$ and for every A

$$\mathsf{P}_x(\Phi_n \in A, \tau = n) = \nu(C \cap A)\mathsf{P}_x(\tau = n). \tag{5.13}$$

To construct τ, let $\boldsymbol{\Phi}$ run until it hits C; from (5.12) this happens eventually with probability one. The time and place of first hitting C will be, say, k and x. Then with probability δ distribute Φ_{k+1} over C according to ν; with probability $(1 - \delta)$ distribute Φ_{k+1} over the whole space with law $Q(x, \cdot)$, where

$$Q(x, A) = [P(x, A) - \delta\nu(A \cap C)]/(1 - \delta);$$

from (5.2) Q is a probability measure, as in (5.6). Repeat this procedure each time Φ enters C; since this happens infinitely often from (5.12) (a fact yet to be proven in Chapter 9), and each time there is an independent probability δ of choosing ν, it is intuitively clear that sooner or later this version of Φ_k is chosen. Let the time when it occurs be τ. Then $\mathsf{P}_x(\tau < \infty) = 1$ and (5.13) clearly holds; and (5.13) says that τ is a regeneration time for the chain.

The two constructions are very close in spirit: if we consider the split chain construction then we can take the random time τ as $\tau_{\check{\alpha}}$, which is identical to the hitting time on the bottom level of the split space.

There are advantages to both approaches, but the Nummelin splitting does not require the recurrence assumption (5.12), and more pertinently, it exploits the rather deep fact that some m-skeleton always obeys the minorization condition when ψ-irreducibility holds, as we now see.

5.2   Small sets

In this section we develop the theory of small sets. These are sets for which the minorization condition holds, at least for the m-skeleton chain. From the splitting construction of Section 5.1.1, then, it is obvious that the existence of small sets is of considerable importance, since they ensure the splitting method is not vacuous.

Small sets themselves behave, in many ways, analogously to atoms, and in particular the conclusions of Proposition 5.1.1 and Proposition 5.1.2 hold. We will find also many cases where we exploit the "pseudo-atomic" properties of small sets without directly using the split chain.

Small sets

A set $C \in \mathcal{B}(\mathsf{X})$ is called a *small set* if there exists an $m > 0$, and a non-trivial measure ν_m on $\mathcal{B}(\mathsf{X})$, such that for all $x \in C$, $B \in \mathcal{B}(\mathsf{X})$,

$$P^m(x, B) \geq \nu_m(B). \qquad (5.14)$$

When (5.14) holds we say that C is ν_m-small.

The central result (Theorem 5.2.2 below), on which a great deal of the subsequent development rests, is that for a ψ-irreducible chain, every set $A \in \mathcal{B}^+(\mathsf{X})$ contains a small set in $\mathcal{B}^+(\mathsf{X})$. As a consequence, every ψ-irreducible chain admits some m-skeleton which can be split, and for which the atomic structure of the split chain can be exploited.

In order to prove this result, we need for the first time to consider the densities of the transition probability kernels. Being a probability measure on $(\mathsf{X}, \mathcal{B}(\mathsf{X}))$ for each individual x and each n, the transition probability kernel $P^n(x, \cdot)$ admits a Lebesgue decomposition into its absolutely continuous and singular parts, with respect to any finite non-trivial measure ϕ on $\mathcal{B}(\mathsf{X})$: we have for any fixed x and $B \in \mathcal{B}(\mathsf{X})$

$$P^n(x, B) = \int_B p^n(x, y)\phi(dy) + P_\perp(x, B). \qquad (5.15)$$

where $p^n(x, y)$ is the density of $P^n(x, \cdot)$ with respect to ϕ and $P_\perp$ is orthogonal to ϕ.

Theorem 5.2.1. *Suppose ϕ is a σ-finite measure on $(\mathsf{X}, \mathcal{B}(\mathsf{X}))$. Suppose A is any set in $\mathcal{B}(\mathsf{X})$ with $\phi(A) > 0$ such that*

$$\phi(B) > 0, \ B \subseteq A \ \Rightarrow \ \sum_{k=1}^{\infty} P^k(x, B) > 0, \quad x \in A.$$

Then, for every n, the function p^n defined in (5.15) can be chosen to be a measurable function on X^2, and there exists $C \subseteq A$, $m > 1$, and $\delta > 0$ such that $\phi(C) > 0$ and

$$p^m(x, y) > \delta, \quad x, y \in C. \qquad (5.16)$$

PROOF   We include a detailed proof because of the central place small sets hold in the development of the theory of ψ-irreducible Markov chains. However, the proof is somewhat complex, and may be omitted without interrupting the flow of understanding at this point.

It is a standard result that the densities $p^n(x, y)$ of $P^n(x, \cdot)$ with respect to ϕ exist for each $x \in \mathsf{X}$, and are unique except for definition on ϕ-null sets. We first need to verify that

(i) the densities $p^n(x, y)$ can be chosen jointly measurable in x and y, for each n;

(ii) the densities $p^n(x, y)$ can be chosen to satisfy an appropriate form of the Chapman–Kolmogorov property, namely for $n, m \in \mathbb{Z}_+$, and all x, z

$$p^{n+m}(x, z) \geq \int_{\mathsf{X}} p^n(x, y)p^m(y, z)\phi(dy). \qquad (5.17)$$

To see (i), we appeal to the fact that $\mathcal{B}(\mathsf{X})$ is assumed countably generated. This means that there exists a sequence $\{\mathcal{B}_i; i \geq 1\}$ of finite partitions of X, such that $\mathcal{B}_{i+1}$ is a refinement of $\mathcal{B}_i$, and which generate $\mathcal{B}(\mathsf{X})$. Fix $x \in \mathsf{X}$, and let $B_i(x)$ denote the element in $\mathcal{B}_i$ with $x \in B_i(x)$.

For each i, the functions

$$p_i^1(x, y) = \begin{cases} 0, & \phi(B_i(y)) = 0, \\ P(x, B_i(y))/\phi(B_i(y)), & \phi(B_i(y)) > 0 \end{cases}$$

are non-negative, and are clearly jointly measurable in x and y. The Basic Differentiation Theorem for measures (cf. Doob [99], Chapter 7, Section 8) now assures us that for y outside a ϕ-null set N,

$$p_\infty^1(x, y) = \lim_{i \to \infty} p_i^1(x, y) \qquad (5.18)$$

exists as a jointly measurable version of the density of $P(x, \cdot)$ with respect to ϕ.

The same construction gives the densities $p_\infty^n(x, y)$ for each n, and so jointly measurable versions of the densities exist as required.

We now define inductively a version $p^n(x, y)$ of the densities satisfying (5.17), starting from $p_\infty^n(x, y)$. Set $p^1(x, y) = p_\infty^1(x, y)$ for all x, y; and set, for $n \geq 2$ and any x, y,

$$p^n(x, y) = p_\infty^n(x, y) \bigvee \max_{1 \leq m \leq n-1} \int P^m(x, dw) p^{n-m}(w, y).$$

One can now check (see Orey [309] p. 6) that the collection $\{p^n(x, y), x, y \in \mathsf{X}, n \in \mathbb{Z}_+\}$ satisfies both (i) and (ii).

We next verify (5.16). The constraints on ϕ in the statement of Theorem 5.2.1 imply that

$$\sum_{n=1}^\infty p^n(x, y) > 0, \quad x \in A, \qquad \text{a.e. } y \in A \ [\phi];$$

and thus we can find integers n, m such that

$$\int_A \int_A \int_A p^n(x, y) p^m(y, z) \phi(dx) \phi(dy) \phi(dz) > 0.$$

Now choose $\eta > 0$ sufficiently small that, writing

$$A_n(\eta) := \{(x, y) \in A \times A : p^n(x, y) \geq \eta\}$$

and ϕ^3 for the product measure $\phi \times \phi \times \phi$ on $\mathsf{X} \times \mathsf{X} \times \mathsf{X}$, we have

$$\phi^3 \left(\{(x, y, z) \in A \times A \times A : (x, y) \in A_n(\eta), (y, z) \in A_m(\eta)\} \right) > 0.$$

We suppress the notational dependence on η from now on, since η is fixed for the remainder of the proof.

For any x, y, set $B_i(x, y) = B_i(x) \times B_i(y)$, where $B_i(x)$ is again the element containing x of the finite partition $\mathcal{B}_i$ above. By the Basic Differentiation Theorem as in (5.18), this time for measures on $\mathcal{B}(\mathsf{X}) \times \mathcal{B}(\mathsf{X})$, there are ϕ^2-null sets $N_k \subseteq \mathsf{X} \times \mathsf{X}$ such that for any k and $(x, y) \in A_k \backslash N_k$,

$$\lim_{i \to \infty} \phi^2(A_k \cap B_i(x, y)) / \phi^2(B_i(x, y)) = 1. \tag{5.19}$$

Now choose a fixed triplet (u, v, w) from the set

$$\{(x, y, z) : (x, y) \in A_n \backslash N_n, (y, z) \in A_m \backslash N_m\}.$$

From (5.19) we can find j large enough that

$$\begin{aligned}
\phi^2(A_n \cap B_j(u, v)) &\geq (3/4)\phi^2(B_j(u, v)), \\
\phi^2(A_m \cap B_j(v, w)) &\geq (3/4)\phi^2(B_j(v, w)).
\end{aligned} \tag{5.20}$$

Let us write $A_n(x) = \{y \in A : (x, y) \in A_n\}$, $A_m^*(z) = \{y \in A : (y, z) \in A_m\}$ for the sections of A_n and A_m in the different directions. If we define

$$E_n = \{x \in A_n \cap B_j(u) : \phi(A_n(x) \cap B_j(v)) \geq (3/4)B_j(v)\} \tag{5.21}$$

$$D_m = \{z \in A_m \cap B_j(w) : \phi(A_m^*(z) \cap B_j(v)) \geq (3/4)B_j(v)\}, \qquad (5.22)$$

then from (5.20) we have that $\phi(E_n) > 0$, $\phi(D_m) > 0$. This then implies, for any pair $(x, z) \in E_n \times D_m$,

$$\phi(A_n(x) \cap A_m^*(z)) \geq (1/2)\phi(B_j(v)) > 0 \qquad (5.23)$$

from (5.21) and (5.22).

Our pieces now almost fit together. We have, from (5.17), that for $(x, z) \in E_n \times D_m$

$$
\begin{aligned}
p^{n+m}(x, z) &\geq \int_{A_n(x) \cap A_m^*(z)} p^n(x, y)p^m(y, z)\phi(dy) \\
&\geq \eta^2 \phi(A_n(x) \cap A_m^*(z)) \\
&\geq [\eta^2/2]\phi(B_j(v)) \\
&\geq \delta_1, \text{ say }. \qquad (5.24)
\end{aligned}
$$

To finish the proof, note that since $\phi(E_n) > 0$, there is an integer k and a set $C \subseteq D_m$ with $P^k(x, E_n) > \delta_2 > 0$, for all $x \in C$. It then follows from the construction of the densities above that for all $x, z \in C$

$$
\begin{aligned}
p^{k+n+m}(x, z) &\geq \int_{E_n} P^k(x, dy)p^{n+m}(y, z) \\
&\geq \delta_1\delta_2,
\end{aligned}
$$

and the result follows with $\delta = \delta_1\delta_2$ and $M = k + n + m$.   □

The key fact proven in this theorem is that we can define a version of the densities of the transition probability kernel such that (5.16) holds uniformly over $x \in C$. This gives us

Theorem 5.2.2. *If Φ is ψ-irreducible, then for every $A \in \mathcal{B}^+(\mathsf{X})$, there exists $m \geq 1$ and a ν_m-small set $C \subseteq A$ such that $C \in \mathcal{B}^+(\mathsf{X})$ and $\nu_m\{C\} > 0$.*

PROOF   When Φ is ψ-irreducible, every set in $\mathcal{B}^+(\mathsf{X})$ satisfies the conditions of Theorem 5.2.1, with the measure $\phi = \psi$. The result then follows immediately from (5.16).   □

As a direct corollary of this result we have

Theorem 5.2.3. *If Φ is ψ-irreducible, then the minorization condition holds for some m-skeleton, and for every K_{a_ε}-chain, $0 < \varepsilon < 1$.*   □

Any Φ which is ψ-irreducible is well endowed with small sets from Theorem 5.2.1, even though it is far from clear from the initial definition that this should be the case. Given the existence of just one small set from Theorem 5.2.2, we now show that it is further possible to cover the whole of X with small sets in the ψ-irreducible case.

Proposition 5.2.4.   (i) *If $C \in \mathcal{B}(\mathsf{X})$ is ν_n-small, and for any $x \in D$ we have $P^m(x, C) \geq \delta$, then D is ν_{n+m}-small, where ν_{n+m} is a multiple of ν_n.*

(ii) *Suppose Φ is ψ-irreducible. Then there exists a countable collection C_i of small sets in $\mathcal{B}(\mathsf{X})$ such that*

$$\mathsf{X} = \bigcup_{i=0}^{\infty} C_i. \tag{5.25}$$

(iii) *Suppose Φ is ψ-irreducible. If $C \in \mathcal{B}^+(\mathsf{X})$ is ν_n-small, then we may find $M \in \mathbb{Z}_+$ and a measure ν_M such that C is ν_M-small, and $\nu_M\{C\} > 0$.*

PROOF   (i)   By the Chapman–Kolmogorov equations, for any $x \in D$,

$$
\begin{aligned}
P^{n+m}(x, B) &= \int_{\mathsf{X}} P^n(x, dy) P^m(y, B) \\
&\geq \int_C P^n(x, dy) P^m(y, B) \\
&\geq \delta \nu_n(B).
\end{aligned}
\tag{5.26}
$$

(ii)   Since Φ is ψ-irreducible, there exists a ν_m-small set $C \in \mathcal{B}^+(\mathsf{X})$ from Theorem 5.2.2. Moreover from the definition of ψ-irreducibility the sets

$$\bar{C}(n, m) := \{y : P^n(y, C) \geq m^{-1}\} \tag{5.27}$$

cover X and each $\bar{C}(n, m)$ is small from (i).

(iii)   Since $C \in \mathcal{B}^+(\mathsf{X})$, we have $K_{a_{\frac{1}{2}}}(x, C) > 0$ for all $x \in \mathsf{X}$. Hence $\nu K_{a_{\frac{1}{2}}}(C) > 0$, and it follows that for some $m \in \mathbb{Z}_+$,

$$\nu_M(C) := \nu P^m(C) > 0.$$

To complete the proof observe that, for all $x \in C$,

$$P^{n+m}(x, B) = \int_{\mathsf{X}} P^n(x, dy) P^m(y, B) \geq \nu P^m(B) = \nu_M(B),$$

which shows that C is ν_M-small, where $M = n + m$.   □

5.3   Small sets for specific models

5.3.1   Random walk on a half line

Random walks on a half line provide a simple example of small sets, regardless of the structure of the increment distribution.

It follows as in the proof of Proposition 4.3.1 that every set $[0, c], c \in \mathbb{R}_+$ is small, provided only that $\Gamma(-\infty, 0) > 0$: in other words, whenever the chain is ψ-irreducible, every compact set is small. Alternatively, we could derive this result by use of Proposition 5.2.4 (i) since $\{0\}$ is, by definition, small.

This makes the analysis of queueing and storage models very much easier than more general models for which there is no atom in the space. We now move on to identify conditions under which these have identifiable small sets.

5.3.2 "Spread-out" random walks

Let us again consider a random walk Φ of the form

$$\Phi_n = \Phi_{n-1} + W_n,$$

satisfying (RW1). We showed in Section 4.3 that, if Γ has a density γ with respect to Lebesgue measure μ^{Leb} on $\mathbb{R}$ with

$$\gamma(x) \geq \delta > 0, \qquad |x| < \beta,$$

then Φ is ψ-irreducible: re-examining the proof shows that in fact we have demonstrated that $C = \{x : |x| \leq \beta/2\}$ is a small set.

Random walks with nonsingular distributions with respect to μ^{Leb}, of which the above are special cases, are particularly well adapted to the ψ-irreducible context. To study them we introduce so-called "spread-out" distributions.

Spread-out random walk

(RW2)   We call the random walk *spread out* (or equivalently, we call Γ spread out) if some convolution power Γ^{n*} is nonsingular with respect to μ^{Leb}.

For spread-out random walks, we find that small sets are in general relatively easy to find.

Proposition 5.3.1. *If Φ is a spread-out random walk, with Γ^{n*} non-singular with respect to μ^{Leb} then there is a neighborhood $C_\beta = \{x : |x| \leq \beta\}$ of the origin which is ν_{2n}-small, where $\nu_{2n} = \varepsilon \mu^{\text{Leb}} \mathbb{I}_{[s,t]}$ for some interval $[s,t]$, and some $\varepsilon > 0$.*

PROOF   Since Γ is spread out, we have for some bounded non-negative function γ with $\int \gamma(x)\,dx > 0$, and some $n > 0$,

$$P^n(0, A) \geq \int_A \gamma(x)\,dx, \qquad A \in \mathcal{B}(\mathbb{R}).$$

Iterating this we have

$$P^{2n}(0, A) \geq \int_A \int_{\mathbb{R}} \gamma(y)\gamma(x-y)\,dy\,dx = \int_A \gamma * \gamma(x)\,dx : \tag{5.28}$$

but since from Lemma D.4.3 the convolution $\gamma * \gamma(x)$ is continuous and not identically zero, there exists an interval $[a, b]$ and a δ with $\gamma * \gamma(x) \geq \delta$ on $[a, b]$. Choose $\beta = [b-a]/4$, and $[s,t] = [a + \beta, b - \beta]$, to prove the result using the translation invariant properties of the random walk. $\square$

For spread out random walks, a far stronger irreducibility result will be provided in Chapter 6 : there we will show that if Φ is a random walk with spread-out increment distribution Γ, with $\Gamma(-\infty, 0) > 0, \Gamma(0, \infty) > 0$, then Φ is μ^{Leb}-irreducible, and every compact set is a small set.

5.3.3   Ladder chains and the GI/G/I queue

Recall from Section 3.5 the Markov chain constructed on $\mathbb{Z}_+ \times \mathbb{R}$ to analyze the GI/G/1 queue, defined by

$$\Phi_n = (N_n, R_n), \quad n \geq 1,$$

where N_n is the number of customers at $T_n'-$ and R_n is the residual service time at $T_n'+$.

This has the transition kernel

$$
\begin{aligned}
P(i, x; j \times A) &= 0, & j &> i+1, \\
P(i, x; j \times A) &= \Lambda_{i-j+1}(x, A), & j &=, 1, \ldots, i+1, \\
P(i, x; 0 \times A) &= \Lambda_i^*(x, A),
\end{aligned}
$$

where

$$\Lambda_n(x, [0, y]) = \int_0^\infty P_n^t(x, y), G(dt), \tag{5.29}$$

$$\Lambda_n^*(x, [0, y]) = \Big[\sum_{n+1}^\infty \Lambda_j(x, [0, \infty)) \Big] H[0, y], \tag{5.30}$$

$$P_n^t(x, y) = \mathsf{P}(S_n' \leq t < S_{n+1}', R_t \leq y \mid R_0 = x); \tag{5.31}$$

here, $R_t = S_{N(t)+1}' - t$, where $N(t)$ is the number of renewals in $[0, t]$ of a renewal process with inter-renewal time H, and if $R_0 = x$ then $S_1' = x$.

At least one collection of small sets for this chain can be described in some detail.

Proposition 5.3.2. *Let* $\boldsymbol{\Phi} = \{N_n, R_n\}$ *be the Markov chain at arrival times of a GI/G/1 queue described above. Suppose* $G(\beta) < 1$ *for all* $\beta < \infty$. *Then the set* $\{0 \times [0, \beta]\}$ *is* ν_1-*small for* $\boldsymbol{\Phi}$, *with* $\nu_1(\cdot)$ *given by* $G(\beta, \infty) H(\cdot)$.

PROOF   We consider the bottom "rung" $\{0 \times \mathbb{R}\}$. By construction

$$\Lambda_0^*(x, [0, \cdot]) = H[0, \cdot][1 - \Lambda_0(x, [0, \infty])],$$

and since

$$
\begin{aligned}
\Lambda_0(x, [0, \infty]) &= \int G(dt) \mathsf{P}(0 \leq t < \sigma_1 \mid R_0 = x) \\
&= \int G(dt) \mathbb{I}\{t < x\} \\
&= G(-\infty, x],
\end{aligned}
$$

we have

$$\Lambda_0^*(x, [0, \cdot]) = H[0, \cdot] G(x, \infty).$$

The result follows immediately, since for $x < \beta, \Lambda_0^*(x, [0, \cdot]) \geq H[0, \cdot] G(\beta, \infty)$. $\square$

5.3.4 The forward recurrence time chain

Consider the forward recurrence time δ-skeleton $\mathbf{V}_\delta^+ = V^+(n\delta), n \in \mathbb{Z}_+$, which was defined in Section 3.5.3: recall that

$$V^+(t) := \inf(Z_n - t : Z_n \geq t), \qquad t \geq 0$$

where $Z_n := \sum_{i=0}^n Y_i$ for $\{Y_1, Y_2, \ldots\}$ a sequence of independent and identical random variables with distribution Γ, and Y_0 a further independent random variable with distribution Γ_0.

We shall prove

Proposition 5.3.3. *When Γ is spread out then for δ sufficiently small the set $[0, \delta]$ is a small set for $\mathbf{V}_\delta^+$.*

PROOF   As in (5.28), since Γ is spread out there exists $n \in \mathbb{Z}_+$, an interval $[a, b]$ and a constant $\beta > 0$ such that

$$\Gamma^{n*}(du) \geq \beta \mu^{\text{Leb}}(du), \qquad du \subseteq [a, b].$$

Hence if we choose small enough δ then we can find $k \in \mathbb{Z}_+$ such that

$$\Gamma^{n*}(du) \geq \beta \mathbb{I}_{[k\delta, (k+4)\delta]}(u) \mu^{\text{Leb}}(du), \qquad du \subseteq [a, b]. \tag{5.32}$$

Now choose $m > 1$ such that $\Gamma[m\delta, (m+1)\delta) = \gamma > 0$; and set $M = k + m + 2$. Then for $x \in [0, \delta)$, by considering the occurrence of the nth renewal where n is the index so that (5.32) holds we find

$$
\begin{aligned}
&\mathsf{P}_x(V^+(M\delta) \in du \cap [0, \delta)) \\
&\geq \quad \mathsf{P}_0(x + Z_{n+1} - M\delta \in du \cap [0, \delta), Y_{n+1} \geq \delta) \\
&= \quad \int_{y \in [\delta, \infty)} \Gamma(dy)\mathsf{P}_0(x + y - M\delta + Z_n \in du \cap [0, \delta)) \\
&\geq \quad \int_{y \in [m\delta, (m+1)\delta)} \Gamma(dy)\mathsf{P}_0(Z_n \in du \cap \{[0, \delta) - x - y + M\delta\}).
\end{aligned}
\tag{5.33}
$$

Now when $y \in [m\delta, (m+1)\delta)$ and $x \in [0, \delta)$, we must have

$$\{[0, \delta) - x - y + M\delta\} \subseteq [k\delta, (k+3)\delta) \tag{5.34}$$

and therefore from (5.33)

$$
\begin{aligned}
\mathsf{P}_x(V^+(M\delta) \in du \cap [0, \delta)) &\geq \quad \beta \mathbb{I}_{[0,\delta)}(u) \mu^{\text{Leb}}(du)\Gamma(m\delta, (m+1)\delta) \\
&\geq \quad \beta\gamma \mathbb{I}_{[0,\delta)}(u)\mu^{\text{Leb}}(du).
\end{aligned}
\tag{5.35}
$$

Hence $[0, \delta)$ is a small set, and the measure ν can be chosen as a multiple of Lebesgue measure over $[0, \delta)$.   □

In this proof we have demanded that (5.32) holds for $u \in [k\delta, (k+4)\delta]$ and in (5.34) we only used the fact that the equation holds for $u \in [k\delta, (k+3)\delta]$. This is not an oversight: we will use the larger range in showing in Proposition 5.4.5 that the chain is also aperiodic.

5.3.5 Linear state space models

For the linear state space LSS(F,G) model we showed in Proposition 4.4.3 that in the Gaussian case when (LSS3) holds, for every initial condition $x_0 \in \mathsf{X} = \mathbb{R}^n$,

$$P^k(x_0, \cdot) = N(F^k x_0, \sum_{i=0}^{k-1} F^i GG^\top F^{i\top}); \tag{5.36}$$

and if (F, G) is controllable then from (4.18) the n-step transition function possesses a smooth density $p_n(x, y)$ which is continuous and everywhere positive on $\mathbb{R}^{2n}$. It follows from continuity that for any pair of bounded open balls B_1 and $B_2 \subset \mathbb{R}^n$, there exists $\varepsilon > 0$ such that

$$p_n(x, y) \geq \varepsilon, \qquad (x, y) \in B_1 \times B_2.$$

Letting ν_n denote the normalized uniform distribution on B_2 we see that B_1 is ν_n-small.

   This shows that for the controllable, Gaussian LSS(F,G) model, all compact subsets of the state space are small.

5.4 Cyclic behavior

5.4.1 The cycle phenomenon

In the previous sections of this chapter we concentrated on the communication structure between states. Here we consider the set of time points at which such communication is possible; for even within a communicating class, it is possible that the chain returns to given states only at specific time points, and this certainly governs the detailed behavior of the chain in any longer term analysis.

   A highly artificial example of cyclic behavior on the finite set $\mathsf{X} = \{1, 2, 3, \ldots, d\}$ is given by the transition probability matrix

$$P(x, x + 1) = 1, \quad x \in \{1, 2, 3, \ldots, d - 1\}, \qquad P(d, 1) = 1.$$

Here, if we start in x then we have $P^n(x, x) > 0$ if and only if $n = 0, d, 2d, \ldots$, and the chain Φ is said to *cycle* through the states of X.

   On a continuous state space the same phenomenon can be constructed equally easily: let $\mathsf{X} = [0, d)$, let U_i denote the uniform distribution on $[i, i + 1)$, and define

$$P(x, \cdot) := \mathbb{I}_{[i-1,i)}(x) U_i(\cdot), \quad i = 0, 1, \ldots, d - 1 \pmod{d}.$$

In this example, the chain again cycles through a fixed finite number of sets. We now prove a series of results which indicate that, no matter how complex the behavior of a ψ-irreducible chain, or a chain on an irreducible absorbing set, the finite cyclic behavior of these examples is typical of the worst behavior to be found.

5.4.2 Cycles for a countable space chain

We discuss this structural question initially for a countable space X.

Let α be a specific state in X, and write

$$d(\alpha) = g.c.d.\{n \geq 1 : P^n(\alpha, \alpha) > 0\}. \tag{5.37}$$

This does not guarantee that $P^{md(\alpha)}(\alpha, \alpha) > 0$ for all m, but it does imply $P^n(\alpha, \alpha) = 0$ unless $n = md(\alpha)$, for some m.

We call $d(\alpha)$ the *period* of α. The result we now show is that the value of $d(\alpha)$ is common to all states y in the class $C(\alpha) = \{y : \alpha \leftrightarrow y\}$, rather than taking a separate value for each y.

Proposition 5.4.1. *Suppose α has period $d(\alpha)$: then for any $y \in C(\alpha)$, $d(\alpha) = d(y)$.*

Proof   Since $\alpha \leftrightarrow y$, we can find m and n such that $P^m(\alpha, y) > 0$ and $P^n(y, \alpha) > 0$. By the Chapman–Kolmogorov equations, we have

$$P^{m+n}(\alpha, \alpha) \geq P^m(\alpha, y)P^n(y, \alpha) > 0, \tag{5.38}$$

and so by definition, $(m + n)$ is a multiple of $d(\alpha)$. Choose k such that k is not a multiple of $d(\alpha)$. Then $(k + m + n)$ is not a multiple of $d(\alpha)$: hence, since

$$P^m(\alpha, y)P^k(y, y)P^n(y, \alpha) \leq P^{k+m+n}(\alpha, \alpha) = 0,$$

we have $P^k(y, y) = 0$, which proves $d(y) \geq d(\alpha)$. Reversing the role of α and y shows $d(\alpha) \geq d(y)$, which gives the result.   □

This result leads to a further decomposition of the transition probability matrix for an irreducible chain; or, equivalently, within a communicating class.

Proposition 5.4.2. *Let Φ be an irreducible Markov chain on a countable space, and let d denote the common period of the states in X. Then there exist disjoint sets $D_1, \ldots, D_d \subseteq X$ such that*

$$X = \bigcup_{i=1}^{d} D_k,$$

and

$$P(x, D_{k+1}) = 1, \qquad x \in D_k, \quad k = 0, \ldots, d-1 \pmod{d}. \tag{5.39}$$

Proof   The proof is similar to that of the previous proposition. Choose $\alpha \in X$ as a distinguished state, and let y be another state, such that for some M

$$P^M(y, \alpha) > 0.$$

Let k be any other integer such that $P^k(\alpha, y) > 0$. Then $P^{k+M}(\alpha, \alpha) > 0$, and thus $k + M = jd$ for some j; equivalently, $k = jd - M$. Now M is fixed, and so we must have $P^k(\alpha, y) > 0$ only for k in the sequence $\{r, r + d, r + 2d, \ldots\}$, where the integer $r = r(y) \in \{1, \ldots, d\}$ is uniquely defined for y.

Call D_r the set of states which are reached with positive probability from α only at points in the sequence $\{r, r + d, r + 2d, \ldots\}$ for each $r \in \{1, 2, \ldots, d\}$. By definition $\alpha \in D_d$, and $P(\alpha, D_1^c) = 0$ so that $P(\alpha, D_1) = 1$. Similarly, for any $y \in D_r$ we have $P(y, D_{r+1}^c) = 0$, giving our result.   □

The sets $\{D_i\}$ covering X and satisfying (5.39) are called *cyclic classes,* or a *d-cycle,* of Φ. With probability one, each sample path of the process Φ "cycles" through values in the sets $D_1, D_2, \ldots, D_d, D_1, D_2, \ldots$.

Diagrammatically, we have shown that we can write an irreducible transition probability matrix in "super-diagonal" form

$$
P = \begin{bmatrix}
0 & P_1 & & & \\
0 & 0 & P_2 & & 0 \\
\vdots & \ddots & 0 & P_3 & \\
\vdots & \vdots & \ddots & 0 & \ddots \\
P_d & \cdots & \cdots & \cdots & 0
\end{bmatrix}
$$

where each block P_i is a square matrix whose dimension may depend upon i.

Aperiodicity

An irreducible chain on a countable space X is called

(i) *aperiodic,* if $d(x) \equiv 1$, $x \in \mathsf{X}$;

(ii) *strongly aperiodic,* if $P(x, x) > 0$ for some $x \in \mathsf{X}$.

Whilst cyclic behavior can certainly occur, as illustrated in the examples at the beginning of this section, and the periodic behavior of the control systems in Theorem 7.3.3 below, most of our results will be given for aperiodic chains. The justification for using such chains is contained in the following, whose proof is obvious.

Proposition 5.4.3. *Suppose Φ is an irreducible chain on a countable space X, with period d and cyclic classes $\{D_1, \ldots, D_d\}$. Then for the Markov chain $\Phi_d = \{\Phi_d, \Phi_{2d}, \ldots\}$ with transition matrix P^d, each D_i is an irreducible absorbing set of aperiodic states.*

5.4.3   Cycles for a general state space chain

The existence of small sets enables us to show that, even on a general space, we still have a finite periodic breakup into cyclic sets for ψ-irreducible chains.

Suppose that C is any ν_M-small set, and assume that $\nu_M(C) > 0$, as we may without loss of generality by Proposition 5.2.4.

We will use the set C and the corresponding measure ν_M to define a cycle for a general irreducible Markov chain. To simplify notation we will suppress the subscript on ν. Hence we have $P^M(x, \cdot) \geq \nu(\cdot)$, $x \in C$, and $\nu(C) > 0$, so that, when the chain starts in C, there is a positive probability that the chain will return to C at time M. Let

$$
E_C = \{n \geq 1 : \text{ the set } C \text{ is } \nu_n\text{-small, with } \nu_n = \delta_n \nu \text{ for some } \delta_n > 0\} \qquad (5.40)
$$

be the set of time points for which C is a small set with minorizing measure proportional to ν. Notice that for $B \subseteq C$, $n, m \in E_C$ implies

$$
\begin{aligned}
P^{n+m}(x, B) &\geq \int_C P^m(x, dy) P^n(y, B) \\
&\geq [\delta_m \delta_n \nu(C)] \nu(B), \quad x \in C;
\end{aligned}
$$

so that E_C is closed under addition. Thus there is a natural "period" for the set C, given by the greatest common divisor of E_C; and from Lemma D.7.4, C is ν_{nd}-small for all large enough n.

We show that this value is in fact a property of the whole chain Φ, and is independent of the particular small set chosen, in the following analogue of Proposition 5.4.2.

Theorem 5.4.4. *Suppose that Φ is a ψ-irreducible Markov chain on X. Let $C \in \mathcal{B}(\mathsf{X})^+$ be a ν_M-small set and let d be the greatest common divisor of the set E_C. Then there exist disjoint sets $D_1 \ldots D_d \in \mathcal{B}(\mathsf{X})$ (a "d-cycle") such that*

(i) *for $x \in D_i$, $P(x, D_{i+1}) = 1$, $i = 0, \ldots, d-1 \pmod{d}$;*

(ii) *the set $N = [\bigcup_{i=1}^d D_i]^c$ is ψ-null.*

The d-cycle $\{D_i\}$ is maximal in the sense that for any other collection $\{d', D_k', k = 1, \ldots, d'\}$ satisfying (i)–(ii), we have d' dividing d; whilst if $d = d'$, then, by reordering the indices if necessary, $D_i' = D_i$ a.e. ψ.

PROOF   For $i = 0, 1, \ldots, d-1$ set

$$
D_i^* = \left\{ y : \sum_{n=1}^{\infty} P^{nd-i}(y, C) > 0 \right\} :
$$

by irreducibility, $\mathsf{X} = \cup D_i^*$.

The D_i^* are in general not disjoint, but we can show that their intersection is ψ-null. For suppose there exists i, k such that $\psi(D_i^* \cap D_k^*) > 0$. Then for some fixed $m, n > 0$, there is a subset $A \subseteq D_i^* \cap D_k^*$ with $\psi(A) > 0$ such that

$$
\begin{aligned}
P^{md-i}(w, C) &\geq \delta_m > 0, \quad w \in A \\
P^{nd-k}(w, C) &\geq \delta_n > 0, \quad w \in A
\end{aligned} \tag{5.41}
$$

and since ψ is the maximal irreducibility measure, we can also find r such that

$$
\int_C \nu(dy) P^r(y, A) = \delta_c > 0. \tag{5.42}
$$

Now we use the fact that C is a ν_M-small set: for $x \in C$, $B \subseteq C$, from (5.41), (5.42),

$$
\begin{aligned}
P^{2M+md-i+r}(x, B) &\geq \int_C P^M(x, dy) \int_A P^r(y, dw) \int_C P^{md-i}(w, dz) P^M(z, B) \\
&\geq [\delta_c \delta_m] \nu(B),
\end{aligned}
$$

so that $[2M+md+r]-i \in E_C$. By identical reasoning, we also have $[2M+nd+r]-k \in E_C$. This contradicts the definition of d, and we have shown that $\psi(D_i^* \cap D_k^*) = 0$, $i \neq k$.

Let $N = \cup_{i,j}(D_i^* \cap D_k^*)$, so that $\psi(N) = 0$. The sets $\{D_i^* \backslash N\}$ form a disjoint class of sets whose union is full. By Proposition 4.2.3, we can find an absorbing set D such that $D_i = D \cap (D_i^* \backslash N)$ are disjoint and $D = \cup D_i$. By the Chapman–Kolmogorov equations again, if $x \in D$ is such that $P(x, D_j) > 0$, then we have $x \in D_{j-1}$, by definition, for $j = 0, \ldots, d-1 \pmod d$. Thus $\{D_i\}$ is a d-cycle.

To prove the maximality and uniqueness result, suppose $\{D_i'\}$ is another cycle with period d', with $N = [\cup D_i']^c$ such that $\psi(N) = 0$. Let k be any index with $\nu(D_k' \cap C) > 0$: since $\psi(N) = 0$ and $\psi \succ \nu$, such a k exists. We then have, since C is a ν_M-small set, $P^M(x, D_k' \cap C) \geq \nu(D_k' \cap C) > 0$ for every $x \in C$. Since $(D_k' \cap C)$ is non-empty, this implies firstly that M is a multiple of d'; since this happens for any $n \in E_C$, by definition of d we have d' divides d as required. Also, we must have $C \cap D_j'$ empty for any $j \neq k$: for if not we would have some $x \in C$ with $P^M(x, C \cap D_k') = 0$, which contradicts the properties of C.

Hence we have $C \subseteq (D_k' \cup N)$, for some particular k. It follows by the definition of the original cycle that each D_j' is a union up to ψ-null sets of (d/d_i) elements of D_i. $\square$

It is obvious from the above proof that the cycle does not depend, except perhaps for ψ-null sets, on the small set initially chosen, and that any small set must be essentially contained inside one specific member of the cyclic class $\{D_i\}$.

Periodic and aperiodic chains

Suppose that $\boldsymbol{\Phi}$ is a φ-irreducible Markov chain.

The largest d for which a d-cycle occurs for $\boldsymbol{\Phi}$ is called the *period* of $\boldsymbol{\Phi}$.

When $d = 1$, the chain $\boldsymbol{\Phi}$ is called *aperiodic*.

When there exists a ν_1-small set A with $\nu_1(A) > 0$, then the chain is called *strongly aperiodic*.

As a direct consequence of these definitions and Theorem 5.2.3 we have

Proposition 5.4.5. *Suppose that $\boldsymbol{\Phi}$ is a ψ-irreducible Markov chain.*

(i) *If $\boldsymbol{\Phi}$ is strongly aperiodic, then the minorization condition (5.2) holds.*

(ii) *The resolvent, or K_{a_ε}-chain, is strongly aperiodic for all $0 < \varepsilon < 1$.*

(iii) *If $\boldsymbol{\Phi}$ is aperiodic, then every skeleton is ψ-irreducible and aperiodic, and some m-skeleton is strongly aperiodic.*

$\square$

This result shows that it is clearly desirable to work with strongly aperiodic chains. Regrettably, this condition is not satisfied in general, even for simple chains; and we will often have to prove results for strongly aperiodic chains and then use special methods to extend them to general chains through the m-skeleton or the K_{a_ε}-chain.

We will however concentrate almost exclusively on aperiodic chains. In practice this is not greatly restrictive, since we have as in the countable case

Proposition 5.4.6. *Suppose Φ is a ψ-irreducible chain with period d and d-cycle $\{D_i, i = 1, \ldots, d\}$. Then each of the sets D_i is an absorbing ψ-irreducible set for the chain Φ_d corresponding to the transition probability kernel P^d, and Φ_d on each D_i is aperiodic.*

PROOF   That each D_i is absorbing and irreducible for Φ_d is obvious: that Φ_d on each D_i is aperiodic follows from the definition of d as the largest value for which a cycle exists. $\qquad\square$

5.4.4 Periodic and aperiodic examples: forward recurrence times

For the forward recurrence time chain on the integers it is easy to evaluate the period of the chain. For let p be the distribution of the renewal variables, and let

$$d = g.c.d.\{n : p(n) > 0\}.$$

It is a simple exercise to check that d is also the g.c.d. of the set of times $\{n : P^n(0,0) > 0\}$ and so d is the period of the chain.

Now consider the forward recurrence time δ-skeleton $\boldsymbol{V}_\delta^+ = V^+(n\delta)$, $n \in \mathbb{Z}_+$ defined in Section 3.5.3. Here, we can find explicit conditions for aperiodicity even though the chain has no atom in the space. We have

Proposition 5.4.7. *If F is spread out, then $\boldsymbol{V}_\delta^+$ is aperiodic for sufficiently small δ.*

PROOF   In Proposition 5.3.3 we showed that for sufficiently small δ, the set $[0, \delta)$ is a ν_M-small set, where ν is a multiple of Lebesgue measure restricted to $[0, \delta)$.

But since the bounds on the densities in (5.35) hold, not just for the range $[k\delta, (k+3)\delta)$ for which they were used, but by construction for the greater range $[k\delta, (k+4)\delta)$, the same proof shows that $[0, \delta)$ is a ν_{M+1}-small set also, and thus aperiodicity follows from the definition of the period of $\boldsymbol{V}_\delta^+$ as the g.c.d. in (5.40). $\qquad\square$

5.5 Petite sets and sampled chains

5.5.1 Sampling a Markov chain

A convenient tool for the analysis of Markov chains is the *sampled chain*, which extends substantially the idea of the m-skeleton or the resolvent chain.

Let $a = \{a(n)\}$ be a distribution, or probability measure, on $\mathbb{Z}_+$, and consider the Markov chain Φ_a with probability transition kernel

$$K_a(x, A) := \sum_{n=0}^{\infty} P^n(x, A)a(n), \qquad x \in \mathsf{X}, A \in \mathcal{B}(\mathsf{X}). \tag{5.43}$$

It is obvious that K_a is indeed a transition kernel, so that Φ_a is well-defined by Theorem 3.4.1.

We will call Φ_a the K_a-chain, with *sampling distribution* a. Probabilistically, Φ_a has the interpretation of being the chain Φ "sampled" at time points drawn successively according to the distribution a, or more accurately, at time points of an independent renewal process with increment distribution a as defined in Section 2.4.1.

There are two specific sampled chains which we have already invoked, and which will be used frequently in the sequel. If $a = \delta_m$ is the Dirac measure with $\delta_m(m) = 1$, then the K_{δ_m}-chain is the m-skeleton with transition kernel P^m. If a_ε is the geometric distribution with

$$a_\varepsilon(n) = [1 - \varepsilon]\varepsilon^n, \qquad n \in \mathbb{Z}_+,$$

then the kernel K_{a_ε} is the resolvent K_ε which was defined in Chapter 3. The concept of sampled chains immediately enables us to develop useful conditions under which one set is uniformly accessible from another. We say that a set $B \in \mathcal{B}(\mathsf{X})$ is *uniformly accessible using* a from another set $A \in \mathcal{B}(\mathsf{X})$ if there exists a $\delta > 0$ such that

$$\inf_{x \in A} K_a(x, B) > \delta; \tag{5.44}$$

and when (5.44) holds we write $A \overset{a}{\rightsquigarrow} B$.

Lemma 5.5.1. *If $A \overset{a}{\rightsquigarrow} B$ for some distribution a, then $A \rightsquigarrow B$.*

PROOF   Since $L(x, B) = \mathsf{P}_x(\tau_B < \infty) = \mathsf{P}_x(\Phi_n \in B$ for some $n \in \mathbb{Z}_+)$ and $K_a(x, B) = \mathsf{P}_x(\Phi_\eta \in B)$ where η has the distribution a, it follows that

$$L(x, B) \geq K_a(x, B) \tag{5.45}$$

for any distribution a, and the result follows.   □

The following relationships will be used frequently.

Lemma 5.5.2.   (i) *If a and b are distributions on $\mathbb{Z}_+$, then the sampled chains with transition laws K_a and K_b satisfy the generalized Chapman–Kolmogorov equations*

$$K_{a*b}(x, A) = \int K_a(x, dy) K_b(y, A) \tag{5.46}$$

where $a * b$ denotes the convolution of a and b.

(ii) *If $A \overset{a}{\rightsquigarrow} B$ and $B \overset{b}{\rightsquigarrow} C$, then $A \overset{a*b}{\rightsquigarrow} C$.*

(iii) *If a is a distribution on $\mathbb{Z}_+$, then the sampled chain with transition law K_a satisfies the relation*

$$U(x, A) \geq \int U(x, dy) K_a(y, A). \tag{5.47}$$

PROOF   To see (i), observe that by definition and the Chapman–Kolmogorov equation

$$
\begin{aligned}
K_{a*b}(x, A) &= \sum_{n=0}^{\infty} P^n(x, A)\, a * b(n) \\
&= \sum_{n=0}^{\infty} P^n(x, A) \sum_{m=0}^{n} a(m)b(n - m) \\
&= \sum_{n=0}^{\infty} \sum_{m=0}^{n} \int P^m(x, dy) P^{n-m}(y, A) a(m)b(n - m) \\
&= \int \sum_{m=0}^{\infty} P^m(x, dy) a(m) \sum_{n=m}^{\infty} P^{n-m}(y, A)b(n - m) \\
&= \int K_a(x, dy) K_b(yA),
\end{aligned}
\tag{5.48}
$$

as required.

The result (ii) follows directly from (5.46) and the definitions.

For (iii), note that for fixed m, n,

$$
P^{m+n}(x, A)a(n) = \int P^m(x, dy) P^n(y, A)a(n)
$$

so that summing over m gives

$$
U(x, A)a(n) \geq \sum_{m > n} P^m(x, A)a(n) = \int U(x, dy) P^n(y, A)a(n);
$$

a second summation over n gives the result since $\sum_n a(n) = 1$.   □

The probabilistic interpretation of Lemma 5.5.2 (i) is simple: if the chain is sampled at a random time $\eta = \eta_1 + \eta_2$, where η_1 has distribution a and η_2 has independent distribution b, then since η has distribution $a*b$, it follows that (5.46) is just a Chapman–Kolmogorov decomposition at the intermediate random time.

5.5.2   The property of petiteness

Small sets always exist in the ψ-irreducible case, and provide most of the properties we need. We now introduce a generalization of small sets, petite sets, which have even more tractable properties, especially in topological analyses.

Petite sets

We will call a set $C \in \mathcal{B}(\mathsf{X})$ ν_a-*petite* if the sampled chain satisfies the bound

$$
K_a(x, B) \geq \nu_a(B),
$$

for all $x \in C$, $B \in \mathcal{B}(\mathsf{X})$, where ν_a is a non-trivial measure on $\mathcal{B}(\mathsf{X})$.

From the definitions we see that a small set is petite, with the sampling distribution a taken as δ_m for some m. Hence the property of being a small set is in general stronger than the property of being petite. We state this formally as

Proposition 5.5.3. *If $C \in \mathcal{B}(\mathsf{X})$ is ν_m-small, then C is ν_{δ_m}-petite.* $\qquad\qquad\qquad$ □

The operation "$\overset{a}{\rightsquigarrow}$" interacts usefully with the petiteness property. We have

Proposition 5.5.4. $\quad$ (i) *If $A \in \mathcal{B}(\mathsf{X})$ is ν_a-petite and $D \overset{b}{\rightsquigarrow} A$, then D is ν_{b*a}-petite, where ν_{b*a} can be chosen as a multiple of ν_a.*

(ii) *If $\mathbf{\Phi}$ is ψ-irreducible and if $A \in \mathcal{B}^+(\mathsf{X})$ is ν_a-petite, then ν_a is an irreducibility measure for $\mathbf{\Phi}$.*

PROOF $\quad$ To prove (i) choose $\delta > 0$ such that for $x \in D$ we have $K_b(x, A) \geq \delta$. By Lemma 5.5.2 (i),

$$
\begin{aligned}
K_{b*a}(x, B) &= \int_{\mathsf{X}} K_b(x, dy) K_a(y, B) \\
&\geq \int_A K_b(x, dy) K_a(y, B) \qquad\qquad (5.49)\\
&\geq \delta \nu_a(B).
\end{aligned}
$$

To see (ii), suppose A is ν_a-petite and $\nu_a(B) > 0$. For $x \in \overline{A}(n, m)$ as in (5.27) we have

$$
P^n K_a(x, B) \geq \int_A P^n(x, dy) K_a(y, B) \geq m^{-1} \nu_a(B) > 0
$$

which gives the result. $\qquad\qquad\qquad\qquad\qquad\qquad\qquad\qquad\qquad\qquad\qquad\qquad$ □

Proposition 5.5.4 provides us with a prescription for generating an irreducibility measure from a petite set A, even if all we know for general $x \in \mathsf{X}$ is that the single petite set A is reached with positive probability. We see the value of this in the examples later in this chapter

The following result illustrates further useful properties of petite sets, which distinguish them from small sets.

Proposition 5.5.5. *Suppose $\mathbf{\Phi}$ is ψ-irreducible.*

(i) *If A is ν_a-petite, then there exists a sampling distribution b such that A is also ψ_b-petite where ψ_b is a maximal irreducibility measure.*

(ii) *The union of two petite sets is petite.*

(iii) *There exists a sampling distribution c, an everywhere strictly positive, measurable function $s \colon \mathsf{X} \to \mathbb{R}$, and a maximal irreducibility measure ψ_c such that*

$$
K_c(x, B) \geq s(x) \psi_c(B), \qquad x \in \mathsf{X}, \ B \in \mathcal{B}(\mathsf{X})
$$

Thus there is an increasing sequence $\{C_i\}$ of ψ_c-petite sets, all with the same sampling distribution c and minorizing measure equivalent to ψ, with $\cup C_i = \mathsf{X}$.

PROOF   To prove (i) we first show that we can assume without loss of generality that ν_a is an irreducibility measure, even if $\psi(A) = 0$.

From Proposition 5.2.4 there exists a ν_b-petite set C with $C \in \mathcal{B}^+(X)$. We have $K_{a_\varepsilon}(y, C) > 0$ for any $y \in X$ and any $\varepsilon > 0$, and hence for $x \in A$,

$$K_{a*a_\varepsilon}(x, C) \geq \int \nu_a(dy) K_{a_\varepsilon}(y, C) > 0.$$

This shows that $A \overset{a*a_\varepsilon}{\rightsquigarrow} C$, and hence from Proposition 5.5.4 we see that A is $\nu_{a*a_\varepsilon *b}$-petite, where $\nu_{a*a_\varepsilon *b}$ is a constant multiple of ν_b. Now, from Proposition 5.5.4 (ii), the measure $\nu_{a*a_\varepsilon *b}$ is an irreducibility measure, as claimed.

We now assume that ν_a is an irreducibility measure, which is justified by the discussion above, and use Proposition 5.5.2 (i) to obtain the bound, valid for any $0 < \varepsilon < 1$,

$$K_{a*a_\varepsilon}(x, B) = K_a K_{a_\varepsilon}(x, B) \geq \nu_a K_{a_\varepsilon}(B), \qquad x \in A, \qquad B \in \mathcal{B}(X).$$

Hence A is ψ_b-petite with $b = a_\varepsilon * a$ and $\psi_b = \nu_a K_{a_\varepsilon}$. Proposition 4.2.2 (iv) asserts that, since ν_a is an irreducibility measure, the measure ψ_b is a maximal irreducibility measure.

To see (ii), suppose that A_1 is ψ_{a_1}-petite, and that A_2 is ψ_{a_2}-petite. Let $A_0 \in \mathcal{B}^+(X)$ be a fixed petite set and define the sampling measure a on $\mathbb{Z}_+$ as $a(i) = \frac{1}{2}[a_1(i) + a_2(i)]$, $i \in \mathbb{Z}_+$.

Since both ψ_{a_1} and ψ_{a_2} can be chosen as maximal irreducibility measures, it follows that for $x \in A_1 \cup A_2$

$$K_a(x, A_0) \geq \frac{1}{2} \min(\psi_{a_1}(A_0), \psi_{a_2}(A_0)) > 0$$

so that $A_1 \cup A_2 \overset{a}{\rightsquigarrow} A_0$. From Proposition 5.5.4 we see that $A_1 \cup A_2$ is petite.

For (iii), first apply Theorem 5.2.2 to construct a ν_n-small set $C \in \mathcal{B}^+(X)$. By (i) above we may assume that C is ψ_b-petite with ψ_b a maximal irreducibility measure. Hence $K_b(y, \cdot) \geq \mathbb{I}_C(y) \psi_b(\cdot)$ for all $y \in X$.

By irreducibility and the definitions we also have $K_{a_\varepsilon}(x, C) > 0$ for all $0 < \varepsilon < 1$, and all $x \in X$. Combining these bounds gives for any $x \in X$, $B \in \mathcal{B}(X)$,

$$K_{b*a_\varepsilon}(x, B) \geq \int_C K_{a_\varepsilon}(y, dz) K_b(z, B) \geq K_{a_\varepsilon}(x, C) \psi_b(B)$$

which shows that (iii) holds with $c = b * a_\varepsilon$, $s(x) = K_{a_\varepsilon}(x, C)$ and $\psi_c = \psi_b$.

The petite sets forming the countable cover can be taken as $C_m := \{x \in X : s(x) \geq m^{-1}\}$, $m \geq 1$.   □

Clearly the result in (ii) is best possible, since the whole space is a countable union of small (and hence petite) sets from Proposition 5.2.4, yet is not necessarily petite itself.

Our next result is interesting of itself, but is more than useful as a tool in the use of petite sets.

Proposition 5.5.6. *Suppose that Φ is ψ-irreducible and that C is ν_a-petite.*

(i) *Without loss of generality we can take a to be either a uniform sampling distribu-*
    tion $a_m(i) = 1/m$, $1 \leq i \leq m$, or a to be the geometric sampling distribution a_ε.
    In either case, there is a finite mean sampling time

$$m_a = \sum_i ia(i).$$

(ii) *If Φ is strongly aperiodic, then the set $C_0 \cup C_1 \subseteq \check{\mathsf{X}}$ corresponding to C is ν_a^*-petite*
     for the split chain $\check{\Phi}$.

PROOF   To see (i), let $A \in \mathcal{B}^+(\mathsf{X})$ be ν_n-small. By Proposition 5.5.5 (i) we have

$$K_b(x, A) \geq \psi_b(A) > 0, \qquad x \in C$$

where ψ_b is a maximal irreducibility measure. Hence $\sum_{k=1}^{N} P^k(x, A) \geq \frac{1}{2}\psi_b(A)$, $x \in C$,
for some N sufficiently large.

Since A is ν_n-small, it follows that for any $B \in \mathcal{B}(\mathsf{X})$,

$$\sum_{k=1}^{N+n} P^k(x, B) \geq \sum_{k=1}^{N} P^{k+n}(x, B) \geq \frac{1}{2}\psi_b(A)\nu_n(B)$$

for $x \in C$. This shows that C is ν_a-petite with $a(k) = (N + n)^{-1}$ for $1 \leq k \leq N + n$.
Since for all ε and m there exists some constant c such that $a_\varepsilon(j) \geq ca_m(j)$, $j \in \mathbb{Z}_+$,
this proves (i).

To see (ii), suppose that the chain is split with the small set $A \in \mathcal{B}^+(\mathsf{X})$. Then
$A_0 \cup \mathsf{X}_1$ is also petite: for X_1 is small, and A_0 is also small since $\check{P}(x, \mathsf{X}_1) \geq \delta$ for
$x_0 \in A_0$, and we know that the union of petite sets is petite, by Proposition 5.5.5.

Since when $x_0 \in A_0^c$ we have for $n \geq 1$, $\check{P}^n(x_0, A_0 \cup \mathsf{X}_1) = \check{P}^n(x_0, A_0 \cup A_1) = P^n(x, A)$
it follows that

$$\check{K}_a(x_0, A_0 \cup \mathsf{X}_1) = \sum_{j=0}^{\infty} a(j)\check{P}^j(x_0, A_0 \cup \mathsf{X}_1)$$

is uniformly bounded from below for $x_0 \in C_0 \setminus A_0$, which shows that $C_0 \setminus A_0$ is petite.
Since the union of petite sets is petite, $C_0 \cup \mathsf{X}_1$ is also petite.   □

5.5.3   Petite sets and aperiodicity

If A is a petite set for a ψ-irreducible Markov chain, then the corresponding minorizing
measure can always be taken to be equal to a maximal irreducibility measure, although
the measure ν_m appropriate to a small set is not as large as this.

We now prove that in the ψ-irreducible aperiodic case, every petite set is also small
for an appropriate choice of m and ν_m.

Theorem 5.5.7. *If Φ is irreducible and aperiodic, then every petite set is small.*

PROOF   Let A be a petite set. From Proposition 5.5.5 we may assume that A is ψ_a-petite, where ψ_a is a maximal irreducibility measure.

Let C denote the small set used in (5.40). Since the chain is aperiodic, it follows from Theorem 5.4.4 and Lemma D.7.4 that for some $n_0 \in \mathbb{Z}_+$, the set C is ν_k-small, with $\nu_k = \delta\nu$ for some $\delta > 0$, for all $n_0/2 - 1 \leq k \leq n_0$.

Since $C \in \mathcal{B}^+(\mathsf{X})$, we may also assume that n_0 is so large that

$$\sum_{k=\lfloor n_0/2 \rfloor}^{\infty} a(k) \leq \tfrac{1}{2}\psi_a(C).$$

With n_0 so fixed, we have for all $x \in A$ and $B \in \mathcal{B}(\mathsf{X})$,

$$
\begin{aligned}
P^{n_0}(x, B) &\geq \sum_{k=0}^{\lceil n_0/2 \rceil} \left\{ \int_C P^k(x, dy) P^{n_0-k}(y, B) \right\} a(k) \\
&\geq \left(\sum_{k=0}^{\lceil n_0/2 \rceil} P^k(x, C)a(k) \right)\left(\delta\nu(B) \right) \\
&\geq \left(\tfrac{1}{2}\psi_a(C) \right)\left(\delta\nu(B) \right)
\end{aligned}
$$

which shows that A is ν_{n_0}-small, with $\nu_{n_0} = \left(\tfrac{1}{2}\delta\psi_a(C) \right)\nu$. $\square$

This somewhat surprising result, together with Proposition 5.5.5, indicates that the class of small sets can be used for different purposes, depending on the choice of sampling distribution we make: if we sample at a fixed finite time we may get small sets with their useful fixed time point properties; and if we extend the sampling as in Proposition 5.5.5, we develop a petite structure with a maximal irreducibility measure. We shall use this duality frequently.

5.6   Commentary

We have already noted that the split chain and the random renewal time approaches to regeneration were independently discovered by Nummelin [301] and Athreya and Ney [13]. The opportunities opened up by this approach are exploited with growing frequency in later chapters.

However, the split chain only works in the generality of φ-irreducible chains because of the existence of small sets, and the ideas for the proof of their existence go back to Doeblin [95], although the actual existence as we have it here is from Jain and Jamison [172]. Our proof is based on that in Orey [309], where small sets are called C-sets. Nummelin [303] Chapter 2 has a thorough discussion of conditions equivalent to that we use here for small sets; Bonsdorff [38] also provides connections between the various small set concepts.

Our discussion of cycles follows that in Nummelin [303] closely. A thorough study of cyclic behavior, expanding on the original approach of Doeblin [95], is given also in Chung [70].

Petite sets as defined here were introduced in Meyn and Tweedie [277]. The "small sets" defined in Nummelin and Tuominen [305] as well as the *petits ensembles* developed in Duflo [102] are also special instances of petite sets, where the sampling distribution a is chosen as $a(i) = 1/N$ for $1 \leq i \leq N$, and $a(i) = (1-\alpha)\alpha^i$ respectively. To a French speaker, the term "petite set" might be disturbing since the gender of *ensemble* is masculine: however, the nomenclature does fit normal English usage since [26] the word "petit" is likened to "puny", while "petite" is more closely akin to "small".

It might seem from Theorem 5.5.7 that there is little reason to consider both petite sets and small sets. However, we will see that the two classes of sets are useful in distinct ways. Petite sets are easy to work with for several reasons: most particularly, they span periodic classes so that we do not have to assume aperiodicity, they are always closed under unions for irreducible chains (Nummelin [303] also finds that unions of small sets are small under aperiodicity), and by Proposition 5.5.5 we may assume that the petite measure is a maximal irreducibility measure whenever the chain is irreducible.

Perhaps most importantly, when in the next chapter we introduce a class of Markov chains with desirable topological properties, we will see that the structure of these chains is closely linked to petiteness properties of compact sets.

Chapter 6

Topology and continuity

The structure of Markov chains is essentially probabilistic, as we have described it so far. In examining the stability properties of Markov chains, the context we shall most frequently use is also a probabilistic one: in Part II, stability properties such as recurrence or regularity will be defined as certain return to sets of positive ψ-measure, or as finite mean return times to petite sets, and so forth.

Yet for many chains, there is more structure than simply a σ-field and a probability kernel available, and the expectation is that any topological structure of the space will play a strong role in defining the behavior of the chain. In particular, we are used thinking of specific classes of sets in $\mathbb{R}^n$ as having intuitively reasonable properties.

When there is a topology, compact sets are thought of in some sense as manageable sets, having the same sort of properties as a finite set on a countable space; and so we could well expect "stable" chains to spend the bulk of their time in compact sets. Indeed, we would expect compact sets to have the sort of characteristics we have identified, and will identify, for small or petite sets.

Conversely, open sets are "non-negligible" in some sense, and if the chain is irreducible we might expect it at least to visit all open sets with positive probability. This indeed forms one alternative definition of "irreducibility".

In this, the first chapter in which we explicitly introduce topological considerations, we will have, as our two main motivations, the desire to link the concept of ψ-irreducibility with that of open set irreducibility and the desire to identify compact sets as petite.

The major achievement of the chapter lies in identifying a topological condition on the transition probabilities which achieves both of these goals, utilizing the sampled chain construction we have just considered in Section 5.5.1.

Assume then that X is equipped with a locally compact, separable, metrizable topology with $\mathcal{B}(\mathsf{X})$ as the Borel σ-field. Recall that a function h from X to $\mathbb{R}$ is *lower semicontinuous* if

$$\liminf_{y \to x} h(y) \geq h(x), \qquad x \in \mathsf{X}:$$

a typical, and frequently used, lower semicontinuous function is the indicator function $\mathbb{I}_O(x)$ of an open set O in $\mathcal{B}(\mathsf{X})$.

We will use the following continuity properties of the transition kernel, couched in terms of lower semicontinuous functions, to define classes of chains with suitable topological properties.

Feller chains, continuous components and T-chains

(i) If $P(\,\cdot\,,O)$ is a lower semicontinuous function for any open set $O \in \mathcal{B}(\mathsf{X})$, then P is called a *(weak) Feller chain.*

(ii) If a is a sampling distribution and there exists a substochastic transition kernel T satisfying

$$K_a(x,A) \geq T(x,A), \qquad x \in \mathsf{X}, \ A \in \mathcal{B}(\mathsf{X}),$$

where $T(\,\cdot\,,A)$ is a lower semicontinuous function for any $A \in \mathcal{B}(\mathsf{X})$, then T is called a *continuous component* of K_a.

(iii) If $\boldsymbol{\Phi}$ is a Markov chain for which there exists a sampling distribution a such that K_a possesses a continuous component T, with $T(x,\mathsf{X}) > 0$ for all x, then $\boldsymbol{\Phi}$ is called a *T-chain.*

We will prove as one highlight of this section

Theorem 6.0.1.   (i) *If $\boldsymbol{\Phi}$ is a T-chain and $L(x,O) > 0$ for all x and all open sets $O \in \mathcal{B}(\mathsf{X})$, then $\boldsymbol{\Phi}$ is ψ-irreducible.*

(ii) *If every compact set is petite, then $\boldsymbol{\Phi}$ is a T-chain; and conversely, if $\boldsymbol{\Phi}$ is a ψ-irreducible T-chain, then every compact set is petite.*

(iii) *If $\boldsymbol{\Phi}$ is a ψ-irreducible Feller chain such that supp ψ has non-empty interior, then $\boldsymbol{\Phi}$ is a ψ-irreducible T-chain.*

PROOF   Proposition 6.2.2 proves (i); (ii) is in Theorem 6.2.5; (iii) is in Theorem 6.2.9. □

In order to have any such links as those in Theorem 6.0.1 between the measure-theoretic and topological properties of a chain, it is vital that there be at least a minimal adaptation of the dynamics of the chain to the topology of the space on which it lives. For consider the chain on $[0,1]$ with transition law for $x \in [0,1]$ given by

$$P(n^{-1}, (n+1)^{-1}) = 1 - \alpha_n, \qquad P(n^{-1}, 0) = \alpha_n, \ n \in \mathbb{Z}_+; \tag{6.1}$$

$$P(x,1) = 1, \qquad x \neq n^{-1}, \qquad n \geq 1. \tag{6.2}$$

This chain fails to visit most open sets, although it is definitely irreducible provided $\alpha_n > 0$ for all n: and although it never leaves a compact set, it is clearly unstable in an obvious way if $\sum_n \alpha_n < \infty$, since then it moves monotonically down the sequence $\{n^{-1}\}$ with positive probability.

Of course, the dynamics of this chain are quite wrong for the space on which we have embedded it: its structure is adapted to the normal topology on the integers, not to that on the unit interval or the set $\{n^{-1}, n \in \mathbb{Z}_+\}$. The Feller property obviously fails at $\{0\}$, as does any continuous component property if $\alpha_n \to 0$.

This is a trivial and pathological example, but one which proves valuable in exhibiting the need for the various conditions we now consider, which do link the dynamics to the structure of the space.

6.1 Feller properties and forms of stability

6.1.1 Weak and strong Feller chains

Recall that the transition probability kernel P acts on bounded functions through the mapping

$$Ph(x) = \int P(x, dy)h(y), \qquad x \in \mathsf{X}. \tag{6.3}$$

Suppose that X is a (locally compact separable metric) topological space, and let us denote the class of bounded continuous functions from X to $\mathbb{R}$ by $\mathcal{C}(\mathsf{X})$.

The (weak) Feller property is frequently defined by requiring that the transition probability kernel P maps $\mathcal{C}(\mathsf{X})$ to $\mathcal{C}(\mathsf{X})$. If the transition probability kernel P maps all bounded measurable functions to $\mathcal{C}(\mathsf{X})$ then P (and also Φ) is called *strong Feller*.

That this is consistent with the definition above follows from

Proposition 6.1.1. (i) *The transition kernel* $P\mathbb{I}_O$ *is lower semicontinuous for every open set* $O \in \mathcal{B}(\mathsf{X})$ *(that is,* Φ *is weak Feller) if and only if* P *maps* $\mathcal{C}(\mathsf{X})$ *to* $\mathcal{C}(\mathsf{X})$; *and* P *maps all bounded measurable functions to* $\mathcal{C}(\mathsf{X})$ *(that is,* Φ *is strong Feller) if and only if the function* $P\mathbb{I}_A$ *is lower semicontinuous for every set* $A \in \mathcal{B}(\mathsf{X})$.

(ii) *If the chain is weak Feller, then for any closed set* $C \subset \mathsf{X}$ *and any non-decreasing function* $m: \mathbb{Z}_+ \to \mathbb{Z}_+$ *the function* $\mathsf{E}_x[m(\tau_C)]$ *is lower semicontinuous in* x. *Hence for any closed set* $C \subset \mathsf{X}$, $r > 1$ *and* $n \in \mathbb{Z}_+$ *the functions*

$$\mathsf{P}_x\{\tau_C \geq n\} \qquad \mathsf{E}_x[\tau_C] \qquad and \qquad \mathsf{E}_x[r^{\tau_C}]$$

are lower semicontinuous.

(iii) *If the chain is weak Feller, then for any open set* $O \subset \mathsf{X}$, *the function* $\mathsf{P}_x\{\tau_O \leq n\}$ *and hence also the functions* $K_a(x, O)$ *and* $L(x, O)$ *are lower semicontinuous.*

PROOF   To prove (i), suppose that Φ is Feller, so that $P\mathbb{I}_O$ is lower semicontinuous for any open set O. Choose $f \in \mathcal{C}(\mathsf{X})$, and assume initially that $0 \leq f(x) \leq 1$ for all x. For $N \geq 1$ define the Nth approximation to f as

$$f_N(x) := \frac{1}{N} \sum_{k=1}^{N-1} \mathbb{I}_{O_k}(x)$$

where $O_k = \{x : f(x) > k/N\}$. It is easy to see that $f_N \uparrow f$ as $N \uparrow \infty$, and by assumption Pf_N is lower semicontinuous for each N. By monotone convergence, $Pf_N \uparrow Pf$ as $N \uparrow \infty$, and hence by Theorem D.4.1 the function Pf is lower semicontinuous. Identical reasoning shows that the function $P(1 - f) = 1 - Pf$, and hence also $-Pf$, is lower semicontinuous. Applying Theorem D.4.1 once more we see that the function Pf is continuous whenever f is continuous with $0 \leq f \leq 1$.

By scaling and translation it follows that Pf is continuous whenever f is bounded and continuous.

Conversely, if P maps $\mathcal{C}(\mathsf{X})$ to itself, and O is an open set then by Theorem D.4.1 there exist continuous positive functions f_N such that $f_N(x) \uparrow \mathbb{I}_O(x)$ for each x as $N \uparrow \infty$. By monotone convergence $P\mathbb{I}_O = \lim Pf_N$, which by Theorem D.4.1 implies that $P\mathbb{I}_O$ is lower semicontinuous.

A similar argument shows that P is strong Feller if and only if the function $P\mathbb{I}_A$ is lower semicontinuous for every set $A \in \mathcal{B}(\mathsf{X})$.

We next prove (ii). By definition of τ_C we have $\mathsf{P}_x\{\tau_C = 0\} = 0$, and hence without loss of generality we may assume that $m(0) = 0$. For each $i \geq 1$ define $\Delta_m(i) := m(i) - m(i-1)$, which is non-negative since m is non-increasing. By a change of summation,

$$
\begin{aligned}
\mathsf{E}[m(\tau_C)] &= \sum_{k=1}^{\infty} m(k)\mathsf{P}_x\{\tau_C = k\} \\
&= \sum_{k=1}^{\infty}\sum_{i=1}^{k} \Delta_m(i)\mathsf{P}_x\{\tau_C = k\} \\
&= \sum_{i=1}^{\infty} \Delta_m(i)\mathsf{P}_x\{\tau_C \geq i\}.
\end{aligned}
$$

Since by assumption $\Delta_m(k) \geq 0$ for each $k > 0$, the proof of (ii) will be complete once we have shown that $\mathsf{P}_x\{\tau_C \geq k\}$ is lower semicontinuous in x for all k.

Since C is closed and hence $\mathbb{I}_{C^c}(x)$ is lower semicontinuous, by Theorem D.4.1 there exist positive continuous functions f_i, $i \geq 1$, such that $f_i(x) \uparrow \mathbb{I}_{C^c}(x)$ for each $x \in \mathsf{X}$.

Extend the definition of the kernel I_A, given by

$$ I_A(x, B) = \mathbb{I}_{A \cap B}(x), $$

by writing for any positive function g

$$ I_g(x, B) := g(x)\mathbb{I}_B(x). $$

Then for all $k \in \mathbb{Z}_+$,

$$ \mathsf{P}_x\{\tau_C \geq k\} = (PI_{C^c})^{k-1}(x, \mathsf{X}) = \lim_{i \to \infty}(PI_{f_i})^{k-1}(x, \mathsf{X}). $$

It follows from the Feller property that $\{(PI_{f_i})^{k-1}(x, \mathsf{X}) : i \geq 1\}$ is an increasing sequence of continuous functions and, again by Theorem D.4.1, this shows that $\mathsf{P}_x\{\tau_C \geq k\}$ is lower semicontinuous in x, completing the proof of (ii).

Result (iii) is similar, and we omit the proof.   □

Many chains satisfy these continuity properties, and we next give some important examples.

Weak Feller chains: the nonlinear state space models

One of the simplest examples of a weak Feller chain is the quite general nonlinear state space model NSS(F).

Suppose conditions (NSS1) and (NSS2) are satisfied, so that $\boldsymbol{X} = \{X_n\}$, where

$$X_k = F(X_{k-1}, W_k),$$

for some smooth (C^∞) function $F \colon \mathsf{X} \times \mathbb{R}^p \to \mathsf{X}$, where X is an open subset of $\mathbb{R}^n$; and the random variables $\{W_k\}$ are a disturbance sequence on $\mathbb{R}^p$.

Proposition 6.1.2. *The NSS(F) model is always weak Feller.*

PROOF   We have by definition that the mapping $x \to F(x, w)$ is continuous for each fixed $w \in \mathbb{R}$. Thus whenever $h \colon \mathsf{X} \to \mathbb{R}$ is bounded and continuous, $h \circ F(x, w)$ is also bounded and continuous for each fixed $w \in \mathbb{R}$. It follows from the Dominated Convergence Theorem that

$$
\begin{aligned}
Ph(x) &= \mathsf{E}[h(F(x, W))] \\
&= \int \Gamma(dw)\, h \circ F(x, w) \qquad\qquad (6.4)
\end{aligned}
$$

is a continuous function of $x \in \mathsf{X}$.   □

This simple proof of weak continuity can be emulated for many models. It implies that this aspect of the topological analysis of many models is almost independent of the random nature of the inputs. Indeed, we could rephrase Proposition 6.1.2 as saying that since the associated control model CM(F) is a continuous function of the state for each fixed control sequence, the stochastic nonlinear state space model NSS(F) is weak Feller.

We shall see in Chapter 7 that this reflection of deterministic properties of CM(F) by NSS(F) is, under appropriate conditions, a powerful and exploitable feature of the nonlinear state space model structure.

Weak and strong Feller chains: the random walk

The difference between the weak and strong Feller properties is graphically illustrated in

Proposition 6.1.3. *The unrestricted random walk is always weak Feller, and is strong Feller if and only if the increment distribution Γ is absolutely continuous with respect to Lebesgue measure μ^{Leb} on $\mathbb{R}$.*

PROOF   Suppose that $h \in \mathcal{C}(\mathsf{X})$: the structure (3.35) of the transition kernel for the random walk shows that

$$
\begin{aligned}
Ph(x) &= \int_{\mathbb{R}} h(y)\Gamma(dy - x) \\
&= \int_{\mathbb{R}} h(y + x)\Gamma(dy) \qquad\qquad (6.5)
\end{aligned}
$$

and since h is bounded and continuous, Ph is also bounded and continuous, again from the Dominated Convergence Theorem. Hence Φ is always weak Feller, as we also know from Proposition 6.1.2.

Suppose next that Γ possesses a density γ with respect to μ^{Leb} on $\mathbb{R}$. Taking h in (6.5) to be any bounded function, we have

$$Ph(x) = \int_{\mathbb{R}} h(y)\gamma(y-x)\,dy; \tag{6.6}$$

but now from Lemma D.4.3 it follows that the convolution $Ph(x) = \gamma * h$ is continuous, and the chain is strong Feller.

Conversely, suppose the random walk is strong Feller. Then for any B such that $\Gamma(B) = \delta > 0$, by the lower semicontinuity of $P(x, B)$ there exists a neighborhood O of $\{0\}$ such that

$$P(x, B) \geq P(0, B)/2 = \Gamma(B)/2 = \delta/2, \qquad x \in O. \tag{6.7}$$

By Fubini's Theorem and the translation invariance of μ^{Leb} we have for any $A \in \mathcal{B}(\mathsf{X})$

$$\begin{aligned}
\int_{\mathbb{R}} \mu^{\mathrm{Leb}}(dy)\Gamma(A-y) &= \int_{\mathbb{R}} \mu^{\mathrm{Leb}}(dy)\int_{\mathbb{R}} \mathbb{I}_{A-y}(x)\Gamma(dx) \\
&= \int_{\mathbb{R}} \Gamma(dx)\int_{\mathbb{R}} \mathbb{I}_{A-x}(y)\mu^{\mathrm{Leb}}(dy) \\
&= \mu^{\mathrm{Leb}}(A)
\end{aligned}$$

since $\Gamma(\mathbb{R}) = 1$. Thus we have in particular from (6.7) and (6.8)

$$\begin{aligned}
\mu^{\mathrm{Leb}}(B) &= \int_{\mathbb{R}} \mu^{\mathrm{Leb}}(dy)\Gamma(B-y) \\
&\geq \int_{O} \mu^{\mathrm{Leb}}(dy)\Gamma(B-y) \\
&\geq \delta\mu^{\mathrm{Leb}}(O)/2
\end{aligned}$$

and hence $\mu^{\mathrm{Leb}} \succ \Gamma$ as required. $\square$

6.1.2   Strong Feller chains and open set irreducibility

Our first interest in chains on a topological space lies in identifying their accessible sets.

Open set irreducibility

(i) A point $x \in \mathsf{X}$ is called *reachable* if for every open set $O \in \mathcal{B}(\mathsf{X})$ containing x (i.e. for every neighborhood of x)

$$\sum_{n} P^n(y, O) > 0, \qquad y \in \mathsf{X}.$$

(ii) The chain Φ is called *open set irreducible* if every point is reachable.

We will use often the following result, which is a simple consequence of the definition of support.

Lemma 6.1.4. *If Φ is ψ-irreducible, then x^* is reachable if and only if $x^* \in \text{supp}\,(\psi)$.*

PROOF   If $x^* \in \text{supp}\,(\psi)$ then, for any open set O containing x^*, we have $\psi(O) > 0$ by the definition of the support. By ψ-irreducibility it follows that $L(x, O) > 0$ for all x, and hence x^* is reachable.

Conversely, suppose that $x^* \notin \text{supp}\,(\psi)$, and let $O = \text{supp}\,(\psi)^c$. The set O is open by the definition of the support, and contains the state x^*. By Proposition 4.2.3 there exists an absorbing, full set $A \subseteq \text{supp}\,(\psi)$. Since $L(x, O) = 0$ for $x \in A$ it follows that x^* is not reachable.   □

It is easily checked that open set irreducibility is equivalent to irreducibility when the state space of the chain is countable and is equipped with the discrete topology.

The open set irreducibility definition is conceptually similar to the ψ-irreducibility definition: they both imply that "large" sets can be reached from every point in the space. In the ψ-irreducible case large sets are those of positive ψ-measure, whilst in the open set irreducible case, large sets are open non-empty sets.

In this book our focus is on the property of ψ-irreducibility as a fundamental structural property. The next result, despite its simplicity, begins to link that property to the properties of open set irreducible chains.

Proposition 6.1.5. *If Φ is a strong Feller chain, and X contains one reachable point x^*, then Φ is ψ-irreducible, with $\psi = P(x^*, \cdot)$.*

PROOF   Suppose A is such that $P(x^*, A) > 0$. By lower semicontinuity of $P(\cdot, A)$, there is a neighborhood O of x^* such that $P(z, A) > 0, z \in O$. Now, since x^* is reachable, for any $y \in X$, we have for some n

$$P^{n+1}(y, A) \geq \int_O P^n(y, dz) P(z, A) > 0 \tag{6.8}$$

which is the result.   □

This gives trivially

Proposition 6.1.6. *If Φ is an open set irreducible strong Feller chain, then Φ is a ψ-irreducible chain.*   □

We will see below in Proposition 6.2.2 that this strong Feller condition, which (as is clear from Proposition 6.1.3) may be unsatisfied for many models, is not needed in full to get this result, and that Proposition 6.1.5 and Proposition 6.1.6 hold for T-chains also.

There are now two different approaches we can take in connecting the topological and continuity properties of Feller chains with the stochastic or measure-theoretic properties of the chain. We can either weaken the strong Feller property by requiring in essence that it only hold partially; or we could strengthen the weak Feller condition whilst retaining its essential flavor.

It will become apparent that the former, T-chain, route is usually far more productive, and we move on to this next. A strengthening of the Feller property to give *e-chains* will then be developed in Section 6.4.

6.2 T-chains

6.2.1 T-chains and open set irreducibility

The calculations for NSS(F) models and random walks show that the majority of the chains we have considered to date have the weak Feller property.

However, we clearly need more than just the weak Feller property to connect measure-theoretic and topological irreducibility concepts: every random walk is weak Feller, and we know from Section 4.3.3 that any chain with increment measure concentrated on the rationals enters every open set but is not ψ-irreducible.

Moving from the weak to the strong Feller property is however excessive. Using the ideas of sampled chains introduced in Section 5.5.1 we now develop properties of the class of T-chains, which we shall find includes virtually all models we will investigate, and which appears almost ideally suited to link the general space attributes of the chain with the topological structure of the space.

The T-chain definition describes a class of chains which are not totally adapted to the topology of the space, in that the strongly continuous kernel T, being only a "component" of P, may ignore many discontinuous aspects of the motion of Φ: but it does ensure that the chain is not completely singular in its motion, with respect to the normal topology on the space, and the strong continuity of T links set properties such as ψ-irreducibility to the topology in a way that is not natural for weak continuity.

We illustrate precisely this point now, with the analogue of Proposition 6.1.5.

Proposition 6.2.1. *If* Φ *is a T-chain, and* X *contains one reachable point* x^*, *then* Φ *is* ψ-*irreducible, with* $\psi = T(x^*, \cdot)$.

PROOF   Let T be a continuous component for K_a: since T is everywhere non-trivial, we must have in particular that $T(x^*, \mathsf{X}) > 0$. Suppose A is such that $T(x^*, A) > 0$. By lower semicontinuity of $T(\cdot, A)$, there is a neighborhood O of x^* such that $T(w, A) > 0$, $w \in O$. Now, since x^* is reachable, for any $y \in \mathsf{X}$, we have from Proposition 5.5.2

$$K_{a_\varepsilon * a}(y, A) \geq \int_O K_{a_\varepsilon}(y, dw) K_a(w, A)$$

$$\geq \int_O K_{a_\varepsilon}(y, dw) T(w, A) > 0$$

which is the result.   □

This result has, as a direct but important corollary

Proposition 6.2.2. *If* Φ *is an open set irreducible T-chain, then* Φ *is a* ψ-*irreducible T-chain.*   □

6.2.2 T-chains and petite sets

When the Markov chain Φ is ψ-irreducible, we know that there always exists at least one petite set. When X is topological, it turns out that there is a perhaps surprisingly direct connection between the existence of petite sets and the existence of continuous components.

In the next two results we show that the existence of sufficient open petite sets implies that Φ is a T-chain.

Proposition 6.2.3. *If an open ν_a-petite set A exists, then K_a possesses a continuous component non-trivial on all of A.*

PROOF   Since A is ν_a-petite, by definition we have

$$K_a(\,\cdot\,,\,\cdot\,) \geq \mathbb{I}_A(\,\cdot\,)\nu\{\,\cdot\,\}.$$

Now set $T(x,B) := \mathbb{I}_A(x)\nu(B)$: this is certainly a component of K_a, non-trivial on A. Since A is an open set its indicator function is lower semicontinuous; hence T is a continuous component of K_a.   □

Using such a construction we can build up a component which is non-trivial everywhere, if the space X is sufficiently rich in petite sets. We need first

Proposition 6.2.4. *Suppose that for each $x \in \mathsf{X}$ there exists a probability distribution a_x on $\mathbb{Z}_+$ such that K_{a_x} possesses a continuous component T_x which is non-trivial at x. Then Φ is a T-chain.*

PROOF   For each $x \in \mathsf{X}$, let O_x denote the set

$$O_x = \{y \in \mathsf{X} : T_x(y, \mathsf{X}) > 0\}.$$

which is open since $T_x(\,\cdot\,, \mathsf{X})$ is lower semicontinuous. Observe that by assumption, $x \in O_x$ for each $x \in \mathsf{X}$.

By Lindelöf's Theorem D.3.1 there exists a countable subcollection of sets $\{O_i : i \in \mathbb{Z}_+\}$ and corresponding kernels T_i and K_{a_i} such that $\bigcup O_i = \mathsf{X}$. Letting

$$T = \sum_{k=1}^{\infty} 2^{-k} T_k \qquad \text{and} \qquad a = \sum_{k=1}^{\infty} 2^{-k} a_k,$$

it follows that $K_a \geq T$, and hence satisfies the conclusions of the proposition.   □

We now get a virtual equivalence between the T-chain property and the existence of compact petite sets.

Theorem 6.2.5.   (i) *If every compact set is petite, then Φ is a T-chain.*

(ii) *Conversely, if Φ is a ψ-irreducible T-chain then every compact set is petite, and consequently if Φ is an open set irreducible T-chain then every compact set is petite.*

PROOF   Since X is σ-compact, there is a countable covering of open petite sets, and the result (i) follows from Proposition 6.2.3 and Proposition 6.2.4.

Now suppose that Φ is ψ-irreducible, so that there exists some petite $A \in \mathcal{B}^+(\mathsf{X})$, and let K_a have an everywhere non-trivial continuous component T.

By irreducibility $K_{a_\varepsilon}(x, A) > 0$, and hence from (5.46)

$$K_{a*a_\varepsilon}(x, A) = K_a K_{a_\varepsilon}(x, A) \geq T K_{a_\varepsilon}(x, A) > 0$$

for all $x \in \mathsf{X}$.

The function $TK_{a_\varepsilon}(\,\cdot\,, A)$ is lower semicontinuous and positive everywhere on X. Hence $K_{a*a_\varepsilon}(x, A)$ is uniformly bounded from below on compact subsets of X. Proposition 5.2.4 completes the proof that each compact set is petite.

The fact that we can weaken the irreducibility condition to open set irreducibility follows from Proposition 6.2.2.   □

The following factorization, which generalizes Proposition 5.5.5, further links the continuity and petiteness properties of T-chains.

Proposition 6.2.6. *If Φ is a ψ-irreducible T-chain, then there is a sampling distribution b, an everywhere strictly positive, continuous function $s' \colon \mathsf{X} \to \mathbb{R}$, and a maximal irreducibility measure ψ_b such that*

$$K_b(x, B) \geq s'(x)\psi_b(B), \qquad x \in \mathsf{X}, \ B \in \mathcal{B}(\mathsf{X}).$$

PROOF   If T is a continuous component of K_a, then we have from Proposition 5.5.5(iii),

$$
\begin{aligned}
K_{a*c}(x, B) &\geq \int K_a(x, dy)s(y)\,\psi_c(B) \\
&\geq T(x, s)\psi_c(B)
\end{aligned}
$$

The function $T(\,\cdot\,, s)$ is positive everywhere and lower semicontinuous, and therefore it dominates an everywhere positive continuous function s'; and we can take $b = a * c$ to get the required properties.   □

6.2.3   Feller chains, petite sets, and T-chains

We now investigate the existence of compact petite sets when the chain satisfies only the (weak) Feller continuity condition. Ultimately this leads to an auxiliary condition, satisfied by very many models in practice, under which a weak Feller chain is also a T-chain.

We first require the following lemma for petite sets for Feller chains.

Lemma 6.2.7. *If Φ is a ψ-irreducible Feller chain, then the closure of every petite set is petite.*

PROOF   By Proposition 5.2.4 and Proposition 5.5.4 and regularity of probability measures on $\mathcal{B}(\mathsf{X})$ (i.e. a set $A \in \mathcal{B}(\mathsf{X})$ may be approximated from within by compact sets), the set A is petite if and only if there exists a probability a on $\mathbb{Z}_+$, $\delta > 0$, and a compact petite set $C \subset \mathsf{X}$ such that

$$K_a(x, C) \geq \delta, \qquad x \in A.$$

By Proposition 6.1.1 the function $K_a(x, C)$ is upper semicontinuous when C is compact. Thus we have

$$\inf_{x \in \bar{A}} K_a(x, C) = \inf_{x \in A} K_a(x, C)$$

and this shows that the closure of a petite set is petite.   □

It is now possible to define auxiliary conditions under which all compact sets are petite for a Feller chain.

Proposition 6.2.8. *Suppose that* Φ *is* ψ-*irreducible. Then all compact subsets of* X *are petite if either:*

(i) Φ *has the Feller property and an open* ψ-*positive petite set exists; or*

(ii) Φ *has the Feller property and* supp ψ *has non-empty interior.*

PROOF   To see (i), let A be an open petite set of positive ψ-measure. Then $K_{a_\varepsilon}(\,\cdot\,, A)$ is lower semicontinuous and positive everywhere, and hence bounded from below on compact sets. Proposition 5.5.4 again completes the proof.

To see (ii), let A be a ψ-positive petite set, and define

$$A_k := \text{closure } \{x : K_{a_\varepsilon}(x, A) \geq 1/k\} \cap \text{supp}\,\psi.$$

By Proposition 5.2.4 and Lemma 6.2.7, each A_k is petite. Since supp ψ has non-empty interior it is of the second category, and hence there exists $k \in \mathbb{Z}_+$ and an open set $O \subset A_k \subset$ supp ψ. The set O is an open ψ-positive petite set, and hence we may apply (i) to conclude (ii).   □

A surprising, and particularly useful, conclusion from this cycle of results concerning petite sets and continuity properties of the transition probabilities is the following result, showing that Feller chains are in many circumstances also T-chains. We have as a corollary of Proposition 6.2.8 (ii) and Proposition 6.2.5 (ii) that

Theorem 6.2.9. *If a* ψ-*irreducible chain* Φ *is weak Feller and if* supp ψ *has nonempty interior then* Φ *is a T-chain.*   □

These results indicate that the Feller property, which is a relatively simple condition to verify in many applications, provides some strong consequences for ψ-irreducible chains.

Since we may cover the state space of a ψ-irreducible Markov chain by a countable collection of petite sets, and since by Lemma 6.2.7 the closure of a petite set is itself petite, it might seem that Theorem 6.2.9 could be strengthened to provide an open covering of X by petite sets without additional hypotheses on the chain. It would then follow by Theorem 6.2.5 that any ψ-irreducible Feller chain is a T-chain.

Unfortunately, this is not the case, as is shown by the following counterexample. Let X $= [0, 1]$ with the usual topology, let $0 < |\alpha| < 1$, and define the Markov transition function P for $x > 0$ by

$$P(x, \{0\}) = 1 - P(x, \{\alpha x\}) = x$$

We set $P(0, \{0\}) = 1$. The transition function P is Feller and δ_0-irreducible. But for any $n \in \mathbb{Z}_+$ we have

$$\lim_{x \to 0} \mathsf{P}_x(\tau_{\{0\}} \geq n) = 1,$$

from which it follows that there does not exist an open petite set containing the point $\{0\}$.

Thus we have constructed a ψ-irreducible Feller chain on a compact state space which is not a T-chain.

6.3 Continuous components for specific models

For a very wide range of the irreducible examples we consider, the support of the irreducibility measure does indeed have non-empty interior under some "spread-out" type of assumption. Hence weak Feller chains, such as the entire class of nonlinear models, will have all of the properties of the seemingly much stronger T-chain models provided they have an appropriate irreducibility structure.

We now identify a number of other examples of T-chains more explicitly.

6.3.1 Random walks

Suppose Φ is random walk on a half line. We have already shown that provided the increment distribution Γ provides some probability of negative increments then the chain is δ_0-irreducible, and moreover all of the sets $[0, c]$ are small sets.

Thus all compact sets are small and we have immediately from Theorem 6.2.5

Proposition 6.3.1. *The random walk on a half line with increment measure Γ is always a ψ-irreducible T-chain provided that $\Gamma(-\infty, 0) > 0$.*   □

Exactly the same argument for a storage model with general state-dependent release rule $r(x)$, as discussed in Section 2.4.4, shows these models to be δ_0-irreducible T-chains when the integral $R(x)$ of (2.32) is finite for all x.

Thus the virtual equivalence of the petite compact set condition and the T-chain condition provides an easy path to showing the existence of continuous components for many models with a real atom in the space.

Assessing conditions for non-atomic chains to be T-chains is not quite as simple in general. However, we can describe exactly what the continuous component condition defining T-chains means in the case of the random walk. Recall that the random walk is called spread-out if some convolution power Γ^{n*} is non-singular with respect to μ^{Leb} on $\mathbb{R}$.

Proposition 6.3.2. *The unrestricted random walk is a T-chain if and only if it is spread out.*

PROOF   If Γ is spread out then for some M, and some positive function γ, we have

$$P^M(x, A) = \Gamma^{M*}(A - x) \geq \int_{A-x} \gamma(y)dy := T(x, A)$$

and exactly as in the proof of Proposition 6.1.3, it follows that T is strong Feller: the spread-out assumption ensures that $T(x, \mathsf{X}) > 0$ for all x, and so by choosing the sampling distribution as $a = \delta_M$ we find that Φ is a T-chain.

The converse is somewhat harder, since we do not know a priori that when Φ is a T-chain, the component T can be chosen to be translation invariant. So let us assume that the result is false, and choose A such that $\mu^{\text{Leb}}(A) = 0$ but $\Gamma^{n*}(A) = 1$ for every n. Then $\Gamma^{n*}(A^c) = 0$ for all n and so for the sampling distribution a associated with the component T,

$$T(0, A^c) \leq K_a(0, A^c) = \sum_n \Gamma^{n*}(A^c)a(n) = 0.$$

The non-triviality of the component T thus ensures $T(0, A) > 0$, and since $T(\tilde{x}, A)$ is lower semicontinuous, there exists a neighborhood O of $\{0\}$ and a $\delta > 0$ such that $T(x, A) \geq \delta > 0$, $x \in O$.

Since T is a component of K_a, this ensures

$$K_a(x, A) \geq \delta > 0, \qquad x \in O.$$

But as in (6.8) by Fubini's Theorem and the translation invariance of μ^{Leb} we have

$$
\begin{aligned}
\mu^{\text{Leb}}(A) &= \int_{\mathbb{R}} \mu^{\text{Leb}}(dy) \Gamma^{n*}(A - y) \\
&= \int_{\mathbb{R}} \mu^{\text{Leb}}(dy) P^n(y, A).
\end{aligned}
\tag{6.9}
$$

Multiplying both sides of (6.9) by $a(n)$ and summing gives

$$
\begin{aligned}
\mu^{\text{Leb}}(A) &= \int_{\mathbb{R}} \mu^{\text{Leb}}(dy) K_a(y, A) \\
&\geq \int_O \mu^{\text{Leb}}(dy) K_a(y, A) \\
&\geq \delta \mu^{\text{Leb}}(O)
\end{aligned}
\tag{6.10}
$$

and since $\mu^{\text{Leb}}(O) > 0$, we have a contradiction. $\qquad\square$

This example illustrates clearly the advantage of requiring only a continuous component, rather than the Feller property for the chain itself.

6.3.2   Linear models as T-chains

Proposition 6.3.2 implies that the random walk model is a T-chain whenever the distribution of the increment variable W is sufficiently rich that, from each starting point, the chain does not remain in a set of zero Lebesgue measure.

This property, that when the set of reachable states is appropriately large the model is a T-chain, carries over to a much larger class of processes, including the linear and nonlinear state space models.

Suppose that X is a LSS(F,G)model, defined as usual by $X_{k+1} = FX_k + GW_{k+1}$. By repeated substitution in (LSS1) we obtain for any $m \in \mathbb{Z}_+$,

$$
X_m = F^m X_0 + \sum_{i=0}^{m-1} F^i GW_{m-i}.
\tag{6.11}
$$

To obtain a continuous component for the LSS(F,G) model, our approach is similar to that in deriving its irreducibility properties in Section 4.4. We require that the set of possible reachable states be large for the associated deterministic linear control system, and we also require that the set of reachable states remain large when the control sequence u is replaced by the random disturbance W. One condition sufficient to ensure this is

---

Non-singularity condition for the LSS(F,G) model

(LSS4)   The distribution Γ of the random variable W is nonsingular with respect to Lebesgue measure, with non-trivial density γ_w.

Using (6.11) we now show that the n-step transition kernel itself possesses a continuous component provided, firstly, Γ is nonsingular with respect to Lebesgue measure and secondly, the chain $\boldsymbol{X}$ can be driven to a sufficiently large set of states in $\mathbb{R}^n$ through the action of the disturbance process $\boldsymbol{W} = \{W_k\}$ as described in the last term of (6.11). This second property is a consequence of the controllability of the associated model LCM(F,G).

In Chapter 7 we will show that this construction extends further to more complex nonlinear models.

Proposition 6.3.3. *Suppose the deterministic control model LCM(F,G) on $\mathbb{R}^n$ satisfies the controllability condition (LCM3), and the associated LSS(F,G) model $\boldsymbol{X}$ satisfies the nonsingularity condition(LSS4).*

Then the n-skeleton possesses a continuous component which is everywhere non-trivial, so that $\boldsymbol{X}$ is a T-chain.

PROOF   We will prove this result in the special case where W is a scalar. The general case with $W \in \mathbb{R}^p$ is proved using the same methods as in the case where $p = 1$, but much more notation is needed for the required change of variables [272].

Let f denote an arbitrary positive function on $\mathsf{X} = \mathbb{R}^n$. From (6.11) together with non-singularity of the disturbance process $\boldsymbol{W}$ we may bound the conditional mean of $f(\Phi_n)$ as follows:

$$P^n f(x_0) \;=\; \mathsf{E}[f(F^n x_0 + \sum_{i=0}^{n-1} F^i G W_{n-i})] \tag{6.12}$$

$$\geq \int \cdots \int f(F^n x_0 + \sum_{i=0}^{n-1} F^i G w_{n-i}) \, \gamma_w(w_1) \cdots \gamma_w(w_n) \, dw_1 \ldots dw_n.$$

Letting C_n denote the controllability matrix in (4.13) and defining the vector valued random variable $\vec{W}_n = (W_1, \ldots, W_n)^\top$, we define the kernel T as

$$T f(x) := \int f(F^n x + C_n \vec{w}_n) \, \gamma_{\vec{w}}(\vec{w}_n) \, d\vec{w}_n.$$

We have $T(x, \mathsf{X}) = \{\int \gamma_w(x) \, dx\}^n > 0$, which shows that T is everywhere non-trivial; and T is a component of P^n since (6.12) may be written in terms of T as

$$P^n f(x_0) \geq \int f(F^n x_0 + C_n \vec{w}_n) \, \gamma_{\vec{w}}(\vec{w}_n) \, d\vec{w}_n = T f(x_0). \tag{6.13}$$

Let $|C_n|$ denote the determinant of C_n, which is non-zero since the pair (F, G) is controllable. Making the change of variables

$$\vec{v}_n = C_n \vec{w}_n, \qquad d\vec{v}_n = |C_n| d\vec{w}_n$$

in (6.13) allows us to write

$$T f(x_0) = \int f(F^n x_0 + \vec{v}_n) \gamma_{\vec{w}}(C_n^{-1} \vec{v}_n) |C_n|^{-1} \, d\vec{v}_n.$$

By Lemma D.4.3 and the Dominated Convergence Theorem, the right hand side of this identity is a continuous function of x_0 whenever f is bounded. This combined with (6.13) shows that T is a continuous component of P^n.   □

In particular this shows that the ARMA process (ARMA1) and any of its variations may be modeled as a T-chain if the noise process W is sufficiently rich with respect to Lebesgue measure, since they possess a controllable realization from Proposition 4.4.2.

In general, we can also obtain a T-chain by restricting the process to a controllable subspace of the state space in the manner indicated after Proposition 4.4.3.

6.3.3   Linear models as ψ-irreducible T-chains

We saw in Proposition 4.4.3 that a controllable LSS(F,G) model is ψ-irreducible (with ψ equivalent to Lebesgue measure) if the distribution Γ of W is Gaussian. In fact, under the conditions of that result, the process is also strong Feller, as we can see from the exact form of (4.18). Thus the controllable Gaussian model is a ψ-irreducible T-chain, with ψ specifically identified and the "component" T given by P itself.

In Proposition 6.3.3 we weakened the Gaussian assumption and still found conditions for the LSS(F,G) model to be a T-chain. We need extra conditions to retain ψ-irreducibility.

Now that we have developed the general theory further we can also use substantially weaker conditions on W to prove the chain possesses a reachable state, and this will give us the required result from Section 6.2.1. We introduce the following condition on the matrix F used in (LSS1):

Eigenvalue condition for the LSS(F,G) model

(LSS5)   The eigenvalues of F fall within the open unit disk in $\mathbb{C}$.

We will use the following lemma to control the growth of the models below.

Lemma 6.3.4. *Let $\rho(F)$ denote the modulus of the eigenvalue of F of maximum modulus, where F is an $n \times n$ matrix. Then for any matrix norm $\|\cdot\|$ we have the limit*

$$\log\big(\rho(F)\big) = \lim_{n \to \infty} \frac{1}{n} \log\big(\|F^n\|\big). \tag{6.14}$$

PROOF   The existence of the limit (6.14) follows from the Jordan Decomposition and is a standard result from linear systems theory: see [57] or Exercises 2.I.2 and 2.I.5 of [102] for details.   □

A consequence of Lemma 6.3.4 is that for any constants $\underline{\rho}$, $\overline{\rho}$ satisfying $\underline{\rho} < \rho(F) < \overline{\rho}$, there exists $c > 1$ such that

$$c^{-1}\underline{\rho}^n \leq \|F^n\| \leq c\overline{\rho}^n. \tag{6.15}$$

Hence for the linear state space model, under the eigenvalue condition (LSS5), the convergence $F^n \to 0$ takes place at a geometric rate. This property is used in the following result to give conditions under which the linear state space model is irreducible.

Proposition 6.3.5. *Suppose that the LSS(F,G) model* $\mathbf{X}$ *satisfies the density condition (LSS4) and the eigenvalue condition (LSS5), and that the associated control system LCM(F,G) is controllable.*

Then $\mathbf{X}$ *is a* ψ*-irreducible T-chain and every compact subset of* X *is small.*

PROOF   We have seen in Proposition 6.3.3 that the linear state space model is a T-chain under these conditions. To obtain irreducibility we will construct a reachable state and use Proposition 6.2.1.

Let $w^\star$ denote any element of the support of the distribution Γ of W, and let

$$x^\star = \sum_{k=0}^{\infty} F^k G w^\star.$$

If in (1.4), the control $u_k = w^\star$ for all k, then the system x_k converges to $x^\star$ uniformly for initial conditions in compact subsets of X.

By (pointwise) continuity of the model, it follows that for any bounded set $A \subset \mathsf{X}$ and open set O containing $x^\star$, there exists $\varepsilon > 0$ sufficiently small and $N \in \mathbb{Z}_+$ sufficiently large such that $x_N \in O$ whenever $x_0 \in A$, and $u_i \in w^\star + \varepsilon B$, for $1 \le i \le N$, where B denotes the open unit ball centered at the origin in X. Since $w^\star$ lies in the support of the distribution of W_k we can conclude that $P^N(x_0, O) \ge \Gamma(w^\star + \varepsilon B)^N > 0$ for $x_0 \in A$.

Hence $x^\star$ is reachable, which by Proposition 6.2.1 and Proposition 6.3.3 implies that Φ is ψ-irreducible for some ψ.

We now show that all bounded sets are small, rather than merely petite. Proposition 6.3.3 shows that P^n possesses a strong Feller component T. By Theorem 5.2.2 there exists a small set C for which $T(x^\star, C) > 0$ and hence, by the Feller property, an open set O containing $x^\star$ exists for which

$$\inf_{x \in O} T(x, C) > 0.$$

By Proposition 5.2.4 O is also a small set. If A is a bounded set, then we have already shown that $A \overset{\delta_M}{\leadsto} O$ for some N, so applying Proposition 5.2.4 once more we have the desired conclusion that A is small.   □

6.3.4   The first-order SETAR model

Results for nonlinear models are not always as easy to establish. However, for simple models similar conditions on the noise variables establish similar results. Here we consider the first-order SETAR models, which are defined as piecewise linear models satisfying

$$X_n = \phi(j) + \theta(j)X_{n-1} + W_n(j), \qquad X_{n-1} \in R_j$$

where $-\infty = r_0 < r_1 < \cdots < r_M = \infty$ and $R_j = (r_{j-1}, r_j]$; for each j, the noise variables $\{W_n(j)\}$ form an i.i.d. zero-mean sequence independent of $\{W_n(i)\}$ for $i \ne j$. Throughout, $W(j)$ denotes a generic variable with distribution Γ_j.

In order to ensure that these models can be analyzed as T-chains we make the following additional assumption, analogous to those above.

> (SETAR2)   For each $j = 1, \ldots, M$, the noise variable $W(j)$ has a density positive on the whole real line.

Even though this model is not Feller, due to the possible presence of discontinuities at the boundary points $\{r_i\}$, we can establish

Proposition 6.3.6. *Under (SETAR1) and (SETAR2), the SETAR model is a φ-irreducible T-process with φ taken as Lebesgue measure μ^{Leb} on $\mathbb{R}$.*

PROOF   The μ^{Leb}-irreducibility is immediate from the assumption of positive densities for each of the $W(j)$. The existence of a continuous component is less simple.

It is obvious from the existence of the densities that at any point in the interior of any of the regions R_i the transition function is strongly continuous. We do not necessarily have this continuity at the boundaries r_i themselves. However, as $x \uparrow r_i$ we have strong continuity of $P(x, \cdot)$ to $P(r_i, \cdot)$, whilst the limits as $x \downarrow r_i$ of $P(x, A)$ always exist giving a limit measure $P'(r_i, \cdot)$ which may differ from $P(r_i, \cdot)$.

If we take $T_i(x, \cdot) = \min(P'(r_i, \cdot), P(r_i, \cdot), P(x, \cdot))$ then T_i is a continuous component of P at least in some neighborhood of r_i; and the assumption that the densities of both $W(i), W(i+1)$ are positive everywhere guarantees that T_i is non-trivial.

But now we may put these components together using Proposition 6.2.4 and we have shown that the SETAR model is a T-chain.   □

Clearly one can weaken the positive density assumption. For example, it is enough for the T-chain result that for each j the supports of $W(j) - \phi(j) - \theta(j)r_j$ and $W(j+1) - \phi(j+1) - \theta(j+1)r_j$ should not be distinct, whilst for the irreducibility one can similarly require only that the densities of $W(j) - \phi(j) - \theta(j)x$ exist in a fixed neighborhood of zero, for $x \in (r_{j-1}, r_j]$. For chains which do not for some structural reason obey (SETAR2) one would need to check the conditions on the support of the noise variables with care to ensure that the conclusions of Proposition 6.3.6 hold.

6.4   e-Chains

Now that we have developed some of the structural properties of T-chains that we will require, we move on to a class of Feller chains which also have desirable structural properties, namely e-chains.

6.4.1   e-Chains and dynamical systems

The stability of weak Feller chains is naturally approached in the context of dynamical systems theory as introduced in the heuristic discussion in Chapter 1. Recall from Section 1.3.2 that the Markov transition function P gives rise to a deterministic map from $\mathcal{M}$, the space of probabilities on $\mathcal{B}(X)$, to itself, and we can construct on this basis a dynamical system $(P, \mathcal{M}, d)$, provided we specify a metric d, and hence also a topology, on $\mathcal{M}$.

To do this we now introduce the topology of weak convergence.

Weak convergence

A sequence of probabilities $\{\mu_k : k \in \mathbb{Z}_+\} \subset \mathcal{M}$ converges *weakly* to $\mu_\infty \in \mathcal{M}$ (denoted $\mu_k \xrightarrow{\text{w}} \mu_\infty$) if

$$\lim_{k \to \infty} \int f \, d\mu_k = \int f \, d\mu_\infty$$

for every $f \in \mathcal{C}(\mathsf{X})$.

Due to our restrictions on the state space X, the topology of weak convergence is induced by a number of metrics on $\mathcal{M}$; see Section D.5. One such metric may be expressed

$$d_m(\mu, \nu) = \sum_{k=0}^{\infty} |\int f_k \, d\mu - \int f_k \, d\nu| 2^{-k}, \qquad \mu, \nu \in \mathcal{M}, \qquad (6.16)$$

where $\{f_k\}$ is an appropriate set of functions in $\mathcal{C}_c(\mathsf{X})$, the set of continuous functions on X with compact support.

For $(P, \mathcal{M}, d_m)$ to be a dynamical system we require that P be a continuous map on $\mathcal{M}$. If P is continuous, then we must have in particular that if a sequence of point masses $\{\delta_{x_k} : k \in \mathbb{Z}_+\} \subset \mathcal{M}$ converge to some point mass $\delta_{x_\infty} \in \mathcal{M}$, then

$$\delta_{x_k} P \xrightarrow{\text{w}} \delta_{x_\infty} P \qquad \text{as } k \to \infty$$

or equivalently, $\lim_{k \to \infty} Pf(x_k) = Pf(x_\infty)$ for all $f \in \mathcal{C}(\mathsf{X})$. That is, if the Markov transition function induces a continuous map on $\mathcal{M}$, then Pf must be continuous for any bounded continuous function f.

This is exactly the weak Feller property. Conversely, it is obvious that for any weak Feller Markov transition function P, the associated operator P on $\mathcal{M}$ is continuous. We have thus shown

Proposition 6.4.1. *The triple $(P, \mathcal{M}, d_m)$ is a dynamical system if and only if the Markov transition function P has the weak Feller property.* $\qquad \square$

Although we do not get further immediate value from this result, since there do not exist a great number of results in the dynamical systems theory literature to be exploited in this context, these observations guide us to stronger and more useful continuity conditions.

Equicontinuity and e-chains

The Markov transition function P is called *equicontinuous* if for each $f \in \mathcal{C}_c(\mathsf{X})$ the sequence of functions $\{P^k f : k \in \mathbb{Z}_+\}$ is equicontinuous on compact sets.

A Markov chain which possesses an equicontinuous Markov transition function will be called an *e-chain*.

There is one striking result which very largely justifies our focus on e-chains, especially in the context of more stable chains.

Proposition 6.4.2. *Suppose that the Markov chain* Φ *has the Feller property, and that there exists a unique probability measure* π *such that for every* x

$$P^n(x, \cdot) \xrightarrow{\text{w}} \pi. \tag{6.17}$$

Then Φ *is an e-chain.*

PROOF   Since the limit in (6.17) is continuous (and in fact constant) it follows from Ascoli's Theorem D.4.2 that the sequence of functions $\{P^k f : k \in \mathbb{Z}_+\}$ is equicontinuous on compact subsets of X whenever $f \in \mathcal{C}(\mathsf{X})$. Thus the chain Φ is an e-chain.   □

Thus chains with good limiting behavior, such as those in Part III in particular, are forced to be e-chains, and in this sense the e-chain assumption is for many purposes a minor extra step after the original Feller property is assumed.

Recall from Chapter 1 that the dynamical system $(P, \mathcal{M}, d_m)$ is called stable in the sense of Lyapunov if for each measure $\mu \in \mathcal{M}$,

$$\lim_{\nu \to \mu} \sup_{k \geq 0} d_m(\nu P^k, \mu P^k) = 0.$$

The following result creates a further link between classical dynamical systems theory, and the theory of Markov chains on topological state spaces. The proof is routine and we omit it.

Proposition 6.4.3. *The Markov chain is an e-chain if and only if the dynamical system* $(P, \mathcal{M}, d_m)$ *is stable in the sense of Lyapunov.*

6.4.2   e-Chains and tightness

Stability in the sense of Lyapunov is a useful concept when a stationary point for the dynamical system exists. If x^* is a stationary point and the dynamical system is stable in the sense of Lyapunov, then trajectories which start near x^* will stay near x^*, and this turns out to be a useful notion of stability.

For the dynamical system $(P, \mathcal{M}, d_m)$, a stationary point is an invariant probability: that is, a probability satisfying

$$\pi(A) = \int \pi(dx)P(x, A), \qquad A \in \mathcal{B}(\mathsf{X}). \tag{6.18}$$

Conditions for such an invariant measure π to exist are the subject of considerable study for ψ-irreducible chains in Chapter 10, and in Chapter 12 we return to this question for weak Feller chains and e-chains.

A more immediately useful concept is that of Lagrange stability. Recall from Section 1.3.2 that $(P, \mathcal{M}, d_m)$ is Lagrange stable if, for every $\mu \in \mathcal{M}$, the orbit of measures μP^k is a precompact subset of $\mathcal{M}$. One way to investigate Lagrange stability for weak Feller chains is to utilize the following concept, which will have much wider applicability in due course.

Chains bounded in probability

The Markov chain Φ is called *bounded in probability* if for each initial condition $x \in X$ and each $\varepsilon > 0$, there exists a compact subset $C \subset X$ such that

$$\liminf_{k \to \infty} \mathsf{P}_x\{\Phi_k \in C\} \geq 1 - \varepsilon.$$

Boundedness in probability is simply tightness for the collection of probabilities $\{P^k(x, \cdot) : k \geq 1\}$. Since it is well known [36] that a set of probabilities $\mathcal{A} \subset \mathcal{M}$ is tight if and only if $\mathcal{A}$ is precompact in the metric space $(\mathcal{M}, d_m)$ this proves

Proposition 6.4.4. *The chain Φ is bounded in probability if and only if the dynamical system $(P, \mathcal{M}, d_m)$ is Lagrange stable.* $\square$

For e-chains, the concepts of boundedness in probability and Lagrange stability also interact to give a useful stability result for a somewhat different dynamical system.

The space $\mathcal{C}(X)$ can be considered as a normed linear space, where we take the norm $|\cdot|_c$ to be defined for $f \in \mathcal{C}(X)$ as

$$|f|_c := \sum_{k=0}^{\infty} 2^{-k} \big(\sup_{x \in C_k} |f(x)| \big)$$

where $\{C_k\}$ is a sequence of open precompact sets whose union is equal to X. The associated metric d_c generates the topology of uniform convergence on compact subsets of X.

If P is a weak Feller kernel, then the mapping P on $\mathcal{C}(X)$ is continuous with respect to this norm, and in this case the triple $(P, \mathcal{C}(X), d_c)$ is a dynamical system.

By Ascoli's Theorem D.4.2, $(P, \mathcal{C}(X), d_c)$ will be Lagrange stable if and only if for each initial condition $f \in \mathcal{C}(X)$, the orbit $\{P^k f : k \in \mathbb{Z}_+\}$ is uniformly bounded, and equicontinuous on compact subsets of X. This fact easily implies

Proposition 6.4.5. *Suppose that Φ is bounded in probability. Then Φ is an e-chain if and only if the dynamical system $(P, \mathcal{C}(X), d_c)$ is Lagrange stable.* $\square$

To summarize, for weak Feller chains boundedness in probability and the equicontinuity assumption are, respectively, exactly the same as Lagrange stability and stability in the sense of Lyapunov for the dynamical system $(P, \mathcal{M}, d_m)$; and these stability conditions are both simultaneously satisfied if and only if the dynamical system $(P, \mathcal{M}, d_m)$ and its dual $(P, \mathcal{C}(X), d_c)$ are simultaneously Lagrange stable.

These connections suggest that equicontinuity will be a useful tool for studying the limiting behavior of the distributions governing the Markov chain Φ, a belief which will be justified in the results in Chapter 12 and Chapter 18.

6.4.3 Examples of e-chains

The easiest example of an e-chain is the simple linear model described by (SLM1) and (SLM2).

If x and y are two initial conditions for this model, and the resulting sample paths are denoted $\{X_n(x)\}$ and $\{X_n(y)\}$ respectively for the same noise path, then by (SLM1) we have

$$X_{n+1}(x) - X_{n+1}(y) = \alpha(X_n(x) - X_n(y)) = \alpha^{n+1}(x - y). \tag{6.19}$$

If $|\alpha| \leq 1$, then this indicates that the sample paths should remain close together if their initial conditions are also close.

From this observation we now show that the simple linear model is an e-chain under the stability condition that $|\alpha| \leq 1$. Since the random walk on $\mathbb{R}$ is a special case of the simple linear model with $\alpha = 1$, this also implies that the random walk is also an e-chain.

Proposition 6.4.6. *The simple linear model defined by (SLM1) and (SLM2) is an e-chain provided that $|\alpha| \leq 1$.*

PROOF   Let $f \in \mathcal{C}_c(\mathsf{X})$. By uniform continuity of f, for any $\varepsilon > 0$ we can find $\delta > 0$ so that $|f(x) - f(y)| \leq \varepsilon$ whenever $|x - y| \leq \delta$. It follows from (6.19) that for any $n \in \mathbb{Z}_+$, and any $x, y \in \mathbb{R}$ with $|x - y| \leq \delta$,

$$
\begin{aligned}
|P^{n+1}f(x) - P^{n+1}f(y)| &= |\mathsf{E}[f(X_{n+1}(x)) - f(X_{n+1}(y))]| \\
&\leq \mathsf{E}[|f(X_{n+1}(x)) - f(X_{n+1}(y))|] \\
&\leq \varepsilon,
\end{aligned}
$$

which shows that $\boldsymbol{X}$ is an e-chain. $\square$

Equicontinuity is rather difficult to verify or rule out directly in general, especially before some form of stability has been established for the process. Although the equicontinuity condition may seem strong, it is surprisingly difficult to construct a natural example of a Feller chain which is not an e-chain. Indeed, our concentration on them is justified by Proposition 6.4.2 and this does provide an indirect way to verify that many Feller examples are indeed e-chains.

One example of a "non-e" chain is, however, provided by a "multiplicative random walk" on $\mathbb{R}_+$, defined by

$$X_{k+1} = \sqrt{X_k}\, W_{k+1}, \qquad k \in \mathbb{Z}_+, \tag{6.20}$$

where $\boldsymbol{W}$ is a disturbance sequence on $\mathbb{R}_+$ whose marginal distribution possesses a finite first moment. The chain is Feller since the right hand side of (6.20) is continuous in X_k. However, $\boldsymbol{X}$ is not an e-chain when $\mathbb{R}$ is equipped with the usual topology.

A complete proof of this fact requires more theory than we have so far developed, but we can give a sketch to illustrate what can go wrong.

When $X_0 \neq 0$, the process $\log X_k$, $k \in \mathbb{Z}_+$, is a version of the simple linear model described in Chapter 2, with $\alpha = \frac{1}{2}$. We will see in Section 10.5.4 that this implies that for any $X_0 = x_0 \neq 0$ and any bounded continuous function f,

$$P^k f(x_0) \to f_\infty, \qquad k \to \infty,$$

where f_∞ is a constant. When $x_0 = 0$ we have that $P^k f(x_0) = f(x_0) = f(0)$ for all k.

From these observations it is easy to see that X is not an e-chain. Take $f \in \mathcal{C}_c(\mathsf{X})$ with $f(0) = 0$ and $f(x) \geq 0$ for all $x > 0$: we may assume without loss of generality that $f_\infty > 0$. Since the one-point set $\{0\}$ is absorbing we have $P^k(0, \{0\}) = 1$ for all k, and it immediately follows that $P^k f$ converges to a discontinuous function. By Ascoli's Theorem the sequence of functions $\{P^k f : k \in \mathbb{Z}_+\}$ cannot be equicontinuous on compact subsets of $\mathbb{R}_+$, which shows that X is not an e-chain.

However by modifying the topology on $\mathsf{X} = \mathbb{R}_+$ we do obtain an e-chain as follows. Define the topology on the strictly positive real line $(0, \infty)$ in the usual way, and define $\{0\}$ to be open, so that X becomes a disconnected set with two open components. Then, in this topology, $P^k f$ converges to a uniformly continuous function which is constant on each component of X. From this and Ascoli's Theorem it follows that X is an e-chain.

It appears in general that such pathologies are typical of "non-e" Feller chains, and this again reinforces the value of our results for e-chains, which constitute the more typical behavior of Feller chains.

6.5   Commentary

The weak Feller chain has been a basic starting point in certain approaches to Markov chain theory for many years. The work of Foguel [121, 123], Jamison [174, 175, 176], Lin [238], Rosenblatt [339] and Sine [356, 357, 358] have established a relatively rich theory based on this approach, and the seminal book of Dynkin [105] uses the Feller property extensively.

We will revisit this in much greater detail in Chapter 12, where we will also take up the consequences of the e-chain assumption: this will be shown to have useful attributes in the study of limiting behavior of chains.

The equicontinuity results here, which relate this condition to the dynamical systems viewpoint, are developed by Meyn [260]. Equicontinuity may be compared to *uniform stability* [174] or *regularity* [115]. Whilst e-chains have also been developed in detail, particularly by Rosenblatt [337], Jamison [174, 175] and Sine [356, 357] they do not have particularly useful connections with the ψ-irreducible chains we are about to explore, which explains their relatively brief appearance at this stage.

The concept of continuous components appears first in Pollard and Tweedie [318, 319], and some practical applications are given in Laslett et al. [237]. The real exploitation of this concept really begins in Tuominen and Tweedie [391], from which we take Proposition 6.2.2. The connections between T-chains and the existence of compact petite sets is a recent result of Meyn and Tweedie [277].

In practice the identification of ψ-irreducible Feller chains as T-chains provided only that supp ψ has non-empty interior is likely to make the application of the results for such chains very much more common. This identification is new. The condition that supp ψ have non-empty interior has however proved useful in a number of associated areas in [319] and in Cogburn [75].

We note in advance here the results of Chapter 9 and Chapter 18, where we will show that a number of stability criteria for general space chains have "topological" analogues which, for T-chains, are exact equivalences. Thus T-chains will prove of on-going interest.

Finding criteria for chains to have continuity properties is a model-by-model exercise, but the results on linear and nonlinear systems here are intended to guide this process in some detail.

The assumption of a spread-out increment process, made in previous chapters for chains such as the unrestricted random walk, may have seemed somewhat arbitrary. It is striking therefore that this condition is both necessary and sufficient for random walk to be a T-chain, as in Proposition 6.3.2 which is taken from Tuominen and Tweedie [391]; they also show that this result extends to random walks on locally compact Haussdorff groups, which are T-chains if and only if the increment measure has some convolution power non-singular with respect to (right) Haar measure. These results have been extended to random walks on semi-groups by Högnas in [162].

In a similar fashion, the analysis carried out in Athreya and Pantula [15] shows that the simple linear model satisfying the eigenvalue condition (LSS5) is a T-chain if and only if the disturbance process is spread out. Chan et al. [64] show in effect that for the SETAR model compact sets are petite under positive density assumptions, but the proof here is somewhat more transparent.

These results all reinforce the impression that even for the simplest possible models it is not possible to dispense with an assumption of positive densities, and we adopt it extensively in the models we consider from here on.

Chapter 7

The nonlinear state space model

In applying the results and concepts of Part I in the domains of times series or systems theory, we have so far analyzed only linear models in any detail, albeit rather general and multidimensional ones. This chapter is intended as a relatively complete description of the way in which nonlinear models may be analyzed within the Markovian context developed thus far. We will consider both the general nonlinear state space model, and some specific applications which take on this particular form.

The pattern of this analysis is to consider first some particular structural or stability aspect of the associated deterministic control, or $CM(F)$, model and then under appropriate choice of conditions on the disturbance or noise process (typically a density condition as in the linear models of Section 6.3.2) to verify a related structural or stability aspect of the stochastic nonlinear state space $NSS(F)$ model.

Highlights of this duality are

(i) if the associated $CM(F)$ model is *forward accessible* (a form of controllability), and the noise has an appropriate density, then the $NSS(F)$ model is a T-chain (Section 7.1);

(ii) a form of irreducibility (the existence of a *globally attracting state* for the $CM(F)$ model) is then equivalent to the associated $NSS(F)$ model being a ψ-irreducible T-chain (Section 7.2);

(iii) the existence of periodic classes for the forward accessible $CM(F)$ model is further equivalent to the associated $NSS(F)$ model being a periodic Markov chain, with the periodic classes coinciding for the deterministic and the stochastic model (Section 7.3).

Thus we can reinterpret some of the concepts which we have introduced for Markov chains in this deterministic setting; and conversely, by studying the deterministic model we obtain criteria for our basic assumptions to be valid in the stochastic case.

In Section 7.4.3 the adaptive control model is considered to illustrate how these results may be applied in specific applications: for this model we exploit the fact that

Φ is generated by a NSS(F) model to give a simple proof that Φ is a ψ-irreducible and aperiodic T-chain.

We will end the chapter by considering the nonlinear state space model without forward accessibility, and showing how c-chain properties may then be established in lieu of the T-chain properties.

7.1 Forward accessibility and continuous components

The nonlinear state space model NSS(F) may be interpreted as a control system driven by a noise sequence exactly as the linear model is interpreted. We will take such a viewpoint in this section as we generalize the concepts used in the proof of Proposition 6.3.3, where we constructed a continuous component for the linear state space model.

7.1.1 Scalar models and forward accessibility

We first consider the scalar model SNSS(F) defined by

$$X_n = F(X_{n-1}, W_n),$$

for some smooth (C^∞) function $F : \mathbb{R} \times \mathbb{R} \to \mathbb{R}$ and satisfying (SNSS1)–(SNSS2).

Recall that in (2.5) we defined the map F_k inductively, for x_0 and w_i arbitrary real numbers, by

$$F_{k+1}(x_0, w_1, \ldots, w_{k+1}) = F(F_k(x_0, w_1, \ldots, w_k), w_{k+1}),$$

so that for any initial condition $X_0 = x_0$ and any $k \in \mathbb{Z}_+$,

$$X_k = F_k(x_0, W_1, \ldots, W_k).$$

Now let $\{u_k\}$ be the associated scalar "control sequence" for CM(F) as in (CM1), and use this to define the resulting state trajectory for CM(F) by

$$x_k = F_k(x_0, u_1, \ldots, u_k), \qquad k \in \mathbb{Z}_+. \tag{7.1}$$

Just as in the linear case, if from each initial condition $x_0 \in \mathsf{X}$ a sufficiently large set of states may be reached from x_0, then we will find that a continuous component may be constructed for the Markov chain $\mathbf{X}$. It is not important that every state may be reached from a given initial condition; the main idea in the proof of Proposition 6.3.3, which carries over to the nonlinear case, is that the set of possible states reachable from a given initial condition is not concentrated in some lower dimensional subset of the state space.

Recall also that we have assumed in (CM1) that for the associated deterministic control model CM(F) with trajectory (7.1), the control sequence $\{u_k\}$ is constrained so that $u_k \in O_w$, $k \in \mathbb{Z}_+$, where the control set O_w is an open set in $\mathbb{R}$.

For $x \in \mathsf{X}$, $k \in \mathbb{Z}_+$, we define $A_+^k(x)$ to be the set of all states reachable from x at time k by CM(F): that is, $A_+^0(x) = \{x\}$, and

$$A_+^k(x) := \Big\{ F_k(x, u_1, \ldots, u_k) : u_i \in O_w, 1 \le i \le k \Big\}, \qquad k \ge 1. \tag{7.2}$$

We define $A_+(x)$ to be the set of all states which are reachable from x at some time in the future, given by

$$A_+(x) := \bigcup_{k=0}^{\infty} A_+^k(x). \tag{7.3}$$

The analogue of controllability that we use for the nonlinear model is called *forward accessibility*.

Forward accessibility

The associated control model $\mathrm{CM}(F)$ is called *forward accessible* if for each $x_0 \in \mathsf{X}$, the set $A_+(x_0) \subset \mathsf{X}$ has non-empty interior.

For general nonlinear models, forward accessibility depends critically on the particular control set O_w chosen. This is in contrast to the linear state space model, where conditions on the driving matrix pair (F, G) sufficed for controllability.

Nonetheless, for the scalar nonlinear state space model we may show that forward accessibility is equivalent to the following "rank condition", similar to (LCM3):

Rank condition for the scalar $\mathrm{CM}(F)$ model

(CM2)   For each initial condition $x_0^0 \in \mathbb{R}$ there exists $k \in \mathbb{Z}_+$ and a sequence $(u_1^0, \ldots, u_k^0) \in O_w^k$ such that the derivative

$$\left[\frac{\partial}{\partial u_1} F_k(x_0^0, u_1^0, \ldots, u_k^0) \mid \cdots \mid \frac{\partial}{\partial u_k} F_k(x_0^0, u_1^0, \ldots, u_k^0) \right] \tag{7.4}$$

is non-zero.

In the scalar linear case the control system (7.1) has the form

$$x_k = Fx_{k-1} + Gu_k,$$

with F and G scalars. In this special case the derivative in (CM2) becomes exactly $[F^{k-1}G \mid \cdots \mid FG \mid G]$, which shows that the rank condition (CM2) is a generalization of the controllability condition (LCM3) for the linear state space model. This connection will be strengthened when we consider multidimensional nonlinear models below.

Theorem 7.1.1. *The control model CM(F) is forward accessible if and only if the rank condition (CM2) is satisfied.*

A proof of this result would take us too far from the purpose of this book. It is similar to that of Proposition 7.1.2, and details may be found in [271, 272].

7.1.2   Continuous components for the scalar nonlinear model

Using the characterization of forward accessibility given in Theorem 7.1.1 we now show how this condition on $\mathrm{CM}(F)$ leads to the existence of a continuous component for the associated $\mathrm{SNSS}(F)$ model.

To do this we need to increase the strength of our assumptions on the noise process, as we did for the linear model or the random walk.

Density for the SNSS(F) model

(SNSS3)   The distribution Γ of W is absolutely continuous, with a density γ_w on $\mathbb{R}$ which is lower semicontinuous.

The control set for the $\mathrm{SNSS}(F)$ model is the open set

$$O_w := \{x \in \mathbb{R} : \gamma_w(x) > 0\}.$$

We know from the definitions that, with probability one, $W_k \in O_w$ for all $k \in \mathbb{Z}_+$. Commonly assumed noise distributions satisfying this assumption include those which possess a continuous density, such as the Gaussian model, or uniform distributions on bounded open intervals in $\mathbb{R}$.

We can now develop an explicit continuous component for such scalar nonlinear state space models.

Proposition 7.1.2. *Suppose that for the SNSS(F) model, the noise distribution satisfies (SNSS3), and that the associated control system CM(F) is forward accessible. Then the SNSS(F) model is a T-chain.*

PROOF   Since $\mathrm{CM}(F)$ is forward accessible we have from Theorem 7.1.1 that the rank condition (CM2) holds. For simplicity of notation, assume that the derivative with respect to the kth disturbance variable is non-zero:

$$\frac{\partial F_k}{\partial w_k}(x_0^0, w_1^0, \ldots, w_k^0) \neq 0 \tag{7.5}$$

with $(w_1^0, \ldots, w_k^0) \in O_w^k$. Define the function $F^k \colon \mathbb{R} \times O_w^k \to \mathbb{R} \times O_w^{k-1} \times \mathbb{R}$ as

$$F^k(x_0, w_1, \ldots, w_k) = (x_0, w_1, \ldots, w_{k-1}, x_k)^\top,$$

where $x_k = F_k(x_0, w_1, \ldots, w_k)$. The total derivative of F^k can be computed as

$$DF^k = \begin{bmatrix} 1 & 0 & \cdots & & 0 \\ 0 & \ddots & & & \vdots \\ \vdots & & & & \\ & & 1 & & 0 \\ \frac{\partial F_k}{\partial x_0} & \frac{\partial F_k}{\partial w_1} & \cdots & & \frac{\partial F_k}{\partial w_k} \end{bmatrix},$$

which is evidently full rank at $(x_0^0, w_1^0, \ldots, w_k^0)$. It follows from the Inverse Function Theorem that there exists an open set

$$B = B_{x_0^0} \times B_{w_1^0} \times \cdots \times B_{w_k^0},$$

containing $(x_0^0, w_1^0, \ldots, w_k^0)$, and a smooth function $G^k : \{F^k\{B\}\} \to \mathbb{R}^{k+1}$ such that

$$G^k(F^k(x_0, w_1, \ldots, w_k)) = (x_0, w_1, \ldots, w_k),$$

for all $(x_0, w_1, \ldots, w_k) \in B$.

Taking G_k to be the final component of G^k, we see that for all $(x_0, w_1, \ldots, w_k) \in B$,

$$G_k(x_0, w_1, \ldots, w_{k-1}, x_k) = G_k(x_0, w_1, \ldots, w_{k-1}, F_k(x_0, w_1, \ldots, w_k)) = w_k.$$

We now make a change of variables, similar to the linear case. For any $x_0 \in B_{x_0^0}$, and any positive function $f : \mathbb{R} \to \mathbb{R}_+$,

$$
\begin{aligned}
P^k f(x_0) &= \int \cdots \int f(F_k(x_0, w_1, \ldots, w_k)) \gamma_w(w_k) \cdots \gamma_w(w_1) \, dw_1 \cdots dw_k \qquad (7.6) \\
&\geq \int_{B_{w_1^0}} \cdots \int_{B_{w_k^0}} f(F_k(x_0, w_1, \ldots, w_k)) \gamma_w(w_k) \cdots \gamma_w(w_1) \, dw_1 \cdots dw_k.
\end{aligned}
$$

We will first integrate over w_k, keeping the remaining variables fixed. By making the change of variables

$$x_k = F_k(x_0, w_1, \ldots, w_k), \qquad w_k = G_k(x_0, w_1, \ldots, w_{k-1}, x_k),$$

so that

$$dw_k = \left| \frac{\partial G_k}{\partial x_k}(x_0, w_1, \ldots, w_{k-1}, x_k) \right| dx_k,$$

we obtain for $(x_0, w_1, \ldots, w_{k-1}) \in B_{x_0^0} \times \cdots \times B_{w_{k-1}^0}$,

$$\int_{B_{w_k^0}} f(F_k(x_0, w_1, \ldots, w_k)) \gamma_w(w_k) \, dw_k = \int_{\mathbb{R}} f(x_k) q_k(x_0, w_1, \ldots, w_{k-1}, x_k) \, dx_k \quad (7.7)$$

where we define, with $\xi := (x_0, w_1, \ldots, w_{k-1}, x_k)$,

$$q_k(\xi) := \mathbb{I}\{G^k(\xi) \in B\} \gamma_w(G_k(\xi)) \left| \frac{\partial G_k}{\partial x_k}(\xi) \right|.$$

Since q_k is positive and lower semicontinuous on the open set $F^k\{B\}$, and zero on $F^k\{B\}^c$, it follows that q_k is lower semicontinuous on $\mathbb{R}^{k+1}$.

Define the kernel T_0 for an arbitrary bounded function f as

$$T_0 f(x_0) := \int \cdots \int f(x_k) q_k(\xi) \gamma_w(w_1) \cdots \gamma_w(w_{k-1}) \, dw_1 \cdots dw_{k-1} dx_k. \qquad (7.8)$$

The kernel T_0 is non-trivial at x_0^0 since

$$q_k(\xi^0) \gamma_w(w_1^0) \cdots \gamma_w(w_{k-1}^0) = \left| \frac{\partial G_k}{\partial x_k}(\xi^0) \right| \gamma_w(w_k^0) \gamma_w(w_1^0) \cdots \gamma_w(w_{k-1}^0) > 0,$$

where $\xi^0 = (x_0^0, w_1^0, \ldots, w_{k-1}^0, x_k^0)$. We will show that $T_0 f$ is lower semicontinuous on $\mathbb{R}$ whenever f is positive and bounded.

Since $q_k(x_0, w_1, \ldots, w_{k-1}, x_k)\gamma_w(w_1)\cdots\gamma_w(w_{k-1})$ is a lower semicontinuous function of its arguments in $\mathbb{R}^{k+1}$, there exists a sequence of positive, continuous functions $r_i: \mathbb{R}^{k+1} \to \mathbb{R}_+$, $i \in \mathbb{Z}_+$, such that for each i, the function r_i has bounded support and, as $i \uparrow \infty$,

$$r_i(x_0, w_1, \ldots, w_{k-1}, x_k) \uparrow q_k(x_0, w_1, \ldots, w_{k-1}, x_k)\gamma_w(w_1)\cdots\gamma_w(w_{k-1})$$

for each $(x_0, w_1, \ldots, w_{k-1}, x_k) \in \mathbb{R}^{k+1}$. Define the kernel T_i using r_i as

$$T_i f(x_0) := \int_{\mathbb{R}^k} f(x_k) r_i(x_0, w_1, \ldots, w_{k-1}, x_k)\, dw_1 \cdots dw_{k-1}\, dx_k.$$

It follows from the dominated convergence theorem that $T_i f$ is continuous for any bounded function f. If f is also positive, then as $i \uparrow \infty$,

$$T_i f(x_0) \uparrow T_0 f(x_0), \qquad x_0 \in \mathbb{R},$$

which implies that $T_0 f$ is lower semicontinuous when f is positive.

Using (7.6) and (7.7) we see that T_0 is a continuous component of P^k which is non-zero at x_0^0. From Theorem 6.2.4, the model is a T-chain as claimed.   □

7.1.3 Simple bilinear model

The forward accessibility of the SNSS(F) model is usually immediate since the rank condition (CM2) is easily checked.

To illustrate the use of Proposition 7.1.2, and in particular the computation of the "controllability vector" (7.4) in (CM2), we consider the scalar example where Φ is the bilinear state space model on $\mathsf{X} = \mathbb{R}$ defined in (SBL1) by

$$X_{k+1} = \theta X_k + b W_{k+1} X_k + W_{k+1}$$

where W is a disturbance process. To place this bilinear model into the framework of this chapter we assume

---

Density for the simple bilinear model

(SBL2)   The sequence W is a disturbance process on $\mathbb{R}$, whose marginal distribution Γ possesses a finite second moment, and a density γ_w which is lower semicontinuous.

Under (SBL1) and (SBL2), the bilinear model X is an SNSS(F) model with F defined in (2.7).

First observe that the one-step transition kernel P for this model cannot possess an everywhere non-trivial continuous component. This may be seen from the fact that

$P(-1/b, \{-\theta/b\}) = 1$, yet $P(x, \{-\theta/b\}) = 0$ for all $x \neq -1/b$. It follows that the only positive lower semicontinuous function which is majorized by $P(\cdot, \{-\theta/b\})$ is zero, and thus any continuous component T of P must be trivial at $-1/b$: that is, $T(-1/b, \mathbb{R}) = 0$.

This could be anticipated by looking at the controllability vector (7.4). The first order controllability vector is

$$\frac{\partial F}{\partial u}(x_0, u_1) = bx_0 + 1,$$

which is zero at $x_0 = -1/b$, and thus the first order test for forward accessibility fails. Hence we must take $k \geq 2$ in (7.4) if we hope to construct a continuous component.

When $k = 2$ the vector (7.4) can be computed using the chain rule to give

$$\left[\frac{\partial F}{\partial x}(x_1, u_2)\frac{\partial F}{\partial u}(x_0, u_1) \mid \frac{\partial F}{\partial u}(x_1, u_2)\right]$$
$$= \ [(\theta + bu_2)(bx_0 + 1) \mid bx_1 + 1]$$
$$= \ [(\theta + bu_2)(bx_0 + 1) \mid \theta bx_0 + b^2 u_1 x_0 + bu_1 + 1]$$

which is non-zero for almost every $\binom{u_1}{u_2} \in \mathbb{R}^2$. Hence the associated control model is forward accessible, and this together with Proposition 7.1.2 gives

Proposition 7.1.3. *If (SBL1) and (SBL2) hold, then the bilinear model is a T-chain.*

7.1.4   Multidimensional models

Most nonlinear processes that are encountered in applications cannot be modeled by a scalar Markovian model such as the SNSS(F) model. The more general NSS(F) model is defined by (NSS1), and we now analyze this in a similar way to the scalar model.

We again call the associated control system CM(F) with trajectories

$$x_k = F_k(x_0, u_1, \ldots, u_k), \qquad k \in \mathbb{Z}_+, \tag{7.9}$$

forward accessible if the set of attainable states $A_+(x)$, defined as

$$A_+(x) := \bigcup_{k=0}^{\infty} \left\{ F_k(x, u_1, \ldots, u_k) : u_i \in O_w, 1 \leq i \leq k \right\}, \qquad k \geq 1, \tag{7.10}$$

has non-empty interior for every initial condition $x \in \mathsf{X}$.

To verify forward accessibility we define a further generalization of the controllability matrix introduced in (LCM3).

For $x_0 \in \mathsf{X}$ and a sequence $\{u_k : u_k \in O_w, k \in \mathbb{Z}_+\}$ let $\{\Xi_k, \Lambda_k : k \in \mathbb{Z}_+\}$ denote the matrices

$$\Xi_{k+1} = \Xi_{k+1}(x_0, u_1, \ldots, u_{k+1}) \quad := \quad \left[\frac{\partial F}{\partial x}\right]_{(x_k, u_{k+1})},$$

$$\Lambda_{k+1} = \Lambda_{k+1}(x_0, u_1, \ldots, u_{k+1}) \quad := \quad \left[\frac{\partial F}{\partial u}\right]_{(x_k, u_{k+1})},$$

where $x_k = F_k(x_0, u_1, \ldots, u_k)$. Let $C_{x_0}^k = C_{x_0}^k(u_1, \ldots, u_k)$ denote the *generalized controllability matrix* (along the sequence $u_1, \ldots, u_k$)

$$C_{x_0}^k := [\Xi_k \cdots \Xi_2 \Lambda_1 \mid \Xi_k \cdots \Xi_3 \Lambda_2 \mid \cdots \mid \Xi_k \Lambda_{k-1} \mid \Lambda_k]. \tag{7.11}$$

If F takes the linear form

$$F(x, u) = Fx + Gu, \tag{7.12}$$

then the generalized controllability matrix again becomes

$$C_{x_0}^k = [F^{k-1}G \mid \cdots \mid G],$$

which is the controllability matrix introduced in (LCM3).

Rank condition for the multidimensional CM(F) model

(CM3)   For each initial condition $x_0 \in \mathbb{R}^n$, there exists $k \in \mathbb{Z}_+$ and a sequence $\vec{u}^0 = (u_1^0, \ldots, u_k^0) \in O_w^k$ such that

$$\operatorname{rank} C_{x_0}^k(\vec{u}^0) = n. \tag{7.13}$$

The controllability matrix C_y^k is the derivative of the state $x_k = F(y, u_1, \ldots, u_k)$ at time k with respect to the input sequence $(u_k^\top, \ldots, u_1^\top)$. The following result is a consequence of this fact together with the Implicit Function Theorem and Sard's Theorem (see [173, 272] and the proof of Proposition 7.1.2 for details).

Proposition 7.1.4. *The nonlinear control model CM(F) satisfying (7.9) is forward accessible if and only the rank condition (CM3) holds.*   □

To connect forward accessibility to the stochastic model (NSS1) we again assume that the distribution of W possesses a density.

Density for the NSS(F) model

(NSS3)   The distribution Γ of W possesses a density γ_w on $\mathbb{R}^p$ which is lower semicontinuous, and the control set for the NSS(F) model is the open set $O_w := \{x \in \mathbb{R} : \gamma_w(x) > 0\}$.

Using an argument which is similar to, but more complicated than the proof of Proposition 7.1.2, we may obtain the following consequence of forward accessibility.

Proposition 7.1.5. *If the NSS(F) model satisfies the density assumption (NSS3), and the associated control model is forward accessible, then the state space* X *may be written as the union of open small sets, and hence the NSS(F) model is a T-chain.*   □

Note that this only guarantees the T-chain property: we now move on to consider the equally needed irreducibility properties of the NNS(F) models.

7.2   Minimal sets and irreducibility

We now develop a more detailed description of reachable states and topological irreducibility for the nonlinear state space NSS(F) model, and exhibit more of the interplay between the stochastic and topological communication structures for NSS(F) models.

Since one of the major goals here is to exhibit further the links between the behavior of the associated deterministic control model and the NSS(F) model, it is first helpful to study the structure of the accessible sets for the control system CM(F) with trajectories (7.9).

A large part of this analysis deals with a class of sets called minimal sets for the control system CM(F). In this section we will develop criteria for their existence and properties of their topological structure. This will allow us to decompose the state space of the corresponding NSS(F) model into disjoint, closed, absorbing sets which are both ψ-irreducible and topologically irreducible.

7.2.1   Minimality for the deterministic control model

We define $A_+(E)$ to be the set of all states attainable by CM(F) from the set E at some time $k \geq 0$, and we let E^0 denote those states which cannot reach the set E:

$$A_+(E) := \bigcup_{x \in E} A_+(x), \qquad E^0 := \{x \in \mathsf{X} : A_+(x) \cap E = \emptyset\}.$$

Because the functions $F_k(\cdot, u_1, \ldots, u_k)$ have the semi-group property

$$F_{k+j}(x_0, u_1, \ldots, u_{k+j}) = F_j(F_k(x_0, u_1, \ldots, u_k), u_{k+1}, \ldots, u_{k+j}),$$

for $x_0 \in \mathsf{X}$, $u_i \in O_w$, $k, j \in \mathbb{Z}_+$, the set maps $\{A_+^k : k \in \mathbb{Z}_+\}$ also have this property: that is,

$$A_+^{k+j}(E) = A_+^k(A_+^j(E)), \qquad E \subset \mathsf{X}, \quad k, j \in \mathbb{Z}_+.$$

If $E \subset \mathsf{X}$ has the property that

$$A_+(E) \subset E,$$

then E is called *invariant*. For example, for all $C \subset \mathsf{X}$, the sets $A_+(C)$ and C^0 are invariant, and since the closure, union, and intersection of invariant sets is invariant, the set

$$\Omega_+(C) := \bigcap_{N=1}^{\infty} \overline{\left\{ \bigcup_{k=N}^{\infty} A_+^k(C) \right\}} \tag{7.14}$$

is also invariant.

The following result summarizes these observations:

Proposition 7.2.1. *For the control system (7.9) we have for any $C \subset \mathsf{X}$,*

(i) *$A_+(C)$ and $\overline{A_+(C)}$ are invariant;*

(ii) *$\Omega_+(C)$ is invariant;*

(iii) *C^0 is invariant, and C^0 is also closed if the set C is open.* □

As a consequence of the assumption that the map F is smooth, and hence continuous, we then have immediately

Proposition 7.2.2. *If the associated CM(F) model is forward accessible, then for the NSS(F) model:*

(i) *a closed subset $A \subset \mathsf{X}$ is absorbing for NSS(F) if and only if it is invariant for CM(F);*

(ii) *if $U \subset \mathsf{X}$ is open, then for each $k \geq 1$ and $x \in \mathsf{X}$,*

$$A_+^k(x) \cap U \neq \emptyset \Longleftrightarrow P^k(x, U) > 0;$$

(iii) *if $U \subset \mathsf{X}$ is open, then for each $x \in \mathsf{X}$,*

$$A_+(x) \cap U \neq \emptyset \Longleftrightarrow K_{a_\varepsilon}(x, U) > 0.$$

□

We now introduce minimal sets for the general CM(F) model.

Minimal sets

We call a set *minimal* for the deterministic control model CM(F) if it is (topologically) closed, invariant, and does not contain any closed invariant set as a proper subset.

For example, consider the LCM(F,G) model introduced in (1.4). The assumption (LCM2) simply states that the control set O_w is equal to $\mathbb{R}^p$.

In this case the system possesses a unique minimal set M which is equal to X_0, the range space of the controllability matrix, as described after Proposition 4.4.3. If the eigenvalue condition (LSS5) holds then this is the only minimal set for the LCM(F,G) model.

The following characterizations of minimality follow directly from the definitions, and the fact that both $\overline{A_+(x)}$ and $\Omega_+(x)$ are closed and invariant.

Proposition 7.2.3. *The following are equivalent for a nonempty set $M \subset \mathsf{X}$:*

(i) *M is minimal for CM(F);*

(ii) *$\overline{A_+(x)} = M$ for all $x \in M$;*

(iii) *$\Omega_+(x) = M$ for all $x \in M$.* □

7.2.2 M-Irreducibility and ψ-irreducibility

Proposition 7.2.3 asserts that any state in a minimal set can be "almost reached" from any other state. This property is similar in flavor to topological irreducibility for a Markov chain. The link between these concepts is given in the following central result for the NSS(F) model.

Theorem 7.2.4. *Let $M \subset \mathsf{X}$ be a minimal set for CM(F). If CM(F) is forward accessible and the disturbance process of the associated NSS(F) model satisfies the density condition (NSS3), then*

(i) *the set M is absorbing for NSS(F);*

(ii) *the NSS(F) model restricted to M is an open set irreducible (and so ψ-irreducible) T-chain.*

PROOF   That M is absorbing follows directly from Proposition 7.2.3, proving $M = \overline{A_+(x)}$ for some x; Proposition 7.2.1, proving $\overline{A_+(x)}$ is invariant; and Proposition 7.2.2, proving any closed invariant set is absorbing for the NSS(F) model.

To see that the process restricted to M is topologically irreducible, let $x_0 \in M$, and let $U \subseteq \mathsf{X}$ be an open set for which $U \cap M \neq \emptyset$. By Proposition 7.2.3 we have $A_+(x_0) \cap U \neq \emptyset$. Hence by Proposition 7.2.2 $K_{a_\varepsilon}(x_0, U) > 0$, which establishes open set irreducibility. The process is then ψ-irreducible from Proposition 6.2.2 since we know it is a T-chain from Proposition 7.1.5.   □

Clearly, under the conditions of Theorem 7.2.4, if X itself is minimal then the NSS(F) model is both ψ-irreducible and open set irreducible. The condition that X be minimal is a strong requirement which we now weaken by introducing a different form of "controllability" for the control system CM(F).

We say that the deterministic control system CM(F) is *indecomposable* if its state space X does not contain two disjoint closed invariant sets. This condition is clearly necessary for CM(F) to possess a unique minimal set. Indecomposability is not sufficient to ensure the existence of a minimal set: take $\mathsf{X} = \mathbb{R}$, $O_w = (0, 1)$, and

$$x_{k+1} = F(x_k, u_{k+1}) = x_k + u_{k+1},$$

so that all proper closed invariant sets are of the form $[t, \infty)$ for some $t \in \mathbb{R}$. This system is indecomposable, yet no minimal sets exist.

Irreducible control models

If CM(F) is indecomposable and also possesses a minimal set M, then CM(F) will be called *M-irreducible*.

If CM(F) is M-irreducible it follows that $M^0 = \emptyset$: otherwise M and M^0 would be disjoint non-empty closed invariant sets, contradicting indecomposability. To establish necessary and sufficient conditions for M-irreducibility we introduce a concept from dynamical systems theory. A state $x^\star \in \mathsf{X}$ is called *globally attracting* if for all $y \in \mathsf{X}$,

$$x^\star \in \Omega_+(y).$$

The following result easily follows from the definitions.

Proposition 7.2.5. (i) *The nonlinear control system (7.9) is M-irreducible if and only if a globally attracting state exists.*

(ii) *If a globally attracting state $x^\star$ exists then the unique minimal set is equal to $A_+(x^\star) = \Omega_+(x^\star)$.* □

We can now provide the desired connection between irreducibility of the nonlinear control system and ψ-irreducibility for the corresponding Markov chain.

Theorem 7.2.6. *Suppose that $CM(F)$ is forward accessible and the disturbance process of the associated $NSS(F)$ model satisfies the density condition (NSS3).*
Then the $NSS(F)$ model is ψ-irreducible if and only if $CM(F)$ is M-irreducible.

PROOF   If the $NSS(F)$ model is ψ-irreducible, let $x^\star$ be any state in $\operatorname{supp}\psi$, and let U be any open set containing $x^\star$. By definition we have $\psi(U) > 0$, which implies that $K_{a_\varepsilon}(x, U) > 0$ for all $x \in \mathsf{X}$. By Proposition 7.2.2 it follows that $x^\star$ is globally attracting, and hence $CM(F)$ is M-irreducible by Proposition 7.2.5.

Conversely, suppose that $CM(F)$ possesses a globally attracting state, and let U be an open petite set containing $x^\star$. Then $A_+(x) \cap U \neq \emptyset$ for all $x \in \mathsf{X}$, which by Proposition 7.2.2 and Proposition 5.5.4 implies that the $NSS(F)$ model is ψ-irreducible for some ψ. □

7.3 Periodicity for nonlinear state space models

We now look at the periodic structure of the nonlinear $NSS(F)$ model to see how the cycles of Section 5.4.3 can be further described, and in particular their topological structure elucidated.

We first demonstrate that minimal sets for the deterministic control model $CM(F)$ exhibit periodic behavior. This periodicity extends to the stochastic framework in a natural way, and under mild conditions on the deterministic control system, we will see that the period is in fact trivial, so that the chain is aperiodic.

7.3.1 Periodicity for control models

To develop a periodic structure for $CM(F)$ we mimic the construction of a cycle for an irreducible Markov chain. To do this we first require a deterministic analogue of small sets: we say that the set C is *k-accessible* from the set B, for any $k \in \mathbb{Z}_+$, if for each $y \in B$,

$$C \subset A_+^k(y).$$

This will be denoted $B \xrightarrow{k} C$. From the Implicit Function Theorem, in a manner similar to the proof of Proposition 7.1.2, we can immediately connect k-accessibility with forward accessibility.

Proposition 7.3.1. *Suppose that the CM(F) model is forward accessible. Then for each $x \in \mathsf{X}$, there exist open sets B_x, $C_x \subset \mathsf{X}$, with $x \in B_x$ and an integer $k_x \in \mathbb{Z}_+$ such that $B_x \xrightarrow{k_x} C_x$.* □

In order to construct a cycle for an irreducible Markov chain, we first constructed a ν_n-small set A with $\nu_n(A) > 0$. A similar construction is necessary for CM(F).

Lemma 7.3.2. *Suppose that the CM(F) model is forward accessible. If M is minimal for CM(F) then there exists an open set $E \subset M$, and an integer $n \in \mathbb{Z}_+$, such that $E \xrightarrow{n} E$.*

PROOF   Using Proposition 7.3.1 we find that there exist open sets B and C, and an integer k with $B \xrightarrow{k} C$, such that $B \cap M \neq \emptyset$. Since M is invariant, it follows that

$$C \subset A_+(B \cap M) \subset M, \tag{7.15}$$

and by Proposition 7.2.1, minimality, and the hypothesis that the set B is open,

$$A_+(x) \cap B \neq \emptyset \tag{7.16}$$

for every $x \in M$.

Combining (7.15) and (7.16) it follows that $A_+^m(c) \cap B \neq \emptyset$ for some $m \in \mathbb{Z}_+$, and $c \in C$. By continuity of the function F we conclude that there exists an open set $E \subset C$ such that

$$A_+^m(x) \cap B \neq \emptyset \qquad \text{for all } x \in E.$$

The set E satisfies the conditions of the lemma with $n = m + k$ since by the semi-group property,

$$A_+^n(x) = A_+^k(A_+^m(x)) \supset A_+^k(A_+^m(x) \cap B) \supset C \supset E$$

for all $x \in E$. □

Call a finite ordered collection of disjoint closed sets $\boldsymbol{G} := \{G_i : 1 \leq i \leq d\}$ a *periodic orbit* if for each i,

$$A_+^1(G_i) \subset G_{i+1}, \qquad i = 1, \ldots, d \pmod{d}.$$

The integer d is called the *period* of $\boldsymbol{G}$.

The cyclic result for CM(F) is given in

Theorem 7.3.3. *Suppose that the function $F \colon \mathsf{X} \times O_w \to \mathsf{X}$ is smooth, and that the system CM(F) is forward accessible.*

If M is a minimal set, then there exists an integer $d \geq 1$, and disjoint closed sets $\boldsymbol{G} = \{G_i : 1 \leq i \leq d\}$ such that $M = \bigcup_{i=1}^d G_i$, and $\boldsymbol{G}$ is a periodic orbit. It is unique in the sense that if $\mathbf{H}$ is another periodic orbit whose union is equal to M with period d', then d' divides d, and for each i the set H_i may be written as a union of sets from $\boldsymbol{G}$.

PROOF   Using Lemma 7.3.2 we can fix an open set E with $E \subset M$, and an integer k such that $E \xrightarrow{k} E$. Define $I \subset \mathbb{Z}_+$ by

$$I := \{n \geq 1 : E \xrightarrow{n} E\}. \tag{7.17}$$

The semi-group property implies that the set I is closed under addition: for if $i, j \in I$, then for all $x \in E$,

$$A_+^{i+j}(x) = A_+^i(A_+^j(x)) \supset A_+^j(E) \supset E.$$

Let d denote g.c.d.(I). The integer d will be called the *period* of M, and M will be called *aperiodic* when $d = 1$.

For $1 \leq i \leq d$ we define

$$G_i := \{x \in M : \bigcup_{k=1}^{\infty} A_+^{kd-i}(x) \cap E \neq \emptyset\}. \tag{7.18}$$

By Proposition 7.2.1 it follows that $M = \bigcup_{i=1}^d G_i$.

Since E is an open subset of M, it follows that for each $i \in \mathbb{Z}_+$, the set G_i is open in the relative topology on M. Once we have shown that the sets $\{G_i\}$ are disjoint, it will follow that they are closed in the relative topology on M. Since M itself is closed, this will imply that for each i, the set G_i is closed.

We now show that the sets $\{G_i\}$ are disjoint. Suppose that on the contrary $x \in G_i \cap G_j$ for some $i \neq j$. Then there exists $k_i, k_j \in \mathbb{Z}_+$ such that

$$A_+^{k_i d-i}(y) \cap E \neq \emptyset \quad \text{and} \quad A_+^{k_j d-j}(y) \cap E \neq \emptyset \tag{7.19}$$

when $y = x$. Since E is open, we may find an open set $O \subset \mathsf{X}$ containing x such that (7.19) holds for all $y \in O$.

By Proposition 7.2.1, there exists $v \in E$ and $n \in \mathbb{Z}_+$ such that

$$A_+^n(v) \cap O \neq \emptyset. \tag{7.20}$$

By (7.20), (7.19), and since $E \xrightarrow{k_0} E$ we have for $\delta = i, j$, and all $z \in E$,

$$\begin{aligned}
A_+^{k_0+k_\delta d-\delta+n+k_0}(z) &\supset A_+^{k_0+k_\delta d-\delta+n}(E) \\
&\supset A_+^{k_0+k_\delta d-\delta}(A_+^n(v) \cap O) \\
&\supset A_+^{k_0}(A_+^{k_\delta d-\delta}(A_+^n(v) \cap O) \cap E) \supset E.
\end{aligned}$$

This shows that

$$2k_0 + k_\delta d - \delta + n \in I$$

for $\delta = i, j$, and this contradicts the definition of d. We conclude that the sets $\{G_i\}$ are disjoint.

We now show that G is a periodic orbit. Let $x \in G_i$, and $u \in O_w$. Since the sets $\{G_i\}$ form a disjoint cover of M and since M is invariant, there exists a unique $1 \leq j \leq d$ such that $F(x, u) \in G_j$. It follows from the semi-group property that $x \in G_{j-1}$, and hence $i = j - 1$.

The uniqueness of this construction follows from the definition given in equation (7.18). $\qquad \square$

The following consequence of Theorem 7.3.3 further illustrates the topological structure of minimal sets.

Proposition 7.3.4. *Under the conditions of Theorem 7.3.3, if the control set O_w is connected, then the periodic orbit G constructed in Theorem 7.3.3 is precisely equal to the connected components of the minimal set M.*

In particular, in this case M is aperiodic if and only if it is connected.

PROOF   First suppose that M is aperiodic. Let $E \xrightarrow{n} E$, and consider a fixed state $v \in E$.

By aperiodicity and Lemma D.7.4 there exists an integer N_0 with the property that

$$e \in A_+^k(v) \qquad\qquad (7.21)$$

for all $k \geq N_0$. Since $A_+^k(v)$ is the continuous image of the connected set $v \times O_w^k$, the set

$$A_+(A_+^{N_0}(v)) = \bigcup_{k=N_0}^{\infty} A_+^k(v) \qquad\qquad (7.22)$$

is connected. Its closure is therefore also connected, and by Proposition 7.2.1 the closure of the set (7.22) is equal to M.

The periodic case is treated similarly. First we show that for some $N_0 \in \mathbb{Z}_+$ we have

$$G_d = \overline{\bigcup_{k=N_0}^{\infty} A_+^{kd}(v)},$$

where d is the period of M, and each of the sets $A_+^{kd}(v)$, $k \geq N_0$, contains v.

This shows that G_d is connected. Next, observe that

$$G_1 = \overline{A_+^1(G_d)},$$

and since the control set O_w and G_d are both connected, it follows that G_1 is also connected. By induction, each of the sets $\{G_i : 1 \leq i \leq d\}$ is connected.   □

7.3.2   Periodicity

All of the results described above dealing with periodicity of minimal sets were posed in a purely deterministic framework. We now return to the stochastic model described by (NSS1)–(NSS3) to see how the deterministic formulation of periodicity relates to the stochastic definition which was introduced for Markov chains in Section 5.4.

As one might hope, the connections are very strong.

Theorem 7.3.5. *If the NSS(F) model satisfies conditions (NSS1)–(NSS3) and the associated control model CM(F) is forward accessible then:*

(i) *if M is a minimal set, then the restriction of the NSS(F) model to M is a ψ-irreducible T-chain, and the periodic orbit $\{G_i : 1 \leq i \leq d\} \subset M$ whose existence is guaranteed by Theorem 7.3.3 is ψ-a.e. equal to the d-cycle constructed in Theorem 5.4.4;*

(ii) *if CM(F) is M-irreducible, and if its unique minimal set M is aperiodic, then the NSS(F) model is a ψ-irreducible aperiodic T-chain.*

PROOF   The proof of (i) follows directly from the definitions, and the observation that by reducing E if necessary, we may assume that the set E which is used in the proof of Theorem 7.3.3 is small. Hence the set E plays the same role as the small set used in the proof of Theorem 5.2.1. The proof of (ii) follows from (i) and Theorem 7.2.4.   □

7.4   Forward accessible examples

We now see how specific models may be viewed in this general context. It will become apparent that without making any unnatural assumptions, both simple models such as the dependent parameter bilinear model, and relatively more complex nonlinear models such as the gumleaf attractor with noise and adaptive control models can be handled within this framework.

7.4.1   The dependent parameter bilinear model

The dependent parameter bilinear model is a simple $\mathrm{NSS}(F)$ model where the function F is given in (2.15) by

$$F\left(\left(\begin{smallmatrix} Y \\ \theta \end{smallmatrix}\right), \left(\begin{smallmatrix} Z \\ W \end{smallmatrix}\right)\right) = \left(\begin{array}{c} \alpha\theta + Z \\ \theta Y + W \end{array}\right). \tag{7.23}$$

Using Proposition 7.1.4 it is easy to see that the associated control model is forward accessible, and then the model is easily analyzed. We have

Proposition 7.4.1. *The dependent parameter bilinear model* $\boldsymbol{\Phi}$ *satisfying assumptions (DBL1)–(DBL2) is a T-chain. If further there exists some one* $z^* \in O_z$ *such that*

$$\left|\frac{z^*}{1-\alpha}\right| < 1, \tag{7.24}$$

then $\boldsymbol{\Phi}$ *is* ψ*-irreducible and aperiodic .*

PROOF   With the noise $\left(\begin{smallmatrix} Z \\ W \end{smallmatrix}\right)$ considered a "control", the first order controllability matrix may be computed to give

$$C^1_{\theta,y} = \frac{\partial \left(\begin{smallmatrix} \theta_1 \\ Y_1 \end{smallmatrix}\right)}{\partial \left(\begin{smallmatrix} Z_1 \\ W_1 \end{smallmatrix}\right)} = \left(\begin{array}{cc} 1 & 0 \\ 0 & 1 \end{array}\right).$$

The control model is thus forward accessible, and hence $\boldsymbol{\Phi} = \left(\begin{smallmatrix} \theta \\ Y \end{smallmatrix}\right)$ is a T-chain.

Suppose now that the bound (7.24) holds for z^* and let w^* denote any element of $O_w \subseteq \mathbb{R}$. If Z_k and W_k are set equal to z^* and w^* respectively in (7.23) then as $k \to \infty$

$$\left(\begin{array}{c} \theta_k \\ Y_k \end{array}\right) \to x^* := \left(\begin{array}{c} z^*(1-\alpha)^{-1} \\ w^*(1-\alpha)(1-\alpha-z^*)^{-1} \end{array}\right).$$

The state x^* is globally attracting, and it immediately follows from Proposition 7.2.5 and Theorem 7.2.6 that the chain is ψ-irreducible. Aperiodicity then follows from the fact that any cycle must contain the state x^*.   □

7.4.2   The gumleaf attractor

Consider the NSS(F) model whose sample paths evolve to create the version of the "gumleaf attractor" illustrated in Figure 2.3. This model is given in (2.12) by

$$X_n = \begin{pmatrix} X_n^a \\ X_n^b \end{pmatrix} = \begin{pmatrix} -1/X_{n-1}^a + 1/X_{n-1}^b \\ X_{n-1}^a \end{pmatrix} + \begin{pmatrix} W_n \\ 0 \end{pmatrix}$$

which is of the form (NSS1), with the associated CM(F) model defined as

$$F\left(\begin{pmatrix} x^a \\ x^b \end{pmatrix}, u \right) = \begin{pmatrix} -1/x^a + 1/x^b \\ x^a \end{pmatrix} + \begin{pmatrix} u \\ 0 \end{pmatrix}. \tag{7.25}$$

From the formulae

$$\frac{\partial F}{\partial x} = \begin{pmatrix} (1/x^a)^2 & -(1/x^b)^2 \\ 1 & 0 \end{pmatrix} \qquad \frac{\partial F}{\partial u} = \begin{pmatrix} 1 \\ 0 \end{pmatrix}$$

we see that the second order controllability matrix is given by

$$C_{x_0}^2(u_1, u_2) = \begin{bmatrix} (1/x_1^a)^2 & 1 \\ 1 & 0 \end{bmatrix}$$

where $x_0 = \begin{pmatrix} x_0^a \\ x_0^b \end{pmatrix}$ and $x_1^a = -1/x_0^a + 1/x_0^b + u_1$. Hence, since $C_{x_0}^2$ is full rank for all x_0, u_1 and u_2, it follows that the control system is forward accessible. Applying Proposition 7.2.6 gives

Proposition 7.4.2. *The NSS(F) model (2.12) is a T-chain if the disturbance sequence W satisfies condition (NSS3).*

7.4.3   The adaptive control model

The adaptive control model described by (2.22)–(2.24) is of the general form of the NSS(F) model and the results of the previous section are well suited to the analysis of this specific example

An apparent difficulty with this model is that the state space X is not an open subset of Euclidean space, so that the general results obtained for the NSS(F) model may not seem to apply directly. However, given our assumptions on the model, the interior of the state space, $(\sigma_z, \frac{\sigma_z}{1-\alpha^2}) \times \mathbb{R}^2$, is absorbing, and is reached in one step with probability one from each initial condition. Hence to obtain a continuous component, and to address periodicity for the adaptive model, we can apply the general results obtained for the nonlinear state space models by first restricting Φ to the interior of X.

Proposition 7.4.3. *If (SAC1) and (SAC2) hold for the adaptive control model defined by (2.22)–(2.24), and if $\sigma_z^2 < 1$, then Φ is a ψ-irreducible and aperiodic T-chain.*

PROOF   To prove the result we show that the associated deterministic control model for the nonlinear state space model defined by (2.22)–(2.24) is forward accessible and, for the associated deterministic control system, a globally attracting point exists.

The second-order controllability matrix has the form

$$
C^2_{\Phi_0}(Z_2, W_2, Z_1, W_1) := \frac{\partial(\Sigma_2, \tilde{\theta}_2, Y_2)^\top}{\partial(Z_2, W_2, Z_1, W_1)^\top} = \begin{bmatrix} 0 & \frac{-2\alpha^2 \sigma_w^2 \Sigma_1^2 Y_1}{(\Sigma_1 Y_1^2 + \sigma_w^2)^2} & 0 & 0 \\ \bullet & \bullet & 1 & \bullet \\ \bullet & \bullet & 0 & 1 \end{bmatrix}
$$

where "$\bullet$" denotes a variable which does not affect the rank of the controllability matrix. It is evident that $C^2_{\Phi_0}$ is full rank whenever $Y_1 = \tilde{\theta}_0 Y_0 + W_1$ is non-zero. This shows that for each initial condition $\Phi_0 \in X$, the matrix $C^2_{\Phi_0}$ is full rank for a.e. $\{(Z_1, W_1), (Z_2, W_2)\} \in \mathbb{R}^4$, and so the associated control model is forward accessible, and hence the stochastic model is a T-chain by Proposition 7.1.5.

It is easily checked that if $\binom{Z}{W}$ is set equal to zero in (2.22)–(2.23) then, since $\alpha < 1$ and $\sigma_z^2 < 1$,

$$
\Phi_k \to \left(\frac{\sigma_z^2}{1 - \alpha^2}, 0, 0\right)^\top \qquad \text{as } k \to \infty.
$$

This shows that the control model associated with the Markov chain Φ is M-irreducible, and hence by Proposition 7.2.6 the chain itself is ψ-irreducible. The limit above also shows that every element of a cycle $\{G_i\}$ for the unique minimal set must contain the point $(\frac{\sigma_z^2}{1-\alpha^2}, 0, 0)$. From Proposition 7.3.4 it follows that the chain is aperiodic.   □

7.5   Equicontinuity and the nonlinear state space model

7.5.1   e-Chain properties of nonlinear state space models

We have seen in this chapter that the NSS(F) model is a T-chain if the noise variable, viewed as a control, can "steer the state process Φ" to a sufficiently large set of states.

If the forward accessibility property does not hold then the chain must be analyzed using different methods. The process is always a Feller Markov chain, because of the continuity of F, as shown in Proposition 6.1.2. In this section we search for conditions under which the process Φ is also an e-chain.

To do this we consider the *sensitivity process* associated with the NSS(F) model, defined by $\nabla^\Phi_0 = I$ and

$$
\nabla^\Phi_{k+1} = [DF(\Phi_k, w_{k+1})]\nabla^\Phi_k, \qquad k \in \mathbb{Z}_+ \tag{7.26}
$$

where ∇^Φ takes values in the set of $n \times n$ matrices, and DF denotes the derivative of F with respect to its first variable.

Since $\nabla^\Phi_0 = I$ it follows from the chain rule and induction that the sensitivity process is in fact the derivative of the present state with respect to the initial state: that is,

$$
\nabla^\Phi_k = \frac{d}{d\Phi_0}\Phi_k \qquad \text{for all } k \in \mathbb{Z}_+.
$$

The main result in this section connects stability of the derivative process with equicontinuity of the transition function for $\mathbf{\Phi}$. Since the system (7.26) is closely related to the system (NSS1), linearized about the sample path $(\Phi_0, \Phi_1, \dots)$, it is reasonable to expect that the stability of $\mathbf{\Phi}$ will be closely related to the stability of ∇^Φ.

Theorem 7.5.1. *Suppose that (NSS1)–(NSS3) hold for the NSS(F) model. Then letting ∇_k^Φ denote the derivative of Φ_k with respect to Φ_0, $k \in \mathbb{Z}_+$, we have*

(i) *if for some open convex set $N \subset \mathsf{X}$,*

$$\mathsf{E}[\sup_{\Phi_0 \in N} \|\nabla_k^\Phi\|] < \infty \qquad\qquad (7.27)$$

   then for all $x \in N$,

$$\frac{d}{dx}\mathsf{E}_x[\Phi_k] = \mathsf{E}_x[\nabla_k^\Phi];$$

(ii) *suppose that (7.27) holds for all sufficiently small neighborhoods N of each $y_0 \in \mathsf{X}$, and further that for any compact set $C \subset \mathsf{X}$,*

$$\sup_{y \in C} \sup_{k \geq 0} \mathsf{E}_y[\|\nabla_k^\Phi\|] < \infty.$$

   Then $\mathbf{\Phi}$ is an e-chain.

Proof   The first result is a consequence of the Dominated Convergence Theorem. To prove the second result, let $f \in \mathcal{C}_c(\mathsf{X}) \cap \mathcal{C}^\infty(\mathsf{X})$. Then

$$\left| \frac{d}{dx} P^k f(x) \right| = \left| \frac{d}{dx} \mathsf{E}_x[f(\Phi_k)] \right| \leq \|f'\|_\infty \mathsf{E}_x[\|\nabla_k^\Phi\|]$$

which by the assumptions of (ii), implies that the sequence of functions $\{P^k f : k \in \mathbb{Z}_+\}$ is equicontinuous on compact subsets of X. Since $C^\infty \cap C_c$ is dense in C_c, this completes the proof.   □

It may seem that the technical assumption (7.27) will be difficult to verify in practice. However, we can immediately identify one large class of examples by considering the case where the i.i.d. process W is uniformly bounded. It follows from the smoothness condition on F that $\sup_{\Phi_0 \in N} \|\nabla_k^\Phi\|$ is almost surely finite for any compact subset $N \subset \mathsf{X}$, which shows that in this case (7.27) is trivially satisfied.

The following result provides another large class of models for which (7.27) is satisfied. Observe that the conditions imposed on W in Proposition 7.5.2 are satisfied for any i.i.d. Gaussian process. The proof is straightforward.

Proposition 7.5.2. *For the Markov chain defined by (NSS1)–(NSS3), suppose that F is a rational function of its arguments, and that for some $\varepsilon_0 > 0$,*

$$\mathsf{E}[\exp(\varepsilon_0 |W_1|)] < \infty.$$

Then letting ∇_k^Φ denote the derivative of Φ_k with respect to Φ_0, we have for any compact set $C \subset \mathsf{X}$, and any $k \geq 0$,

$$\mathsf{E}[\sup_{\Phi_0 \in C} \|\nabla_k^\Phi\|] < \infty.$$

Hence under these conditions,

$$\frac{d}{dx}\mathsf{E}_x[\Phi_k] = \mathsf{E}_x[\nabla_k^{\Phi}].$$

□

7.5.2   Linear state space models

We can easily specialize Theorem 7.5.1 to give conditions under which a linear model is an e-chain.

Proposition 7.5.3. *Suppose the LSS(F,G) model X satisfies (LSS1) and (LSS2), and that the eigenvalue condition (LSS5) also holds. Then Φ is an e-chain.*

PROOF   Using the identity $X_m = F^m X_0 + \sum_{i=0}^{m-1} F^i G W_{m-i}$ we see that

$$\nabla_k^{\Phi} = F^m,$$

which tends to zero exponentially fast, by Lemma 6.3.4. The conditions of Theorem 7.5.1 are thus satisfied, which completes the proof.   □

Observe that Proposition 7.5.3 uses the eigenvalue condition (LSS5), the same assumption which was used in Proposition 4.4.3 to obtain ψ-irreducibility for the Gaussian model, and the same condition that will be used to obtain stability in later chapters.

The analogous Proposition 6.3.3 uses controllability to give conditions under which the linear state space model is a T-chain. Note that controllability is not required here.

Other specific nonlinear models, such as bilinear models, can be analyzed similarly using this approach.

7.6   Commentary*

We have already noted that in the degenerate case where the control set O_w consists of a single point, the NSS(F) model defines a semi-dynamical system with state space X, and in fact many of the concepts introduced in this chapter are generalizations of standard concepts from dynamical systems theory.

Three standard approaches to the qualitative theory of dynamical systems are *topological dynamics* whose principal tool is point set topology; *ergodic theory*, where one assumes (or proves, frequently using a compactness argument) the existence of an ergodic invariant measure; and finally, the *direct method of Lyapunov*, which concerns criteria for stability.

The latter two approaches will be developed in a stochastic setting in Parts II and III. This chapter essentially focused on generalizations of the first approach, which is also based upon, to a large extent, the structure and existence of minimal sets. Two excellent expositions in a purely deterministic and control-free setting are the books by Bhatia and Szegö [34] and Brown [55]. Saperstone [346] considers infinite dimensional spaces so that, in particular, the methods may be applied directly to the dynamical system on the space of probability measures which is generated by a Markov processes.

The connections between control theory and irreducibility described here are taken from Meyn [259] and Meyn and Caines [272, 271]. The dissertations of Chan [61] and Mokkadem [286], and also Diebolt and Guégan [92], treat discrete time nonlinear state space models and their associated control models. Diebolt in [91] considers nonlinear models with additive noise of the form $\Phi_{k+1} = F(\Phi_k) + W_{k+1}$ using an approach which is very different to that described here.

Jakubsczyk and Sontag in [173] present a survey of the results obtainable for forward accessible discrete time control systems in a purely deterministic setting. They give a different characterization of forward accessibility, based upon the rank of an associated Lie algebra, rather than a controllability matrix.

The origin of the approach taken in this chapter lies in the often cited paper by Stroock and Varadhan [378]. There it is shown that the support of the distribution of a diffusion process may be characterized by considering an associated control model. Ichihara and Kunita in [167] and Kliemann in [211] use this approach to develop an ergodic theory for diffusions. The *invariant control sets* of [211] may be compared to minimal sets as defined here.

At this stage, introduction of the e-chain class of models is not well motivated. The reader who wishes to explore them immediately should move to Chapter 12.

In Duflo [102], a condition closely related to the stability condition which we impose on ∇^Φ is used to obtain the Central Limit Theorem for a nonlinear state space model. Duflo assumes that the function F satisfies

$$|F(x, w) - F(y, w)| \leq \alpha(w)|x - y|$$

where α is a function on O_w satisfying, for some sufficiently large m,

$$\mathsf{E}[\alpha(W)^m] < 1.$$

It is easy to see that any process Φ generated by a nonlinear state space model satisfying this bound is an e-chain.

For models more complex than the linear model of Section 7.5.2 it will not be as easy to prove that ∇^Φ converges to zero, so a lengthier stability analysis of this sensitivity process may be necessary. Since ∇^Φ is essentially generated by a random linear system it is therefore likely to either converge to zero or evanesce.

It seems probable that the stochastic Lyapunov function approach of Kushner [232] or Khas'minskii [206], or a more direct analysis based upon limit theorems for products of random matrices as developed in, for instance, Furstenberg and Kesten [134] will be well suited for assessing the stability of ∇^Φ.

Commentary for the second edition: The conjecture voiced in the first edition was confirmed ten years after it was first put into print. A stochastic Lyapunov approach is introduced in [165] for verification of stability of the sensitivity process[1] for a class of Markov models.

A significant omission in the first edition is any discussion of the relationship between stability of the sensitivity process ∇^Φ and Lyapunov exponents (see [212, 255]). For a

---

[1] The sensitivity process was called the *derivative process* in the first edition.

given initial condition x, the top Lyapunov exponent is defined as the random variable

$$\Lambda_x := \limsup_{n \to \infty} \frac{1}{n} \log \| \nabla_n^{\Phi} \|.$$

The choice of norm is arbitrary. There is also a version defined in expectation: for any $p > 0$ denote

$$\Lambda_x(p) := \limsup_{n \to \infty} \frac{1}{n} \log \mathsf{E}_x[\| \nabla_n^{\Phi} \|^p].$$

One approach to establishing the e-chain property is to show that $\Lambda_x(p)$ is independent of x, and negative for all p sufficiently small [165].

Methods for estimating the Lyapunov exponent and conditions for verifying equicontinuity are established for versions of the NSS(F) model, in continuous or discrete time, in several recent papers under a variety of assumptions [370, 371, 22, 165, 20, 323].

A *hidden Markov model* (HMM) is a Markov chain $\boldsymbol{\Phi}$, along with an observation process $\boldsymbol{Y}$ evolving on a state space Y. It is assumed that there is an i.i.d. sequence $\boldsymbol{D}$ evolving on its own state space D, along with a function $G \colon \mathsf{X} \times \mathsf{D} \to \mathsf{Y}$ such that the observation process can be expressed as a noisy function of the chain

$$Y_n = G(\Phi_n, D_n), \qquad n \geq 0.$$

The conditional distribution of X_n given $Y_0, \ldots, Y_n$ is denoted $\hat{\pi}_n$. It is known that $\Upsilon_n := (Y_n, \hat{\pi}_n)$ is itself a Markov chain [106, 107], but one that is rarely ψ-irreducible. Consequently we are forced to consider alternative approaches to address stability of the filtering process $\{\hat{\pi}_n\}$. Lyapunov exponents as well as equicontinuity have proved valuable in the analysis of $\boldsymbol{\Upsilon}$.

Lyapunov exponents for $\boldsymbol{\Upsilon}$ are examined in a series of papers by Zeitouni and coauthors [85, 11]. Under certain conditions on the model the Lyapunov exponent Λ_x is negative and independent of x, which implies that the filter is insensitive to its initial condition. The e-chain property is established directly in [87, 213], under conditions more general than [11]. The recent survey of Chigansky et al. [68] contains an extensive bibliography.

Part II

STABILITY STRUCTURES

Chapter 8

Transience and recurrence

We have developed substantial structural results for ψ-irreducible Markov chains in Part I of this book. Part II is devoted to stability results of ever-increasing strength for such chains.

In Chapter 1, we discussed in a heuristic manner two possible approaches to the stability of Markov chains. The first of these discussed basic ideas of stability and instability, formulated in terms of recurrence and transience for ψ-irreducible Markov chains. The aim of this chapter is to formalize those ideas.

In many ways it is easier to tell when a Markov chain is unstable than when it is stable: it fails to return to its starting point, it eventually leaves any "bounded" set with probability one, it returns only a finite number of times to a given set of "reasonable size". Stable chains are then conceived of as those which do not vanish from their starting points in at least some of these ways. There are many ways in which stability may occur, ranging from weak "expected return to origin" properties, to convergence of all sample paths to a single point, as in global asymptotic stability for deterministic processes. In this chapter we concentrate on rather weak forms of stability, or conversely on strong forms of instability.

Our focus here is on the behavior of the occupation time random variable $\eta_A :=\sum_{n=1}^{\infty} \mathbb{I}\{\Phi_n \in A\}$ which counts the number of visits to a set A. In terms of η_A we study the stability of a chain through the transience and recurrence of its sets.

Uniform transience and recurrence

The set A is called *uniformly transient* if for there exists $M < \infty$ such that $\mathsf{E}_x[\eta_A] \leq M$ for all $x \in A$.

The set A is called *recurrent* if $\mathsf{E}_x[\eta_A] = \infty$ for all $x \in A$.

The highlight of this approach is a solidarity, or dichotomy, theorem of surprising strength.

Theorem 8.0.1. *Suppose that* Φ *is* ψ-*irreducible. Then either*

(i) *every set in* $\mathcal{B}^+(\mathsf{X})$ *is recurrent, in which case we call* Φ *recurrent, or*

(ii) *there is a countable cover of* X *with uniformly transient sets, in which case we call* Φ *transient, and every petite set is uniformly transient.*

PROOF   This result is proved through a splitting approach in Section 8.2.3. We also give a different proof, not using splitting, in Theorem 8.3.4, where the cover with uniformly transient sets is made more explicit, leading to Theorem 8.3.5 where all petite sets are shown to be uniformly transient if there is just one petite set in $\mathcal{B}^+(\mathsf{X})$ which is not recurrent. □

The other high point of this chapter is the first development of one of the themes of the book: the existence of so-called *drift criteria*, couched in terms of the expected change, or drift, defined by the one-step transition function P, for chains to be stable or unstable in the various ways this is defined.

Drift for Markov chains

The (possibly extended valued) drift operator Δ is defined for any non-negative measurable function V by

$$\Delta V(x) := \int P(x, dy)V(y) - V(x), \qquad x \in \mathsf{X}. \tag{8.1}$$

A second goal of this chapter is the development of criteria based on the drift function for both transience and recurrence.

Theorem 8.0.2. *Suppose* Φ *is a* ψ-*irreducible chain.*

(i) *The chain* Φ *is transient if and only if there exists a bounded non-negative function* V *and a set* $C \in \mathcal{B}^+(\mathsf{X})$ *such that for all* $x \in C^c$,

$$\Delta V(x) \geq 0 \tag{8.2}$$

and

$$D = \{V(x) > \sup_{y \in C} V(y)\} \in \mathcal{B}^+(\mathsf{X}). \tag{8.3}$$

(ii) *The chain* Φ *is recurrent if there exists a petite set* $C \subset \mathsf{X}$, *and a function* V *which is unbounded off petite sets in the sense that* $C_V(n) := \{y : V(y) \leq n\}$ *is petite for all* n, *such that*

$$\Delta V(x) \leq 0, \qquad x \in C^c. \tag{8.4}$$

PROOF   The drift criterion for transience is proved in Theorem 8.4.2, whilst the condition for recurrence is in Theorem 8.4.3.   □

Such conditions were developed by Lyapunov as criteria for stability in deterministic systems, by Khas'minskii and others for stochastic differential equations [206, 232], and by Foster as criteria for stability for Markov chains on a countable space: Theorem 8.0.2 is originally due (for countable spaces) to Foster [129] in essentially the form given above.

There is in fact a converse to Theorem 8.0.2 (ii) also, but only for ψ-irreducible Feller chains (which include all countable space chains): we prove this in Section 9.4.2. It is not known whether a converse holds in general.

Recurrence is also often phrased in terms of the hitting time variables $\tau_A = \inf\{k \geq 1 : \Phi_k \in A\}$, with "recurrence" for a set A being defined by $L(x, A) = \mathsf{P}_x(\tau_A < \infty) = 1$ for all $x \in A$. The connections between this condition and recurrence as we have defined it above are simple in the countable state space case: the conditions are in fact equivalent when A is an atom. In general spaces we do not have such complete equivalence. Recurrence properties in terms of τ_A (which we call Harris recurrence properties) are much deeper and we devote much of the next chapter to them. In this chapter we do however give some of the simpler connections: for example, if $L(x, A) = 1$ for all $x \in A$ then $\eta_A = \infty$ a.s. when $\Phi_0 \in A$, and hence A is recurrent (see Proposition 8.3.1).

8.1 Classifying chains on countable spaces

8.1.1 The countable recurrence/transience dichotomy

We turn as before to the countable space to guide and motivate our general results, and to aid in their interpretation.

When $\mathsf{X} = \mathbb{Z}_+$, we initially consider the stability of an *individual state* α. This will lead to a global classification for irreducible chains.

The first, and weakest, stability property involves the *expected number of visits* to α. The random variable $\eta_\alpha = \sum_{n=1}^{\infty} \mathbb{I}\{\Phi_n = \alpha\}$ has been defined in Section 3.4.3 as the number of visits by Φ to α: clearly η_α is a measurable function from Ω to $\mathbb{Z}_+ \cup \{\infty\}$.

Classification of states

The state α is called *transient* if $\mathsf{E}_\alpha(\eta_\alpha) < \infty$, and *recurrent* if $\mathsf{E}_\alpha(\eta_\alpha) = \infty$.

From the definition $U(x, y) = \sum_{n=1}^{\infty} P^n(x, y)$ we have immediately that for any states $x, y \in \mathsf{X}$

$$\mathsf{E}_x[\eta_y] = U(x, y). \tag{8.5}$$

The following result gives a structural dichotomy which enables us to consider, not just the stability of states, but of chains as a whole.

Proposition 8.1.1. *When* X *is countable and* Φ *is irreducible, either* $U(x, y) = \infty$ *for all* $x, y \in \mathsf{X}$ *or* $U(x, y) < \infty$ *for all* $x, y \in \mathsf{X}$.

PROOF   This relies on the definition of irreducibility through the relation $\leftrightarrow$.

If $\sum_n P^n(x,y) = \infty$ for some x, y, then since $u \to x$ and $y \to v$ for any u, v, we have r, s such that $P^r(u,x) > 0$, $P^s(y,v) > 0$ and so

$$\sum_n P^{r+s+n}(u,v) > P^r(u,x)\Big[\sum_n P^n(x,y)\Big] P^s(y,v) = \infty. \qquad (8.6)$$

Hence the series $U(x,y)$ and $U(u,v)$ all converge or diverge simultaneously, and the result is proved. $\square$

Now we can extend these stability concepts for states to the whole chain.

Transient and recurrent chains

If every state is transient, the chain itself is called *transient*.

If every state is recurrent, the chain is called *recurrent*.

The solidarity results of Proposition 8.1.3 and Proposition 8.1.1 enable us to classify irreducible chains by the property possessed by one and then all states.

Theorem 8.1.2. *When $\mathbf{\Phi}$ is irreducible, then either $\mathbf{\Phi}$ is transient or $\mathbf{\Phi}$ is recurrent.*
                                                                                  $\square$

We can say, in the countable case, exactly what recurrence or transience means in terms of the return time probabilities $L(x,x)$. In order to connect these concepts, for a fixed n consider the event $\{\Phi_n = \boldsymbol{\alpha}\}$, and decompose this event over the mutually exclusive events $\{\Phi_n = \boldsymbol{\alpha}, \tau_\alpha = j\}$ for $j = 1, \ldots, n$. Since $\mathbf{\Phi}$ is a Markov chain, this provides the first-entrance decomposition of P^n given for $n \geq 1$ by

$$P^n(x, \boldsymbol{\alpha}) = \mathsf{P}_x\{\tau_\alpha = n\} + \sum_{j=1}^{n-1} \mathsf{P}_x\{\tau_\alpha = j\} P^{n-j}(\boldsymbol{\alpha}, \boldsymbol{\alpha}). \qquad (8.7)$$

If we introduce the generating functions for the series P^n and $_\alpha P^n$ as

$$U^{(z)}(x, \boldsymbol{\alpha}) \ := \ \sum_{n=1}^{\infty} P^n(x, \boldsymbol{\alpha}) z^n, \qquad |z| < 1, \qquad (8.8)$$

$$L^{(z)}(x, \boldsymbol{\alpha}) \ := \ \sum_{n=1}^{\infty} \mathsf{P}_x(\tau_\alpha = n) z^n, \qquad |z| < 1, \qquad (8.9)$$

then multiplying (8.7) by z^n and summing from $n = 1$ to ∞ gives for $|z| < 1$

$$U^{(z)}(x, \boldsymbol{\alpha}) = L^{(z)}(x, \boldsymbol{\alpha}) + L^{(z)}(x, \boldsymbol{\alpha}) U^{(z)}(\boldsymbol{\alpha}, \boldsymbol{\alpha}). \qquad (8.10)$$

From this identity we have

Proposition 8.1.3. *For any $x \in \mathsf{X}$, $U(x,x) = \infty$ if and only if $L(x,x) = 1$.*

PROOF   Consider the first entrance decomposition in (8.10) with $x = \alpha$: this gives

$$U^{(z)}(\alpha, \alpha) = L^{(z)}(\alpha, \alpha) \Big/ \Big[1 - L^{(z)}(\alpha, \alpha)\Big]. \tag{8.11}$$

Letting $z \uparrow 1$ in (8.11) shows that

$$L(\alpha, \alpha) = 1 \Longleftrightarrow U(\alpha, \alpha) = \infty.$$

$\square$

This gives the following interpretation of the transience/recurrence dichotomy of Proposition 8.1.1.

Proposition 8.1.4. *When Φ is irreducible, either $L(x, y) = 1$ for all $x, y \in \mathsf{X}$ or $L(x, x) < 1$ for all $x \in \mathsf{X}$.*

PROOF   From Proposition 8.1.3 and Proposition 8.1.1, we have $L(x, x) < 1$ for all x or $L(x, x) = 1$ for all x. Suppose in the latter case, we have $L(x, y) < 1$ for some pair x, y: by irreducibility, $U(y, x) > 0$ and thus for some n we have $\mathsf{P}_y(\Phi_n = x, \tau_y > n) > 0$, from which we have $L(y, y) < 1$, which is a contradiction. $\square$

In Chapter 9 we will define *Harris recurrence* as the property that $L(x, A) \equiv 1$ for all $x \in A$ and $A \in \mathcal{B}^+(\mathsf{X})$: for countable chains, we have thus shown that recurrent chains are also Harris recurrent, a theme we return to in the next chapter when we explore stability in terms of $L(x, A)$ in more detail.

8.1.2   Specific models: evaluating transience and recurrence

Calculating the quantities $U(x, y)$ or $L(x, x)$ directly for specific models is non-trivial except in the simplest of cases. However, we give as examples two simple models for which this is possible, and then a deeper proof of a result for general random walk.

Renewal processes and forward recurrence time chains

Let the transition matrix of the forward recurrence time chain be given as in Section 3.3. Then it is straightforward to see that for all states $n > 1$,

$$_1P^{n-1}(n, 1) = 1.$$

This gives

$$L(1, 1) = \sum_{n \geq 1} p(n) \, _1P^{n-1}(n, 1) = 1$$

also. Hence the forward recurrence time chain is always recurrent if p is a proper distribution.

The calculation in the proof of Proposition 8.1.3 is actually a special case of the use of the *renewal equation*. Let Z_n be a renewal process with increment distribution p as defined in Section 2.4. By breaking up the event $\{Z_k = n\}$ over the last time before n that a renewal occurred we have

$$u(n) := \sum_{k=0}^{\infty} \mathsf{P}(Z_k = n) = 1 + u * p(n)$$

and multiplying by z^n and summing over n gives the form

$$U(z) = [1 - P(z)]^{-1} \qquad \qquad (8.12)$$

where $U(z) := \sum_{n=0}^{\infty} u(n)z^n$ and $P(z) := \sum_{n=0}^{\infty} p(n)z^n$.

Hence a renewal process is also called recurrent if p is a proper distribution, and in this case $U(1) = \infty$.

Notice that the renewal equation (8.12) is identical to (8.11) in the case of the specific renewal chain given by the return time $\tau_\alpha(n)$ to the state α.

Simple random walk on $\mathbb{Z}_+$

Let P be the transition matrix of random walk on a half line in the simplest irreducible case, namely $P(0,0) = p$ and

$$\begin{aligned}
P(x, x-1) &= p, & x > 0, \\
P(x, x+1) &= q, & x \geq 0.
\end{aligned}$$

where $p + q = 1$. This is known as the simple, or Bernoulli, random walk.

We have that

$$\begin{aligned}
L(0,0) &= p + qL(1,0), \\
L(1,0) &= p + qL(2,0).
\end{aligned}$$

Now we use two tricks specific to chains such as this. Firstly, since the chain is skip-free to the left, it must reach $\{0\}$ from $\{2\}$ only by going through $\{1\}$, so that we have

$$L(2,0) = L(2,1)L(1,0).$$

Secondly, the translation invariance of the chain, which implies $L(j, j-1) = L(1,0)$, $j \geq 1$, gives us

$$L(2,0) = [L(1,0)]^2.$$

Thus from (8.13), we find that

$$L(1,0) = p + q[L(1,0)]^2 \qquad \qquad (8.13)$$

so that $L(1,0) = 1$ or $L(1,0) = p/q$.

This shows that $L(1,0) = 1$ if $p \geq q$, and from (8.13) we derive the well-known result that $L(0,0) = 1$ if $p \geq q$.

Random walk on $\mathbb{Z}$

In order to classify general random walk on the integers we will use the laws of large numbers. Proving these is outside the scope of this book: see, for example, Billingsley [37] or Chung [72] for these results.

Suppose that Φ_n is a random walk such that the increment distribution Γ has a mean which is zero. The form of the Weak Law of Large Numbers that we will use can be stated in our notation as

$$P^n(0, A(\varepsilon n)) \to 1 \qquad \qquad (8.14)$$

for any ε, where the set $A(k) = \{y : |y| \leq k\}$. From this we prove

Theorem 8.1.5. *If Φ is an irreducible random walk on $\mathbb{Z}$ whose increment distribution Γ has mean zero, then Φ is recurrent.*

PROOF   First note that from (8.7) we have for any x

$$\sum_{m=1}^{N} P^m(x,0) = \sum_{k=1}^{N} \sum_{j=0}^{k} \mathsf{P}_x(\tau_0 = k - j) P^j(0,0)$$

$$= \sum_{j=0}^{N} P^j(0,0) \sum_{i=0}^{N-j} \mathsf{P}_x(\tau_0 = i) \qquad (8.15)$$

$$\leq \sum_{j=0}^{N} P^j(0,0).$$

Now using this with the symmetry that $\sum_{m=1}^{N} P^m(x,0) = \sum_{m=1}^{N} P^m(0,-x)$ gives

$$\sum_{m=0}^{N} P^m(0,0) \geq [2M+1]^{-1} \sum_{|x| \leq M} \sum_{j=0}^{N} P^j(0,x)$$

$$\geq [2M+1]^{-1} \sum_{j=0}^{N} P^j(0, A(jM/N)) \qquad (8.16)$$

$$= [2aN+1]^{-1} \sum_{j=0}^{N} P^j(0, A(aj))$$

where we choose $M = Na$ where a is to be chosen later.

But now from the Weak Law of Large Numbers (8.14) we have

$$P^k(0, A(ak)) \to 1$$

as $k \to \infty$; and so from (8.16) we have

$$\liminf_{N \to \infty} \sum_{m=0}^{N} P^m(x,0) \geq \liminf_{N \to \infty} [2aN+1]^{-1} \sum_{j=0}^{N} P^j(0, A(aj))$$

$$= [2a]^{-1}.$$

$$(8.17)$$

Since a can be chosen arbitrarily small, we have $U(0,0) = \infty$ and the chain is recurrent.
$\square$

This proof clearly uses special properties of random walk. If Γ has simpler structure then we shall see that simpler procedures give recurrence in Section 8.4.3.

8.2   Classifying ψ-irreducible chains

The countable case provides guidelines for us to develop solidarity properties of chains which admit a single atom rather than a multiplicity of atoms. These ideas can then be applied to the split chain and carried over through the m-skeleton to the original chain, and this is the agenda in this section.

In order to accomplish this, we need to describe precisely what we mean by recurrence or transience of sets in a general space.

8.2.1   Transience and recurrence for individual sets

For general $A, B \in \mathcal{B}(\mathsf{X})$ recall from Section 3.4.3 the taboo probabilities given by

$$_A P^n(x,B) = \mathsf{P}_x\{\Phi_n \in B, \tau_A \geq n\},$$

and by convention we set $_A P^0(x, A) = 0$. Extending the first entrance decomposition (8.7) from the countable space case, for a fixed n consider the event $\{\Phi_n \in B\}$ for arbitrary $B \in \mathcal{B}(\mathsf{X})$, and decompose this event over the mutually exclusive events $\{\Phi_n \in B, \tau_A = j\}$ for $j = 1, \ldots, n$, where A is any other set in $\mathcal{B}(\mathsf{X})$. The general *first-entrance decomposition* can be written

$$P^n(x, B) = {_A}P^n(x, B) + \sum_{j=1}^{n-1} \int_A {_A}P^j(x, dw) P^{n-j}(w, B) \tag{8.18}$$

whilst the analogous *last-exit decomposition* is given by

$$P^n(x, B) = {_A}P^n(x, B) + \sum_{j=1}^{n-1} \int_A P^j(x, dw) {_A}P^{n-j}(w, B). \tag{8.19}$$

The first-entrance decomposition is clearly a decomposition of the event $\{\Phi_n \in A\}$ which could be developed using the strong Markov property and the stopping time $\zeta = \tau_A \wedge n$. The last-exit decomposition, however, is not an example of the use of the strong Markov property: for, although the first-entrance time τ_A is a stopping time for Φ, the last-exit time is not a stopping time. These decompositions do however illustrate the same principle that underlies the strong Markov property, namely the decomposition of an event over the sub-events on which the random time takes on the (countable) set of values available to it.

We will develop classifications of sets using the generating functions for the series $\{P^n\}$ and $\{_A P^n\}$:

$$U^{(z)}(x, B) := \sum_{n=1}^{\infty} P^n(x, B) z^n, \qquad |z| < 1, \tag{8.20}$$

$$U_A^{(z)}(x, B) := \sum_{n=1}^{\infty} {_A}P^n(x, B) z^n, \qquad |z| < 1. \tag{8.21}$$

The kernel U then has the property

$$U(x, A) = \sum_{n=1}^{\infty} P^n(x, A) = \lim_{z \uparrow 1} U^{(z)}(x, A) \tag{8.22}$$

and as in the countable case, for any $x \in \mathsf{X}, A \in \mathcal{B}(\mathsf{X})$,

$$\mathsf{E}_x(\eta_A) = U(x, A). \tag{8.23}$$

Thus uniform transience or recurrence is quantifiable in terms of the finiteness or otherwise of $U(x, A)$.

The return time probabilities $L(x, A) = \mathsf{P}_x\{\tau_A < \infty\}$ satisfy

$$L(x, A) = \sum_{n=1}^{\infty} {_A}P^n(x, A) = \lim_{z \uparrow 1} U_A^{(z)}(x, A). \tag{8.24}$$

We will prove the solidarity results we require by exploiting the convolution forms in (8.18) and (8.19). Multiplying by z^n in (8.18) and (8.19) and summing, the first entrance and last exit decompositions give, respectively, for $|z| < 1$

$$U^{(z)}(x, B) = U_A^{(z)}(x, B) + \int_A U_A^{(z)}(x, dw) U^{(z)}(w, B), \qquad (8.25)$$

$$U^{(z)}(x, B) = U_A^{(z)}(x, B) + \int_A U^{(z)}(x, dw) U_A^{(z)}(w, B). \qquad (8.26)$$

In classifying the chain $\mathbf{\Phi}$ we will use these relationships extensively.

8.2.2 The recurrence/transience dichotomy: chains with an atom

We can now move to classifying a chain $\mathbf{\Phi}$ which admits an atom in a dichotomous way as either recurrent or transient. Through the splitting techniques of Chapter 5 this will then enable us to classify general chains.

Theorem 8.2.1. *Suppose that $\mathbf{\Phi}$ is ψ-irreducible and admits an atom $\alpha \in \mathcal{B}^+(\mathsf{X})$. Then*

(i) *if α is recurrent, then every set in $\mathcal{B}^+(\mathsf{X})$ is recurrent;*

(ii) *if α is transient, then there is a countable covering of X by uniformly transient sets.*

PROOF   (i)   If $A \in \mathcal{B}^+(\mathsf{X})$ then for any x we have r, s such that $P^r(x, \alpha) > 0$, $P^s(\alpha, A) > 0$ and so

$$\sum_n P^{r+s+n}(x, A) \geq P^r(x, \alpha)\Big[\sum_n P^n(\alpha, \alpha)\Big] P^s(\alpha, A) = \infty. \qquad (8.27)$$

Hence the series $U(x, A)$ diverges for every x, A when $U(\alpha, \alpha)$ diverges.

(ii)   To prove the converse, we first note that for an atom, transience is equivalent to $L(\alpha, \alpha) < 1$, exactly as in Proposition 8.1.3.

Now consider the last exit decomposition (8.26) with $A, B = \alpha$. We have for any $x \in \mathsf{X}$

$$U^{(z)}(x, \alpha) = U_\alpha^{(z)}(x, \alpha) + U^{(z)}(x, \alpha) U_\alpha^{(z)}(\alpha, \alpha)$$

and so by rearranging terms we have for all $z < 1$

$$U^{(z)}(x, \alpha) = U_\alpha^{(z)}(x, \alpha)[1 - U_\alpha^{(z)}(\alpha, \alpha)]^{-1} \leq [1 - L(\alpha, \alpha)]^{-1} < \infty.$$

Hence $U(x, \alpha)$ is bounded for all x.

Now consider the countable covering of X given by the sets

$$\overline{\alpha}(j) = \{y : \sum_{n=1}^{j} P^n(y, \alpha) > j^{-1}\}.$$

Using the Chapman–Kolmogorov equations,

$$U(x, \boldsymbol{\alpha}) \geq j^{-1} U(x, \overline{\boldsymbol{\alpha}}(j)) \inf_{y \in \overline{\boldsymbol{\alpha}}(j)} \sum_{n=1}^{j} P^n(y, \boldsymbol{\alpha}) \geq j^{-2} U(x, \overline{\boldsymbol{\alpha}}(j))$$

and thus $\{\overline{\boldsymbol{\alpha}}(j)\}$ is the required cover by uniformly transient sets.   □

We shall frequently find sets which are not uniformly transient themselves, but which can be covered by a countable number of uniformly transient sets. This leads to the definition

---

Transient sets

If $A \in \mathcal{B}(\mathsf{X})$ can be covered with a countable number of uniformly transient sets, then we call A *transient*.

8.2.3   The general recurrence/transience dichotomy

Now let us consider chains which do not have atoms, but which are strongly aperiodic.

We shall find that the split chain construction leads to a "solidarity result" for the sets in $\mathcal{B}^+(\mathsf{X})$ in the ψ-irreducible case, thus allowing classification of $\boldsymbol{\Phi}$ as a whole. Thus the following definitions will not be vacuous.

Stability classification of ψ-irreducible chains

(i) The chain $\boldsymbol{\Phi}$ is called *recurrent* if it is ψ-irreducible and $U(x, A) \equiv \infty$ for every $x \in \mathsf{X}$ and every $A \in \mathcal{B}^+(\mathsf{X})$.

(ii) The chain $\boldsymbol{\Phi}$ is called *transient* if it is ψ-irreducible and X is transient.

We first check that the split chain and the original chain have mutually consistent recurrent/transient classifications.

Proposition 8.2.2. *Suppose that $\boldsymbol{\Phi}$ is ψ-irreducible and strongly aperiodic. Then either both $\boldsymbol{\Phi}$ and $\check{\boldsymbol{\Phi}}$ are recurrent, or both $\boldsymbol{\Phi}$ and $\check{\boldsymbol{\Phi}}$ are transient.*

PROOF   Strong aperiodicity ensures as in Proposition 5.4.5 that the minorization condition holds, and thus we can use the Nummelin splitting of the chain $\boldsymbol{\Phi}$ to produce a chain $\check{\boldsymbol{\Phi}}$ on $\check{\mathsf{X}}$ which contains an accessible atom $\check{\alpha}$.

We see from (5.9) that for every $x \in \mathsf{X}$, and for every $B \in \mathcal{B}^+(\mathsf{X})$,

$$\sum_{n=1}^{\infty} \int \delta_x^*(dy_i) \check{P}^n(y_i, B) = \sum_{n=1}^{\infty} P^n(x, B). \tag{8.28}$$

If $B \in \mathcal{B}^+(\mathsf{X})$ then since $\psi^*(B_0) > 0$ it follows from (8.28) that if $\check{\Phi}$ is recurrent, so is Φ. Conversely, if $\check{\Phi}$ is transient, by taking a cover of $\check{\mathsf{X}}$ with uniformly transient sets it is equally clear from (8.28) that Φ is transient.

We know from Theorem 8.2.1 that $\check{\Phi}$ is either transient or recurrent, and so the dichotomy extends in this way to Φ. $\qquad \square$

To extend this result to general chains without atoms we first require a link between the recurrence of the chain and its resolvent.

Lemma 8.2.3. *For any $0 < \varepsilon < 1$ the following identity holds:*

$$\sum_{n=1}^{\infty} K_{a_\varepsilon}^n = \frac{1-\varepsilon}{\varepsilon} \sum_{n=0}^{\infty} P^n.$$

Proof   From the generalized Chapman–Kolmogorov equations (5.46) we have

$$\sum_{n=1}^{\infty} K_{a_\varepsilon}^n = \sum_{n=1}^{\infty} K_{a_\varepsilon^{*n}} = \sum_{n=0}^{\infty} b(n) P^n$$

where we define $b(k)$ to be the kth term in the sequence $\sum_{n=1}^{\infty} a_\varepsilon^{*n}$. To complete the proof, we will show that $b(k) = (1-\varepsilon)/\varepsilon$ for all $k \geq 0$.

Let $B(z) = \sum b(k) z^k$, $A_\varepsilon(z) = \sum a_\varepsilon(k) z^k$ denote the power series representation of the sequences b and a_ε. From the identities

$$A_\varepsilon(z) = \left(\frac{1-\varepsilon}{1-\varepsilon z} \right), \qquad B(z) = \sum_{n=1}^{\infty} \Big(A_\varepsilon(z) \Big)^n$$

we see that $B(z) = ((1-\varepsilon)/\varepsilon)(1-z)^{-1}$. By uniqueness of the power series expansion it follows that $b(n) = (1-\varepsilon)/\varepsilon$ for all n, which completes the proof. $\qquad \square$

As an immediate consequence of Lemma 8.2.3 we have

Proposition 8.2.4. *Suppose that Φ is ψ-irreducible.*

(i) *The chain Φ is transient if and only if each K_{a_ε}-chain is transient.*

(ii) *The chain Φ is recurrent if and only if each K_{a_ε}-chain is recurrent.*

$\qquad \square$

We may now prove

Theorem 8.2.5. *If Φ is ψ-irreducible, then Φ is either recurrent or transient.*

Proof   From Proposition 5.4.5 we are assured that the K_{a_ε}-chain is strongly aperiodic. Using Proposition 8.2.2 we know then that each K_{a_ε}-chain can be classified dichotomously as recurrent or transient.

Since Proposition 8.2.4 shows that the K_{a_ε}-chain passes on either of these properties to Φ itself, the result is proved. $\qquad \square$

We also have the following analogue of Proposition 8.2.4:

Theorem 8.2.6. *Suppose that* Φ *is* ψ-*irreducible and aperiodic.*

(i) *The chain* Φ *is transient if and only if one, and then every,* m-*skeleton* Φ^m *is transient.*

(ii) *The chain* Φ *is recurrent if and only if one, and then every,* m-*skeleton* Φ^m *is recurrent.*

PROOF   **(i)**   If A is a uniformly transient set for the m-skeleton Φ^m, with $\sum_j P^{jm}(x, A) \leq M$, then we have from the Chapman–Kolmogorov equations

$$\sum_{j=1}^{\infty} P^j(x, A) = \sum_{r=1}^{m} \int P^r(x, dy) \sum_j P^{jm}(y, A) \leq mM. \tag{8.29}$$

Thus A is uniformly transient for Φ. Hence Φ is transient whenever a skeleton is transient. Conversely, if Φ is transient then every Φ^k is transient, since

$$\sum_{j=1}^{\infty} P^j(x, A) \geq \sum_{j=1}^{\infty} P^{jk}(x, A).$$

(ii)   If the m-skeleton is recurrent then from the equality in (8.29) we again have that

$$\sum P^j(x, A) = \infty, \qquad x \in \mathsf{X}, \ A \in \mathcal{B}^+(\mathsf{X}), \tag{8.30}$$

so that the chain Φ is recurrent.

Conversely, suppose that Φ is recurrent. For any m it follows from aperiodicity and Proposition 5.4.5 that Φ^m is ψ-irreducible, and hence by Theorem 8.2.5, this skeleton is either recurrent or transient. If it were transient we would have Φ transient, from (i). □

It would clearly be desirable that we strengthen the definition of recurrence to a form of Harris recurrence in terms of $L(x, A)$, similar to that in Proposition 8.1.4. The key problem in moving to the general situation is that we do not have, for a general set, the equivalence in Proposition 8.1.3. There does not seem to be a simple way to exploit the fact that the atom in the split chain is not only recurrent but also satisfies $L(\check{\alpha}, \check{\alpha}) = 1$, and the dichotomy in Theorem 8.2.5 is as far as we can go without considerably stronger techniques which we develop in the next chapter.

Until such time as we provide these techniques we will consider various partial relationships between transience and recurrence conditions, which will serve well in practical classification of chains.

8.3   Recurrence and transience relationships

8.3.1   Transience of sets

We next give conditions on hitting times which ensure that a set is uniformly transient, and which commence to link the behavior of τ_A with that of η_A.

Proposition 8.3.1. *Suppose that Φ is a Markov chain, but not necessarily irreducible.*

(i) *If any set $A \in \mathcal{B}(X)$ is uniformly transient with $U(x, A) \leq M$ for $x \in A$, then $U(x, A) \leq 1 + M$ for every $x \in X$.*

(ii) *If any set $A \in \mathcal{B}(X)$ satisfies $L(x, A) = 1$ for all $x \in A$, then A is recurrent. If Φ is ψ-irreducible, then $A \in \mathcal{B}^+(X)$ and we have $U(x, A) \equiv \infty$ for $x \in X$.*

(iii) *If any set $A \in \mathcal{B}(X)$ satisfies $L(x, A) \leq \varepsilon < 1$ for $x \in A$, then we have $U(x, A) \leq 1/[1 - \varepsilon]$ for $x \in X$, so that in particular A is uniformly transient.*

(iv) *Let $\tau_A(k)$ denote the kth return time to A, and suppose that for some m*

$$P_x(\tau_A(m) < \infty) \leq \varepsilon < 1, \qquad x \in A, \tag{8.31}$$

then $U(x, A) \leq 1 + m/[1 - \varepsilon]$ for every $x \in X$.

PROOF **(i)** We use the first-entrance decomposition: letting $z \uparrow 1$ in (8.25) with $A = B$ shows that for all x,

$$U(x, A) \leq 1 + \sup_{y \in A} U(y, A), \tag{8.32}$$

which gives the required bound.

(ii) Suppose that $L(x, A) \equiv 1$ for $x \in A$. The last-exit decomposition (8.26) gives

$$U^{(z)}(x, A) = U_A^{(z)}(x, A) + \int_A U^{(z)}(x, dy) U_A^{(z)}(y, A).$$

Letting $z \uparrow 1$ gives for $x \in A$,

$$U(x, A) = 1 + U(x, A),$$

which shows that $U(x, A) = \infty$ for $x \in A$, and hence that A is recurrent.

Suppose now that Φ is ψ-irreducible. The set $A^\infty = \{x \in X : L(x, A) = 1\}$ contains A by assumption. Hence we have for any x,

$$\int P(x, dy) L(y, A) = P(x, A) + \int_{A^c} P(x, dy) U_A(y, A) = L(x, A).$$

This shows that A^∞ is absorbing, and hence full by Proposition 4.2.3.

It follows from ψ-irreducibility that $K_{a_{\frac{1}{2}}}(x, A) > 0$ for all $x \in X$, and we also have for all x that, from (5.47),

$$U(x, A) \geq \int_A K_{a_{\frac{1}{2}}}(x, dy) U(y, A) = \infty$$

as claimed.

(iii) Suppose on the other hand that $L(x, A) \leq \varepsilon < 1, x \in A$. The last exit decomposition again gives

$$U^{(z)}(x, A) = U_A^{(z)}(x, A) + \int_A U^{(z)}(x, dy) U_A^{(z)}(y, A) \leq 1 + \varepsilon U^{(z)}(x, A)$$

and so $U^{(z)}(x, A) \le [1 - \varepsilon]^{-1}$: letting $z \uparrow 1$ shows that A is uniformly transient as claimed.

(iv)   Suppose now (8.31) holds. This means that for some fixed $m \in \mathbb{Z}_+$, we have $\varepsilon < 1$ with

$$\mathsf{P}_x(\eta_A \ge m) \le \varepsilon, \qquad x \in A; \tag{8.33}$$

by induction in (8.33) we find that

$$
\begin{aligned}
\mathsf{P}_x(\eta_A \ge m(k+1)) &= \int_A \mathsf{P}_x(\Phi_{\tau_A(km)} \in dy)\mathsf{P}_y(\eta_A \ge m) \\
&\le \varepsilon\,\mathsf{P}_x(\tau_A(km) < \infty) \\
&\le \varepsilon\,\mathsf{P}_x(\eta_A \ge km) \\
&\le \varepsilon^{k+1},
\end{aligned}
\tag{8.34}
$$

and so for $x \in A$

$$
\begin{aligned}
U(x, A) &= \sum_{n=1}^{\infty} \mathsf{P}_x(\eta_A \ge n) \\
&\le m[1 + \sum_{k=1}^{\infty} \mathsf{P}_x(\eta_A \ge km)] \\
&\le m/[1 - \varepsilon].
\end{aligned}
\tag{8.35}
$$

We now use (i) to give the required bound over all of X. $\square$

If there is one uniformly transient set then it is easy to identify other such sets, even without irreducibility. We have

Proposition 8.3.2. *If A is uniformly transient, and $B \overset{a}{\rightsquigarrow} A$ for some a, then B is uniformly transient. Hence if A is uniformly transient, there is a countable covering of $\overline{A}$ by uniformly transient sets.*

PROOF   From Lemma 5.5.2 (iii), we have when $B \overset{a}{\rightsquigarrow} A$ that for some $\delta > 0$,

$$U(x, A) \ge \int U(x, dy)K_a(y, A) \ge \delta U(x, B)$$

so that B is uniformly transient if A is uniformly transient. Since $\overline{A}$ is covered by the sets $\overline{A}(m)$, $m \in \mathbb{Z}_+$, and each $\overline{A}(m) \overset{a}{\rightsquigarrow} A$ for some a, the result follows. $\square$

The next result provides a useful condition under which sets are transient even if not uniformly transient.

Proposition 8.3.3. *Suppose D^c is absorbing and $L(x, D^c) > 0$ for all $x \in D$. Then D is transient.*

PROOF   Suppose D^c is absorbing and write $B(m) = \{y \in D : P^m(y, D^c) \ge m^{-1}\}$: clearly, the sets $B(m)$ cover D since $L(x, D^c) > 0$ for all $x \in D$, by assumption.

But since D^c is absorbing, for every $y \in B(m)$ we have

$$\mathsf{P}_y(\eta_{B(m)} \ge m) \le \mathsf{P}_y(\eta_D \ge m) \le [1 - m^{-1}]$$

and thus (8.31) holds for $B(m)$; from (8.35) it follows that $B(m)$ is uniformly transient. $\square$

These results have direct application in the ψ-irreducible case. We next give a number of such consequences.

8.3.2 Identifying transient sets for ψ-irreducible chains

We first give an alternative proof that there is a recurrence/transience dichotomy for general state space chains which is an analogue of that in the countable state space case. Although this result has already been shown through the use of the splitting technique in Theorem 8.2.5, the following approach enables us to identify uniformly transient sets without going through the atom.

Theorem 8.3.4. *If Φ is ψ-irreducible, then Φ is either recurrent or transient.*

PROOF   Suppose Φ is not recurrent: that is, there exists some pair $A \in \mathcal{B}^+(\mathsf{X})$, $x^* \in \mathsf{X}$ with $U(x^*, A) < \infty$. If $A_* = \{y : U(y, A) = \infty\}$, then $\psi(A_*) = 0$: for otherwise we would have $P^m(x^*, A_*) > 0$ for some m, and then

$$
\begin{aligned}
U(x^*, A) &\geq \int_{\mathsf{X}} P^m(x^*, dw) U(w, A) \\
&\geq \int_{A_*} P^m(x^*, dw) U(w, A) = \infty.
\end{aligned}
\tag{8.36}
$$

Set $A_r = \{y \in A : U(y, A) \leq r\}$. Since $\psi(A) > 0$, and $A_r \uparrow A \cap A_*^c$, there must exist some r such that $\psi(A_r) > 0$, and by Proposition 8.3.1 (i) we have for all y,

$$
U(y, A_r) \leq 1 + r.
\tag{8.37}
$$

Consider now $\overline{A}_r(M) = \{y : \sum_{m=0}^M P^m(y, A_r) > M^{-1}\}$. For any x, from (8.37)

$$
\begin{aligned}
M(1 + r) \geq M U(x, A_r) &\geq \sum_{m=1}^M \sum_{n=m}^\infty P^n(x, A_r) \\
&= \sum_{n=0}^\infty \int_{\mathsf{X}} P^n(x, dw) \sum_{m=1}^M P^m(w, A_r) \\
&\geq \sum_{n=0}^\infty \int_{\overline{A}_r(M)} P^n(x, dw) \sum_{m=1}^M P^m(w, A_r) \\
&\geq M^{-1} \sum_{n=0}^\infty P^n(x, \overline{A}_r(M)).
\end{aligned}
\tag{8.38}
$$

Since $\psi(A_r) > 0$ we have $\cup_m \overline{A}_r(m) = \mathsf{X}$, and so the $\{\overline{A}_r(m)\}$ form a partition of X into uniformly transient sets as required. $\square$

The partition of X into uniformly transient sets given in Proposition 8.3.2 and in Theorem 8.3.4 leads immediately to

Theorem 8.3.5. *If Φ is ψ-irreducible and transient, then every petite set is uniformly transient.*

PROOF   If C is petite, then by Proposition 5.5.5 (iii) there exists a sampling distribution a such that $C \overset{a}{\leadsto} B$ for any $B \in \mathcal{B}^+(\mathsf{X})$. If Φ is transient then there exists at least one $B \in \mathcal{B}^+(\mathsf{X})$ which is uniformly transient, so that C is uniformly transient from Proposition 8.3.2. $\square$

Thus petite sets are also "small" within the transience definitions. This gives us a criterion for recurrence which we shall use in practice for many models; we combine it with a criterion for transience in

Theorem 8.3.6. *Suppose that Φ is ψ-irreducible. Then*

(i) *Φ is recurrent if there exists some petite set $C \in \mathcal{B}(\mathsf{X})$ such that $L(x, C) \equiv 1$ for all $x \in C$.*

(ii) *Φ is transient if and only if there exist two sets D, C in $\mathcal{B}^+(\mathsf{X})$ with $L(x, C) < 1$ for all $x \in D$.*

PROOF   **(i)**   From Proposition 8.3.1 (ii) C is recurrent. Since C is petite Theorem 8.3.5 shows Φ is recurrent. Note that we do not assume that C is in $\mathcal{B}^+(\mathsf{X})$, but that this follows also.

(ii)   Suppose the sets C, D exist in $\mathcal{B}^+(\mathsf{X})$. There must exist $D_\varepsilon \subset D$ such that $\psi(D_\varepsilon) > 0$ and $L(x, C) \leq 1 - \varepsilon$ for all $x \in D_\varepsilon$. If also $\psi(D_\varepsilon \cap C) > 0$ then since $L(x, C) \geq L(D_\varepsilon \cap C)$ we have that $D_\varepsilon \cap C$ is uniformly transient from Proposition 8.3.1 and the chain is transient.

Otherwise we must have $\psi(D_\varepsilon \cap C^c) > 0$. The maximal nature of ψ then implies that for some $\delta > 0$ and some $n \geq 1$ the set $C_\delta := \{y \in C : {}_CP^n(y, D_\varepsilon \cap C^c) > \delta\}$ also has positive ψ-measure. Since, for $x \in C_\delta$,

$$1 - L(x, C_\delta) \geq \int_{D_\varepsilon \cap C^c} {}_CP^n(x, dy)[1 - L(y, C_\delta)] \geq \delta\varepsilon$$

the set C_δ is uniformly transient, and again the chain is transient.

To prove the converse, suppose that Φ is transient. Then for some petite set $C \in \mathcal{B}^+(\mathsf{X})$ the set $D = \{y \in C^c : L(y, C) < 1\}$ is non-empty; for otherwise by (i) the chain is recurrent. Suppose that $\psi(D) = 0$. Then by Proposition 4.2.3 there exists a full absorbing set $F \subset D^c$. By definition we have $L(x, C) = 1$ for $x \in F \setminus C$, and since F is absorbing it then follows that $L(x, C) = 1$ for every $x \in F$, and hence also that $L(x, C_0) = 1$ for $x \in F$ where $C_0 = C \cap F$ also lies in $\mathcal{B}^+(\mathsf{X})$.

But now from Proposition 8.3.1 (ii), we see that C_0 is recurrent, which is a contradiction of Theorem 8.3.5; and we conclude that $D \in \mathcal{B}^+(\mathsf{X})$ as required. $\square$

We would hope that ψ-null sets would also have some transience property, and indeed they do.

Proposition 8.3.7. *If Φ is ψ-irreducible, then every ψ-null set is transient.*

PROOF   Suppose that Φ is ψ-irreducible, and D is ψ-null. By Proposition 4.2.3, D^c contains an absorbing set, whose complement can be covered by uniformly transient sets as in Proposition 8.3.3: clearly, these uniformly transient sets cover D itself, and we are finished.   □

As a direct application of Proposition 8.3.7 we extend the description of the cyclic decomposition for ψ-irreducible chains to give

Proposition 8.3.8. *Suppose that Φ is a ψ-irreducible Markov chain on $(X, \mathcal{B}(X))$. Then there exist sets $D_1, \ldots, D_d \in \mathcal{B}(X)$ such that*

(i) *for $x \in D_i$, $P(x, D_{i+1}) = 1$, $i = 0, \ldots, d-1$ (mod d);*

(ii) *the set $N = [\bigcup_{i=1}^{d} D_i]^c$ is ψ-null and transient.*

PROOF   The existence of the periodic sets D_i is guaranteed by Theorem 5.4.4, and the fact that the set N is transient is then a consequence of Proposition 8.3.3, since $\bigcup_{i=1}^{d} D_i$ is itself absorbing.   □

In the main, transient sets and chains are ones we wish to exclude in practice. The results of this section have formalized the situation we would hope would hold: sets which appear to be irrelevant to the main dynamics of the chain are indeed so, in many different ways. But one cannot exclude them all, and for all of the statements where ψ null (and hence transient) exceptional sets occur, one can construct examples to show that the "bad" sets need not be empty.

8.4 Classification using drift criteria

Identifying whether any particular model is recurrent or transient is not trivial from what we have done so far, and indeed, the calculation of the matrix U or the hitting time probabilities L involves in principle the calculation and analysis of all of the P^n, a daunting task in all but the most simple cases such as those addressed in Section 8.1.2.

Fortunately, it is possible to give practical criteria for both recurrence and transience, couched purely in terms of the drift of the one-step transition matrix P towards individual sets, based on Theorem 8.3.6.

8.4.1 A drift criterion for transience

We first give a criterion for transience of chains on general spaces, which rests on finding the minimal solution to a class of inequalities.

Recall that σ_C, the hitting time on a set C, is identical to τ_C on C^c and $\sigma_C = 0$ on C.

Proposition 8.4.1. *For any $C \in \mathcal{B}(X)$, the pointwise minimal non-negative solution to the set of inequalities*

$$\int P(x, dy)h(y) \leq h(x), \qquad x \in C^c,$$

$$(8.39)$$

$$h(x) \geq 1, \qquad x \in C,$$

is given by the function

$$h^*(x) = \mathsf{P}_x(\sigma_C < \infty), \qquad x \in \mathsf{X};$$

and h^* satisfies (8.39) with equality.

PROOF   Since for $x \in C^c$

$$\mathsf{P}_x(\sigma_C < \infty) = P(x, C) + \int_{C^c} P(x, dy)\mathsf{P}_y(\sigma_C < \infty) = Ph^*(x)$$

it is clear that h^* satisfies (8.39) with equality.

Now let h be any solution to (8.39). By iterating (8.39) we have

$$h(x) \;\geq\; \int_C P(x, dy)h(y) + \int_{C^c} P(x, dy)h(y)$$

$$\geq\; \int_C P(x, dy)h(y) + \int_{C^c} P(x, dy)[\int_C P(y, dz)h(z) + \int_{C^c} P(x, dz)h(z)]$$

$$\vdots$$

$$\geq\; \sum_{j=1}^{N} \int_C {}_C P^j(x, dy)h(y) + \int_{C^c} {}_C P^N(x, dy)h(y).$$

$$(8.40)$$

Letting $N \to \infty$ shows that $h(x) \geq h^*(x)$ for all x.   □

This gives the required drift criterion for transience. Recall the definition of the drift operator as $\Delta V(x) = \int P(x, dy)V(y) - V(x)$; obviously Δ is well-defined if V is bounded. We define the sublevel set $C_V(r)$ of any function V for $r \geq 0$ by

$$C_V(r) := \{x : V(x) \leq r\}.$$

Theorem 8.4.2. *Suppose Φ is a ψ-irreducible chain. Then Φ is transient if and only if there exists a bounded function $V : \mathsf{X} \to \mathbb{R}_+$ and $r \geq 0$ such that*

(i) *both $C_V(r)$ and $C_V(r)^c$ lie in $\mathcal{B}^+(\mathsf{X})$;*

(ii) *whenever $x \in C_V(r)^c$,*

$$\Delta V(x) > 0. \qquad (8.41)$$

PROOF   Suppose that V is an arbitrary bounded solution of (i) and (ii), and let M be a bound for V over X. Clearly $M > r$. Set $C = C_V(r)$, $D = C^c$, and

$$h_V(x) = \begin{cases} [M - V(x)]/[M - r] & x \in D \\ 1 & x \in C \end{cases}$$

so that h_V is a solution of (8.39). Then from the minimality of h^* in Proposition 8.4.1, h_V is an upper bound on h^*, and since for $x \in D, h_V(x) < 1$ we must have $L(x, C) < 1$ also for $x \in D$.

Hence Φ is transient as claimed, from Theorem 8.3.6.

Conversely, if Φ is transient, there exists a bounded function V satisfying (i) and (ii). For from Theorem 8.3.6 we can always find $\varepsilon < 1$ and a petite set $C \in \mathcal{B}^+(\mathsf{X})$ such that $\{y \in C^c : L(y, C) < \varepsilon\}$ is also in $\mathcal{B}^+(\mathsf{X})$. Thus from Proposition 8.4.1, the function $V(x) = 1 - \mathsf{P}_x(\sigma_C < \infty)$ has the required properties. $\square$

8.4.2   A drift criterion for recurrence

Theorem 8.4.2 essentially asserts that if Φ "drifts away" in expectation from a set in $\mathcal{B}^+(\mathsf{X})$, as indicated in (8.41), then Φ is transient. Of even more value in assessing stability are conditions which show that "drift toward" a set implies recurrence, and we provide the first of these now. The condition we will use is

---

Drift criterion for recurrence

(V1)   There exists a positive function V and a set $C \in \mathcal{B}(\mathsf{X})$ satisfying

$$\Delta V(x) \le 0, \qquad x \in C^c. \tag{8.42}$$

We will find frequently that, in order to test such drift for the process Φ, we need to consider functions $V : \mathsf{X} \to \mathbb{R}$ such that the set $C_V(M) = \{y \in \mathsf{X} : V(y) \le M\}$ is "finite" for each M. Such a function on a countable space or topological space is easy to define: in this abstract setting we first need to define a class of functions with this property, and we will find that they recur frequently, giving further meaning to the intuitive meaning of petite sets.

Functions unbounded off petite sets

We will call a measurable function $V : \mathsf{X} \to \mathbb{R}_+$ *unbounded off petite sets* for Φ if for any $n < \infty$, the sublevel set $C_V(n)$ is petite, where

$$C_V(n) = \{y : V(y) \le n\}.$$

Note that since, for an irreducible chain, a finite union of petite sets is petite, and since any subset of a petite set is itself petite, a function $V : \mathsf{X} \to \mathbb{R}_+$ will be unbounded off petite sets for Φ if there merely exists a sequence $\{C_j\}$ of petite sets such that, for any $n < \infty$

$$C_V(n) \subseteq \bigcup_{j=1}^{N} C_j \tag{8.43}$$

for some $N < \infty$. In practice this may be easier to verify directly.

We now have a drift condition which provides a test for recurrence.

Theorem 8.4.3. *Suppose Φ is ψ-irreducible. If there exists a petite set $C \subset X$, and a function V which is unbounded off petite sets such that (V1) holds, then $L(x, C) \equiv 1$ and Φ is recurrent.*

PROOF   We will show that $L(x, C) \equiv 1$ which will give recurrence from Theorem 8.3.6. Note that by replacing the set C by $C \cup C_V(n)$ for n suitably large, we can assume without loss of generality that $C \in \mathcal{B}^+(X)$.

Suppose by way of contradiction that the chain is transient, and thus that there exists some $x^* \in C^c$ with $L(x^*, C) < 1$.

Set $C_V(n) = \{y \in X : V(y) \leq n\}$: we know this is petite, by definition of V, and hence it follows from Theorem 8.3.5 that $C_V(n)$ is uniformly transient for any n. Now fix M large enough that

$$M > V(x^*)/[1 - L(x^*, C)]. \tag{8.44}$$

Let us modify P to define a kernel $\widehat{P}$ with entries $\widehat{P}(x, A) = P(x, A)$ for $x \in C^c$ and $\widehat{P}(x, x) = 1, x \in C$. This defines a chain $\widehat{\Phi}$ with C as an absorbing set, and with the property that for all $x \in X$

$$\int \widehat{P}(x, dy)V(y) \leq V(x). \tag{8.45}$$

Since P is unmodified outside C, but $\widehat{\Phi}$ is absorbed in C, we also have

$$\widehat{P}^n(x, C) = \mathsf{P}_x(\tau_C \leq n) \uparrow L(x, C), \qquad x \in C^c, \tag{8.46}$$

whilst for $A \subseteq C^c$

$$\widehat{P}^n(x, A) \leq P^n(x, A), \qquad x \in C^c. \tag{8.47}$$

By iterating (8.45) we thus get, for fixed $x \in C^c$

$$V(x) \;\geq\; \int \widehat{P}^n(x, dy)V(y)$$

$$\geq \int_{C^c \cap [C_V(M)]^c} \widehat{P}^n(x, dy)V(y) \tag{8.48}$$

$$\geq \; M\left[1 - \widehat{P}^n(x, C_V(M) \cup C)\right].$$

Since $C_V(M)$ is uniformly transient, from (8.47) we have

$$\widehat{P}^n(x^*, C_V(M) \cap C^c) \leq P^n(x^*, C_V(M) \cap C^c) \to 0, \qquad n \to \infty. \tag{8.49}$$

Combining this with (8.46) gives

$$[1 - \widehat{P}^n(x^*, C_V(M) \cup C)] \to [1 - L(x^*, C)], \qquad n \to \infty. \tag{8.50}$$

Letting $n \to \infty$ in (8.48) for $x = x^*$ provides a contradiction with (8.50) and our choice of M. Hence we must have $L(x, C) \equiv 1$, and Φ is recurrent, as required.   □

8.4.3 Random walks with bounded range

The drift condition on the function V in Theorem 8.4.3 basically says that, whenever the chain is outside C, it "moves down" towards that part of the space described by the petite sets outside which V tends to infinity.

This condition implies that we know where the petite sets for Φ lie, and can identify those functions which are unbounded off the petite sets. This provides very substantial motivation for the identification of petite sets in a manner independent of Φ; and for many chains we can use the results in Chapter 6 to give such form to the results.

On a countable space, of course, finite sets are petite. Our problem is then to identify the correct test function to use in the criteria.

In order to illustrate the use of the drift criteria we will first consider the simplest case of a random walk on $\mathbb{Z}$ with finite range r. Thus we assume the increment distribution Γ is concentrated on the integers and is such that $\Gamma(x) = 0$ for $|x| > r$. We then have a relatively simple proof of the result in Theorem 8.1.5.

Proposition 8.4.4. *Suppose that Φ is an irreducible random walk on the integers. If the increment distribution Γ has a bounded range and the mean of Γ is zero, then Φ is recurrent.*

PROOF   In Theorem 8.4.3 choose the test function $V(x) = |x|$. Then for $x > r$ we have that

$$\sum_y P(x,y)[V(y) - V(x)] = \sum_y \Gamma(w)w,$$

whilst for $x < -r$ we have that

$$\sum_y P(x,y)[V(y) - V(x)] = -\sum_w \Gamma(w)w.$$

Suppose the "mean drift"

$$\beta = \sum_w \Gamma(w)w = 0.$$

Then the conditions of Theorem 8.4.3 are satisfied with $C = \{-r, \ldots, r\}$ and with (8.42) holding for $x \in C^c$, and so the chain is recurrent.   □

Proposition 8.4.5. *Suppose that Φ is an irreducible random walk on the integers. If the increment distribution Γ has a bounded range and the mean of Γ is non-zero, then Φ is transient.*

PROOF   Suppose Γ has non-zero mean $\beta > 0$. We will establish for some bounded monotone increasing V that

$$\sum_y P(x,y)V(y) = V(x) \tag{8.51}$$

for $x \geq r$.

This time choose the test function $V(x) = 1 - \rho^x$ for $x \geq 0$, and $V(x) = 0$ elsewhere. The sublevel sets of V are of the form $(-\infty, r]$ with $r \geq 0$. This function satisfies (8.51) if and only if for $x \geq r$

$$\sum_y P(x,y)[\rho^y/\rho^x] = 1 \tag{8.52}$$

so that this V can be constructed as a valid test function if (and only if) there is a $\rho < 1$ with

$$\sum_w \Gamma(w)\rho^w = 1. \tag{8.53}$$

Therefore the existence of a solution to (8.53) will imply that the chain is transient, since return to the whole half line $(-\infty, r]$ is less than sure from Proposition 8.4.2. Write $\beta(s) = \sum_w \Gamma(w)s^w$: then β is well defined for $s \in (0,1]$ by the bounded range assumption. By irreducibility, we must have $\Gamma(w) > 0$ for some $w < 0$, so that $\beta(s) \to \infty$ as $s \to 0$. Since $\beta(1) = 1$, and $\beta'(1) = \sum_w w\Gamma(w) = \beta > 0$ it follows that such a ρ exists, and hence the chain is transient.

Similarly, if the mean of Γ is negative, we can by symmetry prove transience because the chain fails to return to the half line $[-r, \infty)$.   □

For random walk on the half line $\mathbb{Z}_+$ with bounded range, as defined by (RWHL1) we find

Proposition 8.4.6. *If the random walk increment distribution Γ on the integers has mean β and a bounded range, then the random walk on $\mathbb{Z}_+$ is recurrent if and only if $\beta \leq 0$.*

PROOF   If β is positive, then the probability of return of the unrestricted random walk to $(-\infty, r]$ is less than one, for starting points above r, and since the probability of return of the random walk on a half line to $[0, r]$ is identical to the return to $(-\infty, r]$ for the unrestricted random walk, the chain is transient.

If $\beta \leq 0$, then we have as for the unrestricted random walk that, for the test function $V(x) = x$ and all $x \geq r$

$$\sum_y P(x,y)[V(y) - V(x)] = \sum_w \Gamma(w)w \leq 0;$$

but since, in this case, the set $\{x \leq r\}$ is finite, we have (8.42) holding and the chain is recurrent.   □

The first part of this proof involves a so-called "stochastic comparison" argument: we use the return time probabilities for one chain to bound the same probabilities for another chain. This is simple but extremely effective, and we shall use it a number of times in classifying random walk. A more general formulation will be given in Section 9.5.1.

Varying the condition that the range of the increment is bounded requires a much more delicate argument, and indeed the known result of Theorem 8.1.5 for a general random walk on $\mathbb{Z}$, that recurrence is equivalent to the mean $\beta = 0$, appears difficult if not impossible to prove by drift methods without some bounds on the spread of Γ.

8.5 Classifying random walk on $\mathbb{R}_+$

In order to give further exposure to the use of drift conditions, we will conclude this chapter with a detailed examination of random walk on $\mathbb{R}_+$.

The analysis here is obviously immediately applicable to the various queueing and storage models introduced in Chapter 2 and Chapter 3, although we do not fill in the details explicitly. The interested reader will find, for example, that the conditions on the increment do translate easily into intuitively appealing statements on the mean input rate to such systems being no larger than the mean service or output rate if recurrence is to hold.

These results are intended to illustrate a variety of approaches to the use of the stability criteria above. Different test functions are utilized, and a number of different methods of ensuring they are applicable are developed. Many of these are used in the sequel where we classify more general models.

As in (RW1) and (RWHL1) we let $\boldsymbol{\Phi}$ denote a chain with

$$\Phi_n = [\Phi_{n-1} + W_n]^+$$

where as usual W_n is a noise variable with distribution Γ and mean β which we shall assume in this section is well defined and finite.

Clearly we would expect from the bounded increments results above that $\beta \leq 0$ is the appropriate necessary and sufficient condition for recurrence of $\boldsymbol{\Phi}$. We now address the three separate cases in different ways.

8.5.1 Recurrence when β is negative

When the inequality is strict it is not hard to show that the chain is recurrent.

Proposition 8.5.1. *If $\boldsymbol{\Phi}$ is random walk on a half line and if*

$$\beta = \int w\,\Gamma(dw) < 0,$$

then $\boldsymbol{\Phi}$ is recurrent.

PROOF   Clearly the chain is φ-irreducible when $\beta < 0$ with $\varphi = \delta_0$, and all compact sets are small as in Chapter 5. To prove recurrence we use Theorem 8.4.3, and show that we can in fact find a suitably unbounded function V and a compact set C satisfying

$$\int P(x, dy)V(y) \leq V(x) - \varepsilon, \qquad x \in C^c, \tag{8.54}$$

for some $\varepsilon > 0$. As in the countable case we note that since $\beta < 0$ there exists $x_0 < \infty$ such that

$$\int_{-x_0}^{\infty} w\,\Gamma(dw) < \beta/2 < 0,$$

and thus if $V(x) = x$, for $x > x_0$

$$\int P(x, dy)[V(y) - V(x)] \leq \int_{-x_0}^{\infty} w\,\Gamma(dw). \tag{8.55}$$

Hence taking $\varepsilon = \beta/2$ and $C = [0, x_0]$ we have the required result.   □

8.5.2   Recurrence when β is zero

When the mean increment $\beta = 0$ the situation is much less simple, and in general the drift conditions cán be verified simply only under somewhat stronger conditions on the increment distribution Γ, such as an assumption of a finite variance of the increments.

We will find it convenient to develop prior to our calculations some detailed bounds on the moments of Γ, which will become relevant when we consider test functions of the form $V(x) = \log(1 + |x|)$.

Lemma 8.5.2. *Let W be a random variable with law Γ, s a positive number and t any real number. Then for any $A \subseteq \{w \in \mathbb{R} : s + tw > 0\}$,*

$$\mathsf{E}[\log(s + tW)\mathbb{I}\{W \in A\}] \;\leq\; \Gamma(A)\log(s) + (t/s)\mathsf{E}[W\mathbb{I}\{W \in A\}]$$

$$- (t^2/(2s^2))\mathsf{E}[W^2\mathbb{I}\{W \in A, tW < 0\}].$$

PROOF   For all $x > -1$, $\log(1 + x) \leq x - (x^2/2)\mathbb{I}\{x < 0\}$. Thus

$$\log(s + tW)\mathbb{I}\{W \in A\} \;=\; [\log(s) + \log(1 + tW/s)]\mathbb{I}\{W \in A\}$$

$$\leq\; [\log(s) + tW/s]\mathbb{I}\{W \in A\}$$

$$- ((tW)^2/(2s^2))\mathbb{I}\{tW < 0, W \in A\}$$

and taking expectations gives the result.   □

Lemma 8.5.3. *Let W be a random variable with law Γ and finite variance. Let s be a positive number and t a real number. Then*

$$\lim_{x\to\infty} -x\mathsf{E}[W\mathbb{I}\{W < t - sx\}] = \lim_{x\to\infty} x\mathsf{E}[W\mathbb{I}\{W > t + sx\}] = 0. \qquad (8.56)$$

Furthermore, if $\mathsf{E}[W] = 0$, then

$$\lim_{x\to\infty} -x\mathsf{E}[W\mathbb{I}\{W > t - sx\}] = \lim_{x\to\infty} x\mathsf{E}[W\mathbb{I}\{W < t + sx\}] = 0. \qquad (8.57)$$

PROOF   This is a consequence of

$$0 \leq \lim_{x\to\infty} (t + sx) \int_{t+sx}^{\infty} w\Gamma(dw) \leq \lim_{x\to\infty} \int_{t+sx}^{\infty} w^2\Gamma(dw) = 0,$$

and

$$0 \leq \lim_{x\to-\infty} (t + sx) \int_{-\infty}^{t+sx} w\Gamma(dw) \leq \lim_{x\to-\infty} \int_{-\infty}^{t+sx} w^2\Gamma(dw) = 0.$$

If $\mathsf{E}[W] = 0$, then $\mathsf{E}[W\mathbb{I}\{W > t + sx\}] = -\mathsf{E}[W\mathbb{I}\{W < t + sx\}]$, giving the second result.   □

We now prove

Proposition 8.5.4. *If W is an increment variable on $\mathbb{R}$ with $\beta = 0$ and*

$$0 < \mathsf{E}[W^2] = \int w^2\, \Gamma(dw) < \infty,$$

then the random walk on $\mathbb{R}_+$ with increment W is recurrent.

PROOF   We use the test function

$$V(x) = \begin{cases} \log(1+x) & x > R \\ 0 & 0 \leq x \leq R \end{cases} \tag{8.58}$$

where R is a positive constant to be chosen. Since $\beta = 0$ and $0 < \mathsf{E}[W^2]$ the chain is δ_0-irreducible, and we have seen that all compact sets are small as in Chapter 5. Hence V is unbounded off petite sets.

For $x > R, 1 + x > 0$, and thus by Lemma 8.5.2,

$$\begin{aligned} \mathsf{E}_x[V(X_1)] &= \mathsf{E}[\log(1 + x + W)\mathbb{I}\{x + W > R\}] \\ &\leq (1 - \Gamma(-\infty, R - x))\log(1+x) + U_1(x) - U_2(x), \end{aligned} \tag{8.59}$$

where in order to bound the terms in the expansion of the logarithms in V, we consider separately

$$\begin{aligned} U_1(x) &= (1/(1+x))\mathsf{E}[W\mathbb{I}\{W > R - x\}] \\ U_2(x) &= (1/(2(1+x)^2))\mathsf{E}[W^2\mathbb{I}\{R - x < W < 0\}]. \end{aligned} \tag{8.60}$$

Since $\mathsf{E}[W^2] < \infty$

$$U_2(x) = (1/(2(1+x)^2))\mathsf{E}[W^2\mathbb{I}\{W < 0\}] - o(x^{-2}),$$

and by Lemma 8.5.3, U_1 is also $o(x^{-2})$.

Thus by choosing R large enough

$$\begin{aligned} \mathsf{E}_x[V(X_1)] &\leq V(x) - (1/(2(1+x)^2))\mathsf{E}[W^2\mathbb{I}\{W < 0\}] + o(x^{-2}) \\ &\leq V(x), \quad x > R. \end{aligned} \tag{8.61}$$

Hence the conditions of Theorem 8.4.3 hold, and chain is recurrent.   □

8.5.3   Transience of skip-free random walk when β is positive

It is possible to verify transience when $\beta > 0$, without any restrictions on the range of the increments of the distribution Γ, thus extending Proposition 8.4.5; but the argument (in Proposition 9.1.2) is a somewhat different one which is based on the Strong Law of Large Numbers and must wait some stronger results on the meaning of recurrence in the next chapter.

Proving transience for random walk without bounded range using drift conditions is difficult in general. There is however one model for which some exact calculations can be made: this is the random walk which is "skip-free to the right" and which models the GI/M/1 queue as in Theorem 3.3.1.

Proposition 8.5.5. *If Φ denotes random walk on a half line $\mathbb{Z}_+$ which is skip-free to the right (so $\Gamma(x) = 0$ for $x > 1$), and if*

$$\beta = \sum w\, \Gamma(w) > 0,$$

then Φ is transient.

PROOF   We can assume without loss of generality that $\Gamma(-\infty, 0) > 0$: for clearly, if $\Gamma[0, \infty) = 1$ then $P_x(\tau_0 < \infty) = 0, x > 0$ and the chain moves inexorably to infinity; hence it is not irreducible, and it is transient in every meaning of the word.

We will show that for a chain which is skip-free to the right the condition $\beta > 0$ is sufficient for transience, by examining the solutions of the equations

$$\sum P(x, y)V(y) = V(x), \qquad x \geq 1, \tag{8.62}$$

and actually constructing a bounded non-constant positive solution if β is positive. The result will then follow from Theorem 8.4.2.

First note that we can assume $V(0) = 0$ by linearity, and write out the equation (8.62) in this case as

$$V(x) = \Gamma(-x+1)V(1) + \Gamma(-x+2)V(2) + \cdots + \Gamma(1)V(1+x). \tag{8.63}$$

Once the first value in the $V(x)$ sequence is chosen, we therefore have the remaining values given by an iterative process. Our goal is to show that we can define the sequence in a way that gives us a non-constant positive bounded solution to (8.63).

In order to do this we first write

$$V^*(z) = \sum_0^\infty V(x)z^x, \qquad \Gamma^*(z) = \sum_{-\infty}^\infty \Gamma(x)z^x,$$

where $V^*(z)$ has yet to be shown to be defined for any z and $\Gamma^*(z)$ is clearly defined at least for $|z| \geq 1$. Multiplying by z^x in (8.63) and summing we have that

$$V^*(z) = \Gamma^*(z^{-1})V^*(z) - \Gamma(1)V(1). \tag{8.64}$$

Now suppose that we can show (as we do below) that there is an analytic expansion of the function

$$z^{-1}[1-z]/[\Gamma^*(z^{-1}) - 1] = \sum_0^\infty b_n z^n \tag{8.65}$$

in the region $0 < z < 1$ with $b_n \geq 0$. Then we will have the identity

$$\begin{aligned}
V^*(z) &= z\Gamma(1)V(1)z^{-1}/[\Gamma^*(z^{-1}) - 1] \\[2mm]
&= z\Gamma(1)V(1)(\sum_0^\infty z^n)z^{-1}[1-z]/[\Gamma^*(z^{-1}) - 1] \tag{8.66} \\[2mm]
&= z\Gamma(1)V(1)(\sum_0^\infty z^n)(\sum_0^\infty b_m z^m).
\end{aligned}$$

From this, we will be able to identify the form of the solution V. Explicitly, from (8.66) we have

$$V^*(z) = z\Gamma(1)V(1) \sum_{n=0}^\infty z^n \sum_{m=0}^n b_m \tag{8.67}$$

so that equating coefficients of z^n in (8.67) gives

$$V(x) = \Gamma(1)V(1) \sum_{m=0}^{x-1} b_m.$$

Clearly then the solution V is bounded and non-constant if

$$\sum_m b_m < \infty. \tag{8.68}$$

Thus we have reduced the question of transience to identifying conditions under which the expansion in (8.65) holds with the coefficients b_j positive and summable.

Let us write $a_j = \Gamma(1 - j)$ so that

$$\Lambda(z) := \sum_0^\infty a_j z^j = z\Gamma^*(z^{-1})$$

and for $0 < z < 1$ we have

$$
\begin{aligned}
B(z) := z[\Gamma^*(z^{-1}) - 1]/[1 - z] \;&=\; [A(z) - z]/[1 - z] \\
&=\; 1 - [1 - A(z)]/[1 - z] \tag{8.69} \\
&=\; 1 - \sum_0^\infty z^j \sum_{n=j+1}^\infty a_n.
\end{aligned}
$$

Now if we have a positive mean for the increment distribution,

$$\Big| \sum_0^\infty z^j \sum_{n=j+1}^\infty a_n \Big| \leq \sum_n n a_n < 1$$

and so $B(z)^{-1}$ is well defined for $|z| < 1$; moreover, by the expansion in (8.69)

$$B(z)^{-1} = \sum b_j z^j$$

with all with all $b_j \geq 0$, and hence by Abel's Theorem,

$$\sum b_j = [1 - \sum_n n a_n]^{-1} = \beta^{-1}$$

which is finite as required.   □

8.6   Commentary*

On countable spaces the solidarity results we generalize here are classical, and thorough expositions are in Feller [114], Chung [71], Çinlar [59] and many more places. Recurrence is called persistence by Feller, but the terminology we use here seems to have become the more standard. The first entrance, and particularly the last exit, decomposition are vital tools introduced and exploited in a number of ways by Chung [71].

There are several approaches to the transience/recurrence dichotomy. A common one which can be shown to be virtually identical with that we present here uses the concept of inessential sets (sets for which η_A is almost surely finite). These play the role of transient parts of the space, with recurrent parts of the space being sets which are not inessential. This is the approach in Orey [309], based on the original methods of Doeblin [95] and Doob [99].

Our presentation of transience, stressing the role of uniformly transient sets, is new, although it is implicit in many places. Most of the individual calculations are in Nummelin [303], and a number are based on the more general approach in Tweedie [394]. Equivalences between properties of the kernel $U(x, A)$, which we have called recurrence and transience properties, and the properties of essential and inessential sets are studied in Tuominen [390].

The uniform transience property is inherently stronger than the inessential property, and it certainly aids in showing that the skeletons and the original chain share the dichotomy between recurrence and transience. For use of the properties of skeleton chains in direct application, see Tjøstheim [386].

The drift conditions we give here are due in the countable case to Foster [129], and the versions for more general spaces were introduced in Tweedie [397, 398] and in Kalashnikov [189]. We shall revisit these drift conditions, and expand somewhat on their implications in the next chapter. Stronger versions of (V1) will play a central role in classifying chains as yet more stable in due course.

The test functions for classifying random walk in the bounded range case are directly based on those introduced by Foster [129]. The evaluation of the transience condition for skip-free walks, given in Proposition 8.5.5, is also due to Foster. The approximations in the case of zero drift are taken from Guo and Petrucelli [149] and are reused in analyzing SETAR models in Section 9.5.2.

The proof of recurrence of random walk in Theorem 8.1.5, using the weak law of large numbers, is due to Chung and Ornstein [73]. It appears difficult to prove this using the elementary drift methods.

The drift condition in the case of negative mean gives, as is well known, a stronger form of recurrence: the concerned reader will find that this is taken up in detail in Chapter 11, where it is a central part of our analysis.

Commentary for the second edition: The drift operator (8.1) is analogous to the *generator* for a Markov process in continuous time. Some of the theory surrounding continuous time models is summarized in Section 20.3, including some foundations of generators and resolvents.

Chapter 9

Harris and topological recurrence

In this chapter we consider stronger concepts of recurrence and link them with the dichotomy proved in Chapter 8. We also consider several obvious definitions of global and local recurrence and transience for chains on topological spaces, and show that they also link to the fundamental dichotomy.

In developing concepts of recurrence for sets $A \in \mathcal{B}(\mathsf{X})$, we will consider not just the first hitting time τ_A, or the expected value $U(\cdot, A)$ of η_A, but also the event that $\Phi \in A$ infinitely often (i.o.), or $\eta_A = \infty$, defined by

$$\{\Phi \in A \text{ i.o.}\} := \bigcap_{N=1}^{\infty} \bigcup_{k=N}^{\infty} \{\Phi_k \in A\}$$

which is well defined as an $\mathcal{F}$-measurable event on Ω. For $x \in \mathsf{X}$, $A \in \mathcal{B}(\mathsf{X})$ we write

$$Q(x, A) := \mathsf{P}_x\{\Phi \in A \text{ i.o.}\} : \tag{9.1}$$

obviously, for any x, A we have $Q(x, A) \leq L(x, A)$, and by the strong Markov property we have

$$Q(x, A) = \mathsf{E}_x[\mathsf{P}_{\Phi_{\tau_A}}\{\Phi \in A \text{ i.o.}\}\mathbb{I}\{\tau_A < \infty\}] = \int_A U_A(x, dy)Q(y, A). \tag{9.2}$$

Harris recurrence

The set A is called *Harris recurrent* if

$$Q(x, A) = \mathsf{P}_x(\eta_A = \infty) = 1, \qquad x \in A.$$

A chain Φ is called *Harris (recurrent)* if it is ψ-irreducible and every set in $\mathcal{B}^+(\mathsf{X})$ is Harris recurrent.

We will see in Theorem 9.1.4 that when $A \in \mathcal{B}^+(\mathsf{X})$ and Φ is Harris recurrent then in fact we have the seemingly stronger and perhaps more commonly used property that $Q(x, A) = 1$ for every $x \in \mathsf{X}$.

It is obvious from the definitions that if a set is Harris recurrent, then it is recurrent. Indeed, in the formulation above the strengthening from recurrence to Harris recurrence is quite explicit, indicating a move from an expected infinity of visits to an almost surely infinite number of visits to a set.

This definition of Harris recurrence appears on the face of it to be stronger than requiring $L(x, A) \equiv 1$ for $x \in A$, which is a standard alternative definition of Harris recurrence. In one of the key results of this section, Proposition 9.1.1, we prove that they are in fact equivalent.

The highlight of the Harris recurrence analysis is

Theorem 9.0.1. *If Φ is recurrent, then we can write*

$$\mathsf{X} = H \cup N \tag{9.3}$$

where H is absorbing and non-empty and every subset of H in $\mathcal{B}^+(\mathsf{X})$ is Harris recurrent; and N is ψ-null and transient.

PROOF   This is proved, in a slightly stronger form, in Theorem 9.1.5.   □

Hence a recurrent chain differs only by a ψ-null set from a Harris recurrent chain. In general we can then restrict analysis to H and derive very much stronger results using properties of Harris recurrent chains.

For chains on a countable space the null set N in (9.3) is empty, so recurrent chains are automatically Harris recurrent.

On a topological space we can also find conditions for this set to be empty, and these also provide a useful interpretation of the Harris property.

We say that a sample path of Φ *converges to infinity* (denoted $\Phi \to \infty$) if the trajectory visits each compact set only finitely often. This definition leads to

Theorem 9.0.2. *For a ψ-irreducible T-chain, the chain is Harris recurrent if and only if $\mathsf{P}_x\{\Phi \to \infty\} = 0$ for each $x \in \mathsf{X}$.*

PROOF   This is proved in Theorem 9.2.2   □

Even without its equivalence to Harris recurrence for such chains this "recurrence" type of property (which we will call *non-evanescence*) repays study, and this occupies Section 9.2.

In this chapter, we also connect local recurrence properties of a chain on a topological space with global properties: if the chain is a ψ-irreducible T-chain, then recurrence of the neighborhoods of any one point in the support of ψ implies recurrence of the whole chain.

Finally, we demonstrate further connections between drift conditions and Harris recurrence, and apply these results to give an increment analysis of chains on $\mathbb{R}$ which generalizes that for the random walk in the previous chapter.

9.1 Harris recurrence

9.1.1 Harris properties of sets

We first develop conditions to ensure that a set is Harris recurrent, based only on the first return time probabilities $L(x, A)$.

Proposition 9.1.1. *Suppose for some one set $A \in \mathcal{B}(\mathsf{X})$ we have $L(x, A) \equiv 1, x \in A$. Then $Q(x, A) = L(x, A)$ for every $x \in \mathsf{X}$, and in particular A is Harris recurrent.*

PROOF   Using the strong Markov property, we have that if $L(y, A) = 1$, $y \in A$, then for any $x \in A$

$$\mathsf{P}_x(\tau_A(2) < \infty) = \int_A U_A(x, dy) L(y, A) = 1;$$

inductively this gives for $x \in A$, again using the strong Markov property,

$$\mathsf{P}_x(\tau_A(k+1) < \infty) = \int_A U_A(x, dy) \mathsf{P}_y(\tau_A(k) < \infty) = 1.$$

For any x we have

$$\mathsf{P}_x(\eta_A \geq k) = \mathsf{P}_x(\tau_A(k) < \infty),$$

and since by monotone convergence

$$Q(x, A) = \lim_k \mathsf{P}_x(\eta_A \geq k)$$

we have $Q(x, A) \equiv 1$ for $x \in A$.

It now follows since

$$Q(x, A) = \int_A U_A(x, dy) Q(y, A) = L(x, A)$$

that the theorem is proved.   □

This shows that the definition of Harris recurrence in terms of Q is identical to a similar definition in terms of L: the latter is often used (see for example Orey [309]) but the use of Q highlights the difference between recurrence and Harris recurrence.

We illustrate immediately the usefulness of the stronger version of recurrence in conjunction with the basic dichotomy to give a proof of transience of random walk on $\mathbb{Z}$.

We showed in Section 8.4.3 that random walk on $\mathbb{Z}$ is transient when the increment has non-zero mean and the range of the increment is bounded.

Using the fact that, on the integers, recurrence and Harris recurrence are identical from Proposition 8.1.3, we can remove this bounded range restriction. To do this we use the strong rather than the weak law of large numbers, as used in Theorem 8.1.5.

The form we require (see again, for example, Billingsley [37]) states that if Φ_n is a random walk such that the increment distribution Γ has a mean β which is not zero, then

$$\mathsf{P}_0(\lim_{n \to \infty} n^{-1} \Phi_n = \beta) = 1.$$

Write C_n for the event $\{|n^{-1}\Phi_n - \beta| > \beta/2\}$. We only use the result, which follows from the strong law, that

$$P_0(\limsup_{n \to \infty} C_n) = 0. \tag{9.4}$$

Now let D_n denote the event $\{\Phi_n = 0\}$, and notice that $D_n \subseteq C_n$ for each n. Immediately from (9.4) we have

$$P_0(\limsup_{n \to \infty} D_n) = 0 \tag{9.5}$$

which says exactly $Q(0,0) = 0$.

Hence we have an elegant proof of the general result

Proposition 9.1.2. *If Φ denotes random walk on $\mathbb{Z}$ and if*

$$\beta = \sum w\,\Gamma(w) > 0,$$

then Φ is transient. □

The most difficult of the results we prove in this section, and the strongest, provides a rather more delicate link between the probabilities $L(x,A)$ and $Q(x,A)$ than that in Proposition 9.1.1.

Theorem 9.1.3.   (i) *Suppose that $D \rightsquigarrow A$ for any sets D and A in $\mathcal{B}(\mathsf{X})$. Then*

$$\{\Phi \in D \text{ i.o.}\} \subseteq \{\Phi \in A \text{ i.o.}\} \qquad a.s. \; [P_*] \tag{9.6}$$

and hence $Q(y,D) \leq Q(y,A)$, for all $y \in \mathsf{X}$.

(ii) *If $\mathsf{X} \rightsquigarrow A$, then A is Harris recurrent, and in fact $Q(x,A) \equiv 1$ for every $x \in \mathsf{X}$.*

PROOF   Since the event $\{\Phi \in A \text{ i.o.}\}$ involves the whole path of Φ, we cannot deduce this result merely by considering P^n for fixed n. We need to consider all the events

$$E_n = \{\Phi_{n+1} \in A\}, \qquad n \in \mathbb{Z}_+$$

and evaluate the probability of those paths such that an infinite number of the E_n hold.

We first show that, if $\mathcal{F}_n^\Phi$ is the σ-field generated by $\{\Phi_0, \ldots, \Phi_n\}$, then as $n \to \infty$

$$P\Big[\bigcup_{i=n}^\infty E_i \mid \mathcal{F}_n^\Phi\Big] \to \mathbb{I}\Big(\bigcap_{m=1}^\infty \bigcup_{i=m}^\infty E_i\Big) \qquad a.s. \quad [P_*]. \tag{9.7}$$

To see this, note that for fixed $k \leq n$

$$P\Big[\bigcup_{i=k}^\infty E_i \mid \mathcal{F}_n^\Phi\Big] \geq P\Big[\bigcup_{i=n}^\infty E_i \mid \mathcal{F}_n^\Phi\Big] \geq P\Big[\bigcap_{m=1}^\infty \bigcup_{i=m}^\infty E_i \mid \mathcal{F}_n^\Phi\Big]. \tag{9.8}$$

Now apply the Martingale Convergence Theorem (see Theorem D.6.1) to the extreme elements of the inequalities (9.8) to give

$$
\begin{aligned}
\mathbb{I}\Big[\bigcup_{i=k}^\infty E_i\Big] &\geq \; \limsup_n P\Big[\bigcup_{i=n}^\infty E_i \mid \mathcal{F}_n^\Phi\Big] \\
&\geq \; \liminf_n P\Big[\bigcup_{i=n}^\infty E_i \mid \mathcal{F}_n^\Phi\Big] \\
&\geq \; \mathbb{I}\Big[\bigcap_{m=1}^\infty \bigcup_{i=m}^\infty E_i\Big].
\end{aligned}
\tag{9.9}
$$

As $k \to \infty$, the two extreme terms in (9.9) converge, which shows the limit in (9.7) holds as required.

By the strong Markov property, $\mathsf{P}_*[\bigcup_{i=n}^{\infty} E_i \mid \mathcal{F}_n^{\Phi}] = L(\Phi_n, A)$ a.s. $[\mathsf{P}_*]$. From our assumption that $D \rightsquigarrow A$ we have that $L(\Phi_n, A)$ is bounded from 0 whenever $\Phi_n \in D$. Thus, using (9.7) we have P_*-a.s,

$$\mathbb{I}\Big(\bigcap_{m=1}^{\infty} \bigcup_{i=m}^{\infty} \{\Phi_i \in D\}\Big) \;\leq\; \mathbb{I}\Big(\limsup_n L(\Phi_n, A) > 0\Big)$$
$$= \;\mathbb{I}\Big(\lim_n L(\Phi_n, A) = 1\Big) \tag{9.10}$$
$$= \;\mathbb{I}\Big(\bigcap_{m=1}^{\infty} \bigcup_{i=m}^{\infty} E_i\Big),$$

which is (9.6).

The proof of (ii) is then immediate, by taking $D = \mathsf{X}$ in (9.6). $\qquad\qquad\square$

As an easy consequence of Theorem 9.1.3 we have the following strengthening of Harris recurrence:

Theorem 9.1.4. *If Φ is Harris recurrent, then $Q(x, B) = 1$ for every $x \in \mathsf{X}$ and every $B \in \mathcal{B}^+(\mathsf{X})$.*

PROOF   Let $\{C_n : n \in \mathbb{Z}_+\}$ be petite sets with $\cup C_n = \mathsf{X}$. Since the finite union of petite sets is petite for an irreducible chain by Proposition 5.5.5, we may assume that $C_n \subset C_{n+1}$ and that $C_n \in \mathcal{B}^+(\mathsf{X})$ for each n.

For any $B \in \mathcal{B}^+(\mathsf{X})$ and any $n \in \mathbb{Z}_+$ we have from Lemma 5.5.1 that $C_n \rightsquigarrow B$, and hence, since C_n is Harris recurrent, we see from Theorem 9.1.3 (i) that $Q(x, B) = 1$ for any $x \in C_n$. Because the sets $\{C_k\}$ cover X, it follows that $Q(x, B) = 1$ for all x as claimed. $\qquad\qquad\square$

Having established these stability concepts, and conditions implying they hold for individual sets, we now move on to consider transience and recurrence of the overall chain in the ψ-irreducible context.

9.1.2   Harris recurrent chains

It would clearly be desirable if, as in the countable space case, every set in $\mathcal{B}^+(\mathsf{X})$ were Harris recurrent for every recurrent Φ. Regrettably this is not quite true.

For consider any chain Φ for which every set in $\mathcal{B}^+(\mathsf{X})$ is Harris recurrent: append to X a sequence of individual points $N = \{x_i\}$, and expand P to P' on $\mathsf{X}' := \mathsf{X} \cup N$ by setting $P'(x, A) = P(x, A)$ for $x \in \mathsf{X}, A \in \mathcal{B}(\mathsf{X})$, and

$$P'(x_i, x_{i+1}) = \beta_i, \qquad P'(x_i, \alpha) = 1 - \beta_i$$

for some one specific $\alpha \in \mathsf{X}$ and all $x_i \in N$.

Any choice of the probabilities β_i which provides

$$1 > \prod_{i=0}^{\infty} \beta_i > 0$$

then ensures that

$$L'(x_i, A) = L'(x_i, \alpha) = 1 - \prod_{n=i}^{\infty} \beta_i < 1, \qquad A \in \mathcal{B}^+(\mathsf{X}),$$

so that no set $B \subset \mathsf{X}'$ with $B \cap \mathsf{X}$ in $\mathcal{B}^+(\mathsf{X})$ and $B \cap N$ non-empty is Harris recurrent: but

$$U'(x_i, A) \geq L'(x_i, \alpha)U(\alpha, A) = \infty, \qquad A \in \mathcal{B}(\mathsf{X}),$$

so that every set in $\mathcal{B}^+(\mathsf{X}')$ is recurrent.

We now show that this example typifies the only way in which an irreducible chain can be recurrent and not Harris recurrent: that is, by the existence of an absorbing set which is Harris recurrent, accompanied by a single ψ-null set on which the Harris recurrence fails.

For any Harris recurrent set D, we write $D^\infty = \{y : L(y, D) = 1\}$, so that $D \subseteq D^\infty$, and D^∞ is absorbing.

We will call D a *maximal absorbing set* if $D = D^\infty$. This will be used, in general, in the following form:

Maximal Harris sets

We call a set H *maximal Harris* if H is a maximal absorbing set such that Φ restricted to H is Harris recurrent.

Theorem 9.1.5. *If Φ is recurrent, then we can write*

$$\mathsf{X} = H \cup N \tag{9.11}$$

where H is a non-empty maximal Harris set and N is transient.

PROOF   Let C be a ψ_a-petite set in $\mathcal{B}^+(\mathsf{X})$, where we choose ψ_a as a maximal irreducibility measure. Set $H = \{y : Q(x, C) = 1\}$ and write $N = H^c$.

Clearly, since $H^\infty = H$, either H is empty or H is maximal absorbing. We first show that H is non-empty.

Suppose otherwise, so that $Q(x, C) < 1$ for all x. We first show this implies the set

$$C_1 := \{x \in C : L(x, C) < 1\} :$$

is in $\mathcal{B}^+(\mathsf{X})$.

For if not, and $\psi(C_1) = 0$, then by Proposition 4.2.3 there exists an absorbing full set $F \subset C_1^c$. We have by definition that $L(x, C) = 1$ for any $x \in C \cap F$, and since F is absorbing we must have $L(x, C \cap F) = 1$ for $x \in C \cap F$. From Proposition 9.1.1 it follows that $Q(x, C \cap F) = 1$ for $x \in C \cap F$, which gives a contradiction, since $Q(x, C) \geq Q(x, C \cap F)$. This shows that in fact $\psi(C_1) > 0$.

But now, since $C_1 \in \mathcal{B}^+(\mathsf{X})$ there exists $B \subseteq C_1, B \in \mathcal{B}^+(\mathsf{X})$ and $\delta > 0$ with $L(x, C_1) \leq \delta < 1$ for all $x \in B$: accordingly

$$L(x, B) \leq L(x, C_1) \leq \delta, \qquad x \in B.$$

Now Proposition 8.3.1 (iii) gives $U(x, B) \leq [1 - \delta]^{-1}, x \in B$ and this contradicts the assumed recurrence of Φ.

Thus H is a non-empty maximal absorbing set, and by Proposition 4.2.3 H is full: from Proposition 8.3.7 we have immediately that N is transient. It remains to prove that H is Harris.

For any set A in $\mathcal{B}^+(\mathsf{X})$ we have $C \rightsquigarrow A$. It follows from Theorem 9.1.3 that if $Q(x, C) = 1$ then $Q(x, A) = 1$ for every $A \in \mathcal{B}^+(\mathsf{X})$. Since by construction $Q(x, C) = 1$ for $x \in H$, we have also that $Q(x, A) = 1$ for any $x \in H$ and $A \in \mathcal{B}^+(\mathsf{X})$: so Φ restricted to H is Harris recurrent, which is the required result. $\qquad\square$

We now strengthen the connection between properties of Φ and those of its skeletons.

Theorem 9.1.6. *Suppose that Φ is ψ-irreducible and aperiodic. Then Φ is Harris if and only if each skeleton is Harris.*

PROOF   If the m-skeleton is Harris recurrent then, since $m\tau_A^m \geq \tau_A$ for any $A \in \mathcal{B}(\mathsf{X})$, where τ_A^m is the first entrance time for the m-skeleton, it immediately follows that Φ is also Harris recurrent.

Suppose now that Φ is Harris recurrent. For any $m \geq 2$ we know from Proposition 8.2.6 that Φ^m is recurrent, and hence a Harris set H_m exists for this skeleton. Since H_m is full, there exists a subset $H \subset H_m$ which is absorbing and full for Φ, by Proposition 4.2.3.

Since Φ is Harris recurrent we have that $\mathsf{P}_x\{\tau_H < \infty\} \equiv 1$, and since H is absorbing we know that $m\tau_H^m \leq \tau_H + m$. This shows that

$$\mathsf{P}_x\{\tau_H^m < \infty\} = \mathsf{P}_x\{\tau_H < \infty\} \equiv 1$$

and hence Φ^m is Harris recurrent as claimed. $\qquad\square$

9.1.3   A hitting time criterion for Harris recurrence

The Harris recurrence results give useful extensions of the results in Theorem 8.3.5 and Theorem 8.3.6.

Proposition 9.1.7. *Suppose that Φ is ψ-irreducible.*

(i) *If some petite set C is recurrent, then Φ is recurrent; and the set $C \cap N$ is uniformly transient, where N is the transient set in the Harris decomposition (9.11).*

(ii) *If there exists some petite set in $\mathcal{B}(\mathsf{X})$ such that $L(x, C) \equiv 1, x \in \mathsf{X}$, then Φ is Harris recurrent.*

PROOF   **(i)**   If C is recurrent then so is the chain, from Theorem 8.3.5. Let $D = C \cap N$ denote the part of C not in H. Since N is ψ-null, and ν is an irreducibility measure we must have $\nu(N) = 0$ by the maximality of ψ; hence (8.33) holds and from (8.35) we have a uniform bound on $U(x, D), x \in \mathsf{X}$ so that D is uniformly transient.

(ii)   If $L(x, C) \equiv 1$, $x \in \mathsf{X}$ for some ψ_a-petite set C, then from Theorem 9.1.3 C is Harris recurrent. Since C is petite we have $C \rightsquigarrow A$ for each $A \in \mathcal{B}^+(\mathsf{X})$. The Harris recurrence of C, together with Theorem 9.1.3 (ii), gives $Q(x, A) \equiv 1$ for all x, so Φ is Harris recurrent. $\square$

This leads to a stronger version of Theorem 8.4.3.

Theorem 9.1.8. *Suppose Φ is a ψ-irreducible chain. If there exists a petite set $C \subset \mathsf{X}$, and a function V which is unbounded off petite sets such that (V1) holds, then Φ is Harris recurrent.*

PROOF   In Theorem 8.4.3 we showed that $L(x, C \cup C_V(n)) \equiv 1$, for some n, so Harris recurrence has already been proved in view of Proposition 9.1.7. $\square$

9.2   Non-evanescent and recurrent chains

9.2.1   Evanescence and transience

Let us now turn to chains on topological spaces. Here, as was the case when considering irreducibility, it is our major goal to delineate behavior on open sets rather than arbitrary sets in $\mathcal{B}(\mathsf{X})$; and when considering questions of stability in terms of sure return to sets, the objects of interest will typically be compact sets.

With probabilistic stability one has "finiteness" in terms of return visits to sets of positive measure of some sort, where the measure is often dependent on the chain; with topological stability the "finite" sets of interest are compact sets which are defined by the structure of the space rather than of the chain. It is obvious from the links between petite sets and compact sets for T-chains that we will be able to describe behavior on compacta directly from the behavior on petite sets described in the previous section, provided there is an appropriate continuous component for the transition law of Φ.

In this section we investigate a stability concept which provides such links between the chain and the topology on the space, and which we touched on in Section 1.3.1.

As we discussed in the introduction of this chapter, a sample path of Φ is said to converge to infinity (denoted $\Phi \to \infty$) if the trajectory visits each compact set only finitely often. Since X is locally compact and separable, it follows from Lindelöf's Theorem D.3.1 that there exists a countable collection of open precompact sets $\{O_n : n \in \mathbb{Z}_+\}$ such that

$$\{\Phi \to \infty\} = \bigcap_{n=0}^{\infty} \{\Phi \in O_n \ \text{i.o.}\}^c.$$

In particular, then, the event $\{\Phi \to \infty\}$ lies in $\mathcal{F}$.

Non-evanescent chains

A Markov chain Φ will be called *non-evanescent* if $\mathsf{P}_x\{\Phi \to \infty\} = 0$ for each $x \in \mathsf{X}$.

We first show that for a T-chain, either sample paths converge to infinity or they enter a recurrent part of the space. Recall that for any A, we have $A^0 = \{y : L(y, A) = 0\}$.

Theorem 9.2.1. *Suppose that Φ is a T-chain. For any $A \in \mathcal{B}(\mathsf{X})$ which is transient, and for each $x \in \mathsf{X}$,*

$$\mathsf{P}_x\Big\{\{\Phi \to \infty\} \cup \{\Phi \text{ enters } A^0\}\Big\} = 1. \tag{9.12}$$

Thus if Φ is a non-evanescent T-chain, then X is not transient.

PROOF   Let $A = \bigcup B_j$, with each B_j uniformly transient; then from Proposition 8.3.2, the sets $\bar{B}_i(M) = \{x \in \mathsf{X} : \sum_{j=1}^{M} P^j(x, B_i) > M^{-1}\}$ are also uniformly transient, for any i, j. Thus $\bar{A} = \bigcup A_i$ where each A_i is uniformly transient.

Since T is lower semicontinuous, the sets $O_{ij} := \{x \in \mathsf{X} : T(x, A_i) > j^{-1}\}$ are open, as is $O_j := \{x \in \mathsf{X} : T(x, A^0) > j^{-1}\}$, $i, j \in \mathbb{Z}_+$. Since T is everywhere non-trivial we have for all $x \in \mathsf{X}$,

$$T(x, (\bigcup A_j) \cup A^0) = T(x, \mathsf{X}) > 0$$

and hence the sets $\{O_{ij}, O_j\}$ form an open cover of X.

Let C be a compact subset of X, and choose M such that $\{O_M, O_{iM} : 1 \le i \le M\}$ is a finite subcover of C. Since each A_i is uniformly transient, and

$$K_a(x, A_i) \ge T(x, A_i) \ge j^{-1}, \qquad x \in O_{ij}, \tag{9.13}$$

we know from Proposition 8.3.2 that each of the sets O_{ij} is uniformly transient. It follows that with probability one, every trajectory that enters C infinitely often must enter O_M infinitely often: that is,

$$\{\Phi \in C \text{ i.o.}\} \subset \{\Phi \in O_M \text{ i.o.}\} \qquad \text{a.s. } [\mathsf{P}_*],$$

But since $L(x, A^0) > 1/M$ for $x \in O_M$ we have by Theorem 9.1.3 that

$$\{\Phi \in O_M \text{ i.o.}\} \subset \{\Phi \in A^0 \text{ i.o.}\} \qquad \text{a.s. } [\mathsf{P}_*]$$

and this completes the proof of (9.12).   □

9.2.2   Non-evanescence and recurrence

We can now prove one of the major links between topological and probabilistic stability conditions.

Theorem 9.2.2. *For a ψ-irreducible T-chain, the space admits a decomposition*

$$\mathsf{X} = H \cup N$$

where H is either empty or a maximal Harris set, and N is transient: and for all $x \in \mathsf{X}$,

$$L(x, H) = 1 - \mathsf{P}_x\{\Phi \to \infty\}. \tag{9.14}$$

Hence we have

  (i) *the chain is recurrent if and only if $\mathsf{P}_x\{\Phi \to \infty\} < 1$ for some $x \in \mathsf{X}$; and*

  (ii) *the chain is Harris recurrent if and only if the chain is non-evanescent.*

PROOF   We have the decomposition $X = H \cup N$ from Theorem 9.1.5 in the recurrent case, and Theorem 8.3.4 otherwise.

We have (9.14) from (9.12), since N is transient and $H = N^0$.

Thus if Φ is a non-evanescent T-chain, then it must leave the transient set N in (9.11) with probability one, from Theorem 9.2.1. By construction, this means N is empty, and Φ is Harris recurrent.

Conversely, if Φ is Harris recurrent (9.14) shows the chain is non-evanescent.   □

This result shows that natural definitions of stability and instability in the topological and in the probabilistic contexts are exactly equivalent, for chains appropriately adapted to the topology.

Before exploring conditions for either recurrence or non-evanescence, we look at the ways in which it is possible to classify individual states on a topological space, and the solidarity between such definitions and the overall classification of the chain which we have just described.

9.3   Topologically recurrent and transient states

9.3.1   Classifying states through neighborhoods

We now introduce some natural stochastic stability concepts for individual states when the space admits a topology. The reader should be aware that uses of terms such as "recurrence" vary across the literature. Our definitions are consistent with those we have given earlier, and indeed will be shown to be identical under appropriate conditions when the chain is an irreducible T-chain or an irreducible Feller process; however, when comparing them with some terms used by other authors, care needs to be taken.

In the general space case, we developed definitions for sets rather than individual states: when there is a topology, and hence a natural collection of sets (the open neighborhoods) associated with each point, it is possible to discuss recurrence and transience of each point even if each point is not itself reached with positive probability.

Topological recurrence concepts

We shall call a point x^* *topologically recurrent* if $U(x^*, O) = \infty$ for all neighborhoods O of x^*, and *topologically transient* otherwise.

We shall call a point x^* *topologically Harris recurrent* if $Q(x^*, O) = 1$ for all neighborhoods O of x^*.

We first determine that this definition of topological Harris recurrence is equivalent to the formally weaker version involving finiteness only of first return times.

Proposition 9.3.1. *The point x^* is topologically Harris recurrent if and only if $L(x^*, O) = 1$ for all neighborhoods O of x^*.*

PROOF   Our assumption is that

$$P_{x^*}(\tau_O < \infty) = 1, \tag{9.15}$$

for each neighborhood O of x^*. We show by induction that if $\tau_O(j)$ is the time of the jth return to O as usual, and for some integer $j \geq 1$,

$$P_{x^*}(\tau_O(j) < \infty) = 1, \tag{9.16}$$

for each neighborhood O of x^*, then for each such neighborhood

$$P_{x^*}(\tau_O(j+1) < \infty) = 1. \tag{9.17}$$

Thus (9.17) holds for all j and the point x^* is by definition topologically Harris recurrent.

Recall that for any $B \subset O$ we have the following probabilistic interpretation of the kernel U_O:

$$U_O(x^*, B) = P_{x^*}(\tau_O < \infty \quad \text{and} \quad \Phi_{\tau_O} \in B).$$

Suppose that $U_O(x^*, \{x^*\}) = q \geq 0$ where $\{x^*\}$ is the set containing the one point x^*, so that

$$U_O(x^*, O\backslash\{x^*\}) = 1 - q. \tag{9.18}$$

The assumption that j distinct returns to O are sure implies that

$$P_{x^*}(\Phi_{\tau_O(1)} = x^*, \Phi_{\tau_O(r)} \in O, r = 2, \dots, j+1) = q. \tag{9.19}$$

Let $O_d \downarrow \{x^*\}$ be a countable neighborhood basis at x^*. The assumption (9.16) applied to each O_d also implies that

$$P_y(\tau_{O_d}(j) < \infty) = 1, \tag{9.20}$$

for almost all y in $O\backslash O_d$ with respect to $U_O(x^*, \cdot)$. But by (9.18) we have

$$U_O(x^*, O\backslash O_d) \uparrow 1 - q,$$

as $O_d \downarrow \{x^*\}$ and so by (9.20),

$$\int_{O\backslash\{x^*\}} U_O(x, dy) P_y(\tau_O(j) < \infty) \geq \lim_{d\downarrow 0} \int_{O\backslash O_d} U_O(x^*, dy) P_y(\tau_{O_d}(j) < \infty)$$
$$= 1 - q. \tag{9.21}$$

This yields the desired conclusion, since by (9.19) and (9.21),

$$P_{x^*}(\tau_O(j+1) < \infty) = \int_O U_O(x^*, dy) P_y(\tau_O(j) < \infty) = 1.$$

$\square$

9.3.2   Solidarity of recurrence for T-chains

For T-chains we can connect the idea of properties of individual states with the properties of the whole space under suitable topological irreducibility conditions.

The key to much of our analysis of chains on topological spaces is the following simple lemma.

Lemma 9.3.2. *If Φ is a T-chain, and $T(x^*, B) > 0$ for some x^*, B, then there is a neighborhood O of x^* and a distribution a such that $O \overset{a}{\rightsquigarrow} B$, and hence from Lemma 5.5.1, $O \rightsquigarrow B$.*

PROOF   Since $\boldsymbol{\Phi}$ is a T-chain, there exists some distribution a such that for all x,

$$K_a(x, B) \geq T(x, B).$$

But since $T(x^*, B) > 0$ and $T(x, B)$ is lower semicontinuous, it follows that for some neighborhood O of x^*,

$$\inf_{x \in O} T(x, B) > 0$$

and thus, as in (5.45),

$$\inf_{x \in O} L(x, B) \geq \inf_{x \in O} K_a(x, B) \geq \inf_{x \in O} T(x, B)$$

and the result is proved.   □

Theorem 9.3.3. *Suppose that $\boldsymbol{\Phi}$ is a ψ-irreducible T-chain, and that x^* is reachable. Then $\boldsymbol{\Phi}$ is recurrent if and only if x^* is topologically recurrent.*

PROOF   If x^* is reachable then $x^* \in \operatorname{supp} \psi$ and so $O \in \mathcal{B}^+(\mathsf{X})$ for every neighborhood of x^*. Thus if $\boldsymbol{\Phi}$ is recurrent then every neighborhood O of x^* is recurrent, and so by definition x^* is topologically recurrent.

If $\boldsymbol{\Phi}$ is transient then there exists a uniformly transient set B such that $T(x^*, B) > 0$, from Theorem 8.3.4, and thus from Lemma 9.3.2 there is a neighborhood O of x^* such that $O \rightsquigarrow B$; and now from Proposition 8.3.2, O is uniformly transient and thus x^* is topologically transient also.   □

We now work towards developing links between topological recurrence and topological Harris recurrence of points, as we did with sets in the general space case.

It is unfortunately easy to construct an example which shows that even for a T-chain, topologically recurrent states need not be topologically Harris recurrent without some extra assumptions. Take $\mathsf{X} = [0, 1] \cup \{2\}$, and define the transition law for $\boldsymbol{\Phi}$ by

$$
\begin{aligned}
P(0, \cdot) &= (\mu + \delta_2)/2, \\
P(x, \cdot) &= \mu, \qquad x \in (0, 1], \\
P(2, \cdot) &= \delta_2,
\end{aligned}
\tag{9.22}
$$

where μ is Lebesgue measure on $[0, 1]$ and δ_2 is the point mass at $\{2\}$. Set the everywhere non-trivial continuous component T of P itself as

$$
\begin{aligned}
T(x, \cdot) &= \mu/2, \qquad x \in [0, 1], \\
T(2, \cdot) &= \delta_2.
\end{aligned}
\tag{9.23}
$$

By direct calculation one can easily see that $\{0\}$ is a topologically recurrent state but is not topologically Harris recurrent.

It is also possible to develop examples where the chain is weak Feller but topological recurrence does not imply topological Harris recurrence of states.

Let $\mathsf{X} = \{0, \pm 1, \pm 2, \ldots, \pm \infty\}$, and choose $0 < p < \frac{1}{2}$ and $q = 1 - p$. Put $P(0, 1) = p, P(0, -1) = q$, and for $n = 1, 2, \ldots$, set

$$
\begin{array}{llll}
P(n, n+1) = p & P(n, n-1) = q & & \\
P(-n, -n-1) = p & P(-n, 0) = \tfrac{1}{2} - p & P(-n, n) = \tfrac{1}{2} & \\
P(-\infty, -\infty) = p & P(-\infty, 0) = \tfrac{1}{2} - p & P(-\infty, \infty) = \tfrac{1}{2} & \\
P(\infty, \infty) = 1. & & &
\end{array}
\tag{9.24}
$$

By comparison with a simple random walk, such as analyzed in Proposition 8.4.4, it is clear that the finite integers are all recurrent states in the countable state space sense.

Now endow the space X with the discrete topology on the integers, and with a countable basis for the neighborhoods at $\infty, -\infty$ given respectively by the two sets $\{n, n+1, \ldots, \infty\}$ and $\{-n, -n-1, \ldots, -\infty\}$ for $n \in \mathbb{Z}_+$. The chain is a Feller chain in this topology, and every neighborhood of $-\infty$ is recurrent so that $-\infty$ is a topologically recurrent state.

But $L(-\infty, \{-\infty, -1\}) < \frac{1}{2}$, so the state at $-\infty$ is not topologically Harris recurrent.

There are however some connections which do hold between recurrence and Harris recurrence.

Proposition 9.3.4. *If Φ is a T-chain and the state x^* is topologically recurrent then $Q(x^*, O) > 0$ for all neighborhoods O of x^*.*

If $P(x^*, \cdot) \cong T(x^*, \cdot)$ then also x^* is topologically Harris recurrent. In particular, therefore, for strong Feller chains topologically recurrent states are topologically Harris recurrent.

PROOF   **(i)**   Assume the state x^* is topologically recurrent but that O is a neighborhood of x^* with $Q(x^*, O) = 0$. Let $O^\infty = \{y : Q(y, O) = 1\}$, so that $L(x^*, O^\infty) = 0$. Since

$$L(x, A) \geq K_a(x, A) \geq T(x, A), \qquad x \in \mathsf{X}, \quad A \in \mathcal{B}(\mathsf{X})$$

this implies $T(x^*, O^\infty) = 0$, and since T is non-trivial, we must have

$$T(x^*, [O^\infty]^c) > 0. \tag{9.25}$$

Let $D_n := \{y : \mathsf{P}_y(\eta_O < n) > n^{-1}\}$: since $D_n \uparrow [O^\infty]^c$, we must have $T(x^*, D_n) > 0$ for some n. The continuity of T now ensures that there exists some δ and a neighborhood $O_\delta \subseteq O$ of x^* such that

$$T(x, D_n) > \delta, \qquad x \in O_\delta. \tag{9.26}$$

Let us take m large enough that $\sum_m^\infty a(j) \leq \delta/2$: then from (9.26) we have

$$\max_{1 \leq j \leq m} P^j(x, D_n) > \delta/2m, \qquad x \in O_\delta, \tag{9.27}$$

which obviously implies

$$\mathsf{P}_x(\tau_{D_n} \leq m) > \delta/2m, \qquad x \in O_\delta. \tag{9.28}$$

It follows that

$$
\begin{aligned}
\mathsf{P}_x(\eta_{O_\delta} \leq m+n) &\geq \mathsf{P}_x(\eta_O \leq m+n) \\
&\geq \textstyle\sum_1^m \int_{D_n} D_n P^k(x, dy) \mathsf{P}_y(\eta_O \leq n) \\
&\geq n^{-1} \mathsf{P}(\tau_{D_n} \leq m) \\
&\geq n^{-1}\delta/2m, \qquad x \in O_\delta.
\end{aligned}
\tag{9.29}
$$

With (9.29) established we can apply Proposition 8.3.1 to see that O_δ is uniformly transient.

This contradicts our assumption that x^* is topologically recurrent, and so in fact $Q(x^*, O) > 0$ for all neighborhoods O.

(ii)   Suppose now that $P(x^*, \cdot)$ and $T(x^*, \cdot)$ are equivalent. Choose x^* topologically recurrent and assume we can find a neighborhood O with $Q(x^*, O) < 1$. Define O^∞ as before, and note that now $P(x^*, [O^\infty]^c) > 0$ since otherwise

$$Q(x^*, O) \geq \int_{O^\infty} P(x^*, dy) Q(y, O) = 1;$$

and so also $T(x^*, [O^\infty]^c) > 0$. Thus we again have (9.25) holding, and the argument in (i) shows that there is a uniformly transient neighborhood of x^*, again contradicting the assumption of topological recurrence. Hence x^* is topologically Harris recurrent.   □

The examples (9.22) and (9.24) show that we do not get, in general, the second conclusion of this proposition if the chain is merely weak Feller or has only a strong Feller component.

In these examples, it is the lack of irreducibility which allows such obvious "pathological" behavior, and we shall see in Theorem 9.3.6 that when the chain is a ψ-irreducible T-chain then this behavior is excluded. Even so, without any irreducibility assumptions we are able to derive a reasonable analogue of Theorem 9.1.5, showing that the non-Harris recurrent states form a transient set.

Theorem 9.3.5. *For any chain Φ there is a decomposition*

$$\mathsf{X} = R \cup N,$$

where R denotes the set of states which are topologically Harris recurrent and N is transient.

PROOF   Let O_i be a countable basis for the topology on X. If $x \in R^c$ then, by Proposition 9.3.1, we have some $n \in \mathbb{Z}_+$ such that $x \in O_n$ with $L(x, O_n) < 1$. Thus the sets $D_n = \{y \in O_n : L(y, O_n) < 1\}$ cover the set of non-topologically Harris recurrent states. We can further partition each D_n into

$$D_n(j) := \{y \in D_n : L(y, O_n) \leq 1 - j^{-1}\}$$

and by this construction, for $y \in D_n(j)$, we have

$$L(y, D_n(j)) \leq L(y, D_n) \leq L(y, O_n) \leq 1 - j^{-1} :$$

it follows from Proposition 8.3.1 that $U(x, D_n(j))$ is bounded above by j, and hence is uniformly transient.   □

Regrettably, this decomposition does not partition X into Harris recurrent and transient states, since the sets $D_n(j)$ in the cover of non-Harris states may not be open. Therefore there may actually be topologically recurrent states which lie in the set which we would hope to have as the "transient" part of the space, as happens in the example (9.22).

We can, for ψ-irreducible T-chains, now improve on this result to round out the links between the Harris properties of points and those of the chain itself.

Theorem 9.3.6. *For a ψ-irreducible T-chain, the space admits a decomposition*

$$\mathsf{X} = H \cup N$$

where H is non-empty or a maximal Harris set and N is transient; the set of Harris recurrent states R is contained in H; and every state in N is topologically transient.

PROOF   The decomposition has already been shown to exist in Theorem 9.2.2. Let $x^* \in R$ be a topologically Harris recurrent state. Then from (9.14), we must have $L(x, H) = 1$, and so $x^* \in H$ by maximality of H.

We can write $N = N_E \cup N_H$ where $N_H = \{y \in N : T(y, H) > 0\}$ and $N_E = \{y \in N : T(y, H) = 0\}$. For fixed $x^* \in N_H$ there exists $\delta > 0$ and an open set O_δ such that $x^* \in O_\delta$ and $T(y, H) > \delta$ for all $y \in O_\delta$, by the lower semicontinuity of $T(\,\cdot\,, H)$.

Hence also the sampled kernel K_a minorized by T satisfies $K_a(y, H) > \delta$ for all $y \in O_\delta$. Now choose M such that $\sum_{n>M} a(n) \le \delta/2$. Then for all $y \in O_\delta$

$$\sum_{n \le M} P^n(y, H) a(n) \ge \delta/2,$$

and since H is absorbing

$$\mathsf{P}_y(\eta_N > M) = \mathsf{P}_y(\tau_H > M) \le 1 - \delta/2,$$

which shows that O_δ is uniformly transient from (8.35).

If on the other hand $x^* \in N_E$ then since T is non-trivial, there exists a uniformly transient set $D \subseteq N$ such $T(x^*, D) > 0$; and now by Lemma 9.3.2, there is again a neighbourhood O of x^* with $O \overset{a}{\leadsto} D$, so that O is uniformly transient by Proposition 8.3.2 as required. $\Box$

The maximal Harris set in Theorem 9.3.6 may be strictly larger than the set R of topologically Harris recurrent states. For consider the trivial example where $\mathsf{X} = [0, 1]$ and $P(x, \{0\}) = 1$ for all x. This is a δ_0-irreducible strongly Feller chain, with $R = \{0\}$ and yet $H = [0, 1]$.

9.4   Criteria for stability on a topological space

9.4.1   A drift criterion for non-evanescence

We can extend the results of Theorem 8.4.3 in a number of ways if we take up the obvious martingale implications of (V1), and in the topological case we can also gain a better understanding of the rather inexplicit concept of functions unbounded off petite sets for a particular chain if we define "coercive" functions.

Coercive functions

A function V is called *coercive* if $V(x) \to \infty$ as $x \to \infty$: this means that the sublevel sets $\{x : V(x) \le r\}$ are precompact for each $r > 0$.

This nomenclature is designed to remind the user that we seek functions which behave like norms: they are large as the distance from the center of the space increases. Typically in practice, a coercive function will be a norm on Euclidean space, or at least a monotone function of a norm. For irreducible T-chains, functions unbounded off petite sets certainly include coercive functions, since compacta are petite in that case; but of course coercive functions are independent of the structure of the chain itself.

Even without irreducibility we get a useful conclusion from applying (V1).

Theorem 9.4.1. *If condition (V1) holds for a coercive function V and a compact set C, then Φ is non-evanescent.*

PROOF   Suppose that in fact $P_x\{\Phi \to \infty\} > 0$ for some $x \in X$. Then, since the set C is compact, there exists $M \in \mathbb{Z}_+$ with

$$P_x\{\{\Phi_k \in C^c, k \geq M\} \cap \{\Phi \to \infty\}\} > 0.$$

Hence letting $\mu = P^M(x, \cdot)$, we have by conditioning at time M,

$$P_\mu\{\{\sigma_C = \infty\} \cap \{\Phi \to \infty\}\} > 0. \tag{9.30}$$

We now show that (9.30) leads to a contradiction.

In order to use the martingale nature of (V1), we write (8.42) as

$$\mathsf{E}[V(\Phi_{k+1}) \mid \mathcal{F}_k^\Phi] \leq V(\Phi_k) \qquad \text{a.s. } [\mathsf{P}_*],$$

when $\sigma_C > k$, $k \in \mathbb{Z}_+$.

Now let $M_i = V(\Phi_i)\mathbb{I}\{\sigma_C \geq i\}$. Using the fact that $\{\sigma_C \geq k\} \in \mathcal{F}_{k-1}^\Phi$, we may show that $(M_k, \mathcal{F}_k^\Phi)$ is a positive supermartingale: indeed,

$$\mathsf{E}[M_k \mid \mathcal{F}_{k-1}^\Phi] = \mathbb{I}\{\sigma_C \geq k\}\mathsf{E}[V(\Phi_k) \mid \mathcal{F}_{k-1}^\Phi] \leq \mathbb{I}\{\sigma_C \geq k\}V(\Phi_{k-1}) \leq M_{k-1}.$$

Hence there exists an almost surely finite random variable M_∞ such that $M_k \to M_\infty$ as $k \to \infty$.

There are two possibilities for the limit M_∞. Either $\sigma_C < \infty$ in which case $M_\infty = 0$, or $\sigma_C = \infty$ in which case $\limsup_{k\to\infty} V(\Phi_k) = M_\infty < \infty$ and in particular $\Phi \not\to \infty$ since V is coercive. Thus we have shown that

$$P_\mu\{\{\sigma_C < \infty\} \cup \{\Phi \to \infty\}^c\} = 1,$$

which clearly contradicts (9.30). Hence Φ is non-evanescent.   □

Note that in general the set C used in (V1) is not necessarily Harris recurrent, and it is possible that the set may not be reached from any initial condition. Consider the example where $X = \mathbb{R}_+$, $P(0, \{1\}) = 1$, and $P(x, \{x\}) \equiv 1$ for $x > 0$. This is non-evanescent, satisfies (V1) with $V(x) = x$, and $C = \{0\}$, but clearly from x there is no possibility of reaching compacta not containing $\{x\}$.

However, from our previous analysis in Theorem 9.1.8 we obviously have that if Φ is ψ-irreducible and condition (V1) holds for C petite, then both C and Φ are Harris recurrent.

9.4.2 A converse theorem for Feller chains

In the topological case we can construct a converse to the drift condition (V1), provided the chain has appropriate continuity properties.

Theorem 9.4.2. *Suppose that Φ is a weak Feller chain, and suppose that there exists a compact set C satisfying $\sigma_C < \infty$ a.s. $[\mathsf{P}_*]$.*

Then there exists a compact set C_0 containing C and a coercive function V, bounded on compacta, such that

$$\Delta V(x) \leq 0, \qquad x \in C_0^c. \tag{9.31}$$

PROOF   Let $\{A_n\}$ be a countable increasing cover of X by open pre-compact sets with $C \subseteq A_0$; and put $D_n = A_n^c$ for $n \in \mathbb{Z}_+$. For $n \in \mathbb{Z}_+$, set

$$V_n(x) = \mathsf{P}_x(\sigma_{D_n} < \sigma_{A_0}). \tag{9.32}$$

For any fixed n and any $x \in A_0^c$ we have from the Markov property that the sequence $V_n(x)$ satisfies, for $x \in A_0^c \cap D_n^c$

$$
\begin{aligned}
\int P(x, dy) V_n(y) &= \mathsf{E}_x[\mathsf{P}_{\Phi_1}\{\sigma_{D_n} < \sigma_{A_0}\}] \\
&= \mathsf{P}_x\{\sigma_{D_n} < \sigma_{A_0}\} \\
&= V_n(x),
\end{aligned}
\tag{9.33}
$$

whilst for $x \in D_n$ we have $V_n(x) = 1$; so that for all $n \in \mathbb{Z}_+$ and $x \in A_0^c$

$$\int P(x, dy) V_n(y) \leq V_n(x). \tag{9.34}$$

We will show that for suitably chosen $\{n_i\}$ the function

$$V(x) = \sum_{i=0}^{\infty} V_{n_i}(x), \tag{9.35}$$

which clearly satisfies the appropriate drift condition by linearity from (9.34) if finitely defined, gives the required converse result.

Since $V_n(x) = 1$ on D_n, it is clear that V is coercive. To complete the proof we must show that the sequence $\{n_i\}$ can be chosen to ensure that V is bounded on compact sets, and it is for this we require the Feller property.

Let $m \in \mathbb{Z}_+$ and take the upper bound

$$
\begin{aligned}
V_n(x) &= \mathsf{P}_x\{\{\sigma_{D_n} < \sigma_{A_0}\} \cap \{\sigma_{A_0} \leq m\} \cup \{\sigma_{D_n} < \sigma_{A_0}\} \cap \{\sigma_{A_0} > m\}\} \\
&\leq \mathsf{P}_x\{\sigma_{D_n} < m\} + \mathsf{P}_x\{\sigma_{A_0} > m\}.
\end{aligned}
\tag{9.36}
$$

Choose the sequence $\{n_i\}$ as follows. By Proposition 6.1.1, the function $\mathsf{P}_x\{\sigma_{A_0} > m\}$ is an upper semicontinuous function of x, which converges to zero as $m \to \infty$ for all x. Hence the convergence is uniform on compacta, and thus we can choose m_i so large that

$$\mathsf{P}_x\{\sigma_{A_0} > m_i\} < 2^{-(i+1)}, \qquad x \in A_i. \tag{9.37}$$

Now for m_i fixed for each i, consider $\mathsf{P}_x\{\sigma_{D_n} < m_i\}$: as a function of x this is also upper semicontinuous and converges to zero as $n \to \infty$ for all x. Hence again we see that the convergence is uniform on compacta, which implies we may choose n_i so large that

$$\mathsf{P}_x\{\sigma_{D_{n_i}} < m_i\} < 2^{-(i+1)}, \qquad x \in A_i. \tag{9.38}$$

Combining (9.36), (9.37) and (9.38) we see that $V_{n_i} \le 2^{-i}$ for $x \in A_i$. From (9.35) this implies, finally, for all $k \in \mathbb{Z}_+$ and $x \in A_k$

$$\begin{aligned} V(x) &\le & k + \sum_{i=k}^{\infty} V_{n_i}(x) \\ &\le & k + \sum_{i=k}^{\infty} 2^{-i} \\ &\le & k+1, \end{aligned} \tag{9.39}$$

which completes the proof.   □

The following somewhat pathological example shows that in this instance we cannot use a strongly continuous component condition in place of the Feller property if we require V to be continuous.

Set $\mathsf{X} = \mathbb{R}_+$ and for every irrational x and every integer x set $P(x, \{0\}) = 1$. Let $\{r_n\}$ be an ordering of the remaining rationals $\mathbb{Q}\setminus\mathbb{Z}_+$, and define P for these states by $P(r_n, 0) = 1/2$, $P(r_n, n) = 1/2$. Then the chain is δ_0-irreducible, and clearly recurrent; and the component $T(x, A) = \frac{1}{2}\delta_0\{A\}$ renders the chain a T-chain. But $PV(r_n) \ge V(n)/2$, so that for any coercive function V, within any open set $\int P(x, dy)V(y)$ is unbounded.

However, for discontinuous V we do get a coercive test function: just take $V(r_n) = n$, and $V(x) = x$, for x not equal to any r_n. Then $PV(r_n) = n/2 < V(r_n)$, and $PV(x) = 0 < V(x)$, for x not equal to any r_n, so that (V1) does hold.

9.4.3 Non-evanescence of random walk

As an example of the use of (V1) we consider in more detail the analysis of the unrestricted random walk

$$\Phi_n = \Phi_{n-1} + W_n.$$

We will show that if W is an increment variable on $\mathbb{R}$ with $\beta = 0$ and

$$\mathsf{E}(W^2) = \int w^2\, \Gamma(dw) < \infty,$$

then the unrestricted random walk on $\mathbb{R}$ with increment W is non-evanescent.

To verify this using (V1) we first need to add to the bounds on the moments of Γ which we gave in Lemma 8.5.2 and Lemma 8.5.3.

Lemma 9.4.3. *Let W be a random variable, s a positive number and t any real number. Then for any $B \subseteq \{w : -s + tw > 0\}$,*

$$\mathsf{E}[\log(-s+tW)\mathbb{I}\{W \in B\}] \le \mathsf{P}(B)(\log(s) - 2) + (t/s)\mathsf{E}[W\mathbb{I}\{W \in B\}].$$

PROOF   For all $x > 1$, $\log(-1 + x) \leq x - 2$. Thus

$$\begin{aligned}
\log(-s + tW)\mathbb{I}\{W \in B\} &= [\log(s) + \log(-1 + tW/s)]\mathbb{I}\{W \in B\} \\
&\leq (\log(s) + tW/s - 2)\mathbb{I}\{W \in B\};
\end{aligned}$$

taking expectations again gives the result.   □

Lemma 9.4.4. *Let W be a random variable with distribution function Γ and finite variance. Let s, c, u_2, and v_2 be positive numbers, and let $t_1 \geq t_2$ and u_1, v_1, t be real numbers. Then*

(i)

$$\lim_{x \to -\infty} x^2[-\Gamma(-\infty, t_1 + sx)\log(u_1 - u_2x) + \Gamma(-\infty, t_2 + sx)(\log(v_1 - v_2x) - c)] \leq 0. \tag{9.40}$$

(ii)

$$\lim_{x \to \infty} x^2[-\Gamma(t_2 + sx, \infty)\log(v_1 + v_2x) + \Gamma(t_1 + sx, \infty)(\log(u_1 + u_2x) - c)] \leq 0. \tag{9.41}$$

PROOF   To see (i), note that from

$$\lim_{x \to \infty} x^2\Gamma(-\infty, t_2 + sx) = 0$$

and

$$\lim_{x \to \infty} \log[(u_1 - u_2x)/(v_1 - v_2x)] = \log(u_2/v_2),$$

we have

$$\begin{aligned}
\lim_{x \to \infty} x^2 &\Big[-\Gamma(-\infty, t_1 + sx)\log(u_1 - u_2x) + \Gamma(-\infty, t_2 + sx)(\log(v_1 - v_2x) - c)\Big] \\
&= \lim_{x \to \infty} \Big[-x^2(\Gamma(-\infty, t_1 + sx) - \Gamma(-\infty, t_2 + sx))\log(u_1 - u_2x)\Big] \\
&\quad \times \Big[-x^2\Gamma(-\infty, t_2 + sx)\log[(u_1 - u_2x)/(v_1 - v_2x)] - cx^2\Gamma(-\infty, t_2 + sx)\Big]
\end{aligned}$$

which is non-positive. The proof of (ii) is similar.   □

We can now prove the most general version of Theorem 8.1.5 using a drift condition that we shall attempt.

Proposition 9.4.5. *If W is an increment variable on $\mathbb{R}$ with $\beta = 0$ and $\mathsf{E}(W^2) < \infty$, then the unrestricted random walk on $\mathbb{R}_+$ with increment W is non-evanescent.*

PROOF   In this situation we use the test function

$$V(x) = \begin{cases} \log(1 + x) & x > R \\ \log(1 - x) & x < -R \end{cases} \tag{9.42}$$

and $V(x) = 0$ in the region $[-R, R]$, where $R > 1$ is again a positive constant to be chosen.

We need to evaluate the behavior of $\mathsf{E}_x[V(X_1)]$ near both ∞ and $-\infty$ in this case, and we write

$$
\begin{aligned}
V_1(x) &= \mathsf{E}_x[\log(1 + x + W)\mathbb{I}\{x + W > R\}] \\
V_2(x) &= \mathsf{E}_x[\log(1 - x - W)\mathbb{I}\{x + W < -R\}]
\end{aligned}
$$

so that

$$
\mathsf{E}_x[V(X_1)] = V_1(x) + V_2(x).
$$

This time we develop bounds using the functions

$$
\begin{aligned}
V_3(x) &= (1/(1 + x))\mathsf{E}[W\mathbb{I}\{W > R - x\}] \\
V_4(x) &= (1/(2(1 + x)^2))\mathsf{E}[W^2\mathbb{I}\{R - x < W < 0\}] \\
V_5(x) &= (1/(1 - x))\mathsf{E}[W\mathbb{I}\{W < -R - x\}].
\end{aligned}
$$

For $x > R, 1 + x > 0$, and thus as in (8.59), by Lemma 8.5.2,

$$
V_1(x) \leq \Gamma(R - x, \infty)\log(1 + x) + V_3(x) - V_4(x),
$$

while $1 - x < 0$, and by Lemma 9.4.3,

$$
V_2(x) \leq \Gamma(-\infty, -R - x)(\log(-1 + x) - 2) - V_5(x).
$$

Since $\mathsf{E}(W^2) < \infty$,

$$
V_4(x) = (1/(2(1 + x)^2))\mathsf{E}[W^2\mathbb{I}\{W < 0\}] - o(x^{-2}),
$$

and by Lemma 8.5.3, both V_3 and V_5 are also $o(x^{-2})$. By Lemma 9.4.4 (i) we also have

$$
-\Gamma(-\infty, R - x)\log(1 + x) + \Gamma(-\infty, -R - x)(\log(-1 + x) - 2) \leq o(x^{-2}).
$$

Thus by choosing R large enough

$$
\begin{aligned}
\mathsf{E}_x[V(X_1)] &\leq V(x) - (1/(2(1 + x)^2))\mathsf{E}[W^2\mathbb{I}\{W < 0\}] + o(x^{-2}) \\
&\leq V(x), \qquad x > R.
\end{aligned} \tag{9.43}
$$

The situation with $x < -R$ is exactly symmetric, and thus we have that V is a coercive function satisfying (V1); and so the chain is non-evanescent from Theorem 9.4.1. $\qquad\square$

9.5   Stochastic comparison and increment analysis

There are two further valuable tools for analyzing specific chains which we will consider in this final section on recurrence and transience. Both have been used implicitly in some of the examples we have looked at in this and the previous chapter, but because they are of wide applicability we will discuss them somewhat more formally here.

The first method analyzes chains through an "increment analysis". Because they consider only expected changes in the one-step position of some function V of the chain, and because expectation is a linear operator, drift criteria such as those in Section 9.4 essentially classify the behavior of the Markov model by a linearization of its increments. They are therefore often relatively easy to use for models where the transitions are already somewhat linear in structure, such as those based on the random walk: we have already seen this in our analysis of random walk on the half line in Section 8.4.3.

Such increment analysis is of value in many models, especially if combined with "stochastic comparison" arguments, which rely heavily on the classification of chains through return time probabilities.

In this section we will further use the stochastic comparison approach to discuss the structure of scalar linear models and general random walk on $\mathbb{R}$, and the special nonlinear SETAR models; we will then consider an increment analysis of general models on $\mathbb{R}_+$ which have no inherent linearity in their structure.

9.5.1   Linear models and the stochastic comparison technique

Suppose we have two φ-irreducible chains $\boldsymbol{\Phi}$ and $\boldsymbol{\Phi}'$ evolving on a common state space, and that for some set C and for all n

$$\mathsf{P}_x(\tau_C \geq n) \leq \mathsf{P}'_x(\tau_C \geq n), \qquad x \in C^c. \tag{9.44}$$

This is not uncommon if the chains have similarly defined structure, as is the case with random walk and the associated walk on a half line.

The stochastic comparison method tells us that a classification of one of the chains may automatically classify the other.

In one direction we have, provided C is a petite set for both chains, that when $\mathsf{P}'_x(\tau_C \geq n) \to 0$ as $n \to \infty$ for $x \in C^c$, then not only is $\boldsymbol{\Phi}'$ Harris recurrent, but $\boldsymbol{\Phi}$ is also Harris recurrent.

This is obvious. Its value arises in cases where the first chain $\boldsymbol{\Phi}'$ has a (relatively) simpler structure so that its analysis is straightforward through, say, drift conditions, and when the validation of (9.44) is also relatively easy.

In many ways stochastic comparison arguments are even more valuable in the transient context: as we have seen with random walk, establishing transience may need a rather delicate argument, and it is then useful to be able to classify "more transient" chains easily.

Suppose that (9.44) holds, and again that C is a φ-irreducible petite set for both chains. Then if $\boldsymbol{\Phi}$ is transient, we know that from Theorem 8.3.6 that there exists $D \subset C^c$ such that $L(x, C) < 1 - \varepsilon$ for $x \in D$ where $\varphi(D) > 0$; it then follows that $\boldsymbol{\Phi}'$ is also transient.

We first illustrate the strengths and drawbacks of this method in proving transience for the general random walk on the half line $\mathbb{R}_+$.

Proposition 9.5.1. *If $\boldsymbol{\Phi}$ is random walk on $\mathbb{R}_+$ and if $\beta > 0$ then $\boldsymbol{\Phi}$ is transient.*

PROOF   Consider the discretized version W_h of the increment variable W with distribution

$$\mathsf{P}(W_h = nh) = \Gamma_h(nh)$$

where $\Gamma_h(nh)$ is constructed by setting, for every n,

$$\Gamma_h(nh) = \int_{nh}^{(n+1)h} \Gamma(dw),$$

and let $\boldsymbol{\Phi}_h$ be the corresponding random walk on the countable half line $\{nh, n \in \mathbb{Z}_+\}$. Then we have firstly that for any starting point nh, the chain $\boldsymbol{\Phi}_h$ is "stochastically smaller" than $\boldsymbol{\Phi}$, in the sense that if τ_0^h is the first return time to zero by $\boldsymbol{\Phi}_h$ then

$$\mathsf{P}_0(\tau_0^h \leq k) \geq \mathsf{P}_0(\tau_0 \leq k).$$

Hence $\boldsymbol{\Phi}$ is transient if $\boldsymbol{\Phi}_h$ is transient.

But now we have that

$$\begin{aligned} \beta_h := \textstyle\sum_n nh\, \Gamma_h(nh) \;\; &\geq \;\; \textstyle\sum_n \int_{nh}^{(n+1)h}(w-h)\Gamma(dw) \\ &= \;\; \int(w-h)\Gamma(dw) \\ &= \;\; \beta - h \end{aligned} \tag{9.45}$$

so that if $h < \beta$ then $\beta_h > 0$.

Finally, for such sufficiently small h we have that the chain $\boldsymbol{\Phi}_h$ is transient from Proposition 9.1.2, as required. $\square$

Let us next consider the use of stochastic comparison methods for the scalar linear model

$$X_n = \alpha X_{n-1} + W_n.$$

Proposition 9.5.2. *Suppose the increment variable W in the scalar linear model is symmetric with density positive everywhere on $[-R, R]$ and zero elsewhere. Then the scalar linear model is Harris recurrent if and only if $|\alpha| \leq 1$.*

PROOF   The linear model is, under the conditions on W, a μ^{Leb}-irreducible chain on $\mathbb{R}$ with all compact sets petite.

Suppose $\alpha > 1$. By stochastic comparison of this model with a random walk $\boldsymbol{\Phi}$ on a half line with mean increment $\alpha - 1$ it is obvious that provided the starting point $x > 1$, then (9.44) holds with $C = (-\infty, 1]$. Since this set is transient for the random walk, as we have just shown, it must therefore be transient for the scalar linear model. Provided the starting point $x < -1$, then by symmetry, the hitting times on the set $C = [-1, \infty)$ are also infinite with positive probability. This argument does not require bounded increments.

If $\alpha < -1$ then the chain oscillates. If the range of W is contained in $[-R, R]$, with $R > 1$, then by choosing $x > R$ we have by symmetry that the hitting time of the chain $X_0, -X_1, X_2, -X_3, \ldots$ on $C = (-\infty, 1]$ is stochastically bounded below by the hitting time of the previous linear model with parameter $|\alpha|$; thus the set $[-R, R]$ is uniformly transient for both models.

Thirdly, suppose that the $0 < \alpha \leq 1$. Then by stochastic comparison with random walk on a half line and mean increment $\alpha - 1$, from $x > R$ we have that the hitting time on $[-R, R]$ of the linear model is bounded above by the hitting time on $[-R, R]$ of the random walk; whilst by symmetry the same is true from $x < -R$. Since we know random walk is Harris recurrent it follows that the linear model is Harris recurrent.

Finally, by considering an oscillating chain we have the same recurrence result for $-1 \leq \alpha \leq 0$. $\square$

The points to note in this example are

(i) without some bounds on W, in general it is difficult to get a stochastic comparison argument for transience to work on the whole real line: on a half line, or equivalently if $\alpha > 0$, the transience argument does not need bounds, but if the chain can oscillate then usually there is insufficient monotonicity to exploit in sample paths for a simple stochastic comparison argument to succeed;

(ii) even with $\alpha > 0$, recurrence arguments on the whole line are also difficult to get to work. They tend to guarantee that the hitting times on half lines such as $C = (-\infty, 1]$ are finite, and since these sets are not compact, we do not have a guarantee of recurrence: indeed, for transient oscillating linear systems such half lines are reached on alternate steps with higher and higher probability.

Thus in the case of unbounded increments more delicate arguments are usually needed, and we illustrate one such method of analysis next.

9.5.2   Unrestricted random walk and SETAR models

Consider next the unrestricted random walk on $\mathbb{R}$ given by

$$\Phi_n = \Phi_{n-1} + W_n.$$

This is easy to analyze in the transient situation using stochastic comparison arguments, given the results already proved.

Proposition 9.5.3. *If the mean increment of an irreducible random walk on $\mathbb{R}$ is non-zero, then the walk is transient.*

PROOF   Suppose that the mean increment of the random walk $\boldsymbol{\Phi}$ is positive. Then the hitting time $\tau_{\{-\infty,0\}}$ on $\{-\infty, 0\}$ from an initial point $x > 0$ is the same as the hitting time on $\{0\}$ itself for the associated random walk on the half line; and we have shown this to be infinite with positive probability. So the unrestricted walk is also transient.

The argument if $\beta < 0$ is clearly symmetric.   □

This model is non-evanescent when $\beta = 0$, as we showed under a finite variance assumption in Proposition 9.4.5.

Now let us consider the more complex SETAR model

$$X_n = \phi(j) + \theta(j)X_{n-1} + W_n(j), \qquad X_{n-1} \in R_j,$$

where $-\infty = r_0 < r_1 < \cdots < r_M = \infty$ and $R_j = (r_{j-1}, r_j]$; recall that for each j, the noise variables $\{W_n(j)\}$ form independent zero-mean noise sequences, and again let $W(j)$ denote a generic variable in the sequence $\{W_n(j)\}$, with distribution Γ_j.

We will see in due course that under a second-order moment condition (SETAR3), we can identify exactly the regions of the parameter space where this nonlinear chain is transient, recurrent and so on.

Here we establish the parameter combinations under which transience will hold: these are extensions of the non-zero mean increment regions of the random walk we have just looked at.

As suggested by Figure B.1–Figure B.3 let us call the *exterior* of the parameter space the area defined by

$$\theta(1) > 1 \tag{9.46}$$

$$\theta(M) > 1 \tag{9.47}$$

$$\theta(1) = 1, \ \theta(M) \leq 1, \ \phi(1) < 0 \tag{9.48}$$

$$\theta(1) \leq 1, \ \theta(M) = 1, \ \phi(M) > 0 \tag{9.49}$$

$$\theta(1) < 0, \ \theta(1)\theta(M) > 1 \tag{9.50}$$

$$\theta(1) < 0, \ \theta(1)\theta(M) = 1, \ \phi(M) + \theta(M)\phi(1) < 0 \tag{9.51}$$

In order to make the analysis more straightforward we will make the following assumption as appropriate.

(SETAR3)   The variances of the noise distributions for the two end intervals are finite; that is,

$$\mathsf{E}(W^2(1)) < \infty, \qquad \mathsf{E}(W^2(M)) < \infty.$$

Proposition 9.5.4. *For the SETAR model satisfying the assumptions (SETAR1)–(SETAR3), the chain is transient in the exterior of the parameter space.*

PROOF   Suppose (9.47) holds. Then the chain is transient, as we show by stochastic comparison arguments. For until the first time the chain enters $(-\infty, -r_{M-1})$ it follows the sample paths of a model

$$X'_n = \phi(M) + \theta(M)X'_{n-1} + W_M$$

and for this linear model $\mathsf{P}_x(\tau_{(-\infty,0)} < \infty) < 1$ for all sufficiently large x, as in the proof of Theorem 9.5.2, by comparison with random walk.

When (9.46) holds, the chain is transient by symmetry: we find $\mathsf{P}_x(\tau_{(0,\infty,)} < \infty) < 1$ for all sufficiently negative x.

When (9.50) holds the same argument can be used, but now for the two step chain: the one-step chain undergoes larger and larger oscillations and thus there is a positive probability of never returning to the set $[r_1, r_{M-1}]$ for starting points of sufficiently large magnitude.

Suppose (9.48) holds and begin the process at $x_o < \min(0, r_1)$. Then until the first time the process exits $(-\infty, \min(0, r_1))$, it has exactly the sample paths of a random walk with negative drift, which we showed to be transient in Section 8.5. The proof of transience when (9.49) holds is similar.

We finally show the chain is transient if (9.51) holds, and for this we need (SETAR3). Here we also need to exploit Theorem 8.4.2 directly rather than construct a stochastic comparison argument.

Let a and b be positive constants such that $-b/a = \theta(1) = 1/\theta(M)$. Since $\phi(M) + \theta(M)\phi(1) < 0$ we can choose u and v such that $-a\phi(1) < au + bv < -b\phi(M)$. Choose c positive such that

$$c/a - u > \max(0, r_{M-1}), \qquad -c/b - v < \min(0, r_1).$$

Consider the function

$$V(x) = \begin{cases} 1 - 1/a(x+u), & x > c/a - u, \\ 1 - 1/c, & -c/b - v < x < c/a - u, \\ 1 + 1/b(x+v), & x < -c/b - v. \end{cases}$$

Suppose $x > R > c/a - u$, where R is to be chosen. Let

$$\lambda(x) = \phi(M) + \theta(M)x + v$$

and

$$\delta(x) = \phi(M) + \theta(M)x + u.$$

If we write

$$\begin{aligned}
V_0(x) &= -a^{-1}\mathsf{E}[(1/(\delta(x) + W(M)))\mathbb{I}_{[W(M) > c/a - \delta(x)]}], \\
V_1(x) &= c^{-1}P(-c/b - \lambda(x) < W(M) < c/a - \delta(x)), \\
V_2(x) &= 1/a(x+u) + b^{-1}\mathsf{E}[(1/(\lambda(x) + W(M)))(_{[W(M) < -c/b - \lambda(x)]}],
\end{aligned}$$

then we get

$$\mathsf{E}_x[V(X_1)] = V(x) + V_0(x) + V_1(x) + V_2(x). \tag{9.52}$$

It is easy to show that both $V_0(x)$ and $V_1(x)$ are $o(x^{-2})$. Since

$$1/(\lambda(x) + W(M)) = 1/\lambda(x) - W(M)/\lambda(x)(\lambda(x) + W(M)),$$

the second summand of $V_2(x)$ equals

$$\Gamma_M(-\infty, -c/b - \lambda(x))/b\lambda(x) - \mathsf{E}[(W(M)/\lambda(x)(\lambda(x) + W(M)))\mathbb{I}_{[W(M) < -c/b - \lambda(x)]}].$$

Since for $0 < W(M) < -c/b - \lambda(x)$,

$$1/(1 + W(M)/\lambda(x)) \le 1 + bW(M)/c,$$

we have in this case that for x large enough

$$\begin{aligned}
0 &\ge -x^2 W(M)/\lambda(x)(\lambda(x) + W(M)) \\
&\ge -x^2 W(M)(1 + bW(M)/c)/\lambda^2(x) \\
&\ge -2W(M)(1 + bW(M)/c)/\theta^2(M); \tag{9.53}
\end{aligned}$$

whilst for $W(M) \le 0$, we have

$$1/(1 + W(M)/\lambda(x)) \le 1$$

and so

$$
\begin{aligned}
0 &\leq -x^2 W(M)/\lambda(x)(\lambda(x) + W(M)) \\
&\leq -x^2 W(M)/\lambda^2(x) \\
&\leq -2W(M)/\theta^2(M).
\end{aligned}
\tag{9.54}
$$

Thus, by the Dominated Convergence Theorem,

$$
\begin{aligned}
\lim x^2 \mathsf{E}[-W(M)/\lambda(x)(\lambda(x) &+ W(M))\mathbb{I}_{[W(M) < -c/b - \lambda(x)]}] \\
&= \mathsf{E}[-W(M)/\theta^2(M)] = 0.
\end{aligned}
\tag{9.55}
$$

From (9.55) we therefore see that V_2 equals

$$
1/a(x+u) + 1/b\lambda(x) - \Gamma_M(-c/b - \lambda(x), \infty)/b\lambda(x) - o(x^{-2})
$$

$$
= (b\phi(M) + bv + au)/ab\lambda(x)(x+u) - o(x^{-2}).
$$

We now have from the breakup (9.52) that by choosing R large enough

$$
\begin{aligned}
\mathsf{E}_x[V(X_1)] &= V(x) + (b\phi(M) + bv + au)/ab\lambda(x)(x+u) - o(x^{-2}) \\
&\geq V(x), \qquad x > R.
\end{aligned}
\tag{9.56}
$$

Similarly, for $x < -R < -c/b - v < r_1$, it can be shown that

$$
\mathsf{E}_x[V(X_1)] \geq V(x).
$$

We may thus apply Theorem 8.4.2 with the set C taken to be $[-R, R]$ and the test function V above to conclude that the process is transient. $\qquad\square$

9.5.3 General chains with bounded increments

One of the more subtle uses of the drift conditions involves a development of the interplay between first and second moment conditions in determining recurrence or transience of a chain.

When the state space is $\mathbb{R}$, then even for a chain $\mathbf{\Phi}$ which is not a random walk it makes obvious sense to talk about the increment at x, defined by the random variable

$$
W_x = \{\Phi_1 - \Phi_0 \mid \Phi_0 = x\}
\tag{9.57}
$$

with probability law

$$
\Gamma_x(A) = \mathsf{P}(\Phi_1 \in A + x \mid \Phi_0 = x).
$$

The defining characteristic of the random walk model is then that the law Γ_x is independent of x, giving the characteristic spatial homogeneity to the model.

In general we can define the "mean drift" at x by

$$
m(x) = \mathsf{E}_x[W_x] = \int w\, \Gamma_x(dw)
$$

so that $m(x) = \Delta V(x)$ for the special choice of $V(x) = x$.

Let us denote the second moment of the drift at x by

$$v(x) = \mathsf{E}_x[W_x^2] = \int w^2\, \Gamma_x(dw).$$

We will now show that there is a threshold or detailed balance effect between these two quantities in considering the stability of the chain.

For ease of exposition let us consider the case where the increments again have uniformly bounded range: that is, for some R and all x,

$$\Gamma_x[-R, R] = 1. \tag{9.58}$$

To avoid somewhat messy calculations such as those for the random walk or SETAR models above we will fix the state space as $\mathbb{R}_+$ and we will make the assumption that the measures Γ_x give sufficient weight to the negative half line to ensure that the chain is a δ_0-irreducible T-chain and also that $v(x)$ is bounded from zero: this ensures that recurrence means that τ_0 is finite with probability one and that transience means that $\mathsf{P}_0(\tau_0 < \infty) < 1$. The δ_0-irreducibility and T-chain properties will of course follow from assuming, for example, that $\varepsilon < \Gamma_x(-\infty, -\varepsilon)$ for some $\varepsilon > 0$.

Theorem 9.5.5. *For the chain Φ with increment (9.57) we have*

(i) *if there exists $\theta < 1$ and x_0 such that for all $x > x_0$*

$$m(x) \leq \theta v(x)/2x, \tag{9.59}$$

then Φ is recurrent;

(ii) *if there exists $\theta > 1$ and x_0 such that for all $x > x_0$*

$$m(x) \geq \theta v(x)/2x, \tag{9.60}$$

then Φ is transient.

PROOF   (i)   We use Theorem 9.1.8, with the test function

$$V(x) = \log(1 + x), \qquad x \geq 0: \tag{9.61}$$

for this test function (V1) requires

$$\int_{-x}^{\infty} \Gamma_x(dw)[\log(w + x + 1) - \log(x + 1)] \leq 0, \tag{9.62}$$

and using the bounded range of the increments, the integral in (9.62) after a Taylor series expansion is, for $x > R$,

$$\int_{-R}^{R} \Gamma_x(dw)[w/(x + 1) - w^2/2(x + 1)^2 + o(x^{-2})]$$

$$= m(x)/(x + 1) - v(x)/2(x + 1)^2 + o(x^{-2}). \tag{9.63}$$

If $x > x_0$ for sufficiently large $x_0 > R$, and $m(x) \leq \theta v(x)/2x$, then

$$\int P(x, dy)V(y) \leq V(x)$$

and hence from Theorem 9.1.8 we have that the chain is recurrent.

(ii)   It is obvious with the assumption of positive mean for Γ_x that for any x the sets $[0, x]$ and $[x, \infty)$ are both in $\mathcal{B}^+(\mathsf{X})$.

In order to use Theorem 9.1.8, we will establish that for some suitable monotonic increasing V

$$\int_y P(x, dy)V(y) \geq V(x) \tag{9.64}$$

for $x \geq x_0$. An appropriate test function in this case is given by

$$V(x) = 1 - [1 + x]^{-\alpha}, \qquad x \geq 0 : \tag{9.65}$$

we can write (9.64) for $x > R$ as

$$\int_{-R}^{R} \Gamma_x(dw)[(w + x + 1)^{-\alpha} - (x + 1)^{-\alpha}] \geq 0. \tag{9.66}$$

Applying Taylor's Theorem we see that for all w we have that the integral in (9.66) equals

$$\alpha m(x)/(x + 1)^{1+\alpha} - \alpha v(x)/2(x + 1)^{2+\alpha} + O(x^{-3-\alpha}). \tag{9.67}$$

Now choose $\alpha < \theta - 1$. For sufficiently large x_0 we have that if $x > x_0$ then from (9.67) we have that (9.66) holds and so the chain is transient.   □

The fact that this detailed balance between first and second moments is a determinant of the stability properties of the chain is not surprising: on the space $\mathbb{R}_+$ all of the drift conditions are essentially linearizations of the motion of the chain, and virtually independently of the test functions chosen, a two-term Taylor series expansion will lead to the results we have described.

One of the more interesting and rather counter-intuitive facets of these results is that it is possible for the first-order mean drift $m(x)$ to be positive and for the chain to still be recurrent: in such circumstances it is the occasional negative jump thrown up by a distribution with a variance large in proportion to its general positive drift which will give recurrence.

Some weakening of the bounded range assumption is obviously possible for these results: the proofs then necessitate a rather more subtle analysis and expansion of the integrals involved. By choosing the iterated logarithm

$$V(x) = \log \log(x + c)$$

as the test function for recurrence, and by more detailed analysis of the function

$$V(x) = 1 - [1 + x]^{-\alpha}$$

as a test for transience, it is in fact possible to develop the following result, whose proof we omit.

Theorem 9.5.6. *Suppose the increment W_x given by (9.57) satisfies*

$$\sup_x \mathsf{E}_x[|W_x|^{2+\varepsilon}] < \infty$$

for some $\varepsilon > 0$. Then

(i) *if there exists $\delta > 0$ and x_0 such that for all $x > x_0$*

$$m(x) \leq v(x)/2x + O(x^{-1-\delta}), \qquad\qquad (9.68)$$

   the chain Φ is recurrent;

(ii) *if there exists $\theta > 1$ and x_0 such that for all $x > x_0$*

$$m(x) \geq \theta v(x)/2x, \qquad\qquad (9.69)$$

   then Φ is transient.   □

The bounds on the spread of Γ_x may seem somewhat artifacts of the methods of proof used, and of course we well know that the zero-mean random walk is recurrent even though a proof using an approach based upon a drift condition has not yet been developed to our knowledge.

We conclude this section with a simple example showing that we cannot expect to drop the higher moment conditions completely.

Let $\mathsf{X} = \mathbb{Z}_+$, and let

$$P(x, x+1) = 1 - c/x, \qquad P(x, 0) = c/x, \qquad x > 0$$

with $P(0, 1) = 1$.

Then the chain is easily shown to be recurrent by a direct calculation that for all $n > 1$

$$\mathsf{P}_0(\tau_0 > n) = \prod_{x=1}^{n}[1 - c/x].$$

But we have $m(x) = -c + 1 - c/x$ and $v(x) = cx + 1 - c/x$ so that

$$2xm(x) - v(x) = (2 - 3c)x^2 - (c+1)x + c,$$

which is clearly positive for $c < 2/3$: hence if Theorem 9.5.6 were applicable we should have the chain transient.

Of course, in this case we have

$$\mathsf{E}_x[|W_x|^{2+\varepsilon}] = x^{2+\varepsilon}c/x + 1 - c/x > x^{1+\varepsilon}$$

and the bound on this higher moment, required in the proof of Theorem 9.5.6, is obviously violated.

9.6   Commentary

Harris chains are named after T. E. Harris who introduced many of the essential ideas in [155]. The important result in Theorem 9.1.3, which enables the properties of Q to be linked to those of L, is due to Orey [308], and our proof follows that in [309]. That recurrent chains are "almost" Harris was shown by Tuominen [390], although the key links between the powerful Harris properties and other seemingly weaker recurrence properties were developed initially by Jain and Jamison [172].

We have taken the proof of transience for random walk on $\mathbb{Z}$ using the Strong Law of Large Numbers from Spitzer [369].

Non-evanescence is a common form of recurrence for chains on $\mathbb{R}^k$: see, for example, Khas'minskii [206]. The links between evanescent and transient chains, and the equivalence between Harris and non-evanescent chains under the T-chain condition, are taken from Meyn and Tweedie [277], who proved Theorem 9.2.2. Most of the connections between neighborhood and global behavior of chains are given by Rosenblatt [338, 339] and Tuominen and Tweedie [391].

The criteria for non-evanescence or Harris recurrence here are of course closely related to those in the previous chapter. The martingale argument for non-evanescence is in [277] and [398], but can be traced back in essentially the same form to Lamperti [234]. The converse to the recurrence criterion under the Feller condition, and the fact that it does not hold in general, are new: the construction of the converse function V is however based on a similar result for countable chains, in Mertens et al. [258].

The term "coercive" to describe functions whose sublevel sets are precompact is new. The justification for the terminology is that coercive functions do, in most of our contexts, measure the distance from a point to a compact "center" of the state space. This will become clearer in later chapters when we see that under a suitable drift condition, the mean time to reach some compact set from $\Phi_0 = x$ is bounded by a constant multiple of $V(x)$. Hence $V(x)$ bounds the mean "distance" to this compact set, measured in units of time. Beneš in [24] uses the term *moment* for these functions. Since "moments" are standard in referring to the expectations of random variables, this terminology is obviously inappropriate here.

Stochastic comparison arguments have been used for far too long to give a detailed attribution. For proving transience, in particular, they are a most effective tool. The analysis we present here of the SETAR model is essentially in Petruccelli et al. [315] and Chan et al. [64].

The analysis of chains via their increments, and the delicate balance required between $m(x)$ and $v(x)$ for recurrence and transience, is found in Lamperti [234]; see also Tweedie [398]. Growth models for which $m(x) \geq \theta v(x)/2x$ are studied by, for example, Kersting (see [205]), and their analysis via suitable renormalization proves a fruitful approach to such transient chains.

It may appear that we are devoting a disproportionate amount of space to unstable chains, and too little to chains with stability properties. This will be rectified in the rest of the book, where we will be considering virtually nothing but chains with ever stronger stability properties.

Chapter 10

The existence of π

In our treatment of the structure and stability concepts for irreducible chains we have to this point considered only the dichotomy between transient and recurrent chains.

For transient chains there are many areas of theory that we shall not investigate further, despite the flourishing research that has taken place in both the mathematical development and the application of transient chains in recent years. Areas which are notable omissions from our treatment of Markovian models thus include the study of potential theory and boundary theory [326], as well as the study of renormalized models approximated by diffusions and the quasi-stationary theory of transient processes [108, 4].

Rather, we concentrate on recurrent chains which have stable properties without renormalization of any kind, and develop the consequences of the concept of recurrence.

In this chapter we further divide recurrent chains into *positive* and *null* recurrent chains, and show here and in the next chapter that the former class provide stochastic stability of a far stronger kind than the latter.

For many purposes, the strongest possible form of stability that we might require in the presence of persistent variation is that the distribution of Φ_n does not change as n takes on different values. If this is the case, then by the Markov property it follows that the finite dimensional distributions of Φ are invariant under translation in time. Such considerations lead us to the consideration of *invariant measures*.

Invariant measures

A σ-finite measure π on $\mathcal{B}(\mathsf{X})$ with the property

$$\pi(A) = \int_{\mathsf{X}} \pi(dx) P(x, A), \qquad A \in \mathcal{B}(\mathsf{X}) \qquad (10.1)$$

will be called *invariant*.

Although we develop a number of results concerning invariant measures, the key conclusion in this chapter is undoubtedly

Theorem 10.0.1. *If the chain Φ is recurrent then it admits a unique (up to constant multiples) invariant measure π, and the measure π has the representation, for any $A \in \mathcal{B}^+(\mathsf{X})$*

$$\pi(B) = \int_A \pi(dw) \mathsf{E}_w \Big[\sum_{n=1}^{\tau_A} \mathbb{I}\{\Phi_n \in B\} \Big], \qquad B \in \mathcal{B}(\mathsf{X}). \tag{10.2}$$

The invariant measure π is finite (rather than merely σ-finite) if there exists a petite set C such that

$$\sup_{x \in C} \mathsf{E}_x[\tau_C] < \infty.$$

PROOF   The existence and representation of invariant measures for recurrent chains is proved in full generality in Theorem 10.4.9: the proof exploits, via the Nummelin splitting technique, the corresponding theorem for chains with atoms as in Theorem 10.2.1, in conjunction with a representation for invariant measures given in Theorem 10.4.9. The criterion for finiteness of π is in Theorem 10.4.10.   □

If an invariant measure is finite, then it may be normalized to a stationary probability measure, and in practice this is the main stable situation of interest. If an invariant measure has infinite total mass, then its probabilistic interpretation is much more difficult, although for recurrent chains, there is at least the interpretation as described in (10.2).

These results lead us to define the following classes of chains.

Positive and null chains

Suppose that Φ is ψ-irreducible, and admits an invariant probability measure π. Then Φ is called a *positive* chain.

If Φ does not admit such a measure, then we call Φ *null*.

10.1 Stationarity and invariance

10.1.1 Invariant measures

Processes with the property that for any k, the marginal distribution of $\{\Phi_n, \ldots, \Phi_{n+k}\}$ does not change as n varies are called *stationary processes*, and whilst it is clear that in general a Markov chain will not be stationary, since in a particular realization we may have $\Phi_0 = x$ with probability one for some fixed x, it is possible that with an appropriate choice of the initial distribution for Φ_0 we may produce a stationary process $\{\Phi_n, n \in \mathbb{Z}_+\}$.

It is immediate that we only need to consider a form of first step stationarity in order to generate an entire stationary process. Given an initial invariant probability measure π such that

$$\pi(A) = \int_X \pi(dw)P(w, A), \tag{10.3}$$

we can iterate to give

$$
\begin{aligned}
\pi(A) &= \int_X \left[\int_X \pi(dx)P(x, dw) \right] P(w, A) \\
&= \int_X \pi(dx) \int_X P(x, dw)P(w, A) \\
&= \int_X \pi(dx)P^2(x, A) \\
&\;\vdots \\
&= \int_X \pi(dx)P^n(x, A) = \mathsf{P}_\pi(\Phi_n \in A),
\end{aligned}
$$

for any n and all $A \in \mathcal{B}(X)$.

From the Markov property, it is clear that Φ is stationary if and only if the distribution of Φ_n does not vary with time. We have immediately

Proposition 10.1.1. *If the chain Φ is positive, then it is recurrent.*

PROOF   Suppose that the chain is positive and let π be a invariant probability measure. If the chain is also transient, let A_j be a countable cover of X with uniformly transient sets, as guaranteed by Theorem 8.3.4, with $U(x, A_j) \le M_j$, say.

Using (10.4) we have for any j, k

$$k\pi(A_j) = \sum_{n=1}^{k} \int \pi(dw)P^n(w, A_j) \le M_j$$

and since the left hand side remains finite as $k \to \infty$, we have $\pi(A_j) = 0$. This implies π is trivial so we have a contradiction. $\square$

Positive chains are often called "positive recurrent" to reinforce the fact that they are recurrent. This also naturally gives the definition

---

Positive Harris chains

If Φ is Harris recurrent and positive, then Φ is called a *positive Harris chain*.

It is of course not yet clear that an invariant probability measure π ever exists, or whether it will be unique when it does exist. It is the major purpose of this chapter to find conditions for the existence of π, and to prove that for any positive (and indeed recurrent) chain, π is essentially unique.

Invariant probability measures are important not merely because they define stationary processes. They will also turn out to be the measures which define the long term or ergodic behavior of the chain. To understand why this should be plausible, consider $P_\mu(\Phi_n \in \cdot)$ for any starting distribution μ. If a limiting measure γ_μ exists in a suitable topology on the space of probability measures, such as

$$P_\mu(X_n \in A) \to \gamma_\mu(A)$$

for all $A \in \mathcal{B}(\mathsf{X})$, then

$$
\begin{aligned}
\gamma_\mu(A) &= \lim_{n \to \infty} \int \mu(dx) P^n(x, A) \\
&= \lim_{n \to \infty} \int_{\mathsf{X}} \mu(dx) \int P^{n-1}(x, dw) P(w, A) \\
&= \int_{\mathsf{X}} \gamma_\mu(dw) P(w, A),
\end{aligned}
\tag{10.4}
$$

since setwise convergence of $\int \mu(dx) P^n(x, \cdot)$ implies convergence of integrals of bounded measurable functions such as $P(w, A)$.

Hence if a limiting distribution exists, it is an invariant probability measure; and obviously, if there is a unique invariant probability measure, the limit γ_μ will be independent of μ whenever it exists.

We will not study the existence of such limits properly until Part III, where our goal will be to develop asymptotic properties of Φ in some detail. However, motivated by these ideas, we will give in Section 10.5 one example, the linear model, where this route leads to the existence of an invariant probability measure.

10.1.2 Subinvariant measures

The easiest way to investigate the existence of π is to consider a yet wider class of measures, satisfying inequalities related to the invariant equation (10.1).

Subinvariant measures

If μ is σ-finite and satisfies

$$\mu(A) \geq \int_{\mathsf{X}} \mu(dx) P(x, A), \quad A \in \mathcal{B}(\mathsf{X}), \tag{10.5}$$

then μ is called *subinvariant*.

The following generalization of the subinvariance equation (10.5) is often useful: we have, by iterating (10.5),

$$\mu(B) \geq \int \mu(dw) P^n(w, B)$$

and hence, multiplying by $a(n)$ and summing,

$$\mu(B) \geq \int \mu(dw) K_a(w, B), \tag{10.6}$$

for any sampling distribution a.

We begin with some structural results for arbitrary subinvariant measures.

Proposition 10.1.2. *Suppose that Φ is ψ-irreducible. If μ is any measure satisfying (10.5) with $\mu(A) < \infty$ for some one $A \in \mathcal{B}^+(\mathsf{X})$, then*

(i) μ *is σ-finite, and thus μ is a subinvariant measure;*

(ii) $\mu \succ \psi$;

(iii) *if C is petite then $\mu(C) < \infty$;*

(iv) *if $\mu(\mathsf{X}) < \infty$ then μ is invariant.*

PROOF   Suppose $\mu(A) < \infty$ for some A with $\psi(A) > 0$. Using $A^*(j) = \{y : K_{a_{1/2}}(y, A) > j^{-1}\}$, we have by (10.6),

$$\infty > \mu(A) \geq \int_{A^*(j)} \mu(dw) K_{a_{1/2}}(w, A) \geq j^{-1} \mu(A^*(j));$$

since $\bigcup A^*(j) = \mathsf{X}$ when $\psi(A) > 0$, such a μ must be σ-finite.

To prove (ii) observe that, by (10.6), if $B \in \mathcal{B}^+(\mathsf{X})$ we have $\mu(B) > 0$, so $\mu \succ \psi$.

Thirdly, if C is ν_a-petite then there exists a set B with $\nu_a(B) > 0$ and $\mu(B) < \infty$, from (i). By (10.6) we have

$$\mu(B) \geq \int \mu(dw) K_a(w, B) \geq \mu(C) \nu_a(B) \tag{10.7}$$

and so $\mu(C) < \infty$ as required.

Finally, if there exists some A such that $\mu(A) > \int \mu(dy) P(y, A)$ then we have

$$
\begin{aligned}
\mu(\mathsf{X}) = \mu(A) + \mu(A^c) \; &> \; \int \mu(dy) P(y, A) + \int \mu(dy) P(y, A^c) \\
&= \int \mu(dy) P(y, \mathsf{X}) \\
&= \mu(\mathsf{X}) \tag{10.8}
\end{aligned}
$$

and if $\mu(\mathsf{X}) < \infty$ we have a contradiction.

$\square$

The major questions of interest in studying subinvariant measures lie with recurrent chains, for we always have

Proposition 10.1.3. *If the chain Φ is transient, then there exists a strictly subinvariant measure for Φ.*

PROOF   Suppose that Φ is transient: then by Theorem 8.3.4, we have that the measures μ_x given by

$$\mu_x(A) = U(x, A), \qquad A \in \mathcal{B}(\mathsf{X}),$$

are σ-finite; and trivially

$$\mu_x(A) = P(x, A) + \int \mu_x(dy) P(y, A) \geq \int \mu_x(dy) P(y, A), \qquad A \in \mathcal{B}(\mathsf{X}) \qquad (10.9)$$

so that each μ_x is subinvariant (and obviously strictly subinvariant, since there is some A with $\mu_x(A) < \infty$ such that $P(x, A) > 0$). $\qquad\qquad\square$

We now move on to study recurrent chains, where the existence of a subinvariant measure is less obvious.

10.2   The existence of π: chains with atoms

Rather than pursue the question of existence of invariant and subinvariant measures on a fully countable space in the first instance, we prove here that the existence of just one atom α in the space is enough to describe completely the existence and structure of such measures.

The following theorem obviously incorporates countable space chains as a special case; but the main value of this presentation will be in the development of a theory for general space chains via the split chain construction of Section 5.1.

Theorem 10.2.1. *Suppose Φ is ψ-irreducible, and X contains an accessible atom α.*

(i) *There is always a subinvariant measure μ_α° for Φ given by*

$$\mu_\alpha^\circ(A) = U_\alpha(\alpha, A) = \sum_{n=1}^{\infty} {}_\alpha P^n(\alpha, A), \qquad A \in \mathcal{B}(\mathsf{X}); \qquad (10.10)$$

and μ_α° is invariant if and only if Φ is recurrent.

(ii) *The measure μ_α° is minimal in the sense that if μ is subinvariant with $\mu(\alpha) = 1$, then*

$$\mu(A) \geq \mu_\alpha^\circ(A), \qquad A \in \mathcal{B}(\mathsf{X}).$$

When Φ is recurrent, μ_α° is the unique (sub)invariant measure with $\mu(\alpha) = 1$.

(iii) *The subinvariant measure μ_α° is a finite measure if and only if*

$$\mathsf{E}_\alpha[\tau_\alpha] < \infty,$$

in which case μ_α° is invariant.

PROOF   **(i)**   By construction we have for $A \in \mathcal{B}(\mathsf{X})$

$$\int_{\mathsf{X}} \mu_\alpha^\circ(dy)P(y,A) \;=\; \mu_\alpha^\circ(\alpha)P(\alpha,A) + \int_{\alpha^c} \sum_{n=1}^{\infty} {}_\alpha P^n(\alpha,dy)P(y,A)$$

$$\leq \;\; {}_\alpha P(\alpha,A) + \sum_{n=2}^{\infty} {}_\alpha P^n(\alpha,A) \qquad\qquad (10.11)$$

$$= \;\; \mu_\alpha^\circ(A),$$

where the inequality comes from the bound $\mu_\alpha^\circ(\alpha) \leq 1$. Thus μ_α° is subinvariant, and is invariant if and only if $\mu_\alpha^\circ(\alpha) = \mathsf{P}_\alpha(\tau_\alpha < \infty) = 1$; that is, from Proposition 8.3.1, if and only if the chain is recurrent.

(ii)   Let μ be any subinvariant measure with $\mu(\alpha) = 1$. By subinvariance,

$$\mu(A) \;\geq\; \int_{\mathsf{X}} \mu(dw)P(w,A)$$

$$\geq \;\; \mu(\alpha)P(\alpha,A) = P(\alpha,A).$$

Assume inductively that $\mu(A) \geq \sum_{m=1}^{n} {}_\alpha P^m(\alpha,A)$, for all A. Then by subinvariance,

$$\mu(A) \;\geq\; \mu(\alpha)P(\alpha,A) + \int_{\alpha^c} \mu(dw)P(w,A)$$

$$\geq \;\; P(\alpha,A) + \int_{\alpha^c} \left[\sum_{m=1}^{n} {}_\alpha P^m(\alpha,dw) \right] P(w,A)$$

$$= \;\; \sum_{m=1}^{n+1} {}_\alpha P^m(\alpha,A).$$

Taking $n \uparrow \infty$ shows that $\mu(A) \geq \mu_\alpha^\circ(A)$ for all $A \in \mathcal{B}(\mathsf{X})$.

Suppose Φ is recurrent, so that $\mu_\alpha^\circ(\alpha) = 1$. If μ_α° differs from μ, there exists A and n such that $\mu(A) > \mu_\alpha^\circ(A)$ and $P^n(w,\alpha) > 0$ for all $w \in A$, since $\psi(\alpha) > 0$. By minimality, subinvariance of μ, and invariance of μ_α°,

$$1 = \mu(\alpha) \;\geq\; \int_{\mathsf{X}} \mu(dw)P^n(w,\alpha)$$

$$> \;\; \int_{\mathsf{X}} \mu_\alpha^\circ(dw)P^n(w,\alpha)$$

$$= \;\; \mu_\alpha^\circ(\alpha) = 1.$$

Hence we must have $\mu = \mu_\alpha^\circ$, and thus when Φ is recurrent, μ_α° is the unique (sub) invariant measure.

(iii)   If μ_α° is finite it follows from Proposition 10.1.2 (iv) that μ_α° is invariant. Finally

$$\mu_\alpha^\circ(\mathsf{X}) = \sum_{n=1}^{\infty} \mathsf{P}_\alpha(\tau_\alpha \geq n) \qquad\qquad (10.12)$$

and so an invariant probability measure exists if and only if the mean return time to α is finite, as stated. $\square$

We shall use π to denote the unique invariant measure in the recurrent case. Unless stated otherwise we will assume π is normalized to be a probability measure when $\pi(\mathsf{X})$ is finite.

The invariant measure μ_α° has an equivalent sample path representation for recurrent chains:

$$\mu_\alpha^\circ(A) = \mathsf{E}_\alpha\left[\sum_{n=1}^{\tau_\alpha} \mathbb{I}\{\Phi_n \in A\}\right], \qquad A \in \mathcal{B}(\mathsf{X}). \tag{10.13}$$

This follows from the definition of the taboo probabilities $_\alpha P^n$.

As an immediate consequence of this construction we have the following elegant criterion for positivity.

Theorem 10.2.2 (Kac's Theorem). *If Φ is ψ-irreducible and admits an atom $\alpha \in \mathcal{B}^+(\mathsf{X})$, then Φ is positive recurrent if and only if $\mathsf{E}_\alpha[\tau_\alpha] < \infty$; and if π is the invariant probability measure for Φ, then*

$$\pi(\alpha) = (\mathsf{E}_\alpha[\tau_\alpha])^{-1}. \tag{10.14}$$

PROOF   If $\mathsf{E}_\alpha[\tau_\alpha] < \infty$, then also $L(\alpha, \alpha) = 1$, and by Proposition 8.3.1 Φ is recurrent; it follows from the structure of π in (10.10) that π is finite so that the chain is positive.

Conversely, $\mathsf{E}_\alpha[\tau_\alpha] < \infty$ when the chain is positive from the structure of the unique invariant measure.

By the uniqueness of the invariant measure normalized to be a probability measure π we have

$$\pi(\alpha) = \frac{\mu_\alpha^\circ(\alpha)}{\mu_\alpha^\circ(\mathsf{X})} = \frac{U_\alpha(\alpha, \alpha)}{U_\alpha(\alpha, \mathsf{X})} = \frac{1}{\mathsf{E}_\alpha[\tau_\alpha]}$$

which is (10.14).   □

The relationship (10.14) is often known as Kac's Theorem. For countable state space models it immediately gives us

Proposition 10.2.3. *For a positive recurrent irreducible Markov chain on a countable space, there is a unique (up to constant multiples) invariant measure π given by*

$$\pi(x) = [\mathsf{E}_x[\tau_x]]^{-1}$$

for every $x \in \mathsf{X}$.   □

We now illustrate the use of the representation of π for a number of countable space models.

10.3   Invariant measures for countable space models*

10.3.1   Renewal chains

Forward recurrence time chains

Consider the forward recurrence time process V^+ with

$$P(1, j) = p(j), \qquad j \geq 1; \qquad P(j, j-1) = 1, \qquad j > 1. \tag{10.15}$$

As noted in Section 8.1.2, this chain is always recurrent since $\sum p(j) = 1$.
By construction we have that

$$_1P^n(1,j) = p(j + n - 1), \qquad j \leq n,$$

and zero otherwise; thus the minimal invariant measure satisfies

$$\pi(j) = U_1(1,j) = \sum_{n \geq j} p(n) \tag{10.16}$$

which is finite if and only if

$$\sum_{j=1}^{\infty} \pi(j) = \sum_{j=1}^{\infty} \sum_{n=j}^{\infty} p(n) = \sum_{n=1}^{\infty} np(n) < \infty: \tag{10.17}$$

that is, if and only if the renewal distribution $\{p(i)\}$ has finite mean.

It is, of course, equally easy to deduce this formula by solving the invariant equations themselves, but the result is perhaps more illuminating from this approach.

Now suppose that the distribution $\{p(j)\}$ is *periodic with period d*: that is, the greatest common divisor of the set $N_p = \{n : p(n) > 0\}$ is d. Let $[N_p]$ denote the span of N_p,

$$[N_p] = \Big\{ \sum m_i r_i : m_i \subset \mathbb{Z}_+, \ r_i \in N_p \Big\}.$$

We have $P^n(j, 1) > 0$ whenever $n - j + 1 \in [N_p]$.

By Lemma D.7.4 there exists an integer $n_0 < \infty$ such that $nd \in [N_p]$ for all $n \geq n_0$. If $d = 1$ it follows that the forward recurrence time process V^+ is aperiodic, since in this case

$$P^n(j, 1) > 0, \qquad n - j + 1 \geq n_0. \tag{10.18}$$

Linked forward recurrence time chains

Consider the forward recurrence time chain with transition law (10.15), and define the bivariate chain $V^* = (V_1^+(n), V_2^+(n))$ on the space $\mathsf{X}^* := \{1, 2, \ldots\} \times \{1, 2, \ldots\}$, with the transition law

$$
\begin{aligned}
P((i,j), (i-1, j-1)) &= 1, & i, j &> 1; \\
P((1,j), (k, j-1)) &= p(k), & k, j &> 1; \\
P((i,1), (i-1, k)) &= p(k), & i, k &> 1; \\
P((1,1), (j, k)) &= p(j)p(k), & j, k &> 1.
\end{aligned}
\tag{10.19}
$$

This chain is constructed by taking the two independent copies $V_1^+(n), V_2^+(n)$ of the forward recurrence time chain and running them independently. It then follows from (10.18) that V^* is ψ-irreducible if $\{p(j)\}$ has period $d = 1$.

Moreover V^* is positive Harris recurrent on X^* provided only $\sum_k kp(k) < \infty$, as was the case for the single copy of the forward recurrence time chain. To prove this we need only note that the product measure $\pi^*(i, j) = \pi(i)\pi(j)$ is invariant for V^*, where

$$\pi(j) = \sum_{k \geq j} p(k) \Big/ \sum_k kp(k)$$

is the invariant probability measure for the forward recurrence time process from (10.16) and (10.17); positive Harris recurrence follows since $\pi^*(X^*) = [\pi(X)]^2 = 1$.

These conditions for positive recurrence of the bivariate forward time process will be of critical use in the development of the asymptotic properties of general chains in Part III.

10.3.2   The number in an M/G/1 queue

Recall from Section 3.3.3 that N^* is a modified random walk on a half line with increment distribution concentrated on the integers $\{\ldots, -1, 0, 1\}$ having the transition probability matrix of the form

$$
P = \begin{pmatrix}
q_0 & q_1 & q_2 & q_3 & \cdots \\
q_0 & q_1 & q_2 & q_3 & \cdots \\
    & q_0 & q_1 & q_2 & \cdots \\
    &   & q_0 & q_1 & \cdots \\
    &   &   & q_0 & \cdots
\end{pmatrix}
$$

where $q_i = \mathsf{P}(Z = i - 1)$ for the increment variable in the chain when the server is busy; that is, for transitions from states other than $\{0\}$. The chain N^* is always ψ-irreducible if $q_0 > 0$, and irreducible in the standard sense if also $q_0 + q_1 < 1$, and we shall assume this to be the case to avoid trivialities.

In this case, we can actually solve the invariant equations explicitly. For $j \geq 1$, (10.1) can be written

$$
\pi(j) = \sum_{k=0}^{j+1} \pi(k) q_{j+1-k} \tag{10.20}
$$

and if we define

$$
\bar{q}_j = \sum_{n=j+1}^{\infty} q_n
$$

we get the system of equations

$$
\begin{aligned}
\pi(1)q_0 &= \pi(0)\bar{q}_0, \\
\pi(2)q_0 &= \pi(0)\bar{q}_1 + \pi(1)\bar{q}_1, \\
\pi(3)q_0 &= \pi(0)\bar{q}_2 + \pi(1)\bar{q}_2 + \pi(2)\bar{q}_1, \\
&\ \ \vdots
\end{aligned}
$$

In this case, therefore, we always get a unique invariant measure, regardless of the transience or recurrence of the chain.

The criterion for positivity follows from (10.21). Note that the mean increment β of Z satisfies

$$
\beta = \sum_{j \geq 0} \bar{q}_j - 1
$$

so that formally summing both sides of (10.21) gives, since $q_0 = 1 - \bar{q}_0$,

$$
(1 - \bar{q}_0) \sum_{j=1}^{\infty} \pi(j) = (\beta + 1)\pi(0) + (\beta + 1 - \bar{q}_0) \sum_{j=1}^{\infty} \pi(j). \tag{10.21}
$$

If the chain is positive, this implies

$$\infty > \sum_{j=1}^{\infty} \pi(j) = -\pi(0)(\beta+1)/\beta,$$

so, since $\beta > -1$, we must have $\beta < 0$. Conversely, if $\beta < 0$, and we take

$$\pi(0) = -\beta,$$

then the same summation (10.21) indicates that the invariant measure π is finite.

Thus we have

Proposition 10.3.1. *The chain N^* is positive if and only if the increment distribution satisfies $\beta = \sum jq_j < 1$.*

This same type of direct calculation can be carried out for any so-called "skip-free" chain with $P(i,j) = 0$ for $j < i-1$, such as the forward recurrence time chain above. For other chains it can be far less easy to get a direct approach to the invariant measure through the invariant equations, and we turn to the representation in (10.10) for our results.

10.3.3   The number in a GI/M/1 queue

We illustrate the use of the structural result in giving a novel interpretation of an old result for the specific random walk on a half line N corresponding to the number in a GI/M/1 queue.

Recall from Section 3.3.3 that N has increment distribution concentrated on the integers $\{\ldots, -1, 0, 1\}$ giving the transition probability matrix of the form

$$P = \begin{pmatrix} \sum_{1}^{\infty} p_i & p_0 & & & \\ \sum_{2}^{\infty} p_i & p_1 & p_0 & \mbox{\Large 0} & \\ \sum_{3}^{\infty} p_i & p_2 & p_1 & p_0 & \cdots \\ \vdots & & \vdots & \vdots & \vdots \end{pmatrix}$$

where $p_i = \mathsf{P}(Z = 1 - i)$. The chain N is ψ-irreducible if $p_0 + p_1 < 1$, and irreducible if $p_0 > 0$ also. Assume these inequalities hold, and let $\{0\} = \alpha$ be our atom.

To investigate the existence of an invariant measure for N, we know from Theorem 10.2.1 that we should look at the quantities $_\alpha P^n(\alpha, j)$.

Write $[k] = \{0, \ldots, k\}$. Because the chain can only move up one step at a time, so the last visit to $[k]$ is at k itself, we have on decomposing over the last visit to $[k]$, for $k \geq 1$

$$_\alpha P^n(\alpha, k+1) = \sum_{r=1}^{n} {}_\alpha P^r(\alpha, k)_{[k]} P^{n-r}(k, k+1). \tag{10.22}$$

Now the translation invariance property of P implies that for $j > k$

$$_{[k]} P^r(k, j) = {}_\alpha P^r(\alpha, j - k). \tag{10.23}$$

Thus, summing (10.22) from 1 to ∞ gives

$$\sum_{n=1}^{\infty} {}_\alpha P^n(\alpha, k+1) = \left[\sum_{n=1}^{\infty} {}_\alpha P^n(\alpha, k)\right]\left[\sum_{n=1}^{\infty} {}_{[k]} P^n(k, k+1)\right]$$

$$= \left[\sum_{n=1}^{\infty} {}_\alpha P^n(\alpha, k)\right]\left[\sum_{n=1}^{\infty} {}_\alpha P^n(\alpha, 1)\right].$$

Using the form (10.10) of μ_α°, we have now shown that

$$\mu_\alpha^\circ(k+1) = \mu_\alpha^\circ(k)\mu_\alpha^\circ(1),$$

and so the minimal invariant measure satisfies

$$\mu_\alpha^\circ(k) = s_\alpha^k \tag{10.24}$$

where $s_\alpha = \mu_\alpha^\circ(1)$.

The chain then has an invariant probability measure if and only if we can find $s_\alpha < 1$ for which the measure μ_α° defined by the geometric form (10.24) is a solution to the subinvariant equations for P: otherwise the minimal subinvariant measure is not summable.

We can go further and identify these two cases in terms of the underlying parameters p_j. Consider the second (that is, the $k = 1$) invariant equation

$$\mu_\alpha^\circ(1) = \sum \mu_\alpha^\circ(k)P(k, 1).$$

This shows that s_α must be a solution to

$$s = \sum_0^{\infty} p_j s^j, \tag{10.25}$$

and since μ_α° is minimal it must be the smallest solution to (10.25). As is well known, there are two cases to consider: since the function of s on the right hand side of (10.25) is strictly convex, a solution $s \in (0, 1)$ exists if and only if

$$\sum_0^{\infty} jp_j > 1,$$

whilst if $\sum_j j\, p_j \le 1$ then the minimal solution to (10.25) is $s_\alpha = 1$.

One can then verify directly that in each of these cases μ_α° solves all of the invariant equations, as required. In particular, if $\sum_j j\, p_j = 1$ so that the chain is recurrent from the remarks following Proposition 9.1.2, the unique invariant measure is $\mu_\alpha(x) \equiv 1, x \in \mathsf{X}$: note that in this case, in fact, the first invariant equation is exactly

$$1 = \sum_{j \ge 0} \sum_{n > j} p_n = \sum_j j\, p_j.$$

Hence for recurrent chains (those for which $\sum_j j\, p_j \ge 1$) we have shown

Proposition 10.3.2. *The unique subinvariant measure for N is given by $\mu_a(k) = s_a^k$, where s_a is the minimal solution to (10.25) in $(0,1]$; and N is positive recurrent if and only if $\sum_j j \, p_j > 1$.* $\qquad\square$

The geometric form (10.24), as a "trial solution" to the equation (10.1), is often presented in an arbitrary way: the use of Theorem 10.2.1 motivates this solution, and also shows that s_a in (10.24) has an interpretation as the expected number of visits to state $k + 1$ from state k, for any k.

10.4 The existence of π: ψ-irreducible chains

10.4.1 Invariant measures for recurrent chains

We prove in this section that a general recurrent ψ-irreducible chain has an invariant measure, using the Nummelin splitting technique.

First we show how subinvariant measures for the split chain correspond with subinvariant measures for Φ.

Proposition 10.4.1. *Suppose that Φ is a strongly aperiodic Markov chain and let $\check{\Phi}$ denote the split chain. Then:*

(i) *If the measure $\check{\pi}$ is invariant for $\check{\Phi}$, then the measure π on $\mathcal{B}(X)$ defined by*

$$\pi(A) = \check{\pi}(A_0 \cup A_1), \qquad A \in \mathcal{B}(X), \tag{10.26}$$

is invariant for Φ, and $\check{\pi} = \pi^*$.

(ii) *If μ is any subinvariant measure for Φ then μ^* is subinvariant for $\check{\Phi}$, and if μ is invariant then so is μ^*.*

PROOF   To prove (i) note that by (5.5), (5.6), and (5.7), we have that the measure $\check{P}(x_i, \cdot)$ is of the form $\mu_{x_i}^*$ for any $x_i \in \check{X}$, where μ_{x_i} is a probability measure on X. By linearity of the splitting and invariance of $\check{\pi}$, for any $\check{A} \in \mathcal{B}(\check{X})$,

$$\check{\pi}(\check{A}) = \int \check{\pi}(dx_i)\check{P}(x_i, \check{A}) = \int \check{\pi}(dx_i)\mu_{x_i}^*(\check{A}) = \left(\int \check{\pi}(dx_i)\mu_{x_i}(\cdot)\right)^*(\check{A}).$$

Thus $\check{\pi} = \pi_0^*$, where $\pi_0 = \int \check{\pi}(dx_i)\mu_{x_i}(\cdot)$.

By (10.26) we have that $\pi(A) = \pi_0^*(A_0 \cup A_1) = \pi_0(A)$, so that in fact $\check{\pi} = \pi^*$. This proves one part of (i), and we now show that π is invariant for Φ. For any $A \in \mathcal{B}(X)$ we have by invariance of π^* and (5.10),

$$\pi(A) = \pi^*(A_0 \cup A_1) = \pi^*\check{P}(A_0 \cup A_1) = \left(\pi P\right)^*(A_0 \cup A_1) = \pi P(A),$$

which shows that π is invariant and completes the proof of (i).

The proof of (ii) also follows easily from (5.10): if the measure μ is subinvariant then

$$\mu^*\check{P} = (\mu P)^* \le \mu^*,$$

which establishes subinvariance of μ^*, and similarly, $\mu^* \check{P} = \mu^*$ if μ is strictly invariant.

$\square$

We can now give a simple proof of

Proposition 10.4.2. *If Φ is recurrent and strongly aperiodic, then Φ admits a unique (up to constant multiples) subinvariant measure which is invariant.*

PROOF   Assume that Φ is strongly aperiodic, and split the chain as in Section 5.1.

If Φ is recurrent then it follows from Proposition 8.2.2 that $\check{\Phi}$ is also recurrent. We have from Theorem 10.2.1 that $\check{\Phi}$ has a unique subinvariant measure $\check{\pi}$ which is invariant. Thus we have from Proposition 10.4.1 that Φ also has an invariant measure.

The uniqueness is equally easy. If Φ has another subinvariant measure μ, then by Proposition 10.4.1 the split measure μ^* is subinvariant for $\check{\Phi}$, and since from Theorem 10.2.1, the invariant measure $\check{\pi}$ is unique (up to constant multiples) for $\check{\Phi}$, we must have for some $c > 0$ that $\mu^* = c\check{\pi}$. By linearity this gives $\mu = c\pi$ as required. $\square$

We can, quite easily, lift this result to the whole chain even in the case where we do not have strong aperiodicity by considering the resolvent chain, since the chain and the resolvent share the same invariant measures.

Theorem 10.4.3. *For any $\varepsilon \in (0,1)$, a measure π is invariant for the resolvent K_{a_ε} if and only if it is invariant for P.*

PROOF   If π is invariant with respect to P then by (10.4) it is also invariant for K_a, for any sampling distribution a.

To see the converse, suppose that π satisfies $\pi K_{a_\varepsilon} = \pi$ for some $\varepsilon \in (0,1)$, and consider the chain of equalities

$$
\begin{aligned}
\pi P &= (1-\varepsilon) \sum_{k=0}^{\infty} \varepsilon^k \pi P^{k+1} \\
&= (1-\varepsilon)\varepsilon^{-1} \left(\sum_{k=0}^{\infty} \varepsilon^k \pi P^k - \pi \right) \\
&= \varepsilon^{-1}(\pi K_{a_\varepsilon} - (1-\varepsilon)\pi) \\
&= \pi.
\end{aligned}
$$

$\square$

This now gives us immediately

Theorem 10.4.4. *If Φ is recurrent then Φ has a unique (up to constant multiples) subinvariant measure which is invariant.*

PROOF   Using Theorem 5.2.3, we have that the K_{a_ε}-chain is strongly aperiodic, and from Theorem 8.2.4 we know that the K_{a_ε}-chain is recurrent. Let π be the unique invariant measure for the K_{a_ε}-chain, guaranteed from Proposition 10.4.2. From Theorem 10.4.3, π is also invariant for Φ.

Suppose that μ is subinvariant for Φ. Then by (10.6) we have that μ is also subinvariant for the K_{a_ε}-chain, and so there is a constant $c > 0$ such that $\mu = c\pi$. Hence we have shown that π is the unique (up to constant multiples) invariant measure for Φ. $\qquad\square$

We may now equate positivity of Φ to positivity for its skeletons as well as the resolvent chains.

Theorem 10.4.5. *Suppose that Φ is ψ-irreducible and aperiodic. Then, for each m, a measure π is invariant for the m-skeleton if and only if it is invariant for Φ.*

Hence, under aperiodicity, the chain Φ is positive if and only if each of the m-skeletons Φ^m is positive.

PROOF   If π is invariant for Φ then it is obviously invariant for Φ^m, by (10.4).

Conversely, if π_m is invariant for the m-skeleton then by aperiodicity the measure π_m is the unique invariant measure (up to constant multiples) for Φ^m. In this case write

$$\pi(A) = \frac{1}{m} \sum_{k=0}^{m-1} \int \pi_m(dw) P^k(w, A), \qquad A \in \mathcal{B}(\mathsf{X}).$$

From the P^m-invariance we have, using operator theoretic notation,

$$\pi P = \frac{1}{m} \sum_{k=0}^{m-1} \pi_m P^{k+1} = \pi$$

so that π is an invariant measure for P. Moreover, since π is invariant for P, it is also invariant for P^m from (10.4), and so by uniqueness of π_m, for some $c > 0$ we have $\pi = c\pi_m$. But as π is invariant for P^j for every j, we have from the definition that

$$\pi = c^{-1} \frac{1}{m} \sum_{k=0}^{m-1} \int \pi P^{k+1} = c^{-1}\pi$$

and so $\pi_m = \pi$. $\qquad\square$

10.4.2   Minimal subinvariant measures

In order to use invariant measures for recurrent chains, we shall study in some detail the structure of the invariant measures we have now proved to exist in Theorem 10.2.1. We do this through the medium of subinvariant measures, and we note that, in this section at least, we do not need to assume any form of irreducibility. Our goal is essentially to give a more general version of Kac's Theorem.

Assume that μ is an arbitrary subinvariant measure, and let $A \in \mathcal{B}(\mathsf{X})$ be such that $0 < \mu(A) < \infty$. Define the measure μ_A° by

$$\mu_A^\circ(B) = \int_A \mu(dy) U_A(y, B), \quad B \in \mathcal{B}(\mathsf{X}). \qquad (10.27)$$

Proposition 10.4.6. *The measure μ_A° is subinvariant, and minimal in the sense that $\mu(B) \geq \mu_A^\circ(B)$ for all $B \in \mathcal{B}(\mathsf{X})$.*

PROOF   If μ is subinvariant, then we have first that

$$\mu(B) \geq \int_A \mu(dw)P(w,B);$$

assume inductively that $\mu(B) \geq \int_A \mu(dw) \sum_{m=1}^n {}_AP^m(w,B)$, for all B. Then, by subinvariance,

$$\begin{aligned}
\mu(B) &\geq \int_{A^c} \left[\int_A \mu(dw) \sum_{m=1}^n {}_AP^m(w,dv) \right] P(v,B) + \int_A \mu(dw)P(w,B) \\
&= \int_A \mu(dw) \sum_{m=1}^{n+1} {}_AP^m(w,B).
\end{aligned}$$

Hence the induction holds for all n, and taking $n \uparrow \infty$ shows that

$$\mu(B) \geq \int_A \mu(dw)U_A(w,B)$$

for all B. Now by this minimality of μ_A°

$$\begin{aligned}
\mu_A^\circ(B) &= \int_A \mu(dw)P(w,B) + \int_A \mu(dw) \sum_{m=2}^\infty {}_AP^m(w,B) \\
&\geq \int_A \mu_A^\circ(dw)P(w,B) + \int_{A^c} [\int_A \mu(dw) \sum_{m=1}^\infty {}_AP^m(w,dv)]P(v,B) \\
&= \int_X \mu_A^\circ(dw)P(w,B).
\end{aligned}$$

Hence μ_A° is subinvariant also.   □

Recall that we define $\overline{A} := \{x : L(x,A) > 0\}$. We now show that if the set A in the definition of μ_A° is Harris recurrent, the minimal subinvariant measure is in fact invariant and identical to μ itself on $\overline{A}$.

Theorem 10.4.7. *If $L(x,A) \equiv 1$ for μ-almost all $x \in A$, then we have*

(i) $\mu(B) = \mu_A^\circ(B)$ *for $B \subset \overline{A}$;*

(ii) μ_A° *is invariant and $\mu_A^\circ(\overline{A}^c) = 0$.*

PROOF   **(i)**   We first show that $\mu(B) = \mu_A^\circ(B)$ for $B \subseteq A$.
   For any $B \subseteq A$, since $L(x,A) \equiv 1$ for μ-almost all $x \in A$, we have from minimality of μ_A°

$$\begin{aligned}
\mu(A) &= \mu(B) + \mu(A \cap B^c) \\
&\geq \mu_A^\circ(B) + \mu_A^\circ(A \cap B^c) \\
&= \int_A \mu(dw)U_A(w,B) + \int_A \mu(dw)U_A(w,A \cap B^c) \\
&= \int_A \mu(dw)U_A(w,A) = \mu(A). \tag{10.28}
\end{aligned}$$

Hence, the inequality $\mu(B) \geq \mu_A^\circ(B)$ must be an equality for all $B \subseteq A$. Thus the measure μ satisfies

$$\mu(B) = \int_A \mu(dw) U_A(w, B) \qquad (10.29)$$

whenever $B \subseteq A$.

We now use (10.29) to prove invariance of μ_A°. For any $B \in \mathcal{B}(\mathsf{X})$,

$$
\begin{aligned}
\int_\mathsf{X} \mu_A^\circ(dy) P(y, B) &= \int_A \mu_A^\circ(dy) P(y, B) \\
&\quad + \int_{A^c} \left[\int_A \mu_A^\circ(dw) U_A(w, dy) \right] P(y, B) \\
&= \int_A \mu_A^\circ(dy) \left[P(y, B) + \sum_2^\infty {}_A P^n(y, B) \right] \\
&= \mu_A^\circ(B) \qquad (10.30)
\end{aligned}
$$

and so μ_A° is invariant for $\boldsymbol{\Phi}$. It follows by definition that $\mu_A^\circ(\overline{A}^c) = 0$, so (ii) is proved.

We now prove (i) by contradiction. Suppose that $B \subseteq \overline{A}$ with $\mu(B) > \mu_A^\circ(B)$. Then we have from invariance of the resolvent chain in Proposition 10.4.3 and minimality of μ_A°, and the assumption that $K_{a_\varepsilon}(x, A) > 0$ for $x \in B$,

$$\mu(A) \geq \int_\mathsf{X} \mu(dy) K_{a_\varepsilon}(y, A) > \int_\mathsf{X} \mu_A^\circ(dy) K_{a_\varepsilon}(y, A) = \mu_A^\circ(A) = \mu(A),$$

and we thus have a contradiction. □

An interesting consequence of this approach is the identity (10.29). This has the following interpretation. Assume A is Harris recurrent, and define the *process on A*, denoted by $\boldsymbol{\Phi}^A = \{\Phi_n^A\}$, by starting with $\Phi_0 = x \in A$, then setting Φ_1^A as the value of $\boldsymbol{\Phi}$ at the next visit to A, and so on. Since return to A is sure for Harris recurrent sets, this is well defined.

Formally, $\boldsymbol{\Phi}^A$ is actually constructed from the transition law

$$U_A(x, B) = \sum_{n=1}^\infty {}_A P^n(x, B) = \mathsf{P}_x\{\Phi_{\tau_A} \in B\},$$

$B \subseteq A$, $B \in \mathcal{B}(\mathsf{X})$. Theorem 10.4.7 thus states that for a Harris recurrent set A, any subinvariant measure restricted to A is actually invariant for the process on A.

One can also go in the reverse direction, starting off with an invariant measure for the process on A. The following result is proved using the same calculations used in (10.30):

Proposition 10.4.8. *Suppose that ν is an invariant probability measure supported on the set A with*

$$\int_A \nu(dx) U_A(x, B) = \nu(B), \qquad B \subseteq A.$$

Then the measure ν° defined as

$$\nu^\circ(B) := \int_A \nu(dx) U_A(x, B), \qquad B \in \mathcal{B}(\mathsf{X}),$$

is invariant for $\boldsymbol{\Phi}$. □

10.4.3   The structure of π for recurrent chains

These preliminaries lead to the following key result.

Theorem 10.4.9. *Suppose Φ is recurrent. Then the unique (up to constant multiples) invariant measure π for Φ is equivalent to ψ and satisfies for any $A \in \mathcal{B}^+(\mathsf{X})$, $B \in \mathcal{B}(\mathsf{X})$,*

$$
\begin{aligned}
\pi(B) &= \int_A \pi(dy) U_A(y, B) \\
&= \int_A \pi(dy) \mathsf{E}_y \Big[\sum_{k=1}^{\tau_A} \mathbb{I}\{\Phi_k \in B\} \Big] \\
&= \int_A \pi(dy) \mathsf{E}_y \Big[\sum_{k=0}^{\tau_A - 1} \mathbb{I}\{\Phi_k \in B\} \Big].
\end{aligned}
\tag{10.31}
$$

PROOF   The construction in Theorem 10.2.1 ensures that the invariant measure π exists. Hence from Theorem 10.4.7 we see that $\pi = \pi_A^\circ$ for any Harris recurrent set A, and π then satisfies the first equality in (10.31) by construction. The second equality is just the definition of U_A. To see the third equality,

$$
\int_A \pi(dy) \mathsf{E}_y \Big[\sum_{k=1}^{\tau_A} \mathbb{I}\{\Phi_k \in B\} \Big] = \int_A \pi(dy) \mathsf{E}_y \Big[\sum_{k=0}^{\tau_A - 1} \mathbb{I}\{\Phi_k \in B\} \Big],
$$

apply (10.29) which implies that

$$
\int_A \pi(dy) \mathsf{E}_y [\mathbb{I}\{\Phi_{\tau_A} \in B\}] = \int_A \pi(dy) \mathsf{E}_y [\mathbb{I}\{\Phi_0 \in B\}].
$$

We finally prove that $\pi \cong \psi$. From Proposition 10.1.2 we need only show that if $\psi(B) = 0$ then also $\pi(B) = 0$. But since $\psi(\bar{B}) = 0$, we have that $B^0 \in \mathcal{B}^+(\mathsf{X})$, and so from the representation (10.31),

$$
\pi(B) = \int_{B^0} \pi(dy) U_{B^0}(y, B) = 0,
$$

which is the required result. $\square$

The interpretation of (10.31) is this: for a fixed set $A \in \mathcal{B}^+(\mathsf{X})$, the invariant measure $\pi(B)$ is proportional to the amount of time spent in B between visits to A, provided the chain starts in A with the distribution π_A which is invariant for the chain Φ^A on A.

When A is a single point, α, with $\pi(\alpha) > 0$ then each visit to α occurs at α. The chain Φ^α is hence trivial, and its invariant measure π_α is just δ_α. The representation (10.31) then reduces to μ_α given in Theorem 10.2.1.

We will use these concepts systematically in building the asymptotic theory of positive chains in Chapter 13 and later work, and in Chapter 11 we develop a number of conditions equivalent to positivity through this representation of π. The next result is a foretaste of that work.

Theorem 10.4.10. *Suppose that Φ is ψ-irreducible, and let μ denote any subinvariant measure.*

(i) *The chain $\mathbf{\Phi}$ is positive if and only if for one, and then every, set with $\mu(A) > 0$*

$$\int_A \mu(dy)\mathsf{E}_y[\tau_A] < \infty. \tag{10.32}$$

(ii) *The measure μ is finite and thus $\mathbf{\Phi}$ is positive recurrent if for some petite set $C \in \mathcal{B}^+(\mathsf{X})$*

$$\sup_{y \in C} \mathsf{E}_y[\tau_C] < \infty. \tag{10.33}$$

The chain $\mathbf{\Phi}$ is positive Harris if also

$$\mathsf{E}_x[\tau_C] < \infty, \qquad x \in \mathsf{X}. \tag{10.34}$$

PROOF   The first result is a direct consequence of (10.27), since we have

$$\mu_A^\circ(\mathsf{X}) = \int_A \mu(dy)U_A(y,\mathsf{X}) = \int_A \mu(dy)\mathsf{E}_y[\tau_A];$$

if this is finite then μ_A° is finite and the chain is positive by definition. Conversely, if the chain is positive then by Theorem 10.4.9 we know that μ must be a finite invariant measure and (10.32) then holds for every A.

The second result now follows since we know from Proposition 10.1.2 that $\mu(C) < \infty$ for petite C; and hence we have positive recurrence from (10.33) and (i), whilst the chain is also Harris if (10.34) holds from the criterion in Theorem 9.1.7.   □

In Chapter 11 we find a variety of usable and useful conditions for (10.33) and (10.34) to hold, based on a drift approach which strengthens those in Chapter 8.

10.5 Invariant measures for general models

The constructive approach to the existence of invariant measures which we have featured so far enables us either to develop results on invariant measures for a number of models, based on the representation in (10.31), or to interpret the invariant measure probabilistically once we have determined it by some other means.

We now give a variety of examples of this.

10.5.1 Random walk

Consider the random walk on the line, with increment measure Γ, as defined in (RW1). Then by Fubini's Theorem and the translation invariance of μ^{Leb} we have for any $A \in \mathcal{B}(\mathsf{X})$

$$
\begin{aligned}
\int_{\mathbb{R}} \mu^{\text{Leb}}(dy)P(y,A) &= \int_{\mathbb{R}} \mu^{\text{Leb}}(dy)\Gamma(A-y) \\
&= \int_{\mathbb{R}} \mu^{\text{Leb}}(dy)\int_{\mathbb{R}} \mathbb{I}_{A-y}(x)\Gamma(dx) \\
&= \int_{\mathbb{R}} \Gamma(dx)\int_{\mathbb{R}} \mathbb{I}_{A-x}(y)\mu^{\text{Leb}}(dy) \tag{10.35} \\
&= \mu^{\text{Leb}}(A)
\end{aligned}
$$

since $\Gamma(\mathbb{R}) = 1$. We have already used this formula in (6.8): here it shows that Lebesgue measure is invariant for unrestricted random walk in either the transient or the recurrent case.

Since Lebesgue measure on $\mathbb{R}$ is infinite, we immediately have from Theorem 10.4.9 that there is no finite invariant measure for this chain: this proves

Proposition 10.5.1. *The random walk on $\mathbb{R}$ is never positive recurrent.* $\square$

If we put this together with the results in Section 9.5, then we have that when the mean β of the increment distribution is zero, then the chain is null recurrent.

Finally, we note that this is one case where the interpretation in (10.31) can be expressed in another way. We have, as an immediate consequence of this interpretation

Proposition 10.5.2. *Suppose Φ is a random walk on $\mathbb{R}$, with spread-out increment measure Γ having zero mean and finite variance.*

Let A be any bounded set in $\mathbb{R}$ with $\mu^{\text{Leb}}(A) > 0$, and let the initial distribution of Φ_0 be the uniform distribution on A. If we let $N_A(B)$ denote the mean number of visits to a set B prior to return to A, then for any two bounded sets B, C with $\mu^{\text{Leb}}(C) > 0$ we have

$$\mathsf{E}[N_A(B)]/\mathsf{E}[N_A(C)] = \mu^{\text{Leb}}(B)/\mu^{\text{Leb}}(C).$$

PROOF   Under the given conditions on Γ we have from Proposition 9.4.5 that the chain is non-evanescent, and hence recurrent.

Using (10.35) we have that the unique invariant measure with $\pi(A) = 1$ is $\pi = \mu^{\text{Leb}}/\pi(A)$, and then the result follows from the form (10.31) of π. $\square$

10.5.2   Forward recurrence time chains

Let us consider the forward recurrence time chain V_δ^+ defined in Section 3.5 for a renewal process on $\mathbb{R}_+$. For any fixed δ consider the expected number of visits to an interval strictly outside $[0, \delta]$. Exactly as we reasoned in the discrete time case studied in Section 10.3, we have

$$F[y, \infty)dy \leq U_{[0,\delta]}(x, dy) \leq F[y - \delta, \infty)dy.$$

Thus, if π_δ is to be the invariant probability measure for V_δ^+, by using the normalized version of the representation (10.31) we obtain

$$\frac{F[y, \infty)dy}{[\int_0^\infty F(w, \infty)dw]} \leq \pi_\delta(dy) \leq \frac{F[y - \delta, \infty)dy}{[\int_\delta^\infty F(w, \infty)dw]}.$$

Now we use uniqueness of the invariant measure to note that, since the chain V_δ^+ is the "two-step" chain for the chain $V_{\delta/2}^+$, the invariant measures π_δ and $\pi_{\delta/2}$ must coincide. Thus letting δ go to zero through the values $\delta/2^n$ we find that for any δ the invariant measure is given by

$$\pi_\delta(dy) = m^{-1}F[y, \infty)dy \tag{10.36}$$

where $m = \int_0^\infty tF(dt)$; and π_δ is a probability measure provided $m < \infty$.

By direct integration it is also straightforward to show that this is indeed the invariant measure for V_δ^+.

This form of the invariant measure thus reinforces the fact that the quantity $F[y, \infty)dy$ is the expected amount of time spent in the infinitesimal set dy on each excursion from the point $\{0\}$, even though in the discretized chain V_δ^+ the point $\{0\}$ is never actually reached.

10.5.3 Ladder chains and GI/G/1 queues

General ladder chains

We will now turn to a more complex structure and see how far the representation of the invariant measure enables us to carry the analysis.

Recall from Section 3.5.4 the Markov chain constructed on $\mathbb{Z}_+ \times \mathbb{R}$ to analyze the GI/G/1 queue, with the "ladder-invariant" transition kernel

$$
\begin{aligned}
P(i, x; j \times A) &= 0, \quad j > i + 1, \\
P(i, x; j \times A) &= \Lambda_{i-j+1}(x, A), \quad j = 1, \ldots, i+1, \\
P(i, x; 0 \times A) &= \Lambda_i^*(x, A).
\end{aligned}
\tag{10.37}
$$

Let us consider the general chain defined by (10.37), where we can treat x and A as general points in and subsets of X, so that the chain Φ now moves on a ladder whose (countable number of) rungs are general in nature. In the special case of the GI/G/1 model the results specialize to the situation where $\mathsf{X} = \mathbb{R}_+$, and there are many countable models where the rungs are actually finite and matrix methods are used to achieve the following results.

Using the representation of π, it is possible to construct an invariant measure for this chain in an explicit way; this then gives the structure of the invariant measure for the GI/G/1 queue also.

Since we are interested in the structure of the invariant probability measure we make the assumption in this section that the chain defined by (10.37) is positive Harris and $\psi([0]) > 0$, where $[0] := \{0 \times \mathsf{X}\}$ is the bottom "rung" of the ladder. We shall explore conditions for this to hold in Chapter 19.

Our assumption ensures we can reach the bottom of the ladder with probability one. Let us denote by π_0 the invariant probability measure for the process on $[0]$, so that π_0 can be thought of as a measure on $\mathcal{B}(\mathsf{X})$.

Our goal will be to prove that the structure of the invariant measure for Φ is an "operator-geometric" one, mimicking the structure of the invariant measure developed in Section 10.3 for skip-free random walk on the integers.

Theorem 10.5.3. *The invariant measure π for Φ is given by*

$$
\pi(k \times A) = \int_\mathsf{X} \pi_0(dy) S^k(y, A),
\tag{10.38}
$$

where

$$
S^k(y, A) = \int_\mathsf{X} S(y, dz) S^{k-1}(z, A)
\tag{10.39}
$$

for a kernel S which is the minimal solution of the operator equation

$$S(y, B) = \sum_{k=0}^{\infty} \int_{\mathsf{X}} S^k(y, dz) \Lambda_k(z, B), \qquad x \in \mathsf{X}, B \in \mathcal{B}(\mathsf{X}). \qquad (10.40)$$

PROOF   Using the structural result (10.31) we have

$$\pi(k \times A) = \int_{[0]} \pi_0(dy) U_{[0]}(0, y; k \times B) \qquad (10.41)$$

so that if we write

$$S^{(k)}(y, A) := U_{[0]}(0, y; k \times A) \qquad (10.42)$$

we have by definition

$$\pi(k \times A) = \int_{[0]} \pi_0(dy) S^{(k)}(y, A). \qquad (10.43)$$

Now if we define the set $[n] = \{0, 1, \dots, n\} \times \mathsf{X}$, by the fact that the chain is translation invariant above the zero level we have that the functions

$$U_{[n]}(n, y; (n+k) \times B) = U_{[0]}(0, y; k \times B) = S^{(k)}(y, A) \qquad (10.44)$$

are independent of n. Using a last-exit decomposition over visits to $[k]$, together with the skip-free property which ensures that the last visit to $[k]$ prior to reaching $(k+1) \times \mathsf{X}$ takes place at the level $k \times \mathsf{X}$, we find

$$\begin{aligned}
{}_{[0]}P^\ell&(0, x; (k+1) \times A) \\
&= \sum_{j=1}^{\ell-1} \int_{\mathsf{X}} {}_{[0]}P^j(0, x; k \times dy)_{[k]} P^{\ell-j}(k, y; (k+1) \times A) \qquad (10.45) \\
&= \sum_{j=1}^{\ell-1} \int_{\mathsf{X}} {}_{[0]}P^j(0, x; k \times dy)_{[0]} P^{\ell-j}(0, y; 1 \times A).
\end{aligned}$$

Summing over ℓ and using (10.44) shows that the operators $S^{(k)}(y, A)$ have the geometric form in (10.39) as stated.

To see that the operator S satisfies (10.40), we decompose ${}_{[0]}P^n$ over the position at time $n - 1$. By construction ${}_{[0]}P^1(0, x; 1 \times B) = \Lambda_0(x, B)$, and for $n > 1$,

$$_{[0]}P^n(0, x; 1 \times B) = \sum_{k \geq 1} \int_{\mathsf{X}} {}_{[0]}P^{n-1}(0, x; k \times dy) \Lambda_k(y, B); \qquad (10.46)$$

summing over n and using (10.39) gives the result (10.40).

To prove minimality of the solution S to (10.40), we first define, for $N \geq 1$, the partial sums

$$S_N(x; k \times B) := \sum_{j=1}^{N} {}_{[0]}P^j(0, x; k \times B) \qquad (10.47)$$

so that as $N \to \infty$, $S_N(x; 1 \times B) \to S(x; B)$.

Using (10.45) these partial sums also satisfy

$$S_{N-1}(x; k+1 \times B) \leq \int S_{N-1}(x; k \times dy) S_{N-1}(y; 1 \times B)$$

so that

$$S_{N-1}(x; k+1 \times B) \leq \int S_{N-1}^k(x; 1 \times dy) S_{N-1}(y; 1 \times B). \tag{10.48}$$

Moreover from (10.46)

$$S_N(x; 1 \times B) = \Lambda_0(x, B) + \sum_{k \geq 1} \int_X S_{N-1}(x; k \times dy) \Lambda_k(y, B). \tag{10.49}$$

Substituting from (10.48) in (10.49) shows that

$$S_N(x; 1, B) \leq \sum_k \int_X S_{N-1}^k(x; 1, dy) \Lambda_k(y, B). \tag{10.50}$$

Now let S^* be any other solution of (10.40). Notice that $S_1(x; 1 \times B) = \Lambda_0(x, B) \leq S^*(x, B)$, from (10.40). Assume inductively that $S_{N-1}(x; 1 \times B) \leq S^*(x, B)$ for all x, B: then we have from (10.50) that

$$S_N(x; 1 \times B) \leq \sum_k \int_X [S^*]^k(x, dy) \Lambda_k(y, B) = S^*(x, B). \tag{10.51}$$

Taking limits as $N \to \infty$ gives $S(x, B) \leq S^*(x, B)$ for all x, B as required.   □

This result is a generalized version of (10.24) and (10.25), where the "rungs" on the ladder were singletons.

The GI/G/1 queue

Note that in the ladder processes above, the returns to the bottom rung of the ladder, governed by the kernels Λ_i^* in (10.37), only appear in the representation (10.38) implicitly, through the form of the invariant measure π_0 for the process on the set [0].

In particular cases it is of course of critical importance to identify this component of the invariant measure also. In the case of a singleton rung, this is trivial since the rung is an atom. This gives the explicit form in (10.24) and (10.25).

We have seen in Section 3.5 that the general ladder chain is a model for the GI/G/1 queue, if we make the particular choice of

$$\Phi_n = (N_n, R_n), \quad n \geq 1$$

where N_n is the number of customers at $T_n'-$ and R_n is the residual service time at $T_n'+$. In this case the representation of $\pi_{[0]}$ can also be made explicit.

For the GI/G/1 chain we have that the chain on [0] has the distribution of R_n at a time point $\{T_n'+\}$ where there were no customers at $\{T_n'-\}$: so at these time points R_n has precisely the distribution of the service brought by the customer arriving at T_n', namely H.

So in this case we have that the process on [0], provided [0] is recurrent, is a process of i.i.d random variables with distribution H, and thus is very clearly positive Harris with invariant probability H.

Theorem 10.5.3 then gives us

Theorem 10.5.4. *The ladder chain* Φ *describing the GI/G/1 queue has an invariant probability if and only if the measure* π *given by*

$$\pi(k \times A) = \int_{\mathsf{X}} H(dy) S^k(y, A) \tag{10.52}$$

is a finite measure, where S *is the minimal solution of the operator equation*

$$S(y, B) = \sum_{k=0}^{\infty} \int_{\mathsf{X}} S^k(y, dz) \Lambda_k(z, B), \qquad x \in \mathsf{X}, B \in \mathcal{B}(\mathsf{X}). \tag{10.53}$$

In this case π *suitably normalized is the unique invariant probability measure for* Φ.

PROOF   Using the proof of Theorem 10.5.3 we have that π is the minimal subinvariant measure for the GI/G/1 queue, and the result is then obvious. $\square$

10.5.4   Linear state space models

We now consider briefly a chain where we utilize the property (10.4) to develop the form of the invariant measure. We will return in much more detail to this approach in Chapter 12.

     We have seen in (10.4) that limiting distributions provide invariant probability measures for Markov chains, provided such limits exist. The linear model has a structure which makes it easy to construct an invariant probability through this route, rather than through the minimal measure construction above.

     Suppose that (LSS1) and (LSS2) are satisfied, and observe that since $\boldsymbol{W}$ is assumed i.i.d. we have for each initial condition $X_0 = x_0 \in \mathbb{R}^n$,

$$
\begin{aligned}
X_k &= F^k x_0 + \sum_{i=0}^{k-1} F^i G W_{k-i} \\
&\sim F^k x_0 + \sum_{i=0}^{k-1} F^i G W_i.
\end{aligned}
$$

This says that for any continuous, bounded function $g \colon \mathbb{R}^n \to \mathbb{R}$,

$$P^k g(x_0) = \mathsf{E}_{x_0}[g(X_k)] = \mathsf{E}[g(F^k x_0 + \sum_{i=0}^{k-1} F^i G W_i)].$$

Under the additional hypothesis that the eigenvalue condition (LSS5) holds, it follows from Lemma 6.3.4 that $F^i \to 0$ as $i \to \infty$ at a geometric rate. Since W has a finite mean then it follows from Fubini's Theorem that the sum

$$X_\infty := \sum_{i=0}^{\infty} F^i G W_i$$

converges absolutely, with $E[\|X_\infty\|] \leq E[\|W\|] \sum_{i=0}^{\infty} \|F^i G\| < \infty$, with $\|\cdot\|$ an appropriate matrix norm. Hence by the Dominated Convergence Theorem, and the assumption that g is continuous,

$$\lim_{k \to \infty} P^k g(x_0) = E[g(X_\infty)].$$

Let us write π_∞ for the distribution of X_∞. Then π_∞ is an invariant probability. For take g bounded and continuous as before, so that using the Feller property for X in Chapter 6 we have that Pg is continuous. For such a function g

$$\begin{aligned}
\pi_\infty(Pg) = E[Pg(X_\infty)] &= \lim_{k \to \infty} P^k(x_0, Pg) \\
&= \lim_{k \to \infty} P^{k+1} g(x_0) \\
&= E[g(X_\infty)] = \pi_\infty(g).
\end{aligned}$$

Since π is determined by its values on continuous bounded functions, this proves that π is invariant.

In the Gaussian case (LSS3) we can express the invariant probability more explicitly. In this case X_∞ itself is Gaussian with mean zero and covariance

$$E[X_\infty X_\infty^\top] = \sum_{k=0}^{\infty} F^i G G^\top F^{i\top}.$$

That is, $\pi = N(0, \Sigma)$ where Σ is equal to the controllability grammian for the linear state space model, defined in (4.17).

The covariance matrix Σ is full rank if and only if the controllability condition (LCM3) holds, and in this case, for any k greater than or equal to the dimension of the state space, $P^k(x, dy)$ possesses the density $p_k(x, y)dy$ given in (4.18). It follows immediately that when (LCM3) holds, the probability π possesses the density p on $\mathbb{R}^n$ given by

$$p(y) = (2\pi|\Sigma|)^{-n/2} \exp\{-\tfrac{1}{2}y^T \Sigma^{-1} y\}, \tag{10.54}$$

while if the controllability condition (LCM3) fails to hold then the invariant probability is concentrated on the controllable subspace $X_0 = \mathcal{R}(\Sigma) \subset X$ and is hence singular with respect to Lebesgue measure.

10.6 Commentary

The approach to positivity given here is by no means standard. It is much more common, especially with countable spaces, to classify chains either through the behavior of the sequence P^n, with null chains being those for which $P^n(x, A) \to 0$ for, say, petite sets A and all x, and positive chains being those for which such limits are not always zero; a limiting argument such as that in (10.4), which we have illustrated in Section 10.5.4, then shows the existence of π in the positive case.

Alternatively, positivity is often defined through the behavior of the expected return times to petite or other suitable sets.

We will show in Chapter 11 and Chapter 18 that even on a general space all of these approaches are identical. Our view is that the invariant measure approach is much more straightforward to understand than the P^n approach, and since one can now develop through the splitting technique a technically simple set of results this gives an appropriate classification of recurrent chains.

The existence of invariant probability measures has been a central topic of Markov chain theory since the inception of the subject. Doob [99] and Orey [309] give some good background. The approach to countable recurrent chains through last-exit probabilities as in Theorem 10.2.1 is due to Derman [86], and has not changed much since, although the uniqueness proofs we give owe something to Vere-Jones [406]. The construction of π given here is of course one of our first serious uses of the splitting method of Nummelin [301]; for strongly aperiodic chains the result is also derived in Athreya and Ney [13]. The fact that one identifies the actual structure of π in Theorem 10.4.9 will also be of great use, and Kac's Theorem [186] provides a valuable insight into the probabilistic difference between positive and null chains: this is pursued in the next chapter in considerably more detail.

Before the splitting technique, verifying conditions for the existence of π had appeared to be a deep and rather difficult task. It was recognized in the relatively early development of general state space Markov chains that one could prove the existence of an invariant measure for Φ from the existence of an invariant probability measure for the "process on A". The approach pioneered by Harris [155] for finding the latter involves using deeper limit theorems for the "process on A" in the special case where A is a ν_n-small set, (called a C-set in Orey [309]) if $a_n = \delta_n$ and $\nu_n\{A\} > 0$. In this methodology, it is first shown that limiting probabilities for the process on A exist, and the existence of such limits then provides an invariant measure for the process on A: by the construction described in this chapter this can be lifted to an invariant measure for the whole chain. Orey [309] remains an excellent exposition of the development of this approach.

This "process on A" method is still the only one available without some regeneration, and we will develop this further in a topological setting in Chapter 12, using many of the constructions above.

We have shown that invariant measures exist without using such deep asymptotic properties of the chain, indicating that the existence and uniqueness of such measures is in fact a result requiring less of the detailed structure of the chain.

The minimality approach of Section 10.4.2 of course would give another route to Theorem 10.4.4, provided we had some method of proving that a "starting" subinvariant measure existed. There is one such approach, which avoids splitting and remains conceptually simple. This involves using the kernels

$$U^{(r)}(x, A) = \sum_{n=1}^{\infty} P^n(x, A) r^n \geq r \int_{\mathsf{X}} U^{(r)}(x, dy) P(y, A) \qquad (10.55)$$

defined for $0 < r < 1$. One can then define a subinvariant measure for Φ as a limit

$$\lim_{r \uparrow 1} \pi_r(\,\cdot\,) := \lim_{r \uparrow 1} [\int_C \nu_n(dy) U^{(r)}(y, \,\cdot\,)] / [\int_C \nu_n(dy) U^{(r)}(y, C)]$$

where C is a ν_n-small set. The key is the observation that this limit gives a non-trivial σ-finite measure due to the inequalities

$$M_j \geq \pi_r(\bar{C}(j)) \qquad (10.56)$$

and

$$\pi_r(A) \geq r^n \nu_n(A), \qquad A \in \mathcal{B}(\mathsf{X}), \qquad\qquad (10.57)$$

which are valid for all r large enough. Details of this construction are in Arjas and Nummelin [7], as is a neat alternative proof of uniqueness.

All of these approaches are now superseded by the splitting approach, but of course only when the chain is ψ-irreducible. If this is not the case then the existence of an invariant measure is not simple. The methods of Section 10.4.2, which are based on Tweedie [402], do not use irreducibility, and in conjunction with those in Chapter 12 they give some ways of establishing uniqueness and structure for the invariant measures from limiting operations, as illustrated in Section 10.5.4.

The general question of existence and, more particularly, uniqueness of invariant measures for non-irreducible chains remains open at this stage of theoretical development.

The invariance of Lebesgue measure for random walk is well known, as is the form (10.36) for models in renewal theory. The invariant measures for queues are derived directly in [59], but the motivation through the minimal measure of the geometric form is not standard. The extension to the operator-geometric form for ladder chains is in [399], and in the case where the rungs are finite, the development and applications are given by Neuts [293, 294].

The linear model is analyzed in Snyders [364] using ideas from control theory, and the more detailed analysis given there allows a generalization of the construction given in Section 10.5.4. Essentially, if the noise does not enter the "unstable" region of the state space then the stability condition on the driving matrix F can be slightly weakened.

Chapter 11

Drift and regularity

Using the finiteness of the invariant measure to classify two different levels of stability is intuitively appealing. It is simple, and it also involves a fundamental stability requirement of many classes of models. Indeed, in time series analysis for example, a standard starting point, rather than an end point, is the requirement that the model be stationary, and it follows from (10.4) that for a stationary version of a model to exist we are in effect requiring that the structure of the model be positive recurrent.

In this chapter we consider two other descriptions of positive recurrence which we show to be equivalent to that involving finiteness of π.

The first is in terms of regular sets.

Regularity

A set $C \in \mathcal{B}(\mathsf{X})$ is called *regular* when $\boldsymbol{\Phi}$ is ψ-irreducible, if

$$\sup_{x \in C} \mathsf{E}_x[\tau_B] < \infty, \qquad B \in \mathcal{B}^+(\mathsf{X}). \tag{11.1}$$

The chain $\boldsymbol{\Phi}$ is called *regular* if there is a countable cover of X by regular sets.

We know from Theorem 10.2.1 that when there is a finite invariant measure and an atom $\alpha \in \mathcal{B}^+(\mathsf{X})$ then $\mathsf{E}_\alpha[\tau_\alpha] < \infty$. A regular set $C \in \mathcal{B}^+(\mathsf{X})$ as defined by (11.1) has the property not only that the return times to C itself, but indeed the mean hitting times on any set in $\mathcal{B}^+(\mathsf{X})$ are bounded from starting points in C.

We will see that there is a second, equivalent, approach in terms of conditions on the one-step "mean drift"

$$\Delta V(x) = \int_\mathsf{X} P(x, dy) V(y) - V(x) = \mathsf{E}_x[V(\Phi_1) - V(\Phi_0)]. \tag{11.2}$$

We have already shown in Chapter 8 and Chapter 9 that for ψ-irreducible chains, drift towards a petite set implies that the chain is recurrent or Harris recurrent, and drift away from such a set implies that the chain is transient. The high points in this chapter are the following much more wide ranging equivalences.

Theorem 11.0.1. *Suppose that* Φ *is a Harris recurrent chain, with invariant measure* π. *Then the following three conditions are equivalent:*

(i) *The measure* π *has finite total mass;*

(ii) *There exists some petite set* $C \in \mathcal{B}(\mathsf{X})$ *and* $M_C < \infty$ *such that*

$$\sup_{x \in C} \mathsf{E}_x[\tau_C] \leq M_C; \tag{11.3}$$

(iii) *There exists some petite set* C *and some extended-real-valued, non-negative test function* V, *which is finite for at least one state in* X, *satisfying*

$$\Delta V(x) \leq -1 + b\mathbb{I}_C(x), \qquad x \in \mathsf{X}. \tag{11.4}$$

When (iii) holds then V *is finite on an absorbing full set* S *and the chain restricted to* S *is regular; and any sublevel set of* V *satisfies (11.3).*

PROOF   That (ii) is equivalent to (i) is shown by combining Theorem 10.4.10 with Theorem 11.1.4, which also shows that some full absorbing set exists on which Φ is regular. The equivalence of (ii) and (iii) is in Theorem 11.3.11, whilst the identification of the set S as the set where V is finite is in Proposition 11.3.13, where we also show that sublevel sets of V satisfy (11.3). $\qquad\square$

Both of these approaches, as well as giving more insight into the structure of positive recurrent chains, provide tools for further analysis of asymptotic properties in Part III.

In this chapter, the equivalence of existence of solutions of the drift condition (11.4) and the existence of regular sets is motivated, and explained to a large degree, by the deterministic results in Section 11.2. Although there are a variety of proofs of such results available, we shall develop a particularly powerful approach via a discrete time form of Dynkin's formula.

Because it involves only the one-step transition kernel, (11.4) provides an invaluable practical criterion for evaluating the positive recurrence of specific models: we illustrate this in Section 11.4.

There exists a matching, although less important, criterion for the chain to be non-positive rather than positive: we shall also prove in Section 11.5.1 that if a test function satisfies the reverse drift condition

$$\Delta V(x) \geq 0, \qquad x \in C^c, \tag{11.5}$$

then provided the increments are bounded in mean, in the sense that

$$\sup_{x \in \mathsf{X}} \int P(x, dy)|V(x) - V(y)| < \infty, \tag{11.6}$$

the mean hitting times $\mathsf{E}_x[\tau_C]$ are infinite for $x \in C^c$.

Prior to considering drift conditions, in the next section we develop through the use of the Nummelin splitting technique the structural results which show why (11.3) holds for some petite set C, and why this "local" bounded mean return time gives bounds on the mean first entrance time to any set in $\mathcal{B}^+(\mathsf{X})$.

11.1   Regular chains

On a countable space we have a simple connection between the concept of regularity and positive recurrence.

Proposition 11.1.1. *For an irreducible chain on a countable space, positive recurrence and regularity are equivalent.*

PROOF   Clearly, from Theorem 10.2.2, positive recurrence is implied by regularity. To see the converse note that, for any fixed states $x, y \in \mathsf{X}$ and any n

$$\mathsf{E}_x[\tau_x] \geq {}_xP^n(x,y)[\mathsf{E}_y[\tau_x] + n].$$

Since the left hand side is finite for any x, and by irreducibility for any y there is some n with ${}_xP^n(x,y) > 0$, we must have $\mathsf{E}_y[\tau_x] < \infty$ for all y also.   □

It will require more work to find the connections between positive recurrence and regularity in general.

It is not implausible that positive chains might admit regular sets. It follows immediately from (10.32) that in the positive recurrent case for any $A \in \mathcal{B}^+(\mathsf{X})$ we have

$$\mathsf{E}_x[\tau_A] < \infty, \qquad \text{a.e. } x \in A \ [\pi]. \tag{11.7}$$

Thus we have from the form of π more than enough "almost-regular" sets in the positive recurrent case.

To establish the existence of true regular sets we first consider ψ-irreducible chains which possess a recurrent atom $\alpha \in \mathcal{B}^+(\mathsf{X})$. Although it appears that regularity may be a difficult criterion to meet since in principle it is necessary to test the hitting time of every set in $\mathcal{B}^+(\mathsf{X})$, when an atom exists it is only necessary to consider the first hitting time to the atom.

Theorem 11.1.2. *Suppose that there exists an accessible atom $\alpha \in \mathcal{B}^+(\mathsf{X})$.*

(i) *If Φ is positive recurrent then there exists a decomposition*

$$\mathsf{X} = S \cup N \tag{11.8}$$

   where the set S is full and absorbing, and Φ restricted to S is regular.

(ii) *The chain Φ is regular if and only if*

$$\mathsf{E}_x[\tau_\alpha] < \infty \tag{11.9}$$

   for every $x \in \mathsf{X}$.

PROOF   Let

$$S := \{x : \mathsf{E}_x[\tau_\alpha] < \infty\};$$

obviously S is absorbing, and since the chain is positive recurrent we have from Theorem 10.4.10 (ii) that $\mathsf{E}_\alpha[\tau_\alpha] < \infty$, and hence $\alpha \in S$. This also shows immediately that S is full by Proposition 4.2.3.

Let B be any set in $\mathcal{B}^+(\mathsf{X})$ with $B \subseteq \boldsymbol{\alpha}^c$, so that for π-almost all $y \in B$ we have $\mathsf{E}_y[\tau_B] < \infty$ from (11.7). From ψ-irreducibility there must then exist amongst these values one w and some n such that $_BP^n(w, \boldsymbol{\alpha}) > 0$. Since

$$\mathsf{E}_w[\tau_B] \geq {}_BP^n(w, \boldsymbol{\alpha})\mathsf{E}_{\boldsymbol{\alpha}}[\tau_B]$$

we must have $\mathsf{E}_{\boldsymbol{\alpha}}[\tau_B] < \infty$.

Let us set

$$S_n = \{y : \mathsf{E}_y[\tau_{\boldsymbol{\alpha}}] \leq n\}. \tag{11.10}$$

We have the obvious inequality for any x and any $B \in \mathcal{B}^+(\mathsf{X})$ that

$$\mathsf{E}_x[\tau_B] \leq \mathsf{E}_x[\tau_{\boldsymbol{\alpha}}] + \mathsf{E}_{\boldsymbol{\alpha}}[\tau_B] \tag{11.11}$$

so that each S_n is a regular set, and since $\{S_n\}$ is a cover of S, we have that $\boldsymbol{\Phi}$ restricted to S is regular.

This proves (i): to see (ii) note that under (11.9) we have $\mathsf{X} = S$, so the chain is regular; whilst the converse is obvious. $\qquad\square$

It is unfortunate that the ψ-null set N in Theorem 11.1.2 need not be empty. For consider a chain on $\mathbb{Z}_+$ with

$$\begin{aligned} P(0,0) &= 1, \\ P(j,0) &= \beta_j > 0, \\ P(j,j+1) &= 1 - \beta_j. \end{aligned} \tag{11.12}$$

Then the chain restricted to $\{0\}$ is trivially regular, and the whole chain is positive recurrent; but if

$$\sum_j \prod_1^j \beta_k = \infty$$

then the chain is not regular, and $N = \{1, 2, \ldots\}$ in (11.8).

It is the weak form of irreducibility we use which allows such null sets to exist: this pathology is of course avoided on a countable space under the normal form of irreducibility, as we saw in Proposition 11.1.1.

However, even under ψ-irreducibility we can extend this result without requiring an atom in the original space.

Let us next consider the case where $\boldsymbol{\Phi}$ is strongly aperiodic, and use the Nummelin splitting to define $\check{\boldsymbol{\Phi}}$ on $\check{\mathsf{X}}$ as in Section 5.1.1.

Proposition 11.1.3. *Suppose that $\boldsymbol{\Phi}$ is strongly aperiodic and positive recurrent. Then there exists a decomposition*

$$\mathsf{X} = S \cup N \tag{11.13}$$

where the set S is full and absorbing, and $\boldsymbol{\Phi}$ restricted to S is regular.

PROOF   We know from Proposition 10.4.2 that the split chain is also positive recurrent with invariant probability measure $\check{\pi}$, and thus for $\check{\pi}$-a.e. $x_i \in \check{X}$, by (11.7) we have that

$$\check{\mathsf{E}}_{x_i}[\tau_{\check{\alpha}}] < \infty. \tag{11.14}$$

Let $\check{S} \subseteq \check{X}$ denote the set where (11.14) holds. Then it is obvious that $\check{S}$ is absorbing, and by Theorem 11.1.2 the chain $\check{\Phi}$ is regular on $\check{S}$. Let $\{\check{S}_n\}$ denote the cover of $\check{S}$ with regular sets.

Now we have $\tilde{N} = \check{X} \backslash \check{S} \subseteq X_0$, and so if we write N as the copy of $\tilde{N}$ and define $S = X \backslash N$, we can cover S with the matching copies S_n. We then have for $x \in S_n$ and any $B \in \mathcal{B}^+(X)$

$$\mathsf{E}_x[\tau_B] \le \check{\mathsf{E}}_{x_0}[\tau_B] + \check{\mathsf{E}}_{x_1}[\tau_B]$$

which is bounded for $x_0 \in \check{S}_n$ and all $x_1 \in \check{\alpha}$, and hence for $x \in S_n$.

Thus S is the required full absorbing set for (11.13) to hold.   □

It is now possible, by the device we have used before of analyzing the m-skeleton, to show that this proposition holds for arbitrary positive recurrent chains.

Theorem 11.1.4. *Suppose that Φ is ψ-irreducible. Then the following are equivalent:*

(i) *The chain Φ is positive recurrent.*

(ii) *There exists a decomposition*

$$X = S \cup N \tag{11.15}$$

   where the set S is full and absorbing, and Φ restricted to S is regular.

PROOF   Assume Φ is positive recurrent. Then the Nummelin splitting exists for some m-skeleton from Proposition 5.4.5, and so we have from Proposition 11.1.3 that there is a decomposition as in (11.15) where the set $S = \cup S_n$ and each S_n is regular for the m-skeleton.

But if τ_B^m denotes the number of steps needed for the m-skeleton to reach B, then we have that

$$\tau_B \le m \, \tau_B^m$$

and so each S_n is also regular for Φ as required.

The converse is almost trivial: when the chain is regular on S then there exists a petite set C inside S with $\sup_{x \in C} \mathsf{E}_x[\tau_C] < \infty$, and the result follows from Theorem 10.4.10.   □

Just as we may restrict any recurrent chain to an absorbing set H on which the chain is Harris recurrent, we have here shown that we can further restrict a positive recurrent chain to an absorbing set where it is regular.

We will now turn to the equivalence between regularity and mean drift conditions. This has the considerable benefit that it enables us to identify exactly the null set on which regularity fails, and thus to eliminate from consideration annoying and pathological behavior in many models. It also provides, as noted earlier, a sound practical approach to assessing stability of the chain.

To motivate and perhaps give more insight into the connections between hitting times and mean drift conditions we first consider deterministic models.

11.2   Drift, hitting times and deterministic models

In this section we analyze a deterministic state space model, indicating the role we might expect the drift conditions (11.4) on ΔV to play. As we have seen in Chapter 4 and Chapter 7 in examining irreducibility structures, the underlying deterministic models for state space systems foreshadow the directions to be followed for systems with a noise component.

Let us then assume that there is a topology on X, and consider the deterministic process known as a semi-dynamical system.

The semi-dynamical system

(DS1)   The process Φ is deterministic, and generated by the nonlinear difference equation, or semi-dynamical system,

$$\Phi_{k+1} = F(\Phi_k), \qquad k \in \mathbb{Z}_+, \tag{11.16}$$

where $F \colon \mathsf{X} \to \mathsf{X}$ is a continuous function.

Although Φ is deterministic, it is certainly a Markov chain (if a trivial one in a probabilistic sense), with Markov transition operator P defined through its operations on any function f on X by

$$Pf(\,\cdot\,) = f(F(\,\cdot\,)).$$

Since we have assumed the function F to be continuous, the Markov chain Φ has the Feller property, although in general it will not be a T-chain.

For such a deterministic system it is standard to consider two forms of stability known as recurrence and ultimate boundedness. We shall call the deterministic system (11.16) *recurrent* if there exists a compact subset $C \subset \mathsf{X}$ such that $\sigma_C(x) < \infty$ for each initial condition $x \in \mathsf{X}$. Such a concept of recurrence here is almost identical to the definition of recurrence for stochastic models. We shall call the system (11.16) *ultimately bounded* if there exists a compact set $C \subset \mathsf{X}$ such that for each fixed initial condition $\Phi_0 \in \mathsf{X}$, the trajectory starting at Φ_0 eventually enters and remains in C. Ultimate boundedness is loosely related to positive recurrence: it requires that the limit points of the process all lie within a compact set C, which is somewhat analogous to the positivity requirement that there be an invariant probability measure π with $\pi(C) > 1 - \varepsilon$ for some small ε.

Drift condition for the semi-dynamical system

(DS2)   There exists a positive function $V: \mathsf{X} \to \mathbb{R}_+$ and a compact set $C \subset \mathsf{X}$ and constant $M < \infty$ such that

$$\Delta V(x) := V(F(x)) - V(x) \leq -1$$

for all x lying outside the compact set C, and

$$\sup_{x \in C} V(F(x)) \leq M.$$

If we consider the sequence $V(\Phi_n)$ on $\mathbb{R}_+$ then this condition requires that this sequence move monotonically downwards at a uniform rate until the first time that Φ enter C. It is therefore not surprising that Φ hits C in a finite time under this condition.

Theorem 11.2.1. *Suppose that Φ is defined by (DS1).*

(i) *If (DS2) is satisfied, then Φ is ultimately bounded.*

(ii) *If Φ is recurrent, then there exists a positive function V such that (DS2) holds.*

(iii) *Hence Φ is recurrent if and only if it is ultimately bounded.*

PROOF   To prove (i), let $\Phi(x, n) = F^n(x)$ denote the deterministic position of Φ_n if the chain starts at $\Phi_0 = x$. We first show that the compact set C' defined as

$$C' := \bigcup \{\Phi(x, i) : x \in C, \, 1 \leq i \leq M + 1\} \cup C$$

where M is the constant used in (DS2), is invariant as defined in Chapter 7.

For any $x \in C$ we have $\Phi(x, i) \in C$ for some $1 \leq i \leq M + 1$ by (DS2) and the hypothesis that V is positive. Hence for an arbitrary $j \in \mathbb{Z}_+$, $\Phi(x, j) = \Phi(y, i)$ for some $y \in C$, and some $1 \leq i \leq M + 1$. This implies that $\Phi(x, j) \in C'$ and hence C' is equal to the invariant set

$$C' = \bigcup_{i=1}^{\infty} \{\Phi(x, i) : x \in C\} \cup C.$$

Because V is positive and decreases on C^c, every trajectory must enter the set C, and hence also C' at some finite time. We conclude that Φ is ultimately bounded.

We now prove (ii). Suppose that a compact set C_1 exists such that $\sigma_{C_1}(x) < \infty$ for each initial condition $x \in \mathsf{X}$. Let O be an open pre-compact set containing C_1, and set $C := cl\, O$. Then the test function

$$V(x) := \sigma_O(x)$$

satisfies (DS2). To see this, observe that if $x \in C^c$, then $V(F(x)) = V(x) - 1$ and hence the first inequality is satisfied. By assumption the function V is everywhere finite, and since O is open it follows that V is upper semicontinuous from Proposition 6.1.1. This implies that the second inequality in (DS2) holds, since a finite-valued upper semicontinuous function is uniformly bounded on compact sets. □

For a semi-dynamical system, this result shows that recurrence is actually equivalent to ultimate boundedness. In this the deterministic system differs from the general NSS(F) model with a non-trivial random component. More pertinently, we have also shown that the semi-dynamical system is ultimately bounded if and only if a test function exists satisfying (DS2).

This test function may always be taken to be the time to reach a certain compact set. As an almost exact analogue, we now go on to see that the expected time to reach a petite set is the appropriate test function to establish positive recurrence in the stochastic framework; and that, as we show in Theorem 11.3.4 and Theorem 11.3.5, the existence of a test function similar to (DS2) is equivalent to positive recurrence.

11.3   Drift criteria for regularity

11.3.1   Mean drift and Dynkin's formula

The deterministic models of the previous section lead us to hope that we can obtain criteria for regularity by considering a drift criterion for positive recurrence based on (11.4).

What is somewhat more surprising is the depth of these connections and the direct method of attack on regularity which we have through this route.

The key to exploiting the effect of mean drift is the following condition, which is stronger on C^c than (V1) and also requires a bound on the drift away from C.

Strict drift towards C

(V2)   For some set $C \in \mathcal{B}(\mathsf{X})$, some constant $b < \infty$, and an extended-real-valued function $V : \mathsf{X} \to [0, \infty]$

$$\Delta V(x) \le -1 + b\mathbb{I}_C(x) \qquad x \in \mathsf{X}. \tag{11.17}$$

This is a portmanteau form of the following two equations:

$$\Delta V(x) \le -1, \qquad x \in C^c, \tag{11.18}$$

for some non-negative function V and some set $C \in \mathcal{B}(\mathsf{X})$; and for some $M < \infty$,

$$\Delta V(x) \le M, \qquad x \in C. \tag{11.19}$$

Thus we might hope that (V2) might have something of the same impact for stochastic models as (DS2) has for deterministic chains.

In essentially the form (11.18) and (11.19) these conditions were introduced by Foster [129] for countable state space chains, and shown to imply positive recurrence. Use of the form (V2) will actually make it easier to show that the existence of everywhere finite solutions to (11.17) is equivalent to regularity and moreover we will identify the sublevel sets of the test function V as regular sets.

The central technique we will use to make connections between one-step mean drifts and moments of first entrance times to appropriate (usually petite) sets hinges on a discrete time version of a result known for continuous time processes as Dynkin's formula.

This formula yields not only those criteria for positive Harris chains and regularity which we discuss in this chapter, but also leads in due course to necessary and sufficient conditions for rates of convergence of the distributions of the process; necessary and sufficient conditions for finiteness of moments; and sample path ergodic theorems such as the Central Limit Theorem and Law of the Iterated Logarithm. All of these are considered in Part III.

Dynkin's formula is a sample path formula, rather than a formula involving probabilistic operators. We need to introduce a little more notation to handle such situations.

Recall from Section 3.4 the definition

$$\mathcal{F}_k^{\Phi} = \sigma\{\Phi_0, \ldots, \Phi_k\}, \tag{11.20}$$

and let $\{Z_k, \mathcal{F}_k^{\Phi}\}$ be an adapted sequence of positive random variables. For each k, Z_k will denote a fixed Borel measurable function of $(\Phi_0, \ldots, \Phi_k)$, although in applications this will usually (although not always) be a function of the last position, so that

$$Z_k(\Phi_0, \ldots, \Phi_k) = Z(\Phi_k)$$

for some measurable function Z. We will somewhat abuse notation and let Z_k denote both the random variable, and the function on X^{k+1}.

For any stopping time τ define

$$\tau^n := \min\{n, \tau, \inf\{k \geq 0 : Z_k \geq n\}\}.$$

The random time τ^n is also a stopping time since it is the minimum of stopping times, and the random variable $\sum_{i=0}^{\tau^n - 1} Z_i$ is essentially bounded by n^2.

Dynkin's formula will now tell us that we can evaluate the expected value of Z_{τ^n} by taking the initial value Z_0 and adding on to this the average increments at each time until τ^n. This is almost obvious, but has widespread consequences: in particular it enables us to use (V2) to control these one-step average increments, leading to control of the expected overall hitting time.

Theorem 11.3.1 (Dynkin's formula). *For each $x \in \mathsf{X}$ and $n \in \mathbb{Z}_+$,*

$$\mathsf{E}_x[Z_{\tau^n}] = \mathsf{E}_x[Z_0] + \mathsf{E}_x\left[\sum_{i=1}^{\tau^n}(\mathsf{E}[Z_i \mid \mathcal{F}_{i-1}^{\Phi}] - Z_{i-1})\right].$$

PROOF   For each $n \in \mathbb{Z}_+$,

$$
\begin{aligned}
Z_{\tau^n} &= Z_0 + \sum_{i=1}^{\tau^n}(Z_i - Z_{i-1}) \\
&= Z_0 + \sum_{i=1}^{n}\mathbb{I}\{\tau^n \geq i\}(Z_i - Z_{i-1}).
\end{aligned}
$$

Taking expectations and noting that $\{\tau^n \geq i\} \in \mathcal{F}_{i-1}^{\Phi}$ we obtain

$$
\begin{aligned}
\mathsf{E}_x[Z_{\tau^n}] &= \mathsf{E}_x[Z_0] + \mathsf{E}_x\Big[\sum_{i=1}^{n}\mathsf{E}_x[Z_i - Z_{i-1} \mid \mathcal{F}_{i-1}^{\Phi}]\mathbb{I}\{\tau^n \geq i\}\Big] \\
&= \mathsf{E}_x[Z_0] + \mathsf{E}_x\Big[\sum_{i=1}^{\tau^n}(\mathsf{E}_x[Z_i \mid \mathcal{F}_{i-1}^{\Phi}] - Z_{i-1})\Big].
\end{aligned}
$$

$\square$

As an immediate corollary we have

Proposition 11.3.2. *Suppose that there exist two sequences of positive functions* $\{s_k, f_k : k \geq 0\}$ *on* X, *such that*

$$
\mathsf{E}[Z_{k+1} \mid \mathcal{F}_k^{\Phi}] \leq Z_k - f_k(\Phi_k) + s_k(\Phi_k).
$$

Then for any initial condition x *and any stopping time* τ

$$
\mathsf{E}_x\Big[\sum_{k=0}^{\tau-1} f_k(\Phi_k)\Big] \leq Z_0(x) + \mathsf{E}_x\Big[\sum_{k=0}^{\tau-1} s_k(\Phi_k)\Big].
$$

PROOF   Fix $N > 0$ and note that

$$
\mathsf{E}[Z_{k+1} \mid \mathcal{F}_k^{\Phi}] \leq Z_k - f_k(\Phi_k) \wedge N + s_k(\Phi_k).
$$

By Dynkin's formula

$$
0 \leq \mathsf{E}_x[Z_{\tau^n}] \leq Z_0(x) + \mathsf{E}_x\Big[\sum_{i=1}^{\tau^n}(s_{i-1}(\Phi_{i-1}) - [f_{i-1}(\Phi_{i-1}) \wedge N])\Big]
$$

and hence by adding the finite term

$$
\mathsf{E}_x\Big[\sum_{k=1}^{\tau^n}[f_{k-1}(\Phi_{k-1}) \wedge N]\Big]
$$

to each side we get

$$
\mathsf{E}_x\Big[\sum_{k=1}^{\tau^n}[f_{k-1}(\Phi_{k-1}) \wedge N]\Big] \leq Z_0(x) + \mathsf{E}_x\Big[\sum_{k=1}^{\tau^n} s_{k-1}(\Phi_{k-1})\Big] \leq Z_0(x) + \mathsf{E}_x\Big[\sum_{k=1}^{\tau} s_{k-1}(\Phi_{k-1})\Big].
$$

Letting $n \to \infty$ and then $N \to \infty$ gives the result by the Monotone Convergence Theorem. $\square$

Closely related to this we have

Proposition 11.3.3. *Suppose that there exists a sequence of positive functions $\{\varepsilon_k : k \geq 0\}$ on X, $c < \infty$, such that*

(i) $\varepsilon_{k+1}(x) \leq c\varepsilon_k(x), \quad k \in \mathbb{Z}_+, x \in A^c;$

(ii) $\mathsf{E}[Z_{k+1} \mid \mathcal{F}_k^{\Phi}] \leq Z_k - \varepsilon_k(\Phi_k), \quad \sigma_A > k.$

Then

$$\mathsf{E}_x \Big[\sum_{i=0}^{\tau_A - 1} \varepsilon_i(\Phi_i) \Big] \leq \begin{cases} Z_0(x), & x \in A^c; \\ \varepsilon_0(x) + cPZ_0(x), & x \in \mathsf{X}. \end{cases}$$

PROOF   Let Z_k and ε_k denote the random variables $Z_k(\Phi_0, \ldots, \Phi_k)$ and $\varepsilon_k(\Phi_k)$ respectively.

By hypothesis $\mathsf{E}[Z_k \mid \mathcal{F}_{k-1}^{\Phi}] - Z_{k-1} \leq -\varepsilon_{k-1}$ whenever $1 \leq k \leq \sigma_A$. Hence for all $n \in \mathbb{Z}_+$ and $x \in \mathsf{X}$ we have by Dynkin's formula

$$0 \leq \mathsf{E}_x[Z_{\tau_A^n}] \leq Z_0(x) - \mathsf{E}_x \Big[\sum_{i=1}^{\tau_A^n} \varepsilon_{i-1}(\Phi_{i-1}) \Big], \qquad x \in A^c.$$

By the Monotone Convergence Theorem it follows that for all initial conditions,

$$\mathsf{E}_x \Big[\sum_{i=1}^{\tau_A} \varepsilon_{i-1}(\Phi_{i-1}) \Big] \leq Z_0(x), \qquad x \in A^c.$$

This proves the result for $x \in A^c$.

For arbitrary x we have

$$\begin{aligned} \mathsf{E}_x \Big[\sum_{i=1}^{\tau_A} \varepsilon_{i-1}(\Phi_{i-1}) \Big] &= \varepsilon_0(x) + \mathsf{E}_x \Big[\mathsf{E}_{\Phi_1} \Big(\sum_{i=1}^{\tau_A} \varepsilon_i(\Phi_{i-1}) \Big) \mathbb{I}(\Phi_1 \in A^c) \Big] \\ &\leq \varepsilon_0(x) + cPZ_0(x). \end{aligned}$$

$\square$

We can immediately use Dynkin's formula to prove

Theorem 11.3.4. *Suppose $C \in \mathcal{B}(\mathsf{X})$, and V satisfies (V2). Then*

$$\mathsf{E}_x[\tau_C] \leq V(x) + b\mathbb{I}_C(x)$$

for all x. Hence if C is petite and V is everywhere finite and bounded on C, then Φ is positive Harris recurrent.

PROOF   Applying Proposition 11.3.3 with $Z_k = V(\Phi_k)$, $\varepsilon_k = 1$ we have the bound

$$\mathsf{E}_x[\tau_C] < \begin{cases} V(x) & \text{for } x \in C^c \\ 1 + PV(x) & x \in C \end{cases}$$

Since (V2) gives $PV \leq V - 1 + b$ on C, we have the required result.

If V is everywhere finite then this bound trivially implies $L(x, C) \equiv 1$ and so, if C is petite, the chain is Harris recurrent from Proposition 9.1.7. Positivity follows from Theorem 10.4.10 (ii). $\square$

We will strengthen Theorem 11.3.4 below in Theorem 11.3.11 where we show that V need not be bounded on C, and moreover that (V2) gives bounds on the mean return time to general sets in $\mathcal{B}^+(\mathsf{X})$.

11.3.2 Hitting times and test functions

The upper bound in Theorem 11.3.4 is a typical consequence of the drift condition. The key observation in showing the actual equivalence of mean drift towards petite sets and regularity is the identification of specific solutions to (V2) when the chain is regular.

For any set $A \in \mathcal{B}(\mathsf{X})$ we define the kernel G_A on $(\mathsf{X}, \mathcal{B}(\mathsf{X}))$ through

$$G_A(x, f) := [I + I_{A^c} U_A](x, f) = \mathsf{E}_x\Big[\sum_{k=0}^{\sigma_A} f(\Phi_k)\Big], \qquad (11.21)$$

where x is an arbitrary state and f is any positive function.

For $f \geq 1$ fixed we will see in Theorem 11.3.5 that the function $V = G_C(\cdot, f)$ satisfies (V2), and also a generalization of this drift condition to be developed in later chapters. In this chapter we concentrate on the special case where $f \equiv 1$ and we will simplify the notation by setting

$$V_C(x) = G_C(x, \mathsf{X}) = 1 + \mathsf{E}_x[\sigma_C]. \qquad (11.22)$$

Theorem 11.3.5. *For any set $A \in \mathcal{B}(\mathsf{X})$ we have*

(i) *The kernel G_A satisfies the identity*

$$PG_A = G_A - I + I_A U_A.$$

(ii) *The function $V_A(\cdot) = G_A(\cdot, \mathsf{X})$ satisfies the identity*

$$PV_A(x) = V_A(x) - 1, \qquad x \in A^c, \qquad (11.23)$$

$$PV_A(x) = \mathsf{E}_x[\tau_A] - 1, \qquad x \in A. \qquad (11.24)$$

Thus if $C \in \mathcal{B}^+(\mathsf{X})$ is regular, V_C is a solution to (11.17).

(iii) *The function $V = V_A - 1$ is the pointwise minimal solution on A^c to the inequalities*

$$PV(x) \leq V(x) - 1, \qquad x \in A^c. \qquad (11.25)$$

PROOF   From the definition

$$U_A := \sum_{k=0}^{\infty} (PI_{A^c})^k P$$

we see that $U_A = P + PI_{A_c} U_A = PG_A$. Since $U_A = G_A - I + I_A U_A$ we have (i), and then (ii) follows.

We have that V_A solves (11.25) from (ii); but if V is any other solution then it is pointwise larger than V_A exactly as in Theorem 11.3.4.   □

We shall use repeatedly the following lemmas, which guarantee finiteness of solutions to (11.17), and which also give a better description of the structure of the most interesting solution, namely V_C.

Lemma 11.3.6. *Any solution of (11.17) is finite ψ-almost everywhere or infinite everywhere.*

PROOF   If V satisfies (11.17), then

$$PV(x) \leq V(x) + b$$

for all $x \in \mathsf{X}$, and it then follows that the set $\{x : V(x) < \infty\}$ is absorbing. If this set is non-empty then it is full by Proposition 4.2.3.   □

Lemma 11.3.7. *If the set C is petite, then the function $V_C(x)$ is unbounded off petite sets.*

PROOF   We have from Chebyshev's inequality that for each of the sublevel sets $C_V(\ell) := \{x : V_C(x) \leq \ell\}$,

$$\sup_{x \in C_V(\ell)} \mathsf{P}_x\{\sigma_C \geq n\} \leq \frac{\ell}{n}.$$

Since the right hand side is less than $\frac{1}{2}$ for sufficiently large n, this shows that $C_V(\ell) \overset{a}{\rightsquigarrow} C$ for a sampling distribution a, and hence, by Proposition 5.5.4, the set $C_V(\ell)$ is petite.   □

Lemma 11.3.7 will typically be applied to show that a given petite set is regular. The converse is always true, as the next result shows:

Proposition 11.3.8. *If the set A is regular, then it is petite.*

PROOF   Again we apply Chebyshev's inequality. If $C \in \mathcal{B}^+(\mathsf{X})$ is petite then

$$\sup_{x \in A} \mathsf{P}_x\{\sigma_C > n\} \leq \frac{1}{n} \sup_{x \in A} \mathsf{E}_x[\tau_C].$$

As in the proof of Lemma 11.3.7 this shows that A is petite if it is regular.   □

11.3.3 Regularity, drifts and petite sets

In this section, using the full force of Dynkin's formula and the form (V2) for the drift condition, we will find we can do rather more than bound the return times to C from states in C. We have first

Lemma 11.3.9. *If (V2) holds, then for each $x \in \mathsf{X}$ and any set $B \in \mathcal{B}(\mathsf{X})$*

$$\mathsf{E}_x[\tau_B] \leq V(x) + b\mathsf{E}_x\Big[\sum_{k=0}^{\tau_B-1} \mathbb{I}_C(\Phi_k)\Big]. \tag{11.26}$$

PROOF   This follows from Proposition 11.3.2 on letting $f_k = 1$, $s_k = b\mathbb{I}_C$.   □

Note that Theorem 11.3.4 is the special case of this result when $B = C$.

In order to derive the central characterization of regularity, we first need an identity linking sampling distributions and hitting times on sets.

Lemma 11.3.10. *For any first entrance time τ_B, any sampling distribution a, and any positive function $f : \mathsf{X} \to \mathbb{R}_+$, we have*

$$\mathsf{E}_x\Big[\sum_{k=0}^{\tau_B-1} K_a(\Phi_k, f)\Big] = \sum_{i=0}^{\infty} a_i \mathsf{E}_x\Big[\sum_{k=0}^{\tau_B-1} f(\Phi_{k+i})\Big].$$

PROOF   By the Markov property and Fubini's Theorem we have

$$\mathsf{E}_x\Big[\sum_{k=0}^{\tau_B-1} K_a(\Phi_k, f)\Big]$$

$$= \sum_{i=0}^{\infty} a_i \mathsf{E}_x\Big[\sum_{k=0}^{\infty} P^i(\Phi_k, f)\mathbb{I}\{k < \tau_B\}\Big]$$

$$= \sum_{i=0}^{\infty}\sum_{k=0}^{\infty} a_i \mathsf{E}_x\Big[\mathsf{E}\big[f(\Phi_{k+i}) \mid \mathcal{F}_k\big]\mathbb{I}\{k < \tau_B\}\Big].$$

But now we have that $\mathbb{I}(k < \tau_B)$ is measurable with respect to $\mathcal{F}_k$ and so by the smoothing property of expectations this becomes

$$\sum_{i=0}^{\infty}\sum_{k=0}^{\infty} a_i \mathsf{E}_x\Big[\mathsf{E}\big[f(\Phi_{k+i})\mathbb{I}\{k < \tau_B\} \mid \mathcal{F}_k\big]\Big]$$

$$= \sum_{i=0}^{\infty}\sum_{k=0}^{\infty} a_i \mathsf{E}_x\Big[f(\Phi_{k+i})\mathbb{I}(k < \tau_B)\Big]$$

$$= \sum_{i=0}^{\infty} a_i \mathsf{E}_x\Big[\sum_{k=0}^{\tau_B-1} f(\Phi_{k+i})\Big].$$

□

We now have a relatively simple task in proving

Theorem 11.3.11. *Suppose that* Φ *is* ψ-*irreducible.*

(i) *If (V2) holds for a function* V *and a petite set* C, *then for any* $B \in \mathcal{B}^+(\mathsf{X})$ *there exists* $c(B) < \infty$ *such that*

$$\mathsf{E}_x[\tau_B] \le V(x) + c(B), \qquad x \in \mathsf{X}.$$

Hence if V *is bounded on* A, *then* A *is regular.*

(ii) *If there exists one regular set* $C \in \mathcal{B}^+(\mathsf{X})$, *then* C *is petite and the function* $V = V_C$ *satisfies (V2), with* V *uniformly bounded on* A *for any regular set* A.

PROOF   To prove (i), suppose that (V2) holds, with V bounded on A and C a ψ_a-petite set. Without loss of generality, from Proposition 5.5.6 we can assume $\sum_{i=0}^{\infty} i\, a_i < \infty$. We also use the simple but critical bound from the definition of petiteness:

$$\mathbb{I}_C(x) \le \psi_a(B)^{-1} K_a(x, B), \qquad x \in \mathsf{X}, B \in \mathcal{B}^+(\mathsf{X}). \tag{11.27}$$

By Lemma 11.3.9 and the bound (11.27) we then have

$$
\begin{aligned}
\mathsf{E}_x[\tau_B] &\le V(x) + b\mathsf{E}_x\left[\sum_{k=0}^{\tau_B-1} \mathbb{I}_C(\Phi_k)\right] \\
&\le V(x) + b\mathsf{E}_x\left[\sum_{k=0}^{\tau_B-1} \psi_a(B)^{-1} K_a(\Phi_k, B)\right] \\
&= V(x) + b\psi_a(B)^{-1} \sum_{i=0}^{\infty} a_i \mathsf{E}_x\left[\sum_{k=0}^{\tau_B-1} \mathbb{I}_B(\Phi_{k+i})\right] \\
&\le V(x) + b\psi_a(B)^{-1} \sum_{i=0}^{\infty} (i+1)a_i
\end{aligned}
$$

for any $B \in \mathcal{B}^+(\mathsf{X})$, and all $x \in \mathsf{X}$. If V is bounded on A, it follows that

$$\sup_{x \in A} \mathsf{E}_x[\tau_B] < \infty,$$

which shows that A is regular.

To prove (ii), suppose that a regular set $C \in \mathcal{B}^+(\mathsf{X})$ exists. By Lemma 11.3.8 the set C is petite. Then $V = V_C$ is clearly positive, and bounded on any regular set A. Moreover, by Theorem 11.3.5 and regularity of C it follows that condition (V2) holds for a suitably large constant b.   □

Boundedness of hitting times from arbitrary initial measures will become important in Part III. The following definition is an obvious one.

Regularity of measures

A probability measure μ is called *regular*, if $\mathsf{E}_\mu[\tau_B] < \infty$ for each $B \in \mathcal{B}^+(\mathsf{X})$.

The proof of the following result for regular measures μ is identical to that of the previous theorem and we omit it.

Theorem 11.3.12. *Suppose that Φ is ψ-irreducible.*

(i) *If (V2) holds for a petite set C and a function V, and if $\mu(V) < \infty$, then the measure μ is regular.*

(ii) *If μ is regular, and if there exists one regular set $C \in \mathcal{B}^+(\mathsf{X})$, then there exists an extended-valued function V satisfying (V2) with $\mu(V) < \infty$.*

$\square$

As an application of Theorem 11.3.11 we obtain a description of regular sets as in Theorem 11.1.4.

Proposition 11.3.13. *If there exists a regular set $C \in \mathcal{B}^+(\mathsf{X})$, then the sets $C_V(\ell) := \{x : V_C(x) \leq \ell, : \ell \in \mathbb{Z}_+\}$ are regular and $S_C = \{y : V_C(y) < \infty\}$ is a full absorbing set such that Φ restricted to S_C is regular.*

PROOF   Suppose that a regular set $C \in \mathcal{B}^+(\mathsf{X})$ exists. Since C is regular it is also ψ_a-petite, and we can assume without loss of generality that the sampling distribution a has a finite mean. By regularity of C we also have, by Theorem 11.3.11 (ii), that (V2) holds with $V = V_C$. From Theorem 11.3.11 each of the sets $C_V(\ell)$ is regular, and by Lemma 11.3.6 the set $S_C = \{y : V_C(y) < \infty\}$ is full and absorbing. $\square$

Theorem 11.3.11 gives a characterization of regular sets in terms of a drift condition. Theorem 11.3.14 now gives such a characterization in terms of the mean hitting times to petite sets.

Theorem 11.3.14. *If Φ is ψ-irreducible, then the following are equivalent:*

(i) *The set $C \in \mathcal{B}(\mathsf{X})$ is petite and $\sup_{x \in C} \mathsf{E}_x[\tau_C] < \infty$.*

(ii) *The set C is regular and $C \in \mathcal{B}^+(\mathsf{X})$.*

PROOF   **(i)**   Suppose that C is petite, and let as before $V_C(x) = 1 + \mathsf{E}_x[\sigma_C]$. By Theorem 11.3.5 and the conditions of the theorem we may find a constant $b < \infty$ such that

$$PV_C \leq V_C - 1 + b\mathbb{I}_C.$$

Since V_C is bounded on C by construction, it follows from Theorem 11.3.11 that C is regular. Since the set C is Harris recurrent it follows from Proposition 8.3.1 (ii) that $C \in \mathcal{B}^+(\mathsf{X})$.

(ii)   Suppose that C is regular. Since $C \in \mathcal{B}^+(\mathsf{X})$, it follows from regularity that $\sup_{x \in C} \mathsf{E}_x[\tau_C] < \infty$, and that C is petite follows from Proposition 11.3.8. $\square$

We can now give the following complete characterization of the case $\mathsf{X} = S$.

Theorem 11.3.15. *Suppose that Φ is ψ-irreducible. Then the following are equivalent:*

(i) *The chain Φ is regular.*

(ii) *The drift condition (V2) holds for a petite set C and an everywhere finite function V.*

(iii) *There exists a petite set C such that the expectation*

$$\mathsf{E}_x[\tau_C]$$

is finite for each x, and uniformly bounded for $x \in C$.

PROOF   If (i) holds, then it follows that a regular set $C \in \mathcal{B}^+(\mathsf{X})$ exists. The function $V = V_C$ is everywhere finite and satisfies (V2), by (11.24), for a suitably large constant b; so (ii) holds. Conversely, Theorem 11.3.11 (i) tells us that if (V2) holds for a petite set C with V finite valued then each sublevel set of V is regular, and so (i) holds.

If the expectation is finite as described in (iii), then by (11.24) we see that the function $V = V_C$ satisfies (V2) for a suitably large constant b. Hence from Theorem 11.3.15 we see that the chain is regular; and the converse is trivial.   □

11.4   Using the regularity criteria

11.4.1   Some straightforward applications

Random walk on a half line

We have already used a drift criterion for positive recurrence, without identifying it as such, in some of our analysis of the random walk on a half line.

Using the criteria above, we have

Proposition 11.4.1. *If Φ is a random walk on a half line with finite mean increment β, then Φ is regular if*

$$\beta = \int w\,\Gamma(dw) < 0;$$

and in this case all compact sets are regular sets.

PROOF   By consideration of the proof of Proposition 8.5.1, we see that this result has already been established, since (11.18) was exactly the condition verified for recurrence in that case, whilst (11.19) is simply checked for the random walk.   □

From the results in Section 8.5, we know that the random walk on $\mathbb{R}_+$ is transient if $\beta > 0$, and that (at least under a second moment condition) it is recurrent in the marginal case $\beta = 0$. We shall show in Proposition 11.5.3 that it is not regular in this marginal case.

Forward recurrence times

We could also use this approach in a simple way to analyze positivity for the forward recurrence time chain.

In this example, using the function $V(x) = x$ we have

$$\sum_y P(x, y)V(y) \; = \; V(x) - 1, \qquad x \geq 1, \qquad (11.28)$$

$$\sum_y P(0, y)V(y) \; = \; \sum_y p(y)\, y. \qquad (11.29)$$

Hence, as we already know, the chain is positive recurrent if $\sum_y p(y)\, y < \infty$.

Since $\mathsf{E}_0[\tau_0] = \sum_y p(y)\, y$ the drift condition with $V(x) = x$ is also necessary, as we have seen.

The forward recurrence time chain thus provides a simple but clear example of the need to include the second bound (11.19) in the criterion for positive recurrence.

Linear models

Consider the simple linear model defined in (SLM1) by

$$X_n = \alpha X_{n-1} + W_n.$$

We have

Proposition 11.4.2. *Suppose that the disturbance variable W for the simple linear model defined in (SLM1), (SLM2) is non-singular with respect to Lebesgue measure, and satisfies $\mathsf{E}[\log(1 + |W|)] < \infty$. Suppose also that $|\alpha| < 1$. Then every compact set is regular, and hence the chain itself is regular.*

PROOF   From Proposition 6.3.5 we know that the chain $\boldsymbol{X}$ is a ψ-irreducible and aperiodic T-chain under the given assumptions.

Let $V(x) = \log(1 + \varepsilon|x|)$, where $\varepsilon > 0$ will be fixed below. We will verify that (V2) holds with this choice of V by applying the following two special properties of this test function:

$$V(x + y) \leq V(x) + V(y), \qquad (11.30)$$

$$\lim_{x \to \infty} [V(x) - V(|\alpha|x)] = \log((|\alpha|^{-1}). \qquad (11.31)$$

From (11.30) and (SLM1),

$$V(X_1) = V(\alpha X_0 + W_1) \leq V(|\alpha|X_0) + V(W_1),$$

and hence from (11.31) there exists $r < \infty$ such that whenever $X_0 \geq r$,

$$V(X_1) \leq V(X_0) - \tfrac{1}{2}\log(|\alpha|^{-1}) + V(W_1).$$

Choosing $\varepsilon > 0$ sufficiently small so that $\mathsf{E}[V(W)] \leq \tfrac{1}{4}\log(|\alpha|^{-1})$ we see that for $x \geq r$,

$$\mathsf{E}_x[V(X_1)] \leq V(x) - \tfrac{1}{4}\log(|\alpha|^{-1}).$$

So we have that (V2) holds with $C = \{x : |x| \leq r\}$ and the result follows.   □

This is part of the recurrence result we proved using a stochastic comparison argument in Section 9.5.1, but in this case the direct proof enables us to avoid any restriction on the range of the increment distribution.

We can extend this simple construction much further, and we shall do so in Chapter 15 in particular, where we show that the geometric drift condition exhibited by the linear model implies much more, including rates of convergence results, than we have so far described.

11.4.2   The GI/G/1 queue with re-entry

In Section 2.4.2 we described models for GI/G/1 queueing systems. We now indicate one class of models where we generalize the conditions imposed on the arrival stream and service times by allowing re-entry to the system, and still find conditions under which the queue is positive Harris recurrent.

As in Section 2.4.2, we assume that customers enter the queue at successive time instants $0 = T_0' < T_1' < T_2' < T_3' < \cdots$. Upon arrival, a customer waits in the queue if necessary, and then is serviced and exits the system. In the G1/G/1 queue, the interarrival times $\{T_{n+1}' - T_n' : n \in \mathbb{Z}_+\}$ and the service times $\{S_i : i \in \mathbb{Z}_+\}$ are i.i.d. and independent of each other with general distributions, and means $1/\lambda$, $1/\mu$ respectively.

After being served, a customer exits the system with probability r and re-enters the queue with probability $1 - r$. Hence the effective rate of customers to the queue is, at least intuitively,

$$\lambda_r := \frac{\lambda}{r}.$$

If we now let N_n denote the queue length (not including the customer which may be in service) at time $T_n'-$, and this time let R_n^+ denote the residual service time (set to zero if the server is free) for the system at time $T_n'-$, then the stochastic process

$$\Phi_n = \begin{pmatrix} N_n \\ R_n^+ \end{pmatrix}, \qquad n \in \mathbb{Z}_+,$$

is a Markov chain with stationary transition probabilities evolving on the ladder-structure space $\mathsf{X} = \mathbb{Z}_+ \times \mathbb{R}_+$.

Now suppose that the load condition

$$\rho_r := \frac{\lambda_r}{\mu} < 1 \tag{11.32}$$

is satisfied. This will be shown to imply positive Harris recurrence for the chain Φ.

Write $[0] = 0 \times 0$ for the state where the queue is empty. Under (11.32), for each $x \in \mathsf{X}$, we may find $m \in \mathbb{Z}_+$ sufficiently large that

$$\mathsf{P}_x\{\Phi_m = [0]\} > 0. \tag{11.33}$$

This follows because under the load constraint, there exists $\delta > 0$ such that with positive probability, each of the first m interarrival times exceeds each of the first m service times by at least δ, and also none of the first m customers re-enter the queue.

For $x, y \in \mathsf{X}$ we say that $x \geq y$ if $x_i \geq y_i$ for $i = 1, 2$. It is easy to see that $\mathsf{P}_x(\Phi_m = [0]) \leq \mathsf{P}_y(\Phi_m = [0])$ whenever $x \geq y$, and hence by (11.33) we have the following result:

Proposition 11.4.3. *Suppose that the load constraint (11.32) is satisfied. Then the Markov chain Φ is $\delta_{[0]}$-irreducible and aperiodic, and every compact subset of X is petite.* $\quad\square$

We let W_n denote the total amount of time that the server will spend servicing the customers which are in the system at time $T_n'+$. Let $V(x) = \mathsf{E}_x[W_0]$. It is easily seen that

$$V(x) = \mathsf{E}[W_n \mid \Phi_n = x],$$

and hence that $P^n V(x) = \mathsf{E}_x[W_n]$.

The random variable W_n is also called the *waiting time* of the nth customer to arrive at the queue. The quantity W_0 may be thought of as the total amount of *work* which is initially present in the system. Hence it is natural that $V(x)$, the expected work, should play the role of a Lyapunov function.

The drift condition we will establish for some $k > 0$ is

$$
\begin{aligned}
\mathsf{E}_x[W_k] &\leq \mathsf{E}_x[W_0] - 1, \qquad x \in A^c, \\
\sup\nolimits_{x \in A} \mathsf{E}_x[W_k] &< \infty;
\end{aligned}
\tag{11.34}
$$

this implies that $V(x)$ satisfies (V2) for the k-skeleton, and hence as in the proof of Theorem 11.1.4 both the k-skeleton and the original chain are regular.

Proposition 11.4.4. *Suppose that $\rho_r < 1$. Then (11.34) is satisfied for some compact set $A \subset \mathsf{X}$ and some $k \in \mathbb{Z}_+$, and hence Φ is a regular chain.*

PROOF   Let $|\cdot|$ denote the Euclidean norm on $\mathbb{R}^2$, and set

$$A_m = \{x \in \mathsf{X} : |x| \leq m\}, \qquad m \in \mathbb{Z}_+.$$

For each $m \in \mathbb{Z}_+$, the set A_m is a compact subset of X.

We first fix k such that $(k/\lambda)(1 - \rho_r) \geq 2$; we can do this since $\rho_r < 1$ by assumption. Let ζ_k then denote the time that the server is active in $[0, T_k']$. We have

$$W_k = W_0 + \sum_{i=1}^{k} \sum_{j=1}^{n_i} S(i, j) - \zeta_k, \tag{11.35}$$

where n_i denotes the number of times that the ith customer visits the system, and the random variables $S(i, j)$ are i.i.d. with mean μ^{-1}.

Now choose m so large that

$$\mathsf{E}_x[\zeta_k] \geq \mathsf{E}_x[T_k'] - 1, \qquad x \in A_m^c.$$

Then by (11.35), and since λ_r/λ is equal to the expected number of times that a customer will re-enter the queue,

$$
\begin{aligned}
\mathsf{E}_x[W_k] &\le \mathsf{E}_x[W_0] + \sum_{i=1}^{k} \mathsf{E}_x[n_i](1/\mu) - (\mathsf{E}[T_k'] - 1) \\
&= \mathsf{E}_x[W_0] + (k\lambda_r/\lambda)(1/\mu) - k/\lambda + 1 \\
&= \mathsf{E}_x[W_0] - (k/\lambda)(1 - \rho_r) + 1,
\end{aligned}
$$

and this completes the proof that (11.34) holds.   □

11.4.3 Regularity of the scalar SETAR model

Let us conclude this section by analyzing the SETAR models defined in (SETAR1) and (SETAR2) by
$$
X_n = \phi(j) + \theta(j)X_{n-1} + W_n(j), \qquad X_{n-1} \in R_j;
$$

these were shown in Proposition 6.3.6 to be φ-irreducible T-chains with φ taken as Lebesgue measure μ^{Leb} on $\mathbb{R}$ under these assumptions.

In Proposition 9.5.4 we showed that the SETAR chain is transient in the "exterior" of the parameter space; we now use Theorem 11.3.15 to characterize the behavior of the chain in the "interior" of the space (see Figure B.1). This still leaves the characterization on the boundaries, which will be done below in Section 11.5.2.

Let us call the *interior* of the parameter space that combination of parameters given by

$$\theta(1) < 1, \ \theta(M) < 1, \ \theta(1)\theta(M) < 1 \tag{11.36}$$

$$\theta(1) = 1, \ \theta(M) < 1, \ \phi(1) > 0 \tag{11.37}$$

$$\theta(1) < 1, \ \theta(M) = 1, \ \phi(M) < 0 \tag{11.38}$$

$$\theta(1) = \theta(M) = 1, \ \phi(M) < 0 < \phi(1) \tag{11.39}$$

$$\theta(1) < 0, \ \theta(1)\theta(M) = 1, \ \phi(M) + \theta(M)\phi(1) > 0. \tag{11.40}$$

Proposition 11.4.5. *For the SETAR model satisfying (SETAR1)–(SETAR2), the chain is regular in the interior of the parameter space.*

PROOF   To prove regularity for this interior set, we use (V2), and show that when (11.36)–(11.40) hold there is a function V and an interval set $[-R, R]$ satisfying the drift condition
$$
\int P(x, dy)V(y) \le V(x) - 1, \qquad |x| > R. \tag{11.41}
$$

First consider the condition (11.36). When this holds it is straightforward to calculate that there must exist positive constants a, b such that

$$1 > \theta(1) > -(b/a),$$

$$1 > \theta(M) > -(a/b).$$

If we now take

$$V(x) = \begin{cases} a\,x & x > 0 \\ b\,|x| & x \le 0 \end{cases}$$

then it is easy to check that (11.41) holds under (11.36) for all $|x|$ sufficiently large.

To prove regularity under (11.37), use the function

$$V(x) = \begin{cases} \gamma\,x & x > 0 \\ 2\,[\phi(1)]^{-1}\,|x| & x \le 0 \end{cases}$$

for which (11.41) is again satisfied provided

$$\gamma > 2\,|\theta(M)|\,[\phi(1)]^{-1}$$

for all $|x|$ sufficiently large. The sufficiency of (11.38) follows by symmetry, or directly by choosing the test function

$$V(x) = \begin{cases} \gamma'\,|x| & x \le 0 \\ -2\,[\phi(M)]^{-1}\,x & x > 0 \end{cases}$$

with

$$\gamma' > -2\,|\theta(1)|\,[\phi(M)]^{-1}.$$

In the case (11.39), the chain is driven by the constant terms and we use the test function

$$V(x) = \begin{cases} 2\,[\phi(1)]^{-1}\,|x| & x \le 0 \\ 2\,[|\phi(M)|]^{-1}\,x & x > 0 \end{cases}$$

to give the result.

The region defined by (11.40) is the hardest to analyze. It involves the way in which successive movements of the chain take place, and we reach the result by considering the two-step transition matrix P^2.

Let f_j denote the density of the noise variable $W(j)$. Fix j and $x \in R_j$ and write

$$R(k,j) = \{y : y + \phi(j) + \theta(j)x \in R_k\},$$

$$\zeta(k,x) = -\phi(k) - \theta(k)\phi(j) - \theta(k)\theta(j)x.$$

If we take the linear test function

$$V(x) = \begin{cases} a\,x & x > 0 \\ b\,|x| & x \le 0 \end{cases}$$

(with a, b to be determined below), then we have

$$\int P^2(x, dy)V(y) = \sum_{k=1}^{M} a \int_{\zeta(k,x)}^{\infty} (u - \zeta(k,x))[\int_{R(k,j)} f_k(u - \theta(k)w)f_j(w)dw]du$$

$$- b \int_{-\infty}^{\zeta(k,x)} (u - \zeta(k,x))[\int_{R(k,j)} f_k(u - \theta(k)w)f_j(w)dw]du.$$

It is straightforward to find from this that for some $R > 0$, we have

$$\int P^2(x, dy)V(y) \leq -bx - (b/2)(\phi(M) + \theta(M)\phi(1)), x \leq -R,$$

$$\int P^2(x, dy)V(y) \leq ax + (a/2)(\phi(1) + \theta(1)\phi(M)), x \geq R.$$

But now by assumption $\phi(M) + \theta(M)\phi(1) > 0$, and the complete set of conditions (11.40) also give $\phi(1) + \theta(1)\phi(M) < 0$. By suitable choice of a, b we have that the drift condition (11.41) holds for the two-step chain, and hence this chain is regular. Clearly, this implies that the one-step chain is also regular, and we are done. □

11.5 Evaluating non-positivity

11.5.1 A drift criterion for non-positivity

Although criteria for regularity are central to analyzing stability, it is also of value to be able to identify unstable models.

Theorem 11.5.1. *Suppose that the non-negative function V satisfies*

$$\Delta V(x) \geq 0, \qquad x \in C^c; \tag{11.42}$$

and

$$\sup_{x \in \mathsf{X}} \int P(x, dy)|V(x) - V(y)| < \infty. \tag{11.43}$$

Then for any $x_0 \in C^c$ such that

$$V(x_0) > V(x), \qquad \text{for all } x \in C \tag{11.44}$$

we have $\mathsf{E}_{x_0}[\tau_C] = \infty$.

PROOF   The proof uses a technique similar to that used to prove Dynkin's formula. Suppose by way of contradiction that $\mathsf{E}_{x_0}[\tau_C] < \infty$, and let $V_k = V(\Phi_k)$. Then we have

$$V_{\tau_C} = V_0 + \sum_{k=1}^{\tau_C}(V_k - V_{k-1})$$

$$= V_0 + \sum_{k=1}^{\infty}(V_k - V_{k-1})\mathbb{I}\{\tau_C \geq k\}.$$

Now from the bound in (11.43) we have for some $B < \infty$

$$\sum_{k=1}^{\infty} \mathsf{E}_{x_0}\left[|\mathsf{E}[(V_k - V_{k-1}) \mid \mathcal{F}_{k-1}^\Phi]\mathbb{I}\{\tau_C \geq k\}|\right] \leq B \sum_{k=1}^{\infty} \mathsf{P}_{x_0}\{\tau_C \geq k\} = B\mathsf{E}_{x_0}[\tau_C]$$

which is finite. Thus the use of Fubini's Theorem is justified, giving

$$\mathsf{E}_{x_0}[V_{\tau_C}] = V_0(x_0) + \sum_{k=1}^{\infty} \mathsf{E}_{x_0}[\mathsf{E}[(V_k - V_{k-1}) \mid \mathcal{F}_{k-1}^{\Phi}]\mathbb{I}\{\tau_C \ge k\}] \ge V_0(x_0).$$

But by (11.44), $V_{\tau_C} < V_0(x_0)$ with probability one, and this contradiction shows that $\mathsf{E}_{x_0}[\tau_C] = \infty$. $\qquad\square$

This gives a criterion for a ψ-irreducible chain to be non-positive. Based on Theorem 11.1.4 we have immediately

Theorem 11.5.2. *Suppose that the chain Φ is ψ-irreducible and that the non-negative function V satisfies (11.42) and (11.43) where $C \in \mathcal{B}^+(\mathsf{X})$. If the set*

$$C_+^c = \{x \in \mathsf{X} : V(x) > \sup_{y \in C} V(y)\}$$

also lies in $\mathcal{B}^+(\mathsf{X})$ then the chain is non-positive.

In practice, one would set C equal to a sublevel set of the function V so that the condition (11.44) is satisfied automatically for all $x \in C^c$.

It is not the case that this result holds without some auxiliary conditions such as (11.43). For take the state space to be $\mathbb{Z}_+$, and define $P(0, i) = 2^{-i}$ for all $i > 0$; if we now choose $k(i) > 2i$, and let

$$P(i, 0) = P(i, k(i)) = 1/2,$$

then the chain is certainly positive Harris, since by direct calculation

$$\mathsf{P}_0(\tau_0 \ge n + 1) \le 2^{-n}.$$

But now if $V(i) = i$ then for all $i > 0$

$$\Delta V(i) = [k(i)/2] - i > 0$$

and in fact we can choose $k(i)$ to give any value of $\Delta V(i)$ we wish.

11.5.2 Applications to random walk and SETAR models

As an immediate application of Theorem 11.5.2 we have

Proposition 11.5.3. *If Φ is a random walk on a half line with mean increment β then Φ is regular if and only if*

$$\beta = \int w\, \Gamma(dw) < 0.$$

PROOF   In Proposition 11.4.1 the sufficiency of the negative drift condition was established. If

$$\beta = \int w\, \Gamma(dw) \ge 0,$$

then using $V(x) = x$ we have (11.42), and the random walk homogeneity properties ensure that the uniform drift condition (11.43) also holds, giving non-positivity.   □

We now give a much more detailed and intricate use of this result to show that the scalar SETAR model is recurrent but not positive on the "margins" of its parameter set, between the regions shown to be positive in Section 11.4.3 and those regions shown to be transient in Section 9.5.2: see Figure B.1–Figure B.3 for the interpretation of the parameter ranges. In terms of the basic SETAR model defined by

$$X_n = \phi(j) + \theta(j)X_{n-1} + W_n(j), \qquad X_{n-1} \in R_j,$$

we call the *margins* of the parameter space the regions defined by

$$\theta(1) < 1, \ \theta(M) = 1, \ \phi(M) = 0 \tag{11.45}$$

$$\theta(1) = 1, \ \theta(M) < 1, \ \phi(1) = 0 \tag{11.46}$$

$$\theta(1) = \theta(M) = 1, \ \phi(M) = 0, \ \phi(1) \geq 0 \tag{11.47}$$

$$\theta(1) = \theta(M) = 1, \ \phi(M) < 0, \ \phi(1) = 0 \tag{11.48}$$

$$\theta(1) < 0, \ \theta(1)\theta(M) = 1, \ \phi(M) + \theta(M)\phi(1) = 0. \tag{11.49}$$

We first establish recurrence; then we establish non-positivity. For this group of parameter combinations, we need test functions of the form $V(x) = \log(u + ax)$ where u, a are chosen to give appropriate drift in (V1). To use these we will need the full force of the approximation results in Lemma 8.5.2, Lemma 8.5.3, Lemma 9.4.3, and Lemma 9.4.4, which we previously used in the analysis of random walk, and to analyze this region we will also need to assume (SETAR3): that is, that the variances of the noise distributions for the two end intervals are finite.

Proposition 11.5.4. *For the SETAR model satisfying (SETAR1)–(SETAR3), the chain is recurrent on the margins of the parameter space.*

PROOF   We will consider the test function

$$V(x) = \begin{cases} \log(u + ax) & x > R > r_{M-1} \\ \log(v - bx) & x < -R < r_1 \end{cases} \tag{11.50}$$

and $V(x) = 0$ in the region $[-R, R]$, where a, b and R are positive constants and u and v are real numbers to be chosen suitably for the different regions (11.45)-(11.49).

We denote the non-random part of the motion of the chain in the two end regions by

$$k(x) = \phi(M) + \theta(M)x$$

and

$$h(x) = \phi(1) + \theta(1)x.$$

We first prove recurrence when (11.45) or (11.46) holds. The proof is similar in style to that used for random walk in Section 9.5, but we need to ensure that the different behavior in each end of the two end intervals can be handled simultaneously.

Consider first the parameter region $\theta(M) = 1$, $\phi(M) = 0$, and $0 \le \theta(1) < 1$, and choose $a = b = u = v = 1$, with $x > R > r_{M-1}$. Write in this case

$$
\begin{aligned}
V_1(x) &= \mathsf{E}[\log(u + ak(x) + aW(M))\mathbb{I}_{[k(x)+W(M)>R]}] \\
V_2(x) &= \mathsf{E}[\log(v - bk(x) - bW(M))\mathbb{I}_{[k(x)+W(M)<-R]}]
\end{aligned}
\tag{11.51}
$$

so that

$$
\mathsf{E}_x[V(X_1)] = V_1(x) + V_2(x).
$$

In order to bound the terms in the expansion of the logarithms in V_1, V_2, we use the further notation

$$
\begin{aligned}
V_3(x) &= (a/(u + ak(x)))\mathsf{E}[W(M)\mathbb{I}_{[W(M)>R-k(x)]}] \\
V_4(x) &= (a^2/(2(u + ak(x))^2))\mathsf{E}[W^2(M)\mathbb{I}_{[R-k(x)<W(M)<0]}] \\
V_5(x) &= (b/(v - bk(x)))\mathsf{E}[W(M)\mathbb{I}_{[W(M)<-R-k(x)]}].
\end{aligned}
\tag{11.52}
$$

Since $\mathsf{E}(W^2(M)) < \infty$

$$
V_4(x) = (a^2/(2(u + ak(x))^2))\mathsf{E}[W^2(M)\mathbb{I}_{[W(M)<0]}] - o(x^{-2}),
$$

and by Lemma 8.5.3 both V_3 and V_5 are also $o(x^{-2})$.

For $x > R, u + ak(x) > 0$, and thus by Lemma 8.5.2,

$$
V_1(x) \le \Gamma_M(R - k(x), \infty) \log(u + ak(x)) + V_3(x) - V_4(x),
$$

while $v - bk(x) < 0$, and thus by Lemma 9.4.3,

$$
V_2(x) \le \Gamma_M(-\infty, -R - k(x))(\log(-v + bk(x)) - 2) - V_5(x).
$$

By Lemma 9.4.4(i) we also have that the terms

$$
-\Gamma_M(-\infty, R - k(x)) \log(u + ak(x)) + \Gamma_M(-\infty, -R - k(x))(\log(-v + bk(x)) - 2)
$$

are $o(x^{-2})$. Thus by choosing R large enough

$$
\begin{aligned}
\mathsf{E}_x[V(X_1)] &\le V(x) - (a^2/(2(u + ak(x))^2))\mathsf{E}[W^2(M)\mathbb{I}_{[W(M)<0]}] + o(x^{-2}) \\
&\le V(x), \qquad x > R.
\end{aligned}
\tag{11.53}
$$

For $x < -R < r_1$ and $\theta(1) = 0$, $\mathsf{E}_x[V(X_1)]$ is a constant and is therefore less than $V(x)$ for large enough R.

For $x < -R < r_1$ and $0 < \theta(1) < 1$, consider

$$
\begin{aligned}
V_6(x) &= \mathsf{E}[\log(u + ah(x) + aW(1))\mathbb{I}_{[h(x)+W(1)>R]}] \\
V_7(x) &= \mathsf{E}[\log(v - bh(x) - bW(1))\mathbb{I}_{[h(x)+W(1)<-R]}] :
\end{aligned}
\tag{11.54}
$$

we have as before

$$
\mathsf{E}_x[V(X_1)] = V_6(x) + V_7(x).
\tag{11.55}
$$

To handle the expansion of terms in this case we use

$$
V_8(x) = (a/(u + ah(x)))\mathsf{E}[W(1)\mathbb{I}_{[W(1)>R-h(x)]}]
$$

$$V_9(x) = (b/v - bh(x)))\mathsf{E}[W(1)\mathbb{I}_{[W(1)<-R-h(x)]}]$$

$$V_{10}(x) = (b^2/(2(v - bh(x))^2))\mathsf{E}[W^2(1)\mathbb{I}_{[-R-h(x)>W(1)>0]}].$$

Since $\mathsf{E}[W^2(1)] < \infty$

$$V_{10}(x) = (b^2/(2(v - bh(x))^2))\mathsf{E}[W^2(1)\mathbb{I}_{[W(1)>0]}] - o(x^{-2}),$$

and by Lemma 8.5.3, both $V_8(x)$ and $V_9(x)$ are $o(x^{-2})$.

For $x < -R$, $u + ah(x) < 0$, we have by Lemma 9.4.3(i),

$$V_6(x) \leq \Gamma_1(R - h(x), \infty)(\log(-u - ah(x)) - 2) - V_8(x),$$

and $v - bh(x) > 0$, so that by Lemma 8.5.2,

$$V_7(x) \leq \Gamma_1(-\infty, -R - h(x)) \log(v - bh(x)) - V_9(x) - V_{10}(x).$$

Hence choosing R large enough that $v - bh(x) \leq v - bx$, we have from (11.55),

$$\Gamma_1(-\infty, -R - h(x)) \log(v - bh(x)) \leq \Gamma_1(-\infty, -R - h(x)) \log(v - bx)$$

$$= V(x) - \Gamma_1(-R - h(x), \infty) \log(v - bx).$$

By Lemma 9.4.4(ii),

$$\Gamma_1(R - h(x), \infty)(\log(-u - ah(x)) - 2) - \Gamma_1(-R - h(x), \infty) \log(v - bx) \leq o(x^{-2}),$$

and thus

$$\begin{aligned} \mathsf{E}_x[V(X_1)] &\leq V(x) - (b^2/(2(v - bh(x))^2))\mathsf{E}[W^2(1)\mathbb{I}_{W(1)>0}] + o(x^{-2}) \\ &\leq V(x), \quad x < -R. \end{aligned} \tag{11.56}$$

Finally consider the region $\theta(M) = 1$, $\phi(M) = 0$, $\theta(1) < 0$, and choose $a = -b\theta(M)$ and $v - u = a\phi(1)$. For $x > R > r_{M-1}$, (11.53) is obtained in a manner similar to the above. For $x < -R < r_1$, we look at

$$V_{11}(x) = (a^2/(2(u + ah(x))^2))\mathsf{E}[W^2(1)\mathbb{I}_{[R-h(x)<W(1)<0]}].$$

By Lemma 9.4.3

$$V_6(x) \leq \Gamma_1(R - h(x), \infty) \log(u + ah(x)) + V_8(x) - V_{11}(x),$$

and

$$V_7(x) \leq \Gamma_1(-\infty, -R - h(x))(\log(-v + bh(x)) - 2) - V_9(x).$$

From the choice of a, b, u and v,

$$\log(u + ah(x)) = \log(v - bx) = V(x),$$

and thus by Lemma 8.5.3 and Lemma 9.4.4(i) for R large enough

$$\begin{aligned} \mathsf{E}_x[V(X_1)] &\leq V(x) - (a^2/(2(u + ah(x))^2))\mathsf{E}[W^2(1)\mathbb{I}_{[W(1)<0]}] + o(x^{-2}) \\ &\leq V(x), \quad x < -R. \end{aligned} \tag{11.57}$$

When (11.46) holds, the recurrence of the SETAR model follows by symmetry from the result in the region (11.45).

(ii)   We now consider the region where (11.47) holds: in (11.48) the result will again follow by symmetry.

Choose $a = b = u = v = 1$ in the definition of V. For $x > R > r_{M-1}$, (11.53) holds as before. For $x < -R < r_1$, since $1 - h(x) \leq 1 - x$,

$$\Gamma_1(-\infty, -R - h(x)) \log(1 - h(x)) \leq \Gamma_1(-\infty, -R - h(x)) \log(1 - x).$$

From this, (11.56) is also obtained as before.

(iii)   Finally we show that the chain is recurrent if the boundary condition (11.49) holds.

Choose $v - u = b\phi(M) = a\phi(1)$, $b = -a\theta(1) = -a/\theta(M)$. For $x > R > r_{M-1}$, consider

$$V_{12}(x) = (b^2/(2(v - bk(x))^2))\mathsf{E}[W^2(M)\mathbb{I}_{[-R-k(x)>W(M)>0]}].$$

By Lemma 9.4.3 we get both

$$V_1(x) \leq \Gamma_M(R - k(x), \infty)(\log(-u - ak(x)) - 2) - V_3(x),$$

$$V_2(x) \leq \Gamma_M(-\infty, -R - k(x)) \log(v - bk(x)) - V_5(x) - V_{12}(x).$$

From the choice of a, b, u and v

$$\Gamma_M(-\infty, -R - k(x)) \log(v - bk(x)) = \log(u + ax) - \Gamma_M(-R - k(x), \infty) \log(u + ax),$$

and thus by Lemma 9.4.4(i) and (iii), for R large enough

$$\begin{aligned} \mathsf{E}_x[V(X_1)]] &\leq V(x) - (b^2/(2(v - bk(x))^2))\mathsf{E}[W^2(M)\mathbb{I}_{[W(M)>0]}] + o(x^{-2}) \\ &\leq V(x), \quad x > R. \end{aligned} \tag{11.58}$$

For $x < -R < r_1$, since

$$\log(u + ah(x)) = \log(v - bx),$$

(11.57) is obtained similarly.

It is obvious that the above test functions V are coercive, and hence (V1) holds outside a compact set $[-R, R]$ in each case. Hence we have recurrence from Theorem 9.1.8.

$\square$

To complete the classification of the model, we need to prove that in this region the model is not positive recurrent.

Proposition 11.5.5. *For the SETAR model satisfying (SETAR1)–(SETAR3), the chain is non-positive on the margins of the parameter space.*

PROOF   We need to show that in the case where

$$\phi(1) < 0, \qquad \phi(1)\phi(M) = 1, \qquad \theta(1)\phi(M) + \theta(M) \leq 0$$

the chain is non-positive. To do this we appeal to the criterion in Section 11.5.1.

As we have $\phi(1)\phi(M) = 1$ we can as before find positive constants a, b such that

$$\phi(1) = -ba^{-1}, \qquad \phi(M) = -ab^{-1}.$$

We will consider the test function

$$V(x) = V_{cd}(x) + \mathbb{I}_{kR}(x) \tag{11.59}$$

where the functions V_{cd} and $\mathbb{I}_{kR}$ are defined for positive c, d, k, R by

$$\mathbb{I}_{kR}(x) = \begin{cases} k & |x| \leq R \\ 0 & |x| > R \end{cases}$$

and

$$V_{cd}(x) = \begin{cases} a\,x + c & x > 0 \\ b\,|x| + d & x \leq 0 \end{cases}.$$

It is immediate that

$$\int P(x, dy)|V(x) - V(y)| \leq a\mathsf{E}[|W_1|] + b\mathsf{E}[|W_M|] + 2(a|\theta(1)| + b|\theta(M)|) + 2|d - c|,$$

whilst V is obviously coercive.

We now verify that indeed the mean drift of $V(\Phi_n)$ is positive. Now for $x \in R_M$, we have

$$\int P(x, dy)V(y) = \int \Gamma_M(dy - \theta(M) - \phi(M)x)V_{cd}(y)$$

$$+ \int \Gamma_M(dy - \theta(M) - \phi(M)x)\mathbb{I}_{kR}(y), \tag{11.60}$$

and the first of these terms can be written as

$$\int \Gamma_M(dy - \theta(M) - \phi(M)x)V_{cd}(y)$$

$$= \int \Gamma_M(dz)[-b(z + \theta(M) + \phi(M)x) + d]$$

$$+ \int_{-\theta(M)-\phi(M)x}^{\infty} \Gamma_M(dz)[(a + b)(z + \theta(M) + \phi(M)x) + c - d]. \tag{11.61}$$

Using this representation we thus have

$$\int P(x, dy)V(y) = ax + d - b\theta(M)$$

$$+ \int_0^{\infty} \Gamma_M(dy - \theta(M) - \phi(M)x)[(a + b)y + c - d]$$

$$+ \int_{-R}^{R} k\Gamma_M(dy - \theta(M) - \phi(M)x). \tag{11.62}$$

A similar calculation shows that for $x \in R_1$,

$$\int P(x, dy)V(y) = -bx + c - a\theta(1)$$

$$- \int_{-\infty}^{0} \Gamma_1(dy - \theta(1) - \phi(1)x)[(a+b)y + c - d]$$

$$+ \int_{-R}^{R} k\Gamma_1(dy - \theta(1) - \phi(1)x). \tag{11.63}$$

Let us now choose the positive constants c, d to satisfy the constraints

$$a\theta(1) \geq d - c \geq b\theta(M) \tag{11.64}$$

(which is possible since $\theta(1)\phi(M) + \theta(M) \leq 0$) and k, R sufficiently large that

$$R \geq \max(|\theta(1)|, |\theta(M)|) \tag{11.65}$$

$$k \geq (a+b)\max(|\theta(1)|, |\theta(M)|). \tag{11.66}$$

It then follows that for all x with $|x|$ sufficiently large

$$\int P(x, dy)V(y) \geq V(x)$$

and the chain is non-positive from Section 11.5.1.   □

11.6   Commentary

For countable space chains, the results of this chapter have been thoroughly explored. The equivalence of positive recurrence and the finiteness of expected return times to each atom is a consequence of Kac's Theorem, and as we saw in Proposition 11.1.1, it is then simple to deduce the regularity of all states. As usual, Feller [114] or Chung [71] or Çinlar [59] provide excellent discussions.

Indeed, so straightforward is this in the countable case that the name "regular chain", or any equivalent term, does not exist as far as we are aware. The real focus on regularity and similar properties of hitting times dates to Isaac [169] and Cogburn [75]; the latter calls regular sets "strongly uniform". Although many of the properties of regular sets are derived by these authors, proving the actual existence of regular sets for general chains is a surprisingly difficult task. It was not until the development of the Nummelin–Athreya–Ney theory of splitting and embedded regeneration occurred that the general result of Theorem 11.1.4, that positive recurrent chains are "almost" regular chains was shown (see Nummelin [302]).

Chapter 5 of Nummelin [303] contains many of the equivalences between regularity and positivity, and our development owes a lot to his approach. The more general f-regularity condition on which he focuses is central to our Chapter 14: it seems worth considering the probabilistic version here first.

For countable chains, the equivalence of (V2) and positive recurrence was developed by Foster [129], although his proof of sufficiency is far less illuminating than the one we have here. The earliest results of this type on a non-countable space appear to be those in Lamperti [235], and the results for general ψ-irreducible chains were developed by Tweedie [397, 398]. The use of drift criteria for continuous space chains, and the use of Dynkin's formula in discrete time, seem to appear for the first time in Kalashnikov [187, 189, 190]. The version used here and later was developed in Meyn and Tweedie [277], although it is well known in continuous time for more special models such as diffusions (see Kushner [232] or Khas'minskii [206]).

There are many rediscoveries of mean drift theorems in the literature. For operations research models (V2) is often known as Pakes' Lemma from [313]: interestingly, Pakes' result rediscovers the original form buried in the discussion of Kendall's famous queueing paper [200], where Foster showed that a sufficient condition for positivity of a chain on $\mathbb{Z}_+$ is the existence of a solution to the pair of equations

$$\sum P(x, y)V(y) \leq V(x) - 1, \qquad x \geq N$$
$$\sum P(x, y)V(y) < \infty, \qquad x < N,$$

although in [129] he only gives the result for $N = 1$. The general N form was also rediscovered by Moustafa [289], and a form for reducible chains given by Mauldon [251]. An interesting state-dependent variation is given by Malyšhev and Men'šikov [243]; we return to this and give a proof based on Dynkin's formula in Chapter 19.

The systematic exploitation of the various equivalences between hitting times and mean drifts, together with the representation of π, is new in the way it appears here. In particular, although it is implicit in the work of Tweedie [398] that one can identify sublevel sets of test functions as regular, the current statements are much more comprehensive than those previously available, and generalize easily to give an appealing approach to f-regularity in Chapter 14.

The criteria given here for chains to be non-positive have a shorter history. The fact that drift away from a petite set implies non-positivity provided the increments are bounded in mean appears first in Tweedie [398], with a different and less transparent proof, although a restricted form is in Doob ([99], p. 308), and a recent version similar to that we give here has been recently given by Fayolle et al. [110]. All proofs we know require bounded mean increments, although there appears to be no reason why weaker constraints may not be as effective.

Related results on the drift condition can be found in Marlin [249], Tweedie [396], Rosberg [336] and Szpankowski [380], and no doubt in many other places: we return to these in Chapter 19.

Applications of the drift conditions are widespread. The first time series application appears to be by Jones [182], and many more have followed. Laslett et al. [237] give an overview of the application of the conditions to operations research chains on the real line. The construction of a test function for the GI/G/1 queue given in Section 11.4.2 is taken from Meyn and Down [273] where this forms a first step in a stability analysis of generalized Jackson networks. A test function approach is also used in Sigman [354] and Fayolle et al. [110] to obtain stability for queueing networks: the interested reader should also note that in Borovkov [43] the stability question is addressed using other means.

The SETAR analysis we present here is based on a series of papers where the SETAR model is analyzed in increasing detail. The positive recurrence and transience results are essentially in Petruccelli et al. [315] and Chan et al. [64], and the non-positivity analysis as we give it here is taken from Guo and Petruccelli [149]. The assumption of finite variances in (SETAR3) is again almost certainly redundant, but an exact condition is not obvious.

We have been rather more restricted than we could have been in discussing specific models at this point, since many of the most interesting examples, both in operations research and in state space and time series models, actually satisfy a stronger version of the drift condition (V2): we discuss these in detail in Chapter 15 and Chapter 16. However, it is not too strong a statement that Foster's criterion (as (V2) is often known) has been adopted as the tool of choice to classify chains as positive recurrent: for a number of applications of interest we refer the reader to the recent books by Tong [388] on nonlinear models and Asmussen [9] on applied probability models. Variations for two-dimensional chains on the positive quadrant are also widespread: the first of these seems to be due to Kingman [207], and ongoing usage is typified by, for example, Fayolle [109].

Chapter 12

Invariance and tightness

In one of our heuristic descriptions of stability, in Section 1.3, we outlined a picture of a chain settling down to a stable regime independent of its initial starting point: we will show in Part III that positive Harris chains do precisely this, and one role of π is to describe the final stochastic regime of the chain, as we have seen.

It is equally possible to approach the problem from the other end: if we have a limiting measure for P^n, then it may well generate a stationary measure for the chain. We saw this described briefly in (10.4): and our main goal now is to consider chains on topological spaces which do not necessarily enjoy the property of ψ-irreducibility, and to show how we can construct invariant measures for such chains through such limiting arguments, rather than through regenerative and splitting techniques.

We will develop the consequences of the following slightly extended form of boundedness in probability, introduced in Chapter 6.

Tightness and boundedness in probability on average

A sequence of probabilities $\{\mu_k : k \in \mathbb{Z}_+\}$ is called *tight* if for each $\varepsilon > 0$, there exists a compact subset $C \subset \mathsf{X}$ such that

$$\liminf_{k \to \infty} \mu_k(C) \geq 1 - \varepsilon. \tag{12.1}$$

The chain Φ will be called *bounded in probability on average* if for each initial condition $x \in \mathsf{X}$ the sequence $\{\overline{P}_k(x, \cdot) : k \in \mathbb{Z}_+\}$ is tight, where we define

$$\overline{P}_k(x, \cdot) := \frac{1}{k} \sum_{i=1}^{k} P^i(x, \cdot). \tag{12.2}$$

We have the following highlights of the consequences of these definitions.

Theorem 12.0.1.   (i) *If* Φ *is a weak Feller chain which is bounded in probability on average, then there exists at least one invariant probability measure.*

(ii) *If* Φ *is an e-chain which is bounded in probability on average, then there exists a weak Feller transition function* Π *such that for each x the measure $\Pi(x, \cdot)$ is invariant, and*

$$\overline{P}_n(x, f) \to \Pi(x, f), \qquad as\ n \to \infty,$$

for all bounded continuous functions f, and all initial conditions $x \in \mathsf{X}$.

PROOF   We prove (i) in Theorem 12.1.2, together with a number of consequents for weak Feller chains. The proof of (ii) essentially occupies Section 12.4, and is concluded in Theorem 12.4.1. $\square$

We will see that for Feller chains, and even more powerfully for e-chains, this approach based upon tightness and weak convergence of probability measures provides a quite different method for constructing an invariant probability measure. This is exemplified by the linear model construction which we have seen in Section 10.5.4.

From such constructions we will show in Section 12.4 that (V2) implies a form of positivity for a Feller chain. In particular, for e-chains, if (V2) holds for a compact set C and an everywhere finite function V then the chain is bounded in probability on average, so that there is a collection of invariant measures as in Theorem 12.0.1 (ii).

In this chapter we also develop a class of kernels, introduced by Neveu in [295], which extend the definition of the kernels U_A. This involves extending the definition of a stopping time to randomized stopping times. These operators have very considerable intuitive appeal and demonstrate one way in which the results of Section 10.4 can be applied to non-irreducible chains.

Using this approach, we will also show that (V1) gives a criterion for the existence of a σ-finite invariant measure for a Feller chain.

12.1   Chains bounded in probability

12.1.1   Weak and vague convergence

It is easy to see that for any chain, being bounded in probability on average is a stronger condition than being non-evanescent.

Proposition 12.1.1. *If* Φ *is bounded in probability on average, then it is non-evanescent.*

PROOF   We obviously have

$$\mathsf{P}_x\{\bigcup_{j=n}^{\infty} \mathbb{I}(\Phi_j \in C)\} \geq P^n(x, C); \tag{12.3}$$

if Φ is evanescent, then for some x there is an $\varepsilon > 0$ such that for every compact C,

$$\limsup_{n \to \infty} \mathsf{P}_x\{\bigcup_{j=n}^{\infty} \mathbb{I}(\Phi_j \in C)\} \leq 1 - \varepsilon$$

and so the chain is not bounded in probability on average.   □

The consequences of an assumption of tightness are well-known (see Billingsley [36]): essentially, tightness ensures that we can take weak limits (possibly through a subsequence) of the distributions $\{\overline{P}_k(x, \cdot) : k \in \mathbb{Z}_+\}$ and the limit will then be a probability measure. In many instances we may apply Fatou's Lemma to prove that this limit is subinvariant for $\mathbf{\Phi}$; and since it is a probability measure it is in fact invariant.

We will then have, typically, that the convergence to the stationary measure (when it occurs) is in the weak topology on the space of all probability measures on $\mathcal{B}(\mathsf{X})$ as defined in Section D.5.

12.1.2 Feller chains and invariant probability measures

For weak Feller chains, boundedness in probability gives an effective approach to finding an invariant measure for the chain, even without irreducibility.

We begin with a general existence result which gives necessary and sufficient conditions for the existence of an invariant probability. From this we will find that the test function approach developed in Chapter 11 may be applied again, this time to establish the existence of an invariant probability measure for a Feller Markov chain.

Recall that the geometrically sampled Markov transition function, or resolvent, K_{a_ε} is defined for $\varepsilon < 1$ as $K_{a_\varepsilon} = (1 - \varepsilon) \sum_{k=0}^{\infty} \varepsilon^k P^k$

Theorem 12.1.2. *Suppose that $\mathbf{\Phi}$ is a Feller Markov chain. Then*

(i) *If an invariant probability does not exist, then for any compact set $C \subset \mathsf{X}$,*

$$\overline{P}_n(x, C) \;\; \to \;\; 0 \qquad as \; n \to \infty \tag{12.4}$$

$$K_{a_\varepsilon}(x, C) \;\; \to \;\; 0 \qquad as \; \varepsilon \uparrow 1 \tag{12.5}$$

uniformly in $x \in \mathsf{X}$.

(ii) *If $\mathbf{\Phi}$ is bounded in probability on average, then it admits at least one invariant probability.*

PROOF   We prove only (12.4), since the proof of (12.5) is essentially identical. The proof is by contradiction: we assume that no invariant probability exists, and that (12.4) does not hold.

Fix $f \in \mathcal{C}_c(\mathsf{X})$ such that $f \geq 0$, and fix $\delta > 0$. Define the open sets $\{A_k : k \in \mathbb{Z}_+\}$ by

$$A_k = \Big\{ x \in \mathsf{X} : \overline{P}_k f > \delta \Big\}.$$

If (12.4) does not hold then for some such f there exists $\delta > 0$ and a subsequence $\{N_i : i \in \mathbb{Z}_+\}$ of $\mathbb{Z}_+$ with $A_{N_i} \neq \emptyset$ for all i. Let $x_i \in A_{N_i}$ for each i, and define

$$\lambda_i := \overline{P}_{N_i}(x_i, \cdot)$$

We see from Proposition D.5.6 that the set of sub-probabilities is sequentially compact with respect to vague convergence. Let λ_∞ be any vague limit point: $\lambda_{n_i} \overset{\mathrm{v}}{\longrightarrow} \lambda_\infty$ for some subsequence $\{n_i : i \in \mathbb{Z}_+\}$ of $\mathbb{Z}_+$. The sub-probability $\lambda_\infty \neq 0$ because, by the definition of vague convergence, and since $x_i \in A_{N_i}$,

$$
\begin{aligned}
\int f \, d\lambda_\infty &\geq \liminf_{i\to\infty} \int f \, d\lambda_i \\
&= \liminf_{i\to\infty} \overline{P}_{N_i}(x_i, f) \\
&\geq \delta > 0.
\end{aligned}
\tag{12.6}
$$

But now λ_∞ is a non-trivial invariant measure. For, letting $g \in \mathcal{C}_c(\mathsf{X})$ satisfy $g \geq 0$, we have by continuity of Pg and (D.6),

$$
\begin{aligned}
\int g \, d\lambda_\infty &= \lim_{i\to\infty} \overline{P}_{N_{n_i}}(x_{n_i}, g) \\
&= \lim_{i\to\infty} [\overline{P}_{N_{n_i}}(x_{n_i}, g) + N_i^{-1}(P^{N_{n_i}+1}(x_{n_i}, g) - Pg)] \\
&= \lim_{i\to\infty} \overline{P}_{N_{n_i}}(x_{n_i}, Pg) \\
&\geq \int (Pg) \, d\lambda_\infty.
\end{aligned}
\tag{12.7}
$$

By regularity of finite measures on $\mathcal{B}(\mathsf{X})$ (cf Theorem D.3.2) this implies that $\lambda_\infty \geq \lambda_\infty P$, which is only possible if $\lambda_\infty = \lambda_\infty P$. Since we have assumed that no invariant probability exists it follows that $\lambda_\infty = 0$, which contradicts (12.6). Thus we have that $A_k = \emptyset$ for sufficiently large k.

To prove (ii), let $\boldsymbol{\Phi}$ be bounded in probability on average. Since we can find $\varepsilon > 0$, $x \in \mathsf{X}$ and a compact set C such that $\overline{P}^j(x, C) > 1 - \varepsilon$ for all sufficiently large j by definition, (12.4) fails and so the chain admits an invariant probability.   □

The following corollary easily follows: notice that the condition (12.8) is weaker than the obvious condition of Lemma D.5.3 for boundedness in probability on average.

Proposition 12.1.3. *Suppose that the Markov chain $\boldsymbol{\Phi}$ has the Feller property, and that a coercive function V exists such that for some initial condition $x \in \mathsf{X}$,*

$$
\liminf_{k\to\infty} \mathsf{E}_x[V(\Phi_k)] < \infty.
\tag{12.8}
$$

Then an invariant probability exists.   □

These results require minimal assumptions on the chain. They do have two drawbacks in practice.

Firstly, there is no guarantee that the invariant probability is unique. Currently, known conditions for uniqueness involve the assumption that the chain is ψ-irreducible. This immediately puts us in the domain of Chapter 10, and if the measure ψ has an open set in its support, then in fact we have the full T-chain structure immediately available, and so we would avoid the weak convergence route.

Secondly, and essentially as a consequence of the lack of uniqueness of the invariant measure π, we do not generally have guaranteed that

$$
P^n(x, \cdot) \xrightarrow{\text{w}} \pi.
$$

However, we do have the result

Proposition 12.1.4. *Suppose that the Markov chain Φ has the Feller property, and is bounded in probability on average.*

If the invariant measure π is unique then for every x

$$\overline{P}_n(x, \cdot) \xrightarrow{\ \mathrm{w}\ } \pi. \tag{12.9}$$

PROOF   Since for every subsequence $\{n_k\}$ the set of probabilities $\{\overline{P}_{n_k}(x, \cdot)\}$ is sequentially compact in the weak topology, then as in the proof of Theorem 12.1.2, from boundedness in probability we have that there is a further subsequence converging weakly to a non-trivial limit which is invariant for P. Since all these limits coincide by the uniqueness assumption on π we must have (12.9).   □

Recall that in Proposition 6.4.2 we came to a similar conclusion. In that result, convergence of the distributions to a unique invariant probability, in a manner similar to (12.9), is given as a condition under which a Feller chain Φ is an e-chain.

12.2   Generalized sampling and invariant measures

In this section we generalize the idea of sampled chains in order to develop another approach to the existence of invariant measures for Φ. This relies on an identity called the resolvent equation for the kernels U_B, $B \in \mathcal{B}(\mathsf{X})$. The idea of the generalized resolvent identity is taken from the theory of continuous time processes, and we shall see that even in discrete time it unifies several concepts which we have used already, and which we shall use in this chapter to give a different construction method for σ-finite invariant measures for a Feller chain, even without boundedness in probability.

To state the resolvent equation in full generality we introduce randomized first entrance times. These include as special cases the ordinary first entrance time τ_A, and also random times which are completely independent of the process: the former have of course been used extensively in results such as the identification of the structure of the unique invariant measure for ψ-irreducible chains, whilst the latter give us the sampled chains with kernel K_{a_ε}.

The more general version involves a function h which will usually be continuous with compact support when the chain is on a topological space, although it need not always be so.

Let $0 \leq h \leq 1$ be a function on X. The random time τ_h which we associate with the function h will have the property that $\mathsf{P}_x\{\tau_h \geq 1\} = 1$, and for any initial condition $x \in \mathsf{X}$ and any time $k \geq 1$,

$$\mathsf{P}_x\{\tau_h = k \mid \tau_h \geq k, \mathcal{F}^\Phi_\infty\} = h(\Phi_k). \tag{12.10}$$

A probabilistic interpretation of this equation is that at each time $k \geq 1$ a weighted coin is flipped with the probability of heads equal to $h(\Phi_k)$. At the first instance k that a head is finally achieved we set $\tau_h = k$. Hence we must have, for any $k \geq 1$,

$$\mathsf{P}_x\{\tau_h = k \mid \mathcal{F}^\Phi_\infty\} \quad = \quad \prod_{i=1}^{k-1}(1 - h(\Phi_i))h(\Phi_k) \tag{12.11}$$

$$\mathsf{P}_x\{\tau_h \geq k \mid \mathcal{F}^\Phi_\infty\} \quad = \quad \prod_{i=1}^{k-1}(1 - h(\Phi_i)) \tag{12.12}$$

where the product is interpreted as one when $k = 1$.

For example, if $h = \mathbb{I}_B$ then we see that $\tau_h = \tau_B$. If $h = \frac{1}{2}\mathbb{I}_B$ then a fair coin is flipped on each visit to B, so that $\Phi_{\tau_h} \in B$, but with probability one half, the random time τ_h will be greater then τ_B.

Note that this is very similar to the Athreya–Ney randomized stopping time construction of an atom, mentioned in Section 5.1.3.

By enlarging the probability space on which Φ is defined, and adjoining an i.i.d. process $Y = \{Y_k, k \in \mathbb{Z}_+\}$ to Φ, we now show that we can explicitly construct the random time τ_h so that it is an ordinary stopping time for the bivariate chain

$$\Psi_k = \begin{pmatrix} \Phi_k \\ Y_k \end{pmatrix}, \qquad k \in \mathbb{Z}_+.$$

Suppose that Y is i.i.d. and independent of Φ, and that each Y_k has distribution μ^{uni}, where μ^{uni} denotes the uniform distribution on $[0,1]$. Then for any sets $A \in \mathcal{B}(\mathsf{X})$, $B \in \mathcal{B}([0,1])$,

$$\mathsf{P}_x\{\Psi_1 \in A \times B \mid \Phi_0 = x, Y_0 = u\} = P(x, A)\mu^{\text{uni}}(B)$$

With this transition probability, Ψ is a Markov chain whose state space is equal to $\mathsf{Y} = \mathsf{X} \times [0,1]$.

Let $A_h \in \mathcal{B}(\mathsf{Y})$ denote the set

$$A_h = \{(x, u) \in \mathsf{Y} : h(x) \geq u\}$$

and define the random time $\tau_h = \min(k \geq 1 : \Psi_k \in A_h)$. Then τ_h is a stopping time for the bivariate chain.

We see at once from the definition and the fact that Y_k is independent of $(\Phi, Y_1, \ldots, Y_{k-1})$ that τ_h satisfies (12.10). For given any $k \geq 1$,

$$\begin{aligned} \mathsf{P}_x\{\tau_h = k \mid \tau_h \geq k, \mathcal{F}_\infty^\Phi\} &= \mathsf{P}_x\{h(\Phi_k) \geq Y_k \mid \tau_h \geq k, \mathcal{F}_\infty^\Phi\} \\ &= \mathsf{P}_x\{h(\Phi_k) \geq Y_k \mid \mathcal{F}_\infty^\Phi\} \\ &= h(\Phi_k), \end{aligned}$$

where in the second equality we used the fact that the event $\{\tau_h \geq k\}$ is measurable with respect to $\{\Phi, Y_1, \ldots, Y_{k-1}\}$, and in the final equality we used independence of Y and Φ.

Now define the kernel U_h on $\mathsf{X} \times \mathcal{B}(\mathsf{X})$ by

$$U_h(x, B) = \mathsf{E}_x\left[\sum_{k=1}^{\tau_h} \mathbb{I}_B(\Phi_k)\right]. \tag{12.13}$$

where the expectation is understood to be on the enlarged probability space. We have

$$U_h(x, B) = \sum_{k=1}^{\infty} \mathsf{E}_x[\mathbb{I}_B(\Phi_k)\mathbb{I}\{\tau_h \geq k\}]$$

and hence from (12.12)

$$U_h(x, B) = \sum_{k=0}^{\infty} P(I_{1-h}P)^k(x, B) \tag{12.14}$$

where I_{1-h} denotes the kernel which gives multiplication by $1-h$. This final expression for U_h defines this kernel independently of the bivariate chain.

In the special cases $h \equiv 0$, $h = \mathbb{I}_B$, and $h \equiv 1$ we have, respectively,

$$U_h = U, \qquad U_h = U_B, \qquad U_h = P.$$

When $h = \frac{1}{2}$ so that τ_h is completely independent of $\boldsymbol{\Phi}$ we have

$$U_{\frac{1}{2}} = \sum_{k=1}^{\infty} (\tfrac{1}{2})^{k-1} P^k = K_{a_{\frac{1}{2}}}.$$

For general functions h, the expression (12.14) defining U_h involves only the transition function P for $\boldsymbol{\Phi}$ and hence allows us to drop the bivariate chain if we are only interested in properties of the kernel U_h. However the existence of the bivariate chain and the construction of τ_h allows a transparent proof of the following resolvent equation.

Theorem 12.2.1 (Resolvent equation). *Let $h \le 1$ and $g \le 1$ be two functions on X with $h \ge g$. Then the resolvent equation holds:*

$$U_g = U_h + U_h I_{h-g} U_g = U_h + U_g I_{h-g} U_h.$$

PROOF   To prove the theorem we will consider the bivariate chain $\boldsymbol{\Psi}$. We will see that the resolvent equation formalizes several relationships between the stopping times τ_g and τ_h for $\boldsymbol{\Psi}$. Note that since $h \ge g$, we have the inclusion $A_g \subseteq A_h$ and hence $\tau_g \ge \tau_h$.

To prove the first resolvent equation we write

$$\sum_{k=1}^{\tau_g} f(\Phi_k) = \sum_{k=1}^{\tau_h} f(\Phi_k) + \mathbb{I}\{\tau_g > \tau_h\} \sum_{k=\tau_h+1}^{\tau_g} f(\Phi_k)$$

so by the strong Markov property for the process $\boldsymbol{\Psi}$,

$$U_g(x, f) = U_h(x, f) + \mathsf{E}_x[\mathbb{I}\{g(\Phi_{\tau_h}) < U_{\tau_h}\} U_g(\Phi_{\tau_h}, f)]. \qquad (12.15)$$

The latter expectation can be computed using (12.12). We have

$$\mathsf{E}_x[\mathbb{I}\{g(\Phi_{\tau_h}) < Y_{\tau_h}\} U_g(\Phi_{\tau_h}, f) \mathbb{I}\{\tau_h = k\} \mid \mathcal{F}_\infty^{\Phi}]$$

$$= \quad \mathsf{E}_x[\mathbb{I}\{g(\Phi_k) < Y_k\} U_g(\Phi_k, f) \mathbb{I}\{\tau_h = k\} \mid \mathcal{F}_\infty^{\Phi}]$$

$$= \quad \mathsf{E}_x[\mathbb{I}\{g(\Phi_k) < Y_k\} \mathbb{I}\{h(\Phi_k) \ge Y_k\} U_g(\Phi_k, f) \mathbb{I}\{\tau_h \ge k\} \mid \mathcal{F}_\infty^{\Phi}]$$

$$= \quad \mathsf{E}_x[\mathbb{I}\{g(\Phi_k) < Y_k \le h(\Phi_k)\} U_g(\Phi_k, f) \mathbb{I}\{\tau_h \ge k\} \mid \mathcal{F}_\infty^{\Phi}]$$

$$= \quad [h(\Phi_k) - g(\Phi_k)] U_g(\Phi_k, f) \prod_{i=1}^{k-1} [1 - h(\Phi_i)].$$

Taking expectations and summing over k gives

$$
\begin{aligned}
&\mathsf{E}_x[\mathbb{I}\{g(\Phi_{\tau_h}) < Y_{\tau_h}\}U_g(\Phi_{\tau_h}, f)]\\
&= \sum_{k=1}^{\infty}\mathsf{E}_x\left[\prod_{i=1}^{k-1}[1 - h(\Phi_i)][h(\Phi_k) - g(\Phi_k)]U_g(\Phi_k, f)\right]\\
&= \sum_{k=0}^{\infty}(PI_{1-h})^k PI_{h-g}U_g\,(x, f).
\end{aligned}
$$

This together with (12.15) gives the first resolvent equation.

To prove the second, break the sum to τ_g into the pieces between consecutive visits to A_h:

$$
\sum_{k=1}^{\tau_g}f(\Phi_k) = \sum_{k=1}^{\tau_h}f(\Phi_k) + \sum_{k=1}^{\tau_g}\mathbb{I}\{\Psi_k \in \{A_h \setminus A_g\}\}\theta^k\Big\{\sum_{i=1}^{\tau_h}f(\Phi_i)\Big\}.
$$

Taking expectations gives

$$
\begin{aligned}
U_g(x, f) =\ & U_h(x, f)\\
& + \mathsf{E}_x\Big[\sum_{k-1}^{\tau_g}\mathbb{I}\{g(\Phi_k) < Y_k \le h(\Phi_k)\}\theta^k\Big\{\sum_{i=1}^{\tau_h}f(\Phi_i)\Big\}\Big]. \quad (12.16)
\end{aligned}
$$

The expectation can be transformed, using the Markov property for the bivariate chain, to give

$$
\begin{aligned}
&\mathsf{E}_x\Big[\sum_{k=1}^{\tau_g}\mathbb{I}\{g(\Phi_k) < Y_k \le h(\Phi_k)\}\theta^k\Big\{\sum_{i=1}^{\tau_h}f(\Phi_i)\Big\}\Big]\\
&= \sum_{k=1}^{\infty}\mathsf{E}_x\Big[\mathbb{I}\{g(\Phi_k) < Y_k \le h(\Phi_k)\}\mathbb{I}\{\tau_g \ge k\}\mathsf{E}_{\Psi_k}\Big[\sum_{i=1}^{\tau_h}f(\Phi_i)\Big]\Big]\\
&= \sum_{k=1}^{\infty}\mathsf{E}_x\Big[[h(\Phi_k) - g(\Phi_k)]\mathbb{I}\{\tau_g \ge k\}U_h(\Phi_k, f)\Big]\\
&= U_g I_{h-g}U_h
\end{aligned}
$$

which together with (12.16) proves the second resolvent equation. □

When τ_h is a.s. finite for each initial condition the kernel P_h defined as

$$
P_h(x, A) = U_h I_h\,(x, A)
$$

is a Markov transition function. This follows from (12.11), which shows that

$$
\begin{aligned}
P_h(x, \mathsf{X}) = U_h(x, h) &= \sum_{k=1}^{\infty}\mathsf{E}_x\Big[\prod_{i=1}^{k-1}(1 - h(\Phi_i))h(\Phi_k)\Big]\\
&= \sum_{k=1}^{\infty}\mathsf{P}_x\{\tau_h = k\} \quad (12.17)
\end{aligned}
$$

and hence $P_h(x, \mathsf{X}) = 1$ if $P_x\{\tau_h < \infty\} = 1$.

It is natural to seek conditions which will ensure that τ_h is finite, since this is of course analogous to the concept of Harris recurrence, and indeed identical to it for $h = \mathbb{I}_C$. The following result answers this question as completely as we will find necessary.

Define $L(x, h) = U_h(x, h)$ and $Q(x, h) = P_x\{\sum_{k=1}^{\infty} h(\Phi_k) = \infty\}$. Theorem 12.2.2 now shows that these functions are extensions of the the functions L and Q which we have used extensively: in the special case where $h = \mathbb{I}_B$ for some $B \in \mathcal{B}(\mathsf{X})$ we have $Q(x, \mathbb{I}_B) = Q(x, B)$ and $L(x, \mathbb{I}_B) = L(x, B)$.

Theorem 12.2.2. *For any $x \in \mathsf{X}$ and function $0 \le h \le 1$,*

(i) $P_x\{\Psi_k \in A_h \quad \text{i.o.}\} = Q(x, h)$;

(ii) $P_x\{\tau_h < \infty\} = L(x, h)$, *and hence* $L(x, h) \ge Q(x, h)$;

(iii) *if for some* $\varepsilon < 1$ *the function* h *satisfies* $h(x) \le \varepsilon$ *for all* $x \in \mathsf{X}$, *then* $L(x, h) = 1$ *if and only if* $Q(x, h) = 1$.

PROOF   **(i)**   We have from the definition of A_h,

$$P_x\{\Psi_k \in A_h \quad \text{i.o.} \mid \mathcal{F}_{\infty}^{\Phi}\} = P_x\{Y_k \le h(\Phi_k) \quad \text{i.o.} \mid \mathcal{F}_{\infty}^{\Phi}\}.$$

Conditioned on $\mathcal{F}_{\infty}^{\Phi}$, the events $\{Y_k \le h(\Phi_k)\}$, $k \ge 1$, are mutually independent. Hence by the Borel-Cantelli Lemma,

$$P_x\{\Psi_k \in A_h \quad \text{i.o.} \mid \mathcal{F}_{\infty}^{\Phi}\} = \mathbb{I}\Big\{\sum_{k=1}^{\infty} P_x\{Y_k \le h(\Phi_k) \mid \mathcal{F}_{\infty}^{\Phi}\} = \infty\Big\}.$$

Since $P_x\{Y_k \le h(\Phi_k) \mid \mathcal{F}_{\infty}^{\Phi}\} = h(\Phi_k)$, taking expectations of each side of this identity completes the proof of (i).

(ii)   This follows directly from the definitions and (12.17).

(iii)   Suppose that $h(x) \le \varepsilon$ for all x, and suppose that $Q(x, h) < 1$ for some x. We will show that $L(x, h) < 1$ also.

If this is the case then by (i), for some $N < \infty$ and $\delta > 0$,

$$P_x\{ \Psi_k \in A_h^c \text{ for all } k > N\} = \delta.$$

But then by the fact that Y is i.i.d. and independent of Φ,

$$\begin{aligned}
1 - L(x, h) &\ge P_x\{ \Psi_k \in A_h^c \text{ for all } k > N, \text{ and } Y_k > \varepsilon \text{ for all } k \le N\} \\
&= P_x\{ \Psi_k \in A_h^c \text{ for all } k > N\}P_x\{ Y_k > \varepsilon \text{ for all } k \le N\} \\
&= \delta(1 - \varepsilon)^N > 0.
\end{aligned}$$

$\square$

We now present an application of Theorem 12.2.2 which gives another representation for an invariant measure, extending the development of Section 10.4.2.

Theorem 12.2.3. *Suppose that $0 \le h \le 1$ with $Q(x, h) = 1$ for all $x \in \mathsf{X}$.*

(i) *If μ is any σ-finite subinvariant measure, then μ is invariant and has the representation*

$$\mu(A) = \int \mu(dx)h(x)U_h(x, A).$$

(ii) *If ν is a finite measure satisfying, for some $A \in \mathcal{B}(X)$,*

$$\nu(B) = \nu U_h I_h(B), \qquad B \subseteq A,$$

then the measure $\mu := \nu U_h$ is invariant for Φ. The sets

$$C_\varepsilon = \{x : K_{a_{\frac{1}{2}}}(x, h) > \varepsilon\}$$

cover X and have finite μ-measure for every $\varepsilon > 0$.

PROOF   We prove (i) by considering the bivariate chain Ψ. The set $A_h \subset Y$ is Harris recurrent and in fact $P_x\{\Psi \in A_h \quad \text{i.o.}\} = 1$ for all $x \in X$ by Theorem 12.2.2. Now define the measure $\overline{\mu}$ on Y by

$$\overline{\mu}(A \times B) = \mu(A)\mu^{\mathrm{uni}}(B), \qquad A \in \mathcal{B}(X), \ B \subset \mathcal{B}([0, 1]). \tag{12.18}$$

Obviously $\overline{\mu}$ is an invariant measure for Ψ and hence by Theorem 10.4.7,

$$\mu(A) = \overline{\mu}(A \times [0, 1]) \; = \; \int_{(x,y) \in A_h} \mu(dx)\mathbf{u}(dy)U_h(x, A)$$

$$= \; \int \mu(dx)h(x)U_h(x, A),$$

which is the first result.

To prove (ii) first extend ν to $\mathcal{B}(Y)$ as μ was extended in (12.18) to obtain a measure $\overline{\nu}$ on $\mathcal{B}(Y)$. Now apply Theorem 10.4.7. The measure $\overline{\mu}'$ defined as

$$\overline{\mu}'(A \times B) = \mathsf{E}_{\overline{\nu}}\Big[\sum_{k=1}^{\tau_h} \mathbb{I}\{\Psi_k \in A \times B\}\Big]$$

is invariant for Ψ, and since the distribution of Φ is the marginal distribution of Ψ, the measure μ defined for $A \in \mathcal{B}(X)$ by $\mu(A) := \overline{\mu}'(A \times [0, 1])$, $A \in \mathcal{B}(X)$, is invariant for Φ.

We now demonstrate that μ is σ-finite. From the assumptions of the theorem and Theorem 12.2.2 (ii) the sets C_ε cover X. We have from the representation of μ,

$$\nu(X) = \mu(h) = \mu K_{a_{\frac{1}{2}}}(h) \geq \varepsilon\mu(C_\varepsilon).$$

Hence for all ε we have the bound $\mu(C_\varepsilon) \leq \mu(h)/\varepsilon < \infty$, which completes the proof of (ii). $\square$

12.3   The existence of a σ-finite invariant measure

12.3.1   The smoothed chain on a compact set

Here we shall give a weak sufficient condition for the existence of a σ-finite invariant measure for a Feller chain. This provides an analogue of the results in Chapter 10 for recurrent chains. The construction we use mimics the construction mentioned in Section 10.4.2: here, though, a function on a compact set plays the part of the petite set A used in the construction of the "process on A", and the fact that there is an invariant measure to play the part of the measure ν in Theorem 10.4.8 is an application of Theorem 12.1.2.

These results will again lead to a test function approach to establishing the existence of an invariant measure for a Feller chain, even without ψ-irreducibility.

We will, however, assume that some one compact set C satisfies a strong form of Harris recurrence: that is, that there exists a compact set $C \subset \mathsf{X}$ with

$$L(x, C) = \mathsf{P}_x\{\Phi \text{ enters } C\} \equiv 1, \qquad x \in \mathsf{X}. \tag{12.19}$$

Observe that by Proposition 9.1.1, (12.19) implies that Φ visits C infinitely often from each initial condition, and hence Φ is at least non-evanescent.

To construct an invariant measure we essentially consider the chain Φ^C obtained by sampling Φ at consecutive visits to the compact set C. Suppose that the resulting sampled chain on C had the Feller property. In this case, since the sampled chain evolves on the compact set C, we could deduce from Theorem 12.1.2 that an invariant probability existed for the sampled chain, and we would then need only a few further steps for an existence proof for the original chain Φ.

However, the transition function P_C for the sampled chain is given by

$$P_C = \sum_{k=0}^{\infty} (PI_{C^c})^k PI_C = U_C I_C,$$

which does not have the Feller property in general. To proceed, we must "smooth around the edges of the compact set C". The kernels P_h introduced in the previous section allow us to do just that.

Let N and O be open subsets of X with compact closure for which $C \subset O \subset \bar{O} \subset N$, where C satisfies (12.19) and let $h \colon \mathsf{X} \to \mathbb{R}$ be a continuous function such as

$$h(x) = \frac{d(x, N^c)}{d(x, N^c) + d(x, \bar{O})}$$

for which

$$\mathbb{I}_O(x) \le h(x) \le \mathbb{I}_N(x). \tag{12.20}$$

The kernel $P_h := U_h I_h$ is a Markov transition function since by (12.19) we have that $Q(x, h) \equiv 1$. Since $P_h(x, \bar{N}) = 1$ for all $x \in \mathsf{X}$, we will immediately have an invariant measure for P_h by Theorem 12.1.2 if P_h has the weak Feller property.

Proposition 12.3.1. *Suppose that the transition function P is weak Feller. If $0 \le h \le 1$ is continuous and if $Q(x, h) \equiv 1$, then P_h is also weak Feller.*

PROOF   By the Feller property, the kernel $(PI_{1-h})^n PI_h$ preserves positive lower semicontinuous functions. Hence if f is positive and lower semicontinuous, then

$$P_h f = \sum_{k=0}^{\infty} (PI_{1-h})^n PI_h f$$

is lower semicontinuous, being the increasing limit of a sequence of lower semicontinuous functions.

Suppose now that f is bounded and continuous, and choose a constant L so large that $L + f$ and $L - f$ are both positive. Then the functions

$$L + f, \qquad L - f, \qquad P_h(L + f), \qquad P_h(L - f),$$

are all positive and lower semicontinuous, from which it follows that $P_h f$ is continuous. Hence P_h is weak Feller as required.   □

We now prove using the generalized resolvent operators

Theorem 12.3.2. *If Φ is Feller and (12.19) is satisfied, then there exists at least one invariant measure which is finite on compact sets.*

PROOF   From Theorem 12.1.2 an invariant probability ν exists which is invariant for $P_h = U_h I_h$. Hence from Theorem 12.2.3, the measure $\mu = \nu U_h$ is invariant for Φ and is finite on the sets $\{x : K_{a_{\frac{1}{2}}}(x, h) > \varepsilon\}$. Since $K_{a_{\frac{1}{2}}}(x, h)$ is a continuous function of x, and is strictly positive everywhere by (12.19), it follows that μ is finite on compact sets.   □

12.3.2   Drift criteria for the existence of invariant measures

We conclude this section by proving that the test function which implies Harris recurrence or regularity for a ψ-irreducible T-chain may also be used to prove the existence of σ-finite invariant measures or invariant probability measures for Feller chains.

Theorem 12.3.3. *Suppose that Φ is Feller and that (V1) is satisfied with a compact set $C \subset X$. Then an invariant measure exists which is finite on compact subsets of X.*

PROOF   If $L(x, C) = 1$ for all $x \in X$, then the proof follows from Theorem 12.3.2.

Consider now the only other possibility, where $L(x, C) \neq 1$ for some x. In this case the adapted process $\{V(\Phi_k) \mathbb{I}\{\tau_C > k\}, \mathcal{F}_k^{\Phi}\}$ is a convergent supermartingale, as in the proof of Theorem 9.4.1, and since by assumption $P_x\{\tau_C = \infty\} > 0$, this shows that

$$P_x\{\limsup_{k \to \infty} V(\Phi_k) < \infty\} \geq 1 - L(x, C) > 0.$$

By Theorem 12.1.2, it follows that an invariant *probability* exists, and this completes the proof.   □

Finally we prove that in the weak Feller case, the drift condition (V2) again provides a criterion for the existence of an invariant probability measure.

Theorem 12.3.4. *Suppose that the chain Φ is weak Feller. If (V2) is satisfied with a compact set C and a positive function V which is finite at one $x_0 \in X$, then an invariant probability measure π exists.*

PROOF   Iterating (V2) n times gives

$$\frac{1}{n}\sum_{k=0}^{n} 1 \le \frac{1}{n}V(x_0) + b\frac{1}{n}\sum_{k=0}^{n}P^k(x_0, C).$$

Letting $n \to \infty$ we see that

$$\liminf_{n\to\infty}\frac{1}{n}\sum_{k=0}^{n}P^k(x_0, C) \ge \frac{1}{b}. \tag{12.21}$$

Theorem 12.3.4 then follows directly from Theorem 12.1.2 (i).   □

12.4   Invariant measures for e-chains

12.4.1   Existence of an invariant measure for e-chains

Up to now we have shown under very mild conditions that an invariant probability measure exists for a Feller chain, based largely on arguments using weak convergence of P^n.

As we have seen, such weak limits will depend in general on the value of x chosen, unless as in Proposition 12.1.4 there is a unique invariant measure. In this section we will explore the properties of the collection of such limiting measures.

Suppose that the chain is weak Feller and we can prove that a Markov transition function Π exists which is itself weak Feller, such that for any $f \in \mathcal{C}(\mathsf{X})$,

$$\lim_{k\to\infty}P^k f(x) = \Pi f(x), \qquad x \in \mathsf{X}. \tag{12.22}$$

In this case, it follows as in Proposition 6.4.2 from Ascoli's Theorem D.4.2 that $\{P^k f : k \in \mathbb{Z}_+\}$ is equicontinuous on compact subsets of X whenever $f \in \mathcal{C}(\mathsf{X})$, and so it is necessary that the chain Φ be an e-chain, in the sense of Section 6.4, whenever we have convergence in the sense of (12.22).

The key to analyzing e-chains lies in the following result:

Theorem 12.4.1. *Suppose that Φ is an e-chain. Then*

(i) *There exists a substochastic kernel Π such that*

$$\overline{P}_k(x, \cdot) \xrightarrow{\text{v}} \Pi(x, \cdot) \qquad as\ k \to \infty \tag{12.23}$$

$$K_{a_\varepsilon}(x, \cdot) \xrightarrow{\text{v}} \Pi(x, \cdot) \qquad as\ \varepsilon \uparrow 1 \tag{12.24}$$

for all $x \in \mathsf{X}$.

(ii) *For each $j,\ k,\ \ell \in \mathbb{Z}_+$ we have*

$$P^j \Pi^k P^\ell = \Pi, \tag{12.25}$$

and hence for all $x \in \mathsf{X}$ the measure $\Pi(x, \cdot)$ is invariant with $\Pi(x, \mathsf{X}) \le 1$.

(iii) *The Markov chain is bounded in probability on average if and only if $\Pi(x, \mathsf{X}) = 1$ for all $x \in \mathsf{X}$.*

PROOF   We prove the result (12.23), the proof of (12.24) being similar. Let $\{f_n\} \subset$ $\mathcal{C}_c(\mathsf{X})$ denote a fixed dense subset. By Ascoli's theorem and a diagonal subsequence argument, there exists a subsequence $\{k_i\}$ of $\mathbb{Z}_+$ and functions $\{g_n\} \subset \mathcal{C}(\mathsf{X})$ such that

$$\lim_{i \to \infty} \overline{P}_{k_i} f_n(x) = g_n(x) \tag{12.26}$$

uniformly for x in compact subsets of X for each $n \in \mathbb{Z}_+$. The set of all subprobabilities on $\mathcal{B}(\mathsf{X})$ is sequentially compact with respect to vague convergence, and any vague limit ν of the probabilities $\overline{P}_{k_i}(x, \cdot)$ must satisfy $\int f_n \, d\nu = g_n(x)$ for all $n \in \mathbb{Z}_+$. Since the functions $\{f_n\}$ are dense in $\mathcal{C}_c(\mathsf{X})$, this shows that for each x there is exactly one vague limit point, and hence a kernel Π exists for which

$$\overline{P}_{k_i}(x, \cdot) \xrightarrow{\text{v}} \Pi(x, \cdot) \qquad \text{as } i \to \infty$$

for each $x \in \mathsf{X}$.

Observe that by equicontinuity, the function Πf is continuous for every function $f \in \mathcal{C}_c(\mathsf{X})$. It follows that Πf is positive and lower semicontinuous whenever f has these properties.

By the Dominated Convergence Theorem we have for all $k, j \in \mathbb{Z}_+$,

$$P^j \Pi^k = \Pi.$$

Next we show that $\Pi P = \Pi$, and hence that

$$\Pi^k P^j = \Pi, \qquad k, j \in \mathbb{Z}_+.$$

Let $f \in \mathcal{C}_c(\mathsf{X})$ be a continuous positive function with compact support. Then, since the function Pf is also positive and continuous, (D.6) implies that

$$\begin{aligned} \Pi(Pf) &\leq \liminf_{i \to \infty} \overline{P}_{k_i}(Pf) \\ &= \Pi f, \end{aligned}$$

which shows that $\Pi P = \Pi$.

We now show that (12.23) holds. Suppose that $\overline{P}_N$ does not converge vaguely to Π. Then there exists a different subsequence $\{m_j\}$ of $\mathbb{Z}_+$, and a distinct kernel Π' such that

$$\overline{P}_{m_j} \xrightarrow{\text{v}} \Pi'(x, \cdot), \qquad j \to \infty.$$

However, for each positive function $f \in \mathcal{C}_c(\mathsf{X})$,

$$\begin{aligned} \Pi f &= \lim_{j \to \infty} \Pi \overline{P}_{m_j} f \\ &= \Pi \Pi' f \qquad \text{by the Dominated Convergence Theorem} \\ &\leq \liminf_{i \to \infty} \overline{P}_{k_i} \Pi' f \qquad \text{since } \Pi' f \text{ is continuous and positive} \\ &= \Pi' f. \end{aligned}$$

Hence by symmetry, $\Pi' = \Pi$, and this completes the proof of (i) and (ii).

The result (iii) follows from (i) and Proposition D.5.6. $\square$

12.4.2   Hitting time and drift criteria for stability of e-chains

We now consider the stability of e-chains. First we show in Theorem 12.4.3 that if the chain hits a fixed compact subset of X with probability one from each initial condition, and if this compact set is positive in a well defined way, then the chain is bounded in probability on average. This is an analogue of the rather more powerful regularity results in Chapter 11.

.This result is then applied to obtain a drift criterion for boundedness in probability using (V2).

To characterize boundedness in probability we use the following weak analogue of Kac's Theorem 10.2.2, connecting positivity of $K_{a_\varepsilon}(x,C)$ with finiteness of the mean return time to C.

Proposition 12.4.2. *For any compact set $C \subset$ X*

$$\liminf_{\varepsilon \uparrow 1} K_{a_\varepsilon}(x,C) \geq \left(\sup_{y \in C} \mathsf{E}_y[\tau_C]\right)^{-1}, \qquad x \in C.$$

PROOF   For the first entrance time τ_C to the compact set C, let θ^{τ_C} denote the τ_C-fold shift on sample space, defined so that $\theta^{\tau_C} f(\Phi_k) = f(\Phi_{k+\tau_C})$ for any function f on X.

Fix $x \in C$, $0 < \varepsilon < 1$, and observe that by conditioning at time τ_C and using the strong Markov property we have for $x \in C$,

$$
\begin{aligned}
K_{a_\varepsilon}(x,C) &= (1-\varepsilon)\mathsf{E}_x\left[\sum_{k=0}^{\infty} \varepsilon^k \mathbb{I}\{\Phi_k \in C\}\right] \\
&= (1-\varepsilon)\mathsf{E}_x\left[1 + \sum_{k=0}^{\infty} \varepsilon^{\tau_C + k}\left(\theta^{\tau_C} \mathbb{I}\{\Phi_k \in C\}\right)\right] \\
&= (1-\varepsilon) + (1-\varepsilon)\mathsf{E}_x\left[\varepsilon^{\tau_C} \mathsf{E}_{\Phi_{\tau_C}}\left[\sum_{k=0}^{\infty} \varepsilon^k \mathbb{I}\{\Phi_k \in C\}\right]\right] \\
&\geq (1-\varepsilon) + \mathsf{E}_x[\varepsilon^{\tau_C}] \inf_{y \in C} K_{a_\varepsilon}(y,C).
\end{aligned}
$$

Taking the infimum over all $x \in C$, we obtain

$$\inf_{y \in C} K_{a_\varepsilon}(y,C) \geq (1-\varepsilon) + \inf_{y \in C} \mathsf{E}_y[\varepsilon^{\tau_C}] \inf_{y \in C} K_{a_\varepsilon}(y,C). \qquad (12.27)$$

By Jensen's inequality we have the bound $\mathsf{E}[\varepsilon^{\tau_C}] \geq \varepsilon^{\mathsf{E}[\tau_C]}$. Hence letting $M_C = \sup_{x \in C} \mathsf{E}_x[\tau_C]$ it follows from (12.27) that for $y \in C$,

$$K_{a_\varepsilon}(y,C) \geq \frac{1-\varepsilon}{1-\varepsilon^{M_C}}.$$

Letting $\varepsilon \uparrow 1$ we have for each $y \in C$,

$$\liminf_{\varepsilon \uparrow 1} K_{a_\varepsilon}(y,C) \geq \lim_{\varepsilon \uparrow 1} \left(\frac{1-\varepsilon}{1-\varepsilon^{M_C}}\right) = \frac{1}{M_C}.$$

$\square$

We saw in Theorem 12.4.1 that Φ is bounded in probability on average if and only if $\Pi(x, \mathsf{X}) = 1$ for all $x \in \mathsf{X}$. Hence the following result shows that compact sets serve as test sets for stability: if a fixed compact set is reachable from all initial conditions, and if Φ is reasonably well behaved from initial conditions on that compact set, then Φ will be bounded in probability on average.

Theorem 12.4.3. *Suppose Φ is an e-chain. Then*

(i) $\max\limits_{x \in \mathsf{X}} \Pi(x, \mathsf{X})$ *exists and is equal to zero or one;*

(ii) *if $\min\limits_{x \in \mathsf{X}} \Pi(x, \mathsf{X})$ exists, then it is equal to zero or one;*

(iii) *if there exists a compact set $C \subset \mathsf{X}$ such that*

$$\mathsf{P}_x\{\tau_C < \infty\} = 1, \qquad x \in \mathsf{X},$$

then $\min\limits_{x \in \mathsf{X}} \Pi(x, \mathsf{X})$ exists and is attained on C, so that

$$\inf_{x \in \mathsf{X}} \Pi(x, \mathsf{X}) = \min_{x \in C} \Pi(x, \mathsf{X});$$

(iv) *if $C \subset \mathsf{X}$ is compact, then*

$$\inf_{x \in C} \Pi(x, \mathsf{X}) \geq \left(\sup_{x \in C} \mathsf{E}_x[\tau_C] \right)^{-1}.$$

PROOF   **(i)**   If $\Pi(x, \mathsf{X}) > 0$ for some $x \in \mathsf{X}$, then an invariant probability π exists. In fact, we may take $\pi = \Pi(x, \cdot)/\Pi(x, \mathsf{X})$.

From the definition of Π and the Dominated Convergence Theorem we have that for any $f \in \mathcal{C}_c(\mathsf{X})$,

$$\pi(f) = \lim_{n \to \infty} [\pi \overline{P}_n(f)] = \pi \Pi(f)$$

which shows that $\pi = \pi \Pi$. Hence $1 = \pi(\mathsf{X}) = \int \pi(dx)\Pi(x, \mathsf{X})$. This shows that $\Pi(y, \mathsf{X}) = 1$ for a.e. $y \in \mathsf{X} [\pi]$, proving (i) of the theorem.

(ii)   Let $\rho = \inf_{x \in \mathsf{X}} \Pi(x, \mathsf{X})$, and let

$$S_\rho = \{x \in \mathsf{X} : \Pi(x, \mathsf{X}) = \rho\}.$$

By the assumptions of (ii), $S_\rho \neq \emptyset$. Letting $u(\cdot) := \Pi(\cdot, \mathsf{X})$, we have $Pu = u$, and this implies that the set S_ρ is absorbing. Since u is lower semicontinuous, the set S_ρ is also a closed subset of X.

Since S_ρ is closed, it follows by vague convergence and (D.6) that for all $x \in \mathsf{X}$,

$$\liminf_{N \to \infty} \overline{P}_N(x, S_\rho^c) \geq \Pi(x, S_\rho^c),$$

and since S_ρ is also absorbing, this shows that for all $x \in S_\rho$

$$\Pi(x, S_\rho^c) = 0. \tag{12.28}$$

Suppose now that $0 \leq \rho < 1$. As in the proof of (i),

$$\pi\{y \in \mathsf{X} : \Pi(y, \mathsf{X}) = 1\} = 1$$

for any invariant probability π, and hence

$$\Pi(x, S_\rho) \leq \Pi(x, \{y \in \mathsf{X} : \Pi(y, \mathsf{X}) < 1\}) = 0. \qquad (12.29)$$

Equations (12.28) and (12.29) show that for any $x \in S_\rho$,

$$\rho = \Pi(x, \mathsf{X}) = \Pi(x, S_\rho) + \Pi(x, S_\rho^c) = 0,$$

and this proves (ii).

(iii)   Since $u(x) := \Pi(x, \mathsf{X})$ is lower semicontinuous we have

$$\inf_{x \in C} u(x) = \min_{x \in C} u(x).$$

That is, the infimum is attained.

Since $Pu = u$, the sequence $\{u(\Phi_k), \mathcal{F}_k^\Phi\}$ is a martingale, which converges to a random variable u_∞ satisfying $\mathsf{E}_x[u_\infty] = u(x)$, $x \in \mathsf{X}$. By Proposition 9.1.1, the assumption that $\mathsf{P}_x\{\tau_C < \infty\} \equiv 1$ implies that

$$\mathsf{P}_x\{\Phi \in C \text{ i.o.}\} = 1, \qquad x \in \mathsf{X}. \qquad (12.30)$$

If $\Phi_k \in C$ for some $k \in \mathbb{Z}_+$, then obviously $u(\Phi_k) \geq \min_{x \in C} u(x)$, which by (12.30) implies that

$$u_\infty = \lim_{k \to \infty} u(\Phi_k) \geq \min_{x \in C} u(x) \qquad \text{a.s.}$$

Taking expectations shows that $u(y) \geq \min_{x \in C} u(x)$ for all $y \in \mathsf{X}$, proving part (iii) of the theorem.

(iv)   Letting $M_C = \sup_{x \in C} \mathsf{E}_x[\tau_C]$ it follows from Proposition 12.4.2 that

$$\inf_{y \in C} \liminf_{\varepsilon \uparrow 1} K_{a_\varepsilon}(y, C) \geq \frac{1}{M_C}.$$

This proves the result since $\limsup_{\varepsilon \uparrow 1} K_{a_\varepsilon}(y, C) \leq \Pi(y, C)$ by Theorem 12.4.1.   □

We have immediately

Proposition 12.4.4. *Let* Φ *be an e-chain, and let* $C \subset \mathsf{X}$ *be compact. If* $\mathsf{P}_x\{\tau_C < \infty\} = 1$, $x \in \mathsf{X}$, *and* $\sup_{x \in C} \mathsf{E}_x[\tau_C] < \infty$, *then* Φ *is bounded in probability on average.*

PROOF   From Theorem 12.4.3 (iii) we see that for all x,

$$\min_{x \in \mathsf{X}} \Pi(x, \mathsf{X}) = \min_{x \in C} \Pi(x, \mathsf{X}) \geq \left(\sup_{x \in C} \mathsf{E}_x[\tau_C]\right)^{-1} > 0.$$

Hence from Theorem 12.4.3 (ii) we have $\Pi(x, \mathsf{X}) = 1$ for all $x \in \mathsf{X}$. Theorem 12.4.1 then implies that the chain is bounded in probability on average.   □

The next result shows that the drift criterion for positive recurrence for ψ-irreducible chains also has an impact on the class of e-chains.

Theorem 12.4.5. *Let* Φ *be an e-chain, and suppose that condition (V2) holds for a compact set* C *and an everywhere finite function* V. *Then the Markov chain* Φ *is bounded in probability on average.*

PROOF   It follows from Theorem 11.3.4 that $\mathsf{E}_x[\tau_C] \leq V(x)$ for $x \in C^c$, so that *a fortiori* we also have $L(x, C) \equiv 1$. As in the proof of Theorem 12.3.4, for any $x \in \mathsf{X}$,

$$\Pi(x, \mathsf{X}) \geq \limsup_{n \to \infty} \frac{1}{n} \sum_{k=0}^{n} P^k(x, C) \geq \frac{1}{b}, \qquad x \in \mathsf{X}.$$

From this it follows from Theorem 12.4.3 (iii) and (ii) that $\Pi(x, \mathsf{X}) \equiv 1$, and hence Φ is bounded in probability on average as claimed. □

12.5   Establishing boundedness in probability

Boundedness in probability is clearly the key condition needed to establish the existence of an invariant measure under a variety of continuity regimes. In this section we illustrate the verification of boundedness in probability for some specific models.

12.5.1   Linear state space models

We show first that the conditions used in Proposition 6.3.5 to obtain irreducibility are in fact sufficient to establish boundedness in probability for the linear state space model. Thus with no extra conditions we are able to show that a stationary version of this model exists.

Recall that we have already seen in Chapter 7 that the linear state space model is an e-chain when (LSS5) holds.

Proposition 12.5.1. *Consider the linear state space model defined by (LSS1) and (LSS2). If the eigenvalue condition (LSS5) is satisfied, then Φ is bounded in probability. Moreover, if the nonsingularity condition (LSS4) and the controllability condition (LCM3) are also satisfied then the model is positive Harris.*

PROOF   Let us take

$$M := I + \sum_{i=1}^{\infty} F^{\top i} F^i,$$

where $F^{\top}$ denotes the transpose of F. If condition (LSS5) holds, then by Lemma 6.3.4 the matrix M is finite and positive definite with $I \leq M$, and for some $\alpha < 1$

$$|Fx|_M^2 \leq \alpha |x|_M^2, \tag{12.31}$$

where $|y|_M^2 := y^{\top} M y$ for $y \in \mathbb{R}^n$.

Let $m = \left(\sum_{i=0}^{\infty} F^i \right) G \, \mathsf{E}[W_1]$, and define

$$V(x) = |x - m|_M^2, \qquad x \in \mathsf{X}. \tag{12.32}$$

Then it follows from (LSS1) that

$$
\begin{aligned}
V(X_{k+1}) &= |F(X_k - m)|_M^2 + |G(W_{k+1} - \mathsf{E}[W_{k+1}])|_M^2 \\
&\quad + (X_k - m)^{\top} F^{\top} M G(W_{k+1} - \mathsf{E}[W_{k+1}]) \\
&\quad + (W_{k+1} - \mathsf{E}[W_{k+1}])^{\top} G^{\top} M F(X_k - m).
\end{aligned}
\tag{12.33}
$$

Since W_{k+1} and X_k are independent, this together with (12.31) implies that

$$\mathsf{E}[V(X_{k+1}) \mid X_0, \ldots, X_k] \leq \alpha V(X_k) + \mathsf{E}[|G(W_{k+1} - \mathsf{E}[W_{k+1}])|_M^2], \qquad (12.34)$$

and taking expectations of both sides gives

$$\limsup_{k \to \infty} \mathsf{E}[V(X_k)] \leq \frac{\mathsf{E}[|G(W_{k+1} - \mathsf{E}[W_{k+1}])|_M^2]}{1 - \alpha} < \infty.$$

Since V is a coercive function on X, Lemma D.5.3 gives a direct proof that the chain is bounded in probability.

We note that (12.34) also ensures immediately that (V2) is satisfied. Under the extra conditions (LSS4) and (LCM3) we have from Proposition 6.3.5 that all compact sets are petite, and it immediately follows from Theorem 11.3.11 that the chain is regular and hence positive Harris.   □

It may be seen that stability of the linear state space model is closely tied to the stability of the deterministic system $x_{k+1} = Fx_k$. For each initial condition $x_0 \in \mathbb{R}^n$ of this deterministic system, the resulting trajectory $\{x_k\}$ satisfies the bound

$$|x_k|_M \leq \alpha^k |x_0|_M$$

and hence is ultimately bounded in the sense of Section 11.2: in fact, in the dynamical systems literature such a system is called *globally exponentially stable*. It is precisely this stability for the deterministic "core" of the linear state space model which allows us to obtain boundedness in probability for the stochastic process Φ.

We now generalize the model (LSS1) to include random variation in the coefficients F and G.

12.5.2   Bilinear models

Let us next consider the scalar example where Φ is the bilinear state space model on $\mathsf{X} = \mathbb{R}$ defined in (SBL1)–(SBL2)

$$X_{k+1} = \theta X_k + bW_{k+1}X_k + W_{k+1}, \qquad (12.35)$$

where W is a zero-mean disturbance process. This is related closely to the linear model above, and the analysis is almost identical.

To obtain boundedness in probability by direct calculation, observe that

$$\mathsf{E}[|X_{k+1}| \mid X_k = x] \leq \mathsf{E}[|\theta + bW_{k+1}|]|x| + \mathsf{E}[|W_{k+1}|]. \qquad (12.36)$$

Hence for every initial condition of the process,

$$\limsup_{k \to \infty} \mathsf{E}[|X_k|] \leq \frac{\mathsf{E}[|W_{k+1}|]}{1 - \mathsf{E}[|\theta + bW_{k+1}|]}$$

provided that

$$\mathsf{E}[|\theta + bW_{k+1}|] < 1. \qquad (12.37)$$

Since $|\cdot|$ is a coercive function on X, this shows that Φ is bounded in probability provided that (12.37) is satisfied.

Again observe that in fact the bound (12.36) implies that the mean drift criterion (V2) holds.

12.5.3   Adaptive control models

Finally we consider the adaptive control model (2.22)–(2.24).

The closed loop system described by (2.25) is a Feller Markov chain, and thus an invariant probability exists if the distributions of the process are tight for some initial condition. We show here that the distributions of Φ are tight when the initial conditions are chosen so that

$$\tilde{\theta}_k = \theta_k - \mathsf{E}[\theta_k \mid \mathcal{Y}_k], \quad \text{and} \quad \Sigma_k = \mathsf{E}[\tilde{\theta}_k^2 \mid \mathcal{Y}_k]. \tag{12.38}$$

For example, this is the case when $y_0 = \tilde{\theta}_0 = \Sigma_0 = 0$. If (12.38) holds then it follows from (2.23) that

$$\mathsf{E}[Y_{k+1}^2 \mid \mathcal{Y}_k] = \Sigma_k Y_k^2 + \sigma_w^2. \tag{12.39}$$

This identity will be used to prove the following result:

Proposition 12.5.2. *For the adaptive control model satisfying (SAC1) and (SAC2), suppose that the process Φ defined in (2.25) satisfies (12.38) and that $\sigma_z^2 < 1$. Then we have*

$$\limsup_{k \to \infty} \mathsf{E}[|\Phi_k|^2] < \infty$$

so that distributions of the chain are tight, and hence Φ is positive recurrent.

PROOF   We note first that since the sequence $\{\Sigma_k\}$ is bounded below and above by $\underline{\Sigma} = \upsilon_z > 0$ and $\overline{\Sigma} = \sigma_z/(1 - \alpha^2) < \infty$, and the process θ clearly satisfies

$$\limsup_{k \to \infty} \mathsf{E}[\theta_k^2] = \frac{\sigma_z^2}{1 - \alpha^2},$$

to prove the proposition it is enough to bound $\mathsf{E}[Y_k^2]$.

From (12.39) and (2.24) we have

$$
\begin{aligned}
\mathsf{E}[Y_{k+1}^2 \Sigma_{k+1} \mid \mathcal{Y}_k] &= \Sigma_{k+1} \mathsf{E}[Y_{k+1}^2 \mid \mathcal{Y}_k] \\
&= \Sigma_{k+1}(\Sigma_k Y_k^2 + \sigma_w^2) \\
&= (\sigma_z^2 + \alpha^2 \sigma_w^2 \Sigma_k (\Sigma_k Y_k^2 + \sigma_w^2)^{-1})(\Sigma_k Y_k^2 + \sigma_w^2) \\
&= \sigma_z^2\Big(Y_k^2 \Sigma_k\Big) + \Big(\sigma_w^2 \sigma_z^2 + \alpha^2 \sigma_w^2 \Sigma_k\Big).
\end{aligned}
\tag{12.40}
$$

Taking total expectations of each side of (12.40), we use the condition $\sigma_z^2 < 1$ to obtain by induction, for all $k \in \mathbb{Z}_+$,

$$\underline{\Sigma}\mathsf{E}[Y_{k+1}^2] \leq \mathsf{E}[Y_{k+1}^2 \Sigma_{k+1}] \leq \frac{\sigma_w^2 \sigma_z^2 + \alpha^2 \sigma_w^2 \overline{\Sigma}}{1 - \sigma_z^2} + \sigma_z^{2k}\mathsf{E}[Y_0^2 \Sigma_0]. \tag{12.41}$$

This shows that the mean of Y_k^2 is uniformly bounded.

Since Φ has the Feller property it follows from Proposition 12.1.3 that an invariant probability exists. Hence from Theorem 7.4.3 the chain is positive recurrent.   □

In fact, we will see in Chapter 16 that not only is the process bounded in probability, but the conditional mean of Y_k^2 converges to the steady state value $\mathsf{E}_\pi[Y_0^2]$ at a geometric rate from every initial condition. These results require a more elaborate stability proof.

Note that equation (12.40) does not obviously imply that there is a solution to a drift inequality such as (V2): the conditional expectation is taken with respect to $\mathcal{Y}_k$, which is strictly smaller than $\mathcal{F}_k^\Phi$.

The condition that $\sigma_z^2 < 1$ cannot be omitted in this analysis: indeed, we have that if $\sigma_z^2 \geq 1$, then

$$\mathsf{E}[Y_k^2] \geq [\sigma_z^2]^k Y_0 + k\sigma_w^2 \to \infty$$

as k increases, so that the chain is unstable in a mean square sense, although it may still be bounded in probability.

It is well worth observing that this is one of the few models which we have encountered where obtaining a drift inequality of the form (V2) is much more difficult than merely proving boundedness in probability. This is due to the fact that the dynamics of this model are extremely nonlinear, and so a direct stability proof is difficult. By exploiting equation (12.39) we essentially linearize a portion of the dynamics, which makes the stability proof rather straightforward. However the identity (12.39) only holds for a restricted class of initial conditions, so in general we are forced to tackle the nonlinear equations directly.

12.6   Commentary

The key result Theorem 12.1.2 is taken from Foguel [121]. Versions of this result have also appeared in papers by Beneš [23, 24] and Stettner [372] which consider processes in continuous time. For more results on Feller chains the reader is referred to Krengel [221], and the references cited therein.

For an elegant operator-theoretic proof of results related to Theorem 12.3.2, see Lin [238] and Foguel [123]. The method of proof based upon the use of the operator $P_h = U_h I_h$ to obtain a σ-finite invariant measure is taken from Rosenblatt [338]. Neveu in [295] promoted the use of the operators U_h, and proved the resolvent equation Theorem 12.2.1 using direct manipulations of the operators. The kernel P_h is often called the *balayage operator* associated with the function h (see Krengel [221] or Revuz [326]). In the Supplement to Krengel's text by Brunel ([221] pp. 301–309) a development of the recurrence structure of irreducible Markov chains is developed based upon these operators. This analysis and much of [326] exploits fully the resolvent equation, illustrating the power of this simple formula although because of our emphasis on ψ-irreducible chains and probabilistic methods, we do not address the resolvent equation further in this book.

Obviously, as with Theorem 12.1.2, Theorem 12.3.4 can be applied to an irreducible Markov chain on countable space to prove positive recurrence. It is of some historical interest to note that Foster's original proof of the sufficiency of (V2) for positivity of such chains is essentially that in Theorem 12.3.4. Rather than showing in any direct way that (V2) gives an invariant measure, Foster was able to use the countable space analogue of Theorem 12.1.2 (i) to deduce positivity from the "non-nullity" of a "compact" finite set of states as in (12.21). We will discuss more general versions of this classification of sets as positive or null further, but not until Chapter 18.

Observe that Theorem 12.3.4 only states that an invariant probability exists. Perhaps surprisingly, it is not known whether the hypotheses of Theorem 12.3.4 imply that the chain is bounded in probability when V is finite valued except for e-chains as in Theorem 12.4.5.

The theory of e-chains is still being developed, although these processes have been the subject of several papers over the past thirty years, most notably by Jamison and Sine [175, 178, 358, 357, 356], Rosenblatt [337], Foguel [121] and the text by Krengel [221]. In most of the e-chain literature, however, the state space is assumed compact so that stability is immediate. The drift criterion for boundedness in probability on average in Theorem 12.4.5 is new. The criterion Theorem 12.3.4 for the existence of an invariant probability for a Feller chain was first shown in Tweedie [402].

The stability analysis of the linear state space model presented here is standard. For an early treatment see Kalman and Bertram [192], while Caines [57] contains a modern and complete development of discrete time linear systems. Snyders [364] treats linear models with a continuous time parameter in a manner similar to the presentation in this book. The bilinear model has been the subject of several papers: see for example Feigin and Tweedie [111], or the discussion in Tong [388]. The stability of the adaptive control model was first resolved in Meyn and Caines [270], and related stability results were described in Solo [365]. The stability proof given here is new, and is far simpler than any previous results.

Part III

CONVERGENCE

Chapter 13

Ergodicity

In Part II we developed the ideas of stability largely in terms of recurrence structures. Our concern was with the way in which the chain returned to the "center" of the space, how sure we could be that this would happen, and whether it might happen in a finite mean time.

Part III is devoted to the perhaps even more important, and certainly deeper, concepts of the chain "settling down", or converging, to a stable or stationary regime.

In our heuristic introduction to the various possible ideas of stability in Section 1.3, such convergence was presented as a fundamental idea, related in the dynamical systems and deterministic contexts to asymptotic stability. We noted briefly, in (10.4) in Chapter 10, that the existence of a finite invariant measure was a necessary condition for such a stationary regime to exist as a limit. In Chapter 12 we explored in much greater detail the way in which convergence of P^n to a limit, on topological spaces, leads to the existence of invariant measures.

In this chapter we begin a systematic approach to this question from the other side. Given the existence of π, when do the n-step transition probabilities converge in a suitable way to π?

We will prove that for positive recurrent ψ-irreducible chains, such limiting behavior takes place with no topological assumptions, and moreover the limits are achieved in a much stronger way than under the tightness assumptions in the topological context. The Aperiodic Ergodic Theorem, which unifies the various definitions of positivity, summarizes this asymptotic theory. It is undoubtedly the outstanding achievement in the general theory of ψ-irreducible Markov chains, even though we shall prove some considerably stronger variations in the next two chapters.

Theorem 13.0.1 (Aperiodic Ergodic Theorem). *Suppose that Φ is an aperiodic Harris recurrent chain, with invariant measure π. The following are equivalent:*

(i) *The chain is positive Harris: that is, the unique invariant measure π is finite.*

(ii) *There exists some ν-small set $C \in \mathcal{B}^+(\mathsf{X})$ and some $P^\infty(C) > 0$ such that as $n \to \infty$, for all $x \in C$*

$$P^n(x, C) \to P^\infty(C). \tag{13.1}$$

313

(iii) *There exists some regular set in $\mathcal{B}^+(\mathsf{X})$: equivalently, there is a petite set $C \in \mathcal{B}(\mathsf{X})$ such that*

$$\sup_{x \in C} \mathsf{E}_x[\tau_C] < \infty. \tag{13.2}$$

(iv) *There exists some petite set C, some $b < \infty$ and a non-negative function V finite at some one $x_0 \in \mathsf{X}$, satisfying*

$$\Delta V(x) := PV(x) - V(x) \leq -1 + b\mathbb{I}_C(x), \qquad x \in \mathsf{X}. \tag{13.3}$$

Any of these conditions is equivalent to the existence of a unique invariant probability measure π such that for every initial condition $x \in \mathsf{X}$,

$$\sup_{A \in \mathcal{B}(\mathsf{X})} |P^n(x, A) - \pi(A)| \to 0 \tag{13.4}$$

as $n \to \infty$, and moreover for any regular initial distributions λ, μ,

$$\sum_{n=1}^{\infty} \int \int \lambda(dx)\mu(dy) \sup_{A \in \mathcal{B}(\mathsf{X})} |P^n(x, A) - P^n(y, A)| < \infty. \tag{13.5}$$

PROOF   That $\pi(\mathsf{X}) < \infty$ in (i) is equivalent to the finiteness of hitting times as in (iii) and the existence of a mean drift test function in (iv) is merely a restatement of the overview Theorem 11.0.1 in Chapter 11.

The fact that any of these positive recurrence conditions imply the uniform convergence over all sets A from all starting points x as in (13.4) is of course the main conclusion of this theorem, and is finally shown in Theorem 13.3.3.

That (ii) holds from (13.4) is obviously trivial by dominated convergence. The cycle is completed by the implication that (ii) implies (13.4), which is in Theorem 13.3.5.

The extension from convergence to summability provided the initial measures are regular is given in Theorem 13.4.4. Conditions under which π itself is regular are also in Section 13.4.2.   □

There are four ideas which should be born in mind as we embark on this third part of the book, especially when coming from a countable space background. The first two involve the types of limit theorems we shall address; the third involves the method of proof of these theorems; and the fourth involves the nomenclature we shall use.

Modes of convergence

The first is that we will be considering, in this and the next three chapters, convergence of a chain in terms of its transition probabilities. Although it is important also to consider convergence of a chain along its sample paths, leading to strong laws, or of normalized variables leading to central limit theorems and associated results, we do not turn to this until Chapter 17.

This is in contrast to the traditional approach in the countable state space case. Typically, there, the search is for conditions under which there exist pointwise limits of the form

$$\lim_{n \to \infty} |P^n(x, y) - \pi(y)| = 0; \tag{13.6}$$

but the results we derive are related to the signed measure $(P^n - \pi)$, and so concern not merely such pointwise or even setwise convergence, but a more global convergence in terms of the total variation norm.

Total variation norm

If μ is a signed measure on $\mathcal{B}(\mathsf{X})$, then the *total variation norm* $\|\mu\|$ is defined as

$$\|\mu\| := \sup_{f:|f|\leq 1} |\mu(f)| = \sup_{A\in\mathcal{B}(\mathsf{X})} \mu(A) - \inf_{A\in\mathcal{B}(\mathsf{X})} \mu(A). \qquad (13.7)$$

The key limit of interest to us in this chapter will be of the form

$$\lim_{n\to\infty} \|P^n(x, \cdot) - \pi\| = 2 \lim_{n\to\infty} \sup_A |P^n(x, A) - \pi(A)| = 0. \qquad (13.8)$$

Obviously when (13.8) holds on a countable space, then (13.6) also holds and indeed holds uniformly in the end point y. This move to the total variation norm, necessitated by the typical lack of structure of pointwise transitions in the general state space, will actually prove exceedingly fruitful rather than restrictive.

When the space is topological, it is also the case that total variation convergence implies weak convergence of the measures in question.

This is clear since (see Chapter 12) the latter is defined as convergence of expectations of functions which are not only bounded but also continuous. Hence the weak convergence of P^n to π as in Proposition 12.1.4 will be subsumed in results such as (13.4) provided the chain is suitably irreducible and positive.

Thus, for example, asymptotic properties of T-chains will be much stronger than those for arbitrary weak Feller chains even when a unique invariant measure exists for the latter.

Independence of initial and limiting distributions

The second point to be made explicitly is that the limits in (13.8), and their refinements and extensions in Chapters 14–16, will typically be found to hold independently of the particular starting point x, and indeed we will be seeking conditions under which this is the case.

Having established this, however, the identification of the class of starting distributions for which particular asymptotic limits hold becomes a question of some importance, and the answer is not always obvious: in essence, if the chain starts with a distribution "too near infinity" then it may never reach the expected stationary distribution.

This is typified in (13.5), where the summability holds only for regular initial measures.

The same type of behavior, and the need to ensure that initial distributions are appropriately "regular" in extended ways, will be a highly visible part of the work in Chapters 14 and 15.

The role of renewal theory and splitting

Thirdly, in developing the ergodic properties of ψ-irreducible chains we will use the splitting techniques of Chapter 5 in a systematic and fruitful way, and we will also need the properties of renewal sequences associated with visits to the atom in the split chain.

Up to now the existence of a "pseudo-atom" has not generated many results that could not have been derived (sometimes with considerable but nevertheless relatively elementary work) from the existence of petite sets: the only real "atom-based" result has been the existence of regular sets in Chapter 11. We have not given much reason for the reader to believe that the atom-based constructions are other than a gloss on the results obtainable through petite sets.

In Part III, however, we will find that the existence of atoms provides a critical step in the development of asymptotic results. This is due to the many limit theorems available for renewal processes, and we will prove such theorems as they fit into the Markov chain development.

We will also see that several generalizations of regular sets also play a key role in such results: the essential equivalence of regularity and positivity, developed in Chapter 11, becomes of far more than academic value in developing ergodic structures.

Ergodic chains

Finally, a word on the term *ergodic*. We will adopt this term for chains where the limit in (13.6) or (13.8) holds as the time sequence $n \to \infty$, rather than as $n \to \infty$ through some subsequence.

Unfortunately, we know that in complete generality Markov chains may be periodic, in which case the limits in (13.6) or (13.8) can hold at best as we go through a periodic sequence nd as $n \to \infty$. Thus by definition, ergodic chains will be aperiodic, and a minor, sometimes annoying but always vital change to the structure of the results is needed in the periodic case.

We will therefore give results, typically, for the aperiodic context and give the required modification for the periodic case following the main statement when this seems worthwhile.

13.1   Ergodic chains on countable spaces

13.1.1   First-entrance last-exit decompositions

In this section we will approach the ergodic question for Markov chains in the countable state space case, before moving on to the general case in later sections. The methods are rather similar: indeed, given the splitting technique there will be a relatively small amount of extra work needed to move to the more general context.

Even in the countable case, the technique of proof we give is simpler and more powerful than that usually presented. One real simplification of the analysis through the use of total variation norm convergence results comes from an extension of the first-entrance and last-exit decompositions of Section 8.2, together with the representation of the invariant probability given in Theorem 10.2.1.

The *first-entrance last-exit decomposition*, for any states $x, y, \alpha \in \mathsf{X}$ is given by

$$P^n(x,y) = {}_\alpha P^n(x,y) + \sum_{j=1}^{n-1} \left[\sum_{k=1}^{j} {}_\alpha P^k(x,\alpha) P^{j-k}(\alpha,\alpha) \right]_\alpha P^{n-j}(\alpha,y), \qquad (13.9)$$

where we have used the notation α to indicate that the specific state being used for the decomposition is distinguished from the more generic states x, y which are the starting and end points of the decomposition.

We will wish in what follows to concentrate on the time variable rather than a particular starting point or end point, and it will prove particularly useful to have notation that reflects this. Let us hold the reference state α fixed and introduce the three forms

$$a_x(n) := \mathsf{P}_x(\tau_\alpha = n), \qquad (13.10)$$

$$u(n) := \mathsf{P}_\alpha(\Phi_n = \alpha), \qquad (13.11)$$

$$t_y(n) := {}_\alpha P^n(\alpha,y). \qquad (13.12)$$

This notation is designed to stress the role of $a_x(n)$ as a delay distribution in the renewal sequence of visits to α, and the "tail properties" of $t_y(n)$ in the representation of π: recall from (10.10) that

$$\begin{aligned}
\pi(y) &= (\mathsf{E}_\alpha[\tau_\alpha])^{-1} \sum_{j=1}^{\infty} {}_\alpha P^j(\alpha,y) \\
&= \pi(\alpha) \sum_{j=1}^{\infty} t_y(j).
\end{aligned} \qquad (13.13)$$

Using this notation the first-entrance and last-exit decompositions become

$$\begin{aligned}
P^n(x,\alpha) &= \sum_{j=0}^{n} \mathsf{P}_x(\tau_\alpha = j) P^{n-j}(\alpha,\alpha) \\
&= \sum_{j=0}^{n} a_x(j) u(n-j), \\
P^n(\alpha,y) &= \sum_{j=0}^{n} P^j(\alpha,\alpha) {}_\alpha P^{n-j}(\alpha,y) \\
&= \sum_{j=0}^{n} u(j) t_y(n-j)
\end{aligned}$$

or, using the convolution notation $a * b(n) = \sum_0^n a(j) b(n-j)$ introduced in Section 2.4.1,

$$P^n(x,\alpha) = a_x * u(n), \qquad (13.14)$$

$$P^n(\alpha,y) = u * t_y(n). \qquad (13.15)$$

The first-exit last-entrance decomposition (13.9) can be written similarly as

$$P^n(x,y) = {}_\alpha P^n(x,y) + a_x * u * t_y(n). \qquad (13.16)$$

The power of these forms becomes apparent when we link them to the representation of the invariant measure given in (13.13). The next decomposition underlies all ergodic theorems for countable space chains.

Proposition 13.1.1. *Suppose that Φ is a positive Harris recurrent chain on a countable space, with invariant probability π. Then for any $x, y, \alpha \in \mathsf{X}$*

$$|P^n(x, y) - \pi(y)| \leq {}_\alpha P^n(x, y) + |a_x * u - \pi(\alpha)| * t_y(n) + \pi(\alpha) \sum_{j=n+1}^{\infty} t_y(j). \quad (13.17)$$

PROOF   From the decomposition (13.16) we have

$$|P^n(x, y) - \pi(y)| \leq {}_\alpha P^n(x, y)$$

$$+ |a_x * u * t_y(n) - \pi(\alpha) \sum_{j=1}^{n} t_y(j)| \quad (13.18)$$

$$+ |\pi(\alpha) \sum_{j=1}^{n} t_y(j) - \pi(y)|.$$

Now we use the representation (13.13) for π and (13.17) is immediate.   □

13.1.2   Solidarity from one ergodic state

If the three terms in (13.17) can all be made to converge to zero, we will have shown that $P^n(x, y) \to \pi(y)$ as $n \to \infty$. The two extreme terms involve the convergence of simple positive expressions, and finding bounds for both of these is at the level of calculation we have already used, especially in Chapters 10 and 11. The middle term involves a deeper limiting operation, and showing that this term does indeed converge is at the heart of proving ergodic theorems.

We can reduce the problem of this middle term entirely to one independent of the initial state x and involving only the reference state α. Suppose we have

$$|u(n) - \pi(\alpha)| \to 0, \qquad n \to \infty. \quad (13.19)$$

Then using Lemma D.7.1 we find

$$\lim_{n \to \infty} a_x * u(n) = \pi(\alpha) \quad (13.20)$$

provided we have (as we do for a Harris recurrent chain) that for all x

$$\sum_j a_x(j) = \mathsf{P}_x(\tau_\alpha < \infty) = 1. \quad (13.21)$$

The convergence in (13.19) will be shown to hold for all states of an aperiodic positive chain in the next section: we first motivate our need for it, and for related results in renewal theory, by developing the ergodic structure of chains with "ergodic atoms".

Ergodic atoms

If $\mathbf{\Phi}$ is positive Harris, an atom $\alpha \in \mathcal{B}^+(\mathsf{X})$ is called *ergodic* if it satisfies

$$\lim_{n \to \infty} |P^n(\alpha, \alpha) - \pi(\alpha)| = 0. \qquad (13.22)$$

In the positive Harris case note that an atom can be ergodic only if the chain is aperiodic.

With this notation, and the prescription for analyzing ergodic behavior inherent in Proposition 13.1.1, we can prove surprisingly quickly the following solidarity result.

Theorem 13.1.2. *If $\mathbf{\Phi}$ is a positive Harris chain on a countable space, and if there exists an ergodic atom α, then for every initial state x*

$$\|P^n(x, \cdot) - \pi\| \to 0, \qquad n \to \infty. \qquad (13.23)$$

PROOF   On a countable space the total variation norm is given simply by

$$\|P^n(x, \cdot) - \pi\| = \sum_y |P^n(x, y) - \pi(y)|$$

and so by (13.17) we have the total variation norm bounded by three terms:

$$\|P^n(x, \cdot) - \pi\| \leq \sum_y {_\alpha}P^n(x, y) + \sum_y |a_x * u - \pi(\alpha)| * t_y(n) + \sum_y \pi(\alpha) \sum_{j=n+1}^{\infty} t_y(j).$$
$$(13.24)$$

We need to show each of these goes to zero. From the representation (13.13) of π and Harris positivity,

$$\infty > \sum_y \pi(y) = \pi(\alpha) \sum_{j=1}^{\infty} \sum_y t_y(j). \qquad (13.25)$$

The third term in (13.24) is the tail sum in this representation and so we must have

$$\pi(\alpha) \sum_{j=n+1}^{\infty} \sum_y t_y(j) \to 0, \qquad n \to \infty. \qquad (13.26)$$

The first term in (13.24) also tends to zero, for we have the interpretation

$$\sum_y {_\alpha}P^n(x, y) = \mathsf{P}_x(\tau_\alpha \geq n) \qquad (13.27)$$

and since $\mathbf{\Phi}$ is Harris recurrent $\mathsf{P}_x(\tau_\alpha \geq n) \to 0$ for every x.

Finally, the middle term in (13.24) tends to zero by a double application of Lemma D.7.1, first using the assumption that α is ergodic so that (13.20) holds and, once we have this, using the finiteness of $\sum_{j=1}^{\infty} \sum_y t_y(j)$ given by (13.25). $\square$

This approach may be extended to give the Ergodic Theorem for a general space chain when there is an ergodic atom in the state space. A first-entrance last-exit decomposition will again give us an elegant proof in this case, and we prove such a result in Section 13.2.3, from which basis we wish to prove the same type of ergodic result for any positive Harris chain. To do this, we must of course prove that the atom $\check{\alpha}$ for the split skeleton chain $\check{\Phi}^m$, which we always have available, is an ergodic atom.

To show that atoms for aperiodic positive chains are indeed ergodic, which is crucial to completing this argument, we need results from renewal theory. This is therefore necessarily the subject of the next section.

13.2   Renewal and regeneration

13.2.1   Coupling renewal processes

When α is a recurrent atom in X, the sequence of return times given by $\tau_\alpha(1) = \tau_\alpha$ and for $n > 1$

$$\tau_\alpha(n) = \min(j > \tau_\alpha(n-1) : \Phi_j = \alpha)$$

is a specific example of a *renewal process*, as defined in Section 2.4.1.

The asymptotic structure of renewal processes has, deservedly, been the subject of a great deal of analysis: such processes have a central place in the asymptotic theory of many kinds of stochastic processes, but nowhere more than in the development of asymptotic properties of general ψ-irreducible Markov chains.

Our goal in this section is to provide essentially those results needed for proving the ergodic properties of Markov chains, and we shall do this through the use of the so-called "coupling approach". We will regrettably do far less than justice to the full power of renewal and regenerative processes, or to the coupling method itself: for more details on renewal and regeneration, the reader should consult Feller [114] or Kingman [208], whilst the more recent flowering of the coupling technique is well covered by the recent book by Lindvall [239].

As in Section 2.4.1 we let $p = \{p(j)\}$ denote the distribution of the increments in a renewal process, whilst $a = \{a(j)\}$ and $b = \{b(j)\}$ will denote possible delays in the first increment variable S_0. For $n = 1, 2, \ldots$ let S_n denote the time of the $(n+1)$st renewal, so that the distribution of S_n is given by $a * p^{n*}$ if S_0 has the delay distribution a.

Recall the standard notation

$$u(n) = \sum_{j=0}^{\infty} p^{j*}(n)$$

for the renewal function for $n \geq 0$. Since $p^{0*} = \delta_0$ we have $u(0) = 1$; by convention we will set $u(-1) = 0$.

If we let $Z(n)$ denote the indicator variables

$$Z(n) = \begin{cases} 1 & S_j = n, \text{ some } j \geq 0 \\ 0 & \text{otherwise,} \end{cases}$$

then we have

$$P_a(Z(n) = 1) = a * u(n),$$

and thus the renewal function represents the probabilities of $\{Z(n) = 1\}$ when there is no delay, or equivalently when $a = \delta_0$.

The coupling approach involves the study of two linked renewal processes with the same increment distribution but different initial distributions, and, most critically, defined on the same probability space.

To describe this concept we define two sets of mutually independent random variables

$$\{S_0, S_1, S_2, \ldots\}, \qquad \{S_0', S_1', S_2', \ldots\}$$

where each of the variables $\{S_1, S_2, \ldots\}$ and $\{S_1', S_2', \ldots\}$ are independent and identically distributed with distribution $\{p(j)\}$; but where the distributions of the independent variables S_0, S_0' are a, b.

The *coupling time* of the two renewal processes is defined as

$$T_{ab} = \min\{j : Z_a(j) = Z_b(j) = 1\}$$

where Z_a, Z_b are the indicator sequences of each renewal process. The random time T_{ab} is the first time that a renewal takes place simultaneously in both sequences, and from that point onwards, because of the loss of memory at the renewal epoch, the renewal processes are identical in distribution.

The key requirement to use this method is that this coupling time be almost surely finite. In this section we will show that if we have an aperiodic *positive recurrent* renewal process with finite mean

$$m_p := \sum_{j=0}^{\infty} jp(j) < \infty, \tag{13.28}$$

then such coupling times are always almost surely finite.

Proposition 13.2.1. *If the increment distribution has an aperiodic distribution p with $m_p < \infty$, then for any initial proper distributions a, b*

$$\mathsf{P}(T_{ab} < \infty) = 1. \tag{13.29}$$

PROOF   Consider the linked forward recurrence time chain $\boldsymbol{V}^*$ defined by (10.19), corresponding to the two independent renewal sequences $\{S_n, S_n'\}$.

Let $\tau_{1,1} = \min(n : V_n^* = (1,1))$. Since the first coupling takes place at $\tau_{1,1} + 1$,

$$T_{ab} = \tau_{1,1} + 1$$

and thus we have that

$$\mathsf{P}(T_{ab} > n) = \mathsf{P}_{a \times b}(\tau_{1,1} \geq n). \tag{13.30}$$

But we know from Section 10.3.1 that, under our assumptions of aperiodicity of p and finiteness of m_p, the chain $\boldsymbol{V}^*$ is $\delta_{1,1}$-irreducible and positive Harris recurrent. Thus for any initial measure μ we have *a fortiori*

$$\mathsf{P}_{\mu}(\tau_{1,1} < \infty) = 1;$$

and hence in particular for the initial measure $a \times b$, it follows that

$$\mathsf{P}_{a \times b}(\tau_{1,1} \geq n) \to 0, \qquad n \to \infty$$

as required.   □

This gives a structure sufficient to prove

Theorem 13.2.2. *Suppose that a, b, p are proper distributions on $\mathbb{Z}_+$, and that u is the renewal function corresponding to p. Then provided p is aperiodic with mean $m_p < \infty$*

$$|a * u(n) - b * u(n)| \to 0, \quad n \to \infty. \tag{13.31}$$

PROOF   Let us define the random variables

$$Z_{ab}(n) = \begin{cases} Z_a(n) & n < T_{ab} \\ Z_b(n) & n \geq T_{ab} \end{cases}$$

so that for any n

$$\mathsf{P}(Z_{ab}(n) = 1) = \mathsf{P}(Z_a(n) = 1). \tag{13.32}$$

We have that

$$
\begin{aligned}
|a * u(n) - b * u(n)| &= |\mathsf{P}(Z_a(n) = 1) - \mathsf{P}(Z_b(n) = 1)| \\
&= |\mathsf{P}(Z_{ab}(n) = 1) - \mathsf{P}(Z_b(n) = 1)| \\
&= |\mathsf{P}(Z_a(n) = 1, T_{ab} > n) + \mathsf{P}(Z_b(n) = 1, T_{ab} \leq n) \\
&\quad - \mathsf{P}(Z_b(n) = 1, T_{ab} > n) - \mathsf{P}(Z_b(n) = 1, T_{ab} \leq n)| \\
&= |\mathsf{P}(Z_a(n) = 1, T_{ab} > n) - \mathsf{P}(Z_b(n) = 1, T_{ab} > n)| \\
&\leq \max\{\mathsf{P}(Z_a(n) = 1, T_{ab} > n), \mathsf{P}(Z_b(n) = 1, T_{ab} > n)\} \\
&\leq \mathsf{P}(T_{ab} > n). \tag{13.33}
\end{aligned}
$$

But from Proposition 13.2.1 we have that $\mathsf{P}(T_{ab} > n) \to 0$ as $n \to \infty$, and (13.31) follows.   □

We will see in Section 18.1.1 that Theorem 13.2.2 holds even without the assumption that $m_p < \infty$. For the moment, however, we will concentrate on further aspects of coupling when we are in the positive recurrent case.

13.2.2   Convergence of the renewal function

Suppose that we have a positive recurrent renewal sequence with finite mean $m_p < \infty$. Then the proper probability distribution $e = e(n)$ defined by

$$e(n) := m_p^{-1} \sum_{j=n+1}^{\infty} p(j) = m_p^{-1}(1 - \sum_{j=0}^{n} p(j)) \tag{13.34}$$

has been shown in (10.16) to be the invariant probability measure for the forward recurrence time chain V^+ associated with the renewal sequence $\{S_n\}$. It also follows that the delayed renewal distribution corresponding to the initial distribution e is given for every $n \geq 0$ by

$$
\begin{aligned}
\mathsf{P}_e(Z(n) = 1) &= e * u(n) \\
&= m_p^{-1}(1 - p * 1) * u(n) \\
&= m_p^{-1}(1 - p * 1) * \left(\sum_{j=0}^{\infty} p^{*j}\right)(n) \\
&= m_p^{-1}\left(1 + 1 * \left(\sum_{j=1}^{\infty} p^{*j}\right)(n) - p * 1 * \left(\sum_{j=0}^{\infty} p^{*j}\right)(n)\right) \\
&= m_p^{-1}.
\end{aligned}
\tag{13.35}
$$

For this reason the distribution e is also called the *equilibrium distribution* of the renewal process.

These considerations show that in the positive recurrent case, the key quantity we considered for Markov chains in (13.22) has the representation

$$
|u(n) - m_p^{-1}| = |\mathsf{P}_{\delta_0}(Z(n) = 1) - \mathsf{P}_e(Z(n) = 1)|
\tag{13.36}
$$

and in order to prove an asymptotic limiting result for an expression of this kind, we must consider the probabilities that $Z(n) = 1$ from the initial distributions δ_0, e.

But we have essentially evaluated this already. We have

Theorem 13.2.3. *Suppose that a, p are proper distributions on $\mathbb{Z}_+$, and that u is the renewal function corresponding to p. Then provided p is aperiodic and has a finite mean m_p*

$$
|a * u(n) - m_p^{-1}| \to 0, \quad n \to \infty.
\tag{13.37}
$$

PROOF   The result follows from Theorem 13.2.2 by substituting the equilibrium distribution e for b and using (13.35).   □

This has immediate application in the case where the renewal process is the return time process to an accessible atom for a Markov chain.

Proposition 13.2.4.   (i) *If Φ is a positive recurrent aperiodic Markov chain, then any atom α in $\mathcal{B}^+(\mathsf{X})$ is ergodic.*

(ii) *If Φ is a positive recurrent aperiodic Markov chain on a countable space, then for every initial state x*

$$
\|P^n(x, \cdot) - \pi\| \to 0, \quad n \to \infty.
\tag{13.38}
$$

PROOF   We know from Proposition 10.2.2 that if Φ is positive recurrent then the mean return time to any atom in $\mathcal{B}^+(\mathsf{X})$ is finite. If the chain is aperiodic then (i) follows directly from Theorem 13.2.3 and the definition (13.22).

The conclusion in (ii) then follows from (i) and Theorem 13.1.2.   □

It is worth stressing explicitly that this result depends on the classification of positive chains in terms of finite mean return times to atoms: that is, in using renewal theory it is the equivalence of positivity and regularity of the chain that is utilized.

13.2.3 The regenerative decomposition for chains with atoms

We now consider general positive Harris chains and use the renewal theorems above to commence development of their ergodic properties.

In order to use the splitting technique for analysis of total variation norm convergence for general state space chains we must extend the first-entrance last-exit decomposition (13.9) to general spaces. For any sets $A, B \in \mathcal{B}(X)$ and $x \in X$ we have, by decomposing the event $\{\Phi_n \in B\}$ over the times of the first and last entrances to A prior to n, that

$$P^n(x, B) = {_A}P^n(x, B) + \sum_{j=1}^{n-1} \int_A \left[\sum_{k=1}^{j} \int_A {_A}P^k(x, dv) P^{j-k}(v, dw) \right]_A P^{n-j}(w, B). \quad (13.39)$$

If we suppose that there is an atom α and take $A = \alpha$ then these forms are somewhat simplified: the decomposition (13.39) reduces to

$$P^n(x, B) = {_\alpha}P^n(x, B) + \sum_{j=1}^{n-1} \left[\sum_{k=1}^{j} {_\alpha}P^k(x, \alpha) P^{j-k}(\alpha, \alpha) \right]_\alpha P^{n-j}(\alpha, B). \quad (13.40)$$

In the general state space case it is natural to consider convergence from an arbitrary initial distribution λ. It is equally natural to consider convergence of the integrals

$$\mathsf{E}_\lambda[f(\Phi_n)] = \int \lambda(dx) \int P^n(x, dy) f(w) \quad (13.41)$$

for arbitrary non-negative functions f. We will use either the probabilistic or the operator-theoretic version of this quantity (as given by the two sides of (13.41)) interchangeably, in what follows.

We explore convergence of $\mathsf{E}_\lambda[f(\Phi_n)]$ for general (unbounded) f in detail in Chapter 14. Here we concentrate on bounded f, in view of the definition (13.7) of the total variation norm.

When α is an atom in $\mathcal{B}^+(X)$, let us therefore extend the notation in (13.10)–(13.12) to the forms

$$a_\lambda(n) = \mathsf{P}_\lambda(\tau_\alpha = n), \quad (13.42)$$

$$t_f(n) = \int {_\alpha}P^n(\alpha, dy) f(y) = \mathsf{E}_\alpha[f(\Phi_n) \mathbb{I}\{\tau_\alpha \geq n\}] : \quad (13.43)$$

these are well defined (although possibly infinite) for any non-negative function f on X and any probability measure λ on $\mathcal{B}(X)$.

As in (13.14) and (13.15) we can use this terminology to write the first-entrance and last-exit formulations as

$$\int \lambda(dx) P^n(x, \alpha) = a_\lambda * u(n), \quad (13.44)$$

$$\int P^n(\alpha, dy) f(y) = u * t_f(n). \quad (13.45)$$

The first-entrance last-exit decomposition (13.40) can similarly be formulated, for any λ, f, as

$$\int \lambda(dx) \int P^n(x, dw)f(w) = \int \lambda(dx) \int {}_\alpha P^n(x, dw)f(w) + a_\lambda * u * t_f(n). \quad (13.46)$$

The general state space version of Proposition 13.1.1 provides the critical bounds needed for our approach to ergodic theorems. Using the notation of (13.41) we have two bounds which we shall refer to as *Regenerative Decompositions*.

Theorem 13.2.5. *Suppose that Φ admits an accessible atom α and is positive Harris recurrent with invariant probability measure π. Then for any probability measure λ and $f \geq 0$,*

$$| \mathsf{E}_\lambda[f(\Phi_n)] - \mathsf{E}_\alpha[f(\Phi_n)] | \leq \mathsf{E}_\lambda[f(\Phi_n)\mathbb{I}\{\tau_\alpha \geq n\}]$$
$$+ |a_\lambda * u - u| * t_f(n), \quad (13.47)$$

$$| \mathsf{E}_\lambda[f(\Phi_n)] - \mathsf{E}_\pi[f(\Phi_n)] | \leq \mathsf{E}_\lambda[f(\Phi_n)\mathbb{I}\{\tau_\alpha \geq n\}]$$
$$+ |a_\lambda * u - \pi(\alpha)| * t_f(n) \quad (13.48)$$
$$+ \pi(\alpha)\sum_{j=n+1}^{\infty} t_f(j).$$

PROOF   The first-entrance last-exit decomposition (13.46), in conjunction with the simple last exit decomposition in the form (13.45), gives the first bound on the distance between $\mathsf{E}_\lambda[f(\Phi_n)]$ and $\mathsf{E}_\alpha[f(\Phi_n)]$ in (13.47).

The decomposition (13.46) also gives

$$| \mathsf{E}_\lambda[f(\Phi_n)] - \mathsf{E}_\pi[f(\Phi_n)] | \leq \mathsf{E}_\lambda[f(\Phi_n)\mathbb{I}\{\tau_\alpha \geq n\}]$$
$$+ \left| a_\lambda * u * t_f(n) - \pi(\alpha)\sum_{j=1}^{n} t_f(j) \right| \quad (13.49)$$
$$+ \left| \pi(\alpha)\sum_{j=1}^{n} t_f(j) - \int \pi(dw)f(w) \right|.$$

Now in the general state space case we have the representation for π given from (10.31) by

$$\int \pi(dw)f(w) = \pi(\alpha)\sum_{1}^{\infty} t_f(y); \quad (13.50)$$

and (13.48) now follows from (13.49). $\qquad\qquad\square$

The Regenerative Decomposition (13.48) in Theorem 13.2.5 shows clearly what is needed to prove limiting results in the presence of an atom. Suppose that f is bounded. Then we must

(E1) control the third term in (13.48), which involves questions of the finiteness of π, but is independent of the initial measure λ: this finiteness is guaranteed for positive chains by definition;

(E2) control the first term in (13.48), which involves questions of the finiteness of the hitting time distribution of τ_α when the chain begins with distribution λ; this is automatically finite as required for a Harris recurrent chain, even without positive recurrence, although for chains which are only recurrent it clearly needs care;

(E3) control the middle term in (13.48), which again involves finiteness of π to bound its last element, but more crucially then involves only the ergodicity of the atom α, regardless of λ: for we know from Lemma D.7.1 that if the atom is ergodic so that (13.19) holds then also

$$\lim_{n \to \infty} a_\lambda * u\,(n) = \pi(\alpha), \tag{13.51}$$

since for Φ a Harris recurrent chain, any probability measure λ satisfies

$$\sum_n a_\lambda(n) = \mathsf{P}_\lambda(\tau_\alpha < \infty) = 1. \tag{13.52}$$

Thus recurrence, or rather Harris recurrence, will be used twice to give bounds: positive recurrence gives one bound; and, centrally, the equivalence of positivity and regularity ensures the atom is ergodic, exactly as in Theorem 13.2.3.

Bounded functions are the only ones relevant to total variation convergence. The Regenerative Decomposition is however valid for all $f \geq 0$. Bounds in this decomposition then involve integrability of f with respect to π, and a non-trivial extension of regularity to what will be called f-regularity. This will be held over to the next chapter, and here we formalize the above steps and incorporate them with the splitting technique, to prove the Aperiodic Ergodic Theorem.

13.3 Ergodicity of positive Harris chains

13.3.1 Strongly aperiodic chains

The prescription (E1)–(E3) above for ergodic behavior is followed in the proof of

Theorem 13.3.1. *If Φ is a positive Harris recurrent and strongly aperiodic chain, then for any initial measure λ*

$$\left\| \int \lambda(dx) P^n(x, \cdot) - \pi \right\| \to 0, \qquad n \to \infty. \tag{13.53}$$

PROOF   (i)   Let us first assume that there is an accessible ergodic atom in the space. The proof is virtually identical to that in the countable case. We have

$$\left\| \int \lambda(dx) P^n(x, \cdot) - \pi \right\| = \sup_{|f| \leq 1} \left| \int \lambda(dx) \int P^n(x, dw) f(w) - \int \pi(dw) f(w) \right|$$

and we use (13.48) to bound these terms uniformly for functions $f \leq 1$.

Since $|f| \leq 1$ the third term in (13.48) is bounded above by

$$\pi(\alpha) \sum_{n+1}^{\infty} t_1(j) \to 0, \qquad n \to \infty \tag{13.54}$$

since it is the tail sum in the representation (13.50) of $\pi(\mathsf{X})$.

The second term in (13.48) is bounded above by

$$|a_\lambda * u - \pi(\alpha)| * t_1(n) \to 0, \qquad n \to \infty, \tag{13.55}$$

by Lemma D.7.1; here we use the fact that α is ergodic and, again, the representation that $\pi(\mathsf{X}) = \pi(\alpha) \sum_1^\infty t_1(j) < \infty$.

We must finally control the first term. To do this, we need only note that, again since $|f| \leq 1$, we have

$$\mathsf{E}_\lambda[f(\Phi_n)\mathbb{I}\{\tau_\alpha \geq n\}] \leq \mathsf{P}_\lambda(\tau_\alpha \geq n) \tag{13.56}$$

and this expression tends to zero by monotone convergence as $n \to \infty$, since α is Harris recurrent and $\mathsf{P}_x(\tau_\alpha < \infty) = 1$ for every x.

Notice explicitly that in (13.54)–(13.56) the bounds which tend to zero are independent of the particular $|f| \leq 1$, and so we have the required supremum norm convergence.

(ii)   Now assume that Φ is strongly aperiodic. Consider the split chain $\check{\Phi}$: we know this is also strongly aperiodic from Proposition 5.5.6 (ii), and positive Harris from Proposition 10.4.2. Thus from Proposition 13.2.4 the atom $\check{\alpha}$ is ergodic. Now our use of total variation norm convergence renders the transfer to the original chain easy. Using the fact that the original chain is the marginal chain of the split chain, and that π is the marginal measure of $\check{\pi}$, we have immediately

$$\begin{aligned}
\left\| \int \lambda(dx)P^n(x, \cdot) - \pi \right\| &= 2 \sup_{A \in \mathcal{B}(\mathsf{X})} \left| \int_\mathsf{X} \lambda(dx)P^n(x, A) - \pi(A) \right| \\
&= 2 \sup_{A \in \mathcal{B}(\mathsf{X})} \left| \int_{\check{\mathsf{X}}} \lambda^*(dx_i)\check{P}^n(x_i, A) - \check{\pi}(A) \right| \\
&\leq 2 \sup_{\check{B} \in \mathcal{B}(\check{\mathsf{X}})} \left| \int_{\check{\mathsf{X}}} \lambda^*(dx_i)\check{P}^n(x_i, \check{B}) - \check{\pi}(\check{B}) \right| \\
&= \left\| \int \lambda^*(dx_i)\check{P}^n(x_i, \cdot) - \check{\pi} \right\|, \tag{13.57}
\end{aligned}$$

where the inequality follows since the first supremum is over sets in $\mathcal{B}(\check{\mathsf{X}})$ of the form $A_0 \cup A_1$ and the second is over all sets in $\mathcal{B}(\check{\mathsf{X}})$.

Applying the result (i) for chains with accessible atoms shows that the total variation norm in (13.57) for the split chain tends to zero, so we are finished.   □

13.3.2   The ergodic theorem for ψ-irreducible chains

We can now move from the strongly aperiodic chain result to arbitrary aperiodic Harris recurrent chains. This is made simpler as a result of another useful property of the total variation norm.

Proposition 13.3.2. *If π is invariant for P, then the total variation norm*

$$\left\| \int \lambda(dx)P^n(x, \cdot) - \pi \right\|$$

is non-increasing in n.

PROOF   We have from the definition of total variation and the invariance of π that

$$\| \int \lambda(dx) P^{n+1}(x, \cdot) - \pi \|$$

$$= \sup_{f:|f| \le 1} | \int \lambda(dx) P^{n+1}(x, dy) f(y) - \int \pi(dy) f(y) |$$

$$= \sup_{f:|f| \le 1} | \int \lambda(dx) P^n(x, dw) \left[\int P(w, dy) f(y) \right] - \int \pi(dw) \left[\int P(w, dy) f(y) \right] |$$

$$\le \sup_{f:|f| \le 1} | \int \lambda(dx) P^n(x, dw) f(w) - \int \pi(dw) f(w) | \qquad (13.58)$$

since whenever $|f| \le 1$ we also have $|Pf| \le 1$.   □

We can now prove the general state space result in the aperiodic case.

Theorem 13.3.3. *If Φ is positive Harris and aperiodic, then for every initial distribution λ*

$$\| \int \lambda(dx) P^n(x, \cdot) - \pi \| \to 0, \qquad n \to \infty. \qquad (13.59)$$

PROOF   Since for some m the skeleton Φ^m is strongly aperiodic, and also positive Harris by Theorem 10.4.5, we know that

$$\| \int \lambda(dx) P^{nm}(x, \cdot) - \pi \| \to 0, \qquad n \to \infty. \qquad (13.60)$$

The result for P^n then follows immediately from the monotonicity in (13.58).   □

As we mentioned in the discussion of the periodic behavior of Markov chains, the results are not quite as simple to state in the periodic as in the aperiodic case; but they can be easily proved once the aperiodic case is understood.

The asymptotic behavior of positive recurrent chains which may not be Harris is also easy to state now that we have analyzed positive Harris chains.

The final formulation of these results for quite arbitrary positive recurrent chains is

Theorem 13.3.4.   (i) *If Φ is positive Harris with period $d \ge 1$, then for every initial distribution λ*

$$\| d^{-1} \int \lambda(dx) \sum_{r=0}^{d-1} P^{nd+r}(x, \cdot) - \pi \| \to 0, \qquad n \to \infty. \qquad (13.61)$$

(ii) *If Φ is positive recurrent with period $d \ge 1$, then there is a π-null set N such that for every initial distribution λ with $\lambda(N) = 0$*

$$\| d^{-1} \int \lambda(dx) \sum_{r=0}^{d-1} P^{nd+r}(x, \cdot) - \pi \| \to 0, \qquad n \to \infty. \qquad (13.62)$$

PROOF   The result (i) is straightforward to check from the existence of cycles in Section 5.4.3, together with the fact that the chain restricted to each cyclic set is aperiodic and positive Harris on the d-skeleton. We then have (ii) as a direct corollary of the decomposition of Theorem 9.1.5.   □

Finally, let us complete the circle by showing the last step in the equivalences in Theorem 13.0.1. Notice that (13.63) is ensured by (13.1), using the Dominated Convergence Theorem, so that our next result is in fact marginally stronger than the corresponding statement of the Aperiodic Ergodic Theorem.

Theorem 13.3.5. *Let Φ be ψ-irreducible and aperiodic, and suppose that there exists some ν-small set $C \in \mathcal{B}^+(\mathsf{X})$ and some $P^\infty(C) > 0$ such that as $n \to \infty$*

$$\int_C \nu_C(dx)(P^n(x,C) - P^\infty(C)) \to 0 \tag{13.63}$$

where $\nu_C(\cdot) = \nu(\cdot)/\nu(C)$ is normalized to a probability on C. Then the chain is positive, and there exists a ψ-null set such that for every initial distribution λ with $\lambda(N) = 0$

$$\left\| \int \lambda(dx)P^n(x,\cdot) - \pi \right\| \to 0, \qquad n \to \infty. \tag{13.64}$$

PROOF   Using the Nummelin splitting via the set C for the m-skeleton, we find that (13.63) taken through the sublattice nm is equivalent to

$$\delta^{-1}(\check{P}^n(\check{\alpha},\check{\alpha}) - \delta P^\infty(C)) \not\to 0. \tag{13.65}$$

Thus the atom $\check{\alpha}$ is ergodic and the results of Section 13.3 all hold, with $P^\infty(C) = \pi(C)$.
                                                                                        □

13.4   Sums of transition probabilities

13.4.1   A stronger coupling theorem

In order to derive bounds such as those in (13.5) on the sums of n-step total variation differences from the invariant measure π, we need to bound sums of terms such as $|P^n(\alpha,\alpha) - \pi(\alpha)|$ rather than the individual terms. This again requires a renewal theory result, which we prove using the coupling method. We have

Proposition 13.4.1. *Suppose that a, b, p are proper distributions on $\mathbb{Z}_+$, and that u is the renewal function corresponding to p. Then provided p is aperiodic and has a finite mean m_p, and a, b also have finite means m_a, m_b, we have*

$$\sum_{n=0}^{\infty} |a * u(n) - b * u(n)| < \infty. \tag{13.66}$$

PROOF   We have from (13.33) that

$$\sum_{n=0}^{\infty} |a * u\,(n) - b * u\,(n)| \le \sum_{n=0}^{\infty} \mathsf{P}(T_{ab} > n) = \mathsf{E}[T_{ab}]. \qquad (13.67)$$

Now we know from Section 10.3.1 that when p is aperiodic and $m_p < \infty$, the linked forward recurrence time chain V^* is positive recurrent with invariant probability

$$e^*(i,j) = e(i)e(j).$$

Hence from any state (i,j) with $e^*(i,j) > 0$ we have as in Proposition 11.1.1

$$\mathsf{E}_{i,j}[\tau_{1,1}] < \infty. \qquad (13.68)$$

Let us consider specifically the initial distributions δ_0 and δ_1: these correspond to the undelayed renewal process and the process delayed by exactly one time unit respectively. For this choice of initial distribution we have for $n > 0$

$$\begin{aligned}
\delta_0 * u\,(n) &= u(n), \\
\delta_1 * u\,(n) &= u(n-1).
\end{aligned}$$

Now $\mathsf{E}[T_{01}] \le \mathsf{E}_{1,2}[\tau_{1,1}] + 1$ and it is certainly the case that $e^*(1,2) > 0$. So from (13.30), (13.67) and (13.68)

$$\mathrm{Var}\,(u) := \sum_{n=0}^{\infty} |u(n) - u(n-1)| \le \mathsf{E}_{1,2}[\tau_{1,1}] + 1 < \infty. \qquad (13.69)$$

We now need to extend the result to more general initial distributions with finite mean. By the triangle inequality it suffices to consider only one arbitrary initial distribution a and to take the other as δ_0. To bound the resulting quantity $|a * u\,(n) - u(n)|$ we write the upper tails of a for $k \ge 0$ as

$$\overline{a}(k) := \sum_{j=k+1}^{\infty} a(j) = 1 - \sum_{j=0}^{k} a(j)$$

and put

$$w(k) = |u(k) - u(k-1)|.$$

We then have the relation

$$
\begin{aligned}
\overline{a} * w\,(n) \;&=\; \sum_{j=0}^{n} \overline{a}(j) w(n-j) \\[2mm]
&\geq\; \Big|\sum_{j=0}^{n}[1 - \sum_{k=0}^{j} a(k)][u(n-j) - u(n-j-1)]\Big| \\[2mm]
&=\; \Big|\sum_{j=0}^{n}[u(n-j) - u(n-j-1)] \\[2mm]
&\qquad - \sum_{j=0}^{n}\sum_{k=0}^{j} a(k)[u(n-j) - u(n-j-1)]\Big| \\[2mm]
&=\; \Big|u(n) - \sum_{k=0}^{n} a(k) \sum_{j=k}^{n}[u(n-j) - u(n-j-1)]\Big| \\[2mm]
&=\; \Big|u(n) - \sum_{k=0}^{n} a(k) u(n-k)\Big| \qquad\qquad (13.70)
\end{aligned}
$$

so that

$$
\sum_{n} |u(n) - a * u\,(n)| \leq \sum_{n} \overline{a} * w\,(n) = [\sum_{n} \overline{a}(n)][\sum_{n} w(n)]. \qquad (13.71)
$$

But by assumption the mean $m_a = \sum \overline{a}(n)$ is finite, and (13.69) shows that the sequence $w(n)$ is also summable; and so we have

$$
\sum_{n} |u(n) - a * u\,(n)| \leq m_a \operatorname{Var}(u) < \infty \qquad (13.72)
$$

as required.   □

It is obviously of considerable interest to know under what conditions we have

$$
\sum_{n} |a * u\,(n) - m_p^{-1}| < \infty; \qquad (13.73)
$$

that is, when this result holds with the equilibrium measure as one of the initial measures.

Using Proposition 13.4.1 we know that this will occur if the equilibrium distribution e has a finite mean; and since we know the exact structure of e it is obvious that $m_e < \infty$ if and only if

$$
s_p := \sum_{n} n^2 p(n) < \infty.
$$

In fact, using the exact form

$$
m_e = [s_p - m_p]/[2m_p]
$$

we have from Proposition 13.4.1 and in particular the bound (13.71) the following pleasing corollary:

Proposition 13.4.2. *If p is an aperiodic distribution with $s_p < \infty$, then*

$$\sum_n |u(n) - m_p^{-1}| \leq \operatorname{Var}(u)[s_p - m_p]/[2m_p] < \infty. \tag{13.74}$$

□

13.4.2 General chains with atoms

We now refine the ergodic theorem Theorem 13.3.3 to give conditions under which sums such as

$$\sum_{n=1}^{\infty} \|P^n(x, \cdot) - P^n(y, \cdot)\|$$

are finite. A result such as this requires regularity of the initial states x, y: recall from Chapter 11 that a probability measure μ on $\mathcal{B}(\mathsf{X})$ is called regular if

$$\mathsf{E}_\mu[\tau_B] < \infty, \qquad B \in \mathcal{B}^+(\mathsf{X}).$$

We will again follow the route of first considering chains with an atom, then translating the results to strongly aperiodic and thence to general chains.

Theorem 13.4.3. *Suppose Φ is an aperiodic positive Harris chain and suppose that the chain admits an atom $\alpha \in \mathcal{B}^+(\mathsf{X})$. Then for any regular initial distributions λ, μ,*

$$\sum_{n=1}^{\infty} \int\int \lambda(dx)\mu(dy)\|P^n(x, \cdot) - P^n(y, \cdot)\| < \infty; \tag{13.75}$$

and in particular, if Φ is regular, then for every $x, y \in \mathsf{X}$

$$\sum_{n=1}^{\infty} \|P^n(x, \cdot) - P^n(y, \cdot)\| < \infty. \tag{13.76}$$

PROOF   By the triangle inequality it will suffice to prove that

$$\sum_{n=1}^{\infty} \int \lambda(dx)\|P^n(x, \cdot) - P^n(\alpha, \cdot)\| < \infty, \tag{13.77}$$

that is, to assume that one of the initial distributions is δ_α.

If we sum the first Regenerative Decomposition (13.47) in Theorem 13.2.5 with $f \leq 1$ we find (13.77) is bounded by two sums: firstly,

$$\sum_{n=1}^{\infty} \int \lambda(dx)_\alpha P^n(x, \mathsf{X}) \;=\; \mathsf{E}_\lambda[\tau_\alpha] \tag{13.78}$$

which is finite since λ is regular; and secondly,

$$\Big\{\sum_{n=1}^{\infty} \int \lambda(dx)|a_x * u(n) - u(n)|\Big\}\Big\{\sum_{n=1}^{\infty} {}_\alpha P^n(\alpha, \mathsf{X})\Big\}. \tag{13.79}$$

To bound this term note that $\sum_{n=1}^{\infty} {}_{\alpha}P^n(\alpha,\mathsf{X}) = \mathsf{E}_{\alpha}[\tau_{\alpha}] < \infty$ since every accessible atom is regular from Theorems 11.1.4 and 11.1.2; and so it remains only to prove that

$$\sum_{n=1}^{\infty} \int \lambda(dx)|a_x * u(n) - u(n)| < \infty. \tag{13.80}$$

From (13.71) we have

$$\sum_{n=1}^{\infty} |a_x * u(n) - u(n)| \leq \left(\sum_{n=1}^{\infty} a_x(n)\right)\left(\sum_{n=1}^{\infty} |u(n) - u(n-1)|\right)$$

$$= \mathsf{E}_x[\tau_{\alpha}]\mathrm{Var}\,(u),$$

and hence the sum (13.80) is bounded by $\mathsf{E}_{\lambda}[\tau_{\alpha}]\mathrm{Var}\,(u)$, which is again finite by Proposition 13.4.1 and regularity of λ. □

13.4.3   General aperiodic chains

The move from the atomic case is by now familiar.

Theorem 13.4.4. *Suppose* Φ *is an aperiodic positive Harris chain. For any regular initial distributions* λ, μ

$$\sum_{n-1}^{\infty} \int\int \lambda(dx)\mu(dy)\|P^n(x,\,\cdot\,) - P^n(y,\,\cdot\,)\| < \infty. \tag{13.81}$$

PROOF   Consider the strongly aperiodic case. The theorem is valid for the split chain, since the split measures λ^*, μ^* are regular for $\check{\Phi}$: this follows from the characterization in Theorem 11.3.12.

Since the result is a total variation result it remains valid when restricted to the original chain, as in (13.57).

In the arbitrary aperiodic case we can apply Proposition 13.3.2 to move to a skeleton chain, as in the proof of Theorem 13.2.5. □

The most interesting special case of this result is given in the following theorem.

Theorem 13.4.5. *Suppose* Φ *is an aperiodic positive Harris chain and that* α *is an accessible atom. If*

$$\mathsf{E}_{\alpha}[\tau_{\alpha}^2] < \infty, \tag{13.82}$$

then for any regular initial distribution λ

$$\sum_{n=1}^{\infty} \|\lambda P^n - \pi\| < \infty. \tag{13.83}$$

□

PROOF   In the case where there is an atom α in the space, we have as in Proposition 13.4.2 that π is a regular measure when the second-order moment (13.82) is finite, and the result is then a consequence of Theorem 13.4.4.

13.5   Commentary*

It is hard to know where to start in describing contributions to these theorems. The countable chain case has an immaculate pedigree: Kolmogorov [215] first proved this result, and Feller [114] and Chung [71] give refined approaches to the single-state version (13.6), essentially through analytic proofs of the lattice renewal theorem.

The general state space results in the positive recurrent case are largely due to Harris [155] and to Orey [308]. Their results and related material, including a null recurrent version in Section 18.1 below, are all discussed in a most readable way in Orey's monograph [309]. Prior to the development of the splitting technique, proofs utilized the concept of the tail σ-field of the chain, which we have not discussed so far, and will only touch on in Chapter 17.

The coupling proofs are much more recent, although they are usually dated to Doeblin [94]. Pitman [317] first exploited the positive recurrent coupling in the way we give it here, and his use of the result in Proposition 13.4.1 was even then new, as was Theorem 13.4.4.

Our presentation of this material has relied heavily on Nummelin [303], and further related results can be found in his Chapter 6. In particular, for results of this kind in a more general setting where the renewal sequence is allowed to vary from the probabilistic structure with $\sum_n p(n) = 1$ which we have used, the reader is referred to Chapters 4 and 6 of [303].

It is interesting to note that the first-entrance last-exit decomposition, which shows so clearly the role of the single ergodic atom, is a relative late-comer on the scene. Although probably used elsewhere, it surfaces in the form given here in Nummelin [301] and Nummelin and Tweedie [307], and appears to be less than well known even in the countable state space case. Certainly, the proof of ergodicity is much simplified by using the Regenerative Decomposition.

We should note, for the reader who is yet again trying to keep stability nomenclature straight, that even the "ergodicity" terminology we use here is not quite standard: for example, Chung [71] uses the word "ergodic" to describe certain ratio limit theorems rather than the simple limit theorem of (13.8). We do not treat ratio limit theorems in this book, except in passing in Chapter 17: it is a notable omission, but one dictated by the lack of interesting examples in our areas of application. Hence no confusion should arise, and our ergodic chains certainly coincide with those of Feller [114], Nummelin [303] and Revuz [326]. The latter two books also have excellent treatments of ratio limit theorems.

We have no examples in this chapter. This is deliberate. We have shown in Chapter 11 how to classify specific models as positive recurrent using drift conditions: we can say little else here other than that we now know that such models converge in the relatively strong total variation norm to their stationary distributions. Over the course of the next three chapters, we will however show that other much stronger ergodic properties hold under other more restrictive drift conditions; and most of the models in which we have been interested will fall into these more strongly stable categories.

Commentary for the second edition: We wrote in Section 13.2 that we *will regrettably do far less than justice to the full power of renewal and regenerative processes, or to the coupling method itself.* It is true that the proof of ergodicity in this chapter and the refinements that follow can be streamlined by using the split chain machinery more fully. In particular, rather than prove a renewal theorem such as (13.31) and then use this to prove an ergodic theorem such as Proposition 13.2.4, it is far simpler to use coupling to prove the ergodic theorem directly as in [127, 128]. See also the aforementioned book by Lindvall on the coupling method [239].

Chapter 14

f-Ergodicity and f-regularity

In Chapter 13 we considered ergodic chains for which the limit

$$\lim_{k \to \infty} \mathsf{E}_x[f(\Phi_k)] = \int f \, d\pi \tag{14.1}$$

exists for every initial condition and every bounded function f on X.

An assumption that f is bounded is often unsatisfactory in applications. For example, f may denote a cost function in an optimal control problem, in which case $f(\Phi_n)$ will typically be a coercive function of Φ_n on X; in queueing applications, the function $f(x)$ might denote buffer levels in a queue corresponding to the particular state $x \in \mathsf{X}$ which is, again, typically an unbounded function on X; in storage models, f may denote penalties for high values of the storage level, which correspond to overflow penalties in reality.

The purpose of this chapter is to relax the boundedness condition by developing more general formulations of regularity and ergodicity. Our aim is to obtain convergence results of the form (14.1) for the mean value of $f(\Phi_k)$, where $f \colon \mathsf{X} \to [1, \infty)$ is an arbitrary fixed function. As in Chapter 13, it will be shown that the simplest approach to ergodic theorems of this kind is to consider simultaneously all functions which are dominated by f: that is, to consider convergence in the f-norm, defined as

$$\|\nu\|_f = \sup_{g:|g| \leq f} |\nu(g)|$$

where ν is any signed measure.

The goals described above are achieved in the following f-Norm Ergodic Theorem for aperiodic chains.

Theorem 14.0.1 (f-Norm Ergodic Theorem). *Suppose that the chain Φ is ψ-irreducible and aperiodic, and let $f \geq 1$ be a function on X. Then the following conditions are equivalent:*

(i) *The chain is positive recurrent with invariant probability measure π and*

$$\pi(f) := \int \pi(dx) f(x) < \infty.$$

336

(ii) *There exists some petite set* $C \in \mathcal{B}(\mathsf{X})$ *such that*

$$\sup_{x \in C} \mathsf{E}_x \Big[\sum_{n=0}^{\tau_C - 1} f(\Phi_n) \Big] < \infty. \tag{14.2}$$

(iii) *There exists some petite set* C *and some extended-valued non-negative function* V *satisfying* $V(x_0) < \infty$ *for some* $x_0 \in \mathsf{X}$, *and*

$$\Delta V(x) \leq -f(x) + b\mathbb{I}_C(x), \qquad x \in \mathsf{X}. \tag{14.3}$$

Any of these three conditions imply that the set $S_V = \{x : V(x) < \infty\}$ *is absorbing and full, where* V *is any solution to (14.3) satisfying the conditions of (iii), and any sublevel set of* V *satisfies (14.2); and for any* $x \in S_V$,

$$\|P^n(x, \cdot) - \pi\|_f \to 0 \tag{14.4}$$

as $n \to \infty$. *Moreover, if* $\pi(V) < \infty$, *then there exists a finite constant* B_f *such that for all* $x \in S_V$,

$$\sum_{n=0}^{\infty} \|P^n(x, \cdot) - \pi\|_f \leq B_f(V(x) + 1). \tag{14.5}$$

PROOF   The equivalence of (i) and (ii) follows from Theorem 14.1.1 and Theorem 14.2.11. The equivalence of (ii) and (iii) is in Theorems 14.2.3 and 14.2.4, and the fact that sublevel sets of V are "self-regular" as in (14.2) is shown in Theorem 14.2.3. The limit theorems are then contained in Theorems 14.3.3, 14.3.4 and 14.3.5.   □

Much of this chapter is devoted to proving this result, and related *f*-regularity properties which follow from (14.2), and the pattern is not dissimilar to that in the previous chapter: indeed, those ergodicity results, and the equivalences in Theorem 13.0.1, can be viewed as special cases of the general *f* results we now develop.

The *f*-norm limit (14.4) obviously implies that the simpler limit (14.1) also holds. In fact, if g is any function satisfying $|g| \leq c(f+1)$ for some $c < \infty$ then $\mathsf{E}_x[g(\Phi_k)] \to \int g \, d\pi$ for states x with $V(x) < \infty$, for V satisfying (14.3). We formalize the behavior we will analyze in

---

f-Ergodicity

We shall say that the Markov chain Φ is *f*-ergodic if $f \geq 1$ and

   (i) Φ is positive Harris recurrent with invariant probability π;

   (ii) the expectation $\pi(f)$ is finite;

   (iii) for every initial condition of the chain,

$$\lim_{k \to \infty} \|P^k(x, \cdot) - \pi\|_f = 0.$$

The f-Norm Ergodic Theorem states that if any one of the equivalent conditions of the Aperiodic Ergodic Theorem holds then the simple additional condition that $\pi(f)$ is finite is enough to ensure that a full absorbing set exists on which the chain is f-ergodic. Typically the way in which finiteness of $\pi(f)$ would be established in an application is through finding a test function V satisfying (14.3): and if, as will typically happen, V is finite everywhere then it follows that the chain is f-ergodic without restriction, since then $S_V = \mathsf{X}$.

14.1 f-Properties: chains with atoms

14.1.1 f-Regularity for chains with atoms

We have already given the pattern of approach in detail in Chapter 13. It is not worthwhile treating the countable case completely separately again: as was the case for ergodicity properties, a single accessible atom is all that is needed, and we will initially develop f-ergodic theorems for chains possessing such an atom.

The generalization from total variation convergence to f-norm convergence given an initial accessible atom α can be carried out based on the developments of Chapter 13, and these also guide us in developing characterizations of the initial measures λ for which general f-ergodicity might be expected to hold. It is in this part of the analysis, which corresponds to bounding the first term in the Regenerative Decomposition of Theorem 13.2.5, that the hard work is needed, as we now discuss.

Suppose that $\boldsymbol{\Phi}$ admits an atom α and is positive Harris recurrent with invariant probability measure π. Let $f \geq 1$ be arbitrary: that is, we place no restrictions on the boundedness or otherwise of f. Recall that for any probability measure λ we have from the Regenerative Decomposition that for arbitrary $|g| \leq f$,

$$|\mathsf{E}_\lambda[g(\Phi_n)] - \pi(g)| \ \leq \ \int \lambda(dx) \int_\alpha P^n(x, dw) f(w) \tag{14.6}$$

$$+ |a_\lambda * u - \pi(\alpha)| * t_f(n) + \pi(\alpha) \sum_{j=n+1}^\infty t_f(j).$$

Using hitting time notation we have

$$\sum_{n=1}^\infty t_f(n) \ = \ \mathsf{E}_\alpha\left[\sum_{j=1}^{\tau_\alpha} f(\Phi_j)\right] \tag{14.7}$$

and thus the finiteness of this expectation will guarantee convergence of the third term in (14.6), as it did in the case of the ergodic theorems in Chapter 13. Also as in Chapter 13, the central term in (14.6) is controlled by the convergence of the renewal sequence u regardless of f, provided the expression in (14.7) is finite.

Thus it is only the first term in (14.6) that requires a condition other than ergodicity and finiteness of (14.7). Somewhat surprisingly, for unbounded f this is a much more troublesome term to control than for bounded f, when it is a simple consequence of recurrence that it tends to zero. This first term can be expressed alternatively as

$$\int \lambda(dx) \int_\alpha P^n(x, dw) f(w) = \mathsf{E}_\lambda\left[f(\Phi_n)\mathbb{I}(\tau_\alpha \geq n)\right] \tag{14.8}$$

and so we have the representation

$$\sum_{n=1}^{\infty} \int \lambda(dx) \int {}_{\alpha}P^n(x, dw)f(w) \;=\; \mathsf{E}_{\lambda}\Big[\sum_{j=1}^{\tau_{\alpha}} f(\Phi_j)\Big]. \tag{14.9}$$

This is similar in form to (14.7), and if (14.9) is finite, then we have the desired conclusion that (14.8) does tend to zero. In fact, it is only the sum of these terms that appears tractable, and for this reason it is in some ways more natural to consider the summed form (14.5) rather than simple *f*-norm convergence.

Given this motivation to require finiteness of (14.7) and (14.9), we introduce the concept of *f-regularity* which strengthens our definition of ordinary regularity.

---

f-Regularity

A set $C \in \mathcal{B}(\mathsf{X})$ is called *f-regular*, where $f\colon \mathsf{X} \to [1, \infty)$ is a measurable function, if for each $B \in \mathcal{B}^+(\mathsf{X})$,

$$\sup_{x \in C} \mathsf{E}_x\Big[\sum_{k=0}^{\tau_B - 1} f(\Phi_k)\Big] < \infty.$$

A measure λ is called *f-regular* if for each $B \in \mathcal{B}^+(\mathsf{X})$,

$$\mathsf{E}_{\lambda}\Big[\sum_{k=0}^{\tau_B - 1} f(\Phi_k)\Big] < \infty.$$

The chain $\boldsymbol{\Phi}$ is called *f-regular* if there is a countable cover of X with *f*-regular sets.

From this definition an *f*-regular state, seen as a singleton set, is a state x for which $\mathsf{E}_x\big[\sum_{k=0}^{\tau_B - 1} f(\Phi_k)\big] < \infty$, $B \in \mathcal{B}^+(\mathsf{X})$.

As with regularity, this definition of *f*-regularity appears initially to be stronger than required since it involves all sets in $\mathcal{B}^+(\mathsf{X})$; but we will show this to be again illusory.

A first consequence of *f*-regularity, and indeed of the weaker "self-*f*-regular" form in (14.2), is

Proposition 14.1.1. *If* $\boldsymbol{\Phi}$ *is recurrent with invariant measure* π *and there exists* $C \in \mathcal{B}(\mathsf{X})$ *satisfying* $\pi(C) < \infty$ *and*

$$\sup_{x \in C} \mathsf{E}_x\Big[\sum_{n=0}^{\tau_C - 1} f(\Phi_n)\Big] < \infty, \tag{14.10}$$

then $\boldsymbol{\Phi}$ *is positive recurrent and* $\pi(f) < \infty$.

PROOF   First of all, observe that under (14.10) the set C is Harris recurrent and hence $C \in \mathcal{B}^+(\mathsf{X})$ by Proposition 9.1.1. The invariant measure π then satisfies, from Theorem 10.4.9,

$$\pi(f) = \int_C \pi(dy)\mathsf{E}_y\left[\sum_{n=0}^{\tau_C-1} f(\Phi_n)\right].$$

If C satisfies (14.10) then the expectation is uniformly bounded on C itself, so that $\pi(f) \leq \pi(C)M_C < \infty$. $\square$

Although f-regularity is a requirement on the hitting times of all sets, when the chain admits an atom it reduces to a requirement on the hitting times of the atom as was the case with regularity.

Proposition 14.1.2. *Suppose Φ is positive recurrent with $\pi(f) < \infty$, and that an atom $\alpha \in \mathcal{B}^+(\mathsf{X})$ exists.*

(i) *Any set $C \in \mathcal{B}(\mathsf{X})$ is f-regular if and only if*

$$\sup_{x\in C}\mathsf{E}_x\left[\sum_{k=0}^{\sigma_\alpha} f(\Phi_k)\right] < \infty.$$

(ii) *There exists an increasing sequence of sets $S_f(n)$ where each $S_f(n)$ is f-regular and the set $S_f = \cup S_f(n)$ is full and absorbing.*

PROOF   Consider the function $G_\alpha(x, f)$ previously defined in (11.21) by

$$G_\alpha(x; f) = \mathsf{E}_x\left[\sum_{k=0}^{\sigma_\alpha} f(\Phi_k)\right]. \tag{14.11}$$

When $\pi(f) < \infty$, by Theorem 11.3.5 the bound $PG_\alpha(x, f) \leq G_\alpha(x, f) + c$ holds for the constant $c = \mathsf{E}_\alpha[\sum_{k=1}^{\tau_\alpha} f(\Phi_k)] = \pi(f)/\pi(\alpha) < \infty$, which shows that the set $\{x : G_\alpha(x, f) < \infty\}$ is absorbing, and hence by Proposition 4.2.3 this set is full.

To prove (i), let B be any sublevel set of the function $G_\alpha(x, f)$ with $\pi(B) > 0$ and apply the bound

$$G_\alpha(x, f) \leq \mathsf{E}_x\left[\sum_{k=0}^{\tau_B-1} f(\Phi_k)\right] + \sup_{y\in B}\mathsf{E}_y\left[\sum_{k=0}^{\sigma_\alpha} f(\Phi_k)\right].$$

This shows that $G_\alpha(x, f)$ is bounded on C if C is f-regular, and proves the "only if" part of (i).

We have from Theorem 10.4.9 that for any $B \in \mathcal{B}^+(\mathsf{X})$,

$$\infty > \int_B \pi(dx)\mathsf{E}_x\left[\sum_{k=0}^{\tau_B} f(\Phi_k)\right]$$

$$\geq \int_B \pi(dx)\mathsf{E}_x\left[\mathbb{I}(\sigma_\alpha < \tau_B)\sum_{k=\sigma_\alpha+1}^{\tau_B} f(\Phi_k)\right]$$

$$= \int_B \pi(dx)\mathsf{P}_x(\sigma_\alpha < \tau_B)\mathsf{E}_\alpha\left[\sum_{k=1}^{\tau_B} f(\Phi_k)\right]$$

where to obtain the last equality we have conditioned at time σ_α and used the strong Markov property.

Since $\alpha \in \mathcal{B}^+(\mathsf{X})$ we have that

$$\pi(\alpha) = \int_B \pi(dx) \mathsf{E}_x \left[\sum_{k=0}^{\tau_B - 1} \mathbb{I}(\Phi_k \in \alpha) \right] > 0,$$

which shows that $\int_B \pi(dx) \mathsf{P}_x(\sigma_\alpha < \tau_B) > 0$. Hence from the previous bounds, we have $\mathsf{E}_\alpha \left[\sum_{k=1}^{\tau_B} f(\Phi_k) \right] < \infty$ for $B \in \mathcal{B}^+(\mathsf{X})$.

Using the bound $\tau_B \leq \sigma_\alpha + \theta^{\sigma_\alpha} \tau_B$, we have for arbitrary $x \in \mathsf{X}$,

$$\mathsf{E}_x \left[\sum_{k=0}^{\tau_B} f(\Phi_k) \right] \leq \mathsf{E}_x \left[\sum_{k=0}^{\sigma_\alpha} f(\Phi_k) \right] + \mathsf{E}_\alpha \left[\sum_{k=1}^{\tau_B} f(\Phi_k) \right] \tag{14.12}$$

and hence C is f-regular if $G_\alpha(x, f)$ is bounded on C, which proves (i).

To prove (ii), observe that from (14.12) we have that the set $S_f(n) := \{x : G_\alpha(x, f) \leq n\}$ is f-regular, and so the proposition is proved. □

14.1.2 *f*-Ergodicity for chains with atoms

As we have foreshadowed, f-regularity is exactly the condition needed to obtain convergence in the f-norm.

Theorem 14.1.3. *Suppose that* Φ *is positive Harris, aperiodic, and that an atom* $\alpha \in \mathcal{B}^+(\mathsf{X})$ *exists.*

(i) *If* $\pi(f) < \infty$, *then the set* S_f *of* f-*regular states is absorbing and full, and for any* $x \in S_f$ *we have*

$$\|P^k(x, \cdot) - \pi\|_f \to 0, \qquad k \to \infty.$$

(ii) *If* Φ *is* f-*regular, then* Φ *is* f-*ergodic.*

(iii) *There exists a constant* $M_f < \infty$ *such that for any two* f-*regular initial distributions* λ *and* μ,

$$\sum_{n=1}^\infty \int \int \lambda(dx) \mu(dy) \| P^n(x, \cdot) - P^n(y, \cdot) \|_f$$

$$\leq M_f \left(\int \lambda(dx) G_\alpha(x, f) + \int \mu(dy) G_\alpha(y, f) \right). \tag{14.13}$$

PROOF   From Proposition 14.1.2 (ii), the set of f-regular states S_f is absorbing and full when $\pi(f) < \infty$. If we can prove $\|P^k(x, \cdot) - \pi\|_f \to 0$, for $x \in S_f$, this will establish both (i) and (ii).

But this f-norm convergence follows from (14.6), where the first term tends to zero since x is f-regular, so that $\mathsf{E}_x[\sum_{n=1}^{\tau_\alpha} f(\Phi_n)] < \infty$; the third term tends to zero since $\sum_{n=1}^\infty t_f(j) = \mathsf{E}_\alpha[\sum_{n=1}^{\tau_\alpha} f(\Phi_n)] = \pi(f)/\pi(\alpha) < \infty$; and the central term converges to zero by Lemma D.7.1 and the fact that α is an ergodic atom.

To prove the result in (iii), we use the same method of proof as for the ergodic case. By the triangle inequality it suffices to assume that one of the initial distributions is δ_α. We again use the first form of the Regenerative Decomposition Theorem to see that for any $|g| \leq f$, $x \in \mathsf{X}$, the sum

$$\sum_{n=1}^{\infty} \int \lambda(dx)|P^n(x,g) - P^n(\alpha,g)|$$

is bounded by the sum of the following two terms:

$$\sum_{n=1}^{\infty} \int \lambda(dx)_\alpha P^n(x,f) \;=\; \mathsf{E}_\lambda\Big[\sum_{n=1}^{\tau_\alpha} f(\Phi_n)\Big], \qquad (14.14)$$

$$\Big\{\sum_{n=1}^{\infty} \int \lambda(dx)|a_x * u(n) - u(n)|\Big\}\Big\{\sum_{n=1}^{\infty} {}_\alpha P^n(\alpha,f)\Big\}. \qquad (14.15)$$

The first of these is again finite since we have assumed λ to be f-regular; and in the second, the right hand term is similarly finite since $\pi(f) < \infty$, whilst the left hand term is independent of f, and since λ is regular (given $f \geq 1$), is bounded by $\mathsf{E}_\lambda[\tau_\alpha]\mathrm{Var}\,(u)$, using (13.72).

Since for some finite M

$$\mathsf{E}_x[\tau_\alpha] \leq \mathsf{E}_x[\sum_{n=1}^{\tau_\alpha} f(\Phi_n)] \leq MG_\alpha(x,f),$$

this completes the proof of (iii).   □

Thus for a chain with an accessible atom, we have very little difficulty moving to f-norm convergence. The simplicity of the results is exemplified in the countable state space case where the f-regularity of all states, guaranteed by Proposition 14.1.2, gives us

Theorem 14.1.4. *Suppose that* Φ *is an irreducible positive Harris aperiodic chain on a countable space. Then if* $\pi(f) < \infty$, *for all* $x, y \in \mathsf{X}$

$$\|P^k(x,\cdot) - \pi\|_f \to 0, \qquad k \to \infty,$$

and

$$\sum_{n=1}^{\infty} \|P^n(x,\cdot) - P^n(y,\cdot)\|_f < \infty.$$

14.2 f-Regularity and drift

It would seem at this stage that all we have to do is move, as we did in Chapter 13, to strongly aperiodic chains; bring the f-properties proved in the previous section above over from the split chain in this case; and then move to general aperiodic chains by using the Nummelin splitting of the m-skeleton.

Somewhat surprisingly, perhaps, this recipe does not work in a trivially easy way. The most difficult step in this approach is that when we go to a split chain it is necessary to consider an m-skeleton, but we do not yet know if the skeletons of an f-regular chain are also f-regular. Such is indeed the case and we will prove this key result in the next section, by exploiting drift criteria.

This may seem to be a much greater effort than we needed for the Aperiodic Ergodic Theorem: but it should be noted that we devoted all of Chapter 11 to the equivalence of regularity and drift conditions in the case of $f \equiv 1$, and the results here actually require rather less effort. In fact, much of the work in this chapter is based on the results already established in Chapter 11, and the duality between drift and regularity established there will serve us well in this more complex case.

14.2.1 The drift characterization of *f*-regularity

In order to establish f-regularity for a chain on a general state space without atoms, we will use the following criterion, which is a generalization of the condition in (V2). As for regular chains, we will find that there is a duality between appropriate solutions to (V3) and f-regularity.

---

f-Modulated drift towards C

(V3)   For a function $f \colon \mathsf{X} \to [1, \infty)$, a set $C \in \mathcal{B}(\mathsf{X})$, a constant $b < \infty$, and an extended-real-valued function $V \colon \mathsf{X} \to [0, \infty]$

$$\Delta V (x) \leq -f(x) + b \mathbb{I}_C (x) \qquad x \in \mathsf{X}. \qquad (14.16)$$

The condition (14.16) is implied by the slightly stronger pair of bounds

$$f(x) + PV (x) \leq \begin{cases} V(x) & x \in C^c \\ b & x \in C \end{cases} \qquad (14.17)$$

with V bounded on C, and it is this form that is often verified in practice.

Those states x for which $V(x)$ is finite when V satisfies (V3) will turn out to be those f-regular states from which the distributions of Φ converge in f-norm. For this reason the following generalization of Lemma 11.3.6 is important: we omit the proof which is similar to that of Lemma 11.3.6 or Proposition 14.1.2.

Lemma 14.2.1. *Suppose that Φ is ψ-irreducible. If (14.16) holds for a positive function V which is finite at some $x_0 \in \mathsf{X}$, then the set $S_f := \{x \in \mathsf{X} : V(x) < \infty\}$ is absorbing and full.*

$\square$

The power of (V3) largely comes from the following

Theorem 14.2.2 (Comparison Theorem). *Suppose that the non-negative functions* V, f, s *satisfy the relationship*

$$PV(x) \leq V(x) - f(x) + s(x), \qquad x \in \mathsf{X}.$$

Then for each $x \in \mathsf{X}$, $N \in \mathbb{Z}_+$, *and any stopping time* τ, *we have*

$$\sum_{k=0}^{N} \mathsf{E}_x[f(\Phi_k)] \leq V(x) + \sum_{k=0}^{N} \mathsf{E}_x[s(\Phi_k)],$$

$$\mathsf{E}_x\left[\sum_{k=0}^{\tau-1} f(\Phi_k)\right] \leq V(x) + \mathsf{E}_x\left[\sum_{k=0}^{\tau-1} s(\Phi_k)\right].$$

PROOF   This follows from Proposition 11.3.2 on letting $f_k = f$, $s_k = s$. □

The first inequality in Theorem 14.2.2 bounds the mean value of $f(\Phi_k)$, but says nothing about the convergence of the mean value. We will see that the second bound is in fact crucial for obtaining f-regularity for the chain, and we turn to this now.

In linking the drift condition (V3) with f-regularity we will consider the extended-real-valued function $G_C(x, f)$ defined in (11.21) as

$$G_C(x, f) = \mathsf{E}_x\left[\sum_{k=0}^{\sigma_C} f(\Phi_k)\right] \tag{14.18}$$

where C is typically f-regular or petite. The following characterization of f-regularity shows that this function is both a solution to (14.16), and can be bounded using any other solution V to (14.16). Together with Lemma 14.2.1, this result proves the equivalence between (ii) and (iii) in the f-Norm Ergodic Theorem.

Theorem 14.2.3. *Suppose that* Φ *is* ψ-*irreducible.*

(i) *If (V3) holds for a petite set* C, *then for any* $B \in \mathcal{B}^+(\mathsf{X})$ *there exists* $c(B) < \infty$ *such that*

$$\mathsf{E}_x\left[\sum_{k=0}^{\tau_B - 1} f(\Phi_k)\right] \leq V(x) + c(B).$$

Hence if V *is bounded on the set* A, *then* A *is* f-*regular.*

(ii) *If there exists one* f-*regular set* $C \in \mathcal{B}^+(\mathsf{X})$, *then* C *is petite and the function* $V(x) = G_C(x, f)$ *satisfies (V3) and is bounded on* A *for any* f-*regular set* A.

PROOF   **(i)**   Suppose that (V3) holds, with C a ψ_a-petite set. By the Comparison Theorem 14.2.2, Lemma 11.3.10, and the bound

$$\mathbb{I}_C(x) \leq \psi_a(B)^{-1} K_a(x, B)$$

in (11.27) we have for any $B \in \mathcal{B}^+(\mathsf{X})$, $x \in \mathsf{X}$,

$$
\begin{aligned}
\mathsf{E}_x\Big[\sum_{k=0}^{\tau_B-1} f(\Phi_k)\Big] &\leq V(x) + b\mathsf{E}_x\Big[\sum_{k=0}^{\tau_B-1} \mathbb{I}_C(\Phi_k)\Big] \\
&\leq V(x) + b\mathsf{E}_x\Big[\sum_{k=0}^{\tau_B-1} \psi_a(B)^{-1} K_a(\Phi_k, B)\Big] \\
&= V(x) + b\psi_a(B)^{-1} \sum_{i=0}^{\infty} a_i \mathsf{E}_x\Big[\sum_{k=0}^{\tau_B-1} \mathbb{I}_B(\Phi_{k+i})\Big] \\
&\leq V(x) + b\psi_a(B)^{-1} \sum_{i=0}^{\infty} ia_i.
\end{aligned}
$$

Since we can choose a so that $m_a = \sum_{i=0}^{\infty} ia_i < \infty$ from Proposition 5.5.6, the result follows with $c(B) = b\psi_a(B)^{-1}m_a$. We then have

$$
\sup_{x \in A} \mathsf{E}_x\Big[\sum_{k=0}^{\tau_B-1} f(\Phi_k)\Big] \leq \sup_{x \in A} V(x) + c(B),
$$

and so if V is bounded on A, it follows that A is f-regular.

(ii)   If an f-regular set $C \in \mathcal{B}^+(\mathsf{X})$ exists, then it is also regular and hence petite from Proposition 11.3.8. The function $G_C(x, f)$ is clearly positive, and bounded on any f-regular set A. Moreover, by Theorem 11.3.5 and f-regularity of C it follows that condition (V3) holds with $V(x) = G_C(x, f)$.   □

14.2.2   *f*-Regular sets

Theorem 14.2.3 gives a characterization of f-regularity in terms of a drift condition. The next result gives such a characterization in terms of the return times to petite sets, and generalizes Proposition 11.3.14: f-regular sets in $\mathcal{B}^+(\mathsf{X})$ are precisely those petite sets which are "self-f-regular".

Theorem 14.2.4. *When Φ is a ψ-irreducible chain, the following are equivalent:*

(i) *$C \in \mathcal{B}(\mathsf{X})$ is petite and*

$$
\sup_{x \in C} \mathsf{E}_x\Big[\sum_{k=0}^{\tau_C-1} f(\Phi_k)\Big] < \infty; \tag{14.19}
$$

(ii) *C is f-regular and $C \in \mathcal{B}^+(\mathsf{X})$.*

PROOF   To see that (i) implies (ii), suppose that C is petite and satisfies (14.19). By Theorem 11.3.5 we may find a constant $b < \infty$ such that (V3) holds for $G_C(x, f)$. It follows from Theorem 14.2.3 that C is f-regular.

The set C is Harris recurrent under the conditions of (i), and hence lies in $\mathcal{B}^+(\mathsf{X})$ by Proposition 9.1.1.

Conversely, if C is f-regular then it is also petite from Proposition 11.3.8, and if $C \in \mathcal{B}^+(\mathsf{X})$ then $\sup_{x \in C} \mathsf{E}_x[\sum_{k=0}^{\tau_C - 1} f(\Phi_k)] < \infty$ by the definition of f-regularity. $\square$

As an easy corollary to Theorem 14.2.3 we obtain the following generalization of Proposition 14.1.2.

Theorem 14.2.5. *If there exists an f-regular set $C \in \mathcal{B}^+(\mathsf{X})$, then there exists an increasing sequence $\{S_f(n) : n \in \mathbb{Z}_+\}$ of f-regular sets whose union is full. Hence there is a decomposition*

$$\mathsf{X} = S_f \cup N \tag{14.20}$$

where the set S_f is full and absorbing and Φ restricted to S_f is f-regular.

PROOF   By f-regularity and positivity of C we have, by Theorem 14.2.3 (ii), that (V3) holds for the function $V(x) = G_C(x, f)$ which is bounded on C, and by Lemma 14.2.1 we have that V is finite π-a.e.

The required sequence of f-regular sets can then be taken as

$$S_f(n) := \{x : V(x) \le n\}, \qquad n \ge 1$$

by Theorem 14.2.3. It is a consequence of Lemma 14.2.1 that $S_f = \cup S_f(n)$ is absorbing. $\square$

We now give a characterization of f-regularity using the Comparison Theorem 14.2.2.

Theorem 14.2.6. *Suppose that Φ is ψ-irreducible. Then the chain is f-regular if and only if (V3) holds for an everywhere finite function V, and every sublevel set of V is then f-regular.*

PROOF   From Theorem 14.2.3 (i) we see that if (V3) holds for a finite-valued V then each sublevel set of V is f-regular. This establishes f-regularity of Φ.

Conversely, if Φ is f-regular then it follows that an f-regular set $C \in \mathcal{B}^+(\mathsf{X})$ exists. The function $V(x) = G_C(x, f)$ is everywhere finite and satisfies (V3), by Theorem 14.2.3 (ii). $\square$

As a corollary to Theorem 14.2.6 we obtain a final characterization of f-regularity of Φ, this time in terms of petite sets:

Theorem 14.2.7. *Suppose that Φ is ψ-irreducible. Then the chain is f-regular if and only if there exists a petite set C such that the expectation*

$$\mathsf{E}_x\Big[\sum_{k=0}^{\tau_C - 1} f(\Phi_k) \Big]$$

is finite for each x and uniformly bounded for $x \in C$.

PROOF   If the expectation is finite as described in the theorem, then by Theorem 11.3.5 the function $G_C(x, f)$ is everywhere finite and satisfies (V3) with the petite set C. Hence from Theorem 14.2.6 we see that the chain is f-regular.

For the converse take C to be any f-regular set in $\mathcal{B}^+(\mathsf{X})$. $\square$

14.2.3   *f*-Regularity and *m*-skeletons

One advantage of the form (V3) over (14.17) is that, once *f*-regularity of Φ is established, we may easily iterate (14.16) to obtain

$$P^m V(x) \le V(x) - \sum_{i=0}^{m-1} P^i f + \sum_{i=0}^{m-1} P^i \mathbb{I}_C(x) \qquad x \in \mathsf{X}. \tag{14.21}$$

This is essentially of the same form as (14.16), and provides an approach to *f*-regularity for the *m*-skeleton which will give us the desired equivalence between *f*-regularity for Φ and its skeletons.

To apply Theorem 14.2.3 and (14.21) to obtain an equivalence between *f*-properties of Φ and its skeletons we must replace the function $\sum_{i=0}^{m-1} P^i \mathbb{I}_C$ with the indicator function of a petite set. The following result shows that this is possible whenever C is petite and the chain is aperiodic.

Let us write for any positive function g on X,

$$g^{(m)} := \sum_{i=0}^{m-1} P^i g. \tag{14.22}$$

Lemma 14.2.8. *If Φ is aperiodic and if $C \in \mathcal{B}(\mathsf{X})$ is a petite set, then for any $\varepsilon > 0$ and $m \ge 1$ there exists a petite set C_ε such that*

$$\mathbb{I}_C^{(m)} \le m \mathbb{I}_{C_\varepsilon} + \varepsilon.$$

PROOF   Since Φ is aperiodic, it follows from the definition of the period given in (5.40) and the fact that petite sets are small, proven in Proposition 5.5.7, that for a non-trivial measure ν and some $k \in \mathbb{Z}_+$, we have the simultaneous bound

$$P^{km-i}(x, B) \ge \mathbb{I}_C(x)\nu(B), \quad x \in \mathsf{X}, \ B \in \mathcal{B}(\mathsf{X}), \qquad 0 \le i \le m-1.$$

Hence we also have

$$P^{km}(x, B) \ge P^i \mathbb{I}_C(x)\nu(B), \qquad x \in \mathsf{X}, \ B \in \mathcal{B}(\mathsf{X}), \qquad 0 \le i \le m-1,$$

which shows that

$$P^{km}(x, \cdot) \ge \mathbb{I}_C^{(m)}(x) m^{-1} \nu.$$

The set $C_\varepsilon = \{x : \mathbb{I}_C^{(m)}(x) \ge \varepsilon\}$ is therefore ν_k-small for the *m*-skeleton, where $\nu_k = \varepsilon m^{-1} \nu$, whenever this set is non-empty. Moreover, $C \subset C_\varepsilon$ for all $\varepsilon < 1$.

Since $\mathbb{I}_C^{(m)} \le m$ everywhere, and since $\mathbb{I}_C^{(m)}(x) < \varepsilon$ for $x \in C_\varepsilon^c$, we have the bound

$$\mathbb{I}_C^{(m)} \le m \mathbb{I}_{C_\varepsilon} + \varepsilon$$

$\qquad\qquad\qquad\qquad\qquad\qquad\qquad\qquad\qquad\qquad\qquad\qquad\qquad\qquad\qquad$ □

We can now put these pieces together and prove the desired solidarity for Φ and its skeletons.

Theorem 14.2.9. *Suppose that Φ is ψ-irreducible and aperiodic. Then $C \in \mathcal{B}^+(\mathsf{X})$ is f-regular if and only if it is $f^{(m)}$-regular for any one, and then every, m-skeleton chain.*

PROOF   If C is $f^{(m)}$-regular for an m-skeleton then, letting τ_B^m denote the hitting time for the skeleton, we have by the Markov property, for any $B \in \mathcal{B}^+(\mathsf{X})$,

$$\mathsf{E}_x\Big[\sum_{k=0}^{\tau_B^m-1}\sum_{i=0}^{m-1} P^i f(\Phi_{km})\Big] = \mathsf{E}_x\Big[\sum_{k=0}^{\tau_B^m-1}\sum_{i=0}^{m-1} f(\Phi_{km+i})\Big]$$

$$\geq \mathsf{E}_x\Big[\sum_{j=0}^{\tau_B-1} f(\Phi_j)\Big].$$

By the assumption of $f^{(m)}$-regularity, the left hand side is bounded over C and hence the set C is f-regular.

Conversely, if $C \in \mathcal{B}^+(\mathsf{X})$ is f-regular then it follows from Theorem 14.2.3 that (V3) holds for a function V which is bounded on C.

By repeatedly applying P to both side of this inequality we obtain as in (14.21)

$$P^m V \leq V - f^{(m)} + b\mathbb{I}_C^{(m)}.$$

By Lemma 14.2.8 we have for a petite set C'

$$P^m V \leq V - f^{(m)} + bm\mathbb{I}_{C'} + \tfrac{1}{2}$$
$$\leq V - \tfrac{1}{2}f^{(m)} + bm\mathbb{I}_{C'},$$

and thus (V3) holds for the m-skeleton. Since V is bounded on C, we see from Theorem 14.2.3 that C is $f^{(m)}$-regular for the m-skeleton.   □

As a simple but critical corollary we have

Theorem 14.2.10. *Suppose that Φ is ψ-irreducible and aperiodic. Then Φ is f-regular if and only if each m-skeleton is $f^{(m)}$-regular.*   □

The importance of this result is that it allows us to shift our attention to skeleton chains, one of which is always strongly aperiodic and hence may be split to form an artificial atom; and this of course allows us to apply the results obtained in Section 14.1 for chains with atoms.

The next result follows this approach to obtain a converse to Proposition 14.1.1, thus extending Proposition 14.1.2 to the non-atomic case.

Theorem 14.2.11. *Suppose that Φ is positive recurrent and $\pi(f) < \infty$. Then there exists a sequence $\{S_f(n)\}$ of f-regular sets whose union is full.*

PROOF   We need only look at a split chain corresponding to the m-skeleton chain, which possess an $f^{(m)}$-regular atom by Proposition 14.1.2. It follows from Proposition 14.1.2 that for the split chain the required sequence of $f^{(m)}$-regular sets exist, and then following the proof of Proposition 11.1.3 we see that for the m-skeleton an increasing sequence $\{S_f(n)\}$ of $f^{(m)}$-regular sets exists whose union is full.

From Theorem 14.2.9 we have that each of the sets $\{S_f(n)\}$ is also f-regular for Φ and the theorem is proved.   □

14.3   *f*-Ergodicity for general chains

14.3.1   The aperiodic *f*-ergodic theorem

We are now, at last, in a position to extend the atom-based *f*-ergodic results of Section 14.1 to general aperiodic chains.

We first give an *f*-ergodic theorem for strongly aperiodic chains. This is an easy consequence of the result for chains with atoms.

Proposition 14.3.1. *Suppose that Φ is strongly aperiodic, positive recurrent, and suppose that $f \geq 1$.*

(i) *If $\pi(f) = \infty$, then $P^k(x, f) \to \infty$ as $k \to \infty$ for all $x \in \mathsf{X}$.*

(ii) *If $\pi(f) < \infty$, then almost every state is f-regular and for any f-regular state $x \in \mathsf{X}$*

$$\|P^k(x, \cdot) - \pi\|_f \to 0, \qquad k \to \infty.$$

(iii) *If Φ is f-regular, then Φ is f-ergodic.*

PROOF   (i) By positive recurrence we have for x lying in the maximal Harris set H, and any $m \in \mathbb{Z}_+$,

$$\liminf_{k \to \infty} P^k(x, f) > \liminf_{k \to \infty} P^k(x, m \wedge f) = \pi(m \wedge f).$$

Letting $m \to \infty$ we see that $P^k(x, f) \to \infty$ for these x. For arbitrary $x \in \mathsf{X}$ we choose n_0 so large that $P^{n_0}(x, H) > 0$. This is possible by ψ-irreducibility. By Fatou's Lemma we then have the bound

$$\liminf_{k \to \infty} P^k(x, f) = \liminf_{k \to \infty} P^{n_0 + k}(x, f) \geq \int_H P^{n_0}(x, dy) \left\{ \liminf_{k \to \infty} P^k(x, f) \right\} = \infty.$$

Result (ii) is now obvious using the split chain, given the results for a chain possessing an atom, and (iii) follows directly from (ii).   □

We again obtain *f*-ergodic theorems for general aperiodic Φ by considering the *m*-skeleton chain. The results obtained in the previous section show that when Φ has appropriate *f*-properties then so does each *m*-skeleton. For aperiodic chains, there always exists some $m \geq 1$ such that the *m*-skeleton is strongly aperiodic, and hence we may apply Theorem 14.3.1 to the *m*-skeleton chain to obtain *f*-ergodicity for this skeleton. This then carries over to the process by considering the m distinct skeleton chains embedded in Φ.

The following lemma allows us to make the desired connections between Φ and its skeletons.

Lemma 14.3.2.   (i) *For any $f \geq 1$ we have for $n \in \mathbb{Z}_+$,*

$$\|P^n(x, \cdot) - \pi\|_f \leq \|P^{km}(x, \cdot) - \pi(\cdot)\|_{f^{(m)}},$$

for k satisfying $n = km + i$ with $0 \leq i \leq m - 1$.

(ii) *If for some $m \geq 1$ and some $x \in \mathsf{X}$ we have $\|P^{km}(x, \cdot) - \pi\|_{f^{(m)}} \to 0$ as $k \to \infty$, then $\|P^k(x, \cdot) - \pi\|_f \to 0$ as $k \to \infty$.*

(iii) *If the m-skeleton is $f^{(m)}$-ergodic, then Φ itself is f-ergodic.*

PROOF   Under the conditions of (i) let $|g| \leq f$ and write any $n \in \mathbb{Z}_+$ as $n = km + i$ with $0 \leq i \leq m - 1$. Then

$$|P^n(x, g) - \pi(g)| = |P^{km}(x, P^i g) - \pi(P^i g)|$$
$$\leq \|P^{km}(x, \cdot) - \pi(\cdot)\|_{f^{(m)}}.$$

This proves (i) and the remaining results then follow.   □

This lemma and the ergodic theorems obtained for strongly aperiodic chains finally give the result we seek.

Theorem 14.3.3. *Suppose that* Φ *is positive recurrent and aperiodic.*

(i) *If* $\pi(f) = \infty$, *then* $P^k(x, f) \to \infty$ *for all* x.

(ii) *If* $\pi(f) < \infty$, *then the set* S_f *of* f-*regular sets is full and absorbing, and if* $x \in S_f$ *then* $\|P^k(x, \cdot) - \pi\|_f \to 0$, *as* $k \to \infty$.

(iii) *If* Φ *is* f-*regular, then* Φ *is* f-*ergodic. Conversely, if* Φ *is* f-*ergodic, then* Φ *restricted to a full absorbing set is* f-*regular.*

PROOF   Result (i) follows as in the proof of Proposition 14.3.1 (i).

If $\pi(f) < \infty$, then there exists a sequence of f-regular sets $\{S_f(n)\}$ whose union is full. By aperiodicity, for some m, the m-skeleton is strongly aperiodic and each of the sets $\{S_f(n)\}$ is $f^{(m)}$-regular. From Proposition 14.3.1 we see that the distributions of the m-skeleton converge in $f^{(m)}$-norm for initial $x \in S_f(n)$.

This and Lemma 14.3.2 proves (ii). The first part of (iii) is then a simple consequence; the converse is also immediate from (ii) since f-ergodicity implies $\pi(f) < \infty$.   □

Note that if Φ is f-ergodic then Φ may not be f-regular: this is already obvious in the case $f = 1$.

14.3.2   Sums of transition probabilities

We now refine the ergodic theorem Theorem 14.3.3 to give conditions under which the sum

$$\sum_{n=1}^{\infty} \|P^n(x, \cdot) - \pi\|_f \tag{14.23}$$

is finite.

The first result of this kind requires f-regularity of the initial probability measures λ, μ. For practical implementation, note that if (V3) holds for a petite set C and a function V, and if $\lambda(V) < \infty$, then from Theorem 14.2.3 (i) we see that the measure λ is f-regular.

Theorem 14.3.4. *Suppose* Φ *is an aperiodic positive Harris chain. If* $\pi(f) < \infty$, *then for any* f-*regular set* $C \in \mathcal{B}^+(\mathsf{X})$ *there exists* $M_f < \infty$ *such that for any* f-*regular initial distributions* λ, μ,

$$\sum_{n=1}^{\infty} \int \int \lambda(dx)\mu(dy)\|P^n(x, \cdot) - P^n(y, \cdot)\|_f \leq M_f(\lambda(V) + \mu(V) + 1) < \infty \tag{14.24}$$

where $V(\cdot) = G_C(\cdot, f)$.

PROOF   Consider first the strongly aperiodic case, and construct a split chain $\check{\Phi}$ using an f-regular set C. The theorem is valid from Theorem 14.1.3 for the split chain, since the split measures μ^*, λ^* are f-regular for $\check{\Phi}$. The bound on the sum can be taken as

$$\sum_{n=1}^{\infty} \int \int \lambda^*(dx)\mu^*(dy)\|\check{P}^n(x, \cdot) - \check{P}^n(y, \cdot)\|_f < M_f(\lambda^*(V) + \mu^*(V) + 1)$$

with $V = \check{G}_{C_0 \cup C_1}(\cdot, f)$, since $C_0 \cup C_1 \in \mathcal{B}^+(\check{X})$ is f-regular for the split chain.

Since the result is a total variation result it is then obviously valid when restricted to the original chain, as in (13.57). Using the identity

$$\int \lambda^*(dx)\check{G}_{C_0 \cup C_1}(x, f) = \int \lambda(dx)G_C(x, f),$$

and the analogous identity for μ, we see that the required bound holds in the strongly aperiodic case.

In the arbitrary aperiodic case we can apply Lemma 14.3.2 to move to a skeleton chain, as in the proof of Theorem 14.3.3.   □

The most interesting special case of this result is given in the following theorem.

Theorem 14.3.5. *Suppose* Φ *is an aperiodic positive Harris chain and that* π *is* f-*regular. Then* $\pi(f) < \infty$ *and for any* f-*regular set* $C \in \mathcal{B}^+(X)$ *there exists* $B_f < \infty$ *such that for any* f-*regular initial distribution* λ

$$\sum_{n=1}^{\infty} \|\lambda P^n - \pi\|_f \leq B_f(\lambda(V) + 1). \tag{14.25}$$

where $V(\cdot) = G_C(\cdot, f)$.   □

Our final f-ergodic result, for quite arbitrary positive recurrent chains is given for completeness in

Theorem 14.3.6.   (i) *If* Φ *is positive recurrent and if* $\pi(f) < \infty$, *then there exists a full set* S_f, *a cycle* $\{D_i : 1 \leq i \leq d\}$ *contained in* S_f, *and probabilities* $\{\pi_i : 1 \leq i \leq d\}$ *such that for any* $x \in D_r$,

$$\|P^{nd+r}(x, \cdot) - \pi_r\|_f \to 0, \qquad n \to \infty. \tag{14.26}$$

(ii) *If* Φ *is* f-*regular, then for all* x,

$$\|d^{-1} \sum_{r=1}^{d} P^{nd+r}(x, \cdot) - \pi\|_f \to 0, \qquad n \to \infty. \tag{14.27}$$

□

14.3.3   A criterion for finiteness of $\pi(f)$

From the Comparison Theorem 14.2.2 and the ergodic theorems presented above we also obtain the following criterion for finiteness of moments.

Theorem 14.3.7. *Suppose that Φ is positive recurrent with invariant probability π, and suppose that V, f and s are non-negative, finite-valued functions on X such that*

$$PV(x) \leq V(x) - f(x) + s(x)$$

for every $x \in X$. Then $\pi(f) \leq \pi(s)$.

PROOF   For π-a.e. $x \in X$ we have from the Comparison Theorem 14.2.2, Theorem 14.3.6 and (if $\pi(f) = \infty$) the aperiodic version of Theorem 14.3.3, whether or not $\pi(s) < \infty$,

$$\pi(f) = \lim_{N \to \infty} \frac{1}{N} \sum_{k=1}^{N} \mathsf{E}_x[f(\Phi_k)] \leq \lim_{N \to \infty} \frac{1}{N} \sum_{k=1}^{N} \mathsf{E}_x[s(\Phi_k)] = \pi(s).$$

$\square$

The criterion for $\pi(X) < \infty$ in Theorem 11.0.1 is a special case of this result. However, it seems easier to prove for quite arbitrary non-negative f, s using these limiting results.

14.4   *f*-Ergodicity of specific models

14.4.1   Random walk on $\mathbb{R}_+$ and storage models

Consider random walk on a half line given by $\Phi_n = [\Phi_{n-1} + W_n]^+$, and assume that the increment distribution Γ has negative first moment and a finite absolute moment $\sigma^{(k)}$ of order k.

Let us choose the test function $V(x) = x^k$. Then using the binomial expansion the drift Δ_V is given for $x > 0$ by

$$\begin{aligned}
\Delta V(x) &= \int_{-x}^{\infty} \Gamma(dy)(x+y)^k - x^k \\
&\leq \left(\int_{-x}^{\infty} \Gamma(dy)y \right) kx^{k-1} + c\sigma^{(k)}x^{k-2} + d
\end{aligned} \tag{14.28}$$

for some finite c, d. We can rewrite (14.28) in the form of (V3); namely for some $c' > 0$, and large enough x

$$\int P(x, dy)y^k \leq x^k - c'x^{k-1}.$$

From this we may prove the following

Proposition 14.4.1. *If the increment distribution Γ has mean $\beta < 0$ and finite $(k+1)^{st}$ moment, then the associated random walk on a half line is $|x|^k$-regular. Hence the process Φ admits a stationary measure π with finite moments of order k; and with $f_k(y) = y^k + 1$,*

(i) *for all* λ *such that* $\int \lambda(dx)x^{k+1} < \infty$,

$$\int \lambda(dx)\|P^n(x,\,\cdot\,) - \pi\|_{f_k} \to 0, \qquad n \to \infty;$$

(ii) *for some* $B_f < \infty$, *and any initial distribution* λ,

$$\sum_{n=0}^{\infty} \int \lambda(dx)\|P^n(x,\,\cdot\,) - \pi\|_{f_{k-1}} \leq B_f\Big(1 + \int x^k\,\lambda(dx)\Big).$$

PROOF   The calculations preceding the proposition show that for some $c_0 > 0$, $d_0 < \infty$, and a compact set $C \subset \mathbb{R}_+$,

$$PV_{i+1}(x) \leq V_{i+1}(x) - c_0 f_i(x) + d_0 \mathbb{1}_C(x) \qquad 0 \leq i \leq k, \tag{14.29}$$

where $V_j(x) = x^j$, $f_j(x) = x^j + 1$. Result (i) is then an immediate consequence of the *f*-Norm Ergodic Theorem.

To prove (ii) apply (14.29) with $i = k$ and Theorem 14.3.7 to conclude that $\pi(V_k) < \infty$. Applying (14.29) again with $i = k - 1$ we see that π is f_{k-1}-regular and then (ii) follows from the *f*-Norm Ergodic Theorem.   □

It is well known that the invariant measure for a random walk on the half line has moments of order one degree lower than those of the increment distribution, but this is a particularly simple proof of this result.

For the Moran dam model or the queueing models developed in Chapter 2, this result translates directly into a condition on the input distribution. Provided the mean input is less than the mean output between input times, then there is a finite invariant measure: and this has a finite k^{th} moment if the input distribution has finite $(k+1)^{st}$ moment.

14.4.2 Bilinear models

The random walk model in the previous section can be generalized in a variety of ways, as we have seen many times in the applications above.

For illustrative purposes we next consider the scalar bilinear model

$$X_{k+1} = \theta X_k + bW_{k+1}X_k + W_{k+1} \tag{14.30}$$

for which we proved boundedness in probability in Section 12.5.2. For simplicity, we take $\mathsf{E}[W] = 0$.

To obtain a solution to (V3), assume that W has finite variance. Then for the test function $V(x) = x^2$, we observe that by independence

$$\mathsf{E}[(X_{k+1})^2 \mid X_k = x] \leq \big[\theta^2 + b^2\mathsf{E}[W_{k+1}^2]\big]x^2 + (2bx+1)\mathsf{E}[W_{k+1}^2]. \tag{14.31}$$

Since this V is a coercive function on $\mathbb{R}$, it follows that (V3) holds with the choice of

$$f(x) = 1 + \delta V(x)$$

for some $\delta > 0$ provided

$$\theta^2 + b^2 \mathsf{E}[W_k^2] < 1. \tag{14.32}$$

Under this condition it follows just as in the LSS(F) model that provided the noise process forces this model to be a T-chain (for example, if the conditions of Proposition 7.1.3 hold) then (14.32) is a condition not just for positive Harris recurrence, but for the existence of a second order stationary model with finite variance: this is precisely the interpretation of $\pi(f) < \infty$ in this case.

A more general version of this result is

Proposition 14.4.2. *Suppose that (SBL1) and (SBL2) hold and*

$$\mathsf{E}[W_n^k] < \infty. \tag{14.33}$$

Then the bilinear model is positive Harris, the invariant measure π also has finite k^{th} moments (that is, satisfies $\int x^k \pi(dx) < \infty$), and

$$\|P^n(x, \cdot) - \pi\|_{x^k} \to 0, \qquad n \to \infty.$$

$\square$

In the next chapter we will show that there is in fact a geometric rate of convergence in this result. This will show that, in essence, the same drift condition gives us finiteness of moments in the stationary case, convergence of time-dependent moments and some conclusion about the rate at which the moments become stationary.

14.5   A key renewal theorem

One of the most interesting applications of the ergodic theorems in these last two chapters is a probabilistic proof of the Key Renewal Theorem.

As in Section 3.5.3, let $Z_n := \sum_{i=0}^n Y_i$, where $\{Y_1, Y_2, \ldots\}$ is a sequence of independent and identical random variables with distribution Γ on $\mathbb{R}_+$, and Y_0 is a further independent random variable with distribution Γ_0 also on $\mathbb{R}_+$; and let $U(\cdot) = \sum_{n=0}^\infty \Gamma^{n*}(\cdot)$ be the associated renewal measure.

Renewal theorems concern the limiting behavior of U; specifically, they concern conditions under which

$$\Gamma_0 * U * f(t) \to \beta^{-1} \int_0^\infty f(s) \, ds \tag{14.34}$$

as $t \to \infty$, where $\beta = \int_0^\infty s\Gamma(ds)$ and f and Γ_0 are an appropriate function and measure respectively.

With minimal assumptions about Γ we have *Blackwell's Renewal Theorem.*

Theorem 14.5.1. *Provided Γ has a finite mean β and is not concentrated on a lattice $nh, n \in \mathbb{Z}_+, h > 0$, then for any interval $[a, b]$ and any initial distribution Γ_0*

$$\Gamma_0 * U[a + t, b + t] \to \beta^{-1}(b - a), \qquad t \to \infty. \tag{14.35}$$

PROOF   This result is taken from Feller ([115], p. 360) and its proof is not one we pursue here. We do note that it is a special case of the general Key Renewal Theorem, which states that under these conditions on Γ, (14.34) holds for all bounded non-negative functions f which are *directly Riemann integrable*, for which again see Feller ([115], p. 361); for then (14.35) is the special case with $f(s) = \mathbb{I}_{[a,b]}(s)$.   □

This result shows us the pattern for renewal theorems: in the limit, the measure U approximates normalized Lebesgue measure.

We now show that one can trade off properties of Γ against properties of f (and to some extent properties of Γ_0) in asserting (14.34). We shall give a proof, based on the ergodic properties we have been considering for Markov chains, of the following Uniform Key Renewal Theorem.

Theorem 14.5.2. *Suppose that Γ has a finite mean β and is spread out (as defined in (RW2)).*

(i) *For any initial distribution Γ_0 we have the uniform convergence*

$$\lim_{t \to \infty} \sup_{|g| \leq f} \left| \Gamma_0 * U * g(t) - \beta^{-1} \int_0^\infty g(s)ds \right| = 0 \qquad (14.36)$$

provided the function $f \geq 0$ satisfies

$$f \quad \text{is bounded;} \qquad (14.37)$$
$$f \quad \text{is Lebesgue integrable;} \qquad (14.38)$$
$$f(t) \quad \to 0, \qquad t \to \infty. \qquad (14.39)$$

(ii) *In particular, for any bounded interval $[a, b]$ and Borel sets B*

$$\lim_{t \to \infty} \sup_{B \subseteq [a,b]} \left| \Gamma_0 * U(t + B) - \beta^{-1} \mu^{\text{Leb}}(B) \right| = 0. \qquad (14.40)$$

(iii) *For any initial distribution Γ_0 which is absolutely continuous, the convergence (14.36) holds for f satisfying only (14.37) and (14.38).*

PROOF   The proof of this set of results occupies the remainder of this section, and contains a number of results of independent interest.   □

Before embarking on this proof, we note explicitly that we have accomplished a number of tradeoffs in this result, compared with the Blackwell Renewal Theorem. By considering spread-out distributions, we have exchanged the direct Riemann integrability condition for the simpler and often more verifiable smoothness conditions (14.37)-(14.39). This is exemplified by the fact that (14.40) allows us to consider the renewal measure of any bounded Borel set, whereas the general Γ version restricts us to intervals as in (14.35). The extra benefits of smoothness of Γ_0 in removing (14.39) as a condition are also in this vein.

Moreover, by moving to the class of spread-out distributions, we have introduced a uniformity into the Key Renewal Theorem which is analogous in many ways to the total variation norm result in Markov chain limit theory. This analogy is not coincidental:

as we now show, these results are all consequences of precisely that total variation convergence for the forward recurrence time chain associated with this renewal process.

Recall from Section 3.5.3 the forward recurrence time process

$$V^+(t) := \inf(Z_n - t : Z_n \geq t), \qquad t \geq 0.$$

We will consider the forward recurrence time δ-skeleton $\boldsymbol{V}_\delta^+ = V^+(n\delta)$, $n \in \mathbb{Z}_+$ for that process, and denote its n-step transition law by $P^{n\delta}(x, \cdot)$. We showed that for sufficiently small δ, when Γ is spread out, then (Proposition 5.3.3) the set $[0, \delta]$ is a small set for $\boldsymbol{V}_\delta^+$, and (Proposition 5.4.7) $\boldsymbol{V}_\delta^+$ is also aperiodic.

It is trivial for this chain to see that (V2) holds with $V(x) = x$, so that the chain is regular from Theorem 11.3.15, and if Γ_0 has a finite mean, then Γ_0 is regular from Theorem 11.3.12.

This immediately enables us to assert from Theorem 13.4.4 that, if Γ_1, Γ_2 are two initial measures both with finite mean, and if Γ itself is spread out with finite mean,

$$\sum_{n=0}^{\infty} \|\Gamma_1 P^{n\delta}(\cdot) - \Gamma_2 P^{n\delta}(\cdot)\| < \infty. \tag{14.41}$$

The crucial corollary to this example of Theorem 13.4.4, which leads to the Uniform Key Renewal Theorem is

Proposition 14.5.3. *If Γ is spread out with finite mean, and if Γ_1, Γ_2 are two initial measures both with finite mean, then*

$$\|\Gamma_1 * U - \Gamma_2 * U\| := \int_0^\infty |\Gamma_1 * U(dt) - \Gamma_2 * U(dt)| < \infty. \tag{14.42}$$

PROOF   By interpreting the measure $\Gamma_0 P^s$ as an initial distribution, observe that for $A \subseteq [t, \infty)$, and fixed $s \in [0, t)$, we have from the Markov property at s the identity

$$\Gamma_0 * U(A) = \Gamma_0 P^s * U(A - s). \tag{14.43}$$

Using this we then have

$$
\begin{aligned}
&\int_0^\infty |\Gamma_1 * U(dt) - \Gamma_2 * U(dt)| \\
&= \quad \sum_{n=0}^\infty \int_{[n\delta,(n+1)\delta)} |\Gamma_1 * U(dt) - \Gamma_2 * U(dt)| \\
&= \quad \sum_{n=0}^\infty \int_{[0,\delta)} |(\Gamma_1 P^{n\delta} - \Gamma_2 P^{n\delta}) * U(dt)| \\
&\leq \quad \sum_{n=0}^\infty \int_{[0,\delta)} \int_{[0,t]} |(\Gamma_1 P^{n\delta} - \Gamma_2 P^{n\delta})(du)| U(dt - u) \\
&\leq \quad \sum_{n=0}^\infty \int_{[0,\delta)} |(\Gamma_1 P^{n\delta} - \Gamma_2 P^{n\delta})(du)| U[0, \delta) \\
&\leq \quad U[0, \delta) \sum_{n=0}^\infty \|\Gamma_1 P^{n\delta} - \Gamma_2 P^{n\delta}\|
\end{aligned}
\tag{14.44}
$$

which is finite from (14.41). $\square$

From this we can prove a precursor to Theorem 14.5.2.

Proposition 14.5.4. *If Γ is spread out with finite mean, and if Γ_1, Γ_2 are two initial measures both with finite mean, then*

$$\sup_{|g| \leq f} |\Gamma_1 * U * g(t) - \Gamma_2 * U * g(t)| \to 0, \qquad t \to \infty \tag{14.45}$$

for any f satisfying (14.37)-(14.39).

PROOF   Suppose that ε is arbitrarily small but fixed. Using Proposition 14.5.3 we can fix T such that

$$\int_T^\infty |(\Gamma_1 * U - \Gamma_2 * U)(du)| \leq \varepsilon. \tag{14.46}$$

If f satisfies (14.39), then for all sufficiently large t,

$$f(t - u) \leq \varepsilon, \qquad u \in [0, T];$$

for such a t, writing $d = \sup f(x) < \infty$ from (14.37), it follows that for any g with $|g| \leq f$,

$$
\begin{aligned}
|\Gamma_1 * U * g(t) - \Gamma_2 * U * g(t)| \ &\leq\ \int_0^T |(\Gamma_1 * U - \Gamma_2 * U(du)| f(t - u) \\
&\quad + \int_T^t |(\Gamma_1 * U - \Gamma_2 * U)(du)| f(t - u) \\
&\leq\ \varepsilon \| \Gamma_1 * U - \Gamma_2 * U \| + \varepsilon d \\
&:=\ \varepsilon'
\end{aligned}
\tag{14.47}
$$

which is arbitrarily small, from (14.44), thus proving the result.   □

This would prove Theorem 14.5.2 (a) if the equilibrium measure

$$\Gamma_e[0, t] = \beta^{-1} \int_0^t \Gamma(u, \infty) du$$

defined in (10.36) were itself regular, since we have that $\Gamma_e * U(\cdot) = \beta^{-1} \mu^{\text{Leb}}(\cdot)$, which gives the right hand side of (14.36). But as can be verified by direct calculation, Γ_e is regular if and only if Γ has a finite second moment, exactly as is the case in Theorem 13.4.5 for general chains with atoms.

However, we can reach the following result, of which Theorem 14.5.2 (a) is a corollary, using a truncation argument.

Proposition 14.5.5. *If Γ is spread out with finite mean, and if Γ_1, Γ_2 are any two initial measures, then*

$$\sup_{|g| \leq f} |\Gamma_1 * U * g(t) - \Gamma_2 * U * g(t)| \to 0, \qquad t \to \infty$$

for any f satisfying (14.37)-(14.39).

PROOF   For fixed v, let $\Gamma^v(A) := \Gamma(A)/\Gamma[0,v]$ for all $A \subseteq [0,v]$ denote the truncation of $\Gamma(A)$ to $[0,v]$.

For any g with $|g| \le f$,

$$|\Gamma_1 * U * g(t) - \Gamma_1^v * U * g(t)| \le \|\Gamma_1 - \Gamma_1^v\| \sup_x U * f(x) \qquad (14.48)$$

which can be made smaller than ε by choosing v large enough, provided $\sup_x U * f(x) < \infty$. But if $t > T$, from (14.47), with $\Gamma_1 = \delta_0$, $\Gamma_2 = \Gamma_e^v$ and $g = f$,

$$
\begin{aligned}
U * f(t) &= \delta_0 * U * f(t) \\[2mm]
&\le \Gamma_e^v * U * f(t) + \varepsilon' \\[2mm]
&\le \left(\Gamma_e[0,v]\right)^{-1} \Gamma_e * U * f(t) + \varepsilon' \\[2mm]
&\le \left(\Gamma_e[0,v]\right)^{-1} \beta^{-1} \int_0^\infty f(u)du + \varepsilon'
\end{aligned}
\qquad (14.49)
$$

which is indeed finite, by (14.38).

The result then follows from Proposition 14.5.4 and (14.48) by a standard triangle inequality argument.   □

Theorem 14.5.2 (b) is a simple consequence of Theorem 14.5.2 (a), but to prove Theorem 14.5.2 (c), we need to refine the arguments above a little.

Suppose that (14.39) does not hold, and write

$$A_\varepsilon(t) := \{u \in [0,T] : f(t-u) \ge \varepsilon\},$$

where ε and T are as in (14.46). We then have

$$
\begin{aligned}
\int_0^T &|(\Gamma_1 * U - \Gamma_2 * U)(du)| f(t-u) \\[2mm]
&\le \int_0^T |(\Gamma_1 * U - \Gamma_2 * U(du)| f(t-u) \mathbb{I}_{[A_\varepsilon(t)]^c}(u) \\[2mm]
&\quad + \int_0^T (\Gamma_1 * U + \Gamma_2 * U)(du) f(t-u) \mathbb{I}_{A_\varepsilon(t)}(u) \\[2mm]
&\le \varepsilon \|\Gamma_1 * U - \Gamma_2 * U\| + d(\Gamma_1 + \Gamma_2) * U(A_\varepsilon(t)).
\end{aligned}
\qquad (14.50)
$$

If we now assume that the measure $\Gamma_1 + \Gamma_2$ to be absolutely continuous with respect to μ^{Leb}, then, so is $(\Gamma_1 + \Gamma_2) * U$ ([115], p. 146).

Now since f is integrable, as $t \to \infty$ for fixed T, ε we must have $\mu^{\text{Leb}}(A_\varepsilon(t)) \to 0$. But since T is fixed, we have that both $\mu^{\text{Leb}}[0,T] < \infty$ and $(\Gamma_1 + \Gamma_2) * U[0,T] < \infty$, and it is a standard result of measure theory ([152], p. 125) that

$$(\Gamma_1 + \Gamma_2) * U(A_\varepsilon(t)) \to 0, \qquad t \to \infty.$$

We can thus make the last term in (14.50) arbitrarily small for large t, even without assuming (14.39); now reconsidering (14.47), we see that Proposition 14.5.4 holds without (14.39), provided we assume the existence of densities for Γ_1 and Γ_2, and then Theorem 14.5.2 (c) follows by the truncation argument of Proposition 14.5.5.

14.6   Commentary*

These results are largely recent. Although the question of convergence of $\mathsf{E}_x[f(\Phi_k)]$ for general f occurs in, for example, Markov reward models [25], most of the literature on Harris chains has concentrated on convergence only for $f \leq 1$ as in the previous chapter. The results developed here are a more complete form of those in Meyn and Tweedie [277], but there the general aperiodic case was not developed: only the strongly aperiodic case is considered in detail. A more embryonic form of the convergence in f-norm, indicating that if $\pi(f) < \infty$ then $\mathsf{E}_x[f(\Phi_k)] \to \pi(f)$, appeared as Theorem 2 of Tweedie [400].

Nummelin [303] considers f-regularity, but does not go on to apply the resulting concepts to f-ergodicity, although in fact there are connections between the two which are implicit through the Regenerative Decomposition in Nummelin and Tweedie [307].

That Theorem 14.1.1 admits a converse, so that when $\pi(f) < \infty$ there exists a sequence of f-regular sets $\{S_f(n)\}$ whose union is full, is surprisingly deep. For general state space chains, the question of the existence of f-regular sets requires the splitting technique as did the existence of regular sets in Chapter 11. The key to their use in analyzing chains which are not strongly aperiodic lies in the duality with the drift condition (V3), and this is given here for the first time.

The fact that (V3) gives a criterion for finiteness of $\pi(f)$ was observed in Tweedie [400]. Its use for asserting the second order stationarity of bilinear and other time series models was developed in Feigin and Tweedie [111], and for analyzing random walk in [401]. Related results on the existence of moments are also in Kalashnikov [188].

The application to the generalized Key Renewal Theorem is particularly satisfying. By applying the ergodic theorems above to the forward recurrence time chain V_δ^+, we have "leveraged" from the discrete time renewal theory results of Section 13.2 to the continuous time ones through the general Markov chain results. This Markovian approach was developed in Arjas et al. [8], and the uniformity in Theorem 14.5.2, which is a natural consequence of this approach, seems to be new there. The simpler form without the uniformity, showing that one can exchange spread-outness of Γ for the weaker conditions on f dates back to the original renewal theorems of Smith [361, 362, 363], whilst Breiman [47] gives a form of Theorem 14.5.2 (b). An elegant and different approach is also possible through Stone's Decomposition of U [374], which shows that when Γ is spread out,

$$U = U_f + U_c$$

where U_f is a finite measure, and U_c has a density p with respect to μ^{Leb} satisfying $p(t) \to \beta^{-1}$ as $t \to \infty$.

The convergence, or rather summability, of the quantities

$$\|P^n(x, \cdot) - \pi\|_f$$

leads naturally to a study of rates of convergence, and this is carried out in Nummelin and Tuominen [306]. Building on this, Tweedie [401] uses similar approaches to those in this chapter to derive drift criteria for more subtle rate of convergence results: the interested reader should note the result of Theorem 3 (iii) of [401]. There it is shown (essentially by using the Comparison Theorem) that if (V3) holds for a function f such that

$$f(x) \geq \mathsf{E}_x[r(\tau_C)], \qquad x \in C^c$$

where $r(n)$ is some function on $\mathbb{Z}_+$, then

$$V(x) \geq \mathsf{E}_x[r^0(\tau_C)], \qquad x \in C^c$$

where $r^0(n) = \sum_1^n r(j)$. If C is petite, then this is (see [306] or Theorem 4 (iii) of [401]) enough to ensure that

$$r(n)\|P^n(x, \cdot) - \pi\| \to 0, \qquad n \to \infty$$

so that (V3) gives convergence at rate $r(n)^{-1}$ in the ergodic theorem.

Applications of these ideas to the Key Renewal Theorem are also contained in [306]. The special case of $r(n) = r^n$ is explored thoroughly in the next two chapters. The rate results above are valuable also in the case of $r(n) = n^k$ since then $r^0(n)$ is asymptotically n^{k+1}. This allows an inductive approach to the level of convergence rate achieved; but this more general topic is not pursued in this book. The interested reader will find the most recent versions, building on those of Nummelin and Tuominen [306], in [393].

Commentary for the second edition: Several topics in this chapter have been extended, or refined in specific applications, since publication of the first edition.

f-Regularity in queueing networks is the subject of [81, 264, 268, 266] – see also the monograph [267]. The Comparison Theorem 14.2.2 is implicit in the stability analysis of Tassiulas's MaxWeight scheduling algorithm, now popular for routing and scheduling in queueing networks [383, 137, 382, 268, 266, 267], and a version of Theorem 14.2.2 is used in [145] in an early "heavy traffic" analysis of a queueing network. The Comparison Theorem is also a component of the approach to network stability and performance approximation developed in [273, 226, 223, 30, 31, 267]. In [81] the assumptions of [393] are verified, provided an associated fluid model for the network is stable. This establishes f-regularity for the network for polynomial f, as well as polynomial rates of convergence in the f-Norm Ergodic Theorem 14.0.1.

Theory surrounding f-regularity is applied in the theory of controlled Markov models (Markov decision processes, or MDPs) in [262, 261, 67, 263, 42, 267]. In particular, [42] characterizes a notion of uniform f-regularity for MDPs.

Recently, Jarner and Roberts introduced a new drift criterion that can be used to simplify the verification of polynomial rates of convergence [180]. Extensions of this approach as well as explicit bounds on the rate of convergence are obtained in [126, 100].

The drift criterion of [180] can be expressed as an intermediate between the drift criteria (V3) and (V4):

Drift criterion of Jarner and Roberts

(V4$^\varrho$)   There exists an extended-real-valued function $V : \mathsf{X} \to [1, \infty]$, a measurable set C, and constants $\beta > 0$, $\varrho > 0$, $b < \infty$, satisfying

$$\Delta V(x) \leq -\beta V^\varrho(x) + b\mathbb{I}_C(x), \quad x \in \mathsf{X}. \tag{14.51}$$

For example, if the inter-arrival times in the GI/M/1 queue possess a finite nth moment, then (V4$^\varrho$) holds with $V(x) = 1 + x^n$ and $\varrho = 1 - n^{-1}$.

We consider the special case $\varrho = \frac{1}{2}$ to illustrate the application of (V4$^\varrho$):

Proposition 14.6.1. *Suppose that the chain* Φ *is* ψ-*irreducible and aperiodic, and that the drift condition (V4$^\varrho$) holds for some extended-real-valued function V satisfying $V(x_0) < \infty$ for some $x_0 \in \mathsf{X}$, with C petite, and $\varrho = \frac{1}{2}$. Then there exists a finite constant B_1 such that for all $x \in S_V$,*

$$\sum_{n=0}^{\infty} \|P^n(x, \cdot) - \pi\| \le B_1 \sqrt{V(x)}. \tag{14.52}$$

PROOF   We establish the assumptions of part (iii) of the f-Norm Ergodic Theorem 14.0.1, with $f \equiv 1$. For this it is sufficient to show that the function $U := 2\beta^{-1}V^{\frac{1}{2}}$ satisfies Foster's criterion, and that $\pi(U) < \infty$.

Finiteness of $\pi(V^{\frac{1}{2}})$ follows from the assumed drift condition and the Comparison Theorem, which gives the explicit bound $\pi(V^{\frac{1}{2}}) \le \beta^{-1}b\pi(C)$.

To show that Foster's criterion is satisfied we begin with an application of Jensen's inequality:

$$PV^{\frac{1}{2}}(x) \le \sqrt{PV(x)} \le \sqrt{V(x) - \beta V^{\frac{1}{2}}(x) + b\mathbb{I}_C(x)}.$$

Concavity of the square root gives the bound $\sqrt{1 + x} \le 1 + \frac{1}{2}x$ for all x. Combining this with the previous bound we obtain

$$PV^{\frac{1}{2}}(x) \le V^{\frac{1}{2}}(x)\sqrt{1 + \frac{-\beta V^{\frac{1}{2}}(x) + b\mathbb{I}_C(x)}{V(x)}}$$

$$\le V^{\frac{1}{2}}(x)\left[1 + \frac{1}{2}\frac{-\beta V^{\frac{1}{2}}(x) + b\mathbb{I}_C(x)}{V(x)}\right]$$

$$= V^{\frac{1}{2}}(x) + \frac{1}{2}\frac{-\beta V^{\frac{1}{2}}(x) + b\mathbb{I}_C(x)}{V^{\frac{1}{2}}(x)}.$$

Multiplying each side by $2\beta^{-1}$ gives Foster's criterion, with Lyapunov function $U = 2\beta^{-1}V^{\frac{1}{2}}$,

$$\Delta U \le -1 + \beta^{-1}\frac{1}{V^{\frac{1}{2}}(x)}b\mathbb{I}_C(x) \le -1 + \beta^{-1}b\mathbb{I}_C(x),$$

where the second inequality follows from the assumption $V \ge 1$. $\square$

Chapter 15

Geometric ergodicity

The previous two chapters have shown that for positive Harris chains, convergence of $\mathsf{E}_x[f(\Phi_k)]$ is guaranteed from almost all initial states x provided only $\pi(f) < \infty$. Strong though this is, for many models used in practice even more can be said: there is often a rate of convergence ρ such that

$$\|P^n(x, \cdot) - \pi\|_f = o(\rho^n)$$

where the rate $\rho < 1$ can be chosen essentially independent of the initial point x.

The purpose of this chapter is to give conditions under which convergence takes place at such a uniform geometric rate. Because of the power of the final form of these results, and the wide range of processes for which they hold (which include many of those already analyzed as ergodic) it is not too strong a statement that this "geometrically ergodic" context constitutes the most useful of all of those we present, and for this reason we have devoted two chapters to this topic.

The following result summarizes the highlights of this chapter, where we focus on bounds such as (15.4) and the strong relationship between such bounds and the drift criterion given in (15.3). In Chapter 16 we will explore a number of examples in detail, and describe techniques for moving from ergodicity to geometric ergodicity. The development there is based primarily on the results of this chapter, and also on an interpretation of the geometric convergence (15.4) in terms of convergence of the kernels $\{P^k\}$ in a certain induced operator norm.

Theorem 15.0.1 (Geometric Ergodic Theorem). *Suppose that the chain Φ is ψ-irreducible and aperiodic. Then the following three conditions are equivalent:*

(i) *The chain Φ is positive recurrent with invariant probability measure π, and there exists some ν-petite set $C \in \mathcal{B}^+(\mathsf{X})$, $\rho_C < 1$, $M_C < \infty$, and $P^\infty(C) > 0$ such that for all $x \in C$*

$$|P^n(x, C) - P^\infty(C)| \le M_C \rho_C^n. \tag{15.1}$$

(ii) *There exists some petite set $C \in \mathcal{B}(\mathsf{X})$ and $\kappa > 1$ such that*

$$\sup_{x \in C} \mathsf{E}_x[\kappa^{\tau_C}] < \infty. \tag{15.2}$$

(iii) *There exists a petite set C, constants $b < \infty$, $\beta > 0$ and a function $V \geq 1$ finite at some one $x_0 \in \mathsf{X}$ satisfying*

$$\Delta V(x) \leq -\beta V(x) + b \mathbb{I}_C(x), \qquad x \in \mathsf{X}. \tag{15.3}$$

Any of these three conditions imply that the set $S_V = \{x : V(x) < \infty\}$ is absorbing and full, where V is any solution to (15.3) satisfying the conditions of (iii), and there then exist constants $r > 1$, $R < \infty$ such that for any $x \in S_V$

$$\sum_n r^n \|P^n(x, \cdot) - \pi\|_V \leq RV(x). \tag{15.4}$$

PROOF   The equivalence of the local geometric rate of convergence property in (i) and the self-geometric recurrence property in (ii) will be shown in Theorem 15.4.3.

The equivalence of the self-geometric recurrence property and the existence of solutions to the drift equation (15.3) is completed in Theorems 15.2.6 and 15.2.4. It is in Theorem 15.4.1 that this is shown to imply the geometric nature of the V-norm convergence in (15.4), while the upper bound on the right hand side of (15.4) follows from Theorem 15.3.3.   □

The notable points of this result are that we can use the same function V in (15.4), which leads to the operator norm results in the next chapter; and that the rate r in (15.4) can be chosen independently of the initial starting point.

We initially discuss conditions under which there exists for some $x \in \mathsf{X}$ a rate $r > 1$ such that

$$\|P^n(x, \cdot) - \pi\|_f \leq M_x r^{-n} \tag{15.5}$$

where $M_x < \infty$. Notice that we have introduced f-norm convergence immediately: it will turn out that the methods are not much simplified by first considering the case of bounded f. We also have another advantage in considering geometric rates of convergence compared with the development of our previous ergodicity results. We can exploit the useful fact that (15.5) is equivalent to the requirement that for some $\bar{r}$, $\bar{M}_x$,

$$\sum_n \bar{r}^n \|P^n(x, \cdot) - \pi\|_f \leq \bar{M}_x. \tag{15.6}$$

Hence it is without loss of generality that we will immediately move also to consider the summed form as in (15.6) rather than the n-step convergence as in (15.5).

---

f-Geometric ergodicity

We shall call Φ *f-geometrically ergodic*, where $f \geq 1$, if Φ is positive Harris with $\pi(f) < \infty$ and there exists a constant $r_f > 1$ such that

$$\sum_{n=1}^{\infty} r_f^n \|P^n(x, \cdot) - \pi\|_f < \infty \tag{15.7}$$

for all $x \in \mathsf{X}$. If (15.7) holds for $f \equiv 1$, then we call Φ *geometrically ergodic*.

The development in this chapter follows a pattern similar to that of the previous two chapters: first we consider chains which possess an atom, then move to aperiodic chains via the Nummelin splitting.

This pattern is now well established: but in considering geometric ergodicity, the extra complexity in introducing both unbounded functions f and exponential moments of hitting times leads to a number of different and sometimes subtle problems. These make the proofs a little harder in the case without an atom than was the situation with either ergodicity or f-ergodicity. However, the final conclusion in (15.4) is well worth this effort.

15.1   Geometric properties: chains with atoms

15.1.1   Using the regenerative decomposition

Suppose in this section that Φ is a positive Harris recurrent chain and that we have an accessible atom α in $\mathcal{B}^+(\mathsf{X})$: as in the previous chapter, we do not consider completely countable spaces separately, as one atom is all that is needed. We will again use the Regenerative Decomposition (13.48) to identify the bounds which will ensure that the chain is f-geometrically ergodic.

Multiplying (13.48) by r^n and summing, we have that

$$\sum_n \|P^n(x, \cdot) - \pi\|_f \, r^n$$

is bounded by the three sums

$$\sum_{n=1}^\infty \int {}_\alpha P^n(x, dw) f(w) \, r^n,$$

$$\pi(\alpha) \sum_{n=1}^\infty \sum_{j=n+1}^\infty t_f(j) \, r^n, \tag{15.8}$$

$$\sum_{n=1}^\infty |a_x * u - \pi(\alpha)| * t_f(n) \, r^n.$$

Now using Lemma D.7.2 and recalling that $t_f(n) = \int {}_\alpha P^n(\alpha, dw) f(w)$, we have that the three sums in (15.8) can be bounded individually through

$$\sum_{n=1}^\infty \int {}_\alpha P^n(x, dw) f(w) r^n \;\leq\; \mathsf{E}_x\Big[\sum_{n=1}^{\tau_\alpha} f(\Phi_n) r^n\Big], \tag{15.9}$$

$$\pi(\alpha) \sum_{n=1}^\infty \sum_{j=n+1}^\infty t_f(j) r^n \;\leq\; \frac{r}{r-1} \mathsf{E}_\alpha\Big[\sum_{n=1}^{\tau_\alpha} f(\Phi_n) r^n\Big], \tag{15.10}$$

$$\sum_{n=1}^{\infty} |a_x * u - \pi(\alpha)| * t_f(n)r^n$$

$$= \left(\sum_{n=1}^{\infty} |a_x * u(n) - \pi(\alpha)|r^n\right)\left(\sum_{n=1}^{\infty} t_f(n)r^n\right) \tag{15.11}$$

$$= \left(\sum_{n=1}^{\infty} |a_x * u(n) - \pi(\alpha)|r^n\right)\left(\mathsf{E}_\alpha\left[\sum_{n=1}^{\tau_\alpha} f(\Phi_n)r^n\right]\right).$$

In order to bound the first two sums (15.9) and (15.10), and the second term in the third sum (15.11), we will require an extension of the notion of regularity, or more exactly of f-regularity. For fixed $r \geq 1$ recall the generating function defined in (8.21) for $r < 1$ by

$$U_\alpha^{(r)}(x, f) := \mathsf{E}_x\left[\sum_{n=1}^{\tau_\alpha} f(\Phi_n)r^n\right]; \tag{15.12}$$

clearly this is defined but possibly infinite for $r \geq 1$. From the inequalities (15.9)–(15.11) above it is apparent that when Φ admits an accessible atom, establishing f-geometric ergodicity will require finding conditions such that $U_\alpha^{(r)}(x, f)$ is finite for some $r > 1$.

The first term in the right hand side of (15.11) can be reduced further. Using the fact that

$$|a_x * u(n) - \pi(\alpha)| = |a_x * (u - \pi(\alpha))(n) - \pi(\alpha) \sum_{j=n+1}^{\infty} a_x(j)|$$

$$< a_x * |(u - \pi(\alpha))|(n) + \pi(\alpha) \sum_{j=n+1}^{\infty} a_x(j)$$

and again applying Lemma D.7.2, we find the bound

$$\sum_{n=1}^{\infty} |a_x * u - \pi(\alpha)|r^n \leq \left(\sum_{n=1}^{\infty} a_x(n)r^n\right)\left(\sum_{n=1}^{\infty} |u(n) - \pi(\alpha)|r^n\right)$$

$$+ \pi(\alpha) \sum_{n=1}^{\infty} \sum_{j=n+1}^{\infty} a_x(j)r^n$$

$$\leq \left(\mathsf{E}_x[r^{\tau_\alpha}]\right)\left(\sum_{n=1}^{\infty} |u(n) - \pi(\alpha)|r^n\right) + \frac{r}{r-1}\mathsf{E}_x[r^{\tau_\alpha}].$$

Thus from (15.9)–(15.11) we might hope to find that convergence of P^n to π takes place at a geometric rate provided

(i) the atom itself is geometrically ergodic, in the sense that

$$\sum_{n=1}^{\infty} |u(n) - \pi(\alpha)|r^n$$

converges for some $r > 1$;

(ii) the distribution of τ_α possess an "f-modulated" geometrically decaying tail from both α and from the initial state x, in the sense that both $U_\alpha^{(r)}(\alpha, f) < \infty$ and

$U_\alpha^{(r)}(x, f) < \infty$ for some $r = r_x > 1$: and if we can choose such an r independent of x then we will be able to assert that the overall rate of convergence in (15.4) is also independent of x.

We now show that as with ergodicity or f-ergodicity, a remarkable degree of solidarity in this analysis is indeed possible.

15.1.2   Kendall's renewal theorem

As in the ergodic case, we need a key result from renewal theory. Kendall's Theorem shows that for atoms, geometric ergodicity and geometric decay of the tails of the return time distribution are actually equivalent conditions.

Theorem 15.1.1 (Kendall's Theorem). *Let $u(n)$ be an ergodic renewal sequence with increment distribution $p(n)$, and write $u(\infty) = \lim_{n \to \infty} u(n)$. Then the following three conditions are equivalent:*

(i) *There exists $r_0 > 1$ such that the series*

$$U_0(z) := \sum_{n=0}^{\infty} |u(n) - u(\infty)| z^n \qquad (15.13)$$

converges for $|z| < r_0$.

(ii) *There exists $r_0 > 1$ such that the function $U(z)$ defined on the complex plane for $|z| < 1$ by*

$$U(z) := \sum_{n=0}^{\infty} u(n) z^n$$

has an analytic extension in the disc $\{|z| < r_0\}$ except for a simple pole at $z = 1$.

(iii) *There exists $\kappa > 1$ such that the series $P(z)$*

$$P(z) := \sum_{n=0}^{\infty} p(n) z^n \qquad (15.14)$$

converges for $\{|z| < \kappa\}$.

PROOF   Assume that (i) holds. Then by construction the function $F(z)$ defined on the complex plane by

$$F(z) := \sum_{n=0}^{\infty} (u(n) - u(n-1)) z^n$$

has no singularities in the disc $\{|z| < r_0\}$, and since

$$F(z) = (1 - z)U(z), \qquad |z| < 1, \qquad (15.15)$$

we have that $U(z)$ has no singularities in the disc $\{|z| < r_0\}$ except a simple pole at $z = 1$, so that (ii) holds.

Conversely suppose that (ii) holds. We can then also extend $F(z)$ analytically in the disc $\{|z| < r_0\}$ using (15.15). As the Taylor series expansion is unique, necessarily $F(z) = \sum_{n=0}^{\infty}(u(n) - u(n-1))z^n$ throughout this larger disc, and so by virtue of Cauchy's inequality

$$\sum_n |u(n) - u(n-1)|r^n < \infty, \qquad r < r_0.$$

Hence from Lemma D.7.2

$$\infty > \sum_n \sum_{m \geq n} |u(m+1) - u(m)|r^n$$

$$\geq \sum_n |\sum_{m \geq n} (u(m+1) - u(m))|r^n$$

$$= \sum_n |u(\infty) - u(n)|r^n$$

so that (i) holds.

Now suppose that (iii) holds. Since $P(z)$ is analytic in the disc $\{|z| < \kappa\}$, for any $\varepsilon > 0$ there are at most finitely many values of z such that $P(z) = 1$ in the smaller disc $\{|z| < \kappa - \varepsilon\}$.

By aperiodicity of the sequence $\{p(n)\}$, we have $p(n) > 0$ for all $n > N$ for some N, from Lemma D.7.4. This implies that for $z \neq 1$ on the unit circle $\{|z| = 1\}$, we have

$$\sum_N^{\infty} p(n)\operatorname{Re}(z^n) < \sum_N^{\infty} p(n),$$

so that

$$\operatorname{Re} P(z) \leq \sum_0^{\infty} p(n)\operatorname{Re}(z^n) < \sum_0^{\infty} p(n) = 1.$$

Consequently only one of these roots, namely $z = 1$, lies on the unit circle, and hence there is some r_0 with $1 < r_0 \leq \kappa$ such that $z = 1$ is the only root of $P(z) = 1$ in the disc $\{|z| < r_0\}$.

Moreover this is a simple root at $z = 1$, since

$$\lim_{z \to 1} \frac{1 - P(z)}{1 - z} = \frac{d}{dz}P(z)|_{z=1} = \sum np(n) \neq 0.$$

Now the renewal equation (8.12) shows that

$$U(z) = [1 - P(z)]^{-1}$$

is valid at least in the disc $\{|z| < 1\}$, and hence

$$F(z) = (1 - z)U(z) = (1 - z)[1 - P(z)]^{-1} \tag{15.16}$$

has no singularities in the disc $\{|z| < r_0\}$; and so (ii) holds.

Finally, to show that (ii) implies (iii) we again use (15.16): writing this as

$$P(z) = [F(z) - 1 + z]/F(z)$$

shows that $P(z)$ is a ratio of analytic functions and so is itself analytic in the disc $\{|z| < \kappa\}$, where now κ is the first zero of $F(z)$ in $\{|z| < r_0\}$; there are only finitely many such zeros and none of them occurs in the closed unit disc $\{|z| \le 1\}$ since $P(z)$ is bounded in this disc, so that $\kappa > 1$ as required. $\square$

It would seem that one should be able to prove this result, not only by analysis but also by a coupling argument as in Section 13.2. Clearly one direction of this is easy: if the renewal times are geometric then one can use coupling to get geometric convergence. The other direction does seem to require analytic tools to the best of our knowledge, and so we have given the classical proof here.

15.1.3   The geometric ergodic theorem

Following this result we formalize some of the conditions that will obviously be required in developing a geometric ergodicity result.

Kendall atoms and geometrically ergodic atoms

An accessible atom is called *geometrically ergodic* if there exists $r_\alpha > 1$ such that

$$\sum_n r_\alpha^n |P^n(\alpha, \alpha) - \pi(\alpha)| < \infty.$$

An accessible atom is called a *Kendall atom* of rate κ if there exists $\kappa > 1$ such that

$$U_\alpha^{(\kappa)}(\alpha, \alpha) = \mathsf{E}_\alpha[\kappa^{\tau_\alpha}] < \infty.$$

Suppose that $f \ge 1$. An accessible atom is called *f-Kendall* of rate κ if there exists $\kappa > 1$ such that

$$\sup_{x \in \alpha} \mathsf{E}_x\Big[\sum_{n=0}^{\tau_\alpha - 1} f(\Phi_n)\kappa^n\Big] < \infty.$$

Equivalently, if f is bounded on the accessible atom α, then α is f-Kendall of rate κ provided

$$U_\alpha^{(\kappa)}(\alpha, f) = \mathsf{E}_\alpha\Big[\sum_{n=1}^{\tau_\alpha} f(\Phi_n)\kappa^n\Big] < \infty.$$

The application of Kendall's Theorem to chains admitting an atom comes from the following, which is straightforward from the assumption that $f \ge 1$, so that $U_\alpha^{(\kappa)}(\alpha, f) \ge \mathsf{E}_\alpha[\kappa^{\tau_\alpha}]$.

Proposition 15.1.2. *Suppose that* Φ *is* ψ-*irreducible and aperiodic, and* α *is an accessible Kendall atom. Then there exists* $r_\alpha > 1$ *and* $R < \infty$ *such that*

$$|P^n(\alpha, \alpha) - \pi(\alpha)| \le Rr_\alpha^{-n}, \qquad n \to \infty.$$

$\square$

This enables us to control the first term in (15.11). To exploit the other bounds in (15.9)–(15.11) we also need to establish finiteness of the quantities $U_\alpha^{(\kappa)}(x, f)$ for values of x other than α.

Proposition 15.1.3. *Suppose that* Φ *is* ψ-*irreducible, and admits an* f-*Kendall atom* $\alpha \in \mathcal{B}^+(\mathsf{X})$ *of rate* κ. *Then the set*

$$S_f^\kappa := \{x : U_\alpha^{(\kappa)}(x, f) < \infty\} \tag{15.17}$$

is full and absorbing.

PROOF   The kernel $U_\alpha^{(\kappa)}(x, \cdot)$ satisfies the identity

$$\int P(x, dy)U_\alpha^{(\kappa)}(y, B) = \kappa^{-1}U_\alpha^{(\kappa)}(x, B) + P(x, \alpha)U_\alpha^{(\kappa)}(\alpha, B)$$

and integrating against f gives

$$PU_\alpha^{(\kappa)}(x, f) = \kappa^{-1}U_\alpha^{(\kappa)}(x, f) + P(x, \alpha)U_\alpha^{(\kappa)}(\alpha, f).$$

Thus the set S_f^κ is absorbing, and since S_f^κ is non-empty it follows from Proposition 4.2.3 that S_f^κ is full. $\square$

We now have sufficient structure to prove the geometric ergodic theorem when an atom exists with appropriate properties.

Theorem 15.1.4. *Suppose that* Φ *is* ψ-*irreducible, with invariant probability measure* π, *and that there exists an* f-*Kendall atom* $\alpha \in \mathcal{B}^+(\mathsf{X})$ *of rate* κ.

Then there exists a decomposition $\mathsf{X} = S^\kappa \cup N$ *where* S^κ *is full and absorbing, such that for all* $x \in S^\kappa$, *some* $R < \infty$, *and some* r *with* $r > 1$

$$\sum_n r^n \|P^n(x, \cdot) - \pi(\cdot)\|_f \le RU_\alpha^{(\kappa)}(x, f) < \infty. \tag{15.18}$$

PROOF   By Proposition 15.1.3 the bounds (15.9) and (15.10), and the second term in the bound (15.11), are all finite for $x \in S^\kappa$; and Kendall's Theorem, as applied in Proposition 15.1.2, gives that for some $r_\alpha > 1$ the other term in (15.11) is also finite. The result follows with $r = \min(\kappa, r_\alpha)$. $\square$

There is an alternative way of stating Theorem 15.1.4 in the simple geometric ergodicity case $f = 1$ which emphasizes the solidarity result in terms of ergodic properties rather than in terms of hitting time properties. The proof uses the same steps as the previous proof, and we omit it.

Theorem 15.1.5. *Suppose that Φ is ψ-irreducible, with invariant probability measure π, and that there is one geometrically ergodic atom $\alpha \in \mathcal{B}^+(\mathsf{X})$. Then there exists $\kappa > 1, r > 1$ and a decomposition $\mathsf{X} = S^\kappa \cup N$ where S^κ is full and absorbing, such that for some $R < \infty$ and all $x \in S^\kappa$*

$$\sum_n r^n \| P^n(x, \cdot) - \pi(\cdot) \| \leq R\mathsf{E}_x[\kappa^{\tau_\alpha}] < \infty, \tag{15.19}$$

so that Φ restricted to S^κ is also geometrically ergodic. $\square$

15.1.4   Some geometrically ergodic chains on countable spaces

Forward recurrence time chains

Consider as in Section 2.4 the forward recurrence time chain V^+.

By construction, we have for this chain that

$$\mathsf{E}_1[r^{\tau_1}] = \sum_n r^n \mathsf{P}_1(\tau_1 = n) = \sum_n r^n p(n)$$

so that the chain is geometrically ergodic if and only if the distribution $p(n)$ has geometrically decreasing tails.

We will see, once we develop a drift criterion for geometric ergodicity, that this duality between geometric tails on increments and geometric rates of convergence to stationarity is repeated for many other models.

A non-geometrically ergodic example

Not all ergodic chains on $\mathbb{Z}_+$ are geometrically ergodic, even if (as in the forward recurrence time chain) the steps to the right are geometrically decreasing. Consider a chain on $\mathbb{Z}_+$ with the transition matrix

$$\begin{aligned}
P(0, j) &= \gamma_j, & j \in \mathbb{Z}_+, \\
P(j, j) &= \beta_j, & j \in \mathbb{Z}_+, \\
P(j, 0) &= 1 - \beta_j, & j \in \mathbb{Z}_+.
\end{aligned} \tag{15.20}$$

where $\sum_j \gamma_j = 1$.

The mean return time from zero to itself is given by

$$\mathsf{E}_0[\tau_0] = \sum_j \gamma_j [1 + (1 - \beta_j)^{-1}]$$

and the chain is thus ergodic if $\gamma_j > 0$ for all j (ensuring irreducibility and aperiodicity), and

$$\sum_j \gamma_j (1 - \beta_j)^{-1} < \infty. \tag{15.21}$$

In this example

$$\mathsf{E}_0[r^{\tau_0}] \geq r \sum_j \gamma_j \mathsf{E}_j[r^{\tau_0}]$$

and
$$P_j(\tau_0 > n) = \beta_j^n.$$

Hence if $\beta_j \to 1$ as $n \to \infty$, then the chain is not geometrically ergodic regardless of the structure of the distribution $\{\gamma_j\}$, even if $\gamma_n \to 0$ sufficiently fast to ensure that (15.21) holds.

Different rates of convergence

Although it is possible to ensure a common rate of convergence in the Geometric Ergodic Theorem, there appears to be no simple way to ensure for a particular state that the rate is best possible. Indeed, in general this will not be the case.

To see this consider the matrix

$$P = \begin{bmatrix} \frac{1}{4} & \frac{1}{2} & \frac{1}{4} \\ 0 & \frac{3}{4} & \frac{1}{4} \\ \frac{3}{4} & 0 & \frac{1}{4} \end{bmatrix}.$$

By direct inspection we find the diagonal elements have generating functions

$$
\begin{aligned}
U^{(z)}(0,0) &= 1 + z/4(1-z), \\
U^{(z)}(1,1) &= 1 + z/2(1-z) + z/4(1-z), \\
U^{(z)}(2,2) &= 1 + z/4(1-z).
\end{aligned}
$$

Thus the best rates for convergence of $P^n(0,0)$ and $\Gamma^n(2,2)$ to their limits $\pi(0) = \pi(2) = \frac{1}{4}$ are $\rho_0 = \rho_2 = 0$: the limits are indeed attained at every step. But the rate of convergence of $P^n(1,1)$ to $\pi(1) = \frac{1}{2}$ is at least $\rho_1 > \frac{1}{4}$.

The following more complex example shows that even on an arbitrarily large finite space $\{1, \ldots, N+1\}$ there may in fact be N different rates of convergence such that

$$|P^n(i,i) - \pi(i)| \le M_i \rho_i^n.$$

Consider the matrix

$$P = \begin{bmatrix}
\beta_1 & \alpha_1 & \alpha_1 & \cdots & \alpha_1 & \alpha_1 & \alpha_1 \\
\alpha_1 & \beta_2 & \alpha_2 & \cdots & \alpha_2 & \alpha_2 & \alpha_2 \\
\alpha_1 & \alpha_2 & \beta_3 & \cdots & \alpha_3 & \alpha_3 & \alpha_3 \\
\vdots & \vdots & \vdots & \cdots & \vdots & \vdots & \vdots \\
\alpha_1 & \alpha_2 & \alpha_3 & \cdots & \beta_{N-1} & \alpha_{N-1} & \alpha_{N-1} \\
\alpha_1 & \alpha_2 & \alpha_3 & \cdots & \alpha_{N-1} & \beta_N & \alpha_N \\
\alpha_1 & \alpha_2 & \alpha_3 & \cdots & \alpha_{N-1} & \alpha_N & \beta_N
\end{bmatrix}$$

so that

$$P(k,k) = \beta_k := 1 - \sum_1^{k-1} \alpha_j - (N+1-k)\alpha_k, \qquad 1 \le k \le N+1,$$

where the off-diagonal elements are ordered by

$$0 < \alpha_N < \alpha_{N-1} < \ldots < \alpha_2 < \alpha_1 \le [N+1]^{-1}.$$

Since P is symmetric it is immediate that the invariant measure is given for all k by

$$\pi(k) = [N+1]^{-1}.$$

For this example it is possible to show [384] that the eigenvalues of P are distinct and are given by $\lambda_1 = 1$ and for $k = 2, \ldots, N+1$

$$\lambda_k = \beta_{N+2-k} - \alpha_{N+2-k}.$$

After considerable algebra it follows that for each k, there are positive constants $s(k,j)$ such that

$$P^m(k,k) - [N+1]^{-1} = \sum_{j=N+2-k}^{N+1} s(k,j)\lambda_j^m$$

and hence k has the exact "self-convergence" rate λ_{N+2-k}.

Moreover, $s(N+1,j) = s(N,j)$ for all $1 \leq j \leq N+1$, and so for the $N+1$ states there are N different "best" rates of convergence.

Thus our conclusion of a common rate parameter is the most that can be said.

15.2   Kendall sets and drift criteria

It is of course now obvious that we should try to move from the results valid for chains with atoms, to strongly aperiodic chains and thence to general aperiodic chains via the Nummelin splitting and the m-skeleton.

We first need to find conditions on the original chain under which the atom in the split chain is an f-Kendall atom. This will give the desired ergodic theorem for the split chain, which is then passed back to the original chain by exploiting a growth rate on the f-norm which holds for "f-geometrically regular chains". This extends the argument used in the proof of Lemma 14.3.2 to prove the f-Norm Ergodic Theorem in Chapter 14.

To do this we need to extend the concepts of Kendall atoms to general sets, and connect these with another and stronger drift condition: this has a dual purpose, for not only will it enable us to move relatively easily between chains, their skeletons, and their split forms, it will also give us a verifiable criterion for establishing geometric ergodicity.

15.2.1 f-Kendall sets and f-geometrically regular sets

The crucial aspect of a Kendall atom is that the return times to the atom from itself have a geometrically bounded distribution. There is an obvious extension of this idea to more general, non-atomic, sets.

Kendall sets and f-geometrically regular sets

A set $A \in \mathcal{B}(\mathsf{X})$ is called a *Kendall set* if there exists $\kappa > 1$ such that

$$\sup_{x \in A} \mathsf{E}_x[\kappa^{\tau_A}] < \infty.$$

A set $A \in \mathcal{B}(\mathsf{X})$ is called an *f-Kendall set* for a measurable $f : \mathsf{X} \to [1, \infty)$ if there exists $\kappa = \kappa(f) > 1$ such that

$$\sup_{x \in A} \mathsf{E}_x \left[\sum_{k=0}^{\tau_A - 1} f(\Phi_k) \kappa^k \right] < \infty. \tag{15.22}$$

A set $A \in \mathcal{B}(\mathsf{X})$ is called *f-geometrically regular* for a measurable $f : \mathsf{X} \to [1, \infty)$ if for each $B \in \mathcal{B}^+(\mathsf{X})$ there exists $r = r(f, B) > 1$ such that

$$\sup_{x \in A} \mathsf{E}_x \left[\sum_{k=0}^{\tau_B - 1} f(\Phi_k) r^k \right] < \infty.$$

Clearly, since we have $r > 1$ in these definitions, an f-geometrically regular set is also f-regular. When a set or a chain is 1-geometrically regular then we will call it geometrically regular.

A Kendall set is, in an obvious way, "self-geometrically regular": return times to the set itself are geometrically bounded, although not necessarily hitting times on other sets.

As in (15.12), for any set C in $\mathcal{B}(\mathsf{X})$ the kernel $U_C^{(r)}(x, B)$ is given by

$$U_C^{(r)}(x, B) = \mathsf{E}_x \left[\sum_{k=1}^{\tau_C} \mathbb{I}_B(\Phi_k) r^k \right]; \tag{15.23}$$

this is again well defined for $r \geq 1$, although it may be infinite. We use this notation in our next result, which establishes that any petite f-Kendall set is actually f-geometrically regular. This is non-trivial to establish, and needs a somewhat delicate "geometric trials" argument.

Theorem 15.2.1. *Suppose that Φ is ψ-irreducible. Then the following are equivalent:*

(i) *The set $C \in \mathcal{B}(\mathsf{X})$ is a petite f-Kendall set.*

(ii) *The set C is f-geometrically regular and $C \in \mathcal{B}^+(\mathsf{X})$.*

PROOF   To prove (ii)$\Rightarrow$(i) it is enough to show that A is petite, and this follows from Proposition 11.3.8, since a geometrically regular set is automatically regular.

To prove (i)$\Rightarrow$(ii) is considerably more difficult, although obviously since a Kendall set is Harris recurrent, it follows from Proposition 9.1.1 that any Kendall set is in $\mathcal{B}^+(\mathsf{X})$.

Suppose that C is an f-Kendall set of rate κ, let $1 < r \leq \kappa$, and define $U^{(r)}(x) = \mathsf{E}_x[r^{\tau_C}]$, so that $U^{(r)}$ is bounded on C. We set $M(r) = \sup_{x \in C} U^{(r)}(x) < \infty$. Put $\varepsilon = \log(r)/\log(\kappa)$: by Jensen's inequality,

$$M(r) = \sup_{x \in C} \mathsf{E}_x[\kappa^{\varepsilon \tau_C}] \leq M(\kappa)^{\varepsilon}.$$

From this bound we see that $M(r) \to 1$ as $r \downarrow 1$.

Let $\tau_C(n)$ denote the nth return time to the set C, where for convenience, we set $\tau_C(0) := 0$. We have by the strong Markov property and induction,

$$
\begin{aligned}
\mathsf{E}_x[r^{\tau_C(n)}] &= \mathsf{E}_x[r^{\tau_C(n-1) + \theta^{\tau_C(n-1)} \tau_C}] \\[2mm]
&= \mathsf{E}_x[r^{\tau_C(n-1)} \mathsf{E}_{\Phi_{\tau_C(n-1)}}[r^{\tau_C}]] \\[2mm]
&\leq M(r) \mathsf{E}_x[r^{\tau_C(n-1)}] \\[2mm]
&\leq (M(r))^{n-1} U^{(r)}(x), \qquad n \geq 1.
\end{aligned}
\tag{15.24}
$$

To prove the theorem we will combine this bound with the sample path bound, valid for any set $B \in \mathcal{B}(\mathsf{X})$,

$$\sum_{i=1}^{\tau_B} r^i f(\Phi_i) \leq \sum_{n=0}^{\infty} \Big(\sum_{j=\tau_C(n)+1}^{\tau_C(n+1)} r^j f(\Phi_j) \Big) \mathbb{I}\{\tau_B > \tau_C(n)\}.$$

Taking expectations and applying the strong Markov property gives

$$
\begin{aligned}
U_B^{(r)}(x,f) &\leq \sum_{n=0}^{\infty} \mathsf{E}_x\Big[\mathbb{I}\{\tau_B > \tau_C(n)\} r^{\tau_C(n)} \mathsf{E}_{\Phi_{\tau_C(n)}}\Big[\sum_{j=1}^{\tau_C} r^j f(\Phi_j) \Big] \Big] \\[2mm]
&\leq \sup_{x \in C} U_C^{(r)}(x,f) \sum_{n=0}^{\infty} \mathsf{E}_x\Big[\mathbb{I}\{\tau_B > \tau_C(n)\} r^{\tau_C(n)} \Big].
\end{aligned}
\tag{15.25}
$$

For any $0 < \gamma < 1$, $n \geq 0$, and positive numbers x and y we have the bound $xy \leq \gamma^n x^2 + \gamma^{-n} y^2$. Applying this bound with $x = r^{\tau_C(n)}$ and $y = \mathbb{I}\{\tau_C(n) < \tau_B\}$ in (15.25), and setting $M_f(r) = \sup_{x \in C} U_C^{(r)}(x,f)$ we obtain for any $B \in \mathcal{B}(\mathsf{X})$,

$$
\begin{aligned}
U_B^{(r)}(x,f) &\leq M_f(r) \sum_{n=0}^{\infty} \Big\{ \gamma^n \mathsf{E}_x[r^{2\tau_C(n)}] + \gamma^{-n} \mathsf{E}_x[\mathbb{I}\{\tau_C(n) < \tau_B\}] \Big\} \\[2mm]
&\leq M_f(r) \Big\{ \sum_{n=0}^{\infty} \gamma^n (M(r^2))^n U^{(r^2)}(x) \\[2mm]
&\qquad\qquad + \sum_{n=0}^{\infty} \gamma^{-n} \mathsf{P}_x\{\tau_C(n) < \tau_B\} \Big\},
\end{aligned}
\tag{15.26}
$$

where we have used (15.24). We still need to prove the right hand side of (15.26) is finite. Suppose now that for some $R < \infty$, $\rho < 1$, and any $x \in \mathsf{X}$,

$$\mathsf{P}_x\{\tau_C(n) < \tau_B\} \leq R\rho^n.\tag{15.27}$$

Choosing $\rho < \gamma < 1$ in (15.26) gives

$$U_B^{(r)}(x,f) \leq M_f(r)\Big\{U^{(r^2)}(x)\sum_{n=0}^{\infty}(\gamma M(r^2))^n + \frac{R}{1-\gamma^{-1}\rho}\Big\}.$$

With γ so fixed, we can now choose $r > 1$ so close to unity that $\gamma M(r^2) < 1$ to obtain

$$U_B^{(r)}(x,f) \leq M_f(r)\Big\{\frac{U^{(r^2)}(x)}{1-\gamma M(r^2)} + \frac{R}{1-\gamma^{-1}\rho}\Big\},$$

and the result holds.

To complete the proof, it is thus enough to bound $\mathsf{P}_x\{\tau_C(n) < \tau_B\}$ by a geometric series as in (15.27). Since C is petite, there exists $n_0 \in \mathbb{Z}_+$, $c < 1$, such that

$$\mathsf{P}_x\{\tau_C(n_0) < \tau_B\} \leq \mathsf{P}_x\{n_0 < \tau_B\} \leq c, \qquad x \in C,$$

and by the strong Markov property it follows that with $m_0 = n_0 + 1$,

$$\mathsf{P}_x\{\tau_C(m_0) < \tau_B\} \leq c, \qquad x \in \mathsf{X}.$$

Hence, using the identity

$$\mathbb{I}\{\tau_C(mm_0) < \tau_B\} = \mathbb{I}\{\tau_C([m-1]m_0) < \tau_B\}\theta^{\tau_C([m-1]m_0)}\mathbb{I}\{\tau_C(m_0) < \tau_B\}$$

we have again by the strong Markov property that for all $x \in \mathsf{X}$, $m \geq 1$,

$$\begin{aligned}
\mathsf{P}_x\{\tau_C(mm_0) < \tau_B\} &= \mathsf{E}_x\Big\{\mathbb{I}\{\tau_C([m-1]m_0) < \tau_B\}\mathsf{P}_{\Phi_{\tau_C([m-1]m_0)}}\{\tau_C(m_0) < \tau_B\}\Big\} \\
&\leq c\mathsf{P}_x\{\tau_C([m-1]m_0) < \tau_B\} \\
&\leq c^m,
\end{aligned}$$

and it now follows easily that (15.27) holds. $\qquad\square$

Notice specifically in this result that there may be a separate rate of convergence r for each of the quantities

$$\sup_{x \in C} U_B^{(r)}(x,f)$$

depending on the quantity ρ in (15.27): intuitively, for a set B "far away" from C it may take many visits to C before an excursion reaches B, and so the value of r will be correspondingly closer to unity.

15.2.2 The geometric drift condition

Whilst for strongly aperiodic chains an approach to geometric ergodicity is possible with the tools we now have directly through petite sets, in order to move from strongly aperiodic to aperiodic chains through skeleton chains and splitting methods an attractive theoretical route is through another set of drift inequalities.

This has, as usual, the enormous practical benefit of providing a set of verifiable conditions for geometric ergodicity. The drift condition appropriate for geometric convergence is:

Geometric drift towards C

(V4)   There exists an extended-real-valued function $V : \mathsf{X} \to [1,\infty]$, a measurable set C, and constants $\beta > 0$, $b < \infty$,

$$\Delta V(x) \leq -\beta V(x) + b\mathbb{I}_C(x), \quad x \in \mathsf{X}. \tag{15.28}$$

We see at once that (V4) is just (V3) in the special case where $f = \beta V$. From this observation we can borrow several results from the previous chapter, and use the approach there as a guide.

We first spell out some useful properties of solutions to the drift inequality in (15.28), analogous to those we found for (14.16).

Lemma 15.2.2. *Suppose that Φ is ψ-irreducible.*

(i) *If V satisfies (15.28), then $\{V < \infty\}$ is either empty or absorbing and full.*

(ii) *If (15.28) holds for a petite set C, then V is unbounded off petite sets.*

PROOF   Since (15.28) implies $PV \leq V + b$ the set $\{V < \infty\}$ is absorbing; hence if it is non-empty it is full, by Proposition 4.2.3.

Since $V \geq 1$, we see that (V4) implies that (V2) holds with $V' = V/(1-\beta)$. From Lemma 11.3.7 it then follows that V' (and hence obviously V) is unbounded off petite sets. $\square$

We now begin a more detailed evaluation of the consequences of (V4). We first give a probabilistic form for one solution to the drift condition (V4), which will prove that (15.2) implies (15.3) has a solution.

Using the kernel $U_C^{(r)}$ we define a further kernel $G_C^{(r)}$ as $G_C^{(r)} = I + I_{C^c} U_C^{(r)}$. For any $x \in \mathsf{X}$, $B \in \mathcal{B}(\mathsf{X})$, this has the interpretation

$$G_C^{(r)}(x,B) = \mathsf{E}_x\Big[\sum_{k=0}^{\sigma_C} \mathbb{I}_B(\Phi_k)r^k\Big]. \tag{15.29}$$

The kernel $G_C^{(r)}(x,B)$ gives us the solution we seek to (15.28).

Lemma 15.2.3. *Suppose that $C \in \mathcal{B}(\mathsf{X})$, and let $r > 1$. Then the kernel $G_C^{(r)}$ satisfies*

$$PG_C^{(r)} = r^{-1}G_C^{(r)} - r^{-1}I + r^{-1}I_C U_C^{(r)}$$

so that in particular for $\beta = 1 - r^{-1}$

$$PG_C^{(r)} - G_C^{(r)} = \Delta G_C^{(r)} \leq -\beta G_C^{(r)} + r^{-1}I_C U_C^{(r)}. \tag{15.30}$$

PROOF   The kernel $U_C^{(r)}$ satisfies the simple identity

$$U_C^{(r)} = rP + rPI_{C^c}U_C^{(r)}. \tag{15.31}$$

Hence the kernel $G_C^{(r)}$ satisfies the chain of identities

$$PG_C^{(r)} = P + PI_{C^c}U_C^{(r)} = r^{-1}U_C^{(r)} = r^{-1}[G_C^{(r)} - I + I_C U_C^{(r)}].$$

$\square$

This now gives us the easier direction of the duality between the existence of f-Kendall sets and solutions to (15.28).

Theorem 15.2.4. *Suppose that $\boldsymbol{\Phi}$ is ψ-irreducible, and admits an f-Kendall set $C \in \mathcal{B}^+(\mathsf{X})$ for some $f \geq 1$. Then the function $V(x) = G_C^{(\kappa)}(x, f) \geq f(x)$ is a solution to (V4).*

PROOF   We have from (15.30) that, by the f-Kendall property, for some $M < \infty$ and $r > 1$,

$$\Delta V \leq -\beta V + r^{-1}M I_C$$

and so the function V satisfies (V4). $\square$

15.2.3   Other solutions of the drift inequalities

We have shown that the existence of f-geometrically regular sets will lead to solutions of (V4). We now show that the converse also holds.

The tool we need in order to consider properties of general solutions to (15.28) is the following "geometric" generalization of the Comparison Theorem.

Theorem 15.2.5. *If (V4) holds, then for any $r \in (1, (1-\beta)^{-1})$ there exists $\varepsilon > 0$ such that for any first entrance time τ_B,*

$$\mathsf{E}_x\left[\sum_{k=0}^{\tau_B-1} V(\Phi_k)r^k\right] \leq \varepsilon^{-1}r^{-1}V(x) + \varepsilon^{-1}b\mathsf{E}_x\left[\sum_{k=0}^{\tau_B-1} I_C(\Phi_k)r^k\right]$$

and hence in particular choosing $B = C$

$$V(x) \leq \mathsf{E}_x\left[\sum_{k=0}^{\tau_C-1} V(\Phi_k)r^k\right] \leq \varepsilon^{-1}r^{-1}V(x) + \varepsilon^{-1}b I_C(x). \tag{15.32}$$

PROOF   We have the bound

$$PV \leq r^{-1}V - \varepsilon V + b I_C$$

where $0 < \varepsilon < \beta$ is the solution to $r = (1 - \beta + \varepsilon)^{-1}$. Defining

$$Z_k = r^k V(\Phi_k)$$

for $k \in \mathbb{Z}_+$, it follows that

$$\mathsf{E}[Z_{k+1} \mid \mathcal{F}_k^{\Phi}] \;=\; r^{k+1} \mathsf{E}[V(\Phi_{k+1}) \mid \mathcal{F}_k^{\Phi}]$$

$$\leq\; r^{k+1}\{r^{-1}V(\Phi_k) - \varepsilon V(\Phi_k) + b\mathbb{I}_C(\Phi_k)\}$$

$$=\; Z_k - \varepsilon r^{k+1}V(\Phi_k) + r^{k+1}b\mathbb{I}_C(\Phi_k).$$

Choosing $f_k(x) = \varepsilon r^{k+1}V(x)$ and $s_k(x) = br^{k+1}\mathbb{I}_C(x)$, we have by Proposition 11.3.2

$$\mathsf{E}_x\Big[\sum_{k=0}^{\tau_B-1} \varepsilon r^{k+1}V(\Phi_k)\Big] \leq Z_0(x) + \mathsf{E}_x\Big[\sum_{k=0}^{\tau_B-1} r^{k+1}b\mathbb{I}_C(\Phi_k)\Big].$$

Multiplying through by $\varepsilon^{-1}r^{-1}$ and noting that $Z_0(x) = V(x)$, we obtain the required bound.

The particular form with $B = C$ is then straightforward.   □

We use this result to prove that in general, sublevel sets of solutions V to (15.28) are V-geometrically regular.

Theorem 15.2.6. *Suppose that Φ is ψ-irreducible, and that (V4) holds for a function V and a petite set C.*

If V is bounded on $A \in \mathcal{B}(\mathsf{X})$, then A is V-geometrically regular.

PROOF   We first show that if V is bounded on A, then $A \subseteq D$ where D is a V-Kendall set.

Assume (V4) holds, let $\rho = 1 - \beta$, and fix $\rho < r^{-1} < 1$. Now consider the set D defined by

$$D := \Big\{x : V(x) \leq \frac{M+b}{r^{-1}-\rho}\Big\}, \tag{15.33}$$

where the integer $M > 0$ is chosen so that $A \subseteq D$ (which is possible because the function V is bounded on A) and $D \in \mathcal{B}^+(\mathsf{X})$, which must be the case for sufficiently large M from Lemma 15.2.2 (i).

Using (V4) we have

$$PV(x) \;\leq\; r^{-1}V(x) - (r^{-1}-\rho)V(x) + b\mathbb{I}_C(x)$$
$$\leq\; r^{-1}V(x) - M, \qquad x \in D^c.$$

Since $PV(x) \leq V(x) + b$, which is bounded on D, it follows that

$$PV \leq r^{-1}V + c\mathbb{I}_D$$

for some $c < \infty$. Thus we have shown that (V4) holds with D in place of C.

Hence using (15.32) there exists $s > 1$ and $\varepsilon > 0$ such that

$$\mathsf{E}_x\Big[\sum_{k=0}^{\tau_D-1} s^k V(\Phi_k)\Big] \;\leq\; \varepsilon^{-1}s^{-1}V(x) + \varepsilon^{-1}c\mathbb{I}_D(x). \tag{15.34}$$

Since V is bounded on D by construction, this shows that D is V-Kendall as required.

By Lemma 15.2.2 (ii) the function V is unbounded off petite sets, and therefore the set D is petite. Applying Theorem 15.2.1 we see that D is V-geometrically regular.

Finally, since by definition any subset of a V-geometrically regular set is itself V-geometrically regular, we have that A inherits this property from D. □

As a simple consequence of Theorem 15.2.6 we can construct, given just one f-Kendall set in $\mathcal{B}^+(\mathsf{X})$, an increasing sequence of f-geometrically regular sets whose union is full: indeed we have a somewhat more detailed description than this.

Theorem 15.2.7. *If there exists an f-Kendall set $C \in \mathcal{B}^+(\mathsf{X})$, then there exists $V \geq f$ and an increasing sequence $\{C_V(i) : i \in \mathbb{Z}_+\}$ of V-geometrically regular sets whose union is full.*

PROOF   Let $V(x) = G_C^{(r)}(x, f)$. Then V satisfies (V4) and by Theorem 15.2.6 the set $C_V(n) := \{x : V(x) \leq n\}$ is V-geometrically regular for each n. Since $S_V = \{V < \infty\}$ is a full absorbing subset of X, the result follows. □

The following alternative form of (V4) will simplify some of the calculations performed later.

Lemma 15.2.8. *The drift condition (V4) holds with a petite set C if and only if V is unbounded off petite sets and*

$$PV \leq \lambda V + L \qquad (15.35)$$

for some $\lambda < 1$, $L < \infty$.

PROOF   If (V4) holds, then (15.35) immediately follows. Lemma 15.2.2 states that the function V is unbounded off petite sets.

Conversely, if (15.35) holds for a function V which is unbounded off petite sets then set $\beta = \frac{1}{2}(1 - \lambda)$ and define the petite set C as

$$C = \{x \in \mathsf{X} : V(x) \leq L/\beta\}$$

It follows that $\Delta V \leq -\beta V + L\mathbb{I}_C$ so that (V4) is satisfied. □

We will find in several examples on topological spaces that the bound (15.35) is obtained for some coercive function V and compact C. If the Markov chain is a ψ-irreducible T-chain it follows from Lemma 15.2.8 that (V4) holds and then that the chain is V-geometrically ergodic.

Although the result that one can use the same function V in both sides of

$$\sum_n r^n \|P^n(x, \cdot) - \pi\|_V \leq RV(x).$$

is an important one, it also has one drawback: as we have larger functions on the left, the bounds on the distance to $\pi(V)$ also increase.

Overall it is not clear when one can have a best common bound on the distance $\|P^n(x, \cdot) - \pi\|_V$ independent of V; indeed, the example in Section 16.2.2 shows that as V increases then one might even lose the geometric nature of the convergence.

However, the following result shows that one can obtain a smaller x-dependent bound in the Geometric Ergodic Theorem if one is willing to use a smaller function V in the application of the V-norm.

Lemma 15.2.9. *If (V4) holds for V, and some petite set C, then (V4) also holds for the function $\sqrt{V}$ and some petite set C.*

PROOF   If (V4) holds for the finite-valued function V then by Lemma 15.2.8 V is unbounded off petite sets and (15.35) holds for some $\lambda < 1$ and $L < \infty$. Letting $V'(x) = \sqrt{V(x)}$, $x \in \mathsf{X}$, we have by Jensen's inequality,

$$
\begin{aligned}
PV'(x) \le \sqrt{PV(x)} &\le \sqrt{\lambda V + L} \\
&\le \sqrt{\lambda}\sqrt{V} + \frac{L}{2\sqrt{\lambda}} \qquad \text{since } V \ge 1 \\
&= \sqrt{\lambda}V' + \frac{L}{2\sqrt{\lambda}},
\end{aligned}
$$

which together with Lemma 15.2.8 implies that (V4) holds with V replaced by $\sqrt{V}$.   □

15.3 f-Geometric regularity of Φ and its skeleton

15.3.1 f-Geometric regularity of chains

There are two aspects to the f-geometric regularity of sets that we need in moving to our prime purpose in this chapter, namely proving the f-geometric convergence part of the Geometric Ergodic Theorem.

The first is to locate sets from which the hitting times on other sets are geometrically fast. For the purpose of our convergence theorems, we need this in a specific way: from an f-Kendall set we will only need to show that the hitting times on a split atom are geometrically fast, and in effect this merely requires that hitting times on a (rather specific) subset of a petite set be geometrically fast. Indeed, note that in the case with an atom we only needed the f-Kendall (or self f-geometric regularity) property of the atom, and there was no need to prove that the atom was fully f-geometrically regular. The other structural results shown in the previous section are an unexpectedly rich by-product of the requirement to delineate the geometric bounds on subsets of petite sets. This approach also gives, as a more directly useful outcome, an approach to working with the m-skeleton from which we will deduce rates of convergence.

Secondly, we can see from the Regenerative Decomposition that we will need the analogue of Proposition 15.1.3: that is, we need to ensure that for some specific set there is a fixed geometric bound on the hitting times of the set from arbitrary starting points. This motivates the next definition.

---

f-Geometric regularity of Φ

The chain Φ is called f-geometrically regular if there exists a petite set C and a fixed constant $\kappa > 1$ such that

$$\mathsf{E}_x\Big[\sum_{/k=0}^{\tau_C-1} f(\Phi_k)\kappa^k \Big] \qquad (15.36)$$

is finite for all $x \in \mathsf{X}$ and bounded on C.

Observe that when κ is taken equal to one, this definition then becomes f-regularity, whilst the boundedness on C implies f-geometric regularity of the set C from Theorem 15.2.1: it is the finiteness from arbitrary initial points that is new in this definition.

The following consequence of f-regularity follows immediately from the strong Markov property and f-geometric regularity of the set C used in (15.36).

Proposition 15.3.1. *If Φ is f-geometrically regular so that (15.36) holds for a petite set C, then for each $B \in \mathcal{B}^+(\mathsf{X})$ there exists $r = r(B) > 1$ and $c(B) < \infty$ such that*

$$U_B^{(r)}(x,f) \leq c(B)U_C^{(r)}(x,f). \qquad (15.37)$$

$\square$

By now the techniques we have developed ensure that f-geometrically regularity is relatively easy to verify.

Proposition 15.3.2. *If there is one petite f-Kendall set C, then there is a decomposition*

$$\mathsf{X} = S_f \cup N$$

where S_f is full and absorbing, and Φ restricted to S_f is f-geometrically regular.

PROOF   We know from Theorem 15.2.1 that when a petite f-Kendall set C exists then C is V-geometrically regular, where $V(x) = G_C^{(r)}(x,f)$ for some $r > 1$. Since V then satisfies (V4) from Lemma 15.2.3, it follows from Lemma 15.2.2 that $S_f = \{V < \infty\}$ is absorbing and full. Now as in (15.32) we have for some $\kappa > 1$

$$V(x) \leq \mathsf{E}_x\Big[\sum_{n=0}^{\tau_C-1} V(\Phi_n)\kappa^n \Big] \leq \varepsilon^{-1}\kappa^{-1}V(x) + \varepsilon^{-1}c\mathbb{I}_C(x) \qquad (15.38)$$

and since the right hand side is finite on S_f the chain restricted to S_f is V-geometrically regular, and hence also f-geometrically regular since $f \leq V$. $\square$

The existence of an everywhere finite solution to the drift inequality (V4) is equivalent to f-geometric regularity, imitating the similar characterization of f-regularity. We have

Theorem 15.3.3. *Suppose that (V4) holds for a petite set C and a function V which is everywhere finite. Then $\mathbf{\Phi}$ is V-geometrically regular, and for each $B \in \mathcal{B}^+(\mathsf{X})$ there exists $c(B) < \infty$ such that*

$$U_B^{(r)}(x, V) \leq c(B)V(x).$$

Conversely, if $\mathbf{\Phi}$ is f-geometrically regular, then there exists a petite set C and a function $V \geq f$ which is everywhere finite and which satisfies (V4).

PROOF   Suppose that (V4) holds with V everywhere finite and C petite. As in the proof of Theorem 15.2.6, there exists a petite set D on which V is bounded, and as in (15.34) there is then $r > 1$ and a constant d such that

$$\mathsf{E}_x\Big[\sum_{k=0}^{\tau_D - 1} V(\Phi_k)r^k\Big] \leq dV(x).$$

Hence $\mathbf{\Phi}$ is V-geometrically regular, and the required bound follows from Proposition 15.3.1.

For the converse, take $V(x) = G_C^{(r)}(x, f)$ where C is the petite set used in the definition of f-geometric regularity. $\square$

This approach, using solutions V to (V4) to bound (15.36), is in effect an extended version of the method used in the atomic case to prove Proposition 15.1.3.

15.3.2   Connections between $\mathbf{\Phi}$ and $\mathbf{\Phi}^n$

A striking consequence of the characterization of geometric regularity in terms of the solution of (V4) is that we can prove almost instantly that if a set C is f-geometrically regular, and if $\mathbf{\Phi}$ is aperiodic, then C is also f-geometrically regular for every skeleton chain.

Theorem 15.3.4. *Suppose that $\mathbf{\Phi}$ is ψ-irreducible and aperiodic.*

  (i) *If V satisfies (V4) with a petite set C, then for any n-skeleton, the function V also satisfies (V4) for some set C' which is petite for the n-skeleton.*

  (ii) *If C is f-geometrically regular, then C is f-geometrically regular for the chain $\mathbf{\Phi}^n$ for any $n \geq 1$.*

PROOF   **(i)**   Suppose $\rho = 1 - \beta$ and $0 < \varepsilon < \rho - \rho^n$. By iteration we have using Lemma 14.2.8 that for some petite set C',

$$P^n V \leq \rho^n V + b \sum_{i=0}^{n-1} P^i \mathbb{I}_C \leq \rho^n V + bm\mathbb{I}_{C'} + \varepsilon.$$

Since $V \geq 1$ this gives

$$P^n V \leq \rho V + bm\mathbb{I}_{C'}, \tag{15.39}$$

and hence (V4) holds for the n-skeleton.

(ii) If C is f-geometrically regular then we know that (V4) holds with $V = G_C^{(r)}(x, f)$. We can then apply Theorem 15.2.6 to the n-skeleton and the result follows.
□

Given this together with Theorem 15.3.3, which characterizes f-geometric regularity, the following result is obvious:

Theorem 15.3.5. *If Φ is f-geometrically regular and aperiodic, then every skeleton is also f-geometrically regular.*
□

We round out this series of equivalences by showing not only that the skeletons inherit f-geometric regularity properties from the chain, but that we can go in the other direction also.

Recall from (14.22) that for any positive function g on X, we write $g^{(m)} = \sum_{i=0}^{m-1} P^i g$. Then we have, as a geometric analogue of Theorem 14.2.9,

Theorem 15.3.6. *Suppose that Φ is ψ-irreducible and aperiodic. Then $C \in \mathcal{B}^+(\mathsf{X})$ is f-geometrically regular if and only if it is $f^{(m)}$-geometrically regular for any one, and then every, m-skeleton chain.*

PROOF   Letting τ_B^m denote the hitting time for the skeleton, we have by the Markov property, for any $B \in \mathcal{B}^+(\mathsf{X})$ and $r > 1$,

$$
\mathsf{E}_x\left[\sum_{k=0}^{\tau_B^m - 1} r^{km} \sum_{i=0}^{m-1} P^i f(\Phi_{km}) \right] \geq r^{-m} \mathsf{E}_x\left[\sum_{k=0}^{\tau_B^m - 1} \sum_{i=0}^{m-1} r^{km+i} f(\Phi_{km+i}) \right]
$$
$$
\geq r^{-m} \mathsf{E}_x\left[\sum_{j=0}^{\tau_B - 1} r^j f(\Phi_j) \right].
$$

If C is $f^{(m)}$-geometrically regular for an m-skeleton then the left hand side is bounded over C for some $r > 1$ and hence the set C is also f-geometrically regular.

Conversely, if $C \in \mathcal{B}^+(\mathsf{X})$ is f-geometrically regular then it follows from Theorem 15.2.4 that (V4) holds for a function $V \geq f$ which is bounded on C.

Thus we have from (15.39) and a further application of Lemma 14.2.8 that for some petite set C'' and $\rho' < 1$

$$
P^m V^{(m)} \leq \rho V^{(m)} + mb \mathbb{I}_{C'}^{(m)} \leq \rho' V^{(m)} + mb \mathbb{I}_{C''}.
$$

and thus (V4) holds for the m-skeleton. Since $V^{(m)}$ is bounded on C by (15.39), we have from Theorem 15.3.3 that C is $V^{(m)}$-geometrically regular for the m-skeleton.   □

This gives the following solidarity result.

Theorem 15.3.7. *Suppose that Φ is ψ-irreducible and aperiodic. Then Φ is f-geometrically regular if and only if each m-skeleton is $f^{(m)}$-geometrically regular.*   □

15.4 f-Geometric ergodicity for general chains

We now have the results that we need to prove the geometrically ergodic limit (15.4). Using the result in Section 15.1.3 for a chain possessing an atom we immediately obtain the desired ergodic theorem for strongly aperiodic chains. We then consider the m-skeleton chain: we have proved that when $\boldsymbol{\Phi}$ is f-geometrically regular then so is each m-skeleton. For aperiodic chains, there always exists some $m \geq 1$ such that the m-skeleton is strongly aperiodic, and hence as in Chapter 14 we can prove geometric ergodicity using this strongly aperiodic skeleton chain.

We follow these steps in the proof of the following theorem.

Theorem 15.4.1. *Suppose that $\boldsymbol{\Phi}$ is ψ-irreducible and aperiodic, and that there is one f-Kendall petite set $C \in \mathcal{B}(\mathsf{X})$.*

Then there exists $\kappa > 1$ and an absorbing full set S_f^κ on which

$$\mathsf{E}_x\Big[\sum_{k=0}^{\tau_C-1} f(\Phi_k)\kappa^k\Big]$$

is finite, and for all $x \in S_f^\kappa$,

$$\sum_n r^n \,\|P^n(x,\cdot) - \pi\|_f \leq R\,\mathsf{E}_x\Big[\sum_{k=0}^{\tau_C} f(\Phi_k)\kappa^k\Big]$$

for some $r > 1$ and $R < \infty$ independent of x.

Proof   This proof is in several steps, from the atomic through the strongly aperiodic to the general aperiodic case. In all cases we use the fact that the seemingly relatively weak f-Kendall petite assumption on C implies that C is f-geometrically regular and in $\mathcal{B}^+(\mathsf{X})$ from Theorem 15.2.1.

Under the conditions of the theorem it follows from Theorem 15.2.4 that

$$V(x) = \mathsf{E}_x\Big[\sum_{k=0}^{\sigma_C} f(\Phi_k)\kappa^k\Big] \geq f(x) \tag{15.40}$$

is a solution to (V4) which is bounded on the set C, and the set $S_f^\kappa = \{x : V(x) < \infty\}$ is absorbing, full, and contains the set C. This will turn out to be the set required for the result.

   (i)   Suppose first that the set C contains an accessible atom α. We know then that the result is true from Theorem 15.1.4, with the bound on the f-norm convergence given from (15.18) and (15.37) by

$$\mathsf{E}_x\Big[\sum_{k=0}^{\tau_\alpha-1} f(\Phi_k)\kappa^k\Big] \leq c(\alpha)\mathsf{E}_x\Big[\sum_{k=0}^{\tau_C-1} f(\Phi_k)\kappa^k\Big]$$

for some $\kappa > 1$ and a constant $c(\alpha) < \infty$.

   (ii)   Consider next the case where the chain is strongly aperiodic, and this time assume that $C \in \mathcal{B}^+(\mathsf{X})$ is a ν_1-small set with $\nu_1(C^c) = 0$. Clearly this will not always be the case, but in part (iii) of the proof we see that this is no loss in generality.

To prove the theorem we abandon the function f and prove V-geometric ergodicity for the chain restricted to S_f^κ and the function (15.40). By Theorem 15.3.3 applied to the chain restricted to S_f^κ we have that for some constants $c < \infty$, $r > 1$,

$$\mathsf{E}_x\Big[\sum_{k=1}^{\tau_C} V(\Phi_k)r^k\Big] \le cV(x). \tag{15.41}$$

Now consider the chain split on C. Exactly as in the proof of Proposition 14.3.1 we have that

$$\check{\mathsf{E}}_{x_i}\Big[\sum_{k=1}^{\tau_{C_0 \cup C_1}} \check{V}(\check{\Phi}_k)r^k\Big] \le c'\check{V}(x_i)$$

where $c' \ge c$ and $\check{V}$ is defined on $\check{\mathsf{X}}$ by $\check{V}(x_i) = V(x)$, $x \in \mathsf{X}$, $i = 0, 1$.

But this implies that $\check{\alpha}$ is a $\check{V}$-Kendall atom, and so from step (i) above we see that for some $r_0 > 1$, $c'' < \infty$,

$$\sum_n r_0^n \|\check{P}^n(x_i, \cdot) - \check{\pi}\|_{\check{V}} \le c''\check{V}(x_i)$$

for all $x_i \in (S_f^\kappa)_0 \cup \mathsf{X}_1$.

It is then immediate that the original (unsplit) chain restricted to S_f^κ is V-geometrically ergodic and that

$$\sum_n r_0^n \|P^n(x, \cdot) - \pi\|_V \le c''V(x).$$

From the definition of V and the bound $V \ge f$ this proves the theorem when C is ν_1-small.

(iii) Now let us move to the general aperiodic case. Choose m so that the set C is itself ν_m-small with $\nu_m(C^c) = 0$: we know that this is possible from Theorem 5.5.7.

By Theorem 15.3.3 and Theorem 15.3.5 the chain and the m-skeleton restricted to S_f^κ are both V-geometrically regular. Moreover, by Theorem 15.3.3 and Theorem 15.3.4 we have for some constants $d < \infty$, $r > 1$,

$$\mathsf{E}_x\Big[\sum_{k=1}^{\tau_C^m} V(\Phi_k)r^k\Big] \le dV(x) \tag{15.42}$$

where as usual τ_C^m denotes the hitting time for the m-skeleton. From (ii), since m is chosen specifically so that C is "ν_1-small" for the m-skeleton, there exists $c < \infty$ with

$$\|P^{nm}(x, \cdot) - \pi\|_V \le cV(x)r_0^{-n}, \qquad n \in \mathbb{Z}_+, \ x \in S_f^\kappa.$$

We now need to compare this term with the convergence of the one-step transition probabilities, and we do not have the contraction property of the total variation norm available to do this. But if (V4) holds for V then we have that

$$PV(x) \le V(x) + b \le (1 + b)V(x),$$

and hence for any $g \leq V$,

$$
\begin{aligned}
|P^{n+1}(x,g) - \pi(g)| &= |P^n(x, Pg) - \pi(Pg)| \\
&\leq \|P^n(x, \cdot) - \pi\|_{(1+b)V} \\
&= (1+b)\|P^n(x, \cdot) - \pi\|_V.
\end{aligned}
$$

Thus we have the bound

$$
\|P^{n+1}(x, \cdot) - \pi\|_V \leq (1+b)\|P^n(x, \cdot) - \pi\|_V. \tag{15.43}
$$

Now observe that for any $k \in \mathbb{Z}_+$, if we write $k = nm + i$ with $0 \leq i \leq m-1$, we obtain from (15.43) the bound, for any $x \in S_f^\kappa$

$$
\begin{aligned}
\|P^k(x, \cdot) - \pi\|_V &\leq (1+b)^m \|P^{nm}(x, \cdot) - \pi\|_V \\
&\leq (1+b)^m c V(x) r_0^{-n} \\
&\leq (1+b)^m c r_0 V(x) (r_0^{1/m})^{-k},
\end{aligned}
$$

and the theorem is proved. $\square$

Intuitively it seems obvious from the method of proof we have used here that f-geometric ergodicity will imply f-geometric regularity for any f, but of course the inequalities in the Regenerative Decomposition are all in one direction, and so we need to be careful in proving this result.

Theorem 15.4.2. *If Φ is f-geometrically ergodic, then there is a full absorbing set S such that Φ is f-geometrically regular when restricted to S.*

PROOF   Let us first assume there is an accessible atom $\alpha \in \mathcal{B}^+(\mathsf{X})$, and that $r > 1$ is such that

$$
\sum_n r^n \|P^n(\alpha, \cdot) - \pi\|_f < \infty.
$$

Using the last exit decomposition (8.19) over the times of entry to α, we have as in the Regenerative Decomposition (13.48)

$$
P^n(\alpha, f) - \pi(f) \geq (u - \pi(\alpha)) * t_f(n) + \pi(\alpha) \sum_{j=n+1}^{\infty} t_f(j). \tag{15.44}
$$

Multiplying by r^n and summing both sides of (15.44) would seem to indicate that α is an f-Kendall atom of rate r, save for the fact that the first term may be negative, so that we could have both positive and negative infinite terms in this sum in principle. We need a little more delicate argument to get around this.

By truncating the last term and then multiplying by $s^n, s \leq r$ and summing to N, we do have

$$
\sum_{n=0}^{N} s^n (P^n(\alpha, f) - \pi(f)) \geq \left[\sum_{n=0}^{N} s^n t_f(n) [\sum_{k=0}^{N-n} s^k (u(k) - \pi(\alpha))] \right]
$$

$$
+ \pi(\alpha) \sum_{n=0}^{N} s^n \sum_{j=n+1}^{N} t_f(j). \tag{15.45}
$$

Let us write $c_N(f,s) = \sum_{n=0}^{N} s^n t_f(n)$, and $d(s) = \sum_{n=0}^{\infty} s^n |u(n) - \pi(\alpha)|$. We can bound the first term in (15.45) in absolute value by $d(s)c_N(f,s)$, so in particular as $s \downarrow 1$, by monotonicity of $d(s)$ we know that the middle term is no more negative than $-d(r)c_N(f,s)$.

On the other hand, the third term is by Fubini's Theorem given by

$$\pi(\alpha)[s-1]^{-1} \sum_{n=0}^{N} t_f(n)(s^n - 1) \geq [s-1]^{-1}[\pi(\alpha)c_N(f,s) - \pi(f) - \pi(\alpha)f(\alpha)]. \quad (15.46)$$

Suppose now that α is not f-Kendall. Then for any $s > 1$ we have that $c_N(f,s)$ is unbounded as N becomes large. Fix s sufficiently small that $\pi(\alpha)[s-1]^{-1} > d(r)$; then we have that the right hand side of (15.45) is greater than

$$c_N(f,s)[\pi(\alpha)[s-1]^{-1} - d(r)] - (\pi(f) + \pi(\alpha)f(\alpha))/(1-s)$$

which tends to infinity as $N \to \infty$. This clearly contradicts the finiteness of the left side of (15.45). Consequently α is f-Kendall of rate s for some $s < r$, and then the chain is f-geometrically regular when restricted to a full absorbing set S from Proposition 15.3.2.

Now suppose that the chain does not admit an accessible atom. If the chain is f-geometrically ergodic, then it is straightforward that for every m-skeleton and every x we have

$$\sum_n r^n |P^{nm}(x,f) - \pi(f)| < \infty,$$

and for the split chain corresponding to one such skeleton we also have $|r^n \check{P}^n(x,f) - \pi(f)|$ summable. From the first part of the proof this ensures that the split chain, and again trivially the m-skeleton is $f^{(m)}$-geometrically regular, at least on a full absorbing set S. We can then use Theorem 15.3.7 to deduce that the original chain is f-geometrically regular on S as required. □

One of the uses of this result is to show that even when $\pi(f) < \infty$ there is no guarantee that geometric ergodicity actually implies f-geometric ergodicity: rates of convergence need not be inherited by the f-norm convergence for "large" functions f. We will see this in the example defined by (16.24) in the next chapter.

However, we can show that local geometric ergodicity does at least give the V-geometric ergodicity of Theorem 15.4.1, for an appropriate V. As in Chapter 13, we conclude with what is now an easy result.

Theorem 15.4.3. *Suppose that Φ is an aperiodic positive Harris chain, with invariant probability measure π, and that there exists some ν-small set $C \in \mathcal{B}^+(\mathsf{X})$, $\rho_C < 1$ and $M_C < \infty$, and $P^\infty(C) > 0$ such that $\nu(C) > 0$ and*

$$\left| \int_C \nu_C(dx)(P^n(x,C) - P^\infty(C)) \right| \leq M_C \rho_C^n \quad (15.47)$$

where $\nu_C(\cdot) = \nu(\cdot)/\nu(C)$ is normalized to a probability measure on C.

Then there exists a full absorbing set S such that the chain restricted to S is geometrically ergodic.

PROOF   Using the Nummelin splitting via the set C for the m-skeleton, we have
exactly as in the proof of Theorem 13.3.5 that the bound (15.47) implies that the atom
in the skeleton chain split at C is geometrically ergodic.

We can then emulate step (iii) of the proof of Theorem 15.4.1 above to reach the
conclusion.   □

Notice again that (15.47) is implied by (15.1), so that we have completed the circle
of results in Theorem 15.0.1.

15.5   Simple random walk and linear models

In order to establish geometric ergodicity for specific models, we will of course use the
drift criterion (V4) as a practical tool to establish the required properties of the chain.

We conclude by illustrating this for three models: the simple random walk on $\mathbb{Z}_+$,
the simple linear model, and a bilinear model. We give many further examples in
Chapter 16, after we have established a variety of desirable and somewhat surprising
consequences of geometric ergodicity.

15.5.1   Bernoulli random walk

Consider the simple random walk on $\mathbb{Z}_+$ with transition law

$$P(x, x+1) = p, \ x \ge 0; \qquad P(x, x-1) = 1 - p, \ x > 0; \qquad P(0,0) = 1 - p.$$

For this chain we can consider directly $\mathsf{P}_x(\tau_0 = n) = a_x(n)$ in order to evaluate the
geometric tails of the distribution of the hitting times. Since we have the recurrence
relations

$$\begin{aligned}
a_x(n) &= (1-p)a_{x-1}(n-1) + pa_{x+1}(n-1), & x > 1; \\
a_x(0) &= 0, & x \ge 1; \\
a_1(n) &= pa_2(n-1), & a_0(0) = 0,
\end{aligned}$$

valid for $n \ge 1$, the generating functions $A_x(z) = \sum_{n=0}^{\infty} a_x(n)z^n$ satisfy

$$\begin{aligned}
A_x(z) &= z(1-p)A_{x-1}(z) + zpA_{x+1}(z), & x > 1; \\
A_1(z) &= z(1-p) + zpA_2(z),
\end{aligned}$$

giving the solution

$$A_x(z) = \left[\frac{1 - (1 - 4pqz^2)^{1/2}}{2pz} \right]^x = \left[A_1(z) \right]^x. \tag{15.48}$$

This is analytic for $z < 2/\sqrt{p(1-p)}$, so that if $p < 1/2$ (that is, if the chain is ergodic)
then the chain is also geometrically ergodic.

Using the drift criterion (V4) to establish this same result is rather easier. Consider
the test function $V(x) = z^x$ with $z > 1$. Then we have, for $x > 0$,

$$\Delta V(x) = z^x[(1-p)z^{-1} + pz - 1]$$

and if $p < 1/2$, then $[(1-p)z^{-1} + pz - 1] = -\beta < 0$ for z sufficiently close to unity, and
so (15.28) holds as desired.

In fact, this same property, that for random walks on the half line ergodic chains are also geometrically ergodic, holds in much wider generality. The crucial property is that the increment distribution have exponentially decreasing right tails, as we shall see in Section 16.1.3.

15.5.2 Autoregressive and bilinear models

Models common in time series, especially those with some autoregressive character, often converge geometrically quickly without the need to assume that the innovation distribution has exponential character. This is because the exponential "drift" of such models comes from control of the autoregressive terms, which "swamp" the linear drift of the innovation terms for large state space values. Thus the linear or quadratic functions used to establish simple ergodicity will satisfy the Foster criterion (V2), not merely in a linear way as is the case of random walk, but in fact in the stronger mode necessary to satisfy (15.28).

We will therefore often find that, for such models, we have already established geometric ergodicity by the steps used to establish simple ergodicity or even boundedness in probability, with no further assumptions on the structure of the model.

Simple linear models

Consider again the simple linear model defined in (SLM1) by

$$X_n = \alpha X_{n-1} + W_n$$

and assume W has an everywhere positive density so the chain is a ψ-irreducible T-chain. Now choosing $V(x) = |x| + 1$ gives

$$\mathsf{E}_x[V(X_1)] \leq |\alpha| V(x) + \mathsf{E}[|W|] + 1. \tag{15.49}$$

We noted in Proposition 11.4.2 that for large enough m, V satisfies (V2) with $C = C_V(m) = \{x : |x| + 1 \leq m\}$, provided that

$$\mathsf{E}[|W|] < \infty, \qquad |\alpha| < 1 :$$

thus $\{X_n\}$ admits an invariant probability measure under these conditions.

But now we can look with better educated eyes at (15.49) to see that V is in fact a solution to (15.28) under precisely these same conditions, and so we can strengthen Proposition 11.4.2 to give the conclusion that such simple linear models are geometrically ergodic.

Scalar bilinear models

We illustrate this phenomenon further by re-considering the scalar bilinear model, and examining the conditions which we showed in Section 12.5.2 to be sufficient for this model to be bounded in probability. Recall that X is defined by the bilinear process on $\mathsf{X} = \mathbb{R}$

$$X_{k+1} = \theta X_k + bW_{k+1}X_k + W_{k+1} \tag{15.50}$$

where W is i.i.d. From Proposition 7.1.3 we know when Φ is a T-chain.

To obtain a geometric rate of convergence, we reinterpret (12.36) which showed that

$$\mathsf{E}[|X_{k+1}| \mid X_k = x] \le \mathsf{E}[|\theta + bW_{k+1}|]|x| + \mathsf{E}[|W_{k+1}|] \tag{15.51}$$

to see that $V(x) = |x| + 1$ is a solution to (V4) provided that

$$\mathsf{E}[|\theta + bW_{k+1}|] < 1. \tag{15.52}$$

Under this condition, just as in the simple linear model, the chain is irreducible and aperiodic and thus again in this case we have that the chain is V-geometrically ergodic with $V(x) = |x| + 1$.

Suppose further that $\boldsymbol{W}$ has finite variance σ_w^2 satisfying

$$\theta^2 + b^2 \sigma_w^2 < 1;$$

exactly as in Section 14.4.2, we see that $V(x) = x^2$ is a solution to (V4) and hence $\boldsymbol{\Phi}$ is V-geometrically ergodic with this V. As a consequence, the chain admits a second order stationary distribution π with the property that for some $r > 1$ and $c < \infty$, and all x and n,

$$\sum_n r^n \left| \int P^n(x, dy) y^2 - \int \pi(dy) y^2 \right| < c(x^2 + 1).$$

Thus not only does the chain admit a second order stationary version, but the time dependent variances converge to the stationary variance.

15.6 Commentary*

Unlike much of the ergodic theory of Markov chains, the history of geometrically ergodic chains is relatively straightforward. The concept was introduced by Kendall in [202], where the existence of the solidarity property for countable space chains was first established: that is, if one transition probability sequence $P^n(i, i)$ converges geometrically quickly, so do all such sequences. In this seminal paper the critical renewal theorem (Theorem 15.1.1) was established.

The central result, the existence of the common convergence rate, is due to Vere-Jones [403] in the countable space case; the fact that no common best bound exists was also shown by Vere-Jones [403], with the more complex example given in Section 15.1.4 being due to Teugels [384]. Vere-Jones extended much of this work to non-negative matrices [405, 406], and this approach carries over to general state space operators [394, 395, 303].

Nummelin and Tweedie [307] established the general state space version of geometric ergodicity, and by using total variation norm convergence, showed that there is independence of A in the bounds on $|P^n(x, A) - \pi(A)|$, as well as an independent geometric rate. These results were strengthened by Nummelin and Tuominen [305], who also show as one important application that it is possible to use this approach to establish geometric rates of convergence in the Key Renewal Theorem of Section 14.5 if the increment distribution has geometric tails. Their results rely on a geometric trials argument to link properties of skeletons and chains: the drift condition approach here is new, as is most of the geometric regularity theory.

The upper bound in (15.4) was first observed by Chan [62]. Meyn and Tweedie [277] developed the f-geometric ergodicity approach, thus leading to the final form of Theorem 15.4.1; as discussed in the next chapter, this form has important operator-theoretic consequences, as pointed out in the case of countable X by Hordijk and Spieksma [163].

The drift function criterion was first observed by Popov [320] for countable chains, with general space versions given by Nummelin and Tuominen [305] and Tweedie [400]. The full set of equivalences in Theorem 15.0.1 is new, although much of it is implicit in Nummelin and Tweedie [307] and Meyn and Tweedie [277].

Initial application of the results to queueing models can be found in Vere-Jones [404] and Miller [284], although without the benefit of the drift criteria, such applications are hard work and restricted to rather simple structures. The bilinear model in Section 15.5.2 is first analyzed in this form in Feigin and Tweedie [111]. Further interpretation and exploitation of the form of (15.4) is given in the next chapter, where we also provide a much wider variety of applications of these results.

In general, establishing exact rates of convergence or even bounds on such rates remains (for infinite state spaces) an important open problem, although by analyzing Kendall's Theorem in detail Spieksma [367] has recently identified upper bounds on the area of convergence for some specific queueing models.

Added in second printing: There has now been a substantial amount of work on this problem, and quite different methods of bounding the convergence rates have been found by Meyn and Tweedie [282], Baxendale [21], Rosenthal [343, 342] and Lund and Tweedie [241]. However, apart from the results in [241] which apply only to stochastically monotone chains, none of these bounds are tight, and much remains to be done in this area.

Commentary for the second edition: This is an evolving research area, and one that is too large to summarize here. Section 20.1 contains a partial survey of the state-of-the-art of geometric ergodicity and its applications. Applications to queueing networks are surveyed in [267].

Chapter 16

V-Uniform ergodicity

In this chapter we introduce the culminating form of the geometric ergodicity theorem, and show that such convergence can be viewed as geometric convergence of an operator norm; simultaneously, we show that the classical concept of uniform (or strong) ergodicity, where the convergence in (13.4) is bounded independently of the starting point, becomes a special case of this operator norm convergence.

We also take up a number of other consequences of the geometric ergodicity properties proven in Chapter 15, and give a range of examples of this behavior. For a number of models, including random walk, time series and state space models of many kinds, these examples have been held back to this point precisely because the strong form of ergodicity we now make available is met as the norm, rather than as the exception. This is apparent in many of the calculations where we verified the ergodic drift conditions (V2) or (V3): often we showed in these verifications that the stronger form (V4) actually held, so that unwittingly we had proved V-uniform or geometric ergodicity when we merely looked for conditions for ergodicity.

To formalize V-uniform ergodicity, let P_1 and P_2 be Markov transition functions, and for a positive function $\infty > V \geq 1$, define the V-norm distance between P_1 and P_2 as

$$\|P_1 - P_2\|_V := \sup_{x \in \mathsf{X}} \frac{\|P_1(x, \cdot) - P_2(x, \cdot)\|_V}{V(x)}. \tag{16.1}$$

The *outer product* of the function 1 and the measure π is denoted

$$[1 \otimes \pi](x, A) = \pi(A), \qquad x \in \mathsf{X}, \ A \in \mathcal{B}(\mathsf{X}).$$

In typical applications we consider the distance $\|P^k - 1 \otimes \pi\|_V$ for large k.

V-uniform ergodicity

An ergodic chain $\boldsymbol{\Phi}$ is called V-*uniformly ergodic* if

$$\|P^n - 1 \otimes \pi\|_V \to 0, \qquad n \to \infty. \tag{16.2}$$

We develop three main consequences of Theorem 15.0.1 in this chapter.

Firstly, we interpret (15.4) in terms of convergence in the operator norm $\|P^k - 1 \otimes \pi\|_V$ when V satisfies (15.3), and consider in particular the uniformity of bounds on the geometric convergence in terms of such solutions of (V4). Showing that the choice of V in the term V-uniformly ergodic is not coincidental, we prove

Theorem 16.0.1. *Suppose that Φ is ψ-irreducible and aperiodic. Then the following are equivalent for any $V \geq 1$:*

(i) *Φ is V-uniformly ergodic.*

(ii) *There exist $r > 1$ and $R < \infty$ such that for all $n \in \mathbb{Z}_+$*

$$\|P^n - 1 \otimes \pi\|_V \leq Rr^{-n}. \tag{16.3}$$

(iii) *There exists some $n > 0$ such that $\|P^i - 1 \otimes \pi\|_V < \infty$ for $i \leq n$ and*

$$\|P^n - 1 \otimes \pi\|_V < 1. \tag{16.4}$$

(iv) *The drift condition (V4) holds for some petite set C and some V_0, where V_0 is equivalent to V in the sense that for some $c \geq 1$,*

$$c^{-1}V \leq V_0 \leq cV. \tag{16.5}$$

PROOF   That (i), (ii) and (iii) are equivalent follows from Proposition 16.1.3. The fact that (ii) follows from (iv) is proven in Theorem 16.1.2, and the converse, that (ii) implies (iv), is Theorem 16.1.4.   □

Secondly, we show that V-uniform ergodicity implies that the chain is *strongly mixing*. In fact, it is shown in Theorem 16.1.5 that for a V-uniformly ergodic chain, there exists R and $\rho < 1$ such that for any $g^2, h^2 \leq V$ and $k, n \in \mathbb{Z}_+$,

$$|\mathsf{E}_x[g(\Phi_k)h(\Phi_{n+k})] - \mathsf{E}_x[g(\Phi_k)]\mathsf{E}_x[h(\Phi_{n+k})]| \leq R\rho^n[1 + \rho^k V(x)].$$

Finally in this chapter, using the form (16.3), we connect concepts of geometric ergodicity with one of the oldest, and strongest, forms of convergence in the study of Markov chains, namely uniform ergodicity (sometimes called strong ergodicity).

Uniform ergodicity

A chain Φ is called *uniformly ergodic* if it is V-uniformly ergodic in the special case where $V \equiv 1$, that is, if

$$\sup_{x \in \mathsf{X}} \|P^n(x, \cdot) - \pi\| \to 0, \qquad n \to \infty. \tag{16.6}$$

There are a large number of stability properties all of which hold uniformly over the whole space when the chain is uniformly ergodic.

Theorem 16.0.2. *For any Markov chain* Φ *the following are equivalent:*

(i) Φ *is uniformly ergodic.*

(ii) *There exist* $r > 1$ *and* $R < \infty$ *such that for all* x

$$\|P^n(x, \cdot) - \pi\| \leq Rr^{-n}; \tag{16.7}$$

that is, the convergence in (16.6) takes place at a uniform geometric rate.

(iii) *For some* $n \in \mathbb{Z}_+$,

$$\sup_{x \in \mathsf{X}} \|P^n(x, \cdot) - \pi(\cdot)\| < 1. \tag{16.8}$$

(iv) *The chain is aperiodic and Doeblin's condition holds: that is, there is a probability measure* ϕ *on* $\mathcal{B}(\mathsf{X})$ *and* $\varepsilon < 1$, $\delta > 0$, $m \in \mathbb{Z}_+$ *such that whenever* $\phi(A) > \varepsilon$

$$\inf_{x \in \mathsf{X}} P^m(x, A) > \delta. \tag{16.9}$$

(v) *The state space* X *is* ν_m*-small for some* m.

(vi) *The chain is aperiodic and there is a petite set* C *with*

$$\sup_{x \in \mathsf{X}} \mathsf{E}_x[\tau_C] < \infty,$$

in which case for every set $A \in \mathcal{B}^+(\mathsf{X})$, $\sup_{x \in \mathsf{X}} \mathsf{E}_x[\tau_A] < \infty$.

(vii) *The chain is aperiodic and there is a petite set* C *and a* $\kappa > 1$ *with*

$$\sup_{x \in \mathsf{X}} \mathsf{E}_x[\kappa^{\tau_C}] < \infty,$$

in which case for every $A \in \mathcal{B}^+(\mathsf{X})$ *we have for some* $\kappa_A > 1$,

$$\sup_{x \in \mathsf{X}} \mathsf{E}_x[\kappa_A^{\tau_A}] < \infty.$$

(viii) *The chain is aperiodic and there is a bounded solution* $V \geq 1$ *to*

$$\Delta V(x) \leq -\beta V(x) + b \mathbb{I}_C(x), \quad x \in \mathsf{X} \tag{16.10}$$

for some $\beta > 0$, $b < \infty$, *and some petite set* C.

Under (v), we have in particular that for any x,

$$\|P^n(x, \cdot) - \pi\| \leq 2\rho^{\lfloor n/m \rfloor} \tag{16.11}$$

where $\rho = 1 - \nu_m(\mathsf{X})$.

PROOF   This cycle of results is proved in Theorem 16.2.1–Theorem 16.2.4.   □

Thus we see that uniform convergence can be embedded as a special case of V-geometric ergodicity, with V bounded; and by identifying the minorization that makes the whole space small we can explicitly bound the rate of convergence.

Clearly then, from these results geometric ergodicity is even richer, and the identification of test functions for geometric ergodicity even more valuable, than the last chapter indicated. This leads us to devote attention to providing a method of moving from ergodicity with a test function V to e^{sV}-geometric convergence, which in practice appears to be a natural tool for strengthening ergodicity to its geometric counterpart.

Throughout this chapter, we provide examples of geometric or uniform convergence for a variety of models. These should be seen as templates for the use of the verification techniques we have given in the theorems of the past several chapters.

16.1   Operator norm convergence

16.1.1   The operator norm $\|\cdot\|_V$

We first verify that $\|\cdot\|_V$ is indeed an operator norm.

Lemma 16.1.1. *Let L_V^∞ denote the vector space of all functions $f\colon \mathsf{X} \to \mathbb{R}_+$ satisfying*

$$|f|_V := \sup_{x\in\mathsf{X}} \frac{|f(x)|}{V(x)} < \infty.$$

If $\|P_1 - P_2\|_V$ is finite then $P_1 - P_2$ is a bounded operator from L_V^∞ to itself, and $\|P_1 - P_2\|_V$ is its operator norm.

PROOF   The definition of $\|\cdot\|_V$ may be restated as

$$
\begin{aligned}
\|P_1 - P_2\|_V &= \sup_{x\in\mathsf{X}}\left\{ \frac{\sup_{|g|\le V} |P_1(x,g) - P_2(x,g)|}{V(x)} \right\} \\
&= \sup_{|g|\le V}\sup_{x\in\mathsf{X}} \frac{|P_1(x,g) - P_2(x,g)|}{V(x)} \\
&= \sup_{|g|\le V} |P_1(\cdot,g) - P_2(\cdot,g)|_V \\
&= \sup_{|g|_V \le 1} |P_1(\cdot,g) - P_2(\cdot,g)|_V
\end{aligned}
$$

which is by definition the operator norm of $P_1 - P_2$ viewed as a mapping from L_V^∞ to itself.   □

We can put this concept together with the results of the last chapter to show

Theorem 16.1.2. *Suppose that Φ is ψ-irreducible and aperiodic and (V4) is satisfied with C petite and V everywhere finite. Then for some $r > 1$,*

$$\sum r^n \|P^n - 1 \otimes \pi\|_V < \infty, \tag{16.12}$$

and hence Φ is V-uniformly ergodic.

PROOF   This is largely a restatement of the result in Theorem 15.4.1. From Theorem 15.4.1 for some $R < \infty$, $\rho < 1$,

$$\|P^n(x, \cdot) - \pi\|_V \leq RV(x)\rho^n, \qquad n \in \mathbb{Z}_+,$$

and the theorem follows from the definition of $\|\cdot\|_V$. $\square$

Because $\|\cdot\|_V$ is a norm it is now easy to show that V-uniformly ergodic chains are always geometrically ergodic, and in fact V-geometrically ergodic.

Proposition 16.1.3. *Suppose that π is an invariant probability and that for some n_0,*

$$\|P - 1 \otimes \pi\|_V < \infty \qquad and \qquad \|P^{n_0} - 1 \otimes \pi\|_V < 1.$$

Then there exists $r > 1$ such that

$$\sum_{n=1}^{\infty} r^n \|P^n - 1 \otimes \pi\|_V < \infty.$$

PROOF   Since $\|\cdot\|_V$ is an operator norm we have for any $m, n \in \mathbb{Z}_+$, using the invariance of π,

$$\|P^{n+m} - 1 \otimes \pi\|_V = \|(P - 1 \otimes \pi)^n (P - 1 \otimes \pi)^m\|_V \leq \|P^n - 1 \otimes \pi\|_V \|P^m - 1 \otimes \pi\|_V.$$

For arbitrary $n \in \mathbb{Z}_+$ write $n = kn_0 + i$ with $1 \leq i \leq n_0$. Then since we have $\|P^{n_0} - 1 \otimes \pi\|_V = \gamma < 1$ and $\|P - 1 \otimes \pi\|_V \leq M < \infty$ this implies that (choosing $M \geq 1$ with no loss of generality)

$$\begin{aligned}
\|P^n - 1 \otimes \pi\|_V &\leq \|P - 1 \otimes \pi\|_V^i \|P^{n_0} - 1 \otimes \pi\|_V^k \\
&\leq M^i \gamma^k \\
&\leq M^{n_0} \gamma^{-1} (\gamma^{1/n_0})^n
\end{aligned}$$

which gives the claimed geometric convergence result. $\square$

Next we conclude the proof that V-uniform ergodicity is essentially equivalent to V solving the drift condition (V4).

Theorem 16.1.4. *Suppose that Φ is ψ-irreducible, and that for some $V \geq 1$ there exist $r > 1$ and $R < \infty$ such that for all $n \in \mathbb{Z}_+$*

$$\|P^n - 1 \otimes \pi\|_V \leq Rr^{-n}. \qquad (16.13)$$

Then the drift condition (V4) holds for some V_0, where V_0 is equivalent to V in the sense that for some $c \geq 1$,

$$c^{-1}V \leq V_0 \leq cV. \qquad (16.14)$$

PROOF   Fix $C \in \mathcal{B}^+(\mathsf{X})$ as any petite set. Then we have from (16.13) the bound

$$P^n(x, C) \geq \pi(C) - R\rho^n V(x)$$

and hence the sublevel sets of V are petite by Proposition 5.5.4 (i), and so V is unbounded off petite sets.

From the bound

$$P^n V \leq R\rho^n V + \pi(V) \qquad (16.15)$$

we see that (15.35) holds for the n-skeleton whenever $R\rho^n < 1$. Fix n with $R\rho^n < e^{-1}$, and set

$$V_0(x) := \sum_{i=0}^{n-1} \exp[i/n] P^i V.$$

We have that $V_0 > V$, and from (16.15),

$$V_0 \leq e^1 nRV + n\pi(V),$$

which shows that V_0 is equivalent to V in the required sense of (16.14).

From the drift (16.15) which holds for the n-skeleton we have

$$\begin{aligned}
PV_0 &= \sum_{i=1}^{n} \exp[i/n - 1/n] P^i V \\
&= \exp[-1/n] \sum_{i=1}^{n-1} \exp[i/n] P^i V + \exp[1 - 1/n] P^n V \\
&\leq \exp[-1/n] \sum_{i=1}^{n-1} \exp[i/n] P^i V + \exp[-1/n] V + \exp[1 - 1/n] \pi(V) \\
&= \exp[-1/n] V_0 + \exp[1 - 1/n] \pi(V).
\end{aligned}$$

This shows that (15.35) also holds for Φ, and hence by Lemma 15.2.8 the drift condition (V4) holds with this V_0, and some petite set C.   □

Thus we have proved the equivalence of (ii) and (iv) in Theorem 16.0.1.

16.1.2 V-geometric mixing and V-uniform ergodicity

In addition to the very strong total variation norm convergence that V-uniformly ergodic chains satisfy by definition, several other ergodic theorems and mixing results may be obtained for these stochastic processes. Much of Chapter 17 will be devoted to proving that the Central Limit Theorem, the Law of the Iterated Logorithm, and an invariance principle holds for V-uniformly ergodic chains. These results are obtained by applying the ergodic theorems developed in this chapter, and by exploiting the V-geometric regularity of these chains. Here we will consider a relatively simple result which is a direct consequence of the operator norm convergence (16.2).

A stochastic process $\boldsymbol{X}$ taking values in X is called *strong mixing* if there exists a sequence of positive numbers $\{\delta(n) : n \geq 0\}$ tending to zero for which

$$\sup |\mathsf{E}[g(X_k)h(X_{n+k})] - \mathsf{E}[g(X_k)]\mathsf{E}[h(X_{n+k})]| \leq \delta(n), \qquad n \in \mathbb{Z}_+,$$

where the supremum is taken over all $k \in \mathbb{Z}_+$, and all g and h such that $|g(x)|, |h(x)| \leq 1$ for all $x \in \mathsf{X}$.

In the following result we show that V-uniformly ergodic chains satisfy a much stronger property. We will call Φ *V-geometrically mixing* if there exists $R < \infty$, $\rho < 1$ such that

$$\sup |\mathsf{E}_x[g(\Phi_k)h(\Phi_{n+k})] - \mathsf{E}_x[g(\Phi_k)]\mathsf{E}_x[h(\Phi_{n+k})]| \leq RV(x)\rho^n, \qquad n \in \mathbb{Z}_+,$$

where we now extend the supremum to include all $k \in \mathbb{Z}_+$, and all g and h such that $g^2(x), h^2(x) \leq V(x)$ for all $x \in \mathsf{X}$.

Theorem 16.1.5. *If Φ is V-uniformly ergodic, then there exists $R < \infty$ and $\rho < 1$ such that for any g^2, $h^2 \leq V$ and $k, n \in \mathbb{Z}_+$,*

$$|\mathsf{E}_x[g(\Phi_k)h(\Phi_{n+k})] - \mathsf{E}_x[g(\Phi_k)]\mathsf{E}_x[h(\Phi_{n+k})]| \leq R\rho^n[1 + \rho^k V(x)],$$

and hence the chain Φ is V-geometrically mixing.

PROOF   For any $h^2 \leq V$, $g^2 \leq V$ let $\overline{h} = h - \pi(h)$, $\overline{g} = g - \pi(g)$. We have by $\sqrt{V}$-uniform ergodicity as in Lemma 15.2.9 that for some $R' < \infty$, $\rho < 1$,

$$\begin{aligned}
|\mathsf{E}_x[\overline{h}(\Phi_k)\overline{g}(\Phi_{k+n})]| &= |\mathsf{E}_x[\overline{h}(\Phi_k)\mathsf{E}_{\Phi_k}[\overline{g}(\Phi_n)]]| \\
&\leq R'\rho^n \mathsf{E}_x\left[|\overline{h}(\Phi_k)|\sqrt{V(\Phi_k)}\right].
\end{aligned}$$

Since $|\overline{h}| \leq \left(1 + \int V^{\frac{1}{2}} d\pi\right) V^{\frac{1}{2}}$ we can set $R'' = R'\left(1 + \int V^{\frac{1}{2}} d\pi\right)$ and apply (15.35) to obtain the bound

$$\begin{aligned}
|\mathsf{E}_x[\overline{h}(\Phi_k)\overline{g}(\Phi_{k+n})]| &\leq R''\rho^n \mathsf{E}_x[V(\Phi_k)] \\
&\leq R''\rho^n \left\{\frac{L}{1-\lambda} + \lambda^k V(x)\right\}.
\end{aligned}$$

Assuming without loss of generality that $\rho \geq \lambda$, and using the bounds

$$\begin{aligned}
|\pi(h) - \mathsf{E}_x[h(\Phi_k)]| &\leq R'''\rho^k \sqrt{V(x)}, \\
|\pi(g) - \mathsf{E}_x[g(\Phi_{k+n})]| &\leq R'''\rho^{k+n} \sqrt{V(x)}
\end{aligned}$$

gives the result for some $R < \infty$. $\square$

It follows from Theorem 16.1.5 that if the chain is V-uniformly ergodic, then for some $R_1 < \infty$,

$$|\mathsf{E}_x[\overline{h}(\Phi_k)\overline{g}(\Phi_{k+n})]| \leq R_1\rho^n[1 + \rho^k V(x)], \qquad k, n \in \mathbb{Z}_+, \tag{16.16}$$

where $\overline{h} = h - \pi(h)$, $\overline{g} = g - \pi(g)$.

By integrating both sides of (16.16) over X, the initial condition x may be replaced with a finite bound for any initial distribution μ with $\mu(V) < \infty$, and a mixing condition will be satisfied for such initial conditions. In the particular case where $\mu = \pi$ we have by stationarity and finiteness of $\pi(V)$ (see Theorem 14.3.7)

$$|\mathsf{E}_\pi[\overline{h}(\Phi_k)\overline{g}(\Phi_{k+n})]| \leq R_2\rho^n, \qquad k, n \in \mathbb{Z}_+, \tag{16.17}$$

for some $R_2 < \infty$; and hence the stationary version of the process satisfies a geometric mixing condition under (V4).

16.1.3 V-uniform ergodicity for regenerative models

In order to establish geometric ergodicity for specific models, we will obviously use the drift criterion (V4) to establish the required convergence. We begin by illustrating this for two regenerative models: we give many further examples later in the chapter.

For many models with some degree of spatial homogeneity, the crucial condition leading to geometric convergence involves exponential bounds on the increments of the process. Let us say that the distribution function G of a random variable is in $\mathcal{G}^+(\gamma)$ if G has a Laplace–Stieltjes transform convergent in $[0, \gamma]$: that is, if

$$\int_0^\infty e^{st} G(dt) < \infty, \qquad 0 < s \le \gamma, \tag{16.18}$$

where $\gamma > 0$.

Forward recurrence time chains

Consider the forward recurrence time δ-skeleton chain V_δ^+ defined by (RT3), based on increments with spread-out distribution Γ.

Suppose that $\Gamma \in \mathcal{G}^+(\gamma)$. By choosing $V(x) = e^{\gamma x}$ we have immediately that (V4) holds for $x \in C$ with $C = [0, \delta]$, and also

$$[V(x)]^{-1} \int P(x, dy) V(y) = e^{\gamma(x-\delta)}/e^{\gamma x} = e^{-\gamma \delta} < 1, \qquad x > \delta.$$

Thus (V4) also holds on C^c, and we conclude that the chain is $e^{\gamma x}$-uniformly ergodic. Moreover, from Theorem 16.0.1 we also have that

$$\int |P^n(x, dy) e^{\gamma y} - \pi(dy) e^{\gamma y}| < e^{\gamma x} r^{-n},$$

so that the moment-generating functions of the model, and moreover all polynomial moments, converge geometrically quickly to their limits with known bounds on the state-dependent constants.

This is the same result we showed in Section 15.1.4 for the forward recurrence time chain on $\mathbb{Z}_+$; here we have used the drift conditions rather than the direct calculation of hitting times to establish geometric ergodicity.

It is obvious from its construction that for this chain the condition $\Gamma \in \mathcal{G}^+(\gamma)$ is also necessary for geometric ergodicity.

The condition for uniform ergodicity for the forward recurrence time chain is also trivial to establish, from the criterion in Theorem 16.0.2 (vi). We will only have this condition holding if Γ is of bounded range so that $\Gamma[0, c] = 1$ for some finite c; in this case we may take the state space X equal to the compact absorbing set $[0, c]$. The existence of such a compact absorbing subset is typical of many uniformly ergodic chains in practice.

Random walk on $\mathbb{R}_+$

Consider now the random walk on $[0, \infty)$, defined by (RWHL1). Suppose that the model has an increment distribution Γ such that

(a) the mean increment $\beta = \int x\,\Gamma(dx) < 0$;

(b) the distribution Γ is in $\mathcal{G}^+(\gamma)$, for some $\gamma > 0$.

Let us choose $V(x) = \exp(sx)$, where $0 < s < \gamma$ is to be selected. Then we have

$$
\begin{aligned}
\int P(x, dy)\Delta V(y)/V(x) \;=\; & \int_{-x}^{\infty} \Gamma(dw)[\exp(sw) - 1] \\
& + \Gamma(-\infty, -x][\exp(-sx) - 1] \\
\leq \; & \int_{-\infty}^{\infty} \Gamma(dw)[\exp(sw) - 1] \\
& + \int_{-\infty}^{-x} \Gamma(dw)[1 - \exp(sw)].
\end{aligned}
\tag{16.19}
$$

But now if we let $s \downarrow 0$, then

$$
s^{-1}\int_{-\infty}^{\infty} \Gamma(dw)[\exp(sw) - 1] \to \beta < 0.
$$

Thus choosing s_0 sufficiently small that $\int_{-\infty}^{\infty} \Gamma(dw)[\exp(s_0 w) - 1] = \xi < 0$, and then choosing c large enough that

$$
\Gamma(-\infty, -x] \leq -\xi/2, \qquad x \geq c,
$$

we have that (V4) holds with $C = [0, c]$. Since C is petite for this chain, the random walk is $\exp(s_0 x)$-uniformly ergodic when (a) and (b) hold.

It is then again a consequence of Theorem 16.0.1 that the moment generating function, and indeed all moments, of the chain converge geometrically quickly.

Thus we see that the behavior of the Bernoulli walk in Section 15.5 is due, essentially, to the bounded and hence exponential nature of its increment distribution.

We will show in Section 16.3 that one can generalize this result to general chains, giving conditions for geometric ergodicity in terms of exponentially decreasing "tails" of the increment distributions.

16.2   Uniform ergodicity

16.2.1   Equivalent conditions for uniform ergodicity

From the definition (16.6), a Markov chain is uniformly ergodic if $\|P^n - 1 \otimes \pi\|_V \to 0$ as $n \to \infty$ when $V \equiv 1$. This simple observation immediately enables us to establish the first three equivalences in Theorem 16.0.2, which relate convergence properties of the chain.

Theorem 16.2.1. *The following are equivalent, without any a priori assumption of ψ-irreducibility or aperiodicity:*

(i) *Φ is uniformly ergodic.*

(ii) *There exists $\rho < 1$ and $R < \infty$ such that for all x*

$$
\|P^n(x, \cdot) - \pi\| \leq R\rho^n.
$$

(iii) *For some $n \in \mathbb{Z}_+$,*

$$\sup_{x \in \mathsf{X}} \| P^n(x, \cdot) - \pi(\cdot) \| < 1.$$

PROOF   Obviously (i) implies (iii); but from Proposition 16.1.3 we see that (iii) implies (ii), which clearly implies (i) as required.   □

Note that uniform ergodicity implies, trivially, that the chain actually is π-irreducible and aperiodic, since for $\pi(A) > 0$ there exists n with $P^n(x, A) \geq \pi(A)/2$ for all x.

We next prove that (v)–(viii) of Theorem 16.0.2 are equivalent to uniform ergodicity.

Theorem 16.2.2. *The following are equivalent for a ψ-irreducible aperiodic chain:*

(i) *Φ is uniformly ergodic.*

(ii) *The state space X is petite.*

(iii) *There is a petite set C with $\sup_{x \in \mathsf{X}} \mathsf{E}_x[\tau_C] < \infty$, in which case for every $A \in \mathcal{B}^+(\mathsf{X})$ we have $\sup_{x \in \mathsf{X}} \mathsf{E}_x[\tau_A] < \infty$.*

(iv) *There is a petite set C and a $\kappa > 1$ with $\sup_{x \in \mathsf{X}} \mathsf{E}_x[\kappa^{\tau_C}] < \infty$ in which case for every $A \subset \mathcal{B}^+(\mathsf{X})$ we have $\sup_{x \in \mathsf{X}} \mathsf{E}_x[\kappa_A^{\tau_A}] < \infty$ for some $\kappa_A > 1$.*

(v) *There is an everywhere bounded solution V to (16.10) for some petite set C.*

PROOF   Observe that the drift inequality (11.17) given in (V2) and the drift inequality (16.10) are identical for bounded V. The equivalence of (iii) and (v) is thus a consequence of Theorem 11.3.11, whilst (iv) implies (iii) trivially and Theorem 15.2.6 shows that (v) implies (iv): such connections between boundedness of τ_A and solutions of (16.10) are by now standard.

To see that (i) implies (ii), observe that if (i) holds, then Φ is π-irreducible and hence there exists a small set $A \in \mathcal{B}^+(\mathsf{X})$. Then, by (i) again, for some $n_0 \in \mathbb{Z}_+$, $\inf_{x \in \mathsf{X}} P^{n_0}(x, A) > 0$ which shows that X is small from Theorem 5.2.4.

The implication that (ii) implies (v) is equally simple. Let $V \equiv 1$, $\beta = b = \frac{1}{2}$, and $C = \mathsf{X}$. We then have

$$\Delta V = -\beta V + b \mathbb{I}_C,$$

giving a bounded solution to (16.10) as required.

Finally, when (v) holds, we immediately have uniform geometric ergodicity by Theorem 16.1.2.   □

Historically, one of the most significant conditions for ergodicity of Markov chains is *Doeblin's condition*.

Doeblin's condition

Suppose there exists a probability measure ϕ with the property that for some m, $\varepsilon < 1, \delta > 0$

$$\phi(A) > \varepsilon \Longrightarrow P^m(x, A) \geq \delta$$

for every $x \in \mathsf{X}$.

From the equivalences in Theorem 16.2.1 and Theorem 16.2.2, we are now in a position to give a very simple proof of the equivalence of uniform ergodicity and this condition.

Theorem 16.2.3. *An aperiodic ψ-irreducible chain Φ satisfies Doeblin's condition if and only if Φ is uniformly ergodic.*

PROOF   Let C be any petite set with $\phi(C) > \varepsilon$ and consider the test function

$$V(x) = 1 + \mathbb{I}_{C^c}(x).$$

Then from Doeblin's condition

$$
\begin{aligned}
P^m V(x) - V(x) = P^m(x, C^c) - \mathbb{I}_{C^c}(x) &\leq 1 - \delta - \mathbb{I}_{C^c}(x) \\
&= -\delta + \mathbb{I}_C(x) \\
&\leq -\tfrac{1}{2}\delta V(x) + \mathbb{I}_C(x).
\end{aligned}
$$

Hence V is a bounded solution to (16.10) for the m-skeleton, and it is thus the case that the m-skeleton and the original chain are uniformly ergodic by the contraction property of the total variation norm.

Conversely, we have from uniform ergodicity in the form (16.7) that for any $\varepsilon > 0$, if $\pi(A) \geq \varepsilon$ then

$$P^n(x, A) \geq \varepsilon - R\rho^n \geq \varepsilon/2$$

for all n large enough that $R\rho^n \leq \varepsilon/2$, and Doeblin's condition holds with $\phi = \pi$.   □

Thus we have proved the final equivalence in Theorem 16.0.2. We conclude by exhibiting the one situation where the bounds on convergence are simply calculated.

Theorem 16.2.4. *If a chain Φ satisfies*

$$P^m(x, A) \geq \nu_m(A) \tag{16.20}$$

for all $x \in \mathsf{X}$ and $A \in \mathcal{B}(\mathsf{X})$, then

$$\|P^n(x, \cdot) - \pi\| \leq 2\rho^{\lfloor n/m \rfloor} \tag{16.21}$$

where $\rho = 1 - \nu_m(\mathsf{X})$.

PROOF   This can be shown using an elegant argument based on the assumption (16.20) that the whole space is small which relies on a coupling method closely connected to the way in which the split chain is constructed.

Write (16.20) as

$$P^m(x, A) \geq (1 - \rho)\nu(A) \tag{16.22}$$

where $\nu = \nu_m/(1 - \rho)$ is a probability measure.

Assume first for simplicity that $m = 1$. Run two copies of the chain, one from the initial distribution concentrated at x and the other from the initial distribution π. At every time point either

(a) with probability $1 - \rho$, choose for both chains the same next position from the distribution ν, after which they will be coupled and then can be run with identical sample paths; or

(b) with probability ρ, choose for each chain an independent position, using the distribution (as in the split chain construction) $[P(x, \cdot) - (1 - \rho)\nu(\cdot)]/\rho$, where x is the current position of the chain.

This is possible because of the minorization in (16.22). The marginal distributions of these chains are identical with the original distributions, for every n. If we let T denote the first time that the chains are chosen using the first option (a), then we have

$$\|P^n(x, \cdot) - \pi\| \leq 2P(T > n) \leq 2\rho^n \tag{16.23}$$

which is (16.21).

When $m > 1$ we can use the contraction property as in Proposition 16.1.3 to give (16.21) in the general case.   □

The optimal use of these many equivalent conditions for uniform ergodicity depends of course on the context of use. In practice, this last theorem, since it identifies the exact rate of convergence, is perhaps the most powerful, and certainly gives substantial impetus to identifying the actual minorization measure which renders the whole space a small set.

It can also be of importance to use these conditions in assessing when uniform convergence does not hold: for example, in the forward recurrence time chain V_δ^+ it is immediate from Theorem 16.2.2 (iii) that, since the mean return time to $[0, \delta]$ from x is of order x, the chain cannot be uniformly ergodic unless the state space can be reduced to a compact set.

Similar remarks apply to random walk on the half line: we see this explicitly in the simple random walk of Section 15.5, but it is a rather deeper result [69] that for general random walk on $[0, \infty)$, $E_x[\tau_0] \sim cx$ so such chains are never uniformly ergodic.

16.2.2   Geometric convergence of given moments

It is instructive to note that, although the concept of uniform ergodicity is a very strong one for convergence of distributions, it need not have any implications for the convergence of moments or other unbounded functionals of the chain at a geometric rate.

This is obviously true in a trivial sense: an i.i.d. sequence Φ_n converges in a uniformly ergodic manner, regardless of whether $\mathsf{E}[\Phi_n]$ is finite or not.

But rather more subtly, we now show that it is possible for us to construct a uniformly ergodic chain with convergence rate ρ such that $\pi(f) < \infty$, so that we know $\mathsf{E}_x[f(\Phi_n)] \to \pi(f)$, but where not only does this convergence not take place at rate ρ, it actually does not take place at any geometric rate at all.

For convenience of exposition we construct this chain on a countable ladder space $\mathsf{X} = \mathbb{Z}_+ \times \mathbb{Z}_+$, even though the example is essentially one-dimensional.

Fix $\beta < 1/4$, and define for the ith rung of the ladder the indices

$$\ell^m(i) := \lfloor (\frac{i-1}{i\beta})^m \rfloor, \qquad i \geq 1, m \geq 0.$$

Note that for $i = 1$ we have $\ell^m(1) = 0$ for all m, but for $i > 1$

$$(\frac{i-1}{i\beta})^{m+1} - (\frac{i-1}{i\beta})^m = (\frac{i-1}{i\beta})^m (\frac{i-1-i\beta}{i\beta}) \geq 1$$

since $(i - 1 - i\beta)/i\beta \geq (3i - 1)/i \geq 2$. Hence from the second rung up, this sequence $\ell^m(i)$ forms a strictly monotone increasing set of states along the rung.

The transition mechanism we consider provides a chain satisfying Doeblin's condition. We suppose P is given by

$$P(i, \ell^m(i); i, \ell^{m+1}(i)) = \beta, \qquad i = 1, 2, \ldots, m = 1, 2, \ldots,$$

$$P(i, \ell^m(i); 0, 0) = 1 - \beta, \qquad i = 1, 2, \ldots, m = 1, 2, \ldots,$$

$$P(i, k; 0, 0) = 1, \qquad i = 1, 2, \ldots, k \neq \ell^m(i), m = 1, 2, \ldots, \quad (16.24)$$

$$P(0, 0; i, j) = \alpha_{ij}, \qquad i, j \in \mathsf{X},$$

$$P(0, k; 0, 0) = 1, \qquad k > 0,$$

where the α_{ij} are to be determined, with $\alpha_{00} > 0$.

In effect this chain moves only on the states $(0, 0)$ and the sequences $\ell^m(i)$, and the whole space is small with

$$P(i, k; \cdot) \geq \min(1 - \beta, \alpha_{00})\delta_{00}(\cdot).$$

Thus the chain is clearly uniformly and hence geometrically ergodic.

Now consider the function f defined by $f(i, k) = k$; that is, f denotes the distance of the chain along the rung independent of the rung in question. We show that the chain is f-ergodic but not f-geometrically ergodic, under suitable choice of the distribution α_{ij}.

First note that we can calculate

$$\mathsf{E}_{i,1}[\textstyle\sum_0^{\tau_{0,0}-1} f(\Phi_n)] = (1-\beta)\sum_{n=0}^{\infty}\beta^n\sum_{m=0}^{n}\ell^m(i)$$

$$\leq (1-\beta)\sum_{n=0}^{\infty}\beta^n\sum_{m=0}^{n}\left(\tfrac{i-1}{i\beta}\right)^m$$

$$= i;$$

$$\mathsf{E}_{i,\ell^m(i)}[\textstyle\sum_0^{\tau_{0,0}-1} f(\Phi_n)] \leq \left(\tfrac{i-1}{i\beta}\right)^m i, \qquad m=1,2,\dots;$$

$$\mathsf{E}_{i,k}[\textstyle\sum_0^{\tau_{0,0}-1} f(\Phi_n)] = k, \qquad k\neq\ell^m(i),\ m=1,2,\dots.$$

Now let us choose

$$\begin{aligned}
\alpha_{ik} &= c2^{-i-k}, & k\neq\ell^m(i),\ m=1,2,\dots;\\
\alpha_{ik} &= c\sum_{m=0}^{\infty}2^{-i-\ell^m(i)}, & k=1,
\end{aligned}$$

and all other values except α_{00} as zero, and where c is chosen to ensure that the α_{ik} form a probability distribution.

With this choice we have

$$\mathsf{E}_{0,0}[\textstyle\sum_0^{\tau_{0,0}-1} f(\Phi_n)] \leq 1 + \sum_{i\geq 1}\sum_{k\neq\ell^m(i),m\geq 0}k2^{-i-k} + \sum_{i\geq 1}\Big[\sum_{m=0}^{\infty}2^{-i-\ell^m(i)}\Big]i$$

$$\leq 1 + 2\sum_{i\geq 1}i2^{-i} < \infty$$

so that the chain is certainly f-ergodic by Theorem 14.0.1. However for any $r\in(1,\beta^{-1})$,

$$\mathsf{E}_{i,1}[\textstyle\sum_0^{\tau_{0,0}-1} f(\Phi_n)r^n] = (1-\beta)\sum_{n=0}^{\infty}\beta^n r^n\sum_{m=0}^{n}\ell^m(i)$$

$$\geq (1-\beta)\sum_{n=0}^{\infty}(\beta r)^n\sum_{m=0}^{n}\Big[\left(\tfrac{i-1}{i\beta}\right)^m - 1\Big]$$

$$= -\left(\tfrac{1-\beta}{1-\beta r}\right) + \sum_{n=0}^{\infty}(\beta r)^n\Big[\tfrac{[(i-1)/i\beta]^{n+1}-1}{[(i-1)/i\beta]-1}\Big]$$

which is infinite if

$$\beta r\big[\tfrac{i-1}{i\beta}\big] > 1;$$

that is, for those rungs i such that $i > r/(r-1)$. Since there is positive probability of reaching such rungs in one step from $(0,0)$ it is immediate that

$$\mathsf{E}_{0,0}\Big[\sum_0^{\tau_{0,0}-1} f(\Phi_n)r^n\Big] = \infty$$

for all $r>1$, and hence from Theorem 15.4.2 for all $r>1$

$$\sum_n r^n\|P^n(0,0;\,\cdot\,)-\pi\|_f = \infty.$$

Since $\{0,0\}\in\mathcal{B}^+(\mathsf{X})$, this implies that $\|P^n(x;\,\cdot\,)-\pi\|_f$ is not $o(\rho^n)$ for any x or any $\rho < 1$.

We have thus demonstrated that the strongest rate of convergence in the simple total variation norm may not be inherited, even by the simplest of unbounded functions; and that one really needs, when considering such functions, to use criteria such as (V4) to ensure that these functions converge geometrically.

16.2.3   Uniform ergodicity: T-chains on compact spaces

For T-chains, we have an almost trivial route to uniform ergodicity, given the results we now have available.

Theorem 16.2.5. *If Φ is a ψ-irreducible and aperiodic T-chain, and if the state space X is compact, then Φ is uniformly ergodic.*

PROOF   If Φ is a ψ-irreducible T-chain, and if the state space X is compact, then it follows directly from Theorem 6.0.1 that X is petite. Applying the equivalence of (i) and (ii) given in Theorem 16.2.2 gives the result.   □

One specific model, the nonlinear state space model, is also worth analyzing in more detail to show how we can identify other conditions for uniform ergodicity.

The NSS(F) model

In a manner similar to the proof of Theorem 16.2.5 we show that the the NSS(F) model defined by (NSS1) and (NSS2) is uniformly ergodic, provided that the associated control model CM(F) is stable in the sense of Lagrange, so that in effect the state space is reduced to a compact invariant subset.

Lagrange stability

The CM(F) model is called *Lagrange stable* if $\overline{A_+(x)}$ is compact for each $x \in$ X.

Typically in applications, when the CM(F) model is Lagrange stable the input sequence will be constrained to lie in a bounded subset of $\mathbb{R}^p$. We stress however that no conditions on the input are made in the general definition of Lagrange stability.

The key to analyzing the NSS(F) corresponding to a Lagrange stable control model lies in the following lemma:

Lemma 16.2.6. *Suppose that the CM(F) model is forward accessible, Lagrange stable, M-irreducible and aperiodic, and suppose that for the NSS(F) model conditions (NSS1)– (NSS3) are satisfied.*

Then for each $x \in$ X the set $\overline{A_+(x)}$ is closed, absorbing, and small.

PROOF   By Lagrange stability it is sufficient to show that any compact and invariant set $C \subset \mathsf{X}$ is small. This follows from Theorem 7.3.5 (ii), which implies that compact sets are small under the conditions of the lemma.   □

Using Lemma 16.2.6 we now establish geometric convergence of the expectation of functions of $\boldsymbol{\Phi}$:

Theorem 16.2.7. *Suppose the NSS(F) model satisfies conditions (NSS1)–(NSS3) and that the associated control model CM(F) is forward accessible, Lagrange stable, M-irreducible and aperiodic.*

Then a unique invariant probability π exists, and the chain restricted to the absorbing set $\overline{A_+(x)}$ is uniformly ergodic for each initial condition.

Hence also for every function $f\colon \mathsf{X} \to \mathbb{R}$ which is uniformly bounded on compact sets, and every initial condition,

$$\mathsf{E}_y[f(\Phi_k)] \to \int f \, d\pi$$

at a geometric rate.

PROOF   When $\mathrm{CM}(F)$ is forward accessible, M-irreducible and aperiodic, we have seen in Theorem 7.3.5 that the Markov chain $\boldsymbol{\Phi}$ is ψ-irreducible and aperiodic.

The result then follows from Lemma 16.2.6: the chain restricted to $\overline{A_+(x)}$ is uniformly ergodic by Theorem 16.0.2.   □

16.3   Geometric ergodicity and increment analysis

16.3.1   Strengthening ergodicity to geometric ergodicity

It is possible to give a "generic" method of establishing that (V4) holds when we have already used the test function approach to establishing simple (non-geometric) ergodicity through Theorem 13.0.1. This method builds on the specific technique for random walks, shown in Section 16.1.3 above, and is an increment-based method similar to that in Section 9.5.1.

Suppose that V is a test function for regularity. We assume that V takes on the "traditional" form due to Foster: V is finite valued, and for some petite set C and some constant $b < \infty$, we have

$$\int P(x, dy)V(y) \leq \begin{cases} V(x) - 1 & \text{for } x \in C^c, \\ b & \text{for } x \in C. \end{cases} \tag{16.25}$$

Recall that $V_C(x) = \mathsf{E}_x[\sigma_C]$ is the minimal solution to (16.25) from Theorem 11.3.5.

Theorem 16.3.1. *If $\boldsymbol{\Phi}$ is a ψ-irreducible ergodic chain and V is a test function satisfying (16.25), and if P satisfies, for some $c, d < \infty$ and $\beta > 0$, and all $x \in \mathsf{X}$,*

$$\int_{V(y) \geq V(x)} P(x, dy) \exp\{\beta(V(y) - V(x))\} \leq c \tag{16.26}$$

and

$$\int_{V(y)<V(x)} P(x,dy)(V(y)-V(x))^2 \le d, \tag{16.27}$$

then Φ is V^*-uniformly ergodic, where $V^*(y) = e^{\delta V(y)}$ for some $\delta < \beta$.

PROOF   For positive $\delta < \beta$ we have

$$[V^*(x)]^{-1} \int P(x,dy)V^*(y) = \int P(x,dy)\exp\{\delta(V(y)-V(x))\}$$

$$= \int P(x,dy)\Big\{1 + \delta(V(y)-V(x))$$

$$+ \tfrac{\delta^2}{2}(V(y)-V(x))^2 \exp\{\delta\theta_x(V(y)-V(x))\}\Big\} \tag{16.28}$$

for some $\theta_x \in [0,1]$, by using a second order Taylor expansion. Since V satisfies (16.25), the right hand side of (16.28) is bounded for $x \in C^c$ by

$$1 - \delta + \tfrac{\delta^2}{2}\Big\{\int_{V(y)<V(x)} P(x,dy)(V(y)-V(x))^2$$

$$+ \int_{V(y)\ge V(x)} P(x,dy)\big((V(y)-V(x))^2 \exp\{\delta(V(y)-V(x))\}\big)\Big\}$$

$$\le 1 - \delta + \tfrac{\delta^2}{2}d + \tfrac{\delta^{2-\xi}}{2}\int_{V(y)\ge V(x)} P(x,dy)\exp\{(\delta+\delta^{\xi/2})(V(y)-V(x))\}$$

$$\le 1 - \delta + \tfrac{\delta^{2-\xi}}{2}(d+c), \tag{16.29}$$

for some $\xi \in (0,1)$ such that $\delta + \delta^{\xi/2} < \beta$ by virtue of (16.26) and (16.27), and the fact that x^2 is bounded by e^x on $\mathbb{R}_+$. This proves the theorem, since we have

$$1 - \delta + \frac{\delta^{2-\xi}}{2}(d+c) < 1$$

for sufficiently small $\delta > 0$, and thus (V4) holds for V^*.   □

The typical example of this behavior, on which this proof is modeled, is the random walk in Section 16.1.3. In that case $V(x) = x$, and (16.26) is the requirement that $\Gamma \in \mathcal{G}^+(\gamma)$. In this case we do not actually need (16.27), which may not in fact hold.

It is often easier to verify the conditions of this theorem than to evaluate directly the existence of a test function for geometric ergodicity, as we shall see in the next section.

How necessary are the conditions of this theorem on the "tails" of the increments? By considering for example the forward recurrence time chain, we see that for some chains $\Gamma \in \mathcal{G}^+(\gamma)$ may indeed be necessary for geometric ergodicity. However, geometric tails are certainly not always necessary for geometric ergodicity: to demonstrate this simply consider any i.i.d. process, which is trivially uniformly ergodic, regardless of its "increment" structure.

It is interesting to note, however, that although they seem somewhat "proof dependent", the uniform bounds (16.26) and (16.27) on P that we have imposed cannot be weakened in general when moving from ergodicity to geometric ergodicity.

We first show that we can ensure lack of geometric ergodicity if the drift to the right is not uniformly controlled in terms of V as in (16.26), even for a chain satisfying all our other conditions. To see this we consider a chain on $\mathbb{Z}_+$ with transition matrix given by, for each $i \in \mathbb{Z}_+$,

$$\begin{aligned}
P(0, i) &= \alpha_i > 0, \\
P(i, i-1) &= \gamma_i > 0, \\
P(i, i+n) &= [1 - \gamma_i][1 - \beta_i]\beta_i^n, \qquad n \in \mathbb{Z}_+.
\end{aligned} \qquad (16.30)$$

where $\sum \alpha_i = 1$ and γ_i, β_i are less than unity for all i.

Provided $\sum i\alpha_i < \infty$ and we choose γ_i sufficiently large that

$$[1 - \gamma_i]\beta_i/[1 - \beta_i] - \gamma_i \leq -\varepsilon$$

for some $\varepsilon > 0$, then the chain is ergodic since $V(x) = x$ satisfies (V2): this can be done if we choose, for example,

$$\gamma_i \geq \beta_i + \varepsilon[1 - \beta_i].$$

And now if we choose $\beta_j \to 1$ as $j \to \infty$ we see that the chain is not geometrically ergodic: we have for any j

$$\mathsf{P}_j(\tau_0 > n) \geq [1 - \gamma_j][1 - \beta_j]\beta_j^n$$

so $\mathsf{P}_0(\tau_0 > n)$ does not decrease geometrically quickly, and the chain is not geometrically ergodic from Theorem 15.4.2 (or directly from Theorem 15.1.1).

In this example we have bounded variances for the left tails of the increment distributions, and exponential tails of the right increments: it is the lack of uniformity in these tails that fails along with the geometric convergence.

To show the need for (16.27), consider the chain on $\mathbb{Z}_+$ with the transition matrix (15.20) given for all $j \in \mathbb{Z}_+$ by $P(0,0) = 0$ and

$$P(0, j) = \gamma_j > 0, \qquad P(j, j) = \beta_j, \qquad P(j, 0) = 1 - \beta_j,$$

where $\sum_j \gamma_j = 1$. We saw in Section 15.1.4 that if $\beta_j \to 1$ as $n \to \infty$, the chain cannot be geometrically ergodic regardless of the structure of the distribution $\{\gamma_j\}$.

If we consider the minimal solution to (16.25), namely

$$V_0(j) = \mathsf{E}_j[\sigma_0] = [1 - \beta_j]^{-1}, \qquad j > 0,$$

then clearly the right hand increments are uniformly bounded in relation to V for $j > 0$: but we find that

$$\sum P(i, j)(V_0(j) - V_0(i))^2 = P(i, 0)[1 - \beta_i]^{-2} = [1 - \beta_i]^{-1} \to \infty, \qquad i \to \infty.$$

Hence (16.27) is necessary in this model for the conclusion of Theorem 16.3.1 to be valid.

16.3.2   Geometric ergodicity and the structure of π

The relationship between spatial and temporal geometric convergence in the previous section is largely a result of the spatial homogeneity we have assumed when using increment analysis.

We now show that this type of relationship extends to the invariant probability measure π also, at least in terms of the "natural" ordering of the space induced by petite sets and test functions.

Let us we write, for any function g,

$$A_{g,n}(x) = \{y : g(y) \le g(x) - n\}.$$

We say that the chain is "g-skip-free to the left" if there is some $k \in \mathbb{Z}_+$, such that for all $x \in \mathsf{X}$,

$$P(x, A_{g,k}(x)) = 0, \tag{16.31}$$

so that the chain can only move a limited amount of "distance" through the sublevel sets of g in one step. Note that such skip-free behavior precludes Doeblin's condition if g is unbounded off petite sets, and requires a more random-walk-like behavior.

Theorem 16.3.2. *Suppose that Φ is geometrically ergodic. Then there exists $\beta > 0$ such that*

$$\int \pi(dy) e^{\beta V_C(y)} < \infty \tag{16.32}$$

where $V_C(y) = \mathsf{E}_y[\sigma_C]$ for any petite set $C \in \mathcal{B}^+(\mathsf{X})$.

If Φ is g-skip-free to the left for a function g which is unbounded off petite sets, then for some $\beta' > 0$

$$\int \pi(dy) e^{\beta' g(y)} < \infty. \tag{16.33}$$

PROOF   From geometric ergodicity, we have from Theorem 15.2.4 that for any petite set $C \in \mathcal{B}^+(\mathsf{X})$ there exists $r > 1$ such that $V(y) = G_C^{(r)}(y, \mathsf{X})$ satisfies (V4). It follows from Theorem 14.3.7 that $\pi(V) < \infty$. Using the interpretation (15.29) we have that

$$\infty > \pi(V) \ge \int \pi(dy) \mathsf{E}_y[r^{\sigma_C}]. \tag{16.34}$$

Now the function $f(j) = z^j$ is convex in $j \in \mathbb{Z}_+$, so that $\mathsf{E}_x[r^{\sigma_C}] \ge r^{\mathsf{E}_x[\sigma_C]}$ by Jensen's inequality. Thus we have (16.32) as desired.

Now suppose that g is such that the chain is g-skip-free to the left, and fix b so that the petite set $C = \{y : g(y) \le b\}$ is in $\mathcal{B}^+(\mathsf{X})$. Because of the left skip-free property (16.31), for $g(x) \ge nk + b$, we have $\mathsf{P}_x(\sigma_C \le n) = 0$ so that $\mathsf{E}_x[r^{\sigma_C}] \ge r^{(g(x)-b)/k}$.

As $\int \pi(dx) \mathsf{E}_x[r^{\sigma_C}] < \infty$ by virtue of (16.34), we have thus proved the second part of the theorem for $e^\beta = \sqrt[k]{r}$. $\square$

This result shows two things; firstly, if we think of V_C (or equivalently $G_C(x, \mathsf{X})$) as providing a natural scaling of the space in some way, then geometrically ergodic chains do have invariant measures with geometric "tails" in this scaling.

Secondly, and in practice more usefully, we have an identifiable scaling for such tails in terms of a "skip-free" condition, which is frequently satisfied by models in queueing applications on $\mathbb{Z}^n$ in particular. For example, if we embed a model at the departure times in such applications, and a limited number of customers leave each time, we get a skip-free condition holding naturally. Indeed, in all of the queueing models of the next section this condition is satisfied, so that this theorem can be applied there.

To see that geometric ergodicity and conditions on π such as (16.33) are not always linked in the given topology on the space, however, again consider any i.i.d. chain. This is always uniformly ergodic, regardless of π: the rescaling through g_C here is too trivial to be useful.

In the other direction, consider again the chain on $\mathbb{Z}_+$ with the transition matrix given for all $j \in \mathbb{Z}_+$ by

$$P(0,j) = \gamma_j, \qquad P(j,j) = \beta_j, \qquad P(j,0) = 1 - \beta_j,$$

where $\sum_j \gamma_j = 1$: we know that if $\beta_j \to 1$ as $n \to \infty$, the chain is not geometrically ergodic. But for this chain, since we know that $\pi(j)$ is proportional to

$$\mathsf{E}_0[\text{Number of visits to } j \text{ before return to } 0],$$

we have

$$\pi(j) \propto \gamma_j [1 - \beta_j]^{-1}$$

and so for suitable choice of γ_j we can clearly ensure that the tails of π are geometric or otherwise in the given topology, regardless of the geometric ergodicity of P.

16.4 Models from queueing theory

We further illustrate the use of these theorems through the analysis of three queueing systems.

These are all models on $\mathbb{Z}_+^n$ and their analysis consists of showing that there exists $\varepsilon_1, \varepsilon_2 > 0$, such that $\varepsilon_1 |i|_1 \le V(i) \le \varepsilon_2 |i|_1$, where V is the minimal solution to (16.25) and $|i|_1$ is the ℓ_1-norm on $\mathbb{Z}_+^n$; we then find that Φ is V^*-uniformly ergodic for $V^*(i) = e^{\delta V(i)}$, so that in particular we conclude that V^* is bounded above and below by exponential functions of $|i|_1$ for these models.

Typically in all of these examples the key extra assumption needed to ensure geometric ergodicity is a geometric tail on the distributions involved: that is, the increment distributions are in $\mathcal{G}^+(\gamma)$ for some γ. Recall that this was precisely the condition used for regenerative models in Section 16.1.3.

16.4.1 The embedded M/G/1 queue N_n

The M/G/1 queue exemplifies the steps needed to apply Theorem 16.3.1 in queueing models.

Theorem 16.4.1. *If Φ the Markov chain N_n defined by (Q4) is ergodic, then Φ is also geometrically ergodic provided the service time distributions are in $\mathcal{G}^+(\gamma)$ for some $\gamma > 0$.*

PROOF   We have seen in Section 11.4 that $V(i) = i$ is a solution to (16.25) with $C = \{0\}$.

Let us now assume that the service time distribution $H \in \mathcal{G}^+(\gamma)$. We prove that (16.26) and (16.27) hold. Application of Theorem 16.3.1 then proves V^*-uniform ergodicity of the embedded Markov chain where $V^*(i) = e^{\delta i}$ for some $\delta > 0$.

Let a_k denote the probability of k arrivals within one service. Note that (16.27) trivially holds, since $\sum_{j \le k} P(k, j)(j - k)^2 \le a_0$. For $l \ge 0$ we have

$$P(k, k + l) = a_{l+1} = \frac{1}{(l+1)!} \int_0^\infty e^{-\lambda t}(\lambda t)^{l+1} dH(t).$$

Let $\delta > 0$, so that

$$\sum_{l \ge 0} e^{\delta(l+1)} P(k, k + l) \le \int_0^\infty \exp\{(e^\delta - 1)\lambda t\} dH(t)$$

which is assumed to be finite for $(e^\delta - 1)\lambda < \gamma$. Thus we have the result.   □

16.4.2   A gated-limited polling system

We next consider a somewhat more complex multidimensional queueing model. Consider a system consisting of K infinite capacity queues and a single server.

The server visits the queues in order (hence the name "polling system") and during a visit to queue k the server serves $\min(x, \ell_k)$ customers, where x is the number of customers present at queue k at the instant the server arrives there: thus ℓ_k is the "gate limit".

To develop a Markovian representation, this system is observed at each instant the server arrives back at queue 1: the queue lengths at the respective queues are then recorded. We thus have a K-dimensional state description $\mathbf{\Phi}_n = \mathbf{\Phi}_n^k$, where $\mathbf{\Phi}_n^k$ stands for the number of customers in queue k at the server's n^{th} visit to queue 1.

The arrival stream at queue k is assumed to be a Poisson stream with parameter λ_k; the amount of service given to a queue k customer is drawn from a general distribution with mean μ_k^{-1}.

To make the process $\mathbf{\Phi}$ a Markov chain we assume that the sequence of service times to queue k are i.i.d. random variables. Moreover, the arrival streams and service times are assumed to be independent of each other.

Theorem 16.4.2. *The gated-limited polling model $\mathbf{\Phi}$ described above is geometrically ergodic provided*

$$1 > \rho := \sum_k \lambda_k / \mu_k \tag{16.35}$$

and the service time distributions are in $\mathcal{G}^+(\gamma)$ for some γ.

PROOF   It is straightforward to show that $\mathbf{\Phi}$ is ergodic for the gated-limited service discipline when (16.35) holds, by identifying a drift function that is linear in the number of customers in the respective queues: specifically $V(i) = \sum_{k=1}^K i_k / \mu_k$, where i is a K-dimensional vector with kth component i_k, can easily be shown to satisfy (16.25).

To apply the results in this section, observe that for this embedded chain there are only finitely many different possible one-step increments, depending on whether Φ_n^k exceeds ℓ_k or equals $x < \ell_k$. Combined with the linearity of V, we conclude that both sums

$$\{ \sum_{j:V(j)\geq V(i)} P(i,j)e^{\lambda(V(j)-V(i))} : i \in \mathsf{X}\}$$

and

$$\{ \sum_{j:V(j)<V(i)} P(i,j)(V(j)-V(i))^2 : i \in \mathsf{X}\}$$

have only finitely many non-zero elements. We must ensure that these expressions are all finite, but it is straightforward to check as in Theorem 16.4.1 that convergence of the Laplace–Stieltjes transforms of the service time distributions in a neighborhood of 0 is sufficient to achieve this, and the theorem follows. $\square$

16.4.3   A queue with phase-type service times

In many cases of ergodic chains there are no closed form expressions for the drift function, even though it follows from Chapter 11 that such functions exist. However, once ergodicity has been established, we do know by minimality that the function $V_C(x) = \mathsf{E}_x[\sigma_C]$ is a finite solution to (16.25). We now consider a queueing model for which we can study properties of this function without explicit calculation: this is the single server queue with phase-type service time distribution.

Jobs arrive at a service facility according to a Poisson process with parameter λ. With probability p_k any job requires k independent exponentially distributed phases of service each with mean ν. The sum of these phases is the "phase-type" service time distribution, with mean service time $\mu^{-1} = \sum_{k=1}^{\infty} kp_k/\nu$.

This process can be viewed as a continuous time Markov process on the state space

$$\mathsf{X} = \{i = (i_1, i_2) \mid i_1, i_2 \in \mathbb{Z}_+\}$$

where i_1 stands for the number of jobs in the queue and i_2 for the remaining number of phases of service the job currently in service is to receive.

We consider an approximating discrete time Markov chain, which has the following transition probabilities for $h < (\lambda + \nu)^{-1}$ and $e_1 = (1,0), e_2 = (0,1)$:

$$
\begin{aligned}
P(0, 0 + e_2) &= \lambda p_l h, \\
P(i, i + e_1) &= \lambda h, \quad i_1, i_2 > 0, \\
P(i, i - e_2) &= \nu h, \quad i_1 > 0, i_2 > 1, \\
P(i, i - e_1 + le_2) &= \nu p_l h, \quad i_1 > 0, i_2 = 1, \\
P(i, i) &= 1 - \sum_{j \neq i} P(i,j).
\end{aligned}
$$

We call this the h-approximation to the M/PH/1 queue.

Although we do not evaluate a drif criterion explicitly for this chain, we will use a coupling argument to show for $V_0(i) = \mathsf{E}_i[\sigma_0]$ that when $i \neq 0$

$$V_0(i + e_2) - V_0(i) = c, \tag{16.36}$$

$$V_0(i + e_1) - V_0(i) = c' := c\sum_{l=1}^{\infty} lp_l \tag{16.37}$$

for some constant $c > 0$, so that $V_0(i) = c'i_1 + ci_2$ is thus linear in both components of the state variable for $i \neq 0$.

Theorem 16.4.3. *The h-approximation of the M/PH/1 queue as in (16.36) is geometrically ergodic whenever it is ergodic, provided the phase distribution of the service times is in $\mathcal{G}^+(\gamma)$ for some $\gamma > 0$.*

In particular if there are a finite number of phases ergodicity is equivalent to geometric ergodicity for the h-approximation.

PROOF   To develop the coupling argument, we first generate sample paths of $\boldsymbol{\Phi}$ drawing from two i.i.d. sequences $U^1 = \{U_n^1\}_n$, $U^2 = \{U_n^2\}_n$ of random variables having a uniform distribution on $(0, 1]$. The first sequence generates arrivals and phase-completions, the second generates the number of phases of service that will be given to a customer starting service. The procedure is as follows. If $U_n^1 \in (0, \lambda h]$ an arrival is generated in $(nh, (n + 1)h]$; if $U_n^1 \in (\lambda h, \lambda h + \nu h]$ a phase completion is generated, and otherwise nothing happens. Similarly, if $U_n^2 \in (\sum_{l=0}^{k-1} p_l, \sum_{l=0}^{k} p_l]$ k phases will be given to the nth job starting service. This stochastic process has the same probabilistic behavior as $\boldsymbol{\Phi}$.

To prove (16.36) we compare two sample paths, say $\phi^k = \{\phi_n^k\}_n$, $k = 1, 2$, with $\phi_1^1 = i$ and $\phi_1^2 = i + e_2$, generated by one realization of U^1 and U^2. Clearly $\phi_n^2 = \phi_n^1 + e_2$, until the first moment that ϕ^1 hits 0, say at time n^*. But then $\phi_{n^*}^2 = (0, 1)$. This holds for all realizations ϕ^1 and ϕ^2 and we conclude that $V_0(i + e_2) = \mathsf{E}_{i+e_2}[\sigma_0] = \mathsf{E}_i[\sigma_0] + \mathsf{E}_{e_2}[\sigma_0] = V_0(i) + c$, for $c = \mathsf{E}_{e_2}[\sigma_0]$.

If ϕ^2 starts in $i + e_1$ then $\phi_{n^*}^2 = (0, l)$ with probability p_l, so that $V_0(i + e_2) = V_0(i) + \sum_l p_l \mathsf{E}_{le_2}[\sigma_0] = V_0(i) + c\sum_l p_l l$.

Hence, (16.37) and (16.36) hold, and the combination of (16.37) and (16.36) proves (16.26) if we assume that the service time distribution is in $\mathcal{G}^+(\gamma)$ for some $\gamma > 0$, again giving sufficiency of this condition for geometric ergodicity.   □

16.5   Autoregressive and state space models

As we saw briefly in Section 15.5.2, models with some autoregressive character may be geometrically ergodic without the need to assume that the innovation distribution is in $\mathcal{G}^+(\gamma)$. We saw this occur for simple linear models, and for scalar bilinear models.

We now consider rather more complex versions of such models and see that the phenomenon persists, even with increasing complexity of space and structure, if there is a multiplicative constant essentially driving the movement of the chain.

16.5.1   Multidimensional RCA models

The model we consider next is a multidimensional version of the RCA model.   The process of n-vector observations $\boldsymbol{\Phi}$ is generated by the Markovian system

$$\Phi_{k+1} = (A + \Gamma_{k+1})\Phi_k + W_{k+1} \tag{16.38}$$

where A is an $n \times n$ non-random matrix, $\boldsymbol{\Gamma}$ is a sequence of random $(n \times n)$ matrices, and $\boldsymbol{W}$ is a sequence of random p-vectors.

Such models are developed in detail in [299], and we will assume familiarity with the Kronecker product "$\otimes$" and the "vec" operations, used in detail there. In particular we use the basic identities

$$
\begin{aligned}
\text{vec}\,(ABC) &= (C^\top \otimes A)\text{vec}\,(B), \\
(A \otimes B)^\top &= (A^\top \otimes B^\top).
\end{aligned}
\tag{16.39}
$$

To obtain a Markov chain and then establish ergodicity we assume:

Random coefficient autoregression

(RCA1)   The sequences $\boldsymbol{\Gamma}$ and $\boldsymbol{W}$ are i.i.d. and also independent of each other.

(RCA2)   The following expectations exist, and have the prescribed values:

$$
\begin{aligned}
\mathsf{E}[W_k] &= 0 & \mathsf{E}[W_k W_k^\top] &= G & (n \times n), \\
\mathsf{E}[\Gamma_k] &= 0 \quad (n \times n) & \mathsf{E}[\Gamma_k \otimes \Gamma_k] &= C & (n^2 \times n^2),
\end{aligned}
$$

and the eigenvalues of $A \otimes A + C$ have moduli less than unity.

(RCA3)   The distribution of $\binom{\Gamma_k}{W_k}$ has an everywhere positive density with respect to μ^{Leb} on $\mathbb{R}^{n^2+p}$.

Theorem 16.5.1. *If the assumptions (RCA1)–(RCA3) hold for the Markov chain defined in (16.38), then $\boldsymbol{\Phi}$ is V-uniformly ergodic, where $V(x) = |x|^2$. Thus these assumptions suffice for a second-order stationary version of $\boldsymbol{\Phi}$ to exist.*

PROOF   Under the assumptions of the theorem the chain is weak Feller and we can take ψ as μ^{Leb} on $\mathbb{R}^n$. Hence from Theorem 6.2.9 the chain is an irreducible T-chain, and compact subsets of the state space are petite. Aperiodicity is immediate from the density assumption (RCA3). We could also apply the techniques of Chapter 7 to conclude that $\boldsymbol{\Phi}$ is a T-chain, and this would allow us to weaken (RCA3).

To prove $|x|^2$-uniform ergodicity we will use the following two results, which are proved in [299]. Suppose that (RCA1) and (RCA2) hold, and let N be any $n \times n$ positive definite matrix.

(i) If M is defined by

$$
\text{vec}\,(M) = (I - A^\top \otimes A^\top - C)^{-1}\text{vec}\,(N),
\tag{16.40}
$$

then M is also positive definite.

(ii) For any x,

$$
\mathsf{E}[\Phi_k^\top (A + \Gamma_{k+1})^\top M (A + \Gamma_{k+1})\Phi_k \mid \Phi_k = x] = x^\top M x - x^\top N x.
\tag{16.41}
$$

Now let N be any positive definite $(n \times n)$-matrix and define M as in (16.40). Then with $V(x) := x^\top M x$,

$$
\begin{aligned}
\mathsf{E}[V(\Phi_{k+1}) \mid \Phi_k = x] &= \mathsf{E}[\Phi_k^\top (A + \Gamma_{k+1})^\top M (A + \Gamma_{k+1}) \Phi_k \mid \Phi_k = x] \\
&\quad + \mathsf{E}[W_{k+1}^\top M W_{k+1}]
\end{aligned}
\tag{16.42}
$$

on applying (RCA1) and (RCA2).

From (16.41) we also deduce that

$$
PV(x) = V(x) - x^\top N x + \operatorname{tr}(VG) < \lambda V(x) + L
\tag{16.43}
$$

for some $\lambda < 1$ and $L < \infty$, from which we see that (V4) follows, using Lemma 15.2.8.

Finally, note that for some constant c we must have $c^{-1}|x|^2 \leq V(x) \leq c|x|^2$ and the result is proved.   □

16.5.2   Adaptive control models

In this section we return to the simple adaptive control model defined by (SAC1)–(SAC2) whose associated Markovian state process Φ is defined by (2.25).

We showed in Proposition 12.5.2 that the distributions of the state process Φ for this adaptive control model are tight whenever stability in the mean square sense is possible, for a certain class of initial distributions. Here we refine the stability proof to obtain V-uniform ergodicity for the model.

Once these stability results are obtained we can further analyze the system equations and find that we can bound the steady state variance of the output process by the mean square tracking error $\mathsf{E}_\pi[|\tilde{\theta}_0|^2]$ and the disturbance intensity σ_w^2.

Let $y \colon \mathsf{X} \to \mathbb{R}$, $\tilde{\theta} \colon \mathsf{X} \to \mathbb{R}$, $\Sigma \colon \mathsf{X} \to \mathbb{R}$ denote the coordinate variables on X so that

$$
Y_k = y(\Phi_k), \quad \tilde{\theta}_k = \tilde{\theta}(\Phi_k), \quad \Sigma_k = \Sigma(\Phi_k), \qquad k \in \mathbb{Z}_+,
$$

and define the coercive function V on X by

$$
V(y, \tilde{\theta}, \Sigma) = \tilde{\theta}^4 + \varepsilon_0 \tilde{\theta}^2 y^2 + \varepsilon_0^2 y^2
\tag{16.44}
$$

where $\varepsilon_0 > 0$ is a small constant which will be specified below.

Letting P denote the Markov transition function for Φ we have by (2.23),

$$
P y^2 = \tilde{\theta}^2 y^2 + \sigma_w^2.
\tag{16.45}
$$

This is far from (V4), but applying the operator P to the function $\tilde{\theta}^2 y^2$ gives

$$
\begin{aligned}
P\tilde{\theta}^2 y^2 &= \mathsf{E}\left[\left(\frac{\alpha \sigma_0^2 \tilde{\theta} - \alpha \Sigma y W_1}{\sigma_0^2 + \Sigma y^2} + Z_1\right)^2 (\tilde{\theta} y + W_1)^2\right] \\
&= \sigma_z^2 \tilde{\theta}^2 y^2 + \sigma_z^2 \sigma_w^2 \\
&\quad + \left(\frac{\alpha}{\sigma_0^2 + \Sigma y^2}\right)^2 \mathsf{E}[(\sigma_0^2 \tilde{\theta} - \Sigma y W_1)^2 (\tilde{\theta} y + W_1)^2]
\end{aligned}
$$

and hence we may find a constant $K_1 < \infty$ such that

$$P\tilde{\theta}^2 y^2 \leq \sigma_z^2 \tilde{\theta}^2 y^2 + K_1(\tilde{\theta}^4 + \tilde{\theta}^2 + 1). \tag{16.46}$$

From (2.22) it is easy to show that for some constant $K_2 > 0$

$$P\tilde{\theta}^4 \leq \alpha^4 \tilde{\theta}^4 + K_2(\tilde{\theta}^2 + 1). \tag{16.47}$$

When $\sigma_z^2 < 1$ we combine (16.45)–(16.47) to find, for any $1 > \rho > \max(\sigma_z^2, \alpha^4)$, constants $R < \infty$ and $\varepsilon_0 > 0$ such that with V defined in (16.44), $PV \leq \rho V + R$. Applying Theorem 16.1.2 and Lemma 15.2.8 we have proved

Proposition 16.5.2. *The Markov chain Φ is V-uniformly ergodic whenever $\sigma_z^2 < 1$, with V given by (16.44); and for all initial conditions $x \in \mathsf{X}$, as $k \to \infty$,*

$$\mathsf{E}_x[Y_k^2] \to \int y^2 \, d\pi \tag{16.48}$$

at a geometric rate.   □

Hence the performance of the closed loop system is characterized by the unique invariant probability π.

From ergodicity of the model it can be shown that in steady state $\tilde{\theta}_k = \theta_k - \mathsf{E}[\theta_k \mid Y_0, \ldots, Y_k]$, and $\Sigma_k = \mathsf{E}[\tilde{\theta}_k^2 \mid Y_0, \ldots, Y_k]$. Using these identities we now obtain bounds on performance of the closed loop system by integrating the system equations with respect to the invariant measure.

Taking expectations in (2.23) and (2.24) under the probability P_π gives

$$\mathsf{E}_\pi[Y_0^2] = \mathsf{E}_\pi[\Sigma_0 Y_0^2] + \sigma_w^2,$$
$$\sigma_z^2 \mathsf{E}_\pi[Y_0^2] = \mathsf{E}_\pi[\Sigma_0 Y_0^2] - \alpha^2 \sigma_w^2 \mathsf{E}_\pi[\Sigma_0].$$

Hence, by subtraction, and using the identity $\mathsf{E}_\pi[|\tilde{\theta}_0|^2] = \mathsf{E}_\pi[\Sigma_0]$, we can evaluate the limit (16.48) as

$$\mathsf{E}_\pi[Y_0^2] = \frac{\sigma_w^2}{1 - \sigma_z^2} \left(1 + \alpha^2 \mathsf{E}_\pi[|\tilde{\theta}_0|^2]\right). \tag{16.49}$$

This shows precisely how the steady state performance is related to the disturbance intensity σ_w^2, the parameter variation intensity σ_z^2, and the mean square parameter estimation error $\mathsf{E}_\pi[|\tilde{\theta}_0|^2]$.

Using obvious bounds on $\mathsf{E}_\pi[\Sigma_0]$ we obtain the following bounds on the steady state performance in terms of the system parameters only:

$$\frac{\sigma_w^2}{1 - \sigma_z^2}(1 + \alpha^2 \sigma_z^2) \leq \mathsf{E}_\pi[Y_0^2] \leq \frac{\sigma_w^2}{1 - \sigma_z^2}\left(1 + \frac{\alpha^2 \sigma_z^2}{1 - \alpha^2}\right).$$

If it were possible to directly observe θ_{k-1} at time k, then the optimal performance would be

$$\mathsf{E}_\pi[Y_0^2] = \frac{\sigma_w^2}{1 - \sigma_z^2}.$$

This shows that the lower bound in the previous chain of inequalities is non-trivial.

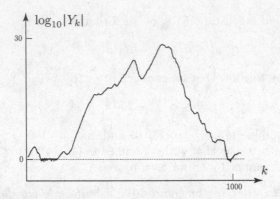

Figure 16.1: The output of the simple adaptive control model when the control U_k is set equal to zero. The resulting process is equivalent to the dependent parameter bilinear model with $\alpha = 0.99$, $W_k \sim N(0, 0.01)$ and $Z_k \sim N(0, 0.04)$.

The performance of the closed loop system is illustrated in Chapter 2.

A sample path of the output Y of the controlled system is given on the left in Figure 2.5, which is comparable to the noise sample path illustrated in Figure 2.6. To see how this compares to the control-free system, a simulation of the simple adaptive control model with the control value U_k set equal to zero for all k is given in Figure 16.1. The resulting process $\binom{\theta}{Y}$ becomes a version of the dependent parameter bilinear model. Even though we will see in Chapter 17 that this process is bounded in probability, the sample paths fluctuate wildly, with the output process Y quickly exceeding 10^{100} in this simulation.

16.6   Commentary*

This chapter brings together some of the oldest and some of the newest ergodic theorems for Markov chains.

Initial results on uniform ergodicity for countable chains under, essentially, Doeblin's condition date to Markov [248]: transition matrices with a column bounded from zero are often called *Markov matrices*. For general state space chains use of the condition of Doeblin is in [93]. These ideas are strengthened in Doob [99], whose introduction and elucidation of Doeblin's condition as Hypothesis D (p. 192 of [99]) still guides the analysis of many models and many applications, especially on compact spaces.

Other areas of study of uniformly ergodic (sometimes called strongly ergodic, or quasi-compact) chains have a long history, much of it initiated by Yosida and Kakutani [412] who considered the equivalence of (iii) and (v) in Theorem 16.0.2, as did Doob [99]. Somewhat surprisingly, even for countable spaces the hitting time criterion of Theorem 16.2.2 for uniformly ergodic chains appears to be as recent as the work of Huang and Isaacson [164], with general-space extensions in Bonsdorff [38]; the obvious value of a bounded drift function is developed in Isaacson and Tweedie [170] in the countable space case. Nummelin ([303], Chapters 5.6 and 6.6) gives a discussion of much of this material.

There is a large subsequent body of theory for quasi-compact chains, exploiting operator-theoretic approaches. Revuz ([326], Chapter 6) has a thorough discussion of uniformly ergodic chains and associated quasi-compact operators when the chain is not irreducible. He shows that in this case there is essentially a finite decomposition into recurrent parts of the space: this is beyond the scope of our work here.

We noted in Theorem 16.2.5 that uniform ergodicity results take on a particularly elegant form when we are dealing with irreducible T-chains: this is first derived in a different way in [391]. It is worth noting that for reducible T-chains there is an appealing structure related to the quasi-compactness above. It is shown by Tuominen and Tweedie [391] that, even for chains which are not necessarily irreducible, if the space is compact then for any T-chain there is also a finite decomposition

$$\mathsf{X} = \bigcup_{k=0}^{n} H_k \cup E$$

where the H_i are disjoint absorbing sets and Φ restricted to any H_k is uniformly ergodic, and E is uniformly transient.

The introduction to uniform ergodicity that we give here appears brief given the history of such theory, but this is a largely a consequence of the fact that we have built up, for ψ-irreducible chains, a substantial set of tools which makes the approach to this class of chains relatively simple.

Much of this simplicity lies in the ability to exploit the norm $\| \cdot \|_V$. This is a very new approach. Although Kartashov [196, 197] has some initial steps in developing a theory of general space chains using the norm $\| \cdot \|_V$, he does not link his results to the use of drift conditions, and the appearance of V-uniform results are due largely to recent observations of Hordijk and Spieksma [366, 163] in the countable space case.

Their methods are substantially different from the general state space version we use, which builds on Chapter 15: the general space version was first developed in [277] for strongly aperiodic chains. This approach shows that for V-uniformly ergodic chains, it is in fact possible to apply the same quasi-compact operator theory that has been exploited for uniformly ergodic chains, at least within the context of the space L_V^∞. This is far from obvious: it is interesting to note Kendall himself ([203], p. 183) saying that " ... the theory of quasi-compact operators is completely useless" in dealing with geometric ergodicity, whilst Vere-Jones [406] found substantial difficulty in relating standard operator theory to geometric ergodicity. This appears to be an area where reasonable further advances may be expected in the theory of Markov chains.

It is shown in Athreya and Pantula [15] that an ergodic chain is always strong mixing. The extension given in Section 16.1.2 for V-uniformly ergodic chains was proved for bounded functions in [92], and the version given in Theorem 16.1.5 is essentially taken from Meyn and Tweedie [277].

Verifying the V-uniform ergodicity properties is usually done through test functions and drift conditions, as we have seen. Uniform ergodicity is generally either a trivial or a more difficult property to verify in applications. Typically one must either take the state space of the chain to be compact (or essentially compact), or be able to apply the Doeblin or small set conditions, in order to gain uniform ergodicity. The identification of the rate of convergence in this last case is a powerful incentive to use such an approach. The delightful proof in Theorem 16.2.4 is due to Rosenthal [341], following the strong stopping time results of Aldous and Diaconis [1, 88], although the result itself is inherent in Theorem 6.15 of Nummelin [303]. An application of this result to Markov chain Monte Carlo methods is given by Tierney [385].

However, as we have shown, V-uniform ergodicity can often be obtained for some V under much more readily obtainable conditions, such as a geometric tail for any i.i.d. random variables generating the process. This is true for queues, general storage models, and other random-walk-related models, as the application of the increment analysis of Section 16.3 shows. Such chains were investigated in detail by Vere-Jones [403] and Miller [284].

The results given in Section 16.3 and Section 16.3.2 are new in the case of general X, but are based on a similar approach for countable spaces in Spieksma and Tweedie [368], which also contains a partial converse to Theorem 16.3.2. There are some precursors to these conditions: one obvious way of ensuring that P has the characteristics in (16.26) and (16.27) is to require that the increments from any state are of bounded range, with the range allowed depending on V, so that for some b

$$|V(j) - V(k)| \geq b \Rightarrow P(k, j) = 0 : \tag{16.50}$$

and in [243] it is shown that under the bounded range condition (16.50) an ergodic chain is geometrically ergodic.

A detailed description of the polling system we consider here can be found in [2]. Note that in [2] the system is modeled slightly differently, with arrivals of the server at each gate defining the times of the embedded process. The coupling construction used to analyze the h-approximation to the phase-service model is based on [350] and clearly is ideal for our type of argument. Further examples are given in [368].

For the adaptive control and linear models, as we have stressed, V-uniform ergodicity is often actually equivalent to simple ergodicity: the examples in this chapter are chosen to illustrate this. The analysis of the bilinear and the vector RCA model given here is taken from Feigin and Tweedie [111]; the former had been previously analyzed by Tong [387]. In a more traditional approach to RCA models through time series methods, Nicholls and Quinn [299] also find (RCA2) appropriate when establishing conditions for strict stationarity of Φ, and also when treating asymptotic results of estimators.

The adaptive model was introduced in [253] and a stability analysis appeared in [270] where the performance bound (16.49) was obtained. Related results appeared in [365, 148, 269, 130]. The stability of the multidimensional adaptive control model was only recently resolved in Rayadurgam et al. [324].

Commentary for the second edition: In the first edition the vector-space setting was credited to work of Kartashov (see preceding text). In fact its origin is the 1969 work of Veinott [185] concerning controlled Markov models. Section 20.1 contains further discussion on the recent evolution of topics in this chapter.

An early application of the skip-free condition is contained in [156], also in the setting of controlled Markov models. Assumption (ii) of this paper is a version of the g-skip-free property, in which the function g represents "reward" in a controlled model.

The implications of Doeblin's condition to large deviations theory and to spectral theory can be found in [140, 218, 408].

Chapter 17

Sample paths and limit theorems

Most of this chapter is devoted to the analysis of the series $S_n(g)$, where we define for any function g on X,

$$S_n(g) := \sum_{k=1}^{n} g(\Phi_k). \tag{17.1}$$

We are concerned primarily with four types of limit theorems for positive recurrent chains possessing an invariant probability π:

(i) those which are based upon the existence of martingales associated with the chain;

(ii) the Strong Law of Large Numbers (LLN), which states that $n^{-1}S_n(g)$ converges to $\pi(g) = \mathsf{E}_\pi[g(\Phi_0)]$, the steady state expectation of $g(\Phi_0)$;

(iii) the Central Limit Theorem (CLT), which states that the sum $S_n(g - \pi(g))$, when properly normalized, is asymptotically normally distributed;

(iv) the Law of the Iterated Logarithm (LIL) which gives precise upper and lower bounds on the limit supremum of the sequence $S_n(g - \pi(g))$, again when properly normalized.

The martingale results (i) provide insight into the structure of irreducible chains, and make the proofs of more elementary ergodic theorems such as the LLN almost trivial. Martingale methods will also prove to be very powerful when we come to the CLT for appropriately stable chains.

The trilogy of the LLN, CLT and LIL provide measures of centrality and variability for Φ_n as n becomes large: these complement and strengthen the distributional limit theorems of previous chapters. The magnitude of variability is measured by the variance given in the CLT, and one of the major contributions of this chapter is to identify the way in which this variance is defined through the autocovariance sequence for the stationary version of the process $\{g(\Phi_k)\}$.

The three key limit theorems which we develop in this chapter using sample path properties for chains which possess a unique invariant probability π are

LLN We say that the *Law of Large Numbers* holds for a function g if

$$\lim_{n \to \infty} \frac{1}{n} S_n(g) = \pi(g) \qquad \text{a.s. } [\mathsf{P}_*].$$ (17.2)

CLT We say that the *Central Limit Theorem* holds for g if there exists a constant $0 < \gamma_g^2 < \infty$ such that for each initial condition $x \in \mathsf{X}$,

$$\lim_{n \to \infty} \mathsf{P}_x \left\{ (n\gamma_g^2)^{-1/2} S_n(\overline{g}) \le t \right\} = \int_{-\infty}^{t} \frac{1}{\sqrt{2\pi}} e^{-x^2/2} \, dx$$

where $\overline{g} = g - \pi(g)$: that is, as $n \to \infty$,

$$(n\gamma_g^2)^{-1/2} S_n(\overline{g}) \xrightarrow{\text{d}} N(0,1).$$

LIL When the CLT holds, we say that the *Law of the Iterated Logarithm* holds for g if the limit infimum and limit supremum of the sequence

$$(2\gamma_g^2 n \log \log(n))^{-1/2} S_n(\overline{g})$$

are respectively -1 and $+1$ with probability one for each initial condition $x \in \mathsf{X}$.

Strictly speaking, of course, the CLT is not a sample path limit theorem, although it does describe the behavior of the sample path averages and these three "classical" limit theorems obviously belong together.

Proofs of all of these results will be based upon martingale techniques involving the path behavior of the chain, and detailed sample path analysis of the process between visits to a recurrent atom.

Much of this chapter is devoted to proving that these limits hold under various conditions. The following set of limit theorems summarizes a large part of this development.

Theorem 17.0.1. *Suppose that Φ is a positive Harris chain with invariant probability π.*

(i) *The LLN holds for any g satisfying $\pi(|g|) < \infty$.*

(ii) *Suppose that Φ is V-uniformly ergodic. Let g be a function on X satisfying $g^2 \le V$, and let $\overline{g}$ denote the centered function $\overline{g} = g - \int g \, d\pi$. Then the constant*

$$\gamma_g^2 := \mathsf{E}_\pi[\overline{g}^2(\Phi_0)] + 2 \sum_{k=1}^{\infty} \mathsf{E}_\pi[\overline{g}(\Phi_0)\overline{g}(\Phi_k)]$$ (17.3)

is well defined, non-negative and finite, and coincides with the asymptotic variance

$$\lim_{n \to \infty} \frac{1}{n} \mathsf{E}_\pi\left[\left(S_n(\overline{g})\right)^2\right] = \gamma_g^2.$$ (17.4)

(iii) *If the conditions of (ii) hold and if $\gamma_g^2 = 0$, then*

$$\lim_{n \to \infty} \frac{1}{\sqrt{n}} S_n(g) = 0 \qquad \text{a.s. } [\mathsf{P}_*].$$

(iv) *If the conditions of (ii) hold and if $\gamma_g^2 > 0$, then the CLT and LIL hold for the function g.*

PROOF   The LLN is proved in Theorem 17.1.7, and the CLT and LIL are proved in Theorem 17.3.6 under conditions somewhat weaker than those assumed here.

It is shown in Lemma 17.5.2 and Theorem 17.5.3 that the asymptotic variance γ_g^2 is given by (17.3) under the conditions of Theorem 17.0.1, and the alternate representation (17.4) of γ_g^2 is given in Theorem 17.5.3. The a.s. convergence in (iii) when $\gamma_g^2 = 0$ is proved in Theorem 17.5.4.   □

While Theorem 17.0.1 summarizes the main results, the reader will find that there is much more to be found in this chapter. We also provide here techniques for proving the LLN and CLT in contexts far more general than given in Theorem 17.0.1. In particular, these techniques lead to a functional CLT for f-regular chains in Section 17.4.

We begin with a discussion of invariant σ-fields, which form the basis of classical ergodic theory.

17.1   Invariant σ-fields and the LLN

Here we introduce the concepts of invariant random variables and σ-fields, and show how these concepts are related to Harris recurrence on the one hand and the LLN on the other.

17.1.1   Invariant random variables and events

For a fixed initial distribution μ, a random variable Y on the sample space $(\Omega, \mathcal{F})$ will be called P_μ-*invariant* if $\theta^k Y = Y$ a.s. $[P_\mu]$ for each $k \in \mathbb{Z}_+$, where θ is the shift operator. Hence Y is P_μ-invariant if there exists a function f on the sample space such that

$$Y = f(\Phi_k, \Phi_{k+1}, \dots) \qquad \text{a.s. } [P_\mu], \quad k \in \mathbb{Z}_+. \tag{17.5}$$

When $Y = \mathbb{I}_A$ for some $A \in \mathcal{F}$, then the set A is called a P_μ-*invariant event*. The set of all P_μ-invariant events is a σ-field, which we denote Σ_μ.

Suppose that an invariant probability measure π exists, and for now restrict attention to the special case where $\mu = \pi$. In this case, Σ_π is equal to the family of invariant events which is commonly used in ergodic theory (see for example Krengel [221]) and is often denoted Σ_I.

For a bounded, P_π-invariant random variable Y we let h_Y denote the function

$$h_Y(x) := E_x[Y], \qquad x \in X. \tag{17.6}$$

By the Markov property and invariance of the random variable Y,

$$h_Y(\Phi_k) = E[\theta^k Y \mid \mathcal{F}_k^\Phi] = E[Y \mid \mathcal{F}_k^\Phi] \qquad \text{a.s. } [P_\pi]. \tag{17.7}$$

This will be used to prove:

Lemma 17.1.1. *If π is an invariant probability measure and Y is a P_π-invariant random variable satisfying $E_\pi[|Y|] < \infty$, then*

$$Y = h_Y(\Phi_0) \qquad \text{a.s. } [P_\pi].$$

PROOF   It follows from (17.7) that the adapted process $(h_Y(\Phi_k), \mathcal{F}_k^\Phi)$ is a convergent martingale for which

$$\lim_{k\to\infty} h_Y(\Phi_k) = Y \qquad \text{a.s. } [\mathsf{P}_\pi].$$

When $\Phi_0 \sim \pi$ the process $h_Y(\Phi_k)$ is also stationary, since Φ is stationary, and hence the limit above shows that its sample paths are almost surely constant. That is, $Y = h_Y(\Phi_k) = h_Y(\Phi_0)$ a.s. $[\mathsf{P}_\pi]$ for all $k \in \mathbb{Z}_+$.   □

It follows from Lemma 17.1.1 that if $X \in L_1(\Omega, \mathcal{F}, \mathsf{P}_\pi)$ then the P_π-invariant random variable $\mathsf{E}[X \mid \Sigma_\pi]$ is a function of Φ_0 alone, which we shall denote $X_\infty(\Phi_0)$, or just X_∞.

The function X_∞ is significant because it describes the limit of the sample path averages of $\{\theta^k X\}$, as we show in the next result.

Theorem 17.1.2. *If Φ is a Markov chain with invariant probability measure π, and $X \in L_1(\Omega, \mathcal{F}, \mathsf{P}_\pi)$, then there exists a set $F_X \in \mathcal{B}(\mathsf{X})$ of full π-measure such that for each initial condition $x \in F_X$,*

$$\lim_{N\to\infty} \frac{1}{N} \sum_{k=1}^{N} \theta^k X = X_\infty(x) \qquad \text{a.s. } [\mathsf{P}_x].$$

PROOF ·   Since Φ is a stationary stochastic process when $\Phi_0 \sim \pi$, the process $\{\theta^k X : k \in \mathbb{Z}_+\}$ is also stationary, and hence the Strong Law of Large Numbers for stationary sequences [99] can be applied:

$$\lim_{N\to\infty} \frac{1}{N} \sum_{k=1}^{N} \theta^k X = \mathsf{E}[X \mid \Sigma_\pi] = X_\infty(\Phi_0) \qquad \text{a.s. } [\mathsf{P}_\pi].$$

Hence, using the definition of P_π, we may calculate

$$\int \mathsf{P}_x \Big\{ \lim_{N\to\infty} \frac{1}{N} \sum_{k=1}^{N} \theta^k X = X_\infty(x) \Big\} \pi(dx) = 1.$$

Since the integrand is always positive and less than or equal to one, this proves the result.   □

This is an extremely powerful result, as it only requires the existence of an invariant probability without any further regularity or even irreducibility assumptions on the chain. As a product of its generality, it has a number of drawbacks. In particular, the set F_X may be very small, may be difficult to identify, and will typically depend upon the particular random variable X.

We now turn to a more restrictive notion of invariance which allows us to deal more easily with null sets such as F_X^c. In particular we will see that the difficulties associated with the general nature of Theorem 17.1.2 are resolved for Harris processes.

17.1.2 Harmonic functions

To obtain ergodic theorems for arbitrary initial conditions, it is helpful to restrict somewhat our definition of invariance.

The concepts introduced in this section will necessitate some care in our definition of a random variable. In this section, a random variable Y must "live on" several different probability spaces at the same time. For this reason we will now stress that Y has the form $Y = f(\Phi_0, \ldots, \Phi_k, \ldots)$ where f is a function which is measurable with respect to $\mathcal{B}(\mathsf{X}^{\mathsf{z}+}) = \mathcal{F}$. We call a random variable Y of this form *invariant* if it is P_μ-invariant for *every* initial distribution μ. The class of invariant events is defined analogously, and is a σ-field which we denote Σ.

Two examples of invariant random variables in this sense are

$$\widetilde{Q}\{A\} = \limsup_{k\to\infty} \mathbb{I}\{\Phi_k \in A\}, \qquad \tilde{\pi}\{A\} = \limsup_{N\to\infty} \frac{1}{N}\sum_{k=1}^{N} \mathbb{I}\{\Phi_k \in A\}$$

with $A \in \mathcal{B}(\mathsf{X})$.

A function $h\colon \mathsf{X} \to \mathbb{R}$ is called *harmonic* if, for all $x \in \mathsf{X}$,

$$\int P(x, dy) h(y) = h(x). \tag{17.8}$$

This is equivalent to the adapted sequence $(h(\Phi_k), \mathcal{F}_k^\Phi)$ possessing the martingale property for each initial condition: that is,

$$\mathsf{E}[h(\Phi_{k+1}) \mid \mathcal{F}_k^\Phi] = h(\Phi_k), \qquad k \in \mathbb{Z}_+, \qquad \text{a.s. } [\mathsf{P}_*].$$

For any measurable set A the function $h_{\widetilde{Q}\{A\}}(x) = Q(x, A)$ is a measurable function of $x \in \mathsf{X}$ which is easily shown to be harmonic. This correspondence is just one instance of the following general result which shows that harmonic functions and invariant random variables are in one-to-one correspondence in a well-defined way.

Theorem 17.1.3.   (i) *If Y is bounded and invariant, then the function h_Y is harmonic, and*

$$Y = \lim_{k\to\infty} h_Y(\Phi_k) \qquad \text{a.s. } [\mathsf{P}_*].$$

 (ii) *If h is bounded and harmonic, then the random variable*

$$H := \limsup_{k\to\infty} h(\Phi_k),$$

is invariant, with $h_H(x) = h(x)$.

PROOF   For (i), first observe that by the Markov property and invariance we may deduce as in the proof of Lemma 17.1.1 that

$$h_Y(\Phi_k) = \mathsf{E}[Y \mid \mathcal{F}_k^\Phi] \qquad \text{a.s. } [\mathsf{P}_*].$$

Since Y is bounded, this shows that $(h_Y(\Phi_k), \mathcal{F}_k^\Phi)$ is a martingale which converges to Y. To see that h_Y is harmonic, we use invariance of Y to calculate

$$P h_Y(x) = \mathsf{E}_x[h_Y(\Phi_1)] = \mathsf{E}_x[\mathsf{E}[Y \mid \mathcal{F}_1^\Phi]] = h_Y(x).$$

To prove (ii), recall that the adapted process $(h(\Phi_k), \mathcal{F}_k^{\Phi})$ is a martingale if h is harmonic, and since h is assumed bounded, it is convergent. The conclusions of (ii) follow.

□

Theorem 17.1.3 shows that there is a one-to-one correspondence between invariant random variables and harmonic functions. From this observation we have as an immediate consequence

Proposition 17.1.4. *The following two conditions are equivalent:*

(i) *All bounded harmonic functions are constant.*

(ii) Σ_μ *and hence* Σ *are* P_μ-*trivial for each initial distribution* μ.

Finally, we show that when Φ is Harris recurrent, all bounded harmonic functions are trivial.

Theorem 17.1.5. *If* Φ *is Harris recurrent, then the constants are the only bounded harmonic functions.*

PROOF   We suppose that Φ is Harris, let h be a bounded harmonic function, and fix a real constant a. If the set $\{x : h(x) \geq a\}$ lies in $\mathcal{B}^+(\mathsf{X})$, then we will show that $h(x) \geq a$ for all $x \in \mathsf{X}$. Similarly, if $\{x : h(x) \leq a\}$ lies in $\mathcal{B}^+(\mathsf{X})$, then we will show that $h(x) \leq a$ for all $x \in \mathsf{X}$. These two bounds easily imply that h is constant, which is the desired conclusion.

If $\{x : h(x) \geq a\} \in \mathcal{B}^+(\mathsf{X})$, then Φ enters this set i.o. from each initial condition, and consequently

$$\limsup_{k \to \infty} h(\Phi_k) \geq a \qquad \text{a.s. } [\mathsf{P}_*].$$

Applying Theorem 17.1.3 we see that $h(x) = \mathsf{E}_x[H] \geq a$ for all $x \in \mathsf{X}$. Identical reasoning shows that $h(x) \leq a$ for all x when $\{x : h(x) \leq a\} \in \mathcal{B}^+(\mathsf{X})$, and this completes the proof.   □

It is of considerable interest to note that in quite another way we have already proved this result: it is indeed a rephrasing of our criterion for transience in Theorem 8.4.2.

In the proof of Theorem 17.1.5 we are not in fact using the full power of the Martingale Convergence Theorem, and consequently the proposition can be extended to include larger classes of functions, extending those which are bounded and harmonic, if this is required.

As an easy consequence we have

Proposition 17.1.6. *Suppose that* Φ *is positive Harris and that any of the LLN, the CLT, or the LIL hold for some* g *and some one initial distribution. Then this same limit holds for every initial distribution.*

PROOF   We will give the proof for the LLN, since the proof of the result for the CLT and LIL is identical.

Suppose that the LLN holds for the initial distribution μ_0, and let $g_\infty(x) = \mathsf{P}_x\{\frac{1}{n}S_n(g) \to \int g \, d\pi\}$. We have by assumption that

$$\int g_\infty \, d\mu_0 = 1.$$

We will now show that g_∞ is harmonic, which together with Theorem 17.1.5 will imply that g_∞ is equal to the constant value 1, and thereby complete the proof. We have by the Markov property and the smoothing property of the conditional expectation,

$$
\begin{aligned}
Pg_\infty\,(x) &= \mathsf{E}_x\Big[\mathsf{P}_{\Phi_1}\Big\{\lim_{n\to\infty}\frac{1}{n}\sum_{k=1}^{n}g(\Phi_k) = \int g\,d\pi\Big\}\Big] \\
&= \mathsf{E}_x\Big[\mathsf{P}_x\Big\{\lim_{n\to\infty}\frac{1}{n}\sum_{k=1}^{n}g(\Phi_{k+1}) = \int g\,d\pi \mid \mathcal{F}_1^{\Phi}\Big\}\Big] \\
&= \mathsf{P}_x\Big\{\lim_{n\to\infty}\Big[\Big(\frac{n+1}{n}\Big)\frac{1}{n+1}\sum_{k=1}^{n+1}g(\Phi_{k+1}) - \frac{g(\Phi_1)}{n}\Big] = \int g\,d\pi\Big\} \\
&= g_\infty(x).
\end{aligned}
$$

$\square$

From these results we may now provide a simple proof of the LLN for Harris chains.

17.1.3   The LLN for positive Harris chains

We present here the LLN for positive Harris chains. In subsequent sections we will prove more general results which are based upon the existence of an atom for the process, or an atom $\check{\alpha}$ for the split version of a general Harris chain.

In the next result we see that when Φ is positive Harris, the null set F_X^c defined in Theorem 17.1.2 is empty:

Theorem 17.1.7. *The following are equivalent when an invariant probability π exists for Φ:*

(i) *Φ is positive Harris.*

(ii) *For each $f \in L_1(\mathsf{X}, \mathcal{B}(\mathsf{X}), \pi)$,*

$$
\lim_{n\to\infty}\frac{1}{n}S_n(f) = \int f\,d\pi \qquad \text{a.s. } [\mathsf{P}_*].
$$

(iii) *The invariant σ-field Σ is P_x-trivial for all x.*

PROOF   (i) $\Rightarrow$ (ii)   If Φ is positive Harris with unique invariant probability π then by Theorem 17.1.2, for each fixed f, there exists a set $G \in \mathcal{B}(\mathsf{X})$ of full π-measure such that the conclusions of (ii) hold whenever the distribution of Φ_0 is supported on G. By Proposition 17.1.6 the LLN holds for every initial condition.

(ii) $\Rightarrow$ (iii)   Let Y be a bounded invariant random variable, and let h_Y be the associated bounded harmonic function defined in (17.6). By the hypotheses of (ii) and Theorem 17.1.3 we have

$$
Y = \lim_{k\to\infty}h_Y(\Phi_k) = \lim_{N\to\infty}\frac{1}{N}\sum_{k=1}^{N}h_Y(\Phi_k) = \int h_Y\,d\pi \qquad \text{a.s. } [\mathsf{P}_*],
$$

which shows that every set in Σ has P_x-measure zero or one.

(iii) $\Rightarrow$ (i)   If (iii) holds, then for any measurable set A the function $Q(\,\cdot\,, A)$ is constant. It follows from Theorem 9.1.3 (ii) that $Q(\,\cdot\,, A) \equiv 0$ or $Q(\,\cdot\,, A) \equiv 1$. When $\pi\{A\} > 0$, Theorem 17.1.2 rules out the case $Q(\,\cdot\,, A) \equiv 0$, which establishes Harris recurrence. $\square$

17.2   Ergodic theorems for chains possessing an atom

In this section we consider chains which possess a Harris recurrent atom α. Under this assumption we can state a self-contained and more transparent proof of the Law of Large Numbers and related ergodic theorems, and the methods extend to general ψ-irreducible chains without much difficulty.

The main step in the proofs of the ergodic theorems considered here is to divide the sample paths of the process into i.i.d. blocks corresponding to pieces of a sample path between consecutive visits to the atom α. This makes it possible to infer most ergodic theorems of interest for the Markov chain from relatively simple ergodic theorems for i.i.d. random variables.

Let $\sigma_\alpha(0) = \sigma_\alpha$, and let $\{\sigma_\alpha(j) : j \geq 1\}$ denote the times of consecutive visits to α so that

$$\sigma_\alpha(k+1) = \theta^{\sigma_\alpha(k)} \tau_\alpha + \sigma_\alpha(k), \qquad k \geq 0.$$

For a function $f \colon \mathsf{X} \to \mathbb{R}$ we let $s_j(f)$ denote the sum of $f(\Phi_i)$ over the jth piece of the sample path of Φ between consecutive visits to α:

$$s_j(f) = \sum_{i=\sigma_\alpha(j)+1}^{\sigma_\alpha(j+1)} f(\Phi_i) \tag{17.9}$$

By the strong Markov property the random variables $\{s_j(f) : j \geq 0\}$ are i.i.d. with common mean

$$\mathsf{E}_\alpha[s_1(f)] = \mathsf{E}_\alpha\left[\sum_{i=1}^{\tau_\alpha} f(\Phi_i)\right] = \int f \, d\mu \tag{17.10}$$

where the definition of μ is self-evident. The measure μ on $\mathcal{B}(\mathsf{X})$ is invariant by Theorem 10.0.1.

By writing the sum of $\{f(\Phi_i)\}$ as a sum of $\{s_i(f)\}$ we may prove the LLN, CLT and LIL for Φ by citing the corresponding ergodic theorem for the i.i.d. sequence $\{s_i(f)\}$. We illustrate this technique first with the LLN.

17.2.1   Ratio form of the law of large numbers

We first present a version of Theorem 17.1.7 for arbitrary recurrent chains.

Theorem 17.2.1. *Suppose that Φ is Harris recurrent with invariant measure π, and suppose that there exists an atom $\alpha \in \mathcal{B}^+(\mathsf{X})$. Then for any $f, g \in L^1(\mathsf{X}, \mathcal{B}(\mathsf{X}), \pi)$ with $\int g \, d\pi \neq 0$,*

$$\lim_{n \to \infty} \frac{S_n(f)}{S_n(g)} = \frac{\pi(f)}{\pi(g)} \qquad \text{a.s. } [\mathsf{P}_*].$$

PROOF   For the proof we assume that each of the functions f and g are positive. The general case follows by decomposing f and g into their positive and negative parts.

We also assume that π is equal to the measure μ defined implicitly in (17.10). This is without loss of generality as any invariant measure is a constant multiple of μ by Theorem 10.0.1.

For $n \geq \sigma_\alpha$ we define

$$\ell_n := \max(k : \sigma_\alpha(k) \leq n) = -1 + \sum_{k=0}^{n} \mathbb{I}\{\Phi_k \in \alpha\} \qquad (17.11)$$

so that from (17.9) we obtain the pair of bounds

$$\sum_{j=0}^{\ell_n-1} s_j(f) \leq \sum_{i=1}^{n} f(\Phi_i) \leq \sum_{j=0}^{\ell_n} s_j(f) + \sum_{i=1}^{\tau_\alpha} f(\Phi_i). \qquad (17.12)$$

Since the same relation holds with f replaced by g we have

$$\frac{\sum_{i=1}^{n} f(\Phi_i)}{\sum_{i=1}^{n} g(\Phi_i)} \leq \frac{\ell_n}{\ell_n - 1} \frac{\left[\frac{1}{\ell_n}\left(\sum_{j=1}^{\ell_n} s_j(f) + \sum_{i=1}^{\tau_\alpha} f(\Phi_i)\right)\right]}{\left[\frac{1}{\ell_n-1}\sum_{j=0}^{\ell_n-1} s_j(g)\right]}.$$

Because $\{s_j(f) : j \geq 1\}$ is i.i.d. and $\ell_n \to \infty$,

$$\frac{1}{\ell_n}\sum_{j=0}^{\ell_n} s_j(f) \to \mathsf{E}[s_1(f)] = \int f\, d\mu$$

and similarly for g. This yields

$$\limsup_{n\to\infty} \frac{\sum_{i=1}^{n} f(\Phi_i)}{\sum_{i=1}^{n} g(\Phi_i)} \leq \frac{\int f\, d\mu}{\int g\, d\mu},$$

and by interchanging the roles of f and g we obtain

$$\liminf_{n\to\infty} \frac{\sum_{i=1}^{n} f(\Phi_i)}{\sum_{i=1}^{n} g(\Phi_i)} \geq \frac{\int f\, d\mu}{\int g\, d\mu}$$

which completes the proof.   □

17.2.2   The CLT and the LIL for chains possessing an atom

Here we show how the CLT and LIL may be proved under the assumption that an atom $\alpha \in \mathcal{B}^+(\mathsf{X})$ exists.

The Central Limit Theorem (CLT) states that the normalized sum

$$(n\gamma_g^2)^{-1/2} S_n(\overline{g})$$

converges in distribution to a standard Gaussian random variable, while the Law of the Iterated Logarithm (LIL) provides sharp bounds on the sequence

$$(2\gamma_g^2 n \log\log(n))^{-1/2} S_n(\overline{g})$$

where $\overline{g}$ is the centered function $\overline{g} := g - \pi(g)$, π is an invariant probability, and γ_g^2 is a normalizing constant.

These results do not hold unless some restrictions are imposed on both the function and the Markov chain: for counterexamples on countable state spaces, the reader is referred to Chung [71]. The purpose of this section is to provide general sufficient conditions for chains which possess an atom.

One might expect that, as in the i.i.d. case, the asymptotic variance γ_g^2 is equal to the variance of the random variable $g(\Phi_k)$ under the invariant probability. Somewhat surprisingly, therefore, we will see below that this is not the case. When an atom α exists we will demonstrate that in fact

$$\gamma_g^2 = \pi\{\alpha\}\mathsf{E}_\alpha\left[\left(\sum_{k=1}^{\tau_\alpha} \overline{g}(\Phi_k)\right)^2\right]. \tag{17.13}$$

The actual variance of $g(\Phi_k)$ in the stationary case is given by Theorem 10.0.1 as

$$\int \overline{g}^2 \, d\pi = \pi\{\alpha\}\mathsf{E}_\alpha\left[\sum_{k=1}^{\tau_\alpha} \left(\overline{g}(\Phi_k)\right)^2\right];$$

thus when Φ is i.i.d., these expressions do coincide, but differ otherwise.

We will need a moment condition to prove the CLT in the case where there is an atom.

CLT moment condition for α

An atom $\alpha \in \mathcal{B}^+(\mathsf{X})$ exists with

$$\mathsf{E}_\alpha[s_0(|g|)^2] < \infty \quad \text{and} \quad \mathsf{E}_\alpha[s_0(1)^2] < \infty. \tag{17.14}$$

This condition will be generalized to obtain the CLT and LIL for general positive Harris chains in Sections 17.3–17.5. We state here the results in the special case where an atom is assumed to exist.

Theorem 17.2.2. *Suppose that Φ is Harris recurrent, $g: \mathsf{X} \to \mathbb{R}$ is a function, and that (17.14) holds so that Φ is in fact positive Harris. Then $\gamma_g^2 < \infty$, and if $\gamma_g^2 > 0$ then the CLT and LIL hold for g.*

PROOF   The proof is a surprisingly straightforward extension of the second proof of the LLN. Using the notation introduced in the proof of Theorem 17.2.1 we obtain the bound

$$\left|\sum_{i=1}^n \overline{g}(\Phi_i) - \sum_{j=0}^{\ell_n-1} s_j(\overline{g})\right| \le s_{\ell_n}(|\overline{g}|). \tag{17.15}$$

By the law of large numbers for the i.i.d. random variables $\{(s_j(|\overline{g}|))^2 : j \geq 1\}$,

$$\lim_{N \to \infty} \frac{1}{N} \sum_{j=1}^{N} (s_j(|\overline{g}|))^2 = \mathsf{E}_\alpha[(s_0(|\overline{g}|))^2] < \infty$$

and hence

$$\lim_{N \to \infty} \frac{1}{N} \sum_{j=1}^{N} (s_j(|\overline{g}|))^2 - \frac{1}{N-1} \sum_{j=1}^{N-1} (s_j(|\overline{g}|))^2 = 0.$$

From these two limits it follows that $(s_n(|\overline{g}|))^2/n \to 0$ as $n \to \infty$, and hence that

$$\limsup_{n \to \infty} \frac{s_{\ell_n}(|\overline{g}|)}{\sqrt{n}} \leq \limsup_{n \to \infty} \frac{s_{\ell_n}(|\overline{g}|)}{\sqrt{\ell_n}} = 0 \qquad \text{a.s. } [\mathsf{P}_*]. \tag{17.16}$$

This and (17.15) show that

$$\left| \frac{1}{\sqrt{n}} \sum_{i=1}^{n} \overline{g}(\Phi_i) - \frac{1}{\sqrt{n}} \sum_{j=0}^{\ell_n - 1} s_j(\overline{g}) \right| \to 0 \qquad \text{a.s. } [\mathsf{P}_*]. \tag{17.17}$$

We now need a more delicate argument to replace the random upper limit in the sum $\sum_{j=0}^{\ell_n - 1} s_j(\overline{g})$ appearing in (17.17) with a deterministic upper bound.

First of all, note that

$$\frac{\ell_n}{\sum_{j=0}^{\ell_n} s_j(1)} \leq \frac{\ell_n}{n} \leq \frac{\ell_n}{\sum_{j=0}^{\ell_n - 1} s_j(1)}.$$

Since $s_0(1)$ is almost surely finite, $s_0(1)/\ell_n \to 0$, and as in (17.16), $s_{\ell_n}(1)/\ell_n \to 0$. Hence by the LLN for i.i.d. random variables,

$$\lim_{n \to \infty} \frac{\ell_n}{n} = \left(\lim_{n \to \infty} \frac{1}{\ell_n} \sum_{j=1}^{\ell_n} s_j(1) \right)^{-1} = \mathsf{E}_\alpha[s_0(1)]^{-1} = \pi\{\alpha\}. \tag{17.18}$$

Let $\varepsilon > 0$, $\underline{n} = \lceil (1 - \varepsilon)\pi\{\alpha\}n \rceil$, $\overline{n} = \lfloor (1 + \varepsilon)\pi\{\alpha\}n \rfloor$, and $n^* = \lceil \pi\{\alpha\}n \rceil$, where $\lceil x \rceil$ ($\lfloor x \rfloor$) denote the smallest integer greater than (greatest integer smaller than) the real number x. Then by the result above, for some n_0

$$\mathsf{P}_x\{\underline{n} \leq \ell_n - 1 \leq \overline{n}\} \geq 1 - \varepsilon, \qquad n \geq n_0. \tag{17.19}$$

Hence for these n we have by Kolmogorov's Inequality (Theorem D.6.3)

$$\mathsf{P}_x\left\{ \left| \frac{1}{\sqrt{n}} \sum_{j=0}^{\ell_n - 1} s_j(\overline{g}) - \frac{1}{\sqrt{n}} \sum_{j=0}^{n^*} s_j(\overline{g}) \right| > \beta \right\} \leq \varepsilon + \mathsf{P}_x\left\{ \max_{\underline{n} \leq l \leq n^*} \left| \sum_{j=l}^{n^*} s_j(\overline{g}) \right| > \beta\sqrt{n} \right\}$$

$$+ \mathsf{P}_x\left\{ \max_{n^* \leq l \leq \overline{n}} \left| \sum_{j=n^*}^{l} s_j(\overline{g}) \right| > \beta\sqrt{n} \right\}$$

$$\leq \varepsilon + \frac{2n\varepsilon \mathsf{E}_\alpha[(s_0(\overline{g}))^2]}{\beta^2 n}.$$

Since $\varepsilon > 0$ is arbitrary, this shows that

$$\left| \frac{1}{\sqrt{n}} \sum_{j=0}^{\ell_n} s_j(\overline{g}) - \frac{1}{\sqrt{n}} \sum_{j=0}^{n^*} s_j(\overline{g}) \right| \to 0$$

in probability. This together with (17.17) implies that also

$$\left| \frac{1}{\sqrt{n}} \sum_{i=1}^{n} \overline{g}(\Phi_i) - \frac{1}{\sqrt{n}} \sum_{j=0}^{n^*} s_j(\overline{g}) \right| \to 0 \qquad (17.20)$$

in probability. By the CLT for i.i.d. sequences, we may let $\sigma^2 = \mathsf{E}_\alpha[(s_0(\overline{g}))^2]$ giving

$$\lim_{n \to \infty} \mathsf{P}_x\left\{ (n\gamma_g^2)^{-1/2} S_n(\overline{g}) \le t \right\} = \lim_{n \to \infty} \mathsf{P}_x\left\{ (n\gamma_g^2)^{-1/2} \sum_{j=0}^{n^*} s_j(\overline{g}) \le t \right\}$$

$$= \lim_{n \to \infty} \mathsf{P}_x\left\{ \sqrt{\frac{\lceil n\pi\{\alpha\}\rceil}{n\pi\{\alpha\}}} \frac{1}{\sqrt{n^* \sigma^2}} \sum_{j=0}^{n^*} s_j(\overline{g}) \le t \right\}$$

$$= \int_{-\infty}^{t} \frac{1}{\sqrt{2\pi}} e^{-1/2\, x^2}\, dx$$

which proves (i).

To prove (ii), observe that (17.17) implies that, as in the proof of the CLT, the analysis can be shifted to the sequence of i.i.d. random variables $\{s_j(\overline{g}) : j \ge 1\}$. By the LIL for this sequence,

$$\limsup_{n \to \infty} \frac{1}{\sqrt{2\sigma^2 \ell_n \log\log(\ell_n)}} \sum_{j=1}^{\ell_n} s_j(\overline{g}) = 1 \qquad \text{a.s. } [\mathsf{P}_*]$$

and the corresponding lim inf is -1. Equation (17.18) shows that $\ell_n/n \to \pi\{\alpha\} > 0$ and hence by a simple calculation $\log\log \ell_n / \log\log n \to 1$ as $n \to \infty$. These relations together with (17.17) imply

$$\limsup_{n \to \infty} \frac{1}{\sqrt{2\gamma_g^2 n \log\log(n)}} \sum_{k=1}^{n} \overline{g}(\Phi_k)$$

$$= \limsup_{n \to \infty} \frac{1}{\sqrt{\pi\{\alpha\}}} \frac{1}{\sqrt{2\sigma^2 n \log\log(n)}} \sum_{k=1}^{\ell_n} s_j(\overline{g})$$

$$= \limsup_{n \to \infty} \frac{1}{\sqrt{\pi\{\alpha\}}} \sqrt{\frac{\ell_n \log\log(\ell_n)}{n \log\log(n)}} \frac{1}{\sqrt{2\sigma^2 \ell_n \log\log(\ell_n)}} \sum_{k=1}^{\ell_n} s_j(\overline{g})$$

$$= 1$$

and the corresponding lim inf is equal to -1 by the same chain of equalities. $\qquad \square$

17.3 General Harris chains

We have seen in the previous section that when $\boldsymbol{\Phi}$ possesses an atom, the sample paths of the process may be divided into i.i.d. blocks to obtain for the Markov chain almost any ergodic theorem that holds for an i.i.d. process.

If $\boldsymbol{\Phi}$ is strongly aperiodic, such ergodic theorems may be established by considering the split chain, which possesses the atom $\mathsf{X} \times \{1\}$. For a general aperiodic chain such a splitting is not possible in such a "clean" form. However, since an m-step skeleton chain is always strongly aperiodic we may split this embedded chain as in Chapter 5 to construct an atom for the split chain. In this section we will show how we can then embed the split chain onto the same probability space as the entire chain $\boldsymbol{\Phi}$. This will again allow us to divide the sample paths of the chain into i.i.d. blocks, and the proofs will be only slightly more complicated than when a genuine atom is assumed to exist.

17.3.1 Splitting general Harris chains

When $\boldsymbol{\Phi}$ is aperiodic, we have seen in Proposition 5.4.5 that every skeleton is ψ-irreducible, and that the minorization condition holds for some skeleton chain. That is, we can find a set $C \in \mathcal{B}^+(\mathsf{X})$, a probability ν, $\delta > 0$, and an integer m such that $\nu(C) = 1$, $\nu(C^c) = 0$ and

$$P^m(x, B) \geq \delta\nu(B), \qquad x \in C, \quad B \in \mathcal{B}(\mathsf{X}).$$

The m-step chain $\{\Phi_{km} : k \in \mathbb{Z}_+\}$ is strongly aperiodic and hence may be split to form a chain which possesses a Harris recurrent atom.

We will now show how the split chain may be put on the same probability space as the entire chain $\boldsymbol{\Phi}$. It will be helpful to introduce some new notation so that we can distinguish between the split skeleton chain, and the original process $\boldsymbol{\Phi}$. We will let $\{Y_n\}$ denote the *level* of the split m-skeleton at time nm; for each n the random variable Y_n may take on the value zero or one. The split chain $\check{\boldsymbol{\Phi}}$ will become the bivariate process $\{\check{\boldsymbol{\Phi}}_n = (\Phi_{mn}, Y_n) : n \in \mathbb{Z}_+\}$, where the equality $\check{\boldsymbol{\Phi}}_n = x_i$ means that $\Phi_{nm} = x$ and $Y_n = i$.

The split chain is constructed by defining the conditional probabilities

$$\check{\mathsf{P}}\{Y_n = 1, \Phi_{nm+1} \in dx_1, \ldots, \Phi_{(n+1)m-1} \in dx_{m-1}, \Phi_{(n+1)m} \in dy$$
$$\mid \Phi_0^{nm}, Y_0^{n-1}; \Phi_{nm} = x\}$$
$$= \check{\mathsf{P}}\{Y_0 = 1, \Phi_1 \in dx_1, \ldots, \Phi_{m-1} \in dx_{m-1}, \Phi_m \in dy \mid \Phi_0 = x\}$$
$$= \delta r(x, y) P(x, dx_1) \cdots P(x_{m-1}, dy) \tag{17.21}$$

where $r \in \mathcal{B}(\mathsf{X}^2)$ is the Radon–Nykodym derivative

$$r(x, y) = \mathbb{I}\{x \in C\}\frac{\nu(dy)}{P^m(x, dy)}.$$

Integrating over $x_1, \ldots, x_{m-1}$ we see that

$$\check{\mathsf{P}}\{Y_n = 1, \Phi_{(n+1)m} \in dy \mid \Phi_0^{nm}, Y_0^{n-1}; \Phi_{nm} = x\}$$
$$= \delta\mathbb{I}(x \in C)\frac{\nu(dy)}{P^m(x, dy)}P^m(x, dy)$$
$$= \delta\mathbb{I}(x \in C)\nu(dy).$$

From Bayes' rule, it follows that

$$\begin{aligned} \check{\mathsf{P}}\{Y_n = 1 \mid \Phi_0^{nm}, Y_0^{n-1}; \Phi_{nm} = x\} &= \delta\mathbb{I}\{x \in C\}, \\ \check{\mathsf{P}}\{\Phi_{(n+1)m} \in dy \mid \Phi_0^{nm}, Y_0^n; \Phi_{nm} = x, Y_n = 1\} &= \nu(dy) \end{aligned}$$

and hence, given that $Y_n = 1$, the pre-nm process and post-$(n+1)m$ process are independent: that is

$$\{\Phi_k, Y_i : k \leq nm, i \leq n\} \text{ is independent of } \{\Phi_k, Y_i : k \geq (n+1)m, i \geq n+1\}.$$

Moreover, the distribution of the post-$(n+1)m$ process is the same as the $\check{\mathsf{P}}_{\nu^*}$-distribution of $\{(\Phi_i, Y_i) : i \geq 0\}$, with the interpretation that ν is "split" to form ν^* as in (5.3) so that

$$\check{\mathsf{P}}_{\nu^*}\{Y_0 = 1, \Phi_0 \in dx\} := \delta\mathbb{I}(x \in C)\nu(dx).$$

For example, for any positive function f on X, we have

$$\check{\mathsf{E}}[f(\Phi_{(n+1)m+k}) \mid \Phi_0^{mn}, Y_0^n; Y_n = 1] = \mathsf{E}_\nu[f(\Phi_k)].$$

Hence the set $\check{\alpha} := C_1 := C \times \{1\}$ behaves very much like an atom for the chain.

We let $\sigma_{\check{\alpha}}(0)$ denote the first entrance time of the split m-step chain to the set $\check{\alpha}$, and $\sigma_{\check{\alpha}}(k)$ the kth entrance time to $\check{\alpha}$ subsequent to $\sigma_{\check{\alpha}}(0)$. These random variables are defined inductively as

$$\begin{aligned} \sigma_{\check{\alpha}}(0) &= \min(k \geq 0 : Y_k = 1), \\ \sigma_{\check{\alpha}}(n) &= \min(k > \sigma_{\check{\alpha}}(n-1) : Y_k = 1), \qquad n \geq 1. \end{aligned}$$

The hitting times $\{\tau_{\check{\alpha}}(k)\}$ are defined in a similar manner:

$$\begin{aligned} \tau_{\check{\alpha}}(1) &= \min(k \geq 1 : Y_k = 1), \\ \tau_{\check{\alpha}}(n) &= \min(k > \tau_{\check{\alpha}}(n-1) : Y_k = 1), \qquad n \geq 1. \end{aligned}$$

For each n define

$$\begin{aligned} s_i(f) &= \sum_{j=m(\sigma_{\check{\alpha}}(i)+1)}^{m\sigma_{\check{\alpha}}(i+1)+m-1} f(\Phi_j) \\ &= \sum_{j=\sigma_{\check{\alpha}}(i)+1}^{\sigma_{\check{\alpha}}(i+1)} Z_j(f) \end{aligned}$$

where

$$Z_j(f) = \sum_{k=0}^{m-1} f(\Phi_{jm+k}).$$

From the remarks above and the strong Markov property we obtain the following result:

Theorem 17.3.1. *The two collections of random variables*

$$\{s_i(f) : 0 \le j \le m-2\}, \qquad \{s_i(f) : j \ge m\}$$

are independent for any $m \ge 2$. The distribution of $s_i(f)$ is, for any i, equal to the $\check{\mathsf{P}}_{\check{\alpha}}$-distribution of the random variable $\sum_{k=m}^{\tau_{\check{\alpha}} m+m-1} f(\Phi_k)$, which is equal to the $\check{\mathsf{P}}_{\nu*}$-distribution of

$$\sum_{k=0}^{\sigma_{\check{\alpha}} m+m-1} f(\Phi_k) = \sum_{k=0}^{\sigma_{\check{\alpha}}} Z_k(f). \tag{17.22}$$

The common mean of $\{s_i(f)\}$ may be expressed

$$\check{\mathsf{E}}[s_i(f)] = \delta^{-1}\pi(C)^{-1}m \int f d\pi. \tag{17.23}$$

PROOF   From the definition of $\{\sigma_{\check{\alpha}}(k)\}$ we have that the distribution of $s_{n+j}(f)$ given $s_0(f), \ldots, s_n(f)$ is equal to the distribution of $s_i(f)$ for all $n \in \mathbb{Z}_+$, $j \ge 1$. This follows from the construction of $\{\sigma_{\check{\alpha}}(k)\}$ which makes the distribution of $\Phi_{\sigma_{\check{\alpha}}(n+j)m+m}$ given $\mathcal{F}^{\Phi}_{\sigma_{\check{\alpha}}(n+j)m} \vee \mathcal{F}^Y_{\sigma_{\check{\alpha}}(n+j)}$ equal to ν.

From this we see that $\{s_n(f) : n \ge 1\}$ is a stationary sequence and, moreover, that $\{s_j(f)\}$ is a one-dependent process: that is, $\{s_0(f), \ldots, s_{n-1}(f)\}$ is independent of $\{s_{n+1}(f), \ldots, \}$ for all $n \ge 1$.

From (17.22) we can express the common mean of $\{s_i(f)\}$ in terms of the invariant mean of f as follows

$$
\begin{aligned}
\check{\mathsf{E}}[s_i(f)] &= \check{\mathsf{E}}_{\check{\alpha}}\left[\sum_{k=1}^{\tau_{\check{\alpha}}} Z_k(f)\right] \\
&= \check{\mathsf{E}}_{\check{\alpha}}\left[\sum_{k=1}^{\infty} Z_k(f)\mathbb{I}\{k \le \tau_{\check{\alpha}}\}\right] \\
&= \check{\mathsf{E}}_{\check{\alpha}}\left[\sum_{k=1}^{\infty} \check{\mathsf{E}}_{\Phi_{mk}}[Z_1(f)]\mathbb{I}\{k \le \tau_{\check{\alpha}}\}\right] \\
&= \delta^{-1}\pi(C)^{-1}\int \pi(dy)\mathsf{E}_y[Z_1(f)] \\
&= \delta^{-1}\pi(C)^{-1}m\int f d\pi
\end{aligned}
$$

where the fourth equality follows from the representation of π given in Theorem 10.0.1 applied to the split m-skeleton chain.   □

Define now, for each $n \in \mathbb{Z}_+$, $\ell_n := \max\{i \ge 0 : m\sigma_{\check{\alpha}}(i) \le n\}$, and write

$$
\begin{aligned}
\sum_{k=1}^n f(\Phi_k) = \; &\sum_{k=1}^{m\sigma_{\check{\alpha}}(0)+m-1} f(\Phi_k) \\
&+ \sum_{i=0}^{\ell_n-1} s_i(f) \\
&+ \sum_{k=m(\sigma_{\check{\alpha}}(\ell_n)+1)}^{n} f(\Phi_k).
\end{aligned} \tag{17.24}
$$

All of the ergodic theorems presented in the remainder of this section are based upon Theorem 17.3.1 and the decomposition (17.24), valid for all $n \ge 1$.

We now apply this construction to give an extension of the Law of Large Numbers.

17.3.2   The LLN for general Harris chains

The following general version of the LLN for Harris chains follows easily by considering the split chain $\check{\Phi}$.

Theorem 17.3.2. *The following are equivalent when a σ-finite invariant measure π exists for Φ:*

(i) *For every $f, g \in L^1(\pi)$ with $\int g \, d\pi \neq 0$,*

$$\lim_{n \to \infty} \frac{S_n(f)}{S_n(g)} = \frac{\pi(f)}{\pi(g)} \qquad \text{a.s. } [\mathsf{P}_*].$$

(ii) *The invariant σ-field Σ is P_x-trivial for all x.*

(iii) *Φ is Harris recurrent.*

PROOF   We just prove the equivalence between (i) and (iii). The equivalence of (i) and (ii) follows from the Chacon–Ornstein Theorem (see Theorem 3.2 of Revuz [326]), and the same argument that was used in the proof of Theorem 17.1.7.

The "if" part is trivial: If $\int f \, d\pi > 0$, then by the ratio limit result which is assumed to hold,

$$\mathsf{P}_x\{f(\Phi_i) > 0 \quad \text{i.o.}\} = 1$$

for all initial conditions, which is seen to be a characterization of Harris recurrence by taking f to be an indicator function.

To prove that (iii) implies (i) we will make use of the decomposition (17.24) and essentially the same proof that was used when an atom was assumed to exist in Theorem 17.2.1.

From (17.24) we have

$$\frac{\sum_{i=1}^n f(\Phi_i)}{\sum_{i=1}^n g(\Phi_i)} \leq \frac{\ell_n}{\ell_n - 1} \frac{\left[\frac{1}{\ell_n} \left(\sum_{j=0}^{\ell_n} s_j(f) + \sum_{k=1}^{m\sigma_{\check{a}}(0)+m-1} f(\Phi_k) \right) \right]}{\left[\frac{1}{\ell_n - 1} \sum_{j=0}^{\ell_n - 1} s_j(f) \right]}.$$

Since by Theorem 17.3.1 the two sequences $\{s_{2k}(f) : k \in \mathbb{Z}_+\}$ and $\{s_{2k+1}(f) : k \in \mathbb{Z}_+\}$ are both i.i.d., we have from (17.23) and the LLN for i.i.d. sequences that

$$\lim_{N \to \infty} \frac{1}{N} \sum_{k=1}^N s_k(f) = \lim_{N \to \infty} \frac{1}{N} \sum_{\substack{k=1 \\ k \text{ odd}}}^N s_k(f) + \lim_{N \to \infty} \frac{1}{N} \sum_{\substack{k=1 \\ k \text{ even}}}^N s_k(f)$$

$$= \frac{1}{2} \left(\delta^{-1} \pi(C)^{-1} m \int f d\pi + \delta^{-1} \pi(C)^{-1} m \int f d\pi \right)$$

$$= \delta^{-1} \pi(C)^{-1} m \int f d\pi.$$

Since $\ell_n \to \infty$ a.s. it follows that

$$\limsup_{n \to \infty} \frac{\sum_{i=1}^n f(\Phi_i)}{\sum_{i=1}^n g(\Phi_i)} \leq \frac{\int f d\pi}{\int g d\pi}.$$

Interchanging the roles of f and g gives an identical lower bound on the limit infimum, and this completes the proof.   □

Observe that this result holds for both positive and null recurrent chains. In the positive case, substituting $g \equiv 1$ gives Theorem 17.2.1.

17.3.3   Applications of the LLN

In this section we will describe two applications of the LLN. The first is a technical result which is generally useful, and will be needed when we prove the functional central limit theorem for Markov chains in Section 17.4.

As a second application of the LLN we will give a proof that the dependent parameter bilinear model is positive recurrent under a weak moment condition on the parameter process.

The running maximum

As a simple application of the Theorem 17.3.2 we will establish here a bound on the running maximum of $g(\Phi_k)$.

Theorem 17.3.3. *Suppose that Φ is positive Harris, and suppose that $\pi(|g|) < \infty$. Then the following limit holds:*

$$\lim_{n \to \infty} \frac{1}{n} \max_{1 \le k \le n} |g(\Phi_k)| = 0 \qquad \text{a.s. } [\mathsf{P}_*].$$

PROOF   We may suppose without loss of generality that $g \ge 0$.

It is easy to verify that the desired limit holds if and only if

$$\lim_{n \to \infty} \frac{1}{n} g(\Phi_n) = 0 \qquad \text{a.s. } [\mathsf{P}_*]. \tag{17.25}$$

It follows from Theorem 17.3.2 and positive Harris recurrence that

$$\lim_{n \to \infty} \left\{ \frac{1}{n} \sum_{k=1}^{n} g(\Phi_k) - \frac{1}{n-1} \sum_{k=1}^{n-1} g(\Phi_k) \right\} = \pi(g) - \pi(g) = 0.$$

The left hand side of this equation is equal to

$$\lim_{n \to \infty} \frac{1}{n} g(\Phi_n) - \frac{1}{n} \frac{1}{n-1} \sum_{k=1}^{n-1} g(\Phi_k).$$

Since by Theorem 17.3.2 we have $\frac{1}{n} \frac{1}{n-1} \sum_{k=1}^{n-1} g(\Phi_k) \to 0$, it follows that (17.25) does hold, and the proof is complete.   □

To illustrate the application of the LLN to the stability of stochastic models we will now consider a linear system with random coefficients.

The dependent parameter bilinear model

Here we revisit the dependent parameter bilinear defined by (DBL1)–(DBL2).

We saw in Proposition 7.4.1 that this model is a Feller T-chain. Since Z is i.i.d., the parameter process θ is itself a Feller T-chain, which is positive Harris by Proposition 11.4.2. Hence the LLN holds for θ, and this fact is the basis of our subsequent analysis of this bilinear model.

Proposition 17.3.4. *If (DBL1) and (DBL2) hold, then θ is positive Harris recurrent with invariant probability π_θ. For any $f \colon \mathbb{R} \to \mathbb{R}$ satisfying*

$$\int_{\mathbb{R}} \{f(x) \vee 0\} \, \pi_\theta(dx) < \infty$$

we have

$$\lim_{N \to \infty} \frac{1}{N} \sum_{k=1}^{N} f(\theta_k) = \int_{\mathbb{R}} f(x) \, \pi_\theta(dx) \qquad \text{a.s. } [\mathsf{P}_*].$$

When $\theta_0 \sim \pi_\theta$ the process is strictly stationary and may be defined on the positive and negative time set $\mathbb{Z}$. For this stationary process, the backwards LLN holds:

$$\lim_{N \to \infty} \frac{1}{N} \sum_{k=1}^{N} f(\theta_{-k}) = \int_{\mathbb{R}} f(x) \, \pi_\theta(dx) \qquad \text{a.s. } [\mathsf{P}_{\pi_\theta}]. \tag{17.26}$$

PROOF   The positivity of θ has already been noted prior to the proposition. The first limit then follows from Theorem 17.1.7 when $\int_{\mathbb{R}} f(x) \, \pi_\theta(dx) > -\infty$. Otherwise, we have from Theorem 17.1.7 and integrability of $f \vee 0$, for any $M > 0$,

$$\limsup_{N \to \infty} \frac{1}{N} \sum_{k=1}^{N} f(\theta_k) \le \limsup_{N \to \infty} \frac{1}{N} \sum_{k=1}^{N} f(\theta_k) \vee (-M) = \int_{\mathbb{R}} \{f(x) \vee (-M)\} \, \pi_\theta(dx),$$

and the right hand side converges to $-\infty = \pi_\theta(f)$ as $M \to \infty$.

The limit (17.26) holds by stationarity, as in the proof of Theorem 17.1.2 (see [99]).
□

We now apply the LLN for θ to obtain stability for the joint process. The bound (17.27) used in Proposition 17.3.5 is analogous to the condition that $|\alpha| < 1$ in the simple linear model. Indeed, suppose that we have the condition that $|\theta_k|$ is less than one only in the mean: $\mathsf{E}_{\pi_\theta}[|\theta_k|] < 1$. Then by Jensen's inequality it follows that the bound (17.27) is also satisfied.

Proposition 17.3.5. *Suppose that (DBL1) and (DBL2) hold, and that*

$$\int_{\mathbb{R}} \log |x| \, \pi_\theta(dx) < 0. \tag{17.27}$$

Then the joint process $\mathbf{\Phi} = \binom{\theta}{Y}$ is positive recurrent and aperiodic.

PROOF   To begin, recall from Theorem 7.4.1 that the joint process $\mathbf{\Phi} = \binom{\theta}{Y}$ is a ψ-irreducible and aperiodic T-chain.

For $y \in \mathbb{R}$ fixed, let $\mu_y = \pi_\theta \times \delta_y$ denote the initial distribution which makes θ a stationary process, and $Y_0 = y$ a.s. We will show that the distributions of Y, and hence of $\mathbf{\Phi}$ are tight whenever $\mathbf{\Phi}_0 \sim \mu_y$. From the Feller property and Theorem 12.1.2, this is sufficient to prove the theorem.

The following equality is obtained by iterating equation (2.13):

$$Y_{k+1} = \sum_{j=1}^{k} (\prod_{i=j}^{k} \theta_i) W_j + (\prod_{i=0}^{k} \theta_i) Y_0 + W_{k+1}. \tag{17.28}$$

Establishing stability is then largely a matter of showing that the product $\prod_{i=j}^{k} \theta_i$ converges to zero sufficiently fast. To obtain such convergence we will apply the LLN Proposition 17.3.4 and (17.27), which imply that as $n \to \infty$,

$$\frac{1}{n} \log \Big(\prod_{i=0}^{n} \theta_{-i}^2 \Big) = 2 \frac{1}{n} \sum_{i=0}^{n} \log |\theta_{-i}| \to 2 \int_{\mathbb{R}} \log |x| \, \pi_\theta(dx) < 0. \tag{17.29}$$

We will see that this limit, together with stationarity of the parameter process, implies exponential convergence of the product $\prod_{i=j}^{k} \theta_i$ to zero. This will give us the desired bounds on Y.

To apply (17.29), fix constants $L < \infty$, $0 < \rho < 1$, let $\Pi_{j,k} = \prod_{i=j}^{k} \theta_i$, and use (17.28) and the inequality $ab \le \frac{1}{2}(a^2 + b^2)$ to obtain the bound

$$P_{\mu_y}\{|Y_{k+1}| \ge L\}$$

$$\le P_{\mu_y}\Big\{ \sum_{j=1}^{k} |\Pi_{j,k}||W_j| + |\Pi_{0,k}||y| + |W_{k+1}| \ge L \Big\}$$

$$\le P_{\mu_y}\Big\{ \sum_{j=0}^{k} \rho^{-(k-j)} \Pi_{j,k}^2 + \sum_{j=0}^{k} \rho^{(k-j)} W_{j+1}^2 \ge 2L - (y^2 + 1) \Big\}$$

$$\le P_{\mu_y}\Big\{ \sum_{j=0}^{k} \rho^{-(k-j)} \Pi_{j,k}^2 \ge L - \frac{1+y^2}{2} \Big\} + P_{\mu_y}\Big\{ \sum_{j=0}^{k} \rho^{(k-j)} W_{j+1}^2 \ge L - \frac{1+y^2}{2} \Big\}.$$

We now use stationarity of θ and independence of W to move the time indices within the probabilities on the right hand side of this bound:

$$P_{\mu_y}\{|Y_{k+1}| \ge L\}$$

$$\le P_{\mu_y}\Big\{ \sum_{j=0}^{k} \rho^{-(k-j)} \Pi_{-(k-j),0}^2 \ge L - \frac{1+y^2}{2} \Big\}$$

$$\quad + P_{\mu_y}\Big\{ \sum_{j=0}^{k} \rho^{(k-j)} W_{k-j}^2 \ge L - \frac{1+y^2}{2} \Big\}$$

$$\le P_{\mu_y}\Big\{ \sum_{\ell=0}^{\infty} \rho^{-\ell} \Pi_{-\ell,0}^2 \ge L - \frac{1+y^2}{2} \Big\}$$

$$\quad + P_{\mu_y}\Big\{ \sum_{\ell=0}^{\infty} \rho^{\ell} W_{\ell}^2 \ge L - \frac{1+y^2}{2} \Big\}. \tag{17.30}$$

From Fubini's Theorem we have, for any $0 < \rho < 1$, that the sum $\sum_{\ell=0}^{\infty} \rho^{\ell} W_{\ell}^2$ converges a.s. to a random variable with finite mean $\sigma_w^2 (1 - \rho)^{-1}$.

We now show that the sum $\sum_{\ell=0}^{\infty} \rho^{-\ell} \Pi_{-\ell,0}^2$ converges a.s. For this we apply the root test. The logarithm of the nth root of the nth term a_n in this series is equal to

$$\log(a_n^{\frac{1}{n}}) := \log(\rho^{-n} \Pi_{-n,0}^2)^{\frac{1}{n}} = -\log(\rho) + 2\frac{1}{n} \sum_{i=0}^{n} \log|\theta_{-i}|.$$

By (17.29) it follows that

$$\lim_{n \to \infty} \log(a_n^{\frac{1}{n}}) = -\log(\rho) + 2 \int_{\mathbb{R}} \log|x|\, \pi_\theta(dx),$$

which is negative for sufficiently large $\rho < 1$. Fixing such a ρ, we have that $\lim_{n \to \infty} a_n^{\frac{1}{n}} < 1$, and thus the root test is positive. Thus the sum $\sum_{\ell=0}^{\infty} \rho^{-\ell} \Pi_{-\ell,0}^2$ converges to a finite limit with probability one.

By (17.30) and finiteness of the sums on the right hand side we conclude that

$$\sup_{k \geq 0} \mathsf{P}_{\mu_y}\{|Y_k| \geq L\} \to 0 \qquad \text{as } L \to \infty,$$

which is the desired tightness property for the process Y. $\qquad\qquad\qquad\square$

This stability result may be surprising given the very weak conditions imposed, and it may be even more surprising to find that these conditions can be substantially relaxed. It is really only the bound (17.27) together with stationarity of the parameter process which was needed in the proof of tightness for the output process Y. The use of the linear model θ was merely a matter of convenience.

This result illustrates the strengths and weaknesses of adopting boundedness in probability, or even positive Harris recurrence as a *stability* condition. Although the dependent parameter bilinear model is positive recurrent under (17.27), the behavior of the sample paths of Y can appear quite explosive. To illustrate this, recall the simulation given in Chapter 16 where we took the simple adaptive control model illustrated in Figure 2.5, but set the control equal to zero for illustrative purposes. This gives the model described in (DBL1)–(DBL2) with Z and W Gaussian $N(0, \sigma_z^2)$ and $N(0, \sigma_w^2)$ respectively, where $\sigma_z = 0.2$ and $\sigma_w = 0.1$. The parameter α is taken as 0.99. These parameter values are identical to those of the simulation given for the simple adaptive control model illustrated on the left in Figure 2.5. The stability condition (17.27) holds in this example since $\int_{\mathbb{R}} \log|x|\, \pi_\theta(dx) \approx -0.3 < 0$.

A sample path of $\log_{10}(|Y_k|)$ is given in Figure 16.1. Note the gross difference in behavior between this model and the simple adaptive control model with the control intact: In less than 700 time points the output of the dependent parameter bilinear model exceeds 10^{100}, while in the controlled case we see in Figures 2.5 and 2.6 that the output is barely distinguishable from the disturbance W when $\sigma_z = 0.2$.

17.3.4   The CLT and LIL for Harris chains

We now give versions of the CLT and LIL without the assumption that a true atom $\alpha \in \mathcal{B}^+(\mathsf{X})$ exists.

We will require the following bounds on the split chain constructed in this section. These conditions will be translated back to a condition on a petite set in Section 17.5.

CLT moment condition for the split chain

For the split chain constructed in this section, $\check{P}_{x_i}\{\sigma_{\check{\alpha}} < \infty\} = 1$ for all $x_i \in \check{X}$, and the function g and the atom $\check{\alpha}$ jointly satisfy the bounds

$$\check{E}_{\nu^*}\left[\left(\sum_{n=0}^{\sigma_{\check{\alpha}}} Z_n(|g|)\right)^2\right] < \infty \quad \text{and} \quad \check{E}_{\nu^*}\left[\sigma_{\check{\alpha}}^2\right] < \infty. \tag{17.31}$$

When these conditions are satisfied we will show that the CLT variance may be written

$$\gamma_g^2 = m^{-1}\check{\pi}(\check{\alpha})\check{E}_{\check{\alpha}}[(s_1(\overline{g}))^2] + 2m^{-1}\check{\pi}(\check{\alpha})\check{E}_{\check{\alpha}}[s_1(\overline{g})s_2(\overline{g})] \tag{17.32}$$

where $\check{\pi}$ is the invariant probability measure for the split chain and $\check{\pi}(\check{\alpha}) = \delta\pi(C)$.

We may now present

Theorem 17.3.6. *Suppose that* Φ *is ergodic and that (17.31) holds. Then* $0 \leq \gamma_g^2 < \infty$, *and if* $\gamma_g^2 > 0$ *then the CLT and LIL hold for* g.

PROOF   The proof is only a minor modification of the previous proof: we recall that $\ell_n := \max(k : m\sigma_{\check{\alpha}}(k) \leq n)$ and observe that in a manner similar to the derivation of (17.17) we may show that

$$\left|\frac{1}{\sqrt{n}}\sum_{j=0}^{n}\overline{g}(\Phi_j) - \frac{1}{\sqrt{n}}\sum_{j=0}^{\ell_n-1}s_j(\overline{g})\right| \to 0 \qquad \text{a.s.} \tag{17.33}$$

From the LLN we have

$$\lim_{n\to\infty}\frac{\ell_n}{n} = \lim_{n\to\infty}\frac{1}{n}\sum_{k=1}^{\lceil n/m\rceil-1}\mathbb{I}\{(\Phi_{mk}, Y_k) \in \check{\alpha}\} = \frac{\check{\pi}(\check{\alpha})}{m} \qquad \text{a.s. } [\mathsf{P}_*]. \tag{17.34}$$

This can be used to replace the upper limit of the second sum in (17.33) by a deterministic bound, just as in the proof of Theorem 17.2.2. Indeed, stationarity and one-dependence of $\{s_j(\overline{g}) : j \geq 1\}$ allow us to apply Kolmogorov's inequality Theorem D.6.3 to obtain the following analogue of (17.20): letting $n^* := \lceil m^{-1}\check{\pi}(\check{\alpha})n\rceil$, we have from (17.34) and (17.33) that

$$\left|\frac{1}{\sqrt{n}}\sum_{i=0}^{n}\overline{g}(\Phi_i) - \frac{1}{\sqrt{n}}\sum_{j=0}^{n^*}s_j(\overline{g})\right| \to 0 \tag{17.35}$$

in probability.

To complete the proof we will obtain a version of the CLT for one-dependent, stationary stochastic processes.

Fix an integer $m \geq 2$ and define $\eta_j = s_{jm+1}(\overline{g}) + \cdots + s_{(j+1)m-1}(\overline{g})$. For all $n \in \mathbb{Z}_+$ we may write

$$\frac{1}{\sqrt{n}} \sum_{j=1}^{n} s_j(\overline{g}) = \frac{1}{\sqrt{n}} \sum_{j=0}^{\lceil n/m \rceil - 1} \eta_j + \frac{1}{\sqrt{n}} \sum_{j=1}^{\lceil n/m \rceil - 1} s_{mj}(\overline{g}) + \frac{1}{\sqrt{n}} \sum_{j=m\lceil n/m \rceil}^{n} s_j(\overline{g}). \quad (17.36)$$

The last term converges to zero in probability, so that it is sufficient to consider the first and second terms on the RHS of (17.36). Since $\{s_i(\overline{g}) : i \geq 1\}$ is stationary and one-dependent, it follows that $\{\eta_j\}$ is an independent and identically distributed process, and also that $\{s_{mj}(\overline{g}) : j \geq 1\}$ is i.i.d.

The common mean of the random variables $\{\eta_j\}$ is zero, and its variance is given by the formula

$$\sigma_m^2 := \check{\mathsf{E}}[\eta_j^2] = (m-1)\check{\mathsf{E}}[s_1(\overline{g})^2] + 2(m-2)\check{\mathsf{E}}[s_1(\overline{g})s_2(\overline{g})].$$

By the CLT for i.i.d. random variables, we have therefore

$$\frac{1}{\sqrt{n}} \sum_{j=0}^{\lceil n/m \rceil - 1} \eta_j \xrightarrow{\mathrm{d}} N(0, m^{-1}\sigma_m^2),$$

and

$$\frac{1}{\sqrt{n}} \sum_{j=0}^{\lceil n/m \rceil} s_{mj}(\overline{g}) \xrightarrow{\mathrm{d}} N(0, m^{-1}\sigma_s^2),$$

where $\sigma_s^2 = \mathsf{E}[s_1(\overline{g})^2]$. Letting $m \to \infty$ we have

$$\begin{aligned} m^{-1}\sigma_m^2 &\to \bar{\sigma}^2 := \check{\mathsf{E}}[s_1(\overline{g})^2] + 2\check{\mathsf{E}}[s_1(\overline{g})s_2(\overline{g})], \\ m^{-1}\sigma_s^2 &\to 0, \end{aligned}$$

from which it can be shown, using (17.36), that

$$\frac{1}{\sqrt{n}} \sum_{j=1}^{n} s_j(\overline{g}) \xrightarrow{\mathrm{d}} N(0, \bar{\sigma}^2) \quad \text{as } n \to \infty.$$

Returning to (17.35) we see that

$$\frac{1}{\sqrt{n}} \sum_{i=0}^{n} \overline{g}(\Phi_i) \to N(0, m^{-1}\check{\pi}(\check{\alpha})\bar{\sigma}^2) \quad \text{as } n \to \infty$$

which establishes the CLT.

We can use Theorem 17.3.1 to prove the LIL, where the details are much simpler. We first write, as in the proof of Theorem 17.2.2,

$$\frac{1}{\sqrt{2n \log \log n}} \left(\sum_{k=1}^{n} \overline{g}(\Phi_k) - \sum_{j=1}^{\ell_n} s_j(\overline{g}) \right) \to 0 \quad \text{a.s.}$$

Using an expression similar to (17.36) together with the LIL for i.i.d. sequences we can easily show that the upper and lower limits of

$$\frac{1}{\sqrt{2n\bar{\sigma}^2 \log\log n}} \sum_{k=1}^{n} s_k(\bar{g})$$

are $+1$ and -1 respectively. Here the proof of Theorem 17.2.2 may be adapted to prove the LIL, which completes the proof of Theorem 17.3.6.   □

17.4   The functional CLT

In this section we show that a sequence of continuous functions obtained by interpolating the values of $S_n(f)$ converge to a standard Brownian motion. The machinery which we develop to prove this result rests heavily on the stability theory developed in Chapters 14 and 15. These techniques are extremely appealing as well as powerful, and can lead to much further insight into asymptotic behavior of the chain. Here we will focus on just one result: a functional central limit theorem, or *invariance principle* for the chain. This will allow us to refine the CLT which was presented in the previous chapter as well as allow us to obtain the expression (17.3) for the asymptotic variance.

We may now drop the aperiodicity assumption which was required in the previous section because of the very different approach taken.

17.4.1   Poisson's equation

Much of this development is based upon the following identity, known as *Poisson's equation*:

$$\hat{g} - P\hat{g} = g - \pi(g). \tag{17.37}$$

The function g is called the *forcing function*. In most cases the forcing function is given, and then $\hat{g}$ is called the *solution to Poisson's equation*.

Given a function g on X with $\pi(|g|) < \infty$ we will require that a finite-valued solution $\hat{g}$ to Poisson's equation (17.37) exist, and we will develop in this section sufficient conditions under which this is the case. The assumption that $\hat{g}$ is finite valued is made without any real loss of generality. If $\hat{g}$ solves Poisson's equation for some finite-valued function g, and if $\hat{g}(x_0)$ is finite for just one $x_0 \in \mathsf{X}$, then the set S_g of all x such that $|\hat{g}(x)| < \infty$ is full and absorbing, and hence the chain may be restricted to the set S_g.

In the special case where $g \equiv 0$, solutions to Poisson's equation are precisely what we have called *harmonic functions* in Section 17.1.2. In general, if $\hat{g}_1$ and $\hat{g}_2$ are two solutions to Poisson's equation then the difference $\hat{g}_1 - \hat{g}_2$ is harmonic. This observation is useful in answering questions regarding the uniqueness of solutions, as we see in Proposition 17.4.1. Integrability of solutions is not guaranteed in general – a much more generally applicable criterion for uniqueness is contained in Theorem 17.7.2.

Proposition 17.4.1. *Suppose that Φ is positive Harris, and suppose that $\hat{g}$ and $\hat{g}_\bullet$ are two solutions to Poisson's equation with $\pi(|\hat{g}| + |\hat{g}_\bullet|) < \infty$. Then for some constant c, $\hat{g}(x) = c + \hat{g}_\bullet(x)$ for a.e. $x \in \mathsf{X} [\pi]$.*

PROOF   We have already remarked that $h := \hat{g} - \hat{g}_\bullet$ is harmonic. To show that h is a constant we will require a strengthening of Theorem 17.1.5.

By iteration of the harmonic equation (17.8) we have $P^k h = h$ for all k, and hence for all n,

$$h = \frac{1}{n} \sum_{k=1}^{n} P^k h.$$

Since by assumption $\pi(|h|) < \infty$, it follows from Theorem 14.3.6 that $h(x) = \pi(h)$ for a.e. x.   □

One approach to the question of existence of solutions to (17.37) when an atom α exists is to let

$$\hat{g}(x) = G_\alpha(x, \overline{g}) = \mathsf{E}_x \Big[\sum_{k=0}^{\sigma_\alpha} \overline{g}(\Phi_k) \Big]. \tag{17.38}$$

The expectation is well defined if the chain is f-regular for some $f \geq |g|$. Since $0 = \pi(\overline{g}) = \pi(\alpha)\mathsf{E}_\alpha[\sum_{k=1}^{\tau_\alpha} \overline{g}(\Phi_k)]$, we have

$$
\begin{aligned}
P\hat{g}(x) &= \mathsf{E}_x \Big[\sum_{k=1}^{\sigma_\alpha} \overline{g}(\Phi_k) \Big] \mathbb{I}(x \notin \alpha) \\
&\quad + \mathsf{E}_\alpha \Big[\sum_{k=1}^{\tau_\alpha} \overline{g}(\Phi_k) \Big] \mathbb{I}(x \in \alpha) \\
&= \mathsf{E}_x \Big[\sum_{k=1}^{\sigma_\alpha} \overline{g}(\Phi_k) \Big] \mathbb{I}(x \notin \alpha).
\end{aligned}
$$

Since $\hat{g}(z) = \overline{g}(z)$ for all $z \in \alpha$, this shows that for all x,

$$P\hat{g}(x) = \mathsf{E}_x \Big[\sum_{k=0}^{\sigma_\alpha} \overline{g}(\Phi_k) \Big] - \overline{g}(x) = \hat{g}(x) - \overline{g}(x),$$

so that Poisson's equation is satisfied.

This approach can be extended to general ergodic chains by considering a split chain. However we will find it more convenient to follow a slightly different approach based upon the ergodic and regularity theorems developed in Chapter 14.

First note the formal similarity between Poisson's equation, which can be written $\Delta\hat{g} = -g + \pi(g)$, and the drift inequality (V3). Poisson's equation and (V3) are closely related, and in fact the inequality implies fairly easily that a solution to Poisson's equation exists. Assume that Φ is f-regular, so that (V3) holds for a function V which is everywhere finite, and a set C which is petite. If Φ is aperiodic, and if $\pi(V) < \infty$, then from the f-Norm Ergodic Theorem 14.0.1 we know that there exists a constant $R < \infty$ such that for any function g satisfying $|g| \leq f$,

$$\sum_{k=0}^{\infty} |P^k(x, g) - \pi(g)| \leq R(V(x) + 1).$$

Hence the function $\hat{g}$ defined as

$$\hat{g}(x) = \sum_{k=0}^{\infty} \{P^k(x,g) - \pi(g)\} \tag{17.39}$$

also satisfies the bound $|\hat{g}| \leq R(V+1)$, and clearly satisfies Poisson's equation. We state a generalization of this important observation as Theorem 17.4.2. The assumption that $\pi(V) < \infty$ is removed in Theorem 17.7.1.

Theorem 17.4.2. *Suppose that Φ is ψ-irreducible, and that (V3) holds with V everywhere finite, $f \geq 1$, and C petite. If $\pi(V) < \infty$, then for some $R < \infty$ and any $|g| \leq f$, Poisson's equation (17.37) admits a solution $\hat{g}$ satisfying the bound $|\hat{g}| \leq R(V+1)$.*

PROOF   The aperiodic case follows from absolute convergence of the sum in (17.39). In the general periodic case it is convenient to consider the K_{a_ε} chain, which is always strongly aperiodic when Φ is ψ-irreducible by Proposition 5.4.5.

To begin, we will show that the resolvent or K_{a_ε}-chain satisfies a version of (V3) with the same function f and a scaled version of the function V used in the theorem. We will on two occasions apply the identity

$$K_{a_\varepsilon} = \varepsilon K_{a_\varepsilon} P + (1-\varepsilon)I, \tag{17.40}$$

whose derivation is straightforward given the definition of the resolvent K_{a_ε}. Hence by (V3) for the kernel P,

$$K_{a_\varepsilon} V \leq \varepsilon K_{a_\varepsilon} (V - f + b\mathbb{I}_C) + (1-\varepsilon)V.$$

Since $f \leq (1-\varepsilon)^{-1} K_{a_\varepsilon} f$ it follows that with V_ε equal to a suitable constant multiple of V we have for some b',

$$K_{a_\varepsilon} V_\varepsilon \leq V_\varepsilon - f + b' K_{a_\varepsilon} \mathbb{I}_C.$$

Since C is petite for Φ and hence also for the K_{a_ε}-chain by Theorem 5.5.6, the set $C_n := \{x : K_{a_\varepsilon}(x,C) \geq 1/n\}$ is petite for the K_{a_ε}-chain for all n. Note that $C \subseteq C_n$ for n sufficiently large. Since C_n is petite we may adopt the proof of Theorem 14.2.9: scaling V_ε as necessary, we may choose n and b_ε so large that

$$K_{a_\varepsilon} V_\varepsilon \leq V_\varepsilon - f + b_\varepsilon \mathbb{I}_{C_n}.$$

Thus the K_{a_ε}-chain is f-regular. By aperiodicity there exists a constant $R_\varepsilon < \infty$ such that for any $|g| \leq f$, we have a solution $\hat{g}_\varepsilon$ to Poisson's equation

$$K_{a_\varepsilon} \hat{g}_\varepsilon = \hat{g}_\varepsilon - \overline{g} \tag{17.41}$$

satisfying $|\hat{g}_\varepsilon| \leq R_\varepsilon(V+1)$.
    To complete the proof let

$$\hat{g} := \frac{\varepsilon}{1-\varepsilon} K_{a_\varepsilon} \hat{g}_\varepsilon = \frac{\varepsilon}{1-\varepsilon} (\hat{g}_\varepsilon - \overline{g}). \tag{17.42}$$

Writing (17.40) in the form

$$\frac{\varepsilon}{1-\varepsilon} P K_{a_\varepsilon} = \frac{1}{1-\varepsilon} K_{a_\varepsilon} - I$$

we have by applying both sides to $\hat{g}_\varepsilon$

$$P\hat{g} = \varepsilon^{-1}\hat{g} - \hat{g}_\varepsilon = \varepsilon^{-1}\hat{g} - (\varepsilon^{-1} - 1)\hat{g} - \overline{g} = \hat{g} - \overline{g}$$

so that Poisson's equation is satisfied. $\qquad\qquad\qquad\qquad\qquad\qquad\square$

The significance of Poisson's equation is that it enables us to apply martingale theory to analyze the series $S_n(\overline{g})$. If $\hat{g}$ solves Poisson's equation, then we may write for any $n \geq 1$,

$$S_n(\overline{g}) = \sum_{k=1}^{n} \overline{g}(\Phi_k) = \sum_{k=1}^{n}[\hat{g}(\Phi_k) - P\hat{g}(\Phi_k)]$$

$$= \sum_{k=1}^{n}[\hat{g}(\Phi_k) - P\hat{g}(\Phi_{k-1})] + \sum_{k=1}^{n}[P\hat{g}(\Phi_{k-1}) - P\hat{g}(\Phi_k)].$$

The second sum on the right hand side is a telescoping series, which telescopes to $P\hat{g}(\Phi_0) - P\hat{g}(\Phi_n)$. We will prove in Theorem 17.4.3 that the first sum is a martingale, which shall be denoted

$$M_n(g) = \sum_{k=1}^{n}[\hat{g}(\Phi_k) - P\hat{g}(\Phi_{k-1})]. \qquad\qquad (17.43)$$

Hence $S_n(\overline{g})$ is equal to a martingale, plus a term which can be easily bounded. We summarize these observations in

Theorem 17.4.3. *Suppose that Φ is positive Harris and that a solution to Poisson's equation (17.37) exists with $\int |\hat{g}|\, d\pi < \infty$. Then when $\Phi_0 \sim \pi$, the series $S_n(\overline{g})$ may be written*

$$S_n(\overline{g}) = M_n(g) + P\hat{g}(\Phi_0) - P\hat{g}(\Phi_n) \qquad\qquad (17.44)$$

where $(M_n(g), \mathcal{F}_n^{\Phi})$ is the martingale defined in (17.43).

PROOF   The expression (17.44) was established prior to the theorem statement. To see that $(M_n(g), \mathcal{F}_n^{\Phi})$ is a martingale, apply the identity

$$\hat{g}(\Phi_k) - P\hat{g}(\Phi_{k-1}) = \hat{g}(\Phi_k) - \mathsf{E}[\hat{g}(\Phi_k) \mid \mathcal{F}_{k-1}^{\Phi}].$$

The integrability condition on $\hat{g}$ is imposed so that

$$\mathsf{E}_\pi[|\hat{g}(\Phi_k) - \mathsf{E}[\hat{g}(\Phi_k) \mid \mathcal{F}_{k-1}^{\Phi}]|] < \infty, \qquad k \geq 1,$$

and hence also $\mathsf{E}_\pi[|M_n|] < \infty$ for all n. $\qquad\qquad\qquad\qquad\qquad\qquad\square$

Theorem 17.4.3 adds a great deal of structure to the problem of analyzing the partial sums $S_n(\overline{g})$ which we may utilize by applying the results of Section D.6.2 for square integrable martingales.

17.4.2 The functional CLT for Markov chains

We now combine the functional CLT for martingales (Theorem D.6.4) and Theorem 17.4.3 to give a functional CLT for Markov chains. In the following main result of this section we consider the function $s_n(t)$ which interpolates the values of the partial sums of $\bar{g}(\Phi_k)$:

$$s_n(t) = S_{\lfloor nt \rfloor}(\bar{g}) + (nt - \lfloor nt \rfloor)\Big[S_{\lfloor nt \rfloor + 1}(\bar{g}) - S_{\lfloor nt \rfloor}(\bar{g})\Big]. \qquad (17.45)$$

Theorem 17.4.4. *Suppose that* Φ *is positive Harris, and suppose that* g *is a function on* X *for which a solution* $\hat{g}$ *to the Poisson equation exists with* $\pi(\hat{g}^2) < \infty$. *If the constant*

$$\gamma_g^2 := \pi(\hat{g}^2 - \{P\hat{g}\}^2) \qquad (17.46)$$

is strictly positive, then as $n \to \infty$,

$$(n\gamma_g^2)^{-1/2} s_n(t) \xrightarrow{\mathrm{d}} B \qquad \text{a.s. } [\mathsf{P}_*]$$

where B *denotes a standard Brownian motion on* $[0,1]$.

PROOF   Using an obvious generalization of Proposition 17.1.6 we see that it is enough to prove the theorem when $\Phi_0 \sim \pi$. From Theorem 17.4.3 we have

$$S_n(\bar{g}) = M_n(g) + P\hat{g}(\Phi_0) - P\hat{g}(\Phi_n).$$

Defining the stochastic process $m_n(t)$ for $t \in [0,1]$ as in (D.7) by

$$m_n(t) = M_{\lfloor nt \rfloor}(g) + (nt - \lfloor nt \rfloor)\Big[M_{\lfloor nt \rfloor + 1}(g) - M_{\lfloor nt \rfloor}(g)\Big], \qquad (17.47)$$

it follows that for all $t \in [0,1]$,

$$(n\gamma_g^2)^{-1/2}|s_n(t) - m_n(t)| \ \leq \ (n\gamma_g^2)^{-1/2}|P\hat{g}(\Phi_0)|$$
$$+ (n\gamma_g^2)^{-1/2} \max_{1 \leq k \leq n} |P\hat{g}(\Phi_k)|. \qquad (17.48)$$

Since $\pi(\hat{g}^2) < \infty$, by Jensen's inequality we also have $\pi(\{P\hat{g}\}^2) < \infty$. Hence by Theorem 17.3.3 it follows that

$$\frac{1}{n} \max_{1 \leq k \leq n} \{P\hat{g}(\Phi_k)\}^2 \to 0 \qquad \text{a.s. } [\mathsf{P}_\pi]$$

as $n \to \infty$, and from (17.48) we have

$$\sup_{0 \leq t \leq 1} (n\gamma_g^2)^{-1/2}|s_n(t) - m_n(t)| \to 0 \qquad \text{a.s. } [\mathsf{P}_\pi]$$

as $n \to \infty$. That is, $|(n\gamma_g^2)^{-1/2}(s_n - m_n)|_c \to 0$ in $\mathcal{C}[0,1]$ with probability one. To prove the theorem, it is therefore sufficient to show that $(n\gamma_g^2)^{-1/2} m_n(t) \xrightarrow{\mathrm{d}} B$.

   We complete the proof by showing that the conditions of Theorem D.6.4 hold for the martingale $M_n(g)$.

To show that (D.8) holds note that

$$\mathsf{E}_\pi[(M_k(g) - M_{k-1}(g))^2 \mid \mathcal{F}^\Phi_{k-1}] = \mathsf{E}_\pi[(\hat{g}(\Phi_k) - P\hat{g}(\Phi_{k-1}))^2 \mid \mathcal{F}^\Phi_{k-1}]$$
$$= P\hat{g}^2(\Phi_{k-1}) - \{P\hat{g}(\Phi_{k-1})\}^2.$$

Since we have assumed that $\hat{g}^2$ is π-integrable, it follows that the function $P\hat{g}^2 - \{P\hat{g}\}^2$ is also π-integrable. Hence the LLN holds:

$$\lim_{n\to\infty} \frac{1}{n} \sum_{k=1}^n \mathsf{E}_\pi[(M_k(g) - M_{k-1}(g))^2 \mid \mathcal{F}^\Phi_{k-1}] = \pi(P\hat{g}^2 - \{P\hat{g}\}^2) = \gamma_f^2 \quad \text{a.s.}$$

We now establish (D.9). Again by the LLN we have for any $b > 0$,

$$\lim_{n\to\infty} \frac{1}{n} \sum_{k=1}^n \mathsf{E}_\pi[(M_k(g) - M_{k-1}(g))^2 \mathbb{I}\{(M_k(g) - M_{k-1}(g))^2 \geq b\} \mid \mathcal{F}^\Phi_{k-1}]$$
$$= \mathsf{E}_\pi[(\hat{g}(\Phi_1) - P\hat{g}(\Phi_0))^2 \mathbb{I}\{(\hat{g}(\Phi_1) - P\hat{g}(\Phi_0))^2 \geq b\}]$$

which tends to zero as $b \to \infty$. It immediately follows that (D.9) holds for any $\varepsilon > 0$, and this completes the proof. □

As an illustration of the implications of Theorem 17.4.4 we state the following corollary, which is an immediate consequence of the fact that both $h(u) = u(1)$ and $h(u) = \max_{0\leq t\leq 1} u(t)$ are continuous functionals on $u \in \mathcal{C}[0,1]$.

Theorem 17.4.5. *Under the conditions of Theorem 17.4.4, the CLT holds for g with γ_g^2 given by (17.46), and as $n \to \infty$,*

$$(n\gamma_g^2)^{-1/2} \max_{1\leq k\leq n} S_k(\overline{g}) \xrightarrow{\text{d}} \max_{0\leq t\leq 1} B(t).$$

17.4.3　The representations of γ_g^2

It is apparent now that the asymptotic variance in the CLT can take on many different forms depending on the context in which this limit theorem is proven. Here we will briefly describe how the various forms may be identified and related.

The CLT variance given in (17.46) can be transformed by substituting in Poisson's equation (17.37), and we thus obtain

$$\gamma_g^2 = \pi(\hat{g}^2 - \{\hat{g} - \overline{g}\}^2) = 2\pi(\hat{g}\overline{g}) - \pi(\overline{g}^2) = \mathsf{E}_\pi[2\hat{g}(\Phi_0)\overline{g}(\Phi_0) - \overline{g}^2(\Phi_0)]. \quad (17.49)$$

Substituting in the particular solution (17.39), which we may write as

$$\hat{g}(x) = \sum_{k=0}^\infty P^k(x, \overline{g}),$$

results in the expression

$$\gamma_g^2 = \pi(\overline{g}^2) + 2\pi\left(\sum_{k=1}^\infty \overline{g}P^k(x, \overline{g})\right). \quad (17.50)$$

This immediately gives the representation (17.3) for γ_g^2 whenever the expectation with respect to π and the infinite sum may be interchanged. We will give such conditions in the next section, under which the identity (17.3) does indeed hold.

Note that if we substituted in a different formula for $\hat{g}$ we would arrive at an entirely different formula. We now show that by taking the specific form (17.38) for $\hat{g}$ we can connect the expression for the asymptotic variance given in Section 17.2 with the formulas given here.

Recall that using the approach of Section 17.2 based upon the existence of an atom we arrived at the identity

$$\gamma_g^2 = \pi(\alpha)\mathsf{E}_\alpha\left[\left(\sum_1^{\tau_\alpha}\overline{g}(\Phi_k)\right)^2\right]. \tag{17.51}$$

It may seem unlikely *a priori* that the two expressions (17.49) and (17.51) coincide. However, as required by the theory, it is of course true that the identity

$$\pi(\alpha)\mathsf{E}_\alpha\left[\left(\sum_{k=1}^{\tau_\alpha}\overline{g}(\Phi_k)\right)^2\right] = \mathsf{E}_\pi[2\hat{g}(\Phi_0)\overline{g}(\Phi_0) - \overline{g}^2(\Phi_0)] \tag{17.52}$$

holds whenever an atom $\alpha \in \mathcal{B}^+(\mathsf{X})$ exists. To see this we will take

$$\hat{g}(x) = \mathsf{E}_x\left[\sum_{j=0}^{\tau_\alpha}\overline{g}(\Phi_j)\right]$$

which is the specific solution (17.38) to Poisson's equation. By the representation of π using the atom α and the formula for the solution $\hat{g}$ to Poisson's equation we then have

$$\begin{aligned}
\mathsf{E}_\pi[2\hat{g}(\Phi_0)\overline{g}(\Phi_0) - \overline{g}^2(\Phi_0)] &= \pi(\alpha)\mathsf{E}_\alpha\left[\sum_{k=1}^{\tau_\alpha}\left(2\overline{g}(\Phi_k)\hat{g}(\Phi_k) - \overline{g}^2(\Phi_k)\right)\right] \\
&= \pi(\alpha)\mathsf{E}_\alpha\left[\sum_{k=1}^{\tau_\alpha}\left(2\overline{g}(\Phi_k)\mathsf{E}_{\Phi_k}\left[\sum_{j=0}^{\sigma_\alpha}\overline{g}(\Phi_j)\right] - \overline{g}^2(\Phi_k)\right)\right] \\
&= \pi(\alpha)\mathsf{E}_\alpha\left[\sum_{k=1}^{\tau_\alpha}\left(2\overline{g}(\Phi_k)\mathsf{E}\left[\theta^k\sum_{j=0}^{\sigma_\alpha}\overline{g}(\Phi_j) \mid \mathcal{F}_k^\Phi\right] - \overline{g}^2(\Phi_k)\right)\right].
\end{aligned}$$

For any $k \geq 1$ we have on the event $\{k \leq \tau_\alpha\}$,

$$\theta^k\sum_{j=0}^{\sigma_\alpha}\overline{g}(\Phi_j) = \sum_{j=k}^{\tau_\alpha}\overline{g}(\Phi_j)$$

and hence the previous equation gives

$$
\begin{aligned}
\mathsf{E}_\pi[2\hat{g}(\Phi_0)\overline{g}(\Phi_0) - \overline{g}^2(\Phi_0)] &= \pi(\alpha)\mathsf{E}_\alpha\Big[\sum_{k=1}^{\tau_\alpha}\Big(2\overline{g}(\Phi_k)\mathsf{E}\Big[\sum_{j=k}^{\tau_\alpha}\overline{g}(\Phi_j)\mid\mathcal{F}_k^\Phi\Big] - \overline{g}^2(\Phi_k)\Big)\Big] \\
&= \pi(\alpha)\mathsf{E}_\alpha\Big[\sum_{k=1}^{\tau_\alpha}\mathsf{E}\Big[\sum_{j=k}^{\tau_\alpha}2\overline{g}(\Phi_k)\overline{g}(\Phi_j) - \overline{g}^2(\Phi_k)\mid\mathcal{F}_k^\Phi\Big]\Big] \\
&= \pi(\alpha)\mathsf{E}_\alpha\Big[\sum_{k=1}^{\tau_\alpha}\Big(\sum_{j=k}^{\tau_\alpha}2\overline{g}(\Phi_k)\overline{g}(\Phi_j) - \overline{g}^2(\Phi_k)\Big)\Big] \\
&= \pi(\alpha)\mathsf{E}_\alpha\Big[\Big(\sum_{k=1}^{\tau_\alpha}\overline{g}(\Phi_k)\Big)^2\Big]
\end{aligned}
$$

which gives (17.52).

We now apply the martingale and atom-based approaches simultaneously to obtain criteria for the CLT and LIL.

17.5   Criteria for the CLT and the LIL

In this section we give more easily verifiable conditions under which the CLT and LIL hold for general Harris chains. Up to now, our assumptions on the chain involve the statistics of the return time to the atom $\check{\alpha}$ for the split chain, or integrability conditions on a solution to Poisson's equation. Neither of these assumptions is easy to interpret, and therefore it is crucial to connect them to verifiable properties of the one-step transition function P. We do this now by proving that a drift property gives a sufficient condition under which the CLT and LIL are valid. Under this condition we will also show that the CLT variance may be written in the form (17.3).

The following conditions will be imposed throughout this section:

CLT moment condition on V, f

The chain Φ is ergodic, and there exists a function $f \geq 1$, a finite-valued function V and a petite set C satisfying (V3).

Letting π denote the unique invariant probability measure for the chain, we assume that $\pi(V^2) < \infty$.

The integrability condition on V^2 can be obtained by applying Theorem 14.3.7, but this condition may be difficult to verify in practice. For this reason we give in the following lemma a stronger condition under which this bound is satisfied automatically.

Lemma 17.5.1. *If Φ is V'-uniformly ergodic, then the CLT moment condition on V, f are satisfied with $V = (1 - \sqrt{1-\beta})^{-1}\sqrt{V'}$ and $f = \sqrt{V'}$.*

PROOF   It follows from Lemma 15.2.9 that the chain is V-uniform, and hence (V3) holds with this V. The finiteness of $\pi(V^2)$ follows from finiteness of $\pi(V')$, which is a consequence of the f-Norm Ergodic Theorem 14.0.1.   □

The following result shows that (V3) provides a sufficient condition under which the assumptions imposed in Section 17.4 and Section 17.3 are satisfied.

Lemma 17.5.2. *Under the CLT moment condition on V, f above we have:*

(i) *there exists a constant $R < \infty$ such that for any function g which satisfies the bound $|g| \leq f$, Poisson's equation (17.37) admits a solution $\hat{g}$ with $|\hat{g}| \leq R(V+1)$;*

(ii) *the split chain satisfies the bound*

$$\check{\mathsf{E}}_{\check{\alpha}}\left[\left(\sum_{\ell=0}^{\tau_{\check{\alpha}}-1} Z_\ell(f)\right)^2\right] < \infty \tag{17.53}$$

and hence the CLT moment condition (17.31) holds for any function g with $|g| \leq f$.

PROOF   Result (i) is simply a restatement of Theorem 17.4.2, so it is enough to prove (ii).

Under the CLT moment condition on V, f above, Φ is f-regular, and hence the m-skeleton is $f^{(m)}$-regular by Theorem 14.2.10. Hence the split chain $\check{\Phi}$ for the m-skeleton is $f^{(m)}$-regular if the set C used in the splitting is a sublevel set of V, and from Theorem 14.2.3 applied to the m-skeleton we have for some $R_0 < \infty$ and any $x_i \in \check{\mathsf{X}}$,

$$\check{\mathsf{E}}_{x_i}\left[\sum_{k=0}^{\tau_{\check{\alpha}}} f^{(m)}(\check{\Phi}_k)\right] \leq R_0(V(x) + 1)$$

where we define $f^{(m)}(\check{\Phi}_k) = f^{(m)}(\Phi_{mk}, Y_k) := f^{(m)}(\Phi_k)$.

Since $\{\tau_{\check{\alpha}} \geq k\} \in \check{\mathcal{F}}_{mk} = \sigma\{Y_i : i \leq k, \Phi_j : j \leq mk\}$, we have for all x_i,

$$\check{\mathsf{E}}_{x_i}\left[\sum_{k=0}^{\tau_{\check{\alpha}}} Z_k(f)\right] = \sum_{k=0}^{\infty} \check{\mathsf{E}}_{x_i}\left[Z_k(f)\mathbb{I}\{\tau_{\check{\alpha}} \geq k\}\right]$$

$$= \sum_{k=0}^{\infty} \check{\mathsf{E}}_{x_i}\left[\check{\mathsf{E}}[Z_k(f) \mid \check{\mathcal{F}}_{mk}]\mathbb{I}\{\tau_{\check{\alpha}} \geq k\}\right].$$

From (17.21) we may find $R_1 < \infty$ such that for $i = 0, 1$,

$$\check{\mathsf{E}}[Z_k(f) \mid \check{\mathcal{F}}_{mk}; \check{\Phi}_k = (\Phi_{mk}, Y_k) = (x, i)] \leq R_1 f^{(m)}(x),$$

and hence

$$\check{\mathsf{E}}_{x_i}\left[\sum_{k=0}^{\tau_{\check{\alpha}}} Z_k(f)\right] \leq R_0 R_1(V(x) + 1), \qquad x_i \in \check{\mathsf{X}}.$$

Under the assumption that $\pi(V^2) < \infty$ we see from the representation of π that

$$\check{\mathsf{E}}_{\check{\alpha}}\left[\sum_{\ell=0}^{\tau_{\check{\alpha}}-1}\left(\check{\mathsf{E}}_{\check{\Phi}_\ell}\left[\sum_{k=0}^{\tau_{\check{\alpha}}} Z_k(f)\right]\right)^2\right] \leq (\check{\pi}(\check{\alpha}))^{-1}(R_0 R_1)^2 \pi([V+1]^2) < \infty. \tag{17.54}$$

Similar arguments give the bound

$$\check{\mathsf{E}}_{\check{\alpha}}\Big[\sum_{\ell=0}^{\tau_{\check{\alpha}}-1}\Big(Z_\ell(f)\Big)^2\Big] = (\check{\pi}(\check{\alpha}))^{-1}\mathsf{E}_\pi\Big[\Big(Z_0(f)\Big)^2\Big] \le (\check{\pi}(\check{\alpha}))^{-1}m^2\pi(f^2) < \infty. \qquad (17.55)$$

Combining (17.54) and (17.55) we obtain

$$\check{\mathsf{E}}_{\check{\alpha}}\Big[\sum_{\ell=0}^{\tau_{\check{\alpha}}-1}\Big(Z_\ell(f) + \check{\mathsf{E}}_{\Phi_{\ell+1}}\Big[\sum_{k=0}^{\tau_{\check{\alpha}}}Z_k(f)\Big]\Big)^2\Big] < \infty.$$

It is now relatively easy to show that the bound (17.53) holds. We may calculate using the ordinary Markov property,

$$\begin{aligned}
\infty \;>\; & \check{\mathsf{E}}_{\check{\alpha}}\Big[\sum_{\ell=0}^{\tau_{\check{\alpha}}-1}\Big(Z_\ell(f) + \check{\mathsf{E}}_{\Phi_{\ell+1}}\Big[\sum_{k=0}^{\tau_{\check{\alpha}}}Z_k(f)\Big]\Big)^2\Big] \\
=\; & \check{\mathsf{E}}_{\check{\alpha}}\Big[\sum_{\ell=0}^{\tau_{\check{\alpha}}-1}\Big(Z_\ell(f) + \check{\mathsf{E}}\Big[\sum_{k=\ell+1}^{\tau_{\check{\alpha}}}Z_k(f) \mid \check{\mathcal{F}}_{m(\ell+1)}\Big]\Big)^2\Big] \\
\ge\; & 2\check{\mathsf{E}}_{\check{\alpha}}\Big[\sum_{\ell=0}^{\tau_{\check{\alpha}}-1}Z_\ell(f)\check{\mathsf{E}}\Big[\sum_{k=\ell+1}^{\tau_{\check{\alpha}}}Z_k(f) \mid \check{\mathcal{F}}_{m(\ell+1)}\Big]\Big] + \check{\mathsf{E}}_{\check{\alpha}}\Big[\sum_{\ell=0}^{\tau_{\check{\alpha}}-1}\Big(Z_\ell(f)\Big)^2\Big] \\
=\; & 2\check{\mathsf{E}}_{\check{\alpha}}\Big[\sum_{\ell=0}^{\tau_{\check{\alpha}}-1}\sum_{k=\ell+1}^{\tau_{\check{\alpha}}}Z_\ell(f)Z_k(f)\Big] + \check{\mathsf{E}}_{\check{\alpha}}\Big[\sum_{\ell=0}^{\tau_{\check{\alpha}}-1}\Big(Z_\ell(f)\Big)^2\Big] \\
=\; & \check{\mathsf{E}}_{\check{\alpha}}\Big[\Big(\sum_{\ell=0}^{\tau_{\check{\alpha}}-1}Z_\ell(f)\Big)^2\Big].
\end{aligned}$$

$$\square$$

Theorem 17.5.3. *Assume the CLT moment condition on V, f, and let g be a function on X with $|g| \le f$. Then the constant γ_g^2 defined as*

$$\gamma_g^2 := \pi(\hat{g}^2 - (P\hat{g})^2)$$

is well defined, non-negative, and finite, and may be written as

$$\gamma_g^2 = \lim_{n\to\infty}\frac{1}{n}\mathsf{E}_\pi\Big[\Big(S_n(\overline{g})\Big)^2\Big] = \mathsf{E}_\pi[\overline{g}^2(\Phi_0)] + 2\sum_{k=1}^{\infty}\mathsf{E}_\pi[\overline{g}(\Phi_0)\overline{g}(\Phi_k)] \qquad (17.56)$$

where the sum converges absolutely.
   If $\gamma_g^2 > 0$, then the CLT and LIL hold for g.

PROOF   To obtain the representation (17.56) for γ_g^2, apply the identity (17.44), from which we obtain

$$\mathsf{E}_\pi[(S_n(\overline{g}) - M_n(g))^2] \le 4\pi(\hat{g}^2).$$

Since $E_\pi[M_n(g)^2] = \sum_1^n E_\pi[(M_k - M_{k-1})^2] = n\gamma_g^2$, it follows that $\frac{1}{n}E_\pi[S_n(\overline{g})^2] \to \gamma_g^2$ as $n \to \infty$.

We now show that $\frac{1}{n}E_\pi[S_n(\overline{g})^2] \to \sum_{-\infty}^{\infty} E_\pi[\overline{g}(\Phi_0)\overline{g}(\Phi_k)]$.

First we show that this sum converges absolutely. By the f-Norm Ergodic Theorem 14.0.1 we have for some $R < \infty$, and each x,

$$\sum_{k=0}^{\infty} |E_x[\overline{g}(\Phi_0)\overline{g}(\Phi_k)]| \leq |\overline{g}(x)| \sum_{k=0}^{\infty} \|P^k(x, \cdot) - \pi\|_f$$

$$\leq |\overline{g}(x)|R(V(x) + 1).$$

Since $|g|$ is bounded by f, which is bounded by a constant times $V + 1$, it follows that for some R',

$$\sum_{k=0}^{\infty} |E_x[\overline{g}(\Phi_0)\overline{g}(\Phi_k)]| \leq R'(V^2(x) + 1)$$

and hence

$$\sum_{k=0}^{\infty} |E_\pi[\overline{g}(\Phi_0)\overline{g}(\Phi_k)]| \leq R'(\pi(V^2) + 1) < \infty.$$

We now compute γ_g^2: For each n we have by invariance,

$$\frac{1}{n}E_\pi[S_n(\overline{g})^2] = E_\pi[\overline{g}(\Phi_0)^2] + 2\frac{1}{n}\sum_{k=1}^{n}\sum_{j-k+1}^{n} E_\pi[\overline{g}(\Phi_k)\overline{g}(\Phi_j)]$$

$$= E_\pi[\overline{g}(\Phi_0)^2] + 2\frac{1}{n}\sum_{k=0}^{n-1}\Big(\sum_{i=1}^{n-1-k} E_\pi[\overline{g}(\Phi_0)\overline{g}(\Phi_i)]\Big),$$

and the right hand side converges to $\sum_{-\infty}^{\infty} E_\pi[\overline{g}(\Phi_0)\overline{g}(\Phi_k)]$ as $n \to \infty$.

To prove that the CLT and LIL hold when $\gamma_g^2 > 0$, observe that by Lemma 17.5.2 under the conditions of this section the hypotheses of both Theorem 17.3.6 and Theorem 17.4.5 are satisfied. Theorem 17.3.6 gives the CLT and LIL, and Theorem 17.4.5 shows that the asymptotic variance is equal to $\pi(\hat{g}^2 - (P\hat{g})^2)$. □

So far we have left open the question of what happens when $\gamma_g^2 = 0$. Under the conditions of Theorem 17.5.3 it may be shown that in this case

$$\frac{1}{\sqrt{n}}S_n(g) \xrightarrow{d} 0.$$

We leave the proof of this general result to the reader. In the next result we give a criterion for the CLT and LIL for V-uniformly ergodic chains, and show that for such chains $\frac{1}{\sqrt{n}}S_n(g)$ converges to zero with probability one when $\gamma_g^2 = 0$.

Theorem 17.5.4. *Suppose that Φ is V-uniformly ergodic. If $g^2 < V$, then the conclusions of Theorem 17.5.3 hold, and if $\gamma_g^2 = 0$, then*

$$\frac{1}{\sqrt{n}}S_n(g) \to 0 \qquad \text{a.s. } [P_*].$$

PROOF   In view of Lemma 17.5.1 and Theorem 17.5.3, the only result which requires
proof is that $(\frac{1}{\sqrt{n}}S_n(\overline{g}) : n \geq 1)$ converges to zero when $\gamma_g^2 = 0$.

Recalling (17.44) we have

$$S_n(\overline{g}) = M_n(g) + P\hat{g}(\Phi_0) - P\hat{g}(\Phi_n).$$

We have shown that $\frac{1}{\sqrt{n}}P\hat{g}(\Phi_n) \to 0$ a.s. in the proof of Theorem 17.4.4. To prove the
theorem we will show that $(M_n(g))$ is a convergent sequence.

We have for all n and x,

$$\mathsf{E}_x[(M_n(g))^2] = \sum_{k=1}^n \mathsf{E}_x[P(\Phi_{k-1}, \hat{g}^2) - P(\Phi_{k-1}, \hat{g})^2].$$

Letting $G(x) = P(x, \hat{g}^2) - P(x, \hat{g})^2$ we have $0 \leq G \leq RV$ for some $R < \infty$, and
$\pi(G) = \gamma_g^2 = 0$. Hence by Theorem 15.0.1,

$$\mathsf{E}_x[(M_n(g))^2] = \sum_{k=1}^n \mathsf{E}_x[G(\Phi_{k-1})] \leq \sum_{k=0}^\infty |P^k(x, G) - \pi(G)| < \infty.$$

By the Martingale Convergence Theorem D.6.1 it follows that $(M_n(g))$ converges to a
finite constant, and is hence bounded in n with probability one.   □

17.6   Applications

From Theorem 17.0.1 we see that any of the V-uniform models which were studied
in the previous chapter satisfy the CLT and LIL as long as the asymptotic variance is
positive. We will consider here two models where moment conditions on the disturbance
process may be given explicitly to ensure that the CLT holds. In the first we avoid
Theorem 17.0.1 since we can obtain a stronger result by using Theorem 17.5.3, which
is based upon the CLT moment condition of the previous section.

17.6.1   Random walks and storage models

Consider random walk on a half line given by $\Phi_n = [\Phi_{n-1} + W_n]^+$, and assume that
the increment distribution Γ is has negative first moment and a finite fifth moment.

We have analyzed this model in Section 14.4 where it was shown in Proposition 14.4.1
that under these conditions the chain is $(x^4 + 1)$-regular.

Let $f(x) = |x|+1$ and $V(x) = cx^2$, with $c > 0$. From (14.29) we have that (V3) holds
for some c, and we have just noted that the chain is V^2-regular. Hence the conditions
imposed in Section 17.5 are satisfied, and applying Theorem 17.5.3 we see that the CLT
and LIL hold for any g satisfying $|g| \leq f$.

In particular, on setting $g(x) = x$ we see that the CLT and LIL hold for Φ itself.

Proposition 17.6.1. *If the increment distribution Γ has mean $\beta < 0$ and finite fifth
moment, then the associated random walk on a half line is positive Harris and the CLT
and LIL hold for the process $\{\Phi_k : k \geq 0\}$.*

The asymptotic variance may be written using (17.3) as $\gamma_g^2 = \sum_{-\infty}^{\infty} \mathsf{E}_\pi[\bar{\Phi}_k \bar{\Phi}_0]$, or using (17.13) with $\alpha = \{0\}$ we have

$$\gamma_g^2 = \pi(0)\mathsf{E}_0\left[\left(\sum_{k=1}^{\tau_0} \Phi_k - \mathsf{E}_\pi[\Phi_k]\right)^2\right].$$

□

17.6.2   Linear state space models

Here we illustrate Theorem 17.0.1. We can easily obtain conditions under which the CLT holds for the linear state space model, and explicitly calculate the asymptotic variance. To avoid unnecessary technicalities we will assume that $\mathsf{E}[W] = 0$.

Let $Y_k = c^\top X_k$, $k \in \mathbb{Z}_+$, where $c \in \mathbb{R}^n$. If the eigenvalue condition (LSS5) holds, then we have seen in Proposition 12.5.1 that a unique invariant probability π exists, and hence a stationary version of the process Y_k also exists, defined for $k \in \mathbb{Z}$. The stationary process can be realized as

$$Y_k = \sum_{\ell=0}^{\infty} h_\ell W_{k-\ell},$$

where $h_\ell = c^\top F^\ell G$ and $(W_k : k \in \mathbb{Z})$ are i.i.d. with mean zero and covariance $\Sigma_W = \mathsf{E}[WW^\top]$, which is assumed to be finite in (LSS2).

Let $R(k)$ denote the autocovariance sequence for the stationary process:

$$R(k) = \mathsf{E}_\pi[Y_k Y_0], \qquad k \in \mathbb{Z}.$$

If the CLT holds for the process Y, then we have seen that the asymptotic variance, which we shall denote γ_c^2, is equal to

$$\gamma_c^2 = \sum_{k=-\infty}^{\infty} R(k). \tag{17.57}$$

The autocovariance sequence can be analyzed through its Fourier series, and this approach gives a simple formula for the limiting variance γ_c^2.

The process Y has a spectral density $D(\omega)$ which is obtained from the autocovariance sequence through the Fourier series

$$D(\omega) = \sum_{m=-\infty}^{\infty} R(m)e^{im\omega},$$

and $R(m)$ can be recovered from $D(\omega)$ by the integral

$$R(m) = \frac{1}{2\pi} \int_{-\pi}^{\pi} e^{-im\omega} D(\omega)\, d\omega.$$

It is a straightforward exercise (see [225], p. 66) to show that the spectral density has the form

$$D(\omega) = H(e^{i\omega})\Sigma_W H(e^{i\omega})^*$$

where

$$H(e^{i\omega}) = \sum_{\ell=0}^{\infty} h_\ell e^{i\ell\omega} = c^\top (I - e^{i\omega} F)^{-1} G.$$

From these calculations we obtain the following CLT for the linear state space model:

Theorem 17.6.2. *Consider the linear state space model defined by (LSS1) and (LSS2). If the eigenvalue condition (LSS5), the nonsingularity condition (LSS4) and the controllability condition (LCM3) are satisfied, then the model is V-uniformly ergodic with $V(x) = |x|^2 + 1$.*

For any vector $c \in \mathbb{R}^n$, the asymptotic variance is given by the formula

$$\gamma_c^2 = c^\top (I - F)^{-1} G\Sigma_W G^\top (I - F^\top)^{-1} c,$$

and the CLT and LIL hold for process Y when $\gamma_c^2 > 0$.

PROOF   We have seen in the proof of Theorem 12.5.1 that (V4) holds for the linear state space model with $V(x) = 1 + x^\top M x$, where M is a positive matrix (see (12.34)). Under the conditions of Theorem 17.6.2 we also have that Φ is a ψ-irreducible and aperiodic T-chain by Proposition 6.3.5. By Lemma 17.5.1 and Theorem 17.5.2 it follows that the CLT and LIL hold for Y and that the asymptotic variance is given by (17.57).

The closed form expression for γ_c follows from the chain of identities

$$\gamma_c^2 = \sum_{k=-\infty}^{\infty} R(k) = D(0) = c^\top (I - F)^{-1} G\Sigma_W G^\top (I - F^\top)^{-1} c.$$

□

Had we proved the CLT for vector-valued functions of the state, it would be more natural in this example to prove directly that the CLT holds for X. In fact, an extension of Theorem D.6.4 to vector-valued processes is possible, and from such a generalization we have under the conditions of Theorem 17.6.2 that

$$\frac{1}{\sqrt{n}} \sum_{k=1}^{n} X_k X_k^\top \xrightarrow{\text{d}} N(0, \Sigma)$$

where $\Sigma = (I - F)^{-1} G\Sigma_W G^\top (I - F^\top)^{-1}$.

17.7   Commentary*

The results of this chapter may appear considerably deeper than those of other chapters, although in truth they are often straightforward from more global stochastic process results, given the embedded regeneration structure of the split chain, or given the existence of a stationary version (that is, of an invariant probability measure) for the chain.

One of the achievements of this chapter is the identification of these links, and in particular the development of a drift-condition approach to the sample path and central limit laws.

These laws are of value for Markov chains exactly as they are for all stochastic processes: the LLN and CLT, in particular, provide the theoretical basis for many results in the statistical analysis of chains as they do in related fields. In particular, the standard proofs of asymptotic efficiency and unbiasedness for maximum likelihood estimators is largely based upon these ergodic theorems. For this and other applications, the reader is referred to [151].

The Law of Large Numbers has a long history whose surface we can only skim here. Theorem 17.1.2 is a result of Doob [99], and the ratio form for Harris chains Theorem 17.3.2 is given in Athreya and Ney [14]. Chapter 3 of Orey [309] gives a good overview of related ratio limit theorems.

The classic text of Chung [71] gives in Section I.16 the CLT and LIL for chains on a countable space from which we adopt many of the proofs of the results in Section 17.2 and Section 17.3. Versions of the Central Limit Theorem for Harris chains may be found in Cogburn [74] and in Nummelin and Niemi [303, 300]. The paper [300] presents an excellent survey of what was the state of the art at that time, and also an excellent development of CLTs in a context more general than we have given.

Neveu remarks in [296] that "the relationship between the theory of martingales and the theory of Markov chains is very deep". At that time he referred mainly to the connections between harmonic functions, martingales, and first hitting probabilities for a Markov chain. In Section III-5 of [296] he develops fairly briefly a remarkably strong classification of a Markov chain as either recurrent or transient, based mainly on martingale limit theory and the existence of harmonic functions. Certainly the connections between martingales and Markov chains are substantial. From the largely martingale-based proof of the functional CLT described in this chapter, and the more general implications of Poisson's equation and its associated martingale to the ergodic theory of Markov chains, it appears that the relationship between Markov chains and martingales is even richer than was thought at the time of Neveu's writing.

The martingale approach via solutions to Poisson's equation which is developed in Section 17.4 is adopted from Duflo [102] and Maigret [242].

For further results on the potential theory of positive kernels we refer the reader to the seminal work of Neveu [295], Revuz [326] and Constantinescu and Cornea [77], and to Nummelin [304] for the most current development. Applications to Markov processes evolving in continuous time are developed in Neveu [295], Kunita [229], and Meyn and Tweedie [278].

For an excellent account of Central Limit Theorems and versions of the Law of the Iterated Logarithm for a variety of processes the reader is referred to Hall and Heyde [151]. Martingale limit theory as presented in, for example, Hall and Heyde [151] allows several obvious extensions of the results given in Section 17.4. For example, a functional Law of the Iterated Logarithm for Markov chains can be proved in a manner similar to the functional Central Limit Theorem given in Theorem 17.4.4. Using the almost sure invariance principle given in Brosamler [54] and Lacey and Philipp [233], it is likely that an almost sure Central Limit Theorem for Markov chains may be obtained under an appropriate drift condition, such as (V4).

In work closely related to the development of Section 17.4, Kurtz [231] considers chains arising in models found in polymer chemistry. These models evolve on the surface of a three-dimensional sphere $X = S^2$, and satisfy a multidimensional version of

Poisson's equation:

$$\int_{\mathsf{X}} P(x, dy)y = \rho x$$

where $|\rho| < 1$. Bhattacharaya [35] also considers the CLT and LIL for Markov processes, using an approach based upon the analogue of Poisson's equation in continuous time.

If a solution to Poisson's equation cannot be found directly as in [231], then a more general approach is needed. This is the main motivation for the development of the drift criteria (V3) and (V4) which is central to this chapter, and all of Part III. Most of these results are either new or very recent in this general state space context. Meyn and Tweedie [277] use a variant of (V4) to obtain the CLT and LIL for ψ-irreducible Markov chains giving Theorem 17.0.1, and the use of (V3) to obtain solutions to Poisson's equation is taken from Glynn and Meyn [139]. Applications to random walks and linear models similar to those given in Section 17.6 are also developed in [139].

Proposition 17.3.5, which establishes stability of the dependent parameter bilinear model, is taken from Brandt et al. [45] where further related results may be found.

The finiteness of the fifth moment of the increment process which is imposed in Proposition 17.6.1 is close to the right condition for guaranteeing that the random walk obey the CLT. Daley [83] shows that for the GI/G/1 queue a fourth moment condition is necessary and sufficient for the absolute convergence of the sum

$$\sum_{-\infty}^{\infty} \mathsf{E}_\pi[\bar{\Phi}_k \bar{\Phi}_0]$$

where $\bar{\Phi}_k = \Phi_k - \mathsf{E}_\pi[\Phi_k]$. Recall that this sum is precisely the asymptotic variance used in Proposition 17.6.1. This strongly suggests that the CLT does not hold for the random walk on the half line when the increment process does not have a finite fourth moment, and also suggests that the CLT may indeed hold when the fourth moment is finite. These subtleties are described further in [139].

Commentary for the second edition: Of all the topics covered in this book, those in this chapter have seen the greatest growth since 1996. The number of recognized open questions has grown at least as quickly as the number of papers providing answers. Section 20.2 contains a survey of advances in simulation methodology based on theory developed in this book.

The CLT for Markov chains is better understood today. Sufficient conditions for the CLT are obtained in [252] under conditions that appear close to minimal, and minimal conditions for chains that are reversible[1] are established in [209].

A future edition of this book will surely draw from Jones's survey [183], which contains many examples along with a streamlined account of the theory. Another survey by Landim [236] develops theory for reversible chains. The rate of convergence in the CLT for geometrically ergodic chains is investigated in [218, 219] – see Section 20.1.5 for results concerning more exotic limit theory, such as large deviations.

Looking back at the first edition, it is a surprise to see how little attention is devoted to Poisson's equation (17.37). This equation is central to many areas in statistics and engineering:

---

[1] See discussion surrounding (20.5) in the new Chapter 20.

(i) Approximate solutions to Poisson's equation are used to obtain performance bounds in Markov models [226, 224, 33, 32, 267].

(ii) This equation emerges in various aspects of statistics and limit theory such as Markov renewal theory [132, 133] and refinements of the CLT [218, 219].

(iii) The martingale property described in Theorem 17.4.3 is central to variance analysis of simulation algorithms. Section 20.2.1 contains a brief survey on the application of Poisson's equation to variance-reduction techniques.

(iv) In controlled Markov models (also called Markov decision processes, or MDPs), a variant of Poisson's equation is known as the (average cost) dynamic programming equation. In this context, the function g appearing in (17.37) is the associated cost function, and the solution $\hat{g}$ is called the *relative value function* [28, 262, 261, 67, 263, 42, 27, 267].

(v) Perturbation theory is typically addressed using Poisson's equation, following the work of Schweitzer [347]. Suppose that $\{P_\alpha : \alpha \in (-1, 1)\}$ is a family of transition kernels, each ergodic with invariant measure π_α. Let c denote a measurable function on X, let $\eta_\alpha = \pi_\alpha(c)$, and let $\hat{c}_\alpha$ denote the solution to Poisson's equation,

$$P_\alpha \hat{c}_\alpha = \hat{c}_\alpha - c + \eta_\alpha.$$

Assuming differentiability of each term, along with suitable regularity conditions, we obtain from the product rule of differentiation

$$P'_\alpha \hat{c}_\alpha + P_\alpha \hat{c}'_\alpha = \hat{c}'_\alpha + \eta'_\alpha.$$

Under mild additional assumptions we obtain the sensitivity formula

$$\eta'_\alpha = \int \int P'_\alpha(x, dy)\hat{c}_\alpha(y)\, \pi_\alpha(dx). \qquad (17.58)$$

This result is a foundation of the theory of singularly perturbed Markov chains [316, 214, 16], it is used in the analysis of numerical integration [351], and it is a major component of the theory of actor-critic algorithms in machine learning [217].

(vi) The 'multiplicative Poisson equation' is central to the theory of large deviations for Markov chains – see Section 20.1.5 – and also risk-sensitive optimal control for MDP models [41]. Closely related techniques are also used in the analysis of change-detection algorithms [131].

The existence of a solution to Poisson's equation is guaranteed without the restrictive assumption $\pi(V) < \infty$ imposed in Theorem 17.4.2. The following improvement is new – it is based on the countable state space result [267, Theorem A.4.5].

Theorem 17.7.1. *Suppose that Φ is ψ-irreducible, and that (V3) holds with V everywhere finite, $f \geq 1$, and C petite. Then, for some $B < \infty$ and any $|g| \leq f$, the Poisson equation (17.37) admits a solution $\hat{g}$ satisfying the bound $|\hat{g}| \leq B(V + 1)$.*

PROOF   Suppose first that the chain is strongly aperiodic. We then consider the split chain – the solution to Poisson's equation is given by $\hat{g}(x) = G_\alpha(x, \bar{g})$, as discussed following (17.38).

In the completely general setting we proceed as in the proof of Theorem 17.4.2. The resolvent kernel K_{a_ε} defined in (3.26) is strongly aperiodic for any $\varepsilon \in (0, 1)$. We can solve Poisson's equation (17.41) for this kernel: the solution satisfies $|\hat{g}_\varepsilon| \le B_\varepsilon(V+1)$ for some fixed constant B_ε. We then recall (17.42), which defines $\hat{g} = \varepsilon(1-\varepsilon)^{-1} K_{a_\varepsilon} \hat{g}_\varepsilon$. The function $\hat{g}$ solves Poisson's equation, and this completes the proof with $B = \varepsilon(1-\varepsilon)^{-1} B_\varepsilon$.
□

Application to performance approximation requires a significant strengthening of the converse result Proposition 17.4.1. Frequently we are given an invariance equation of the form

$$Ph = h - g + \eta \tag{17.59}$$

where g and h are measurable functions and η is constant, and we hope to infer that $\pi(g) = \eta$. We obtain the upper bound $\pi(g) \le \eta$ by the Comparison Theorem when g and h are each non-negative valued.

To strengthen the Comparison Theorem and deduce that $\pi(g) = \eta$ we require bounds on g and h. Suppose that a third function $f \ge 1$ is known to be π-integrable. We have seen in the proof of Theorem 14.2.6 that a solution to (V3) is given by

$$V^*(x) := G_C(x, f) = \mathsf{E}_x \left[\sum_{k=0}^{\sigma_C} f(\Phi_k) \right], \tag{17.60}$$

with $C \in \mathcal{B}^+$ any f-regular set. The following result is adapted from [267, Proposition A.6.2].

Theorem 17.7.2. *Suppose that Φ is a ψ-irreducible, positive recurrent Markov chain on* X. *Assume that the given function $f : $ X $ \to [1, \infty]$ satisfies $\pi(f) < \infty$, and that the given set C satisfies $C \in \mathcal{B}^+$.*

Then the function V^* given in (17.60), defined using this f and C, is finite on a full and absorbing set. If (g, h, η) is any solution to (17.59) satisfying $g \in L_\infty^f$ and $h \in L_\infty^{V^*}$, then $\pi(g) = \eta$, so that h is a solution to Poisson's equation with forcing function g.

PROOF   Under the assumptions of the theorem we have for any $n \ge 1$,

$$n^{-1} P^n h(x) = n^{-1} h(x) + \eta - n^{-1} \sum_{k=0}^{n-1} P^k g(x).$$

The right hand side converges to $\eta - \pi(g)$ for a.e. x by the f-Norm Ergodic Theorem 14.0.1 in the aperiodic case, and by Theorem 14.3.6 in general. The left hand side converges to zero by Lemma 17.7.3, which follows.
□

Lemma 17.7.3. *Under the assumptions of Theorem 17.7.2, there exists a full and absorbing set* X$_f$ *such that*

(i) $V^*(x) < \infty$ *for $x \in$ X$_f$;*

(ii) *Poisson's equation holds on* X_f:

$$PV^* = V^* - f^* \qquad (17.61)$$

where f^* *is the zero-mean function given by*

$$f^*(x) := f(x) - \mathbb{I}_C(x)\mathsf{E}_x\Big[\sum_{k=1}^{\tau_C} f(\Phi_k)\Big]; \qquad (17.62)$$

(iii) *for* $x \in X_f$,

$$\lim_{k \to \infty} k^{-1}\mathsf{E}_x[V^*(\Phi_k)] = \lim_{k \to \infty} \mathsf{E}_x[V^*(\Phi_k)\mathbb{I}\{\tau_C > k\}] = 0.$$

PROOF   For (i) we take X_f equal to the set X_V given in the f-Norm Ergodic Theorem 14.0.1, intersected with the set X_f given in Theorem 14.2.5.

The proof of (ii) is identical to the proof of Theorem 11.3.5. Note that f^* is π-integrable with zero mean by the generalized Kac's Theorem given in (10.2).

To prove the first limit in (iii) we iterate the identity in (ii) to obtain

$$\mathsf{E}_x[V^*(\Phi_n)] = P^n V^*(x) = V^*(x) - \sum_{k=0}^{n-1} P^k f^*(x), \quad n \ge 1.$$

Dividing by n and letting $n \to \infty$ we obtain, whenever $V^*(x) < \infty$,

$$\lim_{n \to \infty} n^{-1}\mathsf{E}_x[V^*(\Phi_n)] = \lim_{n \to \infty} n^{-1}\sum_{k=0}^{n-1} P^k f^*(x).$$

The right hand side is zero for $x \in X_f$ by the f-Norm Ergodic Theorem 14.0.1. The ergodic theorem requires aperiodicity. If this fails, we can apply the theorem to the d-skeleton chain using Theorem 14.3.6.

By the definition of V^* and the Markov property we have for each $m \ge 1$,

$$V^*(X(m)) = \mathsf{E}_{X(m)}\Big[\sum_{k=0}^{\sigma_C} f(\Phi_k)\Big]$$

$$= \mathsf{E}\Big[\sum_{k=m}^{\tau_C} f(\Phi_k) \mid \mathcal{F}_m\Big] \quad \text{on } \{\tau_C \ge m\}. \qquad (17.63)$$

Consequently, by the smoothing property of the conditional expectation,

$$\mathsf{E}_x[V^*(X(m))\mathbb{I}\{\tau_C \ge m\}] = \mathsf{E}\Big[\mathbb{I}\{\tau_C \ge m\}\mathsf{E}\Big[\sum_{k=m}^{\tau_C} f(\Phi_k) \mid \mathcal{F}_m\Big]\Big]$$

$$= \mathsf{E}\Big[\mathbb{I}\{\tau_C \ge m\}\sum_{k=m}^{\tau_C} f(\Phi_k)\Big].$$

If $V^*(x) < \infty$, then the right hand side vanishes as $m \to \infty$ by the Dominated Convergence Theorem. This proves the second limit in (iii).   □

Chapter 18

Positivity

Turning from the sample path and classical limit theorems for normalized sums of the previous chapter, we now return to considering limits of the transition probabilities $P^n(x, A)$.

Our first goal in this chapter is to derive limit theorems for chains which are not positive Harris recurrent. Although some results in this spirit have been derived as ratio limit theorems such as Theorem 17.2.1 and Theorem 17.3.2, we have not to this point considered in any detail the difference between limiting behavior of positive and null recurrent chains.

The last five chapters have amply illustrated the power of ψ-irreducibility in the positive case: that is, in conjunction with the existence of an invariant probability measure. However, even in the non-positive case, powerful and elegant results can be achieved. For Harris recurrent chains we prove a generalized version of the Aperiodic Ergodicity Theorem of Chapter 13, which covers the null recurrent case and actually subsumes the ergodic case also, since it applies to any Harris recurrent chain. We will show

Theorem 18.0.1. *Suppose Φ is an aperiodic Harris recurrent chain. Then for any initial probability distributions λ, μ,*

$$\int \int \lambda(dx)\mu(dy)\|P^n(x, \cdot) - P^n(y, \cdot)\| \to 0, \qquad n \to \infty. \tag{18.1}$$

If Φ is a null recurrent chain with invariant measure π, then for any constant $\varepsilon > 0$, and any initial distribution λ,

$$\lim_{n \to \infty} \sup_{A \in \mathcal{B}(\mathsf{X})} \int \lambda(dx)P^n(x, A)/[\pi(A) + \varepsilon] = 0. \tag{18.2}$$

PROOF   The first result is shown in Theorem 18.1.2 after developing some extended coupling arguments and then applying the splitting technique. The consequence (18.2) is proved in Theorem 18.1.3. □

Our second goal in this chapter is to use these limit results to complete the characterizations of positivity through a positive/null dichotomy of the local behavior of P^n

on suitable sets: not surprisingly, the sets of relevance are petite or compact sets in the general or topological settings respectively.

In classical countable state space analysis, as in Chung [71] or Feller [114] or Çinlar [59], it is standard to first approach positivity as an asymptotic "P^n-property" of individual states. It is not hard to show that when Φ is irreducible, either $\limsup_{n \to \infty} P^n(x, y) > 0$ for all $x, y \in \mathsf{X}$ or $\lim_{n \to \infty} P^n(x, y) = 0$ for all $x, y \in \mathsf{X}$. These classifications then provide different but ultimately equivalent characterizations of positive and null chains in the sense we have defined them, which is through the finiteness or otherwise of $\pi(\mathsf{X})$. In Theorem 18.2.2 we show that for ψ-irreducible chains the positive/null dichotomy as defined in, say, Theorem 13.0.1 is equivalent to similar dichotomous behavior of

$$\limsup_{n \to \infty} P^n(x, C) \tag{18.3}$$

for petite sets, exactly as it is in the countable case.

Hence for irreducible T-chains, positivity of the chain is characterized by positivity of (18.3) for compact sets C. For T-chains we also show in this chapter that positivity is characterized by the behavior of (18.3) for the open neighborhoods of x, and that similar characterizations exist for e-chains. Thus there are, for these two classes of topologically well-behaved chains, descriptions in topological terms of the various concepts embodied in the concept of positivity.

These results are summarized in the following theorem:

Theorem 18.0.2. *Suppose that* Φ *is a chain on a topological space for which a reachable state* $x^* \in \mathsf{X}$ *exists.*

(i) *If the chain is a T-chain, then the following are equivalent:*

   (a) Φ *is positive Harris;*

   (b) Φ *is bounded in probability;*

   (c) Φ *is non-evanescent and* x^* *is "positive".*

   If any of these equivalent conditions hold and if the chain is aperiodic, then for each initial state $x \in \mathsf{X}$,

$$\|P^k(x, \cdot) - \pi\| \to 0 \qquad as\ k \to \infty. \tag{18.4}$$

(ii) *If the chain is an e-chain, then the following are equivalent:*

   (a) *There exists a unique invariant probability* π *and for every initial condition* $x \in \mathsf{X}$ *and each bounded continuous function* $f \in \mathcal{C}(\mathsf{X})$,

$$\lim_{k \to \infty} \overline{P}_k(x, f) = \pi(f),$$

$$\lim_{n \to \infty} \frac{1}{n} \sum_{i=1}^{n} f(\Phi_i) = \pi(f) \qquad in\ probability;$$

   (b) Φ *is bounded in probability on average;*

(c) Φ *is non-evanescent and* x^* *is "positive".*

If any of these equivalent conditions hold and if the reachable state is "aperiodic",
then for each initial state $x \in \mathsf{X}$,

$$P^k(x, \cdot) \xrightarrow{\text{w}} \pi \qquad as \ k \to \infty. \tag{18.5}$$

PROOF   **(i)**   The equivalence of Harris positivity and boundedness in probability
for T-chains is given in Theorem 18.3.2, and the equivalence of (a) and (c) follows from
Proposition 18.3.3.

   (ii)   The equivalences of (a)–(c) follow from Proposition 18.4.2, and the limit result
(18.5) is given in Theorem 18.4.4.   □

Thus we have global convergence properties following from local properties, whether
the local properties are with respect to petite sets as in Theorem 18.0.1 or neighborhoods
of points as in Theorem 18.0.2.

Finally, we revisit the LLN for e-chains in the light of these characterizations and
show that a slight strengthening of the hypotheses of Theorem 18.0.2 are precisely those
needed for such chains to obey such a law.

18.1   Null recurrent chains

Our initial step in examining positivity is to develop, somewhat paradoxically, a limit
result whose main novelty is for null recurrent chains. Orey's Theorem 18.1.2 actually
subsumes some aspects of the ergodic theorem in the positive case, but for us its virtue
lies in ensuring that limits can be also be defined for null chains.

The method of proof is again via a coupling argument and the Regenerative Decomposition.

The coupling in Section 13.2 was made somewhat easier because of the existence of
a finite invariant measure in product form to give positivity of the forward recurrence
time chain. If the mean time between renewals is not finite, then such a coupling of
independent copies of the renewal process may not actually occur with probability one.
To see this, consider the recurrence and transience classification of simple symmetric
random walks in two and four dimensions (see Spitzer [369], Section 8). The former
is known to be recurrent, so the return times to zero form a proper renewal sequence.
Now consider two independent copies of this random walk: this is a four-dimensional
random walk which is equally well known to be transient, so the return time to zero is
infinite with positive probability.

Since this is the coupling time of the two independent renewal processes, we cannot
couple them as we did in the positive recurrent case. It is therefore perhaps surprising
that we can achieve our aims by the following rather different and less obvious coupling
method.

18.1.1   Coupling renewal processes for null chains

As in Section 13.2 we again define two sets of random variables $\{S_0, S_1, S_2, \ldots\}$ and
$\{S_0', S_1', S_2', \ldots\}$, where $\{S_1, S_2, \ldots\}$ are independent and identically distributed with
distribution $\{p(j)\}$, and the distributions of the independent variables S_0, S_0' are a, b.

This time, however, we define the second sequence $\{S'_1, S'_2, \ldots\}$ in a dependent way. Let M be a (typically large, and yet to be chosen) integer. For each j define S'_j as being either exactly S_j if $S_j > M$, or, if $S_j \leq M$, define S'_j as being an independent variable with the same conditional distribution as S_j, namely

$$\mathsf{P}(S'_j = k \mid S_j \leq M) = p(k)/(1 - \overline{p}(M)), \qquad k \leq M,$$

where $\overline{p}(M) = \sum_{j > M} p(j)$.

This construction ensures that for $j \geq 1$ the increments S_j and S'_j are identical in distribution even though they are not independent. By construction, also, the quantities

$$W_j = S_j - S'_j$$

have the properties that they are identically distributed, they are bounded above by M and below by $-M$, and they are symmetric around zero and in particular have zero mean.

Let $\Phi^*_n = \sum_{j=0}^{n} W_j$ denote the random walk generated by this sequence of variables, and let T^*_{ab} denote the first time that the random walk Φ^* returns to zero, when the initial step $W_0 = S_0 - S'_0$ has the distribution induced by choosing a, b as the distributions of S_0, S'_0 respectively.

As in Section 13.2 the coupling time of the two renewal processes is defined as

$$T_{ab} = \min\{j : Z_a(j) = Z_b(j) = 1\}$$

where Z_a, Z_b are the indicator sequences of each renewal process, and since

$$\Phi^*_n = \sum_{j=0}^{n} S_j - \sum_{j=0}^{n} S'_j$$

we have immediately that

$$T_{ab} = T^*_{ab}.$$

But we have shown in Proposition 8.4.4 that such a random walk, with its bounded increments, is recurrent on $\mathbb{Z}$, provided of course that it is ψ-irreducible; and if the random walk is recurrent, $T^*_{ab} < \infty$ with probability one from all initial distributions and we have a successful coupling of the two sequences.

Oddly enough, it is now the irreducibility that causes the problems. Obviously a random walk need not be irreducible if the increment distribution Γ is concentrated on sublattices of $\mathbb{Z}$, and as yet we have no guarantee that Φ^* does not have increments concentrated on such a sublattice: it is clear that it may actually do so without further assumptions.

We now proceed with the proof of the result we require, which is the same conclusion as in Theorem 13.2.2 without the assumption that $m_p < \infty$; and the issues just raised are addressed in that proof.

Theorem 18.1.1. *Suppose that a, b, p are proper distributions on $\mathbb{Z}_+$, and that u is the renewal function corresponding to p. Then provided p is aperiodic*

$$|a * u - b * u|(n) \to 0, \quad n \to \infty, \tag{18.6}$$

and

$$|a * u - b * u| * \overline{p}(n) \to 0, \quad n \to \infty. \tag{18.7}$$

PROOF   We will first assume a stronger form of aperiodicity, namely

$$\text{g.c.d.}\{n - m : m < n,\ p(m) > 0,\ p(n) > 0\} = 1.$$

With this assumption we can choose M sufficiently large that

$$\text{g.c.d.}\{n - m : m < n \leq M,\ p(m) > 0,\ p(n) > 0\} = 1. \qquad (18.8)$$

Let us use this M in the construction of the random walk $\mathbf{\Phi}^*$ above. It is straightforward to check that now $\mathbf{\Phi}^*$ really is irreducible, and so

$$\mathsf{P}(T_{ab} < \infty) = 1$$

for any a, b. In particular, then, (18.6) is true for a, b.

We now move on to prove (18.7), and to do this we will now use the backward recurrence chain rather than the forward recurrence chain.

Let V_a^-, V_b^- be the backward recurrence chains defined for the renewal indicators Z_a^-, Z_b^-: note that the subscripts a, b denote conditional random variables with the initial distributions indicated. It is obvious that the chains V_a^-, V_b^- couple at the same time T_{ab} that Z_a^-, Z_b^- couple.

Now let A be an arbitrary set in $\mathbb{Z}_+$. Since the distributions of V_a^- and V_b^- are identical after the time T_{ab} we have for any $n \geq 1$ by decomposing over the values of T_{ab} and using the Markov or renewal property

$$\mathsf{P}(V_a^-(n) \in A) = \sum_{m=1}^{n} \mathsf{P}(T_{ab} = m)\mathsf{P}(V_a^-(n - m) \in A) + \mathsf{P}(V_a^-(n) \in A, T_{ab} > n),$$

$$\mathsf{P}(V_b^-(n) \in A) = \sum_{m=1}^{n} \mathsf{P}(T_{ab} = m)\mathsf{P}(V_b^-(n - m) \in A) + \mathsf{P}(V_b^-(n) \in A, T_{ab} > n).$$

Using this and the inequality $|x - y| \leq \max(x, y)$, $x, y \geq 0$, we get

$$\sup_{A \subseteq \mathbb{Z}_+} |\mathsf{P}(V_a^-(n) \in A) - \mathsf{P}(V_b^-(n) \in A)| \leq \mathsf{P}(T_{ab} > n). \qquad (18.9)$$

We already know that the right hand side of (18.9) tends to zero. But the left hand side can be written as

$$\sup_{A \subseteq \mathbb{Z}_+} |\mathsf{P}(V_a^-(n) \in A) - \mathsf{P}(V_b^-(n) \in A)|$$

$$= \tfrac{1}{2} \sum_{m=0}^{\infty} |\mathsf{P}(V_a^-(n) = m) - \mathsf{P}(V_b^-(n) = m)|$$

$$= \tfrac{1}{2} \sum_{m=0}^{n} |a * u(n - m)\overline{p}(m) - b * u(n - m)\overline{p}(m)|$$

$$= \tfrac{1}{2}|a * u - b * u| * \overline{p}(n), \qquad (18.10)$$

and so the result (18.7) holds.

It remains to remove the extraneous aperiodicity assumption (18.8).

To do this we use a rather nice trick. Let us modify the distribution $p(j)$ to form another distribution $p^0(j)$ on $\{0, 1, \ldots\}$ defined by setting

$$p^0(0) = p > 0;$$

$$p^0(j) = (1 - p)p(j), \qquad j \geq 1.$$

Let us now carry out all of the above analysis using p^0, noting that even though this is not a standard renewal sequence since $p^0(0) > 0$, all of the operations used above remain valid.

Provided of course that $p(j)$ is aperiodic in the usual way, we certainly have that (18.8) holds for p^0 and we can conclude that as $n \to \infty$,

$$|a * u^0 - b * u^0|(n) \to 0, \tag{18.11}$$

$$|a * u^0 - b * u^0| * \overline{p}^0(n) \to 0. \tag{18.12}$$

Finally, by construction of p^0 we have the two identities

$$\overline{p}^0(n) = (1 - p)\overline{p}(n), \qquad u^0(n) = (1 - p)^{-1}u(n),$$

and consequently, from (18.11) and (18.12), we have exactly (18.6) and (18.7) as required. $\qquad\qquad\Box$

Note that in the null recurrent case, since we do not have $\sum \overline{p}(n) < \infty$, we cannot prove this result from Lemma D.7.1 even though it is a identical conclusion to that reached there in the positive recurrent case.

18.1.2 Orey's convergence theorem

In the positive recurrent case, the asymptotic properties of the chain are interesting largely because of the proper distribution π occurring as the limit of the sequence P^n.

In the null recurrent case we know that no such limiting distribution can exist, since there is no finite invariant measure.

It is therefore remarkable that we can give a strong result on the closeness of the n-step distributions from different initial laws, even for chains which may be null.

Theorem 18.1.2. *Suppose Φ is an aperiodic Harris recurrent chain. Then for any initial probability distributions λ, μ,*

$$\int\int \lambda(dx)\mu(dy)\|P^n(x, \cdot) - P^n(y, \cdot)\| \to 0, \qquad n \to \infty. \tag{18.13}$$

PROOF   Yet again we begin with the assumption that there is an atom α in the space. Then for any x we have from the Regenerative Decomposition (13.47)

$$\|P^n(x, \cdot) - P^n(\alpha, \cdot)\| \leq \mathsf{P}_x(\tau_\alpha \geq n) + |a_x * u - u|(n) + |a_x * u - u| * \overline{p}(n) \tag{18.14}$$

where now $\overline{p}(n) = \mathsf{P}_\alpha(\tau_\alpha > n)$. From Theorem 18.1.1 we know the last two terms in (18.14) tend to zero, whilst the first tends to zero from Harris recurrence.

The result (18.13) then follows for any two specific initial starting points x, y from the triangle inequality; it extends immediately to general initial distributions λ, μ from dominated convergence.

As previously, the extension to strongly aperiodic chains is straightforward, whilst the extension to general aperiodic chains follows from the contraction property of the total variation norm.   □

We conclude with a consequence of this theorem which gives a uniform version of the fact that, in the null recurrent case, we have convergence of the transition probabilities to zero.

Theorem 18.1.3. *Suppose that Φ is aperiodic and null recurrent, with invariant measure π. Then for any initial distribution λ and any constant $\varepsilon > 0$*

$$\lim_{n \to \infty} \sup_{A \in \mathcal{B}(\mathsf{X})} \int \lambda(dx) P^n(x, A)/[\pi(A) + \varepsilon] = 0. \tag{18.15}$$

PROOF   Suppose by way of contradiction that we have a sequence of integers $\{n_k\}$ with $n_k \to \infty$ and a sequence of sets $B_k \in \mathcal{B}(\mathsf{X})$ such that, for some λ, and some $\delta, \varepsilon > 0$,

$$\int \lambda(dx) P^{n_k}(x, B_k) \geq \delta[\pi(B_k) + \varepsilon], \qquad k \in \mathbb{Z}_+. \tag{18.16}$$

Now from (18.13), we know that for every y

$$|\int \lambda(dx) P^{n_k}(x, B_k) - P^{n_k}(y, B_k)| \to 0, \qquad k \to \infty, \tag{18.17}$$

and by Egorov's Theorem and the fact that $\pi(\mathsf{X}) = \infty$ this convergence is uniform on a set with π-measure arbitrarily large.

In particular we can take k and D such that $\pi(D) > \delta^{-1}$ and

$$|\int \lambda(dx) P^{n_k}(x, B_k) - P^{n_k}(y, B_k)| \leq \varepsilon\delta/2, \qquad y \in D. \tag{18.18}$$

Combining (18.16) and (18.18) gives

$$\pi(B_k) = \int \pi(dy) P^{n_k}(y, B_k)$$

$$\geq \int_D \pi(dy) P^{n_k}(y, B_k)$$

$$\geq \pi(D)[\int \lambda(dx) P^{n_k}(x, B_k) - \varepsilon\delta/2]$$

$$\geq \pi(D)[\delta(\pi(B_k) + \varepsilon) - \varepsilon\delta/2] \tag{18.19}$$

which gives

$$\pi(D) \leq \delta^{-1},$$

thus contradicting the definition of D. $\qquad\square$

The two results in Theorem 18.1.2 and Theorem 18.1.3 combine to tell us that, on the one hand, the distributions of the chain are getting closer as n gets large; and that they are getting closer on sets increasingly remote from the "center" of the space, as described by sets of finite π-measure.

18.2 Characterizing positivity using P^n

We have chosen to formulate positive recurrence initially, in Chapter 10, in terms of the finiteness of the invariant measure π. The ergodic properties of such chains are demonstrated in Chapters 13–16 as a consequence of this simple definition.

In contrast to this definition, the classical approach to the classification of irreducible chains as positive or null recurrent uses the transition probabilities rather than the invariant measure: typically, the invariant measure is demonstrated to exist only after a null/positive dichotomy is established in terms of the convergence properties of $P^n(x, A)$ for appropriate sets A. Null chains in this approach are those for which $P^n(x, A) \to 0$ for, say, all x and all small sets A, and almost by default, positive recurrent chains are those which are not null, that is, for which $\limsup P^n(x, A) > 0$.

We now develop a classification of states or of sets as positive recurrent or null using transition probabilities, and show that this approach is consistent with the definitions involving invariant measures in the case of ψ-irreducible chains.

18.2.1 Countable spaces

We will first consider the classical classification of null and positive chains based on P^n in the countable state space case.

When X is countable, recall that recurrence of individual states $x, y \in$ X involves consideration of the finiteness or otherwise of $\mathsf{E}_x(\eta_y) = U(x, y) = \sum_{n=1}^{\infty} P^n(x, y)$. The stronger condition

$$\limsup_{n \to \infty} P^n(x, y) > 0 \qquad (18.20)$$

obviously implies that

$$\mathsf{E}_x(\eta_y) = \infty; \qquad (18.21)$$

and since in general, because of the cyclic behavior in Section 5.4, we may have

$$\liminf_{n \to \infty} P^n(x, y) = 0, \qquad (18.22)$$

the condition (18.20) is often adopted as the next strongest stability condition after (18.21).

This motivates the following definitions.

Null and positive states

(i) The state α is called *null* if $\lim_{n\to\infty} P^n(\alpha,\alpha) = 0$.

(ii) The state α is called *positive* if $\limsup_{n\to\infty} P^n(\alpha,\alpha) > 0$.

When Φ is irreducible, either all states are positive or all states are null, since for any w,z there exist r,s such that $P^r(w,x) > 0$ and $P^s(y,z) > 0$ and

$$\limsup_{n\to\infty} P^{r+s+n}(w,z) > P^r(w,x)[\limsup_{n\to\infty} P^n(x,y)]P^s(y,z). \qquad (18.23)$$

We need to show that these solidarity properties characterize positive and null chains in the sense we have defined them. One direction of this is easy, for if the chain is positive recurrent, with invariant probability π, then we have for any n

$$\pi(y) = \sum_x \pi(x)P^n(x,y);$$

hence if $\lim_{n\to\infty} P^n(w,w) = 0$ for some w then by (18.23) and dominated convergence $\pi(y) \equiv 0$, which is impossible. The other direction is easy only if one knows, not merely that $\limsup_{n\to\infty} P^n(x,y) > 0$, but that (at least through an aperiodic class) this is actually a limit. Theorem 18.1.3 now gives this to us.

Theorem 18.2.1. *If Φ is irreducible on a countable space, then the chain is positive recurrent if and only if some one state is positive. When Φ is positive recurrent, for some $d \geq 1$*

$$\lim_{n\to\infty} P^{nd+r}(x,y) = d\pi(y) > 0$$

for all $x,y \in \mathsf{X}$, and some $0 \leq r(x,y) \leq d-1$; and when Φ is null

$$\lim_{n\to\infty} P^n(x,y) = 0$$

for all $x,y \in \mathsf{X}$.

PROOF   If the chain is transient, then since $U(x,y) < \infty$ for all x,y from Proposition 8.1.1 we have that every state is null; whilst if the chain is null recurrent, then since $\pi(y) < \infty$ for all y, Theorem 18.1.3 shows that every state is null.

Suppose that the chain is positive recurrent, with period d: then the Aperiodic Ergodic Theorem for the chain on the cyclic class D_j shows that for $x,y \in D_j$ we have

$$\lim_{n\to\infty} P^{nr}(x,y) = d\pi(y) > 0$$

whilst for $z \in D_{j-r \pmod d}$ we have $P^{j-r}(z,D_j) = 1$, showing that every state is positive. $\square$

The simple equivalences in this result are in fact surprisingly hard to prove until we have established, not just the properties of the sequences $\limsup P^n$, but the actual existence of the limits of the sequences P^n through the periodic classes. This is why this somewhat elementary result has been reserved until now to establish.

18.2.2   General spaces

We now move on to the equivalent concepts for general chains: here, we must consider properties of sets rather than individual states, but we will see that the results above have completely general analogues.

When X is general, the definitions for sets which we shall use are

---

Null and positive sets

(i) The set A is called *null* if $\lim_{n \to \infty} P^n(x, A) = 0$ for all $x \in A$.

(ii) The set A is called *positive* if $\limsup_{n \to \infty} P^n(x, A) > 0$ for all $x \in A$.

We now prove that these definitions are consistent with the definitions of null and positive recurrence for general ψ-irreducible chains.

Theorem 18.2.2. *Suppose that Φ is ψ-irreducible. Then:*

(i) *the chain Φ is positive recurrent if and only if every set $B \in \mathcal{B}^+(\mathsf{X})$ is positive;*

(ii) *if Φ is null, then every petite set is null and hence there is a sequence of null sets B_j with $\bigcup_j B_j = \mathsf{X}$.*

PROOF   If the chain is null, then either it is transient, in which case each petite set is strongly transient and thus null by Theorem 8.3.5, or it is null and recurrent in which case, since π exists and is finite on petite sets by Proposition 10.1.2, we have that every petite set is again null from Theorem 18.1.3.

Suppose the chain is positive recurrent and we have $A \in \mathcal{B}^+(\mathsf{X})$. For $x \in D_0 \cap H$, where H is the maximal Harris set and D_0 an arbitrary cyclic set, we have for each r

$$\lim_{n \to \infty} P^{nd+r}(x, A) = d\pi(A \cap D_r)$$

which is positive for some r. Since for every x we have $L(x, D_0 \cap H) > 0$ we have that the set A is positive.   □

18.3   Positivity and T-chains

18.3.1   T-chains bounded in probability

In Chapter 12 we showed that chains on a topological space which are bounded in probability admit finite subinvariant measures under a wide range of continuity conditions.

It is thus reasonable to hope that ψ-irreducible chains on a topological space which are bounded in probability will be positive recurrent. Not surprisingly, we will see in this section that such a result is true for T-chains, and indeed we can say considerably more: boundedness in probability is actually equivalent to positive Harris recurrence in this case. Moreover, for T-chains positive or null sets also govern the behavior of the whole chain.

It is easy to see that on a countable space, where the continuous component properties are always satisfied, irreducible chains admit the following connection between boundedness in probability and positive recurrence.

Proposition 18.3.1. *For an irreducible chain on a countable space, positive Harris recurrence is equivalent to boundedness in probability.*

PROOF   In the null case we do not have boundedness in probability since $P^n(x, y) \to 0$ for all x, y from Theorem 18.2.1.

In the positive case we have on each periodic set D_r a finite probability measure π_r such that if $x \in D_0$

$$\lim_{n \to \infty} P^{nd+r}(x, C) = \pi_r(C), \tag{18.24}$$

so by choosing a finite C such that $\pi_r(C) > 1 - \varepsilon$ for all $1 \leq r \leq d$ we have boundedness in probability as required.   □

The identical conclusion holds for T-chains. To get the broadest presentation, recall that a state $x^\star \in \mathsf{X}$ is *reachable* if

$$U(y, O) > 0$$

for every state $y \in \mathsf{X}$ and every open set O containing $x^\star$.

Theorem 18.3.2. *Suppose that Φ is a T-chain and admits a reachable state $x^\star$. Then Φ is a positive Harris chain if and only if it is bounded in probability.*

PROOF   First note from Proposition 6.2.1 that for a T-chain the existence of just one reachable state $x^\star$ gives ψ-irreducibility, and thus Φ is either positive or null.

Suppose that Φ is bounded in probability. Then Φ is non-evanescent from Proposition 12.1.1, and hence Harris recurrent from Theorem 9.2.2.

Moreover, boundedness in probability implies by definition that some compact set is non-null, and hence from Theorem 18.2.2 the chain is positive Harris, since compact sets are petite for T-chains.

Conversely, assume that the chain is positive Harris, with periodic sets D_j each supporting a finite probability measure π_j satisfying (18.24). Choose $\varepsilon > 0$ and compact sets $C_r \subseteq D_r$ such that $\pi_r(C_r) > 1 - \varepsilon$ for each r.

If $x \in D_j$, then with $C := \cup C_r$,

$$\lim_{n \to \infty} P^{nd+r-j}(x, C) = \pi_r(C_r) > 1 - \varepsilon. \tag{18.25}$$

If x is in the non-cyclic set $N = \mathsf{X} \backslash \cup D_j$, then $P^n(x, \cup D_j) \to 1$ by Harris recurrence, and thus from (18.25) we also have $\liminf_n P^n(x, C) > 1 - \varepsilon$, and this establishes boundedness in probability as required.   □

18.3.2 Positive and null states for T-chains

The ideas encapsulated in the definitions of positive and null states in the countable case and positive and null sets in the general state space case find their counterparts in the local behavior of chains on spaces with a topology.

Analogously to the definition of topological recurrence at a point we have

---

Topological positive and null recurrence of states

We shall call a state x^*

(i) *null* if $\lim_{n\to\infty} P^n(x^*, O) = 0$ for some neighborhood O of x^*;

(ii) *positive* if $\limsup_{n\to\infty} P^n(x^*, O) > 0$ for all neighborhoods O of x^*.

We now show that these topological properties for points can be linked to their counterparts for the whole chain when the T-chain condition holds. This completes the series of results begun in Theorem 9.3.3 connecting global properties of T-chains with those at individual points.

Proposition 18.3.3. *Suppose that Φ is a T-chain, and suppose that x^* is a reachable state. Then the chain Φ is positive recurrent if and only if x^* is positive.*

PROOF   From Proposition 6.2.1 the existence of a reachable state ensures the chain is ψ-irreducible. Assume that x^* is positive. Since Φ is a T-chain, there exists an open petite set C containing x^* (take any precompact open neighborhood) and hence by Theorem 18.2.2 the chain is also positive.

Conversely, suppose that Φ has an invariant probability π so that Φ is positive recurrent. Since x^* is reachable it also lies in the support of π, and consequently any neighborhood of x^* is in $\mathcal{B}^+(\mathsf{X})$. Hence x^* is positive as required, from Theorem 18.2.2. □

18.4   Positivity and e-chains

For T-chains we have a great degree of coherence in the concepts of positivity. Although there is not quite the same consistency for weak Feller chains, within the context of chains bounded in probability we can develop several valuable approaches, as we saw in Chapter 12.

In particular, for e-chains we now prove several further positivity results to indicate the level of work needed in the absence of ψ-irreducibility. It is interesting to note that it is the existence of a reachable state that essentially takes over the role of ψ-irreducibility, and that such states then interact well with the e-chain assumption.

18.4.1   Reachability and positivity

To begin we show that for an e-chain which is non-evanescent, the topological irreducibility condition that a reachable state exists is equivalent to the measure-theoretic irreducibility condition that the limiting measure $\Pi(x, \mathsf{X})$ is independent of the starting state x. Boundedness in probability on average is then equivalent to positivity of the reachable state.

We first give a general result for Feller chains:

Lemma 18.4.1. *If Φ is a Feller chain and if a reachable state x^* exists, then for any pre-compact neighborhood O containing x^*,*

$$\{\Phi \to \infty\} = \{\Phi \in O \text{ i.o.}\}^c \qquad \text{a.s. } [\mathsf{P}_*].$$

PROOF   Since $L(x, O)$ is a lower semicontinuous function of x by Proposition 6.1.1, and since by reachability it is strictly positive everywhere, it follows that $L(x, O)$ is bounded from below on compact subsets of X.

Letting $\{O_n\}$ denote a sequence of pre-compact open subsets of X with $O_n \uparrow \mathsf{X}$, it follows that $O_n \rightsquigarrow O$ for each n, and hence by Theorem 9.1.3 we have

$$\{\Phi \in O_n \text{ i.o.}\} \subseteq \{\Phi \in O \text{ i.o.}\} \qquad \text{a.s. } [\mathsf{P}_*].$$

This immediately implies that

$$\{\Phi \to \infty\}^c = \bigcup_{n \geq 1} \{\Phi \in O_n \text{ i.o.}\} \subseteq \{\Phi \in O \text{ i.o.}\} \qquad \text{a.s. } [\mathsf{P}_*],$$

and since it is obvious that $\{\Phi \to \infty\} \subseteq \{\Phi \in O \text{ i.o.}\}^c$, this proves the lemma.   □

Proposition 18.4.2. *Suppose that Φ is an e-chain which is non-evanescent, and suppose that a reachable state $x^* \in \mathsf{X}$ exists. Then the following are equivalent:*

(i) *There exists a unique invariant probability π such that*

$$\overline{P}_k(x, \cdot) \xrightarrow{\text{w}} \pi \qquad as \ k \to \infty.$$

(ii) *Φ is bounded in probability on average.*

(iii) *x^* is positive.*

PROOF   The identity $P\Pi = \Pi$ which is proved in Theorem 12.4.1 implies that for any $f \in \mathcal{C}_c(\mathsf{X})$, the adapted process $(\Pi(\Phi_k, f), \mathcal{F}_k^\Phi)$ is a bounded martingale. Hence by the Martingale Convergence Theorem D.6.1 there exists a random variable $\tilde{\pi}(f)$ for which

$$\lim_{k \to \infty} \Pi(\Phi_k, f) = \tilde{\pi}(f) \qquad \text{a.s. } [\mathsf{P}_*],$$

with $\mathsf{E}_y[\tilde{\pi}(f)] = \Pi(y, f)$ for all $y \in \mathsf{X}$.

Since $\Pi(y, f)$ is a continuous function of y, it follows from Lemma 18.4.1 that

$$\liminf_{k \to \infty} |\Pi(\Phi_k, f) - \Pi(x^*, f)| = 0 \qquad \text{a.s. } [\mathsf{P}_*],$$

which gives $\tilde{\pi}(f) = \Pi(x^\star, f)$ a.s. $[\mathsf{P}_*]$. Taking expectations gives $\Pi(y, f) = \mathsf{E}_y[\tilde{\pi}(f)] = \Pi(x^\star, f)$ for all y.

Since a finite measure on $\mathcal{B}(\mathsf{X})$ is determined by its values on continuous functions with compact support, this shows that the measures $\Pi(y, \cdot)$, $y \in \mathsf{X}$, are identical. Let π denote their common value.

To prove Proposition 18.4.2 we first show that (i) and (iii) are equivalent. To see that (iii) implies (i), observe that under positivity of $x^\star$ we have $\Pi(x^\star, \mathsf{X}) > 0$, and since $\Pi(y, \mathsf{X}) = \pi(\mathsf{X})$ does not depend on y it follows from Theorem 12.4.3 that $\Pi(y, \mathsf{X}) = 1$ for all y. Hence π is an invariant probability, which shows that (i) does hold.

Conversely, if (i) holds, then by reachability of $x^\star$ we have $x^\star \in \operatorname{supp} \pi$ and hence every neighborhood of $x^\star$ is positive. This shows that (iii) also holds.

We now show that (i) is equivalent to (ii).

It is obvious that (i) implies (ii). To see the converse, observe that if (ii) holds, then by Theorem 12.4.1 we have that π is an invariant probability. Moreover, since $x^\star$ is reachable we must have that $\pi(O) > 0$ for any neighborhood of $x^\star$. Since $\Pi(y, O) = \pi(O)$ for every y, this shows that $x^\star$ is positive.

Hence (iii) holds, which implies that (i) also holds. □

18.4.2   Aperiodicity and convergence

The existence of a limit for $\overline{P}_k$ in Proposition 18.4.2 rather than for the individual terms P^n seems to follow naturally in the topology we are using here.

We can strengthen such convergence results using a topological notion of aperiodicity and we turn to such concepts in the this section. It appears to be a particularly difficult problem to find such limits for the terms P^n in the weak Feller situation without an e-chain condition.

In the topological case we use a definition justified by the result in Lemma D.7.4, which is one of the crucial consequences of the definitions in Chapter 5.

Topological aperiodicity of states

A recurrent state x is called *aperiodic* if $P^k(x, O) > 0$ for each open set O containing x and all $k \in \mathbb{Z}_+$ sufficiently large.

The next result justifies this definition of aperiodicity and strengthens Theorem 12.4.1.

Proposition 18.4.3. *Suppose that Φ is an e-chain which is bounded in probability on average. Let $x^\star \in \mathsf{X}$ be reachable and aperiodic, and let $\pi = \Pi(x^\star, \cdot)$. Then for each initial condition y lying in $\operatorname{supp} \pi$,*

$$P^k(y, \cdot) \xrightarrow{\mathrm{w}} \pi \qquad as \ k \to \infty. \tag{18.26}$$

PROOF   For any $f \in \mathcal{C}_c(\mathsf{X})$ we have by stationarity,

$$\int |P^k f| \, d\pi = \int [\int P|P^k f|] d\pi \geq \int |P^{k+1} f| \, d\pi,$$

and hence $v := \lim_{k \to \infty} \int |P^k f| \, d\pi$ exists.

Since $\{P^k f\}$ is equicontinuous on compact subsets of X, there exists a continuous function g, and a subsequence $\{k_i\} \subset \mathbb{Z}_+$ for which $P^{k_i} f \to g$ as $i \to \infty$ uniformly on compact subsets of X. Hence we also have $P^{k_i + \ell} f \to P^\ell g$ as $i \to \infty$ uniformly on compact subsets of X.

By the Dominated Convergence Theorem we have for all $\ell \in \mathbb{Z}_+$,

$$\int P^\ell g \, d\pi = \int f \, d\pi \quad \text{and} \quad \int |P^\ell g| \, d\pi = v. \tag{18.27}$$

We will now show that this implies that the function g cannot change signs on $\operatorname{supp} \pi$.

Suppose otherwise, so that the open sets

$$O_+ := \{x \in \mathsf{X} : g(x) > 0\}, \qquad O_- := \{x \in \mathsf{X} : g(x) < 0\}$$

both have positive π measure.

Because $x^\star \in \operatorname{supp} \pi$, it follows by Proposition 18.4.2 that there exist $k_+, k_- \in \mathbb{Z}_+$ such that

$$P^{k_+}(y, O_+) > 0 \quad \text{and} \quad P^{k_-}(y, O_-) > 0 \tag{18.28}$$

when $y = x^\star$, and since $P^n(\,\cdot\,, O)$ is lower semicontinuous for any open set $O \subset \mathsf{X}$, equation (18.28) holds for all y in an open neighborhood N containing $x^\star$.

We may now use aperiodicity. Since $P^k(x^\star, N) > 0$ for all k sufficiently large, we deduce from (18.28) that there exists $\ell \in \mathbb{Z}_+$ for which

$$P^\ell(y, O_+) > 0 \quad \text{and} \quad P^\ell(y, O_-) > 0$$

when $y = x^\star$, and hence for all y in an open neighborhood N' of $x^\star$. This implies that $|P^\ell g| < P^\ell |g|$ on N', and since $\pi\{N'\} > 0$, that $\int |P^\ell g| \, d\pi < \int |g| \, d\pi$, in contradiction to the second equality in (18.27).

Hence g does not change signs in $\operatorname{supp} \pi$. But by (18.27) it follows that if $\int f \, d\pi = 0$, then

$$0 = |\int g \, d\pi| = \int |g| \, d\pi,$$

so that $g \equiv 0$ on $\operatorname{supp} \pi$. This shows that the limit (18.26) holds for all initial conditions in $\operatorname{supp} \pi$. $\square$

We now show that if a reachable state exists for an e-chain, then the limit in Proposition 18.4.3 holds for each initial condition. A sample path version of Theorem 18.4.4 will be presented below.

Theorem 18.4.4. *Suppose that Φ is an e-chain which is bounded in probability on average. Then:*

(i) *A unique invariant probability π exists if and only if a reachable state $x^\star \in \mathsf{X}$ exists.*

(ii) *If an aperiodic reachable state $x^* \in \mathsf{X}$ exists, then for each initial state $x \in \mathsf{X}$,*

$$P^k(x, \cdot) \xrightarrow{\mathrm{w}} \pi \qquad as \ k \to \infty, \qquad (18.29)$$

where π is the unique invariant probability for $\mathbf{\Phi}$. Conversely, if (18.29) holds for all $x \in \mathsf{X}$, then every state in $\operatorname{supp} \pi$ is reachable and aperiodic.

PROOF   The proof of (i) follows immediately from Proposition 18.4.2, and the converse of (ii) is straightforward.

To prove the remainder, we assume that the state $x^* \in \mathsf{X}$ is reachable and aperiodic, and show that equation (18.29) holds for all initial conditions.

Suppose that $\int f \, d\pi = 0$, $|f(x)| \le 1$ for all x, and for fixed $\varepsilon > 0$ define the set

$$O_\varepsilon := \{x \in \mathsf{X} : \limsup_{k \to \infty} |P^k f| < \varepsilon\}.$$

Because the Markov transition function P is equicontinuous, and because Proposition 18.4.3 implies that (18.29) holds for all initial conditions in $\operatorname{supp} \pi$, the set O_ε is an open neighborhood of $\operatorname{supp} \pi$.

Hence $\pi\{O_\varepsilon\} = 1$, and since O_ε is open, it follows from Theorem 18.4.4 (i) that

$$\lim_{N \to \infty} \overline{P}_N(x, O_\varepsilon) = 1.$$

Fix $x \in \mathsf{X}$, and choose $N_0 \in \mathbb{Z}_+$ such that $P^{N_0}(x, O_\varepsilon) \ge 1 - \varepsilon$. We then have by the definition of O_ε and Fatou's Lemma,

$$\limsup_{k \to \infty} |P^{N_0+k} f(x)| \ \le \ P^{N_0}(x, O_\varepsilon^c) + \limsup_{k \to \infty} \int_{O_\varepsilon} P^{N_0}(x, dy) |P^k f(y)|$$

$$\le \ 2\varepsilon.$$

Since ε is arbitrary, this completes the proof. $\qquad\qquad\qquad\qquad\qquad\qquad \square$

18.5   The LLN for e-chains

As a final result, illustrating both these methods and the sample path methods developed in Chapter 17, we now give a sample path version of Proposition 18.4.2 for e-chains.

Define the *occupation probabilities* as

$$\tilde{\mu}_n\{A\} := S_n(\mathbb{I}_A) = \frac{1}{n} \sum_{k=1}^{n} \mathbb{I}\{\mathbf{\Phi}_k \in A\}, \qquad n \in \mathbb{Z}_+, \quad A \in \mathcal{B}(\mathsf{X}). \qquad (18.30)$$

Observe that $\{\tilde{\mu}_k\}$ are not probabilities in the usual sense, but are probability-valued random variables.

The Law of Large Numbers (Theorem 17.1.2) states that if an invariant probability measure π exists, then the occupation probabilities converge with probability one for each initial condition lying in a set of full π-measure. We now present two versions of the law of large numbers for e-chains where the null set appearing in Theorem 17.1.2 is removed by restricting consideration to continuous, bounded functions. The first is a Weak Law of Large Numbers, since the convergence is only in probability, while the second is a Strong Law with convergence occurring almost surely.

Theorem 18.5.1. *Suppose that Φ is an e-chain bounded in probability on average, and suppose that a reachable state exists. Then a unique invariant probability π exists and the following limits hold.*

(i) *For any $f \in \mathcal{C}(\mathsf{X})$, as $k \to \infty$*

$$\int f \, d\tilde{\mu}_k \to \int f \, d\pi$$

in probability for each initial condition.

(ii) *If for each initial condition of the Markov chain the occupation probabilities are almost surely tight, then as $k \to \infty$*

$$\tilde{\mu}_k \xrightarrow{\text{w}} \pi \qquad \text{a.s. } [\mathsf{P}_*]. \tag{18.31}$$

PROOF   Let $f \in \mathcal{C}(\mathsf{X})$ with $0 \le f(x) \le 1$ for all x, let $C \subset \mathsf{X}$ be compact and choose $\varepsilon > 0$. Since $\overline{P}_k f \to \int f \, d\pi$ as $k \to \infty$, uniformly on compact subsets of X, there exists M sufficiently large for which

$$\left| \frac{1}{N} \sum_{k=1}^{N} \overline{P}_M f(\Phi_k) - \int f \, d\pi \right| \le \varepsilon + \frac{1}{N} \sum_{i=1}^{N} \mathbb{I}\{\Phi_i \in C^c\}. \tag{18.32}$$

Now for any $M \in \mathbb{Z}_+$, we will show that

$$\left| \frac{1}{N} \sum_{k=1}^{N} f(\Phi_k) - \int f \, d\pi \right| = \left| \frac{1}{N} \sum_{k=1}^{N} \overline{P}_M f(\Phi_k) - \int f \, d\pi \right| + o(1) \tag{18.33}$$

where the term $o(1)$ converges to zero as $n \to \infty$ with probability one.

For each $N, n \in \mathbb{Z}_+$ we have

$$\frac{1}{N} \sum_{k=1}^{N} f(\Phi_k) - \int f \, d\pi = \sum_{i=0}^{n-1} \frac{1}{N} \sum_{k=1}^{N} \left(P^i f(\Phi_{k-i}) - P^{i+1} f(\Phi_{k-i-1}) \right)$$

$$+ \frac{1}{N} \sum_{k=1}^{N} P^n f(\Phi_k) - \int f \, d\pi$$

$$+ \frac{1}{N} \sum_{k=1}^{N} \left(P^n f(\Phi_{k-n}) - P^n f(\Phi_k) \right)$$

where we adopt the convention that $\Phi_k = \Phi_0$ for $k \le 0$. For each $M \in \mathbb{Z}_+$ we may average the right hand side of this equality from $n = 1$ to M to obtain

$$\frac{1}{N}\sum_{k=1}^{N} f(\Phi_k) - \int f\, d\pi = \frac{1}{M}\sum_{n=1}^{M}\left(\sum_{i=0}^{n-1}\frac{1}{N}\sum_{k=1}^{N}\left(P^i f(\Phi_{k-i}) - P^{i+1} f(\Phi_{k-i-1})\right)\right)$$

$$+ \frac{1}{N}\sum_{k=1}^{N}\left(\frac{1}{M}\sum_{n=1}^{M} P^n f(\Phi_k)\right) - \int f\, d\pi$$

$$+ \frac{1}{M}\sum_{n=1}^{M}\left(\frac{1}{N}\sum_{k=1}^{N} P^n f(\Phi_{k-n}) - P^n f(\Phi_k)\right).$$

The fourth term is a telescoping series, and hence recalling our definition of the transition function $\overline{P}_M$ we have

$$\left|\frac{1}{N}\sum_{k=1}^{N} f(\Phi_k) - \int f\, d\pi\right| \leq \sum_{i=0}^{M-1}\left|\frac{1}{N}\sum_{k=1}^{N}\left(P^i f(\Phi_{k-i}) - P^{i+1} f(\Phi_{k-i-1})\right)\right|$$

$$+ \left|\frac{1}{N}\sum_{k=1}^{N}\left(\overline{P}_M f(\Phi_k) - \int f\, d\pi\right)\right|$$

$$+ \frac{2M}{N}. \tag{18.34}$$

For each fixed $0 < i \leq M - 1$ the sequence

$$\left(P^i f(\Phi_{k-i}) - P^{i+1} f(\Phi_{k-i-1}), \mathcal{F}_{k-i}^{\Phi}\right), \qquad k > i,$$

is a bounded martingale difference process. Hence by Theorem 5.2 of Chapter 4 of [99], the first summand converges to zero almost surely for every $M \in \mathbb{Z}_+$, and thus (18.33) is proved.

Hence for any $\gamma > \varepsilon$, it follows from (18.33) and (18.32) that

$$\limsup_{N\to\infty} P_x\left\{\left|\frac{1}{N}\sum_{k=1}^{N} f(\Phi_k) - \int f\, d\pi\right| \geq \gamma\right\}$$

$$\leq \limsup_{N\to\infty} P_x\left\{\frac{1}{N}\sum_{i=1}^{N}\mathbb{I}\{\Phi_i \in C^c\} \geq \gamma - \varepsilon\right\}$$

$$\leq \frac{1}{\gamma - \varepsilon}\limsup_{N\to\infty} E_x\left[\frac{1}{N}\sum_{i=1}^{N}\mathbb{I}\{\Phi_i \in C^c\}\right].$$

Since Φ is bounded in probability on average, the right hand side decreases to zero as $C \uparrow \mathsf{X}$, which completes the proof of (i).

To prove (ii), suppose that the occupation probabilities $\{\tilde{\mu}_k\}$ are tight along some sample path. Then we may choose the compact set C in (18.32) so that along this sample path

$$\limsup_{N\to\infty}\left|\frac{1}{N}\sum_{k=1}^{N}\overline{P}_M f(\Phi_k) - \int f\, d\pi\right| \leq 2\varepsilon.$$

Since $\varepsilon > 0$ is arbitrary, (18.33) shows that

$$\lim_{N \to \infty} \frac{1}{N} \sum_{k=1}^{N} f(\Phi_k) = \int f \, d\pi \qquad \text{a.s. } [\mathsf{P}_*],$$

so that the Strong Law of Large Numbers holds for all $f \in \mathcal{C}(\mathsf{X})$ and all initial conditions $x \in \mathsf{X}$.

Let $\{f_n\}$ be a sequence of continuous functions with compact support which is dense in $\mathcal{C}_c(\mathsf{X})$ in the uniform norm. Such a sequence exists by Proposition D.5.1. Then by the preceding result,

$$\mathsf{P}_x \left\{ \lim_{k \to \infty} \int f_n \, d\tilde{\mu}_k = \int f_n \, d\pi \qquad \text{for each } n \in \mathbb{Z}_+ \right\} = 1,$$

which implies that $\tilde{\mu}_k \xrightarrow{\text{v}} \pi$ as $k \to \infty$. Since π is a probability, this shows that in fact $\tilde{\mu}_k \xrightarrow{\text{w}} \pi$ a.s. $[\mathsf{P}_*]$, and this completes the proof. $\qquad\qquad \square$

We conclude by stating a result which, combined with Theorem 18.5.1, provides a test function approach to establishing the Law of Large Numbers for Φ. For a proof see [259].

Theorem 18.5.2. *If a coercive function V and a compact set C satisfy condition (V4), then Φ is bounded in probability, and the occupation probabilities are almost surely tight for each initial condition. Hence, if Φ is an e-chain, and if a reachable state exists,*

$$\tilde{\mu}_k \xrightarrow{\text{w}} \pi \quad \text{as } k \to \infty \qquad \text{a.s. } [\mathsf{P}_*]. \tag{18.35}$$

18.6   Commentary

Theorem 18.1.2 for positive recurrent chains is first proved in Orey [308], and the null recurrent version we give here is in Jamison and Orey [177]. The dependent coupling which we use to prove this result for null recurrent chains is due to Ornstein [310], [311], and is also developed in Berbee [25]. Our presentation of this material has relied heavily on Nummelin [303], and further related results can be found in his Chapter 6.

Theorem 18.1.3 is due to Jain [171], and our proof is taken from Orey [309].

The links between positivity of states, boundedness in probability, and positive Harris recurrence for T-chains are taken from Meyn [259], Meyn and Tweedie [277] and Tuominen and Tweedie [391]. In [277] analogues of Theorem 18.3.2 and Proposition 18.3.3 are obtained for non-irreducible chains.

The convergence result Theorem 18.4.4 for chains possessing an aperiodic reachable state is based upon Theorem 8.7.2 of Feller [115].

The use of the martingale property of $\Pi(\Phi_k, f)$ to obtain uniqueness of the invariant probability in Proposition 18.4.2 is originally in [175]. This is a powerful technique which is perhaps even more interesting in the absence of a reachable state.

For suppose that the chain is bounded in probability but a reachable state does not exist, and define an equivalence relation on X as follows: $x \leftrightarrow y$ if and only if $\Pi(x, \cdot) = \Pi(y, \cdot)$. It follows from the same techniques which were used in the proof of Proposition 18.4.2 that if x is recurrent, then the set of all states $\overline{E}_x$ for which $y \leftrightarrow x$

is closed. Since $x \in \overline{E}_x$ for every recurrent point $x \in R$, $F = \mathsf{X} - \sum \overline{E}_x$ consists entirely of non-recurrent points. It then follows from Proposition 3.3 of Tuominen and Tweedie [392] that F is transient.

From this decomposition and Proposition 18.4.3 it is straightforward to generalize Theorem 18.4.4 to chains which do not possess a reachable state. The details of this decomposition are spelled out in Meyn and Tweedie [281].

Such decompositions have a large literature for Feller chains and e-chains: see for example Jamison [175] and also Rosenblatt [337] for e-chains, and Jamison and Sine [178], Sine [358, 357, 356] and Foguel [121, 123] for Feller chains and the detailed connections between the Feller property and the stronger e-chain property. All of these papers consider exclusively compact state spaces. The results for non-compact state spaces appear here for the first time.

The LLN for e-chains is originally due to Breiman [46] who considered Feller chains on a compact state space. Also on a compact state space is Jamison's extension of Breiman's result [174] where the LLN is obtained without the assumption that a unique invariant probability exists.

One of the apparent difficulties in establishing this result is finding a candidate limit $\tilde{\pi}(f)$ of the sample path averages $\frac{1}{n} S_n(f)$. Jamison resolved this by considering the transition function Π, and the associated convergent martingale $(\Pi(\Phi_k, A), \mathcal{F}_k^{\Phi})$. If the chain is bounded in probability on average, then we define the *random probability* $\tilde{\pi}$ as

$$\tilde{\pi}\{A\} := \lim_{k \to \infty} \Pi(\Phi_k, A), \qquad A \in \mathcal{B}(\mathsf{X}). \tag{18.36}$$

It is then easy to show by modifying (18.34) that Theorem 18.5.1 continues to hold with $\int f \, d\pi$ replaced by $\int f \, d\tilde{\pi}$, even when no reachable state exists for the chain. The proof of Theorem 18.5.1 can be adopted after it is appropriately modified using the limit (18.36).

Chapter 19

Generalized classification criteria

We have now developed a number of simple criteria, solely involving the one-step transition function, which enable us to classify quite general Markov chains. We have seen, for example, that the equivalences in Theorem 11.0.1, Theorem 13.0.1, or Theorem 15.0.1 give an effective approach to the analysis of many systems.

For more complex models, however, the analysis of the simple one-step drift

$$\Delta V(x) = \int P(x, dy)[V(y) - V(x)]$$

towards petite sets may not be straightforward, or indeed may even be impracticable. Even though we know from the powerful converse theory in the theorems just mentioned that for most forms of stability, there must be at least one V with the one-step drift ΔV suitably negative, finding such a function may well be non-trivial.

In this chapter we conclude our approach to stochastic stability by giving a number of more general drift criteria which enable the classification of chains where the one-step criteria are not always straightforward to construct. All of these variations are within the general framework described previously. The steps to be used in practice are, we hope, clear from the preceding chapters, and follow the route reiterated in Appendix A.

There are three generalizations of the drift criteria which we consider here.

(a) *State-dependent drift conditions,* which allow for negative drifts after a number of steps $n(x)$ depending on the state x from which the chain starts.

(b) *Path- or history-dependent drift conditions,* which allow for functions of the whole past of the process to show a negative drift.

(c) *Mixed or "average" drift conditions,* which allow for functions whose drift varies in direction, but which is negative in a suitably "averaged" way.

For each of these we also indicate the application of the method by example. The state-dependent drift technique is used to analyze random walk on $\mathbb{R}^2_+$ and a model of invasion/defense, where simple one-step drift conditions seem almost impossible to construct; the history-dependent methods are shown to be suited to bilinear models with random coefficients, where again one-step drift conditions seem to fail; and, finally, the mixed drift analysis gives us a criterion for ladder processes, and in particular the Markovian representation of the full GI/G/1 queue, to be ergodic.

19.1 State-dependent drifts

19.1.1 The state-dependent drift criteria

In this section we consider consequences of state-dependent drift conditions of the form

$$\int P^{n(x)}(x, dy)V(y) \leq g[V(x), n(x)], \qquad x \in C^c, \tag{19.1}$$

where $n(x)$ is a function from X to $\mathbb{Z}_+$, g is a function depending on which type of stability we seek to establish, and C is an appropriate petite set.

The function $n(x)$ here provides the state dependence of the drift conditions, since from any x we must wait $n(x)$ steps for the drift to be negative.

In order to develop results in this framework we work with an "embedded" chain $\hat{\mathbf{\Phi}}$. Using $n(x)$ we define the new transition law $\{\hat{P}(x, A)\}$ by

$$\hat{P}(x, A) = P^{n(x)}(x, A), \qquad x \in \mathsf{X}, \ A \in \mathcal{B}(\mathsf{X}), \tag{19.2}$$

and let $\hat{\mathbf{\Phi}}$ be the corresponding Markov chain. This Markov chain may be constructed explicitly as follows. The time $n(x)$ is a (trivial) stopping time. Let $s(k)$ denote its iterates: that is, along any sample path, $s(0) = 0$, $s(1) = n(x)$ and

$$s(k + 1) = s(k) + n(\mathbf{\Phi}_{s(k)}).$$

Then it follows from the strong Markov property that

$$\hat{\mathbf{\Phi}}_k = \mathbf{\Phi}_{s(k)}, \qquad k \geq 0, \tag{19.3}$$

is a Markov chain with transition law $\hat{P}$.

Let $\hat{\mathcal{F}}_k = \mathcal{F}_{s(k)}$ be the σ-field generated by the events "before $s(k)$": that is,

$$\hat{\mathcal{F}}_k := \{A : A \cap \{s(k) \leq n\} \in \mathcal{F}_n, n \geq 0\}.$$

We let $\hat{\tau}_A, \hat{\sigma}_A$ denote the first-return and first-entry index to A respectively for the chain $\hat{\mathbf{\Phi}}$. Clearly $s(k)$ and the events $\{\hat{\sigma}_A \geq k\}$, $\{\hat{\tau}_A \geq k\}$ are $\hat{\mathcal{F}}_{k-1}$-measurable for any $A \in \mathcal{B}(\mathsf{X})$.

Note that $s(\hat{\tau}_C)$ denotes the time of first return to C by the original chain $\mathbf{\Phi}$ along an embedded path, defined by

$$s(\hat{\tau}_C) := \sum_0^{\hat{\tau}_C - 1} n(\hat{\mathbf{\Phi}}_k). \tag{19.4}$$

From (19.3) we have

$$s(\hat{\tau}_C) \geq \tau_C, \qquad s(\hat{\sigma}_C) \geq \sigma_C, \qquad \text{a.s. } [\mathsf{P}_*]. \tag{19.5}$$

These relations will enable us to use the drift equations (19.1), with which we will bound the index at which $\hat{\boldsymbol{\Phi}}$ reaches C, to bound the hitting times on C by the original chain.

We first give a state-dependent criterion for Harris recurrence.

Theorem 19.1.1. *Suppose that $\boldsymbol{\Phi}$ is a ψ-irreducible chain on X, and let $n(x)$ be a function from X to $\mathbb{Z}_+$. The chain is Harris recurrent if there exists a non-negative function V unbounded off petite sets and some petite set C satisfying*

$$\int P^{n(x)}(x, dy) V(y) \;\; \leq \;\; V(x), \qquad x \in C^c. \tag{19.6}$$

PROOF   The proof is an adaptation of the proof of Theorem 9.4.1.

Let $C_0 = C$, and let $C_n = \{x \in \mathsf{X} : V(x) \leq n\}$. By assumption, the sets C_n, $n \in \mathbb{Z}_+$, are petite.

Now suppose by way of contradiction that $\boldsymbol{\Phi}$ is not Harris recurrent. By Theorem 8.0.1 the chain is either recurrent, but not Harris recurrent, or the chain is transient. In either case, we show that there exists an initial condition x_0 such that

$$\mathsf{P}_{x_0}\{(\boldsymbol{\Phi} \in C \text{ i.o.})^c \cap (V(\Phi_k) \to \infty)\} > 0. \tag{19.7}$$

Firstly, if the chain is transient, then by Theorem 8.3.5 each C_n is uniformly transient, and hence $V(\Phi_k) \to \infty$ as $k \to \infty$ a.s. $[\mathsf{P}_*]$, and so (19.7) holds.

Secondly, if $\boldsymbol{\Phi}$ is recurrent, then the state space may be written as

$$\mathsf{X} = H \cup N \tag{19.8}$$

where $H = N^c$ is a maximal Harris set and $\psi(N) = 0$; this follows from Theorem 9.0.1. Since for each n the set C_n is petite we have $C_n \rightsquigarrow H$, and hence by Theorem 9.1.3,

$$\{\boldsymbol{\Phi} \in C_n \text{ i.o.}\} \subset \{\boldsymbol{\Phi} \in H \text{ i.o.}\} \qquad \text{a.s. } [\mathsf{P}_*].$$

It follows that the inclusion $\{\liminf V(\Phi_n) < \infty\} \subset \{\boldsymbol{\Phi} \in H \text{ i.o.}\}$ holds with probability one. Thus (19.7) holds for any $x_0 \in N$, and if the chain is not Harris, we know N is non-empty.

Now from (19.7) there exists $M \in \mathbb{Z}_+$ with

$$\mathsf{P}_{x_0}\{(\Phi_k \in C^c, k \geq M) \cap (V(\Phi_k) \to \infty)\} > 0 :$$

letting $\mu = P^M(x_0, \cdot)$, we have by conditioning at time M,

$$\mathsf{P}_\mu\{(\sigma_C = \infty) \cap (V(\Phi_k) \to \infty)\} > 0. \tag{19.9}$$

We now show that (19.9) leads to a contradiction when (19.6) holds.

Define the chain $\hat{\boldsymbol{\Phi}}$ as in (19.3). We can write (19.6), for every k, as

$$\mathsf{E}[V(\hat{\Phi}_{k+1}) \mid \hat{\mathcal{F}}_k] \leq V(\hat{\Phi}_k) \qquad \text{a.s. } [\mathsf{P}_*]$$

when $\hat{\sigma}_C > k$, $k \in \mathbb{Z}_+$.

Let $M_i = V(\hat{\Phi}_i)\mathbb{I}\{\hat{\sigma}_C \geq i\}$. Using the fact that $\{\hat{\sigma}_C \geq k\} \in \hat{\mathcal{F}}_{k-1}$, we have that

$$\mathsf{E}[M_k \mid \hat{\mathcal{F}}_{k-1}] = \mathbb{I}\{\hat{\sigma}_C \geq k\}\mathsf{E}[V(\hat{\Phi}_k) \mid \hat{\mathcal{F}}_{k-1}] \leq \mathbb{I}\{\hat{\sigma}_C \geq k\}V(\hat{\Phi}_{k-1}) \leq M_{k-1}.$$

Hence $(M_k, \hat{\mathcal{F}}_k)$ is a positive supermartingale, so that from Theorem D.6.2 there exists an almost surely finite random variable M_∞ such that $M_k \to M_\infty$ a.s. as $k \to \infty$. From the construction of M_i, either $\hat{\sigma}_C < \infty$ in which case $M_\infty = 0$, or $\hat{\sigma}_C = \infty$ in which case $\limsup_{k\to\infty} V(\hat{\Phi}_k) = M_\infty < \infty$ a.s.

Since $\sigma_C < \infty$ whenever $\hat{\sigma}_C < \infty$, this shows that for any initial distribution μ,

$$\mathsf{P}_\mu\big\{\{\sigma_C < \infty\} \cup \{\liminf_{n\to\infty} V(\Phi_n) < \infty\}^c\big\} = 1.$$

This contradicts (19.9), and hence the chain is Harris recurrent. □

We next prove a state-dependent criterion for positive recurrence.

Theorem 19.1.2. *Suppose that Φ is a ψ-irreducible chain on X, and let $n(x)$ be a function from X to $\mathbb{Z}_+$. The chain is positive Harris recurrent if there exists some petite set C, a non-negative function V bounded on C, and a positive constant b satisfying*

$$\int P^{n(x)}(x, dy)V(y) \leq V(x) - n(x) + b\mathbb{I}_C(x), \qquad x \in \mathsf{X}, \tag{19.10}$$

in which case for all x

$$\mathsf{E}_x[\tau_C] \leq V(x) + b. \tag{19.11}$$

PROOF   The state-dependent drift criterion for positive recurrence is a direct consequence of the f-ergodicity results of Theorem 14.2.2, which tell us that without any irreducibility or other conditions on Φ, if f is a non-negative function and

$$\int P(x, dy)V(y) \leq V(x) - f(x) + b\mathbb{I}_C(x), \qquad x \in \mathsf{X}, \tag{19.12}$$

for some set C, then for all $x \in \mathsf{X}$

$$\mathsf{E}_x\Big[\sum_{k=0}^{\tau_C - 1} f(\Phi_k)\Big] \leq V(x) + b. \tag{19.13}$$

Again define the chain $\hat{\Phi}$ as in (19.3). From (19.10) we can use (19.13) for $\hat{\Phi}$, with $f(x)$ taken as $n(x)$, to deduce that

$$\mathsf{E}_x\Big[\sum_{k=0}^{\hat{\tau}_C - 1} n(\hat{\Phi}_k)\Big] \leq V(x) + b. \tag{19.14}$$

But we have by adding the lengths of the embedded times $n(x)$ along any sample path that from (19.4)

$$\sum_{k=0}^{\hat{\tau}_C - 1} n(\hat{\Phi}_k) = s(\hat{\tau}_C) \geq \tau_C.$$

Thus from (19.14) and the fact that V is bounded on the petite set C, we have that Φ is positive Harris using the one-step criterion in Theorem 13.0.1, and the bound (19.11) follows also from (19.14). $\qquad\qquad\square$

We conclude the section with a state-dependent criterion for geometric ergodicity.

Theorem 19.1.3. *Suppose that Φ is a ψ-irreducible chain on X, and let $n(x)$ be a function from X to $\mathbb{Z}_+$. The chain is geometrically ergodic if it is aperiodic and there exists some petite set C, a non-negative function $V \geq 1$ and bounded on C, and positive constants $\lambda < 1$ and $b < \infty$ satisfying*

$$\int P^{n(x)}(x, dy)V(y) \;\leq\; \lambda^{n(x)}[V(x) + b\mathbb{I}_C(x)]. \qquad (19.15)$$

When (19.15) holds,

$$\sum_n r^n \| P^n(x, \cdot) - \pi \| \leq RV(x), \qquad x \in \mathsf{X} \qquad (19.16)$$

for some constants $R < \infty$ and $r > 1$.

PROOF   Suppose that (19.15) holds, and define

$$V'(x) = 2(V(x) - 1/2) \geq 1.$$

Then we can write (19.15) as

$$\begin{aligned}
\int \hat{P}(x, dy)V'(y) &\leq \lambda^{n(x)}[2V(x) + 2b\mathbb{I}_C(x)] - 1 \\
&= \lambda^{n(x)}[V'(x) + 1 + 2b\mathbb{I}_C(x)] - 1.
\end{aligned} \qquad (19.17)$$

Without loss of generality we will therefore assume that V itself satisfies the inequality

$$\int \hat{P}(x, dy)V(y) \;\leq\; \lambda^{n(x)}[V(x) + 1 + b\mathbb{I}_C(x)] - 1. \qquad (19.18)$$

We now adapt the proof of Theorem 15.2.5. Define the random variables

$$Z_k = \kappa^{s(k)}V(\hat{\Phi}_k)$$

for $k \in \mathbb{Z}_+$. It follows from (19.18) that for $\kappa = \lambda^{-1}$, since $\kappa^{s(k+1)}$ is $\hat{\mathcal{F}}_k$-measurable,

$$\begin{aligned}
\mathsf{E}[Z_{k+1} \mid \hat{\mathcal{F}}_k] &= \kappa^{s(k+1)}\mathsf{E}[V(\hat{\Phi}_{k+1}) \mid \hat{\mathcal{F}}_k] \\[2mm]
&\leq \kappa^{s(k+1)}\{\kappa^{-n(\Phi_k)}[V(\Phi_k) + 1 + b\mathbb{I}_C(\Phi_k)] - 1\} \\[2mm]
&= Z_k - \kappa^{s(k+1)} + \kappa^{s(k)} + \kappa^{s(k)}b\mathbb{I}_C(\Phi_k).
\end{aligned}$$

Using Proposition 11.3.2 we have

$$\mathsf{E}_x\Big[\sum_{k=0}^{\hat{\tau}_C - 1}[\kappa^{s(k+1)} - \kappa^{s(k)}]\Big] \;\leq\; Z_0(x) + \mathsf{E}_x\Big[\sum_{k=0}^{\hat{\tau}_C - 1}\kappa^{s(k)}b\mathbb{I}_C(\hat{\Phi}_k)\Big].$$

Collapsing the sum on the left and using the fact that only the first term in the sum on the right is non-zero, we get

$$\mathsf{E}_x[\kappa^{s(\hat{\tau}_C)} - 1] \leq V(x) + b\mathbb{I}_C(x). \tag{19.19}$$

Since $V < \infty$ and V is assumed bounded on C, and again using the fact that $s(\hat{\tau}_C) > \tau_C$, we have from Theorem 15.0.1 (ii) that the chain is geometrically ergodic.

The final bound in (19.16) comes from the fact that for some r, an upper bound on the state-dependent constant term in (19.16) is shown in Theorem 15.4.1 to be given by

$$R(x) = \mathsf{E}_x[\kappa^{\tau_C}] \leq \mathsf{E}_x[\kappa^{s(\hat{\tau}_C)}] \leq (2 + b)V(x)$$

since $V \geq 1$. □

19.1.2 Models on $\mathbb{R}^2_+$

State-dependent criteria appear to be of most use in analyzing multidimensional models, especially those on the positive orthant of Euclidean space. This is because, although the normal one-step drift conditions may work in the interior of such spaces, the constraints on the faces of the orthant can imply that drift is not negative in this part of the space.

We illustrate this in a simple case when the space is $\mathbb{R}^2_+ = \{(x, y), x \geq 0, y \geq 0\}$.

Consider the case of random walk restricted to the positive orthant. Let $Z_k = (Z_k(1), Z_k(2))$ be a sequence of i.i.d. random variables in $\mathbb{R}^2$ and define the chain Φ by

$$(\Phi_n(1), \Phi_n(2)) = ([\Phi_{n-1}(1) + Z_n(1)]^+, [\Phi_{n-1}(2) + Z_n(2)]^+). \tag{19.20}$$

Let us assume that for each coordinate we have negative increments: that is,

$$\mathsf{E}[Z_k(1)] < 0, \qquad \mathsf{E}[Z_k(2)] < 0.$$

This assumption ensures that the chain is a $\delta_{(0,0)}$-irreducible chain with all compact sets petite. To see this note that there exists $h > 0$ such that

$$\mathsf{P}(Z_k(1) < -h) > h, \qquad \mathsf{P}(Z_k(2) < -h) > h,$$

and so for any square $S_w = \{x \leq w, y \leq w\}$ we have that, choosing $m \geq w/h$,

$$P^m((x, y), (0, 0)) > h^{2m} > 0, \qquad (x, y) \in S_w.$$

This provides $\delta_{(0,0)}$-irreducibility, and moreover shows that S_w is small, with $\nu = \delta_{0,0}$ in (5.14).

We will also assume that the second moments of the increments are finite:

$$\mathsf{E}[Z_k^2(1)] < \infty, \qquad \mathsf{E}[Z_k^2(2)] < \infty.$$

Thus it follows from Proposition 14.4.1 that each of the marginal random walks on $[0, \infty)$ is positive Harris with stationary measures π_1, π_2 satisfying

$$\beta_1 := \int z\pi_1(dz) < \infty, \qquad \beta_2 := \int z\pi_2(dz) < \infty. \tag{19.21}$$

Of course, from this we could establish positivity merely by noting that $\pi = \pi_1 \times \pi_2$ is invariant for the bivariate chain. However, in order to illustrate the methods of this section we will establish that Φ is positive Harris by considering the test function $V(x,y) = x + y$: this also gives us a bound on the hitting times of rectangles that the more indirect result does not provide.

By choosing M large enough we can ensure that the truncated versions of the increments are also negative, so that for some $\varepsilon > 0$

$$\mathsf{E}[Z_k(1)\mathbb{I}\{Z_k(1) \geq -M\}] < -\varepsilon, \qquad \mathsf{E}[Z_k(2)\mathbb{I}\{Z_k(2) \geq -M\}] < -\varepsilon.$$

This ensures that on the set $A(M) = \{x \geq M, y \geq M\}$, we have that (19.10) holds with $n(x,y) = 1$ in the usual manner.

Now consider the strip $A_1(M,m) = \{x \leq M, y \geq m\}$, and fix $(x,y) \in A_1(M,m)$.

Let us choose a given fixed number of steps n, and choose $m > (M+1)n$. At each step in the time period $\{0, \ldots, n\}$ the expected value of $\Phi_n(2)$ decreases in expectation by at least ε. Moreover, from (19.21) and the f-norm ergodic result (14.5) we have that by convergence there is a constant c_0 such that for all n

$$\mathsf{E}_{(0,y)}[\Phi_n(1)] \leq c_0 \tag{19.22}$$

independent of y. From stochastic monotonicity we also have that for all $x \leq M$, if τ_0 denotes the first hitting time on $\{0\}$ for the marginal chain $\Phi_n(1)$,

$$\begin{aligned} \mathsf{E}_{(x,y)}[\Phi_n(1)\mathbb{I}\{\tau_0 > n\}] &\leq \mathsf{E}_{(M,y)}[\Phi_n(1)\mathbb{I}\{\tau_0 > n\}] \\ &:= \zeta_M(n) \end{aligned} \tag{19.23}$$

which is finite and tends to zero as $n \to \infty$, from Theorem 14.2.7, independent of y. Let us choose n large enough that $\zeta_M(n) \leq \varepsilon_0$.

We thus have from the Markov property

$$\begin{aligned} \mathsf{E}_{(x,y)}[\Phi_n(1) + \Phi_n(2)] &= \mathsf{E}_{(x,y)}[\Phi_n(2)] + \mathsf{E}_{(x,y)}[\Phi_n(1)\mathbb{I}\{\tau_0 > n\}] \\ &\quad + \mathsf{E}_{(x,y)}[\Phi_n(1)\mathbb{I}\{\tau_0 \leq n\}] \\ &\leq y - n\varepsilon + \varepsilon_0 + c_0. \end{aligned} \tag{19.24}$$

Thus for $x \leq M$, we have uniform negative n-step drift in the region $A_1(M,m)$ provided

$$n\varepsilon > M + \varepsilon_0 + c_0$$

as required.

A similar construction enables us to find that for fixed large n the n-step drift in the region $A_2(m,M)$ is negative also. Thus we have shown

Theorem 19.1.4. *If the bivariate random walk on $\mathbb{R}_+^2$ has negative mean increments and finite second moments in both coordinates, then it is positive Harris recurrent, and for sets $A(m) = \{x \geq m, y \geq m\}$ with m large, and some constant c,*

$$\mathsf{E}_{(x,y)}[\tau_{A(m)}] \leq c(x+y). \tag{19.25}$$

In this example, we do not use the full power of the results of Section 19.1. Only three values of $n(x, y)$ are used, and indeed it is apparent from the construction in (19.24) that we could have treated the whole chain on the region

$$\{x \geq M + n\} \cup \{y \geq M + n\}$$

for the same n. In this case the n-skeleton $\{\Phi_{nk}\}$ would be shown to be positive recurrent, and it follows from the fact that the invariant measure for $\{\Phi_k\}$ is also invariant for $\{\Phi_{nk}\}$ that the original chain is positive Harris: see Chapter 10. This example does, however, indicate the steps that we could go through to analyze less homogeneous models, and also indicates that it is easier to analyze the boundaries or non-standard regions independently of the interior or standard region of the space without the need to put the results together for a single fixed skeleton.

19.1.3 An invasion/antibody model

We conclude this section with the analysis of an invasion/antibody model on a countable space, illustrating another type of model where control of state-dependent drift is useful.

Models for competition between two groups can be modeled as bivariate processes on the integer-valued quadrant $\mathbb{Z}_+^2 = \{i, j \in \mathbb{Z}_+\}$. Consider such a process in discrete time with the first coordinate process $\Phi_n(1)$ denoting the numbers of invaders and the second coordinate process $\Phi_n(2)$ denoting the numbers of defenders.

(A1) Suppose first that the defenders and invaders mutually tend to reduce the opposition numbers when both groups are present, even though "reinforcements" may join either side. Thus on the interior of the space, denoted $I = \{i, j \geq 1\}$, we assume that for some $\varepsilon_i, \varepsilon_j \geq \varepsilon > 1/2$

$$\mathsf{E}_{i,j}[\Phi_1(1) + \Phi_1(2)] \leq (i - \varepsilon_i) + (j - \varepsilon_j) \leq i + j - 2\varepsilon, \qquad i, j > 1. \qquad (19.26)$$

Such behavior might model, for example, antibody action against invasive bodies where there is physical attachment of at least one antibody to each invader and then both die: in such a context we would have $\varepsilon_i = \varepsilon_j = 1$.

(A2) On one boundary, when the defender numbers reach the level 0, if the invaders are above a threshold level d the body dies in which case the invaders also die and the chain drops to $(0, 0)$, so that

$$P((i, 0), (0, 0)) = 1, \qquad i > d; \qquad (19.27)$$

otherwise a new population of antibodies or defenders of finite mean size is generated. These assumptions are of course somewhat unrealistic and clearly with more delicate arguments can be made much more general if required.

(A3) Much more critically, on the other boundary, when the invader numbers fall to level 0, and the defenders are of size $j > 0$, a new "invading army" is raised to bring the invaders to size N, where N is a random variable concentrated on $\{j + 1, j + 2, \ldots, j + d\}$ for the same threshold d, so that

$$\sum_{k=1}^{d} P((0, j), (j + k, j)) = 1 : \qquad (19.28)$$

this distribution being concentrated above j represents the physically realistic concept that a new invasion will fail instantly if the invading population is not at least the size of the defending population. The bounded size of the increment is purely for convenience of exposition.

Note that the chain is $\delta_{(0,0)}$-irreducible under assumptions (A1)–(A3), regardless of the behavior at zero. Thus the model can be formulated to allow for a stationary distribution at $(0,0)$ (i.e., extinction) or for rebirth and a more generally distributed stationary distribution over the whole of $\mathbb{Z}_2^+$. The only restriction we place in general is that the increments from $(0,0)$ have finite mean: here we will not make this more explicit as it does not affect our analysis.

Let us, to avoid unrewarding complexities, add to (19.26) the additional condition that the model is "left-continuous," that is, has bounded negative increments defined by

$$P((i,j),(i-l,j-k)) = 0, \qquad i,j > 0, \quad k,l > 1 : \tag{19.29}$$

this would be appropriate if the chain were embedded at the jumps of a continuous time process, for example.

To evaluate positive recurrence of the model, we use the test function $V(i,j) = [i+j]/\beta$, where $\beta < \varepsilon$ is to be chosen.

Analysis of this model in the interior of the space is not difficult: by using (V2) with $V(i,j)$ on $I = \{i,j \geq 1\}$, we have that $\mathsf{E}_{i,j}[\tau_{I^c}] < (i+j)/\beta$ from assumption (A1). The difficulty with such multidimensional models is that even though they reach I^c in a finite mean time, they may then "escape" along one or both of the boundaries. It is in this region that the tools of Section 19.1 are useful in assisting with the classification of the model.

Starting at $B_1(c) = \{(i,0), i > c\}$, the infinite boundary edge above c, we have that the value of $V(\Phi_1)$ is zero if $c > d$, so that (19.10) also holds with $n = 1$ provided we choose $c > \max(d, \beta^{-1})$.

On the other infinite boundary edge, denoted $B_2(c) = \{(0,j), j > c\}$, however, we have positive one-step drift of the function V. Now from the starting point $(0,j)$, let us consider the $(j+1)$-step drift. This is bounded above by $[j + d - 2j\varepsilon]/\beta$ and so we have (19.10) also holds with $n(j) = j + 1$ provided

$$[j + d - 2j\varepsilon]/\beta < -j - 1,$$

which will hold provided $\beta < 2\varepsilon - 1$ and we then choose $c > (d+\beta)/(2\varepsilon - 1 - \beta)$.

Consequently we can assert that, writing $C = I \cup B_2(c) \cup B_1(c)$ with c satisfying both these constraints, the mean time is bounded as

$$\mathsf{E}_{(i,j)}[\tau_C] \leq [i+j]/\beta$$

regardless of the threshold level d, and so the invading strategy is successful in overcoming the antibody defense.

Note that in this model there is no fixed time at which the drift from all points on the boundary $B_2(c)$ is uniformly negative, no matter what the value of c chosen. Thus, state-dependent drift conditions appear needed to analyze this model.

To test for geometric ergodicity we use the function $V(i,j) = \exp(\alpha i) + \exp(\alpha j)$ and adopt the approach in Section 16.3.

We assume that the increments in the model have uniformly geometrically decreasing tails and bounded second moments: specifically, we assume each coordinate process satisfies, for some $\gamma > 0$,

$$\theta_i(\gamma) := \sum_{k \geq i-1} \exp(\gamma k) \mathsf{P}_{i,j}(\Phi_1(1) = i + k) < \infty, \qquad j \geq 1,$$

$$\theta'_j(\gamma) := \sum_{k \geq j-1} \exp(\gamma k) \mathsf{P}_{i,j}(\Phi_1(2) = j + k) < \infty, \qquad i \geq 1,$$

(19.30)

and

$$\sum_{k \geq i-1} k^2 \mathsf{P}_{i,j}(\Phi_1(1) = i + k) \quad < \quad D_1, \qquad j \geq 1,$$

$$\sum_{k \geq j-1} k^2 \mathsf{P}_{i,j}(\Phi_1(2) = j + k) \quad < \quad D_2, \qquad i \geq 1.$$

(19.31)

Then on the interior set I we have, for $\alpha < \gamma$,

$$\sum_j P((r,s),(i,j))V(i,j) \quad \leq \quad \exp(\alpha r)[\theta_i(\alpha) - 1]$$

$$+ \exp(\alpha s)[\theta'_j(\alpha) - 1]$$

$$\leq \quad \alpha \exp(\alpha r)(-\varepsilon_r/2)$$

$$+ \alpha \exp(\alpha s)(-\varepsilon_s/2)$$

for small enough α, using a Taylor series expansion and the uniform conditions (19.30) and (19.31). Thus (19.15) holds with $n = 1$ and $\lambda = 1 - \alpha\varepsilon/2$.

Starting at $B_1(c)$, (19.15) also obviously holds provided we choose c large enough. On the other infinite boundary edge $B_2(c) = \{(0,j), j > c\}$ we have a similar construction for the $(j + 1)$-step drift. We have, using the uniform bounds (19.31) assumed on the variances,

$$\sum_j P^{j+1}((0,s),(i,j))V(i,j) \quad \leq \quad \exp(\alpha(j+d))[1 - \varepsilon/2]^j$$

$$+ \exp(\alpha s)[1 - \varepsilon/2]^j$$

and so, for α suitably small, we have (19.15) holding again as required. $\square$

19.2   History-dependent drift criteria

The approach through Dynkin's formula to obtaining bounds on hitting times of appropriate sets allows a straightforward generalization to more complex, history-dependent, test functions with very little extra effort above that expended already.

Rather than considering a fixed function V of the state Φ_k, we will now let $\{V_k : k \in \mathbb{Z}_+\}$ denote a family of non-negative Borel measurable functions $V_k : \mathsf{X}^{k+1} \to \mathbb{R}_+$. By imposing the appropriate "drift condition" on the stochastic process $\{V_k = V_k(\Phi_0, \ldots, \Phi_k)\}$, we will obtain generalized criteria for stability and non-stability. The value of this generalization will be illustrated below in an application to an autoregressive model with random coefficients.

19.2.1   Generalized criteria for positivity and nullity

We first consider, in the time-varying context, drift conditions on such a family $\{V_k : k \in \mathbb{Z}_+\}$ for chains to be positive or to be null. We call a sequence $\{V_k, \mathcal{F}_k^{\Phi}\}$ *adapted* if V_k is measurable with respect to $\mathcal{F}_k^{\Phi}$ for each k.

The following condition generalizes (V2).

Generalized negative drift condition

There exists a set $C \in \mathcal{B}(\mathsf{X})$, and an adapted sequence $\{V_k, \mathcal{F}_k^{\Phi}\}$, such that for some $\varepsilon > 0$,

$$\mathsf{E}[V_{k+1} \mid \mathcal{F}_k^{\Phi}] \leq V_k - \varepsilon \qquad \text{a.s. } [\mathsf{P}_*] \tag{19.32}$$

when $\sigma_C > k$, $k \in \mathbb{Z}_+$.

As usual the condition that $\sigma_C > k$ means that $\Phi_i \in C^c$ for each i between 0 and k. Since C will usually be assumed "small" in some sense (either petite or compact), (19.32) implies that there is a drift towards the "center" of the state space when Φ is "large" in exactly the same way that (V2) does.

From these generalized drift conditions and Dynkin's formula we find

Theorem 19.2.1. *If $\{V_k\}$ satisfies (19.32), then*

$$\mathsf{E}_x[\tau_C] \leq \begin{cases} \varepsilon^{-1} V_0(x) & x \in C^c \\ 1 + \varepsilon^{-1} P V_0(x) & x \in C. \end{cases}$$

Hence if C is petite and $\sup_{x \in C} \mathsf{E}_x[V_0(\Phi_1)] < \infty$, then Φ is regular.

PROOF   The proof follows immediately from Proposition 11.3.3 by letting $Z_k = V_k$, $\varepsilon_k = \varepsilon$, exactly as in Theorem 11.3.4.   □

There is a similar generalization of the drift criterion for determining whether a given chain is null.

Generalized positive drift condition

There exists a set $C \in \mathcal{B}(\mathsf{X})$, and an adapted sequence $\{V_k, \mathcal{F}_k^{\Phi}\}$ with

$$\mathsf{E}[V_{k+1} \mid \mathcal{F}_k^{\Phi}] \geq V_k \qquad \text{a.s. } [\mathsf{P}_*], \tag{19.33}$$

when $\sigma_C > k$, $k \in \mathbb{Z}_+$.

Clearly the process $V_k \equiv 1$ satisfies (19.33), so we will need some auxiliary conditions to prove anything specific when (19.33) holds.

Theorem 19.2.2. *Suppose that $\{V_k\}$ satisfies (19.33), and let $x_0 \in C^c$ be such that*

$$V_0(x_0) > V_k(x_0, \ldots, x_k), \qquad x_k \in C, \ k \in \mathbb{Z}_+. \tag{19.34}$$

Suppose moreover the conditional absolute increments have bounded means: that is, for some constant $B < \infty$,

$$\mathsf{E}[|V_k - V_{k-1}| \mid \mathcal{F}_{k-1}^\Phi] \leq B. \tag{19.35}$$

Then $\mathsf{E}_{x_0}[\tau_C] = \infty$.

Proof   The proof of Theorem 11.5.1 goes through without change, although in this case the functions V_k in that proof are not taken simply as $V(\Phi_k)$ but as $V_k(\Phi_0, \ldots, \Phi_k)$.   □

19.2.2   Generalized criteria for geometric ergodicity

We can extend the results of Chapter 15 in a similar way when the space admits a topology. In order to derive such criteria we need to adapt the sequence $\{V_k\}$ appropriately to the topology. Let us call the whole sequence $\{V_k\}$ coercive if there exists a coercive function $V \colon \mathsf{X} \to \mathbb{R}_+$ with the property

$$V_k(x_0, \ldots, x_k) \geq V(x_k) \geq 0 \tag{19.36}$$

for all $k \in \mathbb{Z}_+$ and all $x_i \in \mathsf{X}$.

The criterion for such a family $\{V_k\}$ generalizes (15.35), which we showed in Lemma 15.2.8 to be equivalent to (V4).

Generalized geometric drift condition

There exist $\lambda < 1$, $L < \infty$ and an adapted coercive sequence $\{V_k, \mathcal{F}_k^\Phi\}$ such that

$$\mathsf{E}_x[V_{k+1} \mid \mathcal{F}_k^\Phi] \leq \lambda V_k + L \qquad \text{a.s. } [\mathsf{P}_*], \quad k \in \mathbb{Z}_+. \tag{19.37}$$

Theorem 19.2.3. *Suppose that Φ is an irreducible aperiodic T-chain. If the generalized geometric drift condition (19.37) holds, and if V_0 is uniformly bounded on compact subsets of X, then there exist $R < \infty$ and $r > 1$ such that*

$$\sum_{n=1}^{\infty} r^n \|P^n(x, \cdot) - \pi\|_f \leq R(V_0(x) + 1), \qquad n \in \mathbb{Z}_+, \ x \in \mathsf{X},$$

where $f = V + 1$ and V is as defined in (19.36). In particular, Φ is then f-geometrically ergodic.

PROOF   Let $\lambda < \rho < 1$, and define the pre-compact set C and the constant $\varepsilon > 0$ by

$$C = \{x \in \mathsf{X} : V(x) \leq \frac{2L}{\rho - \lambda} + 1\}, \qquad \varepsilon = \frac{\rho - \lambda}{2}.$$

Then for all $k \in \mathbb{Z}_+$,

$$\mathsf{E}[V_{k+1} \mid \mathcal{F}_k^\Phi] \leq \rho V_k + \left\{ [L + (\rho - \lambda)] - \frac{\rho - \lambda}{2}(V(\Phi_k) + 1) \right\} - \frac{\rho - \lambda}{2}(V(\Phi_k) + 1).$$

Hence $\mathsf{E}[V_{k+1} \mid \mathcal{F}_k^\Phi] \leq \rho V_k - \varepsilon f(\Phi_k)$ when $\Phi_k \in C^c$. Letting $Z_k = r^k V_k$, where $r = \rho^{-1}$, we then have $\mathsf{E}[Z_k \mid \mathcal{F}_{k-1}^\Phi] - Z_{k-1} \leq -\varepsilon r^k f(\Phi_{k-1})$ when $\Phi_{k-1} \in C^c$. We now use Dynkin's formula to deduce that for all $x \in \mathsf{X}$,

$$0 \leq \mathsf{E}_x[z_{\tau_C^m}] = \mathsf{E}_x[Z_1] + \mathsf{E}_x\left[\left(\sum_{k=2}^{\tau_C^m} \mathsf{E}[Z_k \mid \mathcal{F}_{k-1}^\Phi] - Z_{k-1} \right) \mathbb{I}(\tau_C \geq 2) \right]$$

$$\leq \mathsf{E}_x[Z_1] - \mathsf{E}_x\left[\sum_{k=2}^{\tau_C^m} \varepsilon r^k f(\Phi_{k-1}) \mathbb{I}(\tau_C \geq 2) \right].$$

This and the Monotone Convergence Theorem shows that for all $x \in \mathsf{X}$,

$$\mathsf{E}_x\left[\sum_{k=1}^{\tau_C} r^k f(\Phi_{k-1}) \right] \leq \varepsilon^{-1} r \mathsf{E}_x[V_1] + r V(x).$$

This completes the proof, since $\mathsf{E}_x[V_1] + V(x) \leq \lambda V_0(x) + L + V_0(x)$ by (19.37) and (19.36).   □

19.2.3 Generalized criteria for non-evanescence and transience

A general criterion for Harris recurrence on a topological space can be obtained from the following history-dependent drift condition, which generalizes (V1).

Generalized non-positive drift condition

There exists a compact set $C \subset \mathsf{X}$, and an adapted coercive sequence $\{V_k, \mathcal{F}_k^\Phi\}$ such that

$$\mathsf{E}[V_{k+1} \mid \mathcal{F}_k^\Phi] \leq V_k \qquad \text{a.s. } [\mathsf{P}_*], \tag{19.38}$$

when $\sigma_C > k$, $k \in \mathbb{Z}_+$.

Theorem 19.2.4. *If (19.38) holds, then Φ is non-evanescent. Hence if Φ is a ψ-irreducible T-chain and (19.38) holds for a coercive sequence and a compact C, then Φ is Harris recurrent.*

PROOF   The proof is almost identical to that of Theorem 9.4.1. If $P_x\{\Phi \to \infty\} > 0$ for some $x \in X$, then (9.30) holds, so that for some M

$$P_\mu\{\{\sigma_C = \infty\} \cap \{\Phi \to \infty\}\} > 0, \qquad (19.39)$$

where $\mu = P^M(x, \cdot)$.

This time let $M_i = V_i \mathbb{I}\{\sigma_C \geq i\}$. Again we have that $(M_k, \mathcal{F}_k^\Phi)$ is a positive supermartingale, since

$$E[M_k \mid \mathcal{F}_{k-1}^\Phi] = \mathbb{I}\{\sigma_C \geq k\} E[V_k \mid \mathcal{F}_{k-1}^\Phi] \leq \mathbb{I}\{\sigma_C \geq k\} V_{k-1} \leq M_{k-1}.$$

Hence there exists an almost surely finite random variable M_∞ such that $M_k \to M_\infty$ as $k \to \infty$.

But as in Theorem 9.4.1, either $\sigma_C < \infty$ in which case $M_\infty = 0$, or $\sigma_C = \infty$ which contradicts (19.39). Hence Φ is again non-evanescent.

The Harris recurrence when Φ is a T-chain follows as usual by Theorem 9.2.2.   □

Finally, we give a criterion for transience using a time-varying test function.

Generalized non-negative drift condition

There exists a set $A \in \mathcal{B}(X)$, and a uniformly bounded, adapted sequence $\{V_k, \mathcal{F}_k^\Phi\}$ such that

$$E[V_{k+1} \mid \mathcal{F}_k^\Phi] \geq V_k \qquad \text{a.s. } [P_*], \qquad (19.40)$$

when $\sigma_A > k$, $k \in \mathbb{Z}_+$.

Theorem 19.2.5. *Suppose that the process V_k satisfies (19.40) for a set A, and suppose that for deterministic constants $L > M$,*

$$V_k \leq L, \qquad \mathbb{I}\{\sigma_A = k\} V_k \leq M, \qquad k \in \mathbb{Z}_+.$$

Then for all $x \in X$

$$P_{x_0}\{\sigma_A = \infty\} \geq \frac{V_0(x) - M}{L - M}.$$

Hence if both A and $\{x : V_0(x) > M\}$ lie in $\mathcal{B}^+(X)$, then Φ is transient.

PROOF   Define the sequence $\{M_k\}$ by

$$M_{k+1} = V_{k+1} \mathbb{I}\{\sigma_A > k\} + M\mathbb{I}\{\sigma_A \leq k\}.$$

Then, since $\{\sigma_A \leq k\} \in \mathcal{F}_k^\Phi$, we have

$$\begin{aligned}
E[M_{k+1} \mid \mathcal{F}_k^\Phi] &\geq V_k \mathbb{I}\{\sigma_A > k\} + M\mathbb{I}\{\sigma_A \leq k\} \\
&\geq V_k \mathbb{I}\{\sigma_A > k\} + V_k \mathbb{I}\{\sigma_A = k\} + M\mathbb{I}\{\sigma_A \leq k-1\} \\
&= M_k
\end{aligned}$$

and the adapted process $(M_k, \mathcal{F}_k^\Phi)$ is thus a submartingale. Hence $(L - M_k, \mathcal{F}_k^\Phi)$ is a positive supermartingale. By Kolmogorov's Inequality (Theorem D.6.3) it follows that for any $T > 0$

$$P_x\{\sup_{k \geq 0}(L - M_k) \geq T\} \leq \frac{L - M_0(x)}{T}.$$

Letting $T = L - M$, and noting that $M_0(x) \geq V_0(x)$, gives

$$P_x\{\inf_{k \geq 0} M_k \leq M\} \leq \frac{L - V_0(x)}{L - M}.$$

Finally, since $M_k = M$ for all k sufficiently large whenever $\sigma_A < \infty$, it follows that

$$P_x\{\sigma_A = \infty\} \geq P_x\{\inf_{k \geq 0} M_k > M\} \geq \frac{V_0(x) - M}{L - M}$$

which is the desired bound.   □

19.2.4   The dependent parameter bilinear model

To illustrate the general results described above we will analyze the dependent parameter bilinear model defined as in (7.23) by the pair of equations

$$\begin{aligned} \theta_{k+1} &= \alpha\theta_k + Z_{k+1}, \qquad |\alpha| < 1, \\ Y_{k+1} &= \theta_k Y_k + W_{k+1}. \end{aligned}$$

This model is just the simple adaptive control model with the control set to zero; but while the model is somewhat simpler to define than the adaptive control model, we will see that the lack of control makes it much more difficult to show that the model is geometrically ergodic. One of the difficulties with this model is that to date a test function of the form (V4) has not been explicitly computed, though we will show here that a time-varying test function of the form (19.37) can be constructed.

The proof will require a substantially more stringent bound on the parameter process than that which was used in the proof of Proposition 17.3.5. We will assume that

$$\zeta_z^2 := \mathsf{E}\Big[\exp\Big\{\frac{2}{1 - |\alpha|}|Z_1| - 2\Big\}\Big] < 1. \tag{19.41}$$

Using a history-dependent test function of the form (19.37) we will prove the following:

Theorem 19.2.6. *Suppose that conditions (DBL1)–(DBL2) hold, and (19.41) is satisfied. Then Φ is geometrically ergodic, and hence possesses a unique invariant probability π. The CLT and LIL hold for the processes Y and θ, and for each initial condition $x \in \mathsf{X}$,*

$$\lim_{N \to \infty} \frac{1}{N}\sum_{k=1}^N Y_k^2 = \int y^2\, d\pi < \infty \qquad a.s.\ [P_x],$$

$$|\mathsf{E}_x[Y_k^2] - \int y^2 d\pi| \leq M(x)\rho^k, \quad k \geq 0,$$

where M is a continuous function on X and $0 < \rho < 1$.   □

PROOF   It follows as in the proof of Proposition 17.3.5 that the joint process $\Phi_k = \binom{\theta_k}{Y_k}$, $k \geq 0$, is an aperiodic, ψ-irreducible T-chain.

In view of Theorem 19.2.3 it is enough to show that the history-dependent drift (19.37) holds for an adapted process $\{V_k\}$. We now indicate how to construct such a process.

First use the estimate $x \leq e^{-1} e^x$ to show

$$|\prod_{i=j}^{k} \theta_i| \leq e^{-(k-j+1)} \left(\prod_{i=j}^{k} \exp|\theta_i|\right) = e^{-(k-j+1)} \exp\left(\sum_{i=j}^{k} |\theta_i|\right). \qquad (19.42)$$

But since by (2.14),

$$\sum_{i=j}^{k} |\theta_i| \leq |\alpha| \sum_{i=j}^{k} |\theta_i| + |\alpha||\theta_{j-1}| + \sum_{i=j}^{k} |Z_i|,$$

we have

$$\sum_{i=j}^{k} |\theta_i| \leq \frac{|\alpha|}{1-|\alpha|} |\theta_{j-1}| + \frac{1}{1-|\alpha|} \sum_{i=j}^{k} |Z_i|, \qquad (19.43)$$

and (19.42) and (19.43) imply the bound, for $j \geq 1$,

$$|\prod_{i=j}^{k} \theta_i| \leq e^{-(k-j+1)} \exp\{\frac{|\alpha|}{1-|\alpha|} |\theta_{j-1}|\} \times \exp\{\frac{1}{1-|\alpha|} \sum_{i=j}^{k} |Z_i|\}. \qquad (19.44)$$

Squaring both sides of (17.28) and applying (19.44), we obtain the bound

$$Y_{k+1}^2 \leq 3A_k + 3B_k + 3W_{k+1}^2 \qquad (19.45)$$

for all $k \in \mathbb{Z}_+$, where

$$A_k = \{\sum_{j=1}^{k} |W_j| \exp\{\frac{|\alpha|}{1-|\alpha|} |\theta_{j-1}|\} \prod_{i=j}^{k} \exp\{\frac{1}{1-|\alpha|} |Z_i| - 1\}\}^2,$$

$$B_k = \theta_0^2 Y_0^2 \exp\{\frac{2|\alpha|}{1-|\alpha|} |\theta_0|\} \prod_{i=1}^{k} \exp\{\frac{2}{1-|\alpha|} |Z_i| - 2\}.$$

If we define

$$C_k = \exp\{\frac{2|\alpha|}{1-|\alpha|} |\theta_k|\},$$

we have the three bounds, valid for any $\varepsilon > 0$,

$$\mathsf{E}[A_{k+1} \mid \mathcal{F}_k^\Phi] \leq \zeta_z^2 \{(1+\varepsilon)A_k + (1+\varepsilon^{-1})\mathsf{E}[W^2]C_k\},$$
$$\mathsf{E}[B_{k+1} \mid \mathcal{F}_k^\Phi] \leq \zeta_z^2 B_k,$$
$$\mathsf{E}[C_{k+1} \mid \mathcal{F}_k^\Phi] \leq |\alpha|C_k + (1-|\alpha|)(\mathsf{E}[\exp\{\frac{2|\alpha|}{1-|\alpha|} |Z_1|\}])^{\frac{1}{1-|\alpha|}}.$$

This is shown in [275] and we omit the details which are too lengthy for this exposition. The constant ε will be assumed small, but we will keep it free until we have performed one more calculation. For $k \geq 0$ we make the definition

$$V_k = \varepsilon^3 Y_k^2 + \varepsilon^2 A_k + B_k + C_k.$$

We have for any $k \geq 0$,

$$\varepsilon^3 Y_k^2 + \exp\{\frac{2|\alpha|}{1-|\alpha|}|\theta_k|\} \leq V_k,$$

and since $V(y, \theta) = \varepsilon^3 y^2 + \exp\{\frac{2|\alpha|}{1-|\alpha|}|\theta|\}$ is a coercive function on X, it follows that the sequence $\{V_k : k \in \mathbb{Z}_+\}$ is coercive.

Using the bounds above we have for some $R < \infty$,

$$\begin{aligned}
\mathsf{E}[V_{k+1} \mid \mathcal{F}_k^\Phi] \quad \leq \quad & 3\varepsilon^3 A_k^2 + 3\varepsilon^3 B_k + \zeta_z^2 \varepsilon^2 (1+\varepsilon) A_k + \zeta_z^2 \varepsilon^2 (1+\varepsilon^{-1}) \mathsf{E}[W^2] C_k \\
& + \zeta_z^2 B_k + |\alpha| C_k + R.
\end{aligned}$$

Rearranging terms gives

$$\begin{aligned}
\mathsf{E}[V_{k+1} \mid \mathcal{F}_k^\Phi] \quad \leq \quad & \{3\varepsilon + \zeta_z^2(1+\varepsilon)\}\varepsilon^2 A_k + \{3\varepsilon^3 + \zeta_z^2\}B_k \\
& + \{|\alpha| + \varepsilon^2(1+\varepsilon^{-1})\mathsf{E}[W^2]\zeta_z^2\}C_k + R.
\end{aligned}$$

Hence (19.37) holds with

$$\lambda = \max(|\alpha| + \zeta_z^2 \varepsilon^2 (1+\varepsilon^{-1})\mathsf{E}[W^2], \zeta_z^2 + 3\varepsilon^3, \zeta_z^2(1+\varepsilon) + 3\varepsilon),$$

and for ε sufficiently small, we have $\lambda < 1$ as required. $\square$

19.3 Mixed drift conditions

One of the themes of this book has been the interplay between the various stability concepts, and the existence of test functions which give appropriate and consistent drift towards the center of the space.

We conclude with a section which considers chains where the drift is mixed: that is, inward in some parts of the space, and outward in other parts. Of course, it again follows from all we have done to date that for some functions (and in particular the expected hitting time functions V_C) the one-step drift will always be towards the set C from initial conditions outside of C. However, it is of considerable intuitive interest to consider the drift when the function V is relatively arbitrary, in which case there is no reason *a priori* to expect that the drift will be consistent in any useful way.

We will find in this section that for a large class of functions, an appropriately averaged drift over the state space is indeed "inwards" when the chain is positive, and "outwards" when the chain is null. This accounts in yet another way for the success of the seemingly simple drift criteria as tools for classifying general chains.

19.3.1 The limiting-average drift

Suppose that V is an everywhere finite non-negative function satisfying

$$\int P(x,dy)|V(y) - V(x)| \le d < \infty, \qquad x \in \mathsf{X}. \tag{19.46}$$

Then we have, for all $n \in \mathbb{Z}_+$, $x \in \mathsf{X}$,

$$\int P^n(x,dy)|\Delta V(y)| \le d,$$

and thus the functions

$$n^{-1} \sum_{k=1}^n \int_{C^c} P^k(x,dy)\Delta V(y) \tag{19.47}$$

are all well defined and finite everywhere. Obviously we need a little less than (19.46) to guarantee this, but (19.46) will also be a convenient condition elsewhere.

Theorem 19.3.1. *Suppose that Φ is ψ-irreducible and that $V \ge 0$ satisfies (19.46). A sufficient condition for the chain to be positive is that for some one $x \in \mathsf{X}$ and some petite set C*

$$\liminf_{n \to \infty} n^{-1} \sum_{k=1}^n \int_{C^c} P^k(x,dy)\Delta V(y) < 0. \tag{19.48}$$

PROOF   By definition we have

$$\int P^{n+1}(x,dy)V(y) = \int P^n(x,dw) \int P(w,dy)V(y)$$
$$= \int P^n(x,dy)\Delta V(y) + \int P^n(x,dy)V(y) \tag{19.49}$$

where all the terms in (19.49) are finite by induction and (19.46). By iteration, we then get

$$n^{-1}\int P^{n+1}(x,dy)V(y) = n^{-1}\sum_{k=1}^n \int P^k(x,dy)\Delta V(y) + n^{-1}[\Delta V(x) + V(x)]$$

so that as $n \to \infty$

$$\liminf n^{-1}\sum_{k=1}^n \int P^n(x,dy)\Delta V(y) \ge 0. \tag{19.50}$$

Now suppose by way of contradiction that Φ is null; then from Theorem 18.2.2 we have that the petite set C is null, and so for every x we have by the bound in (19.46)

$$\lim_{n \to \infty} \int_C P^n(x,dy)\Delta V(y) = 0.$$

This, together with (19.50), cannot be true when we have assumed (19.48); so the chain is indeed positive. $\square$

There is a converse to this result. We first show that for positive chains and suitable functions V, the drift ΔV, π-averaged over the whole space, is in fact zero.

Theorem 19.3.2. *Suppose that* Φ *is* ψ-*irreducible, positive with invariant probability measure* π, *and that* $V \geq 0$ *satisfies (19.46). Then*

$$\int_{\mathsf{X}} \pi(dy)\Delta V(y) = 0. \tag{19.51}$$

PROOF   Consider the function $M_z(x)$ defined for $z \in (0,1)$ by

$$M_z(x) = \int P(x,dy)[z^{V(x)} - z^{V(y)}]/[1-z].$$

We first show that $|M_z(x)|$ is uniformly bounded for $x \in \mathsf{X}$ and $z \in (\frac{1}{2},1)$ under the bound (19.46).

By the Mean Value Theorem and non-negativity of V we have for any $0 < z < 1$,

$$
\begin{aligned}
|z^{V(x)} - z^{V(y)}| &\leq |V(x) - V(y)| \sup_{t \geq 0} |\frac{d}{dt}z^t| \\
&= |V(x) - V(y)||\log(z)|. \tag{19.52}
\end{aligned}
$$

Hence under (19.46), for all $x \in \mathsf{X}$ and $z \in (0,1)$,

$$|M_z(x)| \leq \frac{|\log(z)|}{1-z}\int P(x,dy)|V(x) - V(y)| \leq \frac{d}{z} \tag{19.53}$$

which establishes the claimed boundedness of $|M_z(x)|$.

Moreover, by (19.52) and dominated convergence,

$$\lim_{z\uparrow 1} M_z(x) = \int P(x,dy)\left\{\lim_{z\uparrow 1}\frac{z^{V(x)} - z^{V(y)}}{1-z}\right\} = \Delta V(x). \tag{19.54}$$

Since $\int \pi(dx)z^{V(x)} < \infty$ for fixed $z \in (0,1)$, we can interchange the order of integration and find

$$\int \pi(dx)M_z(x) = \int \pi(dx)\int P(x,dy)[z^{V(x)} - z^{V(y)}]/[1-z] = 0.$$

Hence by the Dominated Convergence Theorem once more we have

$$
\begin{aligned}
0 &= \lim_{z\uparrow 1}\int \pi(dx)M_z(x) \\
&= \int \pi(dx)\left[\lim_{z\uparrow 1} M_z(x)\right] \tag{19.55} \\
&= \int \pi(dx)\Delta V(x)
\end{aligned}
$$

as required.   □

Intuitively, one might expect from stationarity that the balance equation (19.51) will hold in complete generality. But we know that this is not the case without some auxiliary conditions such as (19.46): we saw this in Section 11.5.1, where we showed an example of a positive chain with everywhere strictly positive drift.

We now see that the balanced drift of (19.51) occurs, as one might expect from (19.48), from the inward drift towards suitable sets C, combined with an outward drift from such sets. This gives us the converse to Theorem 19.3.1.

Theorem 19.3.3. *Suppose that Φ is ψ-irreducible and that $V \geq 0$ satisfies (19.46). If C is a sublevel set of V with $C^c, C \in \mathcal{B}^+(\mathsf{X})$, then a necessary condition for the chain to be positive is that*

$$\int_{C^c} \pi(dw)\Delta V(w) < 0, \tag{19.56}$$

in which case for almost all $x \in \mathsf{X}$

$$\lim_{n \to \infty} n^{-1} \sum_{k=1}^{n} \int_{C^c} P^k(x, dy)\Delta V(y) < 0. \tag{19.57}$$

Thus, under these conditions, (19.48) is necessary and sufficient for positivity.

PROOF   Suppose the chain is positive, and that $C = \{x : V(x) \leq b\} \in \mathcal{B}^+(\mathsf{X})$ is a sublevel set of the function V, so that obviously

$$V(y) > \sup_{x \in C} V(x), \qquad y \in C^c. \tag{19.58}$$

From (19.46) we certainly have that drift off C is bounded, so that

$$|\Delta V(x)| \leq B' < \infty, \qquad x \in C, \tag{19.59}$$

and in particular $\int_C \pi(dw)\Delta V(w) \leq B'$.
    Using the invariance of π,

$$
\begin{aligned}
\int_C \pi(dw)\Delta V(w) &= \int_C \pi(dx) \int P(x, dw)V(w) - \int_C \pi(dw)V(w) \\
&= \int_C \pi(dx)[\int_{C^c} P(x, dw)V(w) + \int_C P(x, dw)V(w)] \\
&\quad - \int_C [\int_{\mathsf{X}} \pi(dx)P(x, dw)]V(w) \\
&= \int_C \pi(dx) \int_{C^c} P(x, dw)V(w) \\
&\quad + \int_C \pi(dx) \int_C P(x, dw)V(w) \\
&\quad - \int_{\mathsf{X}} \pi(dx) \int_C P(x, dw)V(w).
\end{aligned}
\tag{19.60}
$$

Now provided the set C^c is in $\mathcal{B}^+(\mathsf{X})$, we show the right hand side of (19.60) is strictly positive. To see this requires two steps.
    First observe that $\int_C \pi(dx)P(x, C^c) > 0$ since $C, C^c \in \mathcal{B}^+(\mathsf{X})$. Since $V(y) > \sup_{w \in C} V(w)$ for $y \in C^c$ we have

$$\int_C \pi(dx) \int_{C^c} P(x, dw)V(w) > \left(\sup_{w \in C} V(w)\right) \int_C \pi(dx)P(x, C^c), \tag{19.61}$$

showing from (19.60) that

$$\int_C \pi(dw)\Delta V(w) > \left(\sup_{w \in C} V(w)\right)\left[\int_C \pi(dx)P(x, C^c) - \int_{C^c} \pi(dx)P(x, C)\right]. \tag{19.62}$$

Secondly, we have the balanced-flow equation

$$
\begin{aligned}
\int_C \pi(dx) P(x, C^c) &= \int_C \pi(dx)[1 - P(x, C)] \\
&= \pi(C) - \int_C \pi(dx) P(x, C) \\
&= \int_{\mathsf{X}} \pi(dx) P(x, C) - \int_C \pi(dx) P(x, C) \\
&= \int_{C^c} \pi(dx) P(x, C).
\end{aligned}
\tag{19.63}
$$

Putting this into the strict inequality in (19.62), we have that

$$
\int_C \pi(dw) \Delta V(w) > 0
\tag{19.64}
$$

provided that V does not vanish on C. If V does vanish on C, then (19.64) holds automatically.

But now, under (19.46) we have $\int \pi(dx)\Delta V(x) = 0$ from (19.51), and so (19.56) is a consequence of this and (19.64). Since $\Delta V(y)$ is bounded under (19.46), (19.57) is actually identical to (19.56) and the theorem is proved. $\qquad\square$

These results show that for a wide class of functions, our criteria for positivity and nullity, given respectively in Section 11.3 and Section 11.5.1, are essentially the two extreme cases of this mixed drift result. We conclude with an example where similar mixed behavior may be exhibited quite explicitly.

19.3.2   A mixed drift criterion for stability of the ladder chain

We return to the ladder chain defined by (10.37). Recall that the structure of the stationary measure, when it exists, is known to have an operator-geometric form as in Section 10.5.3. Here we consider conditions under which such a stationary measure exists.

If we assume that the zero-level transitions have the form

$$
\Lambda_i^*(x, A) = P(i, x; 0, A) = \sum_{j=k+1}^{\infty} \Lambda_j(x, A)
\tag{19.65}
$$

so that there is a greater degree of homogeneity than in the general model, then the operator

$$
\Lambda(x, A) := \sum_{j=0}^{\infty} \Lambda_j(x, A)
$$

is stochastic.

Thus $\Lambda(x, A)$ defines a Markov chain $\mathbf{\Phi}^\Lambda$, which is the marginal position of $\mathbf{\Phi}$ ignoring the actual rung: by direct calculation we can check that for any B

$$
P^n(i, x; \mathbb{Z}_+ \times B) = \Lambda^n(x, B).
\tag{19.66}
$$

Moreover, (19.66) immediately gives that if $\mathbf{\Phi}$ is ψ-irreducible, then $\mathbf{\Phi}^\Lambda$ is ψ^*-irreducible, where $\psi^*(B) = \psi(\mathbb{Z}_+ \times B)$.

Now define, for any $w \in \mathsf{X}$, the expected change in ladder height by

$$\beta(w) = \sum_{j=0}^{\infty} j\Lambda_j(x,\mathsf{X}) : \tag{19.67}$$

if $\beta(w) > 1 + \delta$ for all w then, exactly as in our analysis of the random walk on a half line, we have that

$$\mathsf{E}_{(i,w)}[\tau_C] < \infty$$

for all $i > M, w \in \mathsf{X}$, where $C = \cup_0^M \{j \times \mathsf{X}\}$ is the "bottom end" of the ladder.

But one might not have such downwards drift uniform across the rungs. The result we prove is thus an average drift criterion.

Theorem 19.3.4. *Suppose that the chain $\boldsymbol{\Phi}$ is ψ-irreducible and has the structure (19.65). If the marginal chain $\boldsymbol{\Phi}^\Lambda$ admits an invariant probability measure ν such that*

$$\int \nu(dw)\beta(w) > 1, \tag{19.68}$$

then $\boldsymbol{\Phi}$ admits an invariant probability measure π.

PROOF   The proof is similar to that of Theorem 19.3.1, but we do not assume boundedness of the drifts so we must be a little more delicate. Choosing $V(i,w) = i$, we have first that

$$\Delta V(i,w) = 1 - \sum_{j=0}^{i} j\Lambda_j(x,\mathsf{X}) - (i+1) \sum_{j=i+1}^{\infty} \Lambda_j(x,\mathsf{X});$$

note that in particular for $i > d$ this gives

$$\Delta V(i,w) \le \Delta V(d,w), \qquad w \in \mathsf{X}. \tag{19.69}$$

Now even though (19.46) is not assumed, because $|\Delta V(i,w)| \le d+1$ for $i \le d$ and because, starting at level i, after k steps the chain cannot be above level $i+k$, we see exactly as in proving (19.50) that

$$\liminf n^{-1} \sum_{k=1}^{n} \int \sum_j P^k(i,x;j \times dy)\Delta V(j;y) \ge 0. \tag{19.70}$$

We now show that this average non-negative drift is not possible under (19.68), unless the chain is positive.

From (19.68) we have

$$0 > \lim_{k \to \infty} \int \nu(dw)\Delta V(k,w). \tag{19.71}$$

Choose d sufficiently large that

$$0 > \int \nu(dw)\Delta V(d,w). \tag{19.72}$$

Further truncate by choosing $v \geq 1$ large enough that if $D_v = \{y : \Delta V(d, y) \geq -v\}$ then, using (19.72),

$$0 > \int_{D_v} \nu(dw)\Delta V(d, w). \tag{19.73}$$

Now decompose the left hand side of (19.70) as

$$n^{-1}\sum_{k=1}^{n}\int_{\mathsf{X}}\sum_{j}P^k(i, x; j \times dy)\Delta V(j, y)$$

$$= n^{-1}\sum_{k=1}^{n}\int_{\mathsf{X}}\sum_{j=0}^{d-1}P^k(i, x; j \times dy)\Delta V(j, y)$$

$$+ n^{-1}\sum_{k=1}^{n}\int_{\mathsf{X}}\sum_{j\geq d}P^k(i, x; j \times dy)\Delta V(j, y)$$

$$\leq n^{-1}\sum_{k=1}^{n}d\sum_{j=0}^{d-1}P^k(i, x; j \times \mathsf{X})$$

$$+ n^{-1}\sum_{k=1}^{n}\int_{D_v}\sum_{j\geq d}P^k(i, x; j \times dy)\Delta V(j, y) \tag{19.74}$$

since on D_v^c we have $\Delta V(d, y) \leq -1$.

Assume the chain is not positive: we now show that (19.74) is strictly negative, and this provides the required contradiction of (19.70).

We know from Theorem 18.2.2 that there exists a sequence C_n of null sets with $C_n \uparrow \mathbb{Z}_+ \times \mathsf{X}$.

In fact, in this model we now show that every rung is such a null set. Fix a rung $j \times \mathsf{X}$, and let $C_n(j) = C_n \cap j \times \mathsf{X}$. Since Φ is assumed ψ^*-irreducible with an invariant probability measure ν, we have from the ergodic theorem (13.62) that for ψ^*-a.e. x and any M,

$$\lim n^{-1}\sum_{k=1}^{n}\Lambda^k(x, C_M(j)) = \nu(C_M(j)).$$

Choose M so large that $\nu(C_M(j)) \geq 1 - \varepsilon$ for a given $\varepsilon > 0$. Then we have

$$\lim n^{-1}\sum_{k=1}^{n}P^k(i, x; j \times \mathsf{X}) = \lim n^{-1}\sum_{k=1}^{n}P^k(i, x; j \times C_M(j))$$

$$+ \lim n^{-1}\sum_{k=1}^{n}P^k(i, x; j \times [C_M(j)]^c)$$

$$\leq \lim n^{-1}\sum_{k=1}^{n}P^k(i, x; C_M)$$

$$+ \lim n^{-1}\sum_{k=1}^{n}\Lambda^k(x, [C_M(j)]^c)$$

$$\leq \varepsilon \tag{19.75}$$

which shows the rung $j \times \mathsf{X}$ to be null as claimed.

Using (19.75) we have in particular that for any B, and d as above,

$$
\begin{aligned}
\nu(B) &= \lim n^{-1} \sum_{k=1}^{n} \Lambda^k(x, B) \\
&= \lim n^{-1} \sum_{k=1}^{n} \sum_{j=0}^{d-1} P^k(i, x; j \times B) \\
&\quad + \lim n^{-1} \sum_{k=1}^{n} \sum_{j=d}^{\infty} P^k(i, x; j \times B) \\
&= \lim n^{-1} \sum_{k=1}^{n} \sum_{j=d}^{\infty} P^k(i, x; j \times B).
\end{aligned}
\tag{19.76}
$$

We now use (19.75) and (19.76) in (19.74). This gives, successively,

$$
\liminf_{n \to \infty} n^{-1} \sum_{k=1}^{n} \int_{\times} \sum_{j} P^k(i, x; j \times dy) \Delta V(j, y)
$$

$$
\leq \quad \liminf_{n \to \infty} n^{-1} \sum_{k=1}^{n} \int_{D_v} \sum_{j \geq d} P^k(i, x; j \times dy) \Delta V(j, y)
$$

$$
= \quad \int_{D_v} \nu(dy) \Delta V(j, y) < 0
$$

from the construction in (19.73).

This is the required contradiction of (19.70) and we are finished. $\square$

It is obviously of interest to know whether the same average drift condition suffices for positivity when (19.65) does not hold.

In general, this is a subtle question. Writing as before $[0] = 0 \times \mathsf{X}$, we obviously have that under (19.68)

$$
\mathsf{E}_{0,y}[\tau_{[0]}] < \infty
\tag{19.77}
$$

for ν-a.e. y, since this quantity does not depend on the detailed hitting distribution on $[0]$. But although this ensures that the process on $[0]$ is well defined, it does not even ensure that it is recurrent.

As an example of the range of behaviors possible, let us take $\mathsf{X} = \mathbb{Z}_+$ also, and consider a chain that can move only up one rung or down one rung: specifically, choose $0 < p, q < 1$ and

$$
\begin{aligned}
\Lambda_0(x, x-1) &= pq, & x &\geq 1, \\
\Lambda_0(x, x+1) &= (1-p)q, & x &\geq 0, \\
\Lambda_2(x, x-1) &= p(1-q), & x &\geq 1, \\
\Lambda_2(x, x+1) &= (1-p)(1-q), & x &\geq 0,
\end{aligned}
\tag{19.78}
$$

with the transitions on the boundary given by

$$
\begin{aligned}
\Lambda_0(0,0) &= pq, \\
\Lambda_2(0,0) &= p(1-q).
\end{aligned}
\tag{19.79}
$$

The marginal chain $\mathbf{\Phi}^\Lambda$ is a random walk on the half line $\{0, 1, \ldots\}$ with an invariant measure ν if and only if $p > 1/2$. On the other hand, $\beta(x) > 1$ if and only if $q < 1/2$. Thus (19.68) holds if $q < 1/2 < p$.

This chain falls into the class that we have considered in Theorem 19.3.4; but other behaviors follow if we vary the structure at the bottom rung.

Let us then specify the boundary conditions in a manner other than (19.65): put $\Lambda_1^*(x, x-1) = p(1-q)$ and $\Lambda_1^*(x, x+1) = (1-p)(1-q)$ but

$$
\begin{array}{rclcl}
\Lambda_0^*(x, x-1) & = & r(1-q), & x & \geq & 1, \\
\Lambda_0^*(x, x+1) & = & (1-r)(1-q), & x & \geq & 1,
\end{array}
\tag{19.80}
$$

where $0 < r < 1$.

Consider now the expected increments in the chain $\mathbf{\Phi}^{[0]}$ on $[0]$. By considering whether the chain leaves $[0]$ or not we have for all $x \geq 1$

$$
\mathsf{E}[\Phi_n^{[0]} \mid \Phi_{n-1}^{[0]} = x] - x \geq (1-2r)(1-q) + (1-2p)\Big(\frac{1-q}{1-2q}+1\Big)q :
\tag{19.81}
$$

here the second term follows since, on an excursion from $[0]$, the expected drift to the left at every step is no more than $(1-2p)$ independent of level change, and the expected number of steps to return to $[0]$ from $1 \times \mathsf{X}$ is $(1-q)/(1-2q)$.

From (19.81) we therefore have that the chain $\mathbf{\Phi}^{[0]}$ is transient if r and q are small enough, and $p - 1/2$ is not too large.

This example shows the critical need to identify petite sets and the return times to them in classifying any chain: here we have an example where the set $[0]$ is not petite, although it has many of the properties of a petite set. Yet even though we have (19.77) proven, we do not even have enough to guarantee the chain is recurrent.

19.3.3　Stability of the GI/G/1 queue

We saw in Section 3.5 that with appropriate choice of kernels the ladder chain serves as a model for the GI/G/1 queue. We will use the average drift condition of Theorem 19.3.4 to derive criteria for stability of this model.

Of course, in this case we do not have (19.65), and the example at the end of the last section shows that we cannot necessarily deduce anything from (19.68).

In this case, however, we have as in Section 10.5.3 that $[0]$ is petite, and that the process on $[0]$, if honest, has invariant measure H where H is the service time distribution. If we can satisfy (19.68), then it follows from (19.77) that the process on $[0]$ is indeed honest, and we only have to check further that

$$
\int H(dy)\mathsf{E}_{0,y}[\tau_{[0]}] < \infty
\tag{19.82}
$$

to derive positivity.

We conclude by proving through this approach a result complementing the result found in quite another way in Proposition 11.4.4.

Theorem 19.3.5. *The GI/G/1 queue with mean inter-arrival time λ and mean service time μ satisfies (19.68) if and only if $\lambda > \mu$, and in this case the chain has an invariant measure given by (10.52).*

PROOF   From the representations (3.42) and (3.43), we have the kernel

$$\Lambda(x, [0, y]) = \int_0^\infty G(dt) P^t(x, y)$$

where $P^t(x, y) = \mathsf{P}(R_t \leq y \mid R_0 \leq y)$ is the forward recurrence time process in a renewal process $N(t)$ generated by increments with distribution H.

Since H has finite mean μ, we know from (10.36) that $P^\delta(x, y)$ has invariant measure

$$\nu[0, x] = \mu^{-1} \int_0^x [1 - H(x)] dx$$

for every δ: thus ν is also invariant for Λ.

On the other hand, from (3.42),

$$\begin{aligned}
\beta(x) &= \sum_{n=0}^\infty n \Lambda_n(x, [0, \infty)) \\
&= \sum_{n=0}^\infty n \int G(dt) P_n^t(x, \infty) \\
&= \int G(dt) \mathsf{E}[N(t) \mid R_0 = x].
\end{aligned}$$

The stationarity of ν for the renewal process $N(t)$ shows that

$$\int_0^\infty \nu(dx) \mathsf{E}[N(t) \mid R_0 = x] = t/\mu$$

and so by Fubini's Theorem, we therefore have

$$\begin{aligned}
\int \nu(dx) \beta(x) &= \int_0^\infty \left[\int_0^\infty \nu(dx) \mathsf{E}[N(t) \mid R_0 = x] \right] G(dt) \\
&= \int_0^\infty [t/\mu] G(dt) \qquad\qquad\qquad\qquad (19.83) \\
&= \lambda/\mu
\end{aligned}$$

which proves the first part of the theorem.

To conclude, we note that in this particular case, we know more about the structure of $\mathsf{E}_{0,y}[\tau_{[0]}]$, and this enables us to move from the case where (19.65) holds. Given the starting configuration $(0, y)$, let n_y denote the number of customers arriving in the first service time y: if $\eta(\leq \infty)$ denotes the expected number of customers in a busy period of the queue, then by using the trick of rearranging the order of service to deal with each of the identical n_y "busy periods" generated by these customers separately, we have the linear structure

$$\mathsf{E}_{0,y}[\tau_{[0]}] = 1 + \mathsf{E}_{0,y}[n_y \eta] = 1 + \eta \sum_{n=0}^\infty G^{n*}[0, y]. \qquad (19.84)$$

As in (19.77), we at least know that since (19.68) holds, the left hand side of this equation is finite, so that $\eta < \infty$. Moreover, from the Blackwell Renewal Theorem

(Theorem 14.5.1) we have for any ε and large y

$$\sum_{n=0}^{\infty} G^{n*}[0, y] \leq y[\lambda^{-1} + \varepsilon] \qquad (19.85)$$

so that, finally, (19.82) follows from (19.84), (19.85), and the fact that the mean of H is finite. □

19.4　Commentary*

Despite the success of the simple drift, or Foster–Lyapunov, approach there is a growing need for more subtle variations such as those we present here.

There are several cases in the literature where the analysis of state-dependent (or at least not simple one-step) drift appears unavoidable: see Tjøstheim [386] or Chen and Tsay [66], where m-step skeletons $\{\Phi_{mk}\}$ are analyzed. Analysis of this kind is simplified if the various parts of the space can be considered separately as in Section 19.1.2.

In the countable space context, Theorem 19.1.1 was first shown as Theorem 1.3 and Theorem 19.1.2 as Theorem 1.4 of Malyšev and Men'šikov [243]. Their proofs, especially of Theorem 19.1.2, are more complex than those based on sample path arguments, which were developed along with Theorem 19.1.3 in [283]. As noted there, the result can be extended by choosing $n(x)$ as a random variable, conditionally independent of the process, on $\mathbb{Z}_+$. In the special case where $n(x)$ has a uniform distribution on $[1, n]$ independent of x, we get a time-averaged result used by Meyn and Down [273] in analyzing stability of queueing networks. If the variable has a point mass at $n(x)$ we get the results given here.

Models of random walk on the orthant in Section 19.1.2 have been analyzed in numerous different ways on the integer quadrant $\mathbb{Z}_+^2$ by, for example, [244, 257, 243, 340, 109]. Much of their work pertains to more general models which assume different drifts on the boundary, thus leading to more complex conditions. In [244, 257, 243] it is assumed that the increments are bounded (although they also analyze higher dimensional models), whilst in [340, 109] it is shown that one can actually choose $n = 1$ if a quadratic function is used for a test function, whilst weakening the bounded increments assumption to a second moment condition: this method appears to go back to Kingman [207].

As we have noted, positive recurrence in the simple case illustrated here could be established more easily given the independence of the two components. However, the bound using linear functions in (19.25) seems to be new, as does the continuous space methodology we use here.

The antibody model here is based on that in [283]. The attack pattern of the "invaders" is modeled to a large extent on the rabies model developed in Bartoszyński [19], although the need to be the same order of magnitude as the antibody group is a weaker assumption than that implicit in the continuous time continuous space model there.

The results in Section 19.2 are largely taken from Meyn and Tweedie [277]: they appear to give a fruitful approach to more complex models, and the seeming simplicity of the presentation here is largely a function of the development of the methods based on Dynkin's formula for the non-time-varying case. An application to adaptive control is given in Meyn and Guo [274], where drift functions which depend on the whole history of the chain are used systematically. Regrettably, examples using this approach are typically too complex to present here.

The dependent parameter bilinear time series model is analyzed in [275], from which we adopt the proof of Theorem 19.2.6. In Karlsen [195] a decoupling inequality of [210] is used to obtain a second order stationary solution in the Gaussian parameter case, and Brandt [44] provides a simple argument, similar to the proof of Proposition 17.3.4, to obtain boundedness in probability for general bilinear time series models with stationary coefficients.

Results on mixed drifts, such as those in Section 19.3.1, have been discovered independently several times.

Although Neuts [292] analyzed a two-drift chain in detail, on a countable space the first approach to classifying chains with different drifts appears to be due to Marlin [249]. He considered the special case of $V(x) = x$ and assumed a fixed finite number of different drifts. The form given here was developed for countable spaces by Tweedie [396] (although the proof there is incomplete) and Rosberg [336], who gives a slightly different converse statement. A general state space form is in Tweedie [398].

The condition (19.53) for the converse result to hold, and which also suffices to ensure that $\Delta V(w) \geq 0$ on C^c implies non-positivity, is known as Kaplan's condition [193]: the general state space version sketched here is adapted from a countable space version in [349]. Related results are in [380].

The average mean drift criterion for the ladder process in Section 19.3.2 is due to Neuts [293] when the rungs are finite, and is proved there by matrix methods: the general result is in [399], and (19.68) is also shown there to be necessary for positivity under reasonable assumptions.

The final criterion for stability of the GI/G/1 queue produced by this analysis is of course totally standard [9]: that the very indirect Markovian approach reproduces this result exactly brings us to a remarkably reassuring conclusion.

Added in second printing: In the past year, Dai has shown in [80] that the state-dependent drift criterion Theorem 19.1.2 leads to a new approach to the stability of stochastic queueing network models via the analysis of a simpler deterministic fluid model. Related work has been developed by Chen [65] and Stolyar [373], and these results have been strengthened in Dai and Weiss [82] and Dai and Meyn [81].

Commentary for the second edition: Over the past ten years there have been many further improvements in the theory surrounding the multi-step drift criterion for stability within specific applications. Applications include stochastic approximation [40, 39], Markov chain Monte Carlo (MCMC) [100], as well as stochastic networks [267], which was the original motivation for the technique in [243].

Chapter 20

Epilogue to the second edition

Following publication of the "Big Red Book" in the early nineties, Richard and I devoted more attention to applications. Each of us became interested in simulation, albeit in entirely different contexts. In addition, Richard spent more of his time on topics in statistics, and I became increasingly involved in topics surrounding control and performance evaluation for networks.

Personally, I thought that I would abandon Markov chains as a research topic. This, fortunately, has turned out to be an impossible task!

The three sections that follow can be regarded as proposals for future monographs that will never be written by either of us. The first section comprises our biggest thrust shortly after the book was complete, along with my own view of geometric ergodicity and spectral theory. The second section describes how methods in this book can be applied to construct and analyze simulation algorithms. The final section explains how theory in continuous time can be generated from discrete time counterparts.

20.1   Geometric ergodicity and spectral theory

The weighted supremum norm $\| \cdot \|_V$ that forms a foundation of this subject was brought to our attention by Arie Hordijk. In his paper [163], co-authored with Floske Spieksma, they establish a version of the Geometric Ergodic Theorem for countable state space models. This technique had tremendous influence on our research during the writing of the first edition, and in subsequent research.

The Geometric Ergodic Theorem and refinements developed in Chapter 16 have found application in many fields. Examples include numerical integration [250], statistics [60], machine learning [217, 135], Markov decision theory [41], and economics [154, 254]. This ergodic theorem and other ideas in Chapter 16 have proved valuable in the development of various theoretical aspects of Markov chains. Some of these ideas are surveyed in the discussion that follows.

20.1.1   The spectrum of P

One topic missing from the first edition is the relationship between ergodic theory and the spectral properties of the transition kernel P that defines the chain (see (1.1)).

Among the many applications of spectral theory is the identification of the rate of convergence in the Geometric Ergodic Theorem. One elegant bound of Diaconis and Stroock is described below in (20.6). Spectral theory and surrounding techniques are also used to construct finite-rank approximations of a transition kernel. These take the form

$$\widehat{P} = \sum_{i=1}^{n} s_i \otimes \mu_i, \qquad (20.1)$$

where for a function r and measure μ we define $[r \otimes \mu](x, dy) := r(x)\mu(dy)$. In most cases we restrict to an L_∞^V setting. In this case it is assumed that $\widehat{P}$ is a bounded linear operator on L_∞^V, which amounts to the inclusions $\{s_i\} \subset L_\infty^V$ and $\{\mu_i\} \subset \mathcal{M}_1^V$ (i.e., V is μ_i-integrable for each i).

For any complex number $z \in \mathbb{C}$ we denote

$$T_z := [Iz - P]^{-1}, \qquad Z_z := [Iz - P + 1 \otimes \pi]^{-1}, \qquad (20.2)$$

provided the inverse exists as a bounded linear operator on L_∞^V. This is true whenever $|z| > 1$ and $\|P^n\|_V$ is uniformly bounded in n since we can express the inverses as power series

$$T_z = \sum_{n=0}^{\infty} z^{-n-1} P^n, \qquad Z_z = \sum_{n=0}^{\infty} z^{-n-1} [P - 1 \otimes \pi]^n.$$

Hence these kernels generalize the resolvents K_{a_ε} and U_h, defined in (3.26) and (12.13), respectively.

We omit the subscript on Z_z when $z = 1$. In this special case, $Z = Z_1$ is known as the *fundamental kernel*. In the proof of Theorem 17.4.2 we noted that the function $\hat{g} := Zg$ is a solution to Poisson's equation (17.37) for any $g \in L_\infty^V$.

This kernel T_z defined in (20.2) is used to define the spectrum of the kernel P:

Spectrum for a Markov chain

(i) The *spectrum* $S_V(P)$ is the set of all $z \in \mathbb{C}$ such that the operator T_z defined in (20.2) *does not exist* as a bounded linear operator on L_∞^V.

(ii) $z_0 \in S_V(P)$ is a *pole of (finite) multiplicity* n if it is an isolated point in $S_V(P)$ and the associated projection operator

$$\widehat{P}_{z_0} := \lim_{\varepsilon \to 0} \frac{1}{2\pi i} \int_{\{z:|z-z_0|=\varepsilon\}} [Iz - P]^{-1} dz$$

can be expressed as a finite-rank operator on L_∞^V of the form (20.1).

(iii) There is a *spectral gap* in L_∞^V if 1 is an isolated point in the spectrum with finite multiplicity.

(iv) The spectrum is said to be *discrete* in L_∞^V if the set $\{z : z \in S_V(P), |z| \geq \varepsilon\}$ is finite for each $\varepsilon > 0$ and contains only poles of finite multiplicity.

Spectral theory for Markov chains arises in a finer analysis of the Geometric Ergodic Theorem and in the theory of large deviations. Just as in the theory of linear systems and finite state space Markov chains, the dynamics of the chain can be understood through an analysis of the spectrum of P. We survey these ideas next.

20.1.2   Rates of convergence and eigenvalues

The rate of convergence in the Geometric Ergodic Theorem is intimately related to the spectrum of P. To see this, recall that a (possibly complex) number $\lambda \in S_V(P)$ is called an *eigenvalue* if there exists $h \in L_\infty^V$ satisfying $Ph = \lambda h$ and $\pi(|h|) \neq 0$. In this case h is called an *eigenfunction*.   Provided $|\lambda| < 1$ and $\pi(h) = 0$, we obtain an exact bound on the rate of convergence to steady state when h is an eigenfunction:

$$|P^n h - \pi(h)| = |\lambda|^n |h|.$$

This observation can be strengthened. If the state space is finite, it is known that the rate of convergence to equilibrium is determined by the second largest eigenvalue, and this result can be generalized to obtain bounds on the rate of convergence in the Geometric Ergodic Theorem.

We have the following general result that follows from ideas in [282, 218].   The constant $\lambda_* \in (0, 1)$ appearing in (20.3) is called the *spectral radius* of $P^n - 1 \otimes \pi$.

Theorem 20.1.1. *Suppose that* Φ *is a Markov chain on a general state space.*

(i) *If Φ is V-uniformly ergodic, then there is a spectral gap in L_∞^V, and there exists $\varepsilon_0 < 1$ such that the inverse operator Z_z exists as a bounded linear operator on L_∞^V for every $z \in \mathbb{C}$ satisfying $|z| > \varepsilon_0$.*

(ii) *Conversely, suppose that there exists $\varepsilon_0 < 1$ such that $\|Z_z\|_V < \infty$ for every $z \in \mathbb{C}$ satisfying $|z| > \varepsilon_0$. Then Φ is V-uniformly ergodic, and the rate of convergence is bounded as follows:*

$$\lim_{n \to \infty} n^{-1} \log(\|P^n - 1 \otimes \pi\|_V) \leq \log(\lambda_*) \tag{20.3}$$

where $\lambda_* < 1$ is the minimum value of ε_0, which coincides with the minimal bound on the spectrum within the open unit disk: $\lambda_* = \max\{|z| : z \neq 1, z \in \mathcal{S}_V(P)\}$.

PROOF   For (i) we begin with the proof that Z_z is defined for this range of z.
Under the assumptions of (i) we have for some $R < \infty$, $r > 1$,

$$\|P^n - 1 \otimes \pi\|_V \leq Rr^{-n}, \qquad n \geq 0. \tag{20.4}$$

In this case the power series representation is justified,

$$Z_z = \sum_{n=0}^\infty z^{-n-1}[P - 1 \otimes \pi]^n = \sum_{n=0}^\infty z^{-n-1}[P^n \quad 1 \otimes \pi].$$

The triangle inequality gives the bound

$$\|Z_z\|_V \leq \sum_{n=0}^\infty |z|^{-n-1} \|P^n - 1 \otimes \pi\|_V,$$

which when combined with the sequence of bounds given in (20.4) gives

$$\|Z_z\|_V \leq R \sum_{n=0}^\infty |z|^{-n-1} r^{-n} = R|z|^{-1} \frac{1}{1 - |z|^{-1} r^{-1}}$$

for $|z| > r^{-1}$.
We now turn to T_z. The two inverses are related: whenever Z_z is defined we can express T_z in terms of Z_z,

$$[Iz - P]^{-1} = [(Iz - P + 1 \otimes \pi) - 1 \otimes \pi]^{-1} = Z_z[I - 1 \otimes \pi Z_z]^{-1}.$$

From the definition (20.2) we also have $\pi Z_z = Z_z \pi = z^{-1} \pi$. Consequently, provided $z \neq 1$,

$$T_z = [Iz - P]^{-1} = Z_z[I - z^{-1} 1 \otimes \pi]^{-1} = Z_z[I + (z-1)^{-1} 1 \otimes \pi].$$

This representation completes the proof of (i).
To prove (ii) we first observe that the norm $\|Z_z\|_V$ is continuous on the set $\{z : |z| > \lambda_*\}$. Consequently, for each $\varepsilon > \lambda_*$ there exists $B_\varepsilon < \infty$ such that $\|Z_z\|_V \leq B_\varepsilon$ when $|z| = \varepsilon$. For any $f \in L_\infty^V$ and $n \geq 1$ we denote

$$f_n(x) := \frac{1}{2\pi i} \int_0^{2\pi} \left(Z_{\varepsilon e^{i\theta}} f(x)\right) e^{-in\theta} \, dz, \qquad x \in \mathsf{X}.$$

The bound $\|Z_z\|_V \leq B_\varepsilon$ implies that $\|f_n\|_V \leq B_\varepsilon \|f\|_V$. In fact we can identify this function as

$$f_n(x) = \varepsilon^{-n-1}(P_.^n f(x) - \pi(f)).$$

We conclude that $\|P^n f - \pi(f)\|_V \leq B_\varepsilon \|f\|_V \varepsilon^{n+1}$, and hence

$$\lim_{n \to \infty} n^{-1} \log(\|P^n - 1 \otimes \pi\|_V) \leq \log(\varepsilon).$$

This establishes (20.3) since $\varepsilon > \lambda_*$ was arbitrary.   □

While elegant, the limit (20.3) tells us nothing about the actual error $|P^n f(x) - \pi(f)|$ for a given finite n. Elegant bounds are available for chains that are *reversible*.

A Markov chain with invariant measure π is called reversible if the statistics of the stationary process on the two-sided time interval are invariant under time reversal:

$$\{\Phi_t : t \in \mathbb{R}\} \overset{\text{dist}}{=} \{\Phi_{-t} : t \in \mathbb{R}\}.$$

Applying the Markov property, it can be shown that this invariance holds if and only if the bivariate distributions are insensitive to time reversal in steady state:

$$(\Phi_t, \Phi_{t+1}) \overset{\text{dist}}{=} (\Phi_{t+1}, \Phi_t).$$

The bivariate distributions can be identified, leading to the more standard definition: Φ is reversible if the *detailed balance equations* hold

$$\pi(dx)P(x, dy) = \pi(dy)P(y, dx). \tag{20.5}$$

For a reversible chain with finite state space each of the eigenvalues is real. Diaconis and Stroock in [90] obtain bounds on the second largest eigenvalue in this setting. A striking conclusion is the following explicit bound on the rate of convergence

$$\|P^n(x, \cdot) - \pi\|_V \leq \sqrt{\tfrac{1-\pi(x)}{\pi(x)}} \lambda_*^n, \tag{20.6}$$

where λ_* is the magnitude of the second largest eigenvalue, $\lambda_* = \max\{|\lambda| : \lambda \neq 1\}$, and $V \equiv 1$. Bounds on the rate of convergence for chains that are not necessarily reversible are obtained in [119], again in the finite state space case. The bounds are based on spectral theory, but the spectrum of the symmetrized kernel $P\widetilde{P}$ is considered, where $\widetilde{P}$ is the transition kernel for the time-reversed chain.

See Diaconis and Saloff-Coste [89], Rosenthal's survey [344], and Baxendale [21] for bibliographies and further generalizations and improvements since 1996.

Spectral theory is often cast in a Hilbert space setting in the space $L_2(\pi)$, defined as the set of all measurable functions f on X satisfying $\pi(f^2) < \infty$. For arbitrary $p \geq 1$, the $L_p(\pi)$ norm of a function $f \colon \mathsf{X} \to \mathbb{R}$ is defined by $\|f\|_p = \sqrt[p]{\pi(|f|^p)}$. It is natural to extend the definitions of spectrum and spectral gap to the $L_p(\pi)$ norm. In particular, the chain is called *geometrically ergodic in* $L_p(\pi)$ if there exist $r > 1$ and $R < \infty$ such that for all $n \in \mathbb{Z}_+$, $f \in L_p(\pi)$,

$$\|P^n f - \pi(f)\|_p^p := \int |P^n f(x) - \pi(f)|^p \, \pi(dx) \leq R r^{-n} \|f\|_p^p. \tag{20.7}$$

In operator-theoretic language, (20.7) means that the spectral radius of $P - 1 \otimes \pi$ (in the induced operator norm) is strictly less than unity.

An application of Theorem 15.0.1 shows that, for any p, if the chain is geometrically ergodic in $L_p(\pi)$ then it is also V-uniformly ergodic for some $V : \mathsf{X} \to [1, \infty]$, finite a.e. $[\pi]$. A few other relationships between these different notions of ergodicity are summarized in the following.

Proposition 20.1.2. *Each of the following statements refers to a positive recurrent ψ-irreducible Markov chain with invariant probability measure π:*

(i) *There is a reversible Markov chain that is V-uniformly ergodic but not geometrically ergodic in $L_1(\pi)$.*

(ii) *There is a Markov chain that is V-uniformly ergodic but not geometrically ergodic in $L_2(\pi)$.*

(iii) *If the chain is reversible, then it is geometrically ergodic in $L_2(\pi)$ if and only if it is V-uniformly ergodic for some $V : \mathsf{X} \to [1, \infty]$, finite a.e. $[\pi]$.*

PROOF   The M/M/1 queue provides an example in (i). Proposition 20.1.3 that follows demonstrates that this model is V-uniformly ergodic and reversible provided a "load condition" holds. It is shown that there is a constant $\underline{\lambda} \in (0, 1)$ such that *every* $\lambda \in [\underline{\lambda}, 1)$ is an eigenvalue, with corresponding eigenfunction $h \in L_\infty^V$, and hence also $h \in L_1(\pi)$. The eigenfunction property gives $\pi(h) = 0$ and $\|P^n h - \pi(h)\|_1 = \lambda^n \|h\|_1$ for each n. This rules out (20.7) for a *fixed* $r > 1$.

To prove (ii), note first that if the chain is geometrically ergodic in $L_2(\pi)$, then (20.7) combined with the definition (17.4) gives the following bound on the asymptotic variance:

$$\gamma_f^2 \leq R \frac{r+1}{r-1} \|f\|_2^2 < \infty, \qquad f \in L_2(\pi).$$

In particular, the asymptotic variance is finite whenever the ordinary variance is finite. Häggström in [150] gives an example of a V-uniformly ergodic Markov chain and a function f such that $f \in L_2(\pi)$, yet the asymptotic variance is not finite.

Part (iii) is established by Roberts and Rosenthal in [329].   □

20.1.3   The spectrum of the M/M/1 queue

The M/M/1 queue is the simplest queueing model in which service times and inter-arrival times are exponentially distributed. Following sampling, as described in Section 2.4.2, the discrete time model is identical to the random walk on the half line (1.7) in which the marginal distribution of W is supported on the two points ± 1. The following discussion is based on discussion in [282].

The Markov transition matrix can be expressed for any function h by

$$Ph(n) = ph(n+1) + (1-p)h(n-1), \qquad n \in \mathsf{X} := \{0, 1, 2, \ldots\}, \tag{20.8}$$

where p denotes the probability that W_k is equal to one. In the special case $n = 0$ we set $h(n-1) = h(-1) = h(0)$ to make this formula consistent with the dynamics

(1.7). We have seen in Section 16.1.3 that this chain is V-uniformly ergodic provided $\rho := p/(1-p) < 1$. We can take $V(n) = r_0^n$, $n \geq 0$, for any $r_0 \in (1, \rho^{-1})$. It follows from Theorem 20.1.1 that there is a spectral gap. The unique invariant measure is geometric, $\pi(n) = (1 - \rho)\rho^n$, $n \geq 0$. The detailed balance equations (20.5) are easily verified, so that we can conclude that the M/M/1 queue is reversible.

We next consider the spectrum in L_∞^V with this V, where $r_0 \in (\rho^{-\frac{1}{2}}, \rho^{-1})$ is fixed. We find in Proposition 20.1.3 that the spectrum is not discrete even when the model admits a spectral gap.

The structure of eigenfunctions can be identified through the form of the transition law (20.8). This expression suggests the application of transform techniques: define for any complex z,

$$H(z) = \sum z^{-n} h(n).$$

This is defined for $|z| > r_0^{-1}$ whenever $h \in L_\infty^V$. If h is an eigenfunction, then on taking transforms of each side of the eigenfunction equation $Ph = \lambda h$, and applying (20.8), we find that H can be expressed as the ratio of quadratic functions. If the roots of the denominator are distinct, then H can be expressed as the sum of two simpler rational functions

$$H(z) = c_1 \frac{z}{z - \beta_1} + c_2 \frac{z}{z - \beta_2}$$

for constants c_1, c_2, where β_1, β_2 are the poles of H.

Proposition 20.1.3. *Suppose that $\rho < 1$, fix $r_0 \in (\rho^{-\frac{1}{2}}, \rho^{-1})$, and define $V(n) = r_0^n$ for $n \geq 0$. Then*

(i) *The queue is V-uniformly ergodic.*

(ii) *For each $\beta \in (\rho^{-\frac{1}{2}}, r_0)$ an element λ of $\mathcal{S}_V(P)$ is given by $\lambda = p\beta + (1-p)\beta^{-1} \in (0, 1)$. This is also an eigenvalue for P in L_∞^V with eigenfunction given by the difference of scaled geometric series*

$$h(n) = (1 - \beta^{-1})\beta^n - (1 - \beta_-^{-1})\beta_-^{\,n}, \qquad n \in \mathsf{X}, \tag{20.9}$$

where $\beta_- \in (1, \rho^{-\frac{1}{2}})$ is the second solution to the equation $\lambda = p\beta_- + (1-p)\beta_-^{-1}$.

(iii) *As $\beta \downarrow \rho^{-\frac{1}{2}}$ the eigenvalues converge, with $\lambda \downarrow 2\sqrt{p(1-p)}$. This limiting value is also an eigenvalue, with eigenfunction*

$$\underline{h}(n) = (1 - (p^{-1} - 2)n)\rho^{-\frac{1}{2}n}, \qquad n \in \mathsf{X}.$$

PROOF   We have already established (i). To see (ii) and (iii), first observe that the eigenfunction equation $Ph(n) = \lambda h(n)$ holds for $n \geq 1$ for the function $h(n) = \beta^n$, regardless of the value of β, with $\lambda = p\beta + (1-p)\beta^{-1}$. However, except for the special case $\beta = 1$, no single geometric series defines an eigenfunction.

To cope with the special case $n = 0$ we consider a pair of geometric series to cancel an "error term." Consider $h(n) = \beta^n - c\beta_-^{\,n}$, where β and β_- are given in the proposition. From the foregoing we do have $Ph(n) = \lambda h(n)$ for $n \geq 1$. We choose c to ensure that this also holds with $n = 0$, which gives the unique value $c = (1 - \beta_-^{-1})/(1 - \beta^{-1})$. The resulting function is a scalar multiple of (20.9).

Finally, we note that a unique solution $\beta_- \in (1, \rho^{-\frac{1}{2}})$ exists since the function $g(x) = px + (1-p)x^{-1}$ is convex, tends to infinity for $x \sim 0$ and for $x \sim \infty$ is equal to unity for $x = 1$ and $x = \rho^{-1}$, and attains a unique minimum at $x = \rho^{-\frac{1}{2}}$. $\square$

In conclusion, the M/M/1 queue admits a spectral gap, but the spectrum is not discrete. Moreover, the set of eigenvalues depends on the choice of V, with $\sup\{\lambda : \lambda < 1\} = pr_0 + (1-p)r_0^{-1}$ approaching unity as r_0 increases to the upper bound ρ^{-1}.

We next demonstrate that a suitably strong drift condition implies a discrete spectrum in $L_2(\pi)$ and L_∞^V simultaneously.

20.1.4   Drift criterion for a discrete spectrum

We saw in Theorem 20.1.1 that V-uniform ergodicity implies a spectral gap in L_∞^V. Hence the drift condition (V4) can be regarded as a sufficient condition for a spectral gap for an aperiodic chain.

There is a stronger drift condition that provides a minimal sufficient condition for a discrete spectrum. It is most conveniently expressed in terms of the *nonlinear generator*, defined for measurable functions $G: \mathsf{X} \to \mathbb{R}$ via

$$\mathcal{H}(G) := \log\left(\frac{Pe^G}{e^G}\right). \tag{20.10}$$

The nonlinear generator was introduced by Fleming for Markov models in continuous time [120, 104, 116, 117, 219], following Donsker and Varadhan [96, 97].

The following drift condition is analogous to (V3). A closely related bound is one component of the assumptions used in Donsker and Varadhan's classic papers on the large deviations theory for Markov models [96, 97, 98]. Condition (DV3), together with techniques surveyed in [89], is applied in [58] to bound rates of convergence for a Markov chain.

Drift criterion of Donsker and Varadhan

(DV3)   For a function $f: \mathsf{X} \to [1, \infty)$, a set $C \in \mathcal{B}(\mathsf{X})$, constants $\delta > 0$ and $b < \infty$, and an extended-real-valued function $V: \mathsf{X} \to [0, \infty]$

$$\mathcal{H}(V) \leq -\delta f + b\mathbb{I}_C. \tag{20.11}$$

From the definition of the nonlinear generator the bound (20.11) can be expressed as

$$Pe^V \leq \exp(-\delta f + b\mathbb{I}_C)e^V.$$

If V is bounded on the set C, then this implies a version of (V4):

$$\Delta e^V(x) \leq -\beta e^V(x) + b'\mathbb{I}_C(x), \quad x \in \mathsf{X},$$

where $b' = b + \sup_{x \in C} e^V(x)$ and $1 - \beta = \sup_{x \in \mathsf{X}} e^{-\delta f(x)}$. We have $\beta \leq 1 - e^{-\delta} < 1$ under the assumption that $f \geq 1$.

Consider for example the LSS(F, G) model under the assumptions of Proposition 7.5.3. Assume in addition that the distribution of the disturbance has a "Gaussian tail": $\mathsf{E}[\exp(\varepsilon\|W_k\|^2)] < \infty$ for some $\varepsilon > 0$. One solution to (20.11) is obtained using the quadratic $V(x) = 1 + \varepsilon_0 |x|_M^2$ in which $\varepsilon_0 > 0$ is chosen sufficiently small and the norm $|y|_M^2 := y^\top My$, $y \in \mathbb{R}^n$, is defined in (12.31). In this case the function f can be chosen with linear growth, $f(x) = 1 + |x|_M$.

For the purposes of spectral analysis, (DV3) is used to justify truncation of the transition kernel. Define for $n \geq 1$,

$$\widehat{P}_n := I_{\{C_f(n)\}}P$$

where $C_f(n) := \{x : f(x) \leq n\}$.

Proposition 20.1.4. *Suppose that (20.11) holds with f unbounded. Assume moreover that $C \subset C_f(n)$ for all $n \geq 1$ sufficiently large. Then*

$$\|P - \widehat{P}_n\|_v \to 0, \qquad n \to \infty,$$

where $v := e^V$.

PROOF   Under the assumptions of the proposition we have $Pv \leq e^{-\delta f}v$ on $\{C_f(n)\}^c$ for n sufficiently large. From the definition of the sublevel set this gives

$$I_{\{C_f(n)\}^c}Pv \leq e^{-\delta n}v,$$

so that $\|P - \widehat{P}_n\|_v = \|I_{\{C_f(n)\}^c}P\|_v \leq e^{-\delta n}$. $\square$

Wu, beginning with his 1995 work [409], has developed this truncation technique for establishing large deviations limit theory, as well as a spectral gap in the L_p norm. For recent bibliographies on these methods and other applications, see [141, 147].

The L_∞ setting of Proposition 20.1.4 is the foundation of research with Kontoyiannis in [219, 220]. To illustrate its application consider the discrete state space setting:

Proposition 20.1.5. *Suppose that X is countable and that (20.11) holds for a coercive function f and a finite set C. Then P has a discrete spectrum in L_∞^v.*

The proof of Proposition 20.1.5 follows from Theorem 3.5 of [219]. The idea is that $\widehat{P}_n$ can be expressed as a finite-rank operator, of the form (20.1), and hence its spectrum is finite.

Proposition 20.1.4 implies that P can be approximated by $\widehat{P}_n$ in norm. From the proof we obtain the explicit bound $\|P - \widehat{P}_n\|_v \leq e^{-\delta n}$. It is shown in [219] that the spectrum of P is discrete if it can be approximated by finite-rank kernels in this fashion.

20.1.5   Multiplicative ergodic theory and large deviations

Suppose that $F \colon \mathsf{X} \to \mathbb{R}$ is a measurable function, denote $S_n(F) := \sum_{k=0}^{n-1} F(\Phi_k)$, and consider the expectation

$$\lambda_{n,x}(F) = \mathsf{E}_x\big[\exp\big(S_n(F)\big)\big]. \tag{20.12}$$

Multiplicative ergodic theory is the study of the asymptotics of this quantity for large n. Under suitable conditions on the chain and the function F the multiplicative ergodic theorem holds,

$$\lim_{n \to \infty} \frac{1}{n} \log(\lambda_{n,x}(F)) = \Lambda(F),$$

where the *limiting log-moment generating function* $\Lambda(F)$ is independent of the initial condition x.

To place this problem within the context of spectral theory, introduce the positive kernel $P_f(x, dy) = f(x)P(x, dy)$, with $f = e^F$. The iterates P_f^n are defined in the usual way and have the representation

$$P_f^n(x, A) = \mathsf{E}_x\big[\exp(S_n(F))\mathbb{I}\{\Phi_n \in A\}\big]. \tag{20.13}$$

Setting $A = \mathsf{X}$ gives $\lambda_{n,x}(F) = P_f^n(x, \mathsf{X})$. This is known as the *Feynman–Kac semigroup*, though this terminology is usually reserved for processes in continuous time. Recall that the kernel U_h was defined based on a power series with respect to this semi-group with $f = 1 - h \geq 0$ (see (12.14)).

When the limit defining $\Lambda(F)$ exists, we typically have $\Lambda(F) = \log(\lambda)$, where λ is the largest eigenvalue of P_f (known as the Perron–Frobenius eigenvalue) [303, 297, 298]. Foundations of Perron–Frobenius theory go back to Tweedie's earliest work on positive operators [394, 395], following the work of Vere-Jones and Seneta for positive matrices and Markov chains on a countable or finite state space [405, 348].

The spectrum of P_f is defined precisely as for a probabilistic kernel. Criteria for a spectral gap are developed in [17, 218, 219], along with multiplicative ergodic theorems. These take the form

$$\lim_{n \to \infty} \lambda^{-n} \lambda_{n,x}(F) = \lim_{n \to \infty} \mathsf{E}_x\big[\exp(S_n(F) - n\Lambda(F))\big] = \check{f}(x), \qquad x \in \mathsf{X}, \tag{20.14}$$

where the limit $\check{f}$ is a Perron–Frobenius eigenfunction associated with the eigenvalue λ, so that $P_f \check{f} = \lambda \check{f}$. In terms of the nonlinear generator (20.10), the eigenvector equation is a multiplicative version of Poisson's equation,

$$\mathcal{H}(\check{F}) = \check{F} - F + \Lambda(F),$$

where $F = \log(f)$ and $\check{F} = \log(\check{f})$.

The multiplicative ergodic theorem holds for bounded functions with sufficiently small L_∞ norm if the chain is V-uniformly ergodic [218], and for a class of unbounded functions under the drift criterion (DV3) of Donsker and Varadhan [17, 219]. The proof of the multiplicative ergodic theorem (20.14) is by reduction to the probabilistic setting. First the eigenfunction is constructed, and from this a *twisted kernel* is defined by

$$\check{P}(x, A) := \frac{f(x)}{\lambda \check{f}(x)} \int_{y \in A} P(x, dy)\check{f}(y).$$

In several different settings, it is shown in [218, 17, 219] that the chain with this transition kernel is geometrically ergodic, and this implies that the convergence in (20.14) holds at a geometric rate.

The convergence of the log-moment generating functions is used in [218, 17, 219] to prove large deviations estimates for the partial sums S_n – see also [84, 265, 141] and their references. The simplest estimates take the following form: for $c > \pi(F)$,

$$\lim_{n \to \infty} \log P_\pi \{S_n \geq nc\} = -I_F(c), \qquad (20.15)$$

where I_F is the Fenchel–Legendre transform

$$I_F(c) = \sup_{\theta \in \mathbb{R}} (\theta c - \Lambda(\theta F)).$$

The limit (20.15) is established for geometrically ergodic models in [218] provided F is bounded and $c > \pi(F)$ is close enough to the mean $\pi(F)$. An elegant *bound* on the error probability $P_\pi \{S_n \geq nc\}$ is obtained in [140] for uniformly ergodic chains, similar to the coupling bound in Theorem 16.2.4.

Another approach to obtaining bounds on the rate of convergence as well as large deviations asymptotics for Markov chains is based on *Sobolev inequalities* and their generalizations [89, 407]. The relationship between these conditions and (DV3) is explored in [58].

20.1.6 Quasi-stationarity

Metastability refers to the presence of near stationary behavior of a process during a time period in which it remains in some restricted region of the state space. Quasi-stationarity has the following precise definition: a set M is quasi-stationary if there exists a probability measure π_M satisfying

$$\lim_{n \to \infty} P_x \{\Phi_n \in A \mid \tau_M > n\} = \pi_M \{A\}, \qquad x \in M, \ A \in \mathcal{B}.$$

A special case of the Feynman–Kac semi-group is obtained with $f = \mathbb{I}_M$, in which case we denote $P_f = P_M$. The semi-group has a different interpretation in this case: for any time n, state x, and set A we have

$$P_M^n(x, A) = P_x \{\Phi_n \in A \ and \ \tau_M > n\}. \qquad (20.16)$$

It follows that quasi-stationarity is equivalent to the ratio limit theorem

$$\lim_{n \to \infty} \frac{P_M^n(x, A)}{P_M^n(x, \mathsf{X})} = \pi_M \{A\}, \qquad x \in M, \ A \in \mathcal{B}.$$

The existence of a limit can be established exactly as in the proof of (20.14), provided a twisted kernel is ergodic [118].

The analysis carries over to processes in continuous time. It is shown in [166] that the twisted semi-group is exponentially ergodic for a diffusion process when the set M is taken to be a connected component of $\{x : h(x) \neq 0\}$, with h an eigenfunction. Further analysis reveals that the exit time from M is approximately exponentially distributed with mean $|\Lambda|^{-1}$, where $\Lambda < 0$ is the corresponding eigenvalue for the Markovian generator. Generalizations to the case in which Λ is complex are contained in [276].

20.2   Simulation and MCMC

Suppose that Φ is a Markov chain that we can simulate on a computer. A function $f\colon \mathsf{X} \to \mathbb{R}$ is given, and we wish to compute estimates of the steady state mean $\pi(f)$.

The usual Monte Carlo method constructs estimates recursively through the sample path averages

$$\hat{\pi}_n(f) := \frac{1}{n} \sum_{k=0}^{n-1} f(\Phi_k). \tag{20.17}$$

In some applications the probability measure π is *given*, and the question is how to construct a Markov chain with invariant distribution π. Answers to this question are contained in the Markov chain Monte Carlo (MCMC) literature.

20.2.1   Variance reduction

Performance of the estimator (20.17) is naturally addressed through the CLT, which can be expressed

$$\hat{\pi}_n(f) \approx \pi(f) + n^{-\frac{1}{2}} \gamma_g W, \tag{20.18}$$

where W is an $N(0,1)$ random variable and the approximation is in distribution. The asymptotic variance γ_g^2 is the subject of Section 17.4.3.

There are other estimators of $\pi(f)$ that satisfy a CLT with lower variance. One class of estimators is based on the control variate method [10], in which a zero-mean process $\{\mathcal{W}_k\}$ is introduced in (20.17):

$$\hat{\pi}_n^{\mathrm{CV}}(f) := \frac{1}{n} \sum_{k=0}^{n-1} [f(\Phi_k) - \mathcal{W}_k]. \tag{20.19}$$

The question then is how to construct a process with zero mean, and one for which the asymptotic variance is reduced.

Henderson and Glynn introduced a collection of techniques for this purpose in [158, 160] – see also the recent monograph [10]. Suppose that the chain is V-uniformly ergodic, fix a function $g \in L_\infty^V$, and define

$$\mathcal{W}_t := g(\Phi_t) - Pg(\Phi_t). \tag{20.20}$$

Although the mean of $\mathcal{W}_t$ may not be zero for arbitrary initial conditions, its steady state mean is always zero:

$$\mathsf{E}_\pi[\mathcal{W}_t] = -\int (\Delta g(x)) \pi(dx) = \pi(g - Pg) = 0.$$

See Theorem 17.7.2 for relaxed assumptions on g under which this limit holds.

With proper choice of g the variance of the resulting estimate is reduced. In fact, under mild conditions the control variate can be constructed so that the asymptotic variance of the resulting estimator is *zero*. Suppose that $f^2 \in L_\infty^V$, and set $g = \hat{f}$ equal to a solution to Poisson's equation, $Ph = h - f + \pi(f)$. In this case we have $\mathcal{W}_t = f(\Phi_t) - \pi(f)$, so that $\hat{\pi}_n^{\mathrm{CV}}(f) = \pi(f)$ for each $n \geq 1$.

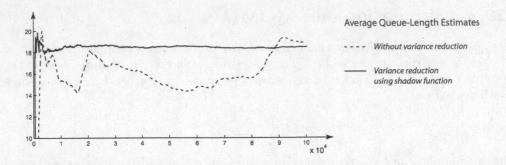

Figure 20.1: Results for a simulation run of length 100,000 steps in a network model. The dashed line represents the running average cost using the standard estimator. The solid line represents the running average cost for the estimator (20.19).

Of course, if we can solve Poisson's equation then we have computed the steady state mean, so there is no reason to simulate. Henderson and Glynn propose *approximate* solutions to reduce the variance in the standard estimator. See the survey by Glynn and Szechtman [138], and results specialized to network models in [159, 265, 267]. The function $g - Pg$ appearing in (20.20) is called a *shadow function* in [159, 267] since it is meant to eclipse the function f to be simulated.

Figure 20.1 shows a comparison of the standard estimator and the estimator (20.19) for a network model. Details on the model and the construction of the control variate can be found in [159] and [267, Chapter 11]. The main idea is to use the control variate (20.20) in which g is a *fluid value function* that approximates the solution to Poisson's equation. The introduction of the zero-mean term $\mathcal{W}_t$ results in a 100-fold reduction in variance over the standard estimator in this example.

20.2.2 Markov chain Monte Carlo (MCMC)

Given a *target* density π on X, the Metropolis–Hastings algorithm constructs a Markov chain $\boldsymbol{X}$ with stationary distribution π. To simplify the description of this technique we restrict to the case in which $\mathsf{X} = \mathbb{R}$, and π possesses a density which we also denote by π.

At each iteration, and conditional on $X_{k-1} = x$, it *proposes* a candidate new value Y_k according to a transition density $q(x, \cdot)$. The new value of the Markov chain X_k is chosen via the following mechanism. The value $X_k = Y_k$ is accepted with conditional probability given by

$$\alpha(x, y) = \min\left\{1, \frac{\pi(y)q(y, x)}{\pi(x)q(x, y)}\right\}; \qquad (20.21)$$

otherwise, the previous state value $X_k = X_{k-1}$ is retained. The resulting Markov chain satisfies the detailed balance equations (20.5) with invariant measure π.

Virtually any transition density $q(\cdot, \cdot)$ can be used in this construction, although some may be more useful then others [327]. There are two generic and popular choices: a) the *independence sampler*, where $q(x, \cdot) \sim p(\cdot)$ for some fixed probability distribution p on $\mathbb{R}$, independent of x; and b) the *Metropolis–Hastings algorithm*, where $q(\cdot, \cdot)$

is the transition density of a symmetric random walk, $q(x,y) = p(|y - x|)$, $x, y \in \mathbb{R}$, where p is again fixed.

The papers [256, 179] set out general conditions relating properties of the distribution used in the simulation with geometric ergodicity of the algorithm. It is shown that:

(i) The independence sampler is uniformly ergodic if and only if $\sup_y (\pi(y)/p(y)) < \infty$. Where this condition fails, the independence sampler fails to be geometrically ergodic. This gives clear, practical guidance about the construction of proposal distributions. It also leads to a powerful and general method for constructing perfect simulation algorithms [49].

(ii) The Metropolis–Hastings algorithm is rarely uniformly ergodic for unbounded state spaces. The algorithm is geometrically ergodic if and only if the tails of the target density π are bounded by $ae^{-b|x|}$ for positive constants a and b.

The Geometric Ergodic Theorem is used to analyze the Metropolis–Hastings algorithm in [330, 331, 79]. These results are generalized to multidimensional models in [335], with surprising consequences for apparently innocuous target densities. The paper [332] contains a survey of MCMC for general state space Markov chains.

A new approach to MCMC analysis was introduced in [125]. This work is based on a fluid model constructed as an approximation to the dynamics of the chain for a large initial condition, similar to the way in which fluid models are used in the theory of stochastic networks (see the commentary for Chapter 19 and [267]). The MCMC algorithm is stable, in the sense that the algorithm is ergodic, provided the fluid model is stable. The structure of the fluid model also provides insight regarding the dynamics of the MCMC algorithm.

20.2.3 Machine learning

Much of the machine learning literature is based on Markovian models [29, 379, 217, 246, 389]. A theme in this literature is the approximation of value functions such as Poisson's equation, which we have already noted arises as the *relative value function* in average-cost optimal control. The final chapter of [267] explains how the "TD learning" technique for value function approximation can be analyzed within the framework of this book.

Many of the recursive algorithms found in this literature can be regarded as instances of stochastic approximation [39], which is itself a generalization of the Monte Carlo method for estimation. It would be worthwhile to search for generalizations of the control variate technique to accelerate algorithms found in the machine learning literature.

20.3 Continuous time models

We wrote in the preface to the first edition that we had *not yet adjusted to the fact that a similar development of the continuous time theory clearly needs to be written next.* Although our interest in continuous time models grew in the years following publication of the book, we never found the time to write this sequel.

One of our earliest contributions to Markov processes in continuous time was at a workshop honoring the contributions of Wolfgang Doeblin to the field of stochastic processes. Entitled *50 Years after Doeblin: Developments in the Theory of Markov Chains, Markov Processes and Sums of Random Variables*, it was held in 1991, just before our book went to print. It is remarkable that before his death in 1940, at the age of just 25, he was able to make such influential contributions to these fields. We can thank Doeblin for the coupling method [94], his minorization condition [93, 95], as well as the Doeblin decomposition [95].

Even more remarkable is that in 1991 his most important achievements remained sealed and unknown to any living person. Quoting from the abstract of a recent movie on Doeblin's life [153], *the full measure of his mathematical stature became apparent only in 2000 when the sealed envelope containing his construction of diffusion processes in terms of a time change of Brownian motion was finally opened, 60 years after it was sent to the Academy of Sciences in Paris.* Among the ideas contained in these notes, written by Doeblin on the front lines of France during World War II, is the one-dimensional equation

$$X(t) = x + \int_0^t a(s, X(s)) \, ds + \beta \Big(\int_0^t \sigma^2(s, X(s)) \, ds \Big) \qquad (20.22)$$

with β Brownian motion on $\mathbb{R}$. This is a sample path representation of the solution to a stochastic differential equation

$$dX(t) = a(t, X(t)) \, dt + \sigma^2(t, X(t)) \, d\beta(t), \qquad (20.23)$$

where the stochastic differential equation is defined in the usual L_2 sense, introduced by Itô several years after Doeblin's early death. More on this history can be found in [153], and scientific details are contained in [411].

In this section we provide highlights of the general theory of ψ-irreducible Markov models in continuous time, without examples and without discussion of specific model classes such as the Doeblin–Itô stochastic differential equation. In particular, left out are the fruits of Richard's collaboration with Gareth Roberts on Langevin algorithms [334], which led to fundamental results on the stability of Langevin diffusions and their discretizations [375, 376].

The focus here is on methods for translating theory from discrete to continuous time. This is made possible through the resolvent kernels, and associated resolvent equations, as described in one of our contributions to the 1991 Blaubeuren meeting [278].

20.3.1   Structure and sampled chains

If time is continuous, then we can still define a Markovian semi-group $\{P^t : t \in \mathbb{R}\}$ via conditional probabilities, exactly as in the discrete time setting. There are only a few technicalities that must be addressed. The first is to rule out "explosion", and for this we restrict to a topological state space. The process is called *non-explosive* if there exists a sequence of compact sets $K_1 \subset K_2 \subset \cdots \subset K_n$ such that $\mathsf{X} = \cup K_n$, and for each initial condition and each time T,

$$\lim_{n \to \infty} \mathsf{P}_x \big\{ \sigma_{K_n^c} < T \big\} = 0,$$

where the hitting times are defined as in discrete time by

$$\sigma_A = \inf\{t \geq 0 : \Phi_t \in A\}, \qquad \tau_A = \inf\{t \geq 1 : \Phi_t \in A\}.$$

This minimal condition is assumed throughout this section, and throughout most of the literature.

When considering sample path properties of the process, it is often desirable to have $\Phi_{\tau_K} \in K$ when K is closed. This is true provided the sample paths of Φ are right continuous. The Markov process is called CADLAG if this condition holds and in addition the sample paths have left hand limits with probability one at any point of discontinuity. The acronym is taken from the French *continue à droite, limite à gauche*. The CADLAG assumption will be assumed throughout.

Translation of results from discrete time to the continuous time domain is made possible through sampling. We consider the δ-skeleton $\{\Phi_{\delta n} : n \geq 0\}$ for arbitrary $\delta > 0$, and we also sample randomly as in Section 5.5.1.

For a given $\alpha > 0$, the resolvent is defined as the Laplace transform

$$U_\alpha := \int_0^\infty e^{-\alpha t} P^t \, dt. \tag{20.24}$$

Random sampling provides a representation for this kernel. Let $\{T_k\}$ denote a Poisson process with rate α, independent of the process Φ. That is, $T_0 = 0$, and $T_{k+1} = T_k + \alpha^{-1} A_{k+1}$ for $k \geq 0$, where A is an i.i.d. sequence, independent of Φ, with standard exponential distribution. The sampled chain is defined exactly as in discrete time by

$$\Phi_k^\alpha := \Phi_{T_k}, \qquad k \geq 0.$$

The transition kernel for the Markov chain Φ^α coincides with the normalized resolvent αU_α.

The definitions of irreducibility and the petiteness property of sets are then based on properties of the sampled chains:

ψ-Irreducibility for a Markov process

We say that the Markov process is

(i) *ψ-irreducible* if Φ^α is ψ-irreducible for some $\alpha > 0$;

(ii) *positive recurrent* if some Φ^α is positive recurrent.

(iii) A ψ-irreducible Markov process is called *aperiodic* if a δ-skeleton is ψ-irreducible for some $\delta > 0$.

The definition of aperiodicity deserves explanation. For the sake of illustration, suppose that some δ-skeleton is Harris ergodic: there is a unique invariant measure π satisfying, for any initial distribution μ,

$$\lim_{n \to \infty} \|\mu P^{n\delta} - \pi\| = 0. \tag{20.25}$$

For an initial condition x and $t \in (0, \delta)$ we can set $\mu = P^t(x, \cdot)$, giving $\mu P^{n\delta} = P^{t+n\delta}(x, \cdot)$. Hence $\|P^{t+n\delta}(x, \cdot) - \pi\| \to 0$, $n \to \infty$, as well. The total variation norm $\|P^t(x, \cdot) - \pi\|$ is non-increasing in t, which shows that Φ is ergodic:

$$\lim_{t \to \infty} \|P^t(x, \cdot) - \pi\| = 0. \tag{20.26}$$

20.3.2   Resolvent equations and drift

To express Lyapunov criteria for stability, a natural analogue of the drift operator Δ defined in (8.1) is the *generator*, defined for a function V by

$$\mathcal{D}V(x) := \lim_{t \to 0} \frac{1}{t}\Big(\int P^t(x, dy)V(y) - V(x)\Big), \qquad x \in \mathsf{X}. \tag{20.27}$$

The existence of the limit is guaranteed under various conditions on the model and the function V.

It is more convenient to work with a relaxed definition. Recall that in the development of limit theory in Section 17.4 we relied upon the construction of martingales, such as $\{M_n(g)\}$ defined in (17.43). For a given function $h \colon \mathsf{X} \to \mathbb{R}$ define $g = \Delta h$, where $\Delta = P - I$ is the drift operator, and consider the stochastic process

$$M_n := h(\Phi_n) - h(\Phi_0) - \sum_{k=1}^{n} g(\Phi_k), \qquad n \geq 1.$$

This process is a martingale under general conditions on the function h. A sufficient condition is finiteness of the n-step expectation $P^n|h|(x)$ for each n and x.

The drift operator is nothing but a discrete time analogue of the generator (20.27). This suggests the following definition: a function V is in the domain of the *extended generator* if there exists a function g such that the process below is a *local martingale* for each initial condition of Φ,

$$M_T := h(\Phi_T) - h(\Phi_0) - \int_0^T g(\Phi_s)\, ds, \qquad T \geq 0. \tag{20.28}$$

This means that there exists a sequence of stopping times $\{\tau_n\}$ such that the stopped process $\{M_{t \wedge \tau_n} : t \geq 0\}$ is a martingale, for each $n \geq 1$, and $\tau_n \uparrow \infty$ a.s. as $n \to \infty$. We let $\mathcal{A}$ denote the extended generator, and denote $\mathcal{A}f = g$ when M is a local martingale.

Two resolvent equations are given in the following. The second equation (20.30) implies that the domain of the extended generator includes the range of the resolvent.

Theorem 20.3.1 (Resolvent equations in continuous time). *If $\{P^t\}$ is a Markovian semi-group, then the following hold:*

(i) *For any pair of positive constants β and α*

$$U_\alpha = U_\beta + (\beta - \alpha)U_\beta U_\alpha = U_\beta + (\beta - \alpha)U_\alpha U_\beta. \tag{20.29}$$

(ii) *For each $\alpha > 0$ and bounded measurable function $g \colon \mathsf{X} \to \mathbb{R}$, the function $U_\alpha g$ is in the domain of the extended generator with*

$$\mathcal{A}U_\alpha g = \alpha U_\alpha g - g. \tag{20.30}$$

$\square$

20.3.3   Ergodic theory

We have already seen how to translate ergodic theorems from discrete to continuous time in the implication (20.25) $\Rightarrow$ (20.26). We now ask, *how can we establish ergodicity of a skeleton based on primitive properties of the continuous time process?*

By applying the differential resolvent equation (20.30) we can translate a Lyapunov function for the sampled chain Φ^{α} to the original process, and vice versa. The martingale characterization of the generator (20.28) can be used to translate a Lyapunov function from the continuous time process to a δ skeleton. These ideas are the basis of the development in [278, 279, 280, 101, 218, 219].

Theorem 20.3.2 is one such result – the extension of the Geometric Ergodic Theorem to continuous time. The drift condition (V4) is defined exactly as in discrete time:

Exponential drift towards C

There exist an extended-real-valued function $V : \mathsf{X} \to [1, \infty]$, a measurable set C, and constants $\beta > 0$, $b < \infty$, such that

$$\mathcal{A}V(x) \le -\beta V(x) + b\mathbb{I}_C(x), \quad x \in \mathsf{X}. \tag{20.31}$$

A set C is called petite if it is petite for some sampled chain: for a probability distribution a on $\mathbb{R}_+$, all x, and all $B \in \mathcal{B}$,

$$K_a(x, B) := \int_0^{\infty} P^t(x, B)\, a(dt) \ge \nu_a(B),$$

where ν_a is a non-trivial measure on $\mathcal{B}(\mathsf{X})$. In (20.31) we typically take C to be petite.

A proof of Theorem 20.3.2 can be found in [101] – see [218] for refinements and extensions.

Theorem 20.3.2 (Exponential Ergodic Theorem). *Suppose that the process Φ is ψ-irreducible and aperiodic. Then the following three conditions are equivalent:*

(i) *The chain Φ is positive recurrent with invariant probability measure π, and there exists some ν-petite set $C \in \mathcal{B}^+(\mathsf{X})$, $\rho_C < 1$, $M_C < \infty$, and $P^{\infty}(C) > 0$ such that for all $x \in C$*

$$|P^t(x, C) - P^{\infty}(C)| \le M_C \rho_C^t. \tag{20.32}$$

(ii) *There exists a closed petite set $C \in \mathcal{B}(\mathsf{X})$ and $\kappa > 1$ such that*

$$\sup_{x \in C} \mathsf{E}_x[\kappa^{\tau_C}] < \infty. \tag{20.33}$$

(iii) *There exists a closed petite set C, constants $b < \infty$, $\beta > 0$ and a function $V \ge 1$ finite at some one $x_0 \in \mathsf{X}$ satisfying (20.31).*

Any of these three conditions imply that the set $S_V = \{x : V(x) < \infty\}$ is absorbing and full, where V is any solution to (20.31) satisfying the conditions of (iii), and there then exist constants $r > 1$, $R < \infty$ such that for any $x \in S_V$,

$$\|P^t - 1 \otimes \pi\|_V \leq R r^{-t}, \qquad t \geq 0. \tag{20.34}$$

□

20.3.4 Multiplicative ergodic theory and large deviations

The generalized resolvent U_h defined in (12.13) also has a continuous time analogue. Let H be a measurable function on X, and set $h = e^H$. The *Feynman–Kac semi-group* is defined for $x \in \mathsf{X}$, $A \in \mathcal{B}(\mathsf{X})$, and $t \in \mathbb{R}_+$ in analogy with the discrete time formula (20.13),

$$P_h^t(x, A) := \mathsf{E}_x\left[\exp\left\{\int_0^t H(\Phi(s))\, ds\right\} \mathbb{I}_A(\Phi(t))\right]. \tag{20.35}$$

Integrating the semi-group defines the resolvent kernel,

$$U_h := \int_0^\infty P_h^t\, dt. \tag{20.36}$$

This reduces to the simple resolvent kernel $U_h = U_\alpha$ when $h = e^{-\alpha}$ is independent of x. The differential resolvent equation (20.30) is often expressed $[\alpha I - \mathcal{A}]U_\alpha g = g$, or

$$U_\alpha = [\alpha I - \mathcal{A}]^{-1}.$$

The generalized resolvent has an analogous formal representation,

$$U_h = [I_{-H} - \mathcal{A}]^{-1},$$

where I_F is the multiplication operator defined by $I_F g = Fg$ for any function g on X. See [278, 295] for a formal treatment, and [218, 219] for application to the theory of large deviations.

Part IV

APPENDICES

Appendices

Despite our best efforts, we understand that the scope of this book inevitably leads to the potential for confusion in readers new to the subject, especially in view of the variety of approaches to stability which we have given, the many related and perhaps (until frequently used) forgettable versions of the "Foster–Lyapunov" drift criteria, and the sometimes widely separated conditions on the various models which are introduced throughout the book.

At the risk of repetition, we therefore gather together in this Appendix several discussions which we hope will assist in giving both the big picture, and a detailed illustration of how the structural results developed in this book may be applied in different contexts.

We first give a succinct series of equivalences between and implications of the various classifications we have defined, as a quick "mud map" to where we have been. In particular, this should help to differentiate between those stability conditions which are "almost" the same.

Secondly, we list together the drift conditions, in slightly abbreviated form, together with references to their introduction and the key theorems which prove that they are indeed criteria for different forms of stability and instability. As a guide to their usage we then review the analysis of one specific model (the scalar threshold autoregression, or SETAR model).

This model incorporates a number of sub-models (specifically, random walks and scalar linear models) which we have already analyzed individually: thus, although not the most complex model available, the SETAR model serves to illustrate many of the technical steps needed to convert elegant theory into practical use in a number of fields of application. The scalar SETAR model also has the distinct advantage that under the finite second moment conditions we impose, it can be analyzed fully, with a complete categorization of its parameter space to place each model into an appropriate stability class.

Thirdly, we give a glossary of the assumptions employed in each of the various models we have analyzed. This list is not completely self-contained: to do this would extend repetition beyond reasonable bounds. However, our experience is that, when looking at a multiply analyzed model, one can run out of hands with which to hold pages open, so we trust that this recapitulation will serve our readers well.

We conclude with a short collection of mathematical results which underpin and are used in proving results throughout the book: these are intended to render the book self-contained, but make no pretence at giving any more comprehensive overview of the areas of measure theory, analysis, topology and even number theory which contribute to the overall development of the theory of general Markov chains.

Appendix A

Mud maps

The wide variety of approaches to and definitions of stability can be confusing. Unfortunately, if one insists on non-countable spaces there is little that can be done about the occasions when two definitions are "almost the same" except to try and delineate the differences.

Here then is an overview of the structure of Markov chains we have developed, at least for the class of chains on which we have concentrated, namely

$$\mathcal{I} := \{ \boldsymbol{\Phi} : \boldsymbol{\Phi} \text{ is } \psi\text{-irreducible for some } \psi \}.$$

We have classified chains in $\mathcal{I}$ using three different but (almost) equivalent properties:

P^n-*properties*: that is, direct properties of the transition laws P^n;

τ-*properties*: properties couched in terms of the hitting times τ_A for appropriate sets A;

drift properties: properties using one-step increments of the form of ΔV for some function V.

A.1 Recurrence versus transience

The first fundamental dichotomy (Chapter 8) is

$$\mathcal{I} = \mathcal{T} + \mathcal{R}$$

where $\mathcal{T}$ denotes the class of *transient chains* and $\mathcal{R}$ denotes the class of *recurrent chains*. This is defined as a dichotomy through a P^n-property in Theorem 8.0.1:

P^n-definition of recurrent and transient chains

$$\Phi \in \mathcal{R} \iff \sum_n P^n(x, A) = \infty, \qquad x \in \mathsf{X},\ A \in \mathcal{B}^+(\mathsf{X}),$$

$$\Phi \in \mathcal{T} \iff \sum_n P^n(x, A_j) \le M_j < \infty, \qquad x \in \mathsf{X},\ \mathsf{X} = \cup A_j.$$

A recurrent chain is "almost" a Harris chain (Chapter 9). Define $\mathcal{H} \subseteq \mathcal{R}$ by the *Harris τ-property*

$$\Phi \in \mathcal{H} \iff \mathsf{P}_x(\tau_A < \infty) \equiv 1, \qquad x \in \mathsf{X},\ A \in \mathcal{B}^+(\mathsf{X}).$$

If $\Phi \in \mathcal{R}$, then (Theorem 9.0.1) there is a full absorbing set (a maximal Harris set) H such that

$$\mathsf{X} = H \cup N$$

and Φ can be restricted in a unique way to a chain $\Phi \in \mathcal{H}$ on the set H.

The τ-classification of $\mathcal{T}$ and $\mathcal{R}$ can be made stronger in terms of

$$Q(x, A) = \mathsf{P}_x(\Phi \in A\ \text{i.o.}).$$

We have from Theorem 8.0.1 and Theorem 9.0.1:

τ-Classification of recurrent and transient chains

$$\Phi \in \mathcal{R} \iff Q(x, A) = 1, \qquad x \in H,\ A \in \mathcal{B}^+(\mathsf{X}),$$

$$\Phi \in \mathcal{T} \iff Q(x, A) = 0, \qquad x \subset \mathsf{X},\ A \text{ petite}.$$

If indeed $\Phi \in \mathcal{H}$, then the first of these holds for all x since $H = \mathsf{X}$.

The drift classification we have derived is then (Theorem 9.1.8 and Theorem 8.0.2):

Drift classification of recurrent and transient chains

$$\Phi \in \mathcal{H} \impliedby \Delta V(x) \leq 0, \qquad x \in C^c,$$
$$C \text{ petite, } V \text{ unbounded off petite sets;}$$

$$\Phi \in \mathcal{T} \iff \Delta V(x) \geq 0, \qquad x \in C^c,$$
$$C \text{ petite, } V \text{ bounded and increasing off } C.$$

There is thus only one gap in these classifications, namely the actual equivalence of the drift condition for recurrence. We have shown (Theorem 9.4.2) that such equivalence holds for Feller (including countable space) chains.

Finally, it is valuable in practice in a topological context to recall that for T-chains, which (Proposition 6.2.8) include all Feller chains in $\mathcal{I}$ such that supp ψ has non-empty interior,

(i) if Φ is in $\mathcal{I}$, then (Theorem 6.2.5)

$$\Phi \text{ is a T-chain} \iff \text{every compact set is petite;}$$

(ii) if Φ is a T-chain in $\mathcal{I}$, then (Theorem 9.2.2)

$$\Phi \in \mathcal{H} \iff \Phi \text{ is non-evanescent;}$$

that is, Harris chains in this case do not leave compact sets forever.

A.2   Positivity versus nullity

The second fundamental dichotomy (Chapter 10) is

$$\mathcal{I} = \mathcal{P} + \mathcal{N}$$

where $\mathcal{N}$ denotes the set of *null chains* and $\mathcal{P} \subseteq \mathcal{R}$ denotes the set of *positive chains*. Since every transient chain is *a fortiori* null, this is in any real sense a breakup of $\mathcal{R}$ rather than the complete set $\mathcal{I}$, and is defined in Chapter 10 through a P^n-property:

First P^n-definition of positive and null chains

$$\Phi \in \mathcal{P} \iff \pi(A) = \int \pi(dy) P^n(y, A), \qquad A \in \mathcal{B}(\mathsf{X}),$$

where π is a probability measure with $\pi(\mathsf{X}) = 1$;

$$\Phi \in \mathcal{N} \iff \mu(A) \geq \int \mu(dy) P^n(y, A), \qquad A \in \mathcal{B}(\mathsf{X}),$$

where μ is a measure with $\mu(\mathsf{X}) = \infty$.

A positive chain is again "almost" a *regular chain*. Define the collection $\mathcal{S} \subseteq \mathcal{P}$ by the *τ-property of regularity*

$$\Phi \in \mathcal{S} \iff \sup_{x \in C_j} \mathsf{E}_x[\tau_A] < \infty, \qquad A \in \mathcal{B}^+(\mathsf{X}), \ \mathsf{X} = \cup C_j.$$

If $\Phi \in \mathcal{P}$, then (Theorem 11.0.1) there is a full absorbing set S such that

$$\mathsf{X} = S \cup N$$

and Φ can be restricted in a unique way to a regular chain $\Phi \in \mathcal{S}$ on the set S.

The τ-classification of $\mathcal{P}$ and $\mathcal{N}$ can be made stronger, in almost exact analogy to the recurrence classification above. Theorem 11.0.1 shows:

τ-Classification of positive and null chains

$$\Phi \in \mathcal{P} \iff \sup_{x \in C_j} \mathsf{E}_x[\tau_A] < \infty, \qquad A \in \mathcal{B}^+(\mathsf{X}), \ S = \cup C_j,$$

$$\Phi \in \mathcal{N} \iff \int_C \pi(dx) \mathsf{E}_x[\tau_C] = \infty, \qquad C \in \mathcal{B}^+(\mathsf{X}).$$

Again, if $\Phi \in \mathcal{S}$, then the first of these holds with $S = \mathsf{X}$. We might expect that

$$\Phi \in \mathcal{N} \iff \inf_{x \in C} \mathsf{E}_x[\tau_C] = \infty, \qquad \text{some } C \in \mathcal{B}^+(\mathsf{X}) :$$

clearly the infinite expected hitting times will imply the chain is not positive, but the converse appears to be so far unknown except when C is an atom.

The drift classification is:

Drift classification of positive and null chains

$$\Phi \in \mathcal{S} \iff \Delta V(x) \leq -1 + b\mathbb{I}_C, \ x \in \mathsf{X}, \ C \text{ petite};$$

$$\Phi \in \mathcal{N} \impliedby \begin{cases} \Delta V(x) \geq 0, & x \in C^c, \\ \int P(x, dy)|V(y) - V(x)| \text{ bounded}, \\ C \text{ petite}, \ V \text{ increasing off } C. \end{cases}$$

There is again one open question in these classifications, namely that of the equivalence or otherwise of the drift condition for nullity. We do not know how close this is to complete.

In a topological context we know again (see Chapter 18) that for T-chains, there is a further stability property completely equivalent to positivity: if Φ is an aperiodic T-chain in $\mathcal{R}$ then

$$\Phi \in \mathcal{P} \iff \{P^n(x, \cdot)\} \text{ is tight, a.e. } x \in \mathsf{X}.$$

Both the P^n and τ properties are essentially properties involving the whole trajectory of the chain. The drift conditions, and in particular their sufficiency for classification, are powerful practical tools of analysis because they involve only the one-step movement of the chain: this is summarized further in Section B.1.

A.3   Convergence properties

There is a further P^n-description of $\mathcal{P}$ and $\mathcal{N}$, closer to the recurrence/transience dichotomy, which is developed in Chapter 18, and which is the classical starting point in countable chain theory.

Second P^n-definition of positive and null chains

$$\Phi \in \mathcal{P} \iff \limsup_{n \to \infty} P^n(x, A) > 0, \qquad x \in \mathsf{X}, \ A \in \mathcal{B}^+(\mathsf{X});$$

$$\Phi \in \mathcal{N} \iff \lim_{n \to \infty} P^n(x, B_j) = 0, \qquad x \in \mathsf{X}, \ \mathsf{X} = \bigcup B_j.$$

However, these are weak categorizations of the types of convergence which hold for these chains. For *aperiodic* chains we have (Theorem 13.0.1)

$$\mathcal{H} \cap \mathcal{P} = \mathcal{E}$$

where the class $\mathcal{E}$ is the set of *ergodic chains* such that

$$\Phi \in \mathcal{E} \iff \lim_{n \to \infty} \|P^n(x, \cdot) - \pi\| = 0, \qquad x \in \mathsf{X}.$$

The properties of $\mathcal{E}$ are delineated further in Part III, and in particular in our next appendix we summarize criteria (drift conditions) for classifying sub-classes of $\mathcal{E}$.

Appendix B

Testing for stability

B.1 Glossary of drift conditions

In this section we collect together the various "Foster–Lyapunov" or "drift" criteria which we have developed for the testing of various forms of stability described in Appendix A.

In using each of these drift conditions, one is required to find two chain-related characteristics:

(i) a suitable non-negative "test function" which is always denoted V;

(ii) a suitable "test set" which is always denoted C.

Typically, for well-behaved chains we are able without great difficulty to give conditions showing a set C to be a "test set"; these sets are usually petite, or for T-chains, compact. The choice of V, on the other hand, is an art form and depends strongly on intuition regarding the movement of the chain.

The recurrence criterion (V1)

The weakest stability condition was introduced on page 189. Its use in general requires the existence of a function V, unbounded off petite sets, or coercive on topological spaces, and a petite or compact set C, with

$$\Delta V(x) \leq 0, \qquad x \in C^c. \tag{8.42}$$

Several theorems show this to be an appropriate condition for various forms of recurrence, including Theorem 8.4.3, Theorem 9.4.1, and Theorem 12.3.3.

The positivity/regularity criterion (V2)

The second condition (often known as Foster's condition) was introduced on page 263. We require for some constant $b < \infty$

$$\Delta V(x) \leq -1 + b\mathbb{I}_C(x), \qquad x \in \mathsf{X}, \tag{11.17}$$

where V is allowed to be an extended-real-valued function $V: \mathsf{X} \to [0, \infty]$ provided it is finite at some point in X, and C is typically petite or compact. Theorems which show this to be an appropriate condition for various forms of regularity, existence of invariant measures, positive recurrence and ergodicity are Theorem 11.3.4, Theorem 11.3.11, Theorem 11.3.15, Theorem 12.3.4, Theorem 12.4.5 and Theorem 13.0.1.

The f-positivity/f-regularity criterion (V3)

The third condition was introduced on page 343. Here again V is an extended-real-valued function $V: \mathsf{X} \to [0, \infty]$ finite at some point in X and C is typically petite or compact; and we require for some function $f: \mathsf{X} \to [1, \infty)$, and a constant $b < \infty$,

$$\Delta V (x) \le -f(x) + b\mathbb{I}_C (x), \qquad x \in \mathsf{X}. \tag{14.16}$$

Various theorems which show this to be an appropriate condition for various forms of f-regularity, existence of f-moments of π and f-ergodicity and even sample path results such as the Central Limit Theorem and the Law of the Iterated Logarithm include Theorem 14.2.3, Theorem 14.2.6, Theorem 14.3.7 and Theorem 17.5.3.

See also the drift criterion of Jarner and Roberts described on page 360.

The V-uniform/V-geometric ergodicity criterion (V4)

The strongest stability condition was introduced on page 376. Again V is an extended-real-valued function $V: \mathsf{X} \to [1, \infty]$ finite at some point in X, and for constants $\beta > 0$ and $b < \infty$,

$$\Delta V(x) \le -\beta V(x) + b\mathbb{I}_C (x), \quad x \in \mathsf{X}. \tag{15.28}$$

Critical theorems which show this to be an appropriate condition for various forms of V-geometric regularity, geometric ergodicity, V-uniform ergodicity are Theorem 15.2.6 and Theorem 16.1.2. We also showed in Lemma 15.2.8 that (V4) holds with a petite set C if and only if V is unbounded off petite sets and

$$PV \le \lambda V + L \tag{15.35}$$

holds for some $\lambda < 1$, $L < \infty$, and this is a frequently used alternative form.

Results in Section 20.1 show that (V4) characterizes the existence of a spectral gap for the transition kernel. The stronger drift criterion of Donsker and Varadhan introduced on page 517 is closely related to a discrete spectrum for P.

The transience/nullity criterion

Finally, we introduced conditions for instability. These involve the relation

$$\Delta V(x) \ge 0, \qquad x \in C^c, \tag{8.41}$$

which was introduced on both page 278 and page 188.

Theorems which show this to be an appropriate condition for various forms of non-positivity or nullity include Theorem 11.5.1: typically these require V to have bounded increments in expectation, and C to be a sublevel set of V.

Exactly the same drift criterion can also be shown to give an appropriate condition for various forms of transience, as in Theorem 8.4.2: these require, typically, that V be bounded, and C be a sublevel set of V with both C and C^c in $\mathcal{B}^+(\mathsf{X})$.

These criteria form the basis for classification of the chains we have considered into the various stability classes, and despite their simplicity they appear to work well across a great range of cases. It is our experience that in the use of the two commonest criteria (V2) and (V4) for models on $\mathbb{R}^k$, quadratic forms are the most useful to use, although the choice of a suitable form is not always trivial.

Finally, we mention that in some cases where identifying the test function is difficult we may need greater subtlety: the generalizations in Chapter 19 then provide a number of other methods of approach.

B.2   The scalar SETAR model: a complete classification

In this section we summarize, for illustration, the use of these drift conditions in practice for scalar first order SETAR models: recall that these are piecewise linear models satisfying

$$X_n = \phi(j) + \theta(j)X_{n-1} + W_n(j), \qquad X_{n-1} \in R_j,$$

where $-\infty = r_0 < r_1 < \cdots < r_M = \infty$ and $R_j = (r_{j-1}, r_j]$; for each j, the noise variables $\{W_n(j)\}$ form an i.i.d. zero-mean sequence independent of $\{W_n(i)\}$ for $i \neq j$.

We assume (for convenience of exposition) that the following conditions hold on the noise distributions:

(i) each $\{W_n(i)\}$ has a density positive on the whole real line, and

(ii) the variances of the noise distributions for the two end intervals are finite.

Neither of these conditions is necessary for what follows, although weakening them makes proofs rather more difficult.

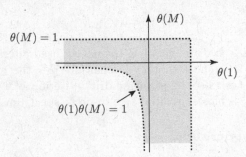

Figure B.1: The SETAR model: stability classification of $(\theta(1), \theta(M))$-space. The model is regular in the shaded "interior" area (11.36), and transient in the unshaded "exterior" (9.46), (9.47) and (9.50). The boundaries are in the figures below.

In Figure B.1, Figure B.2 and Figure B.3 we depict the parameter space in terms of $\phi(1), \theta(1), \phi(M)$, and $\theta(M)$. The results we have proved show that in the "interior" and "boundary" areas, the SETAR model is Harris recurrent; and it is transient in the "exterior" of the parameter space. In accordance with intuition, the model is null on the boundaries themselves, and regular (and indeed, in this case, geometrically ergodic) in the strict interior of the parameter space.

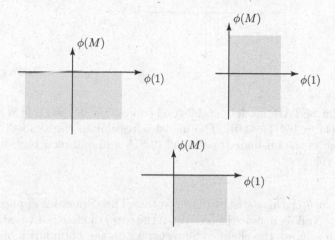

Figure B.2: The SETAR model: stability classification of $(\phi(1), \phi(M))$-space in the regions $(\theta(M) = 1; \theta(1) \leq 1)$ and $(\theta(M) \leq 1; \theta(1) = 1)$. The model is regular in the shaded "interior" areas, which are (clockwise starting with the plot on the far left) (11.38), (11.37) and (11.39); transient in the unshaded "exterior" (9.49), (9.48); and null recurrent on the "margins" described clockwise by (11.45), (11.46) and (11.47)–(11.48).

The steps taken to carry out this classification form a template for analyzing many models, which is our reason for reproducing them in summary form here.

(STEP 1)   As a first step, we show in Theorem 6.3.6 that the SETAR model is a φ-irreducible T-process with φ taken as Lebesgue measure μ^{Leb} on $\mathbb{R}$. Thus compact sets are test sets in all of the criteria above.

(STEP 2)   In the "interior" of the parameter space we are able to identify geometric ergodicity in Proposition 11.4.5, by using (V4) with linear test functions of the form

$$V(x) = \begin{cases} a\,x, & x > 0, \\ b\,|x|, & x \leq 0, \end{cases}$$

and suitable choice of the coefficients a, b, related to the parameters of the model. Note that we only indicated that V satisfied (V2), but the stronger form is actually proved in that result.

(STEP 3)   We establish transience on the "exterior" of the parameter space as in Proposition 9.5.4 using the bounded function

$$V(x) = \begin{cases} 1 - 1/a(x + u), & x > c/a - u, \\ 1 - 1/c, & -c/b - v < x < c/a - u, \\ 1 + 1/b(x + v), & x < -c/b - v, \end{cases}$$

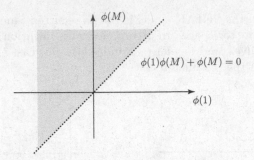

Figure B.3: The SETAR model: stability classification of $(\phi(1), \phi(M))$-space in the region $(\theta(M)\,\theta(1) = 1; \theta(1) \le 0)$. The model is regular in the shaded "interior" area (11.40); transient in the unshaded "exterior" (9.51); and null recurrent on the "margin" described by (11.49).

for suitable u, v, a, b, c: this satisfies (8.41) so that Theorem 8.4.2 applies.

(STEP 4)   Null recurrence is, as is often the case, the hardest to establish. Firstly, Proposition 11.5.4 shows the chain to be recurrent on the boundaries of the parameter space. This is done by applying (V1) with a logarithmic test function

$$V(x) = \begin{cases} \log(u + ax), & x > R > r_{M-1}, \\ \log(v - bx), & x < -R < r_1, \end{cases}$$

and $V(x) = 0$ in the region $[-R, R]$, where a, b, R, u and v are constants chosen suitably for different regions of the parameter space.

To complete the classification of the model, we need to prove that in this region the model is not positive recurrent. In Proposition 11.5.5 we show that the chain is indeed null on the margins of the parameter space, using essentially linear test functions in (11.42).

This model, although not linear, is sufficiently so that the methods applied to the random walk or the simple autoregressive models work here also. In this sense the SETAR model is an example of greater complexity but not of a step change in type. Indeed, the fact that the drift conditions only have to hold outside a compact set means that for this model we really only have to consider the two linear models one each of the end intervals, rendering its analysis even more straightforward.

For more detail on this model, see Tong [388]; and for some of the complications in moving to multidimensional versions, see Brockwell, Liu and Tweedie [52].

Other generalized random coefficient models or completely nonlinear models with which we have dealt are in many ways more difficult to classify. Nevertheless, steps similar to those above are frequently the only ones available, and in practice linearization to enable use of test functions of these forms will often be the approach taken.

Appendix C

Glossary of model assumptions

Here we gather together the assumptions used for the classes of models we have analyzed as continuing examples. The equation numbering and assumption item labels (such as (RT1)) coincide with those used in the main body of the book.

C.1 Regenerative models

We first consider the class of models loosely defined as "regenerative". Such models are usually addressed in applied probability or operations research contexts.

C.1.1 Recurrence time chains

Both discrete time and continuous time renewal processes have served as examples as well as tools in our analysis.

(RT1) If $\{Z_n\}$ is a discrete time renewal process, then the *forward recurrence time chain* $\boldsymbol{V}^+ = V^+(n), n \in \mathbb{Z}_+$ is given by

$$V^+(n) := \inf(Z_m - n : Z_m > n), \qquad n \geq 0.$$

(RT2) The *backward recurrence time chain* $\boldsymbol{V}^- = V^-(n), n \in \mathbb{Z}_+$ is given by

$$V^-(n) := \inf(n - Z_m : Z_m \leq n), \qquad n \geq 0.$$

(RT3) If $\{Z_n\}$ is a renewal process in continuous time with no delay, then we call the process

$$V^+(t) := \inf(Z_n - t : Z_n > t, \, n \geq 1), \qquad t \geq 0$$

the *forward recurrence time process*; and for any $\delta > 0$, the discrete time chain $\boldsymbol{V}_\delta^+ = V^+(n\delta), n \in \mathbb{Z}_+$ is called the *forward recurrence time δ-skeleton*.

(RT4) We call the process

$$V^-(t) := \inf(t - Z_n : Z_n \leq t, \, n \geq 1), \qquad t \geq 0$$

the *backward recurrence time process*; and for any $\delta > 0$, the discrete time chain $\boldsymbol{V}_\delta^- = V^-(n\delta), n \in \mathbb{Z}_+$ is called the *backward recurrence time δ-skeleton*.

C.1.2　Random walk

We have analyzed both random walk on the real line and random walk on the half line, and many models based on these.

(RW1) Suppose that $\boldsymbol{\Phi} = \{\Phi_n; n \in \mathbb{Z}_+\}$ is a collection of random variables defined by choosing an arbitrary distribution for Φ_0 and setting for $k \geq 1$

$$\Phi_k = \Phi_{k-1} + W_k$$

where the W_k are i.i.d. random variables taking values in $\mathbb{R}$ with

$$\Gamma(-\infty, y] = \mathsf{P}(W_n \leq y). \tag{1.6}$$

Then $\boldsymbol{\Phi}$ is called *random walk* on $\mathbb{R}$.

(RW2) We call the random walk spread out (or equivalently, we call Γ spread out) if some convolution power Γ^{n*} is non-singular with respect to μ^{Leb}.

(RWHL1) Suppose $\boldsymbol{\Phi} = \{\Phi_n; n \in \mathbb{Z}_+\}$ is defined by choosing an arbitrary distribution for Φ_0 and taking
$$\Phi_n = [\Phi_{n-1} + W_n]^+ \tag{1.7}$$

where $[\Phi_{n-1} + W_n]^+ := \max(0, \Phi_{n-1} + W_n)$ and again the W_n are i.i.d. random variables taking values in $\mathbb{R}$ with $\Gamma(-\infty, y] = \mathsf{P}(W \leq y)$.

Then $\boldsymbol{\Phi}$ is called *random walk on a half line*.

C.1.3　Storage models and queues

Random walks provide the underlying structure for both queueing and storage models, and we have assumed several specializations for these physical systems.

Queueing models and storage models are closely related in formal structure, although the physical interpretation of the quantities of interest are somewhat different.

We have analyzed *GI/G/1 queueing models* under the following assumptions:

(Q1) Customers arrive into a service operation at time points $T_0 = 0$, $T_0 + T_1$, $T_0 + T_1 + T_2, \ldots$ where the inter-arrival times T_i, $i \geq 1$, are i.i.d. random variables, distributed as a random variable T with $G(-\infty, t] = \mathsf{P}(T \leq t)$.

(Q2) The nth customer brings a job requiring service S_n where the service times are independent of each other and of the inter-arrival times, and are distributed as a variable S with distribution $H(-\infty, t] = \mathsf{P}(S \leq t)$.

(Q3) There is one server and customers are served in order of arrival.

In such a general situation we have often considered the countable space chain consisting of the number of customers in the queue either at arrival or at departure times. Under some exponential assumptions these give the GI/M/1 and M/G/1 queueing systems:

(Q4) If the distribution $H(-\infty, t]$ of service times is exponential with

$$H(-\infty, t] = 1 - e^{-\mu t}, \quad t \geq 0$$

then the queue is called a *GI/M/1* queue.

(Q5) If the distribution $G(-\infty, t]$ of inter-arrival times is exponential with

$$G(-\infty, t] = 1 - e^{-\lambda t}, \quad t \geq 0$$

then the queue is called an *M/G/1 queue.*

In storage models we have a special case of random walk on a half line, but here we consider the model at the times of input and break the increment into the input and output components.

The simple storage model has the following assumptions:

(SSM1) For each $n \geq 0$ let S_n and T_n be i.i.d. random variables on $\mathbb{R}$ with distributions H and G.

(SSM2) Define the random variables

$$\Phi_{n+1} = [\Phi_n + S_n - J_n]^+$$

where the variables J_n are i.i.d., with

$$\mathsf{P}(J_n \leq x) = G(-\infty, x/r] \tag{2.31}$$

for some $r > 0$.

Then the chain $\Phi = \{\Phi_n\}$ represents the contents of a storage system at the times $\{T_n -\}$ immediately before each input and is called the *simple storage model* with release rate r.

More complex content-dependent storage models have the following assumptions:

(CSM1) For each $n \geq 0$ let $S_n(x)$ and T_n be i.i.d. random variables on $\mathbb{R}$ with distributions H_x and G.

(CSM2) Define the random variables

$$\Phi_{n+1} = [\Phi_n - J_n + S_n(\Phi_n - J_n)]^+$$

where the variables J_n are independently distributed, with

$$\mathsf{P}(J_n \leq y \mid \Phi_n = x) = \int G(dt)\mathsf{P}(J_x(t) \leq y). \tag{2.33}$$

The chain $\Phi = \{\Phi_n\}$ can be interpreted as the content of the storage system at the times $\{T_n -\}$ immediately before each input and is called the *content-dependent storage model*.

We also note that these models can be used to represent a number of state-dependent queueing systems where the rate of service depends on the actual state of the system rather than being independent.

C.2   State space models

The other broad class of models we have considered are loosely described as "state space models" and occur in communication and control engineering, other areas of systems analysis, and in time series.

C.2.1   Linear models

The process $\boldsymbol{X} = \{X_n, n \in \mathbb{Z}_+\}$ is called the *simple linear model* if

(SLM1) X_n and W_n are random variables on $\mathbb{R}$ satisfying, for some $\alpha \in \mathbb{R}$,

$$X_n = \alpha X_{n-1} + W_n, \qquad n \geq 1;$$

(SLM2) the random variables $\{W_n\}$ are an i.i.d. sequence with distribution Γ on $\mathbb{R}$.

Next suppose $\boldsymbol{X} = \{X_k\}$ is a stochastic process for which

(LSS1) There exists an $n \times n$ matrix F and an $n \times p$ matrix G such that for each $k \in \mathbb{Z}_+$, the random variables X_k and W_k take values in $\mathbb{R}^n$ and $\mathbb{R}^p$, respectively, and satisfy inductively for $k \geq 1$, and arbitrary W_0,

$$X_k = FX_{k-1} + GW_k;$$

(LSS2) The random variables $\{W_n\}$ are i.i.d. with common finite mean, taking values on $\mathbb{R}^p$, with distribution $\Gamma(A) = \mathsf{P}(W_j \in A)$.

Then $\boldsymbol{X}$ is called the *linear state space model driven by* F, G, or the LSS(F,G) model, with associated control model LCM(F,G) (defined below).

Further assumptions are required for the stability analysis of this model. These include, at different times

(LSS3) The noise variable W has a Gaussian distribution on $\mathbb{R}^p$ with zero mean and unit variance: that is, $W \sim N(0, I)$.

(LSS4) The distribution Γ of the random variable W is non-singular with respect to Lebesgue measure, with non-trivial density γ_w.

(LSS5) The eigenvalues of F fall within the open unit disk in $\mathbb{C}$.

The associated *(linear) control model* LCM(F,G) is defined by the following two sets of assumptions.

Suppose $\boldsymbol{x} = \{x_k\}$ is a deterministic process on $\mathbb{R}^n$ and $\boldsymbol{u} = \{u_n\}$ is a deterministic process on $\mathbb{R}^p$, for which x_0 is arbitrary; then $\boldsymbol{x}$ is called the *linear control model driven by* F, G, or the LCM(F,G) model, if for $k \geq 1$

(LCM1) there exists an $n \times n$ matrix F and an $n \times p$ matrix G such that for each $k \in \mathbb{Z}_+$,

$$x_{k+1} = Fx_k + Gu_{k+1}; \tag{1.4}$$

(LCM2) the sequence $\{u_n\}$ on $\mathbb{R}^p$ is chosen deterministically.

A process $\boldsymbol{Y} = \{Y_n\}$ is called a (scalar) *autoregression of order k*, or AR(k) model, if it satisfies

(AR1) for each $n \geq 0$, Y_n and W_n are random variables on $\mathbb{R}$ satisfying, inductively for $n \geq k$,
$$Y_n = \alpha_1 Y_{n-1} + \alpha_2 Y_{n-2} + \cdots + \alpha_k Y_{n-k} + W_n,$$
for some $\alpha_1, \ldots, \alpha_k \in \mathbb{R}$;

(AR2) the sequence $\boldsymbol{W}$ is an error or innovation sequence on $\mathbb{R}$.

The process $\boldsymbol{Y} = \{Y_n\}$ is called an *autoregressive moving-average process of order (k, ℓ)*, or ARMA(k, ℓ) model, if it satisfies

(ARMA1) for each $n \geq 0$, Y_n and W_n are random variables on $\mathbb{R}$ satisfying, inductively for $n \geq k$,
$$Y_n = \sum_{j=1}^{k} \alpha_j Y_{n-j} + \sum_{j=1}^{\ell} \beta_j W_{n-j} + W_n,$$
for some $\alpha_1, \ldots, \alpha_k, \beta_1, \ldots, \beta_\ell \in \mathbb{R}$;

(ARMA2) the sequence $\boldsymbol{W}$ is an error or innovation sequence on $\mathbb{R}$.

C.2.2   Nonlinear models

The stochastic nonlinear systems we analyze have a deterministic analogue in *semi-dynamical systems*, defined by:

(DS1) The process $\boldsymbol{\Phi}$ is deterministic and generated by the nonlinear difference equation, or semi-dynamical system,
$$\Phi_{k+1} = F(\Phi_k), \qquad k \in \mathbb{Z}_+, \tag{11.16}$$
where $F \colon \mathsf{X} \to \mathsf{X}$ is a continuous function.

(DS2) There exists a positive function $V \colon \mathsf{X} \to \mathbb{R}_+$ and a compact set $C \subset \mathsf{X}$ and constant $M < \infty$ such that
$$\Delta V(x) := V(F(x)) - V(x) \leq -1$$
for all x lying outside the compact set C, and
$$\sup_{x \in C} V(F(x)) \leq M.$$

The chain $\boldsymbol{X} = \{X_n\}$ is called a *scalar nonlinear state space model on $\mathbb{R}$ driven by F*, or SNSS(F) model, if it satisfies

(SNSS1) for each $n \geq 0$, X_n and W_n are random variables on $\mathbb{R}$ satisfying, inductively for $n \geq 1$,

$$X_n = F(X_{n-1}, W_n),$$

for some smooth (C^∞) function $F : \mathbb{R} \times \mathbb{R} \to \mathbb{R}$.

We also use, for various results at various times,

(SNSS2) The sequence $\boldsymbol{W}$ is a disturbance sequence on $\mathbb{R}$, whose marginal distribution Γ possesses a density γ_w supported on an open set O_w, called the *control set*.

(SNSS3) The distribution Γ of W is absolutely continuous, with a density γ_w on $\mathbb{R}$ which is lower semicontinuous.

Suppose $\boldsymbol{X} = \{X_k\}$, where

(NSS1) for each $k \geq 0$, X_k and W_k are random variables on $\mathbb{R}^n$, $\mathbb{R}^p$ respectively satisfying, inductively for $k \geq 1$,

$$X_k = F(X_{k-1}, W_k),$$

for some smooth (C^∞) function $F \colon \mathsf{X} \times O_w \to \mathsf{X}$, where X is an open subset of $\mathbb{R}^n$ and O_w is an open subset of $\mathbb{R}^p$.

Then $\boldsymbol{X}$ is called a *nonlinear state space model driven by F, or NSS(F) model*, with *control set O_w*.

Again for various properties to hold we require

(NSS2) The random variables $\{W_k\}$ are a disturbance sequence on $\mathbb{R}^p$, whose marginal distribution Γ possesses a density γ_w which is supported on an open set O_w.

(NSS3) The distribution Γ of W possesses a density γ_w on $\mathbb{R}^p$ which is lower semicontinuous, and the *control set* is the open set

$$O_w := \{x \in \mathbb{R} : \gamma_w(x) > 0\}.$$

The *associated control model* $\mathrm{CM}(F)$ is defined as follows:

(CM1) The deterministic system

$$x_k = F_k(x_0, u_1, \ldots, u_k), \qquad k \in \mathbb{Z}_+, \tag{2.8}$$

where the sequence of maps $\{F_k : \mathsf{X} \times O_w^k \to \mathsf{X} : k \geq 0\}$ is defined by (2.5), is called the associated control system for the NSS(F) model (denoted $\mathrm{CM}(F)$) provided the deterministic control sequence $\{u_1, \ldots, u_k, k \in \mathbb{Z}_+\}$ lies in the control set $O_w \subseteq \mathbb{R}^p$.

To obtain a T-chain, we assume for the SNSS(F) model,

(CM2) For each initial condition $x_0^0 \in \mathbb{R}$ there exists $k \in \mathbb{Z}_+$ and a sequence $(u_1^0, \ldots, u_k^0) \in O_w^k$ such that the derivative

$$\left[\frac{\partial}{\partial u_1} F_k\left(x_0^0, u_1^0, \ldots, u_k^0\right) \mid \cdots \mid \frac{\partial}{\partial u_k} F_k\left(x_0^0, u_1^0, \ldots, u_k^0\right) \right] \tag{7.4}$$

is non-zero.

For the multidimensional NSS(F) model we often assume

(CM3) For each initial condition $x_0^0 \in \mathbb{R}$ there exists $k \in \mathbb{Z}_+$ and a sequence $\vec{u}^0 = (u_1^0, \ldots, u_k^0) \in O_w^k$ such that

$$\operatorname{rank} C_x^k(\vec{u}^0) = n. \tag{7.13}$$

A specific example of the NSS(F) model is the nonlinear autoregressive moving-average, or NARMA, model.

The process $Y = \{Y_n\}$ is called a *nonlinear autoregressive moving-average process of order* (k, ℓ) if the values $Y_0, \ldots, Y_{k-1}$ are arbitrary and

(NARMA1) for each $n \geq 0$, Y_n and W_n are random variables on $\mathbb{R}$ satisfying, inductively for $n \geq k$,

$$Y_n = G(Y_{n-1}, Y_{n-2}, \ldots, Y_{n-k}, W_n, W_{n-1}, W_{n-2}, \ldots, W_{n-\ell})$$

where the function $G \colon \mathbb{R}^{k+\ell+1} \to \mathbb{R}$ is smooth (C^∞).

(NARMA2) the sequence W is an error sequence on $\mathbb{R}$.

C.2.3 Particular examples

The *simple adaptive control model* is a triple Y, U, θ where

(SAC1) the output sequence Y and parameter sequence θ are defined inductively for any input sequence U by

$$Y_{k+1} = \theta_k Y_k + U_k + W_{k+1} \tag{2.20}$$

$$\theta_{k+1} = \alpha \theta_k + Z_{k+1}, \qquad k \geq 1 \tag{2.21}$$

where α is a scalar with $|\alpha| < 1$;

(SAC2) the bivariate disturbance process $\binom{Z}{W}$ is Gaussian and satisfies

$$\mathsf{E}[\binom{Z_n}{W_n}] = \binom{0}{0}$$

$$\mathsf{E}[\binom{Z_n}{W_n}(Z_k, W_k)] = \begin{pmatrix} \sigma_z^2 & 0 \\ 0 & \sigma_w^2 \end{pmatrix} \delta_{n-k}, \qquad n \geq 1$$

with $\sigma_z < 1$;

(SAC3) the input process satisfies $U_k \in \mathcal{Y}_k$, $k \in \mathbb{Z}_+$, where $\mathcal{Y}_k = \sigma\{Y_0, \ldots, Y_k\}$.

With the control U_k chosen as $U_k = -\hat{\theta}_k Y_k$, $k \in \mathbb{Z}_+$, the closed loop system equations for the simple adaptive control model are

$$\tilde{\theta}_{k+1} = \alpha \tilde{\theta}_k - \alpha \Sigma_k Y_{k+1} Y_k (\Sigma_k Y_k^2 + \sigma_w^2)^{-1} + Z_{k+1} \tag{2.22}$$

$$Y_{k+1} = \tilde{\theta}_k Y_k + W_{k+1} \tag{2.23}$$

$$\Sigma_{k+1} = \sigma_z^2 + \alpha^2 \sigma_w^2 \Sigma_k (\Sigma_k Y_k^2 + \sigma_w^2)^{-1}, \qquad k \geq 1 \tag{2.24}$$

where the triple $\Sigma_0, \tilde{\theta}_0, Y_0$ is given as an initial condition.

The closed loop system gives rise to a Markovian system of the form (NSS1), so that $\Phi_k = (\Sigma_k, \tilde{\theta}_k, Y_k)^\top$ is a Markov chain with state space $\mathsf{X} = [\sigma_z^2, \frac{\sigma_z^2}{1-\alpha^2}] \times \mathbb{R}^2$.

A chain $X = \{X_n\}$ is called a *scalar self-exciting threshold autoregression (or SE-TAR) model* if it satisfies

(SETAR1) for each $1 \leq j \leq M$, X_n and $W_n(j)$ are random variables on $\mathbb{R}$ satisfying, inductively for $n \geq 1$,

$$X_n = \phi(j) + \theta(j)X_{n-1} + W_n(j), \qquad r_{j-1} < X_{n-1} \leq r_j,$$

where $-\infty = r_0 < r_1 < \cdots < r_M = \infty$ and $\{W_n(j)\}$ forms an i.i.d. zero-mean error sequence for each j, independent of $\{W_n(i)\}$ for $i \neq j$.

For stability classification we often use

(SETAR2) For each $j = 1, \ldots, M$, the noise variable $W(j)$ has a density positive on the whole real line.

(SETAR3) The variances of the noise distributions for the two end intervals are finite; that is,

$$\mathsf{E}(W^2(1)) < \infty, \qquad \mathsf{E}(W^2(M)) < \infty.$$

A chain $X = \{X_n\}$ is called a *simple (first order) bilinear process* if it satisfies

(SBL1) For each $n \geq 0$, X_n and W_n are random variables on $\mathbb{R}$ satisfying, for $n \geq 1$,

$$X_n = \theta X_{n-1} + bX_{n-1}W_n + W_n$$

where θ and b are scalars and the sequence W is an error sequence on $\mathbb{R}$.

(SBL2) The sequence W is a disturbance process on $\mathbb{R}$, whose marginal distribution Γ possesses a finite second moment, and a density γ_w which is lower semicontinuous.

The process $\Phi = \binom{\theta}{Y}$ is called the *dependent parameter bilinear model* if it satisfies

(DBL1) For some $|\alpha| < 1$ and all $k \in \mathbb{Z}_+$,

$$Y_{k+1} = \theta_k Y_k + W_{k+1} \tag{2.13}$$

$$\theta_{k+1} = \alpha\theta_k + Z_{k+1}. \tag{2.14}$$

We often also require

(DBL2) The joint process $(Z, W)^\top$ is a disturbance sequence on $\mathbb{R}^2$, Z and W are mutually independent, and the distributions Γ_w and Γ_z of W, Z respectively possess densities which are lower semicontinuous. It is assumed that W has a finite second moment, and that $\mathsf{E}[\log(1 + |Z|)] < \infty$.

The chain $X = \{X_k\}$ is called a *random coefficient autoregression (RCA) process* if it satisfies, for each $k \geq 0$,

$$X_{k+1} = (A + \Gamma_{k+1})X_k + W_{k+1}$$

where X_k, Γ_k and W_k are random variables satisfying the following:

(RCA1) The sequences $\boldsymbol{\Gamma}$ and $\boldsymbol{W}$ are i.i.d. and also independent of each other.

Conditions which lead to stability are then

(RCA2) The following expectations exist, and have the prescribed values:

$$
\begin{aligned}
\mathsf{E}[W_k] &= 0 & \mathsf{E}[W_k W_k^\top] &= G & (n \times n), \\
\mathsf{E}[\Gamma_k] &= 0 & (n \times n) \quad \mathsf{E}[\Gamma_k \otimes \Gamma_k] &= C & (n^2 \times n^2),
\end{aligned}
$$

and the eigenvalues of $A \otimes A + C$ have moduli less than unity.

(RCA3) The distribution of $\binom{\Gamma_k}{W_k}$ has an everywhere positive density with respect to μ^{Leb} on $\mathbb{R}^{n^2+p}$

Appendix D

Some mathematical background

In this final section we collect together, for ease of reference, many of those mathematical results which we have used in developing our results on Markov chains and their applications: these come from probability and measure theory, topology, stochastic processes, the theory of probabilities on topological spaces, and even number theory.

We have tried to give results at a relevant level of generality for each of the types of use: for example, since we assume that the leap from countable to general spaces or topological spaces is one that this book should encourage, we have reviewed (even if briefly) the simple aspects of this theory; conversely, we assume that only a relatively sophisticated audience will wish to see details of sample path results, and the martingale background provided requires some such sophistication.

Readers who are unfamiliar with any particular concepts and who wish to delve further into them should consult the standard references cited, although in general a deep understanding of many of these results is not vital to follow the development in this book itself.

D.1 Some measure theory

We assume throughout this book that the reader has some familiarity with the elements of measure and probability theory. The following sketch of key concepts will serve only as a reminder of terms, and perhaps as an introduction to some non-elementary concepts; anyone who is unfamiliar with this section must take much in the general state space part of the book on trust, or delve into serious texts such as Billingsley [37], Chung [72] or Doob [99] for enlightenment.

D.1.1 Measurable spaces and σ-fields

A general *measurable space* is a pair $(\mathsf{X}, \mathcal{B}(\mathsf{X}))$ with

X: an abstract set of points;

$\mathcal{B}(X)$: a σ-field of subsets of X; that is,

(a) $X \in \mathcal{B}(X)$;

(b) if $A \in \mathcal{B}(X)$, then $A^c \in \mathcal{B}(X)$;

(c) if $A_k \in \mathcal{B}(X)$, $k = 1, 2, 3, \ldots$, then $\bigcup_{k=1}^{\infty} A_k \in \mathcal{B}(X)$.

A σ-field $\mathcal{B}$ is *generated* by a collection of sets $\mathcal{A}$ in $\mathcal{B}$ if $\mathcal{B}$ is the smallest σ-field containing the sets $\mathcal{A}$, and then we write $\mathcal{B} = \sigma(\mathcal{A})$; a σ-field $\mathcal{B}$ is *countably generated* if it is generated by a countable collection $\mathcal{A}$ of sets in $\mathcal{B}$. The σ-fields $\mathcal{B}(X)$ we use are always assumed to be countably generated.

On the real line $\mathbb{R} := (-\infty, \infty)$ the Borel σ-field $\mathcal{B}(\mathbb{R})$ is generated by the countable collection of sets $\mathcal{A} = (a, b]$ where a, b range over the rationals $\mathbb{Q}$.

When our state space is $\mathbb{R}$, then we always assume it is equipped with the Borel σ-field.

If $(X_1, \mathcal{B}(X_1))$ is a measurable space and $(X_2, \mathcal{B}(X_2))$ is another measurable space, then a mapping $h \colon X_1 \to X_2$ is called a *measurable function* if

$$h^{-1}\{B\} := \{x : h(x) \in B\} \in \mathcal{B}(X_1)$$

for all sets $B \in \mathcal{B}(X_2)$.

As a convention, functions on $(X, \mathcal{B}(X))$ which we use are always assumed to be measurable, and in general this is omitted from theorem statements and the like.

D.1.2 Measures

A *(signed) measure* μ on the space $(X, \mathcal{B}(X))$ is a function from $\mathcal{B}(X)$ to $(-\infty, \infty]$ which is countably additive: if $A_k \in \mathcal{B}(X)$, $k = 1, 2, 3, \ldots$, and $A_i \cap A_j = \emptyset$, $i \neq j$, then

$$\mu(\bigcup_{i=1}^{\infty} A_i) = \sum_{i=1}^{\infty} \mu(A_i).$$

We say that μ is *positive* if $\mu(A) \geq 0$ for any A. The measure μ is called a *probability* (or *subprobability*) measure if it is positive and $\mu(X) = 1$ (or $\mu(X) < 1$).

A positive measure μ is σ-finite if there is a countable collection of sets $\{A_k\}$ such that $X = \cup A_k$ and $\mu(A_k) < \infty$ for each k.

On the real line $(\mathbb{R}, \mathcal{B}(\mathbb{R}))$ *Lebesgue measure* μ^{Leb} is a positive measure defined for intervals $(a, b]$ by $\mu^{\text{Leb}}(a, b] = b - a$, and for the other sets in $\mathcal{B}(\mathbb{R})$ by an obvious extension technique. Lebesgue measure on higher dimensional Euclidean space $\mathbb{R}^p$ is constructed similarly using the area of rectangles as a basic definition.

The *total variation norm* of a signed measure is defined as $\|\mu\| := \sup \int f \, d\mu$, where the supremum is taken over all measurable functions f from $(X, \mathcal{B}(X))$ to $(\mathbb{R}, \mathcal{B}(\mathbb{R}))$, such that $|f(x)| \leq 1$ for all $x \in X$.

For a signed measure μ, the state space X may be written as the union of disjoint sets X_+ and X_- where

$$\mu(X_+) - \mu(X_-) = \|\mu\|.$$

This is known as the *Hahn decomposition*.

D.1.3   Integrals

Suppose that h is a non-negative measurable function from $(X, \mathcal{B}(X))$ to $(\mathbb{R}, \mathcal{B}(\mathbb{R}))$. The *Lebesgue integral* of h with respect to a positive finite measure μ is defined in three steps.

Firstly, for $A \in \mathcal{B}(X)$ define $\mathbb{I}_A(x) = 1$ if $x \in A$, and 0 otherwise: $\mathbb{I}_A$ is called the *indicator function* of the set A. In this case we define

$$\int_X \mathbb{I}_A(x)\mu(dx) := \mu(A).$$

Next consider *simple functions* h such that there exist sets $\{A_1, \ldots A_N\} \subset \mathcal{B}(X)$ and positive numbers $\{b_1, \ldots b_N\} \subset \mathbb{R}_+$ with $h = \sum_{k=1}^N b_k \mathbb{I}_{A_k}$.

If h is a simple function we can unambiguously define

$$\int_X h(x)\mu(dx) := \sum_{k=1}^N b_k \mu\{A_k\}.$$

Finally, since it is possible to show that given any non-negative measurable h, there exists a sequence of simple functions $\{h_k\}_{k=1}^\infty$ such that for each $x \in X$,

$$h_k(x) \uparrow h(x),$$

we can take

$$\int_X h(x)\mu(dx) := \lim_k \int_X h_k(x)\mu(dx)$$

which always exists, though it may be infinite.

This approach works if h is non-negative. If not, write

$$h = h^+ - h^-$$

where h^+ and h^- are both non-negative measurable functions, and define

$$\int_X h(x)\mu(dx) := \int_X h^+(x)\mu(dx) - \int_X h^-(x)\mu(dx),$$

if both terms on the right are finite. Such functions are called μ-*integrable*, or just integrable if there is no possibility of confusion; and we frequently denote the integral by

$$\int h \, d\mu := \int_X h(x)\mu(dx).$$

The extension to σ-finite measures is then straightforward.

Convergence of sequences of integrals is central to much of this book. There are three results which we use regularly:

Theorem D.1.1 (Monotone Convergence Theorem). *If μ is a σ-finite positive measure on $(X, \mathcal{B}(X))$ and $\{f_i : i \in \mathbb{Z}_+\}$ are measurable functions from $(X, \mathcal{B}(X))$ to $(\mathbb{R}, \mathcal{B}(\mathbb{R}))$ which satisfy $0 \leq f_i(x) \uparrow f(x)$ for μ-almost every $x \in X$, then*

$$\int_X f(x)\mu(dx) = \lim_i \int_X f_i(x)\mu(dx). \tag{D.1}$$

Note that in this result the monotone limit f may not be finite even μ-almost everywhere, but the result continues to hold in the sense that both sides of (D.1) will be finite or infinite together.

Theorem D.1.2 (Fatou's Lemma). *If μ is a σ-finite positive measure on $(\mathsf{X}, \mathcal{B}(\mathsf{X}))$ and $\{f_i : i \in \mathbb{Z}_+\}$ are non-negative measurable functions from $(\mathsf{X}, \mathcal{B}(\mathsf{X}))$ to $(\mathbb{R}, \mathcal{B}(\mathbb{R}))$, then*

$$\int_\mathsf{X} \liminf_i f_i(x)\mu(dx) \le \liminf_i \int_\mathsf{X} f_i(x)\mu(dx). \tag{D.2}$$

Theorem D.1.3 (Dominated Convergence Theorem). *Suppose that μ is a σ-finite positive measure on $(\mathsf{X}, \mathcal{B}(\mathsf{X}))$ and $g \ge 0$ is a μ-integrable function from $(\mathsf{X}, \mathcal{B}(\mathsf{X}))$ to $(\mathbb{R}, \mathcal{B}(\mathbb{R}))$.*

If f and $\{f_i : i \in \mathbb{Z}_+\}$ are measurable functions from $(\mathsf{X}, \mathcal{B}(\mathsf{X}))$ to $(\mathbb{R}, \mathcal{B}(\mathbb{R}))$ satisfying $|f_i(x)| \le g(x)$ for μ-almost every $x \in \mathsf{X}$, and if $f_i(x) \to f(x)$ as $i \to \infty$ for μ-a.e. $x \in \mathsf{X}$, then each f_i is μ-integrable and

$$\int_\mathsf{X} f(x)\mu(dx) = \lim_i \int_\mathsf{X} f_i(x)\mu(dx).$$

D.2   Some probability theory

A general *probability space* is an ordered triple $(\Omega, \mathcal{F}, \mathsf{P})$ with Ω an abstract set of points, $\mathcal{F}$ a σ-field of subsets of Ω, and P a probability measure on $\mathcal{F}$.

If $(\Omega, \mathcal{F}, \mathsf{P})$ is a probability space and $(\mathsf{X}, \mathcal{B}(\mathsf{X}))$ is a measurable space, then a mapping $X : \Omega \to \mathsf{X}$ is called a *random variable* if

$$X^{-1}\{B\} := \{\omega : X(\omega) \in B\} \in \mathcal{F}$$

for all sets $B \in \mathcal{B}(\mathsf{X})$: that is, if X is a measurable mapping from Ω to X.

Given a random variable X on the probability space $(\Omega, \mathcal{F}, \mathsf{P})$, we define the σ-field *generated by* X, denoted $\sigma\{X\} \subseteq \mathcal{F}$, to be the smallest σ-field on which X is measurable.

If X is a random variable from a probability space $(\Omega, \mathcal{F}, \mathsf{P})$ to a general measurable space $(\mathsf{X}, \mathcal{B}(\mathsf{X}))$, and h is a real-valued measurable mapping from $(\mathsf{X}, \mathcal{B}(\mathsf{X}))$ to the real line $(\mathbb{R}, \mathcal{B}(\mathbb{R}))$ then the composite function $h(X)$ is a real-valued random variable on $(\Omega, \mathcal{F}, \mathsf{P})$: note that some authors reserve the term "random variable" for such real-valued mappings. For such functions, we define the *expectation* as

$$\mathsf{E}[h(X)] = \int_\Omega h(X(\omega))\mathsf{P}(dw).$$

The set of real-valued random variables Y for which the expectation is well defined and finite is denoted $L^1(\Omega, \mathcal{F}, \mathsf{P})$. Similarly, we use $L^\infty(\Omega, \mathcal{F}, \mathsf{P})$ to denote the collection of essentially bounded real-valued random variables Y, that is, those for which there is a bound M and a set $A_M \subset \mathcal{F}$ with $\mathsf{P}(A_M) = 0$ such that $\{\omega : |Y(\omega)| > M\} \subseteq A_M$.

Suppose that $Y \in L^1(\Omega, \mathcal{F}, \mathsf{P})$ and $\mathcal{G} \subset \mathcal{F}$ is a sub-σ-field of $\mathcal{F}$. If $\hat{Y} \in L^1(\Omega, \mathcal{G}, \mathsf{P})$ and satisfies

$$\mathsf{E}[YZ] = \mathsf{E}[\hat{Y}Z] \quad \text{for all } Z \in L^\infty(\Omega, \mathcal{G}, \mathsf{P}),$$

then $\hat{Y}$ is called the *conditional expectation* of Y given $\mathcal{G}$, and denoted $\mathsf{E}[Y \mid \mathcal{G}]$. The conditional expectation defined in this way exists and is unique (modulo P-null sets) for any $Y \in L^1(\Omega, \mathcal{F}, \mathsf{P})$ and any sub σ-field $\mathcal{G}$.

Suppose now that we have another σ-field $\mathcal{D} \subset \mathcal{G} \subset \mathcal{F}$. Then

$$\mathsf{E}[Y \mid \mathcal{D}] = \mathsf{E}[\mathsf{E}[Y \mid \mathcal{G}] \mid \mathcal{D}]. \tag{D.3}$$

The identity (D.3) is often called "the smoothing property of conditional expectations".

D.3   Some topology

We summarize in this section several concepts needed for chains on topological spaces, and for the analysis of some of the applications on such spaces. The classical texts of Kelley [198] and Halmos [152] are excellent references for details at the level we require, as is the more introductory but very readable exposition of Simmons [355].

D.3.1   Topological spaces

On any abstract space X a topology $\mathcal{T} := \{$open subsets of $\mathsf{X}\}$ is a collection of sets containing

(i) arbitrary unions of members of $\mathcal{T}$,

(ii) finite intersections of members of $\mathcal{T}$,

(iii) the whole space X and the empty set $\emptyset$.

Those members of $\mathcal{T}$ containing a point x are called the *neighborhoods* of x, and the complements of open sets are called *closed*.

A set C is called *compact* if any cover of C with open sets admits a finite subcover, and a set D is *dense* if the smallest closed set containing D (the *closure* of D) is the whole space. A set is called *pre-compact* if it has a compact closure.

When there is a topology assumed on the state spaces for the Markov chains considered in this book, it is always assumed that these render the space locally compact and separable metric: a *locally compact* space is one for which each open neighborhood of a point contains a compact neighborhood, and a *separable* space is one for which a countable dense subset of X exists. A metric space is such that there is a metric d on X which generates its topology.

For the topological spaces we consider, Lindelöf's Theorem holds:

Theorem D.3.1 (Lindelöf's Theorem)**.** *If X is a separable metric space, then every cover of an open set by open sets admits a countable subcover.*

If X is a topological space with topology $\mathcal{T}$, then there is a natural σ-field on X containing $\mathcal{T}$. This σ-field $\mathcal{B}(\mathsf{X})$ is defined as

$$\mathcal{B}(\mathsf{X}) := \bigcap \{G : \mathcal{T} \subset G, \ G \text{ a } \sigma\text{-field on } \mathsf{X}\}$$

so that $\mathcal{B}(\mathsf{X})$ is generated by the open subsets of X.

Extending the terminology from $\mathbb{R}$, this is often called the *Borel σ-field* of X: throughout this book, we have assumed that on a topological space the Borel σ-field is being addressed, and so our general notation $\mathcal{B}(X)$ is consistent in the topological context with the conventional notation.

A measure μ is called *regular* if for any set $E \in \mathcal{B}(X)$,

$$\mu(E) = \inf\{\mu(O) : E \subseteq O, \ O \text{ open}\} = \sup\{\mu(C) : C \subseteq E, \ C \text{ compact}\}.$$

For the topological spaces we consider, measures on $\mathcal{B}(X)$ are regular: we have ([345] p. 49).

Theorem D.3.2. *If* X *is locally compact and separable, then every σ-finite measure on* $\mathcal{B}(X)$ *is regular.*

D.4 Some real analysis

A function $f: X \to \mathbb{R}$ on a space X with a metric d is called *continuous* if for each $\varepsilon > 0$ there exists $\delta > 0$ such that for any two $x, y \in X$, if $d(x, y) < \delta$ then $|f(x) - f(y)| < \varepsilon$. The set of all bounded continuous functions on the locally compact and separable metric space X forms a metric space denoted $\mathcal{C}(X)$, whose metric is generated by the supremum norm

$$|f|_c := \sup_{x \in X} |f(x)|.$$

A function $f: X \to \mathbb{R}$ is called *lower semicontinuous* if the sublevel set $\{x : f(x) \leq c\}$ is closed for any constant c, and *upper semicontinuous* if $\{x : f(x) < c\}$ is open for any constant c.

Theorem D.4.1. *A real-valued function f on* X *is continuous if and only if it is simultaneously upper semicontinuous and lower semicontinuous.*

If the function f is positive, then it is lower semicontinuous if and only if there exists a sequence of continuous bounded positive functions $\{f_n : n \in \mathbb{Z}_+\} \subset \mathcal{C}(X)$, each with compact support, such that for all $x \in X$,

$$f_n(x) \uparrow f(x) \qquad \text{as } n \to \infty.$$

A sequence of functions $\{f_i : i \in \mathbb{Z}_+\} \subset \mathcal{C}(X)$ is called *equicontinuous* if for each $\varepsilon > 0$ there exists $\delta > 0$ such that for any two $x, y \in X$, if $d(x, y) < \delta$ then $|f_i(x) - f_i(y)| < \varepsilon$ for all i.

Theorem D.4.2 (Ascoli's Theorem). *Suppose that the topological space* X *is compact. A collection of functions $\{f_i : i \in \mathbb{Z}_+\} \subset \mathcal{C}(X)$ is pre-compact as a subset of $\mathcal{C}(X)$ if and only if the following two conditions are satisfied:*

(i) *The sequence is uniformly bounded: that is, for some $M < \infty$, and all $i \in \mathbb{Z}_+$,*

$$|f_i|_c = \sup_{x \in X} |f_i(x)| \leq M.$$

(ii) *The sequence is equicontinuous.*

Finally, in our context one of the most frequently used of all results on continuous functions is that which assures us that the convolution operation applied to any pair of $L^1(\mathbb{R}, \mathcal{B}(\mathbb{R}), \mu^{\text{Leb}})$ and $L^\infty(\mathbb{R}, \mathcal{B}(\mathbb{R}), \mu^{\text{Leb}})$ functions is continuous.

For two functions $f, g \colon \mathbb{R} \to \mathbb{R}$, the *convolution* $f * g$ is the function on $\mathbb{R}$ defined for $t \in \mathbb{R}$ by

$$f * g(t) = \int_{-\infty}^{\infty} f(s)g(t-s)\,ds.$$

This is well defined if, for example, both f and g are positive. We have (see [345], p. 196):

Theorem D.4.3. *Suppose that f and g are measurable functions on $\mathbb{R}$, that f is bounded, and that $\int |g|\,dx < \infty$. Then the convolution $f * g$ is a bounded continuous function.*

D.5 · Convergence concepts for measures

In this section we summarize various forms of convergence of probability measures which are used throughout the book. For further information the reader is referred to Parthasarathy [314] and Billingsley [36].

Assume X to be a locally compact and separable metric space. Letting $\mathcal{M}$ denote the set of probability measures on $\mathcal{B}(\mathsf{X})$, we can construct a number of natural topologies on $\mathcal{M}$.

As is obvious in Part III of this book, we are frequently concerned with the very strong topology of convergence in total variation norm. However, for individual sequences of measures, the topologies of weak or vague convergence prove more natural in many respects.

D.5.1 Weak convergence

In the topology of *weak convergence* a sequence $\{\nu_k : k \in \mathbb{Z}_+\}$ of elements of $\mathcal{M}$ converges to ν if and only if

$$\lim_{k \to \infty} \int f\,d\nu_k = \int f\,d\nu \tag{D.4}$$

for every $f \in \mathcal{C}(\mathsf{X})$.

In this case we say that $\{\nu_k\}$ converges *weakly* to ν as $k \to \infty$, and this will be denoted $\nu_k \xrightarrow{\text{w}} \nu$.

The following key result is given as Theorem 6.6 in [314]:

Proposition D.5.1. *There exists a sequence of uniformly continuous, uniformly bounded functions $\{g_n : n \in \mathbb{Z}_+\} \subset \mathcal{C}(\mathsf{X})$ with the property that*

$$\mu_k \xrightarrow{\text{w}} \mu_\infty \quad \Longleftrightarrow \quad \forall n \in \mathbb{Z}_+, \ \lim_{k \to \infty} \int g_n\,d\mu_k = \int g_n\,d\mu_\infty. \tag{D.5}$$

It follows that $\mathcal{M}$ can be considered as a metric space with metric $|\cdot|_w$ defined for $\nu, \mu \in \mathcal{M}$ by

$$|\nu - \mu|_w := \sum_{k=1}^{\infty} 2^{-k} \left| \int g_k \, d\nu - \int g_k \, d\mu \right|.$$

Other metrics relevant to weak convergence are summarized in, for example, [191].

A set of probability measures $\mathcal{A} \subset \mathcal{M}$ is called *tight* if for every $\varepsilon \geq 0$ there exists a compact set $C \subset X$ for which

$$\nu\{C\} \geq 1 - \varepsilon \qquad \text{for every } \nu \in \mathcal{A}.$$

The following result, which characterizes tightness with $\mathcal{M}$ viewed as a metric space, follows from Proposition D.5.6 below.

Proposition D.5.2. *The set of probabilities $A \subset \mathcal{M}$ is pre-compact if and only if it is tight.*

A function $V \colon X \to \mathbb{R}_+$ is called *coercive* if there exists a sequence of compact sets, $C_n \subset X$, $C_n \uparrow X$ such that

$$\lim_{n \to \infty} \left(\inf_{x \in C_n^c} V(x) \right) = \infty$$

where we adopt the convention that the infimum of a function over the empty set is infinity. If X is a closed and unbounded subset of $\mathbb{R}^k$, it is evident that $V(x) = |x|^p$ is coercive for any $p > 0$. If X is *compact*, then our convention implies that any positive function V is coercive because we may set $C_n = X$ for all $n \in \mathbb{Z}_+$.

It is easily verified that a collection of probabilities $\mathcal{A} \subset \mathcal{M}$ is tight if and only if a coercive function V exists such that

$$\sup_{\nu \in \mathcal{A}} \int V \, d\nu < \infty.$$

The following simple lemma will often be needed.

Lemma D.5.3.   (i) *A sequence of probabilities $\{\nu_k : k \in \mathbb{Z}_+\}$ is tight if and only if there exists a coercive function V such that*

$$\limsup_{k \to \infty} \nu_k(V) < \infty.$$

(ii) *If for each $x \in X$ there exists a coercive function $V_x(\cdot)$ on X such that*

$$\limsup_{k \to \infty} \mathsf{E}_x[V_x(\Phi_k)] < \infty,$$

then the chain is bounded in probability. $\square$

The next result can be found in [36] and [314].

Theorem D.5.4. *The following are equivalent for a sequence of probabilities $\{\nu_k : k \in \mathbb{Z}_+\} \subset \mathcal{M}$:*

(i) $\nu_k \xrightarrow{\text{w}} \nu$;

(ii) *for all open sets* $O \subset \mathsf{X}$, $\liminf\limits_{k\to\infty} \nu_k\{O\} \geq \nu\{O\}$;

(iii) *for all closed sets* $C \subset \mathsf{X}$, $\limsup\limits_{k\to\infty} \nu_k\{C\} \leq \nu\{C\}$;

(iv) *for every uniformly bounded and equicontinuous family of functions* $\mathcal{C} \subset \mathcal{C}(\mathsf{X})$,

$$\lim_{k\to\infty} \sup_{f\in\mathcal{C}} \left| \int f d\nu_k - \int f d\nu \right| = 0.$$

$\square$

D.5.2   Vague convergence

Vague convergence is less stringent than weak convergence. Let $\mathcal{C}_0(\mathsf{X}) \subset \mathcal{C}(\mathsf{X})$ denote the set of continuous functions on X which converge to zero on the "boundary" of X: that is, $f \in \mathcal{C}_0(\mathsf{X})$ if for some (and hence any) sequence $\{C_k : k \in \mathbb{Z}_+\}$ of compact sets which satisfy

$$C_k \subset C_{k+1} \qquad \text{and} \qquad \bigcup_{k=0}^{\infty} C_k = \mathsf{X}$$

we have

$$\lim_{k\to\infty} \sup_{x\in C_k^c} |f(x)| = 0.$$

The space $\mathcal{C}_0(\mathsf{X})$ is simply the closure of $\mathcal{C}_c(\mathsf{X})$, the space of continuous functions with compact support, in the uniform norm.

A sequence of subprobability measures $\{\nu_k : k \in \mathbb{Z}_+\}$ is said to converge *vaguely* to a subprobability measure ν if for all $f \in \mathcal{C}_0(\mathsf{X})$

$$\lim_{k\to\infty} \int f \, d\nu_k = \int f \, d\nu,$$

and in this case we will write

$$\nu_k \xrightarrow{\text{v}} \nu \qquad \text{as } k \to \infty.$$

In this book we often apply the following result, which follows from the observation that positive lower semicontinuous functions on X are the pointwise supremum of a collection of positive, continuous functions with compact support (see Theorem D.4.1).

Lemma D.5.5. *If* $\nu_k \xrightarrow{\text{v}} \nu$, *then*

$$\liminf_{k\to\infty} \int f \, d\nu_k \geq \int f \, d\nu \tag{D.6}$$

for any positive lower semicontinuous function f on X.

It is obvious that weak convergence implies vague convergence. On the other hand, a sequence of probabilities converges weakly if and only if it converges vaguely and is tight.

The use and direct verification of boundedness in probability will often follow from the following results: the first of these is a consequence of our assumption that the state space is locally compact and separable (see Billingsley [36] and Revuz [326]).

Proposition D.5.6.   (i) *For any sequence of subprobabilities* $\{\nu_k : k \in \mathbb{Z}_+\}$ *there exists a subsequence* $\{n_k\}$ *and a subprobability* ν_∞ *such that*

$$\nu_{n_k} \xrightarrow{\text{v}} \nu_\infty, \qquad k \to \infty.$$

(ii) *If* $\{\nu_k\}$ *is tight and each* ν_k *is a probability measure, then* $\nu_{n_k} \xrightarrow{\text{w}} \nu_\infty$ *and* ν_∞ *is a probability measure.*

$\square$

D.6   Some martingale theory

D.6.1   The martingale convergence theorem

A sequence of integrable random variables $\{M_n : n \in \mathbb{Z}_+\}$ is called *adapted* to an increasing family of σ-fields $\{\mathcal{F}_n : n \in \mathbb{Z}_+\}$ if M_n is $\mathcal{F}_n$-measurable for each n. The sequence is called a *martingale* if $\mathsf{E}[M_{n+1} \mid \mathcal{F}_n] = M_n$ for all $n \in \mathbb{Z}_+$, and a *supermartingale* if $\mathsf{E}[M_{n+1} \mid \mathcal{F}_n] \leq M_n$ for $n \in \mathbb{Z}_+$.

A *martingale difference sequence* $\{Z_n : n \in \mathbb{Z}_+\}$ is an adapted sequence of random variables such that the sequence $M_n = \sum_{k=0}^{n} Z_k$ is a martingale.

The following result is basic:

Theorem D.6.1 (Martingale Convergence Theorem). *Let* M_n *be a supermartingale, and suppose that*

$$\sup_n \mathsf{E}[|M_n|] < \infty.$$

Then $\{M_n\}$ *converges to a finite limit with probability one.*

If $\{M_n\}$ is a positive, real-valued supermartingale, then by the smoothing property of conditional expectations (D.3),

$$\mathsf{E}[|M_n|] = \mathsf{E}[M_n] \leq \mathsf{E}[M_0] < \infty, \qquad n \in \mathbb{Z}_+.$$

Hence we have as a direct corollary to the Martingale Convergence Theorem:

Theorem D.6.2. *A positive supermartingale converges to a finite limit with probability one.*

Since a positive supermartingale is convergent, it follows that its sample paths are bounded with probability one. The following result gives an upper bound on the magnitude of variation of the sample paths of both positive supermartingales, and general martingales.

Theorem D.6.3 (Kolmogorov's Inequality).   (i) *If M_n is a martingale, then for each $c > 0$ and $p \geq 1$,*

$$P\{\max_{0 \leq k \leq n} |M_k| \geq c\} \leq \frac{1}{c^p} E[|M_n|^p].$$

(ii) *If M_n is a positive supermartingale, then for each $c > 0$*

$$P\{\sup_{0 \leq k \leq \infty} M_k \geq c\} \leq \frac{1}{c} E[M_0].$$

These results, and related concepts, can be found in Billingsley [37], Chung [72], Hall and Heyde [151], and of course Doob [99].

D.6.2 The functional CLT for martingales

Consider a general martingale $(M_n, \mathcal{F}_n)$. Our purpose is to analyze the following sequence of continuous functions on $[0, 1]$:

$$m_n(t) := M_{\lfloor nt \rfloor} + (nt - \lfloor nt \rfloor)\left[M_{\lfloor nt \rfloor + 1} - M_{\lfloor nt \rfloor}\right], \qquad 0 \leq t \leq 1. \tag{D.7}$$

The function $m_n(t)$ is piecewise linear and is equal to M_i when $t = i/n$ for $0 \leq t \leq 1$. In Theorem D.6.4 below we give conditions under which the normalized sequence $\{n^{-1/2} m_n(t) : n \in \mathbb{Z}_+\}$ converges to a continuous process (Brownian motion) on $[0, 1]$. This result requires some care in the definition of convergence for a sequence of stochastic processes.

Let $\mathcal{C}[0, 1]$ denote the normed space of all continuous functions $\phi : [0, 1] \to \mathbb{R}$ under the uniform norm, which is defined as

$$|\phi|_c = \sup_{0 \leq t \leq 1} |\phi(t)|.$$

The vector space $\mathcal{C}[0, 1]$ is a complete, separable metric space, and hence the theory of weak convergence may be applied to analyze measures on $\mathcal{C}[0, 1]$.

The stochastic process $m_n(t)$ possesses a distribution μ_n, which is a probability measure on $\mathcal{C}[0, 1]$. We say that $m_n(t)$ *converges in distribution* to a stochastic process $m_\infty(t)$ as $n \to \infty$, which is denoted $m_n \xrightarrow{d} m_\infty$, if the sequence of measures μ_n converges weakly to the distribution μ_∞ of m_∞. That is, for any bounded continuous functional h on $\mathcal{C}[0, 1]$,

$$E[h(m_n)] \to E[h(m_\infty)] \qquad \text{as } n \to \infty.$$

The limiting process, standard Brownian motion on $[0, 1]$, which we denote by B, is defined as follows:

Standard Brownian motion

Brownian motion $B(t)$ is a real-valued stochastic process on $[0,1]$ with $B(0) = 0$, satisfying

(i) the sample paths of B are continuous with probability one;

(ii) the increment $B(t) - B(s)$ is independent of $\{B(r) : r \le s\}$ for each $0 \le s \le t \le 1$;

(iii) the distribution of $B(t) - B(s)$ is Gaussian $N(0, |t - s|)$.

To prove convergence we use the following key result which is a consequence of Theorem 4.1 of [151].

Theorem D.6.4. *Let $(M_n, \mathcal{F}_n)$ be a square integrable martingale, so that for all $n \in \mathbb{Z}_+$*

$$\mathsf{E}[M_n^2] = \mathsf{E}[M_0^2] + \sum_{k=1}^{n} \mathsf{E}[(M_k - M_{k-1})^2] < \infty,$$

and suppose that the following conditions hold:

(i) *For some constant $0 < \gamma^2 < \infty$,*

$$\lim_{n \to \infty} \frac{1}{n} \sum_{k=1}^{n} \mathsf{E}[(M_k - M_{k-1})^2 | \mathcal{F}_{k-1}] = \gamma^2 \qquad \text{a.s.} \tag{D.8}$$

(ii) *For all $\varepsilon > 0$,*

$$\lim_{n \to \infty} \frac{1}{n} \sum_{k=1}^{n} \mathsf{E}[(M_k - M_{k-1})^2 \mathbb{I}\{(M_k - M_{k-1})^2 \ge \varepsilon n\} | \mathcal{F}_{k-1}] = 0 \qquad \text{a.s.} \tag{D.9}$$

Then $(\gamma^2 n)^{-1/2} m_n \xrightarrow{d} B$. □

Function space limits of this kind are often called *invariance principles*, though we have avoided this term because *functional CLT* seems more descriptive.

D.7 Some results on sequences and numbers

We conclude with some useful lemmas on sequences and convolutions. The first gives an interaction between convolutions and limits. Recall that for two series a, b on $\mathbb{Z}_+$, the convolution is defined as

$$a * b\,(n) := \sum_{j=0}^{n} a(j)b(n-j).$$

Lemma D.7.1. *If $\{a(n)\}, \{b(n)\}$ are non-negative sequences such that $b(n) \to b(\infty) < \infty$ as $n \to \infty$ and $\sum a(j) < \infty$, then*

$$a * b\,(n) \to b(\infty) \sum_{j=0}^{\infty} a(j) < \infty, \qquad n \to \infty. \tag{D.10}$$

PROOF   Set $b(n) = 0$ for $n < 0$. Since $b(n)$ converges it is bounded, and so by the Dominated Convergence Theorem

$$\lim_{n \to \infty} a * b\,(n) = \sum_{j=0}^{\infty} a(j) \lim_{n \to \infty} b(n - j) = b(\infty) \sum_{j=0}^{\infty} a(j) \tag{D.11}$$

as required.   □

The next lemma contains two valuable summation results for series.

Lemma D.7.2.   (i) *If $c(n)$ is a non-negative sequence, then for any $r > 1$,*

$$\sum_{n \geq 0} \Big[\sum_{m \geq n} c(m) \Big] r^n \leq \frac{r}{r - 1} \sum_{m \geq 0} c(m) r^m$$

and hence the two series

$$\sum_{n \geq 0} c(n) r^n, \qquad \sum_{n \geq 0} \Big[\sum_{m \geq n} c(m) \Big] r^n$$

converge or diverge together.

(ii) *If a, b are two non-negative sequences and $r \geq 0$, then*

$$\sum a * b\,(n) r^n = \Big[\sum a(n) r^n \Big] \Big[\sum b(n) r^n \Big].$$

PROOF   By Fubini's Theorem we have

$$\sum_{n \geq 0} [\sum_{m \geq n} c(m)] r^n = \sum_{m \geq 0} c(m) \sum_{n \leq m} r^n$$

$$= \sum_{m \geq 0} c(m)[r^{m+1} - 1]/[r - 1]$$

which gives the first result. Similarly, we have

$$\sum_{n \geq 0} a * b\,(n) r^n = \sum_{n \geq 0} [\sum_{m \leq n} a(m) b(n - m)] r^n$$

$$= \sum_{m \geq 0} a(m) r^m \sum_{n \geq m} b(n - m) r^{n-m}$$

$$= \sum_{m \geq 0} a(m) r^m \sum_{n \geq 0} b(n) r^n$$

which gives the second result.   □

An elementary result on the greatest common divisor is useful for periodic chains.

Lemma D.7.3. *Let d denote the greatest common divisor (g.c.d.) of the numbers m, n. Then there exist integers a, b such that*

$$am + bn = d.$$

For a proof, see the corollary to Lemma 1.31 in Herstein [161].

Finally, in analyzing the periodic behavior of Markov chains, the following lemma is invaluable on very many occasions in ensuring positivity of transition probabilities. A proof can be found in Billingsley [37].

Lemma D.7.4. *Suppose that $\mathcal{N} \subset \mathbb{Z}_+$ is a subset of the integers which is closed under addition: for each $j, k \in \mathcal{N}$, $j + k \in \mathcal{N}$. Let d denote the greatest common divisor of the set $\mathcal{N}$. Then there exists $n_0 < \infty$ such that $nd \in \mathcal{N}$ for all $n \geq n_0$.*

Bibliography

[1] D. Aldous and P. Diaconis. Strong uniform times and finite random walks. *Adv. Applied Maths.*, 8:69–97, 1987.

[2] E. Altman, P. Konstantopoulos, and Z. Liu. Stability, monotonicity and invariant quantities in general polling systems. *Queueing Syst. Theory Appl.*, 11:35–57, 1992.

[3] B. D. O. Anderson and J. B. Moore. *Optimal Control: Linear Quadratic Methods.* Prentice-Hall, Englewood Cliffs, N.J., 1990.

[4] W.J. Anderson. *Continuous Time Markov Chains: An Applications-Oriented Approach.* Springer-Verlag, New York, 1991.

[5] M. Aoki. *State Space Modeling of Time Series.* Springer-Verlag, Berlin, 1990.

[6] A. Arapostathis, V. S. Borkar, E. Fernandez-Gaucherand, M. K. Ghosh, and S. I. Marcus. Discrete-time controlled Markov processes with average cost criterion: a survey. *SIAM J. Control Optim.*, 31:282–344, 1993.

[7] E. Arjas and E. Nummelin. A direct construction of the R-invariant measure for a Markov chain on a general state space. *Ann. Probab.*, 4:674–679, 1976.

[8] E. Arjas, E. Nummelin, and R. L. Tweedie. Uniform limit theorems for non-singular renewal and Markov renewal processes. *J. Appl. Probab.*, 15:112–125, 1978.

[9] S. Asmussen. *Applied Probability and Queues.* John Wiley & Sons, New York, 1987.

[10] S. Asmussen and P. W. Glynn. *Stochastic Simulation: Algorithms and Analysis*, volume 57 of *Stochastic Modelling and Applied Probability.* Springer-Verlag, New York, 2007.

[11] R. Atar and O. Zeitouni. Lyapunov exponents for finite state nonlinear filtering. *SIAM J. Control Optim.*, 35(1):36–55, 1997.

[12] K. B. Athreya and P. Ney. *Branching Processes.* Springer-Verlag, New York, 1972.

[13] K. B. Athreya and P. Ney. A new approach to the limit theory of recurrent Markov chains. *Trans. Amer. Math. Soc.*, 245:493–501, 1978.

[14] K. B. Athreya and P. Ney. Some aspects of ergodic theory and laws of large numbers for Harris recurrent Markov chains. *Colloquia Mathematica Societatis János Bolyai. Nonparametric Statistical Inference*, 32:41–56, 1980. Budapest, Hungary.

[15] K. B. Athreya and S. G. Pantula. Mixing properties of Harris chains and autoregressive processes. *J. Appl. Probab.*, 23:880–892, 1986.

[16] K. E. Avrachenkov and J. B. Lasserre. The fundamental matrix of singularly perturbed Markov chains. *Adv. Appl. Probab.*, 31(3):679–697, 1999.

[17] S. Balaji and S. P. Meyn. Multiplicative ergodicity and large deviations for an irreducible Markov chain. *Stoch. Proc. Applns.*, 90(1):123–144, 2000.

[18] D. J. Bartholomew and D. J. Forbes. *Statistical Techniques for Manpower Planning*. John Wiley & Sons, New York, 1979.

[19] R. Bartoszyński. On the risk of rabies. *Math. Biosci.*, 24:355–377, 1975.

[20] P. Baxendale. Stochastic averaging and asymptotic behavior of the stochastic Duffing-van der Pol equation. *Stoch. Proc. Applns.*, 113:235–272, 2004.

[21] P. H. Baxendale. Renewal theory and computable convergence rates for geometrically ergodic Markov chains. *Adv. Appl. Probab.*, 15(1B):700–738, 2005.

[22] P. H. Baxendale and L. Goukasian. Lyapunov exponents for small random perturbations of Hamiltonian systems. *Ann. Probab.*, 30(1):101–134, 2002.

[23] V. E. Beneš. Existence of finite invariant measures for Markov processes. *Proc. Amer. Math. Soc.*, 18:1058–1061, 1967.

[24] V. E. Beneš. Finite regular invariant measures for Feller processes. *J. Appl. Probab.*, 5:203–209, 1968.

[25] H. C. P. Berbee. *Random walks with stationary increments and renewal theory*. PhD thesis, Vrije University, Amsterdam, 1979.

[26] J. R. L. Bernard, editor. *Macquarie English Thesaurus*. Macquarie Library, 1984.

[27] D. P. Bertsekas. *Dynamic Programming and Optimal Control*. Athena Scientific, Cambridge, Mass., third edition, 2007.

[28] D. P. Bertsekas and J. N. Tsitsiklis. An analysis of stochastic shortest path problems. *Math. Oper. Res.*, 16(3):580–595, 1991.

[29] D. P. Bertsekas and J. N. Tsitsiklis. *Neuro-Dynamic Programming*. Athena Scientific, Cambridge, Mass., 1996.

[30] D. Bertsimas, D. Gamarnik, and J. N. Tsitsiklis. Stability conditions for multiclass fluid queueing networks. *IEEE Trans. Automat. Control*, 41(11):1618–1631, 1996.

[31] D. Bertsimas, D. Gamarnik, and J. N. Tsitsiklis. Correction to: "Stability conditions for multiclass fluid queueing networks" [IEEE Trans. Automat. Control. **41** (1996), no. 11, 1618–1631; MR1419686 (97f:90028)]. *IEEE Trans. Automat. Control*, 42(1):128, 1997.

[32] D. Bertsimas, D. Gamarnik, and J. N. Tsitsiklis. Performance of multiclass Markovian queueing networks via piecewise linear Lyapunov functions. *Ann. Appl. Probab.*, 11(4):1384–1428, 2001.

[33] D. Bertsimas, I. Paschalidis, and J. N. Tsitsiklis. Optimization of multiclass queueing networks: polyhedral and nonlinear characterizations of achievable performance. *Ann. Appl. Probab.*, 4:43–75, 1994.

[34] N. P. Bhatia and G. P. Szegö. *Stability Theory of Dynamical Systems*. Springer-Verlag, New York, 1970.

[35] R. N. Bhattacharaya. On the functional Central Limit Theorem and the law of the iterated logarithm for Markov processes. *Z. Wahrscheinlichkeitstheorie und Verw. Geb.*, 60:185–201, 1982.

[36] P. Billingsley. *Convergence of Probability Measures*. John Wiley & Sons, New York, 1968.

[37] P. Billingsley. *Probability and Measure*. John Wiley & Sons, New York, 1995.

[38] H. Bonsdorff. Characterisation of uniform recurrence for general Markov chains. *Ann. Acad. Scientarum Fennicae Ser. A, I. Mathematica, Dissertationes*, 32, 1980.

[39] V. S. Borkar. *Stochastic Approximation: A Dynamical Systems Viewpoint.* Hindustan Publishing Agency and Cambridge University Press (jointly), Delhi and Cambridge, 2008.

[40] V. S. Borkar and S. P. Meyn. The O.D.E. method for convergence of stochastic approximation and reinforcement learning. *SIAM J. Control Optim.*, 38(2):447–469, 2000. (also presented at the IEEE CDC, December 1998).

[41] V. S. Borkar and S. P. Meyn. Risk-sensitive optimal control for Markov decision processes with monotone cost. *Math. Oper. Res.*, 27(1):192–209, 2002.

[42] Vivek S. Borkar. Uniform stability of controlled Markov processes. In *System Theory: Modeling, Analysis and Control (Cambridge, MA, 1999)*, volume 518 of *Kluwer Internat. Ser. Engrg. Comput. Sci.*, pages 107–120. Kluwer Acad. Publ., Boston, 2000.

[43] A. A. Borovkov. Limit theorems for queueing networks. *Theory Probab. Appl.*, 31:413–427, 1986.

[44] A. Brandt. The stochastic equation $y_{n+1} = a_n y_n + b_n$ with stationary coefficients. *Adv. Appl. Probab.*, 18:211–220, 1986.

[45] A. Brandt, P. Franken, and B. Lisek. *Stationary Stochastic Models.* Akademie-Verlag, Berlin, 1990.

[46] L. Breiman. The strong law of large numbers for a class of Markov chains. *Ann. Math. Statist.*, 31:801–803, 1960.

[47] L. Breiman. Some probabilistic aspects of the renewal theorem. In *Trans. 4th Prague Conf. on Inf. Theory, Statist. Dec. Functions and Random Procs.*, pages 255–261. Academia, Prague, 1967.

[48] L. Breiman. *Probability.* Addison-Wesley, Reading, Mass., 1968.

[49] L. A. Breyer and G. O. Roberts. Catalytic perfect simulation. *Methodology and Computing in Applied Probability*, 3(2):161–177, 2001.

[50] P. J. Brockwell, 1992. Personal communication.

[51] P. J. Brockwell and R. A. Davis. *Time Series: Theory and Methods.* Springer-Verlag, New York, second edition, 1991.

[52] P. J. Brockwell, J. Liu, and R. L. Tweedie. On the existence of stationary threshold autoregressive moving-average processes. *J. Time Ser. Anal.*, 13:95–107, 1992.

[53] P. J. Brockwell, S. J. Resnick, and R. L. Tweedie. Storage processes with general release rule and additive inputs. *Adv. Appl. Probab.*, 14:392–433, 1982.

[54] G. A. Brosamler. An almost everywhere Central Limit Theorem. *Math. Proc. Camb. Phil. Soc.*, 104:561–574, 1988.

[55] J. B. Brown. *Ergodic Theory and Topological Dynamics.* Academic Press, New York, 1976.

[56] S. Browne and K. Sigman. Work-modulated queues with applications to storage processes. *J. Appl. Probab.*, 29:699–712, 1992.

[57] P. E. Caines. *Linear Stochastic Systems.* John Wiley & Sons, New York, 1988.

[58] P. Cattiaux, A. Guillin, F.-Y. Wang, and L. Wu. Lyapunov conditions for logarithmic Sobolev and super Poincaré inequality. ArXiv 0712.0235 [math.PR], 2007.

[59] E. Çinlar. *Introduction to Stochastic Processes.* Prentice-Hall, Englewood Cliffs, N.J., 1975.

[60] H. P. Chan and T. L. Lai. Saddlepoint approximations and nonlinear boundary crossing probabilities of Markov random walks. *Ann. Appl. Probab.*, 13(2):395–429, 2003.

[61] K. S. Chan. *Topics in Nonlinear Time Series Analysis.* PhD thesis, Princeton University, 1986.

[62] K. S. Chan. A note on the geometric ergodicity of a Markov chain. *Adv. Appl. Probab.*, 21:702–704, 1989.

[63] K. S. Chan. Asymptotic behaviour of the Gibbs sampler. *J. Amer. Statist. Assoc.*, 88:320–326, 1993.

[64] K. S. Chan, J. Petruccelli, H. Tong, and S. W. Woolford. A multiple threshold AR(1) model. *J. Appl. Probab.*, 22:267–279, 1985.

[65] H. Chen. Fluid approximations and stability of multiclass queueing networks: work-conserving disciplines. *Ann. Appl. Probab.*, 5(3):637–665, 1995.

[66] R. Chen and R. S. Tsay. On the ergodicity of TAR(1) processes. *Ann. Appl. Probab.*, 1:613–634, 1991.

[67] R-R. Chen and S. P. Meyn. Value iteration and optimization of multiclass queueing networks. *Queueing Syst. Theory Appl.*, 32(1-3):65–97, 1999.

[68] P. Chigansky, R. Liptser, and R. van Handel. Intrinsic methods in filter stability. In *Handbook of Nonlinear Filtering.* Oxford University Press, 2008. To appear.

[69] Y. S. Chow and H. Robbins. A renewal theorem for random variables which are dependent or non-identically distributed. *Ann. Math. Statist.*, 34:390–395, 1963.

[70] K. L. Chung. The general theory of Markov processes according to Doeblin. *Z. Wahrscheinlichkeitstheorie und Verw. Geb.*, 2:230–254, 1964.

[71] K. L. Chung. *Markov Chains with Stationary Transition Probabilities.* Springer-Verlag, Berlin, second edition, 1967.

[72] K. L. Chung. *A Course in Probability Theory.* Academic Press, New York, second edition, 1974.

[73] K. L. Chung and D. Ornstein. On the recurrence of sums of random variables. *Bull. Amer. Math. Soc.*, 68:30–32, 1962.

[74] R. Cogburn. The Central Limit Theorem for Markov processes. In L. M. Le Cam, J. Neyman, and E. L. Scott, editors, *Proceedings of the 6th Berkeley Symposium on Mathematical Statistics and Probability*, pages 485–512. University of California Press, Berkeley, 1972.

[75] R. Cogburn. A uniform theory for sums of Markov chain transition probabilities. *Ann. Probab.*, 3:191–214, 1975.

[76] J. W. Cohen. *The Single Server Queue.* North-Holland, Amsterdam, second edition, 1982.

[77] C. Constantinescu and A. Cornea. *Potential Theory on Harmonic Spaces.* Springer-Verlag, Berlin, 1972.

[78] P. C. Consul. Evolution of surnames. *Int. Statist. Rev.*, 59:271–278, 1991.

[79] J. N. Corcoran and R. L. Tweedie. Perfect sampling of ergodic Harris chains. *Ann. Appl. Probab.*, 11(2):438–451, 2001.

[80] J. G. Dai. On positive Harris recurrence of multiclass queueing networks: a unified approach via fluid limit models. *Ann. Appl. Probab.*, 5(1):49–77, 1995.

[81] J. G. Dai and S. P. Meyn. Stability and convergence of moments for multiclass queueing networks via fluid limit models. *IEEE Trans. Automat. Control*, 40:1889–1904, 1995.

[82] J. G. Dai and G. Weiss. Stability and instability of fluid models for reentrant lines. *Math. Oper. Res.*, 21(1):115–134, 1996.

[83] D. Daley. The serial correlation coefficients of waiting times in a stationary single server queue. *J. Austral. Math. Soc.*, 8:683–699, 1968.

[84] A. de Acosta and P. Ney. Large deviation lower bounds for arbitrary additive functionals of a Markov chain. *Ann. Probab.*, 26(4):1660–1682, 1998.

[85] B. Delyon and O. Zeitouni. Lyapunov exponents for filtering problems. In *Applied stochastic analysis (London, 1989)*, volume 5 of *Stochastics Monogr.*, pages 511–521. Gordon and Breach, New York, 1991.

[86] C. Derman. A solution to a set of fundamental equations in Markov chains. *Proc. Amer. Math. Soc.*, 5:332–334, 1954.

[87] G. B. Di Masi and Ł. Stettner. Ergodicity of hidden Markov models. *Math. Control Signals Systems*, 17(4):269–296, 2005.

[88] P. Diaconis. *Group Representations in Probability and Statistics.* Institute of Mathematical Statistics, Hayward, Calif., 1988.

[89] P. Diaconis and L. Saloff-Coste. Logarithmic sobolev inequalities for finite Markov chains. *Ann. Appl. Probab.*, 6(3):695–750, 1996.

[90] P. Diaconis and D. Stroock. Geometric bounds for eigenvalues of Markov chains. *Ann. Appl. Probab.*, 1:36–61, 1991.

[91] J. Diebolt. Loi stationnaire et loi des fluctuations pour le processus autorégressif général d'ordre un. *C. R. Acad. Sci.*, 310:449–453, 1990.

[92] J. Diebolt and D. Guégan. Probabilistic properties of the general nonlinear Markovian process of order one and applications to time series modeling. Technical report 125, Laboratoire de Statistique Théorique et Appliquée, Université Paris, 1990.

[93] W. Doeblin. Sur les propriétés asymptotiques de mouvement régis par certain types de chaînes simples. *Bull. Math. Soc. Roum. Sci.*, 39(1):57–115; 39(2), 3–61, 1937.

[94] W. Doeblin. Exposé de la théorie des chaînes simples constantes de Markov à un nombre fini d'états. *Revue Mathematique de l'Union Interbalkanique*, 2:77–105, 1938.

[95] W. Doeblin. Eléments d'une théorie générale des chaînes simples constantes de Markoff. *Ann. Sci. Ec. Norm. Sup.*, 57:61–111, 1940.

[96] M. D. Donsker and S. R. S. Varadhan. Asymptotic evaluation of certain Markov process expectations for large time. I. II. *Comm. Pure Appl. Math.*, 28:1–47; 28 (1975), 279–301, 1975.

[97] M. D. Donsker and S. R. S. Varadhan. Asymptotic evaluation of certain Markov process expectations for large time. III. *Comm. Pure Appl. Math.*, 29(4):389–461, 1976.

[98] M. D. Donsker and S. R. S. Varadhan. Asymptotic evaluation of certain Markov process expectations for large time. IV. *Comm. Pure Appl. Math.*, 36(2):183–212, 1983.

[99] J. L. Doob. *Stochastic Processes.* John Wiley & Sons, New York, 1953.

[100] R. Douc, G. Fort, E. Moulines, and P. Soulier. Practical drift conditions for subgeometric rates of convergence. *Ann. Appl. Probab.*, 14(3):1353–1377, 2004.

[101] D. Down, S. P. Meyn, and R. L. Tweedie. Exponential and uniform ergodicity of Markov processes. *Ann. Probab.*, 23(4):1671–1691, 1995.

[102] M. Duflo. *Méthodes Récursives Aléatoires*. Masson, Paris, 1990.

[103] W. T. M. Dunsmuir, S. P. Meyn, and G. Roberts. Obituary: Richard Lewis Tweedie. *J. Appl. Probab.*, 39(2):441–454, 2002.

[104] P. Dupuis and R. S. Ellis. *A Weak Convergence Approach to the Theory of Large Deviations*. John Wiley & Sons Inc., New York, 1997.

[105] E. B. Dynkin. *Markov Processes I, II*. Academic Press, New York, 1965.

[106] R. J. Elliott, L. Aggoun, and J. B. Moore. *Hidden Markov Models*. Springer-Verlag, New York, 1995.

[107] Y. Ephraim and N. Merhav. Hidden Markov processes. *IEEE Trans. Inform. Theory*, 48(6):1518–1569, 2002.

[108] S. N. Ethier and T. G. Kurtz. *Markov Processes: Characterization and Convergence*. John Wiley & Sons, New York, 1986.

[109] G. Fayolle. On random walks arising in queueing systems: ergodicity and transience via quadratic forms as Lyapounov functions. I. *Queueing Systems*, 5:167–183, 1989.

[110] G. Fayolle, V. A. Malyshev, M. V. Menshikov, and A. F. Sidorenko. Lyapunov functions for Jackson networks. *Math. Oper. Res.*, 18(4):916–927, 1993.

[111] P. D. Feigin and R. L. Tweedie. Random coefficient autoregressive processes: a Markov chain analysis of stationarity and finiteness of moments. *J. Time Ser. Anal.*, 6:1–14, 1985.

[112] P. D. Feigin and R. L. Tweedie. Linear functionals and Markov chains associated with the Dirichlet process. *Math. Proc. Camb. Phil. Soc.*, 105:579–585, 1989.

[113] E. Feinberg and A. Shwartz, editors. *Markov Decision Processes: Models, Methods, Directions, and Open Problems*. Kluwer Acad. Publ., Holland, 2001.

[114] W. Feller. *An Introduction to Probability Theory and Its Applications. I.* John Wiley & Sons, New York, third edition, 1968.

[115] W. Feller. *An Introduction to Probability Theory and Its Applications. II.* John Wiley & Sons, New York, second edition, 1971.

[116] J. Feng. Martingale problems for large deviations of Markov processes. *Stoch. Proc. Applns.*, 81:165–212, 1999.

[117] J. Feng and T. G. Kurtz. *Large Deviations for Stochastic Processes*, volume 131 of *Mathematical Surveys and Monographs*. American Mathematical Society, Providence, R.I., 2006.

[118] P. A. Ferrari, H. Kesten, and S. Martínez. R-positivity, quasi-stationary distributions and ratio limit theorems for a class of probabilistic automata. *Ann. Appl. Probab.*, 6:577–616, 1996.

[119] J. A. Fill. Eigenvalue bounds on convergence to stationarity for nonreversible Markov chains, with an application to the exclusion process. *Ann. Appl. Probab.*, 1(1):62–87, 1991.

[120] W. H. Fleming. Exit probabilities and optimal stochastic control. *App. Math. Optim.*, 4:329–346, 1978.

[121] S. R. Foguel. Positive operators on $C(X)$. *Proc. Amer. Math. Soc.*, 22:295–297, 1969.

[122] S. R. Foguel. *The Ergodic Theory of Markov Processes*. Van Nostrand Reinhold, New York, 1969.

[123] S. R. Foguel. The ergodic theory of positive operators on continuous functions. *Ann. Scuola Norm. Sup. Pisa*, 27:19–51, 1973.

[124] S. R. Foguel. *Selected Topics in the Study of Markov Operators*. Carolina Lecture Series. Dept. of Mathematics, University of North Carolina at Chapel Hill, 1980.

[125] G. Fort, S. Meyn, E. Moulines, and P. Priouret. ODE methods for skip-free Markov chain stability with applications to MCMC. *Ann. Appl. Probab.*, 18(2):664–707, 2008.

[126] G. Fort and E. Moulines. Polynomial ergodicity of Markov transition kernels. *Stoch. Proc. Applns.*, 103(1):57–99, 2003.

[127] S. G. Foss and R. L. Tweedie. Perfect simulation and backward coupling. *Comm. Statist. Stochastic Models*, 14(1-2):187–203, 1998. Special issue in honor of Marcel F. Neuts.

[128] S. G. Foss, R. L. Tweedie, and J. N. Corcoran. Simulating the invariant measures of Markov chains using backward coupling at regeneration times. *Probab. Engrg. Inform. Sci.*, 12(3):303–320, 1998.

[129] F. G. Foster. On the stochastic matrices associated with certain queuing processes. *Ann. Math. Statist.*, 24:355–360, 1953.

[130] J. J. Fuchs and B. Delyon. Adaptive control of a simple time-varying system. *IEEE Trans. Automat. Control*, 37:1037–1040, 1992.

[131] Cheng-Der Fuh. Asymptotic operating characteristics of an optimal change point detection in hidden Markov models. *Ann. Statist.*, 32(5):2305–2339, 2004.

[132] Cheng-Der Fuh and Tze Leung Lai. Asymptotic expansions in multidimensional Markov renewal theory and first passage times for Markov random walks. *Adv. in Appl. Probab.*, 33(3):652–673, 2001.

[133] Cheng-Der Fuh and Cun-Hui Zhang. Poisson equation, moment inequalities and quick convergence for Markov random walks. *Stoch. Proc. Applns.*, 87(1):53–67, 2000.

[134] H. Furstenberg and H. Kesten. Products of random matrices. *Ann. Math. Statist.*, 31:457–469, 1960.

[135] D. Gamarnik. Extension of the PAC framework to finite and countable Markov chains. *IEEE Trans. Inform. Theory*, 49(1):338–345, 2003.

[136] J. M. Gani and I. W. Saunders. Some vocabulary studies of literary texts. *Sankhyā Ser. B*, 38:101–111, 1976.

[137] L. Georgiadis, W. Szpankowski, and L. Tassiulas. A scheduling policy with maximal stability region for ring networks with spatial reuse. *Queueing Syst. Theory Appl.*, 19(1-2):131–148, 1995.

[138] P. Glynn and R. Szechtman. Some new perspectives on the method of control variates. In K.T. Fang, F.J. Hickernell, and H. Niederreiter, editors, *Monte Carlo and Quasi-Monte Carlo Methods 2000: Proceedings of a Conference held at Hong Kong Baptist University, Hong Kong SAR, China*, pages 27–49. Springer-Verlag, Berlin, 2002.

[139] P. W. Glynn and S. P. Meyn. A Liapounov bound for solutions of the Poisson equation. *Ann. Probab.*, 24(2):916–931, 1996.

[140] P. W. Glynn and D. Ormoneit. Hoeffding's inequality for uniformly ergodic Markov chains. *Statistics and Probability Letters*, 56:143–146, 2002.

[141] F. Z. Gong and L. M. Wu. Spectral gap of positive operators and applications. *J. Math. Pure Appl.*, 85:151–191, 2006.

[142] G. C. Goodwin, P. J. Ramadge, and P. E. Caines. Discrete time stochastic adaptive control. *SIAM J. Control Optim.*, 19:829–853, 1981.

[143] C. W. J. Granger and P. Andersen. *An Introduction to Bilinear Time Series Models.* Vandenhoeck and Ruprecht, Göttingen, 1978.

[144] R. M. Gray. *Entropy and Information Theory.* Springer-Verlag, New York, 1990.

[145] J. A. Gubner, B. Gopinath, and S. R. S. Varadhan. Bounding functions of Markov processes and the shortest queue problem. *Adv. in Appl. Probab.*, 21(4):842–860, 1989.

[146] D. Guégan. Different representations for bilinear models. *J. Time Ser. Anal.*, 8:389–408, 1987.

[147] A. Guillin, C. Leonard, L. Wu, and N. Yao. Transportation inequalities for Markov processes. Unpublished manuscript. Preprint available at http://www.latp.univ-mrs.fr/~guillin/index3.html, 2007.

[148] L. Guo and S. P. Meyn. Adaptive control for time-varying systems: a combination of martingale and Markov chain techniques. *Int. J. Adaptive Control and Signal Processing*, 3:1–14, 1989.

[149] M. Guo and J. Petruccelli. On the null recurrence and transience of a first-order SETAR model. *J. Appl. Probab.*, 28:584–592, 1991.

[150] Olle Häggström. On the Central Limit Theorem for geometrically ergodic Markov chains. *Prob. Theory Related Fields*, 132(1):74–82, 2005.

[151] P. Hall and C. C. Heyde. *Martingale Limit Theory and Its Application.* Academic Press, New York, 1980.

[152] P. R. Halmos. *Measure Theory.* Van Nostrand, Princeton, 1950.

[153] A. Handwerk and H. Willems. *Wolfgang Doeblin: A mathematician rediscovered.* Video-MATH. Springer, Berlin, 2007.

[154] L. P. Hansen and J. Scheinkman. Long term risk: an operator approach. In preparation, 2008.

[155] T. E. Harris. The existence of stationary measures for certain Markov processes. In *Proceedings of the 3rd Berkeley Symposium on Mathematical Statistics and Probability*, volume 2, pages 113–124. University of California Press, Berkeley, 1956.

[156] J. M. Harrison. Discrete dynamic programming with unbounded rewards. *Ann. Math. Statist.*, 43:636–644, 1972.

[157] J. M. Harrison and S. I. Resnick. The stationary distribution and first exit probabilities of a storage process with general release rule. *Math. Oper. Res.*, 1:347–358, 1976.

[158] S. G. Henderson and P. W. Glynn. Approximating martingales for variance reduction in Markov process simulation. *Math. Oper. Res.*, 27(2):253–271, 2002.

[159] S. G. Henderson, S. P. Meyn, and V. B. Tadić. Performance evaluation and policy selection in multiclass networks. *Discrete Event Dynamic Systems: Theory and Applications*, 13(1-2):149–189, 2003.

[160] S.G. Henderson. *Variance Reduction via an Approximating Markov Process.* PhD thesis, Stanford University, Stanford, Calif., 1997.

[161] I. N. Herstein. *Topics in Algebra.* John Wiley & Sons, New York, second edition, 1975.

[162] G. Högnas. On random walks with continuous components. Preprint no 26, Aarhus Universitet, 1977.

[163] A. Hordijk and F.M. Spieksma. On ergodicity and recurrence properties of a Markov chain with an application. *Adv. Appl. Probab.*, 24:343–376, 1992.

[164] C. Huang and D. Isaacson. Ergodicity using mean visit times. *J. Lond. Math. Soc.*, 14:570–576, 1976.

[165] J. Huang, I. Kontoyiannis, and S. P. Meyn. The ODE method and spectral theory of Markov operators. In T. E. Duncan and B. Pasik-Duncan, editors, *Proceedings of the workshop held at the University of Kansas, Lawrence, October 18–20, 2001*, pages 205–222. Springer-Verlag, Berlin, 2002.

[166] W. Huisinga, S. Meyn, and C. Schütte. Phase transitions and metastability in Markovian and molecular systems. *Ann. Appl. Probab.*, 14(1):419–458, 2004.

[167] K. Ichihara and H. Kunita. A classification of the second order degenerate elliptic operator and its probabilistic characterization. *Z. Wahrscheinlichkeitstheorie und Verw. Geb.*, 30:235–254, 1974.

[168] N. Ikeda and S. Watanabe. *Stochastic Differential Equations and Diffusion Processes.* North-Holland, Amsterdam, 1981.

[169] R. Isaac. Some topics in the theory of recurrent Markov processes. *Duke Math. J.*, 35:641–652, 1968.

[170] D. Isaacson and R. L. Tweedie. Criteria for strong ergodicity of Markov chains. *J. Appl. Probab.*, 15:87–95, 1978.

[171] N. Jain. Some limit theorem for a general Markov process. *Z. Wahrscheinlichkeitstheorie und Verw. Geb.*, 6:206–223, 1966.

[172] N. Jain and B. Jamison. Contributions to Doeblin's theory of Markov processes. *Z. Wahrscheinlichkeitstheorie und Verw. Geb.*, 8:19–40, 1967.

[173] B. Jakubczyk and E. D. Sontag. Controllability of nonlinear discrete-time systems: a lie-algebraic approach. *SIAM J. Control Optim.*, 28:1–33, 1990.

[174] B. Jamison. Asymptotic behavior of successive iterates of continuous functions under a Markov operator. *J. Math. Anal. Appl.*, 9:203–214, 1964.

[175] B. Jamison. Ergodic decomposition induced by certain Markov operators. *Trans. Amer. Math. Soc.*, 117:451–468, 1965.

[176] B. Jamison. Irreducible Markov operators on $C(S)$. *Proc. Amer. Math. Soc.*, 24:366–370, 1970.

[177] B. Jamison and S. Orey. Markov chains recurrent in the sense of Harris. *Z. Wahrscheinlichkeitstheorie und Verw. Geb.*, 8:41–48, 1967.

[178] B. Jamison and R. Sine. Sample path convergence of stable Markov processes. *Z. Wahrscheinlichkeitstheorie und Verw. Geb.*, 28:173–177, 1974.

[179] S. F. Jarner and S. Hansen. Geometric ergodicity of Metropolis algorithms. *Stoch. Proc. Applns.*, 85(2):341–361, 2000.

[180] S. F. Jarner and G. O. Roberts. Polynomial convergence rates of Markov chains. *Ann. Appl. Probab.*, 12(1):224–247, 2002.

[181] A. A. Johnson and G. L. Jones. Gibbs sampling for a Bayesian hierarchical general linear model. ArXiv:0712.3056 [math.PR], 2007.

[182] D. A. Jones. Non-linear autoregressive processes. *Proc. Roy. Soc. A*, 360:71–95, 1978.

[183] G. L. Jones. On the Markov chain Central Limit Theorem. *Probab. Surv.*, 1:299–320 (electronic), 2004.

[184] G. L. Jones and J. P. Hobert. Honest exploration of intractable probability distributions via Markov chain Monte Carlo. *Statist. Sci.*, 16(4):312–334, 2001.

[185] A.F. Veinott Jr. Discrete dynamic programming with sensitive discount optimality criteria. *Ann. Math. Statist.*, 40(5):1635–1660, 1969.

[186] M. Kac. On the notion of recurrence in discrete stochastic processes. *Bull. Amer. Math. Soc.*, 53:1002–1010, 1947.

[187] V. V. Kalashnikov. Analysis of ergodicity of queueing systems by Lyapunov's direct method (in russian). *Avtomatica i Telemechanica*, 4:46–54, 1971.

[188] V. V. Kalashnikov. The property of gamma-reflexivity for Markov sequences. *Soviet Math. Dokl.*, 14:1869–1873, 1973.

[189] V. V. Kalashnikov. Stability analysis in queueing problems by the method of test functions. *Theory Probab. Appl.*, 22:86–103, 1977.

[190] V. V. Kalashnikov. *Qualitative Analysis of Complex Systems Behaviour by the Test Functions Method (in Russian)*. Nauka, Moscow, 1978. [in Russian].

[191] V. V. Kalashnikov and S. T. Rachev. *Mathematical Methods for Construction of Queueing Models*. Wadsworth and Brooks/Cole, New York, 1990.

[192] R. E. Kalman and J. E. Bertram. Control system analysis and design by the second method of Lyapunov. *Trans. ASME Ser. D: J. Basic Eng.*, 82:371–400, 1960.

[193] M. Kaplan. A sufficient condition for nonergodicity of a Markov chain. *IEEE Trans. Inform. Theory*, 25:470–471, 1979.

[194] S. Karlin and H. M. Taylor. *A First Course in Stochastic Processes*. Academic Press, New York, second edition, 1975.

[195] H. A. Karlsen. Existence of moments in a stationary stochastic difference equation. *Adv. Appl. Probab.*, 22:129–146, 1990.

[196] N. V. Kartashov. Criteria for uniform ergodicity and strong stability of Markov chains with a common phase space. *Theory Probab. Appl.*, 30:71–89, 1985.

[197] N.V. Kartashov. Inequalities in theorems of ergodicity and stability for Markov chains with a common phase space. *Theory Probab. Appl.*, 30:247–259, 1985.

[198] J. L. Kelley. *General Topology*. Van Nostrand, Princeton, N.J., 1955.

[199] F. P. Kelly. *Reversibility and Stochastic Networks*. John Wiley & Sons, Chichester, U.K., 1979.

[200] D. G. Kendall. Some problems in the theory of queues. *J. Roy. Statist. Soc. Ser. B*, 13:151–185, 1951.

[201] D. G. Kendall. Stochastic processes occurring in the theory of queues and their analysis by means of the imbedded Markov chain. *Ann. Math. Statist.*, 24:338–354, 1953.

[202] D. G. Kendall. Unitary dilations of Markov transition operators and the corresponding integral representation for transition-probability matrices. In U. Grenander, editor, *Probability and Statistics*, pages 139–161. Almqvist and Wiksell, Stockholm, 1959.

[203] D. G. Kendall. Geometric ergodicity in the theory of queues. In K. J. Arrow, S. Karlin, and P. Suppes, editors, *Mathematical Methods in the Social Sciences*, pages 176–195. Stanford University Press, Stanford, 1960.

[204] D. G. Kendall. Kolmogorov as I remember him. *Statist. Sci.*, 6:303–312, 1991.

[205] G. Kersting. On recurrence and transience of growth models. *J. Appl. Probab.*, 23:614–625, 1986.

[206] R. Z. Khas'minskii. *Stochastic Stability of Differential Equations*. Sijthoff & Noordhoff, Netherlands, 1980.

[207] J. F. C. Kingman. The ergodic behaviour of random walks. *Biometrika*, 48:391–396, 1961.

[208] J. F. C. Kingman. *Regenerative Phenomena*. John Wiley & Sons, London, 1972.

[209] C. Kipnis and S. R. S. Varadhan. Central limit theorem for additive functionals of reversible Markov processes and applications to simple exclusions. *Comm. Math. Phys.*, 104(1):1–19, 1986.

[210] A. Klein, L. J. Landau, and D. S. Shucker. Decoupling inequalities for stationary Gaussian processes. *Ann. Probab.*, 10:702–708, 1981.

[211] W. Kliemann. Recurrence and invariant measures for degenerate diffusions. *Ann. Probab.*, 15:690–707, 1987.

[212] W. Kliemann and N. Sri Namachchivaya, editors. *Nonlinear Dynamics and Stochastic Mechanics*. CRC Press, Boca Raton, Fla., 1995.

[213] F. Kochman and J. Reeds. A simple proof of Kaijser's unique ergodicity result for hidden Markov α-chains. *Ann. Appl. Probab.*, 16(4):1805–1815, 2006.

[214] P. V. Kokotovic, J. O'Reilly, and J. K. Khalil. *Singular Perturbation Methods in Control: Analysis and Design*. Academic Press, Orlando, Fla., 1986.

[215] A. N. Kolmogorov. Über die analytischen methoden in der wahrscheinlichkeitsrechnung. *Math. Ann.*, 104:415–458, 1931.

[216] A. N. Kolmogorov. Anfangsgründe der theorie der Markoffschen ketten mit unendlichen vielen möglichen zuständen. *Mat. Sbornik N.S. Ser*, pages 607–610, 1936.

[217] V. R. Konda and J. N. Tsitsiklis. On actor-critic algorithms. *SIAM J. Control Optim.*, 42(4):1143–1166 (electronic), 2003.

[218] I. Kontoyiannis and S. P. Meyn. Spectral theory and limit theorems for geometrically ergodic Markov processes. *Ann. Appl. Probab.*, 13:304–362, 2003.

[219] I. Kontoyiannis and S. P. Meyn. Large deviations asymptotics and the spectral theory of multiplicatively regular Markov processes. *Electron. J. Probab.*, 10(3):61–123 (electronic), 2005.

[220] I. Kontoyiannis and S. P. Meyn. Finite state-space Markov chain approximations for diffusions. In preparation, 2008.

[221] U. Krengel. *Ergodic Theorems*. Walter de Gruyter, Berlin, New York, 1985.

[222] P. Krugman and M. Miller, editors. *Exchange Rate Targets and Currency Bands*. Cambridge University Press, Cambridge, 1992.

[223] P. R. Kumar and S. P. Meyn. Stability of queueing networks and scheduling policies. *IEEE Trans. Automat. Control*, 40(2):251–260, 1995.

[224] P. R. Kumar and S. P. Meyn. Duality and linear programs for stability and performance analysis queueing networks and scheduling policies. *IEEE Trans. Automat. Control*, 41(1):4–17, 1996.

[225] P. R. Kumar and P. P. Varaiya. *Stochastic Systems: Estimation, Identification and Adaptive Control*. Prentice-Hall, Englewood Cliffs, N.J., 1986.

[226] S. Kumar and P. R. Kumar. Performance bounds for queueing networks and scheduling policies. *IEEE Trans. Automat. Control*, AC-39:1600–1611, 1994.

[227] H. Kunita. Diffusion processes and control systems. Course at the University of Paris VI, 1974.

[228] H. Kunita. Supports of diffusion processes and controllability problems. In K. Itô, editor, *Proceedings of the International Symposium on Stochastic Differential Equations*, pages 163–185. John Wiley & Sons, New York, 1978.

[229] H. Kunita. *Stochastic Flows and Stochastic Differential Equations*. Cambridge University Press, Cambridge, 1990.

[230] B. C. Kuo. *Automatic Control Systems*. Prentice-Hall, Englewood Cliffs, N.J., sixth edition, 1990.

[231] T. G. Kurtz. The Central Limit Theorem for Markov chains. *Ann. Probab.*, 9:557–560, 1981.

[232] H. J. Kushner. *Stochastic Stability and Control*. Academic Press, New York, 1967.

[233] M. T. Lacey and W. Philipp. A note on the almost sure Central Limit Theorem. *Statistics and Probability Letters*, 9:201–205, 1990.

[234] J. Lamperti. Criteria for the recurrence or transience of stochastic processes I. *J. Math. Anal. Appl.*, 1:314–330, 1960.

[235] J. Lamperti. Criteria for stochastic processes II: passage time moments. *J. Math. Anal. Appl.*, 7:127–145, 1963.

[236] C. Landim. Central limit theorem for Markov processes. In *From Classical to Modern Probability*, volume 54 of *Progr. Probab.*, pages 145–205. Birkhäuser, Basel, 2003.

[237] G. M. Laslett, D. B. Pollard, and R. L. Tweedie. Techniques for establishing ergodic and recurrence properties of continuous-valued Markov chains. *Nav. Res. Log. Quart.*, 25:455–472, 1978.

[238] M. Lin. Conservative Markov processes on a topological space. *Israel J. Math.*, 8:165–186, 1970.

[239] T. Lindvall. *Lectures on the Coupling Method*. John Wiley & Sons, New York, 1992.

[240] R. S. Liptster and A. N. Shiryayev. *Statistics of Random Processes, II: Applications*. Springer-Verlag, New York, 1978.

[241] R. Lund and R. L. Tweedie. Geometric convergence rates for stochastically ordered Markov chains. *Math. Oper. Res.*, 20:182–194, 1996.

[242] N. Maigret. Théorème de limite centrale pour une chaine de Markov récurrente Harris positive. *Ann. Inst. Henri Poincaré Ser. B*, 14:425–440, 1978.

[243] V. A. Malyšev and M. V. Men′šikov. Ergodicity, continuity and analyticity of countable Markov chains. *Trudy Moskov. Mat. Obshch.*, 39:3–48, 235, 1979. *Trans. Moscow Math. Soc.*, pp. 1-48, 1981.

[244] V. A. Malyšhev. Classification of two-dimensional positive random walks and almost linear semi-martingales. *Soviet Math. Dokl.*, 13:136–139, 1972.

[245] R. S. Mamon and R. J. Elliott. *Hidden Markov Models in Finance*, volume 104 of *International Series in Operations Research & Management Science*. Springer-Verlag, New York, 2007.

[246] P. Marbach and J. N. Tsitsiklis. Simulation-based optimization of Markov reward processes. *IEEE Trans. Automat. Control*, 46(2):191–209, 2001.

[247] I. M. Y. Mareels and R. R. Bitmead. Bifurcation effects in robust adaptive control. *IEEE Trans. Circuits and Systems*, 35:835–841, 1988.

[248] A. A. Markov. Extension of the law of large numbers to dependent quantities (in Russian). *Izv. Fiz.-Matem. Obsch. Kazan Univ. (2nd Ser.)*, 15:135–156, 1906.

[249] P. G. Marlin. On the ergodic theory of Markov chains. *Operations Res.*, 21:617–622, 1973.

[250] P. Mathé. Numerical integration using V-uniformly ergodic Markov chains. *J. Appl. Probab.*, 41(4):1104–1112, 2004.

[251] J. G. Mauldon. On non-dissipative Markov chains. *Math. Proc. Camb. Phil. Soc.*, 53:825–835, 1958.

[252] M. Maxwell and M. Woodroofe. Central limit theorems for additive functionals of Markov chains. *Ann. Probab.*, 28(2):713–724, 2000.

[253] D. Q. Mayne. Optimal nonstationary estimation of the parameters of a linear system with Gaussian inputs. *J. Electron. Contr.*, 14:101, 1963.

[254] A. Medio. Invariant probability distributions in economic models: a general result. *Macroeconomic Dynamics*, 8(2):162–187, 2004. Available at http://ideas.repec.org/a/cup/macdyn/v8y2004i02p162-187_03.html.

[255] A. I. Mees, editor. *Nonlinear Dynamics and Statistics.* Birkhäuser, Boston, 2001. Selected papers from the workshop held at Cambridge University, Cambridge, September 1998.

[256] K.L. Mengersen and R.L. Tweedie. Rates of convergence of the Hastings and Metropolis algorithms. *Ann. Statist.*, 24:101–121, 1996.

[257] M. V. Men'šikov. Ergodicity and transience conditions for random walks in the positive octant of space. *Soviet Math. Dokl.*, 15:1118–1121, 1974.

[258] J.-F. Mertens, E. Samuel-Cahn, and S. Zamir. Necessary and sufficient conditions for recurrence and transience of Markov chains, in terms of inequalities. *J. Appl. Probab.*, 15:848–851, 1978.

[259] S. P. Meyn. Ergodic theorems for discrete time stochastic systems using a stochastic Lyapunov function. *SIAM J. Control Optim.*, 27:1409–1439, 1989.

[260] S. P. Meyn. Stability of Markov chains on topological spaces with applications to adaptive control and time series analysis. In L. Gerencsér and P. E. Caines, editors, *Topics in Stochastic Systems: Modelling, Estimation and Adaptive Control*, pages 369–401. Springer-Verlag, New York, 1991.

[261] S. P. Meyn. The policy iteration algorithm for average reward Markov decision processes with general state space. *IEEE Trans. Automat. Control*, 42(12):1663–1680, 1997.

[262] S. P. Meyn. Stability and optimization of queueing networks and their fluid models. In *Mathematics of Stochastic Manufacturing Systems (Williamsburg, VA, 1996)*, pages 175–199. American Mathematical Society, Providence, R.I., 1997.

[263] S. P. Meyn. Algorithms for optimization and stabilization of controlled Markov chains. *Sādhanā*, 24(4-5):339–367, 1999.

[264] S. P. Meyn. Sequencing and routing in multiclass queueing networks I: feedback regulation. *SIAM J. Control Optim.*, 40(3):741–776, 2001.

[265] S. P. Meyn. Large deviation asymptotics and control variates for simulating large functions. *Ann. Appl. Probab.*, 16(1):310–339, 2006.

[266] S. P. Meyn. Myopic policies and MaxWeight policies for stochastic networks. In *Proc. of the 46th Conf. on Dec. and Control*, pages 639–646, 2007.

[267] S. P. Meyn. *Control Techniques for Complex Networks.* Cambridge University Press, Cambridge, 2008.

[268] S. P. Meyn. Stability and asymptotic optimality of generalized MaxWeight policies. To appear in *SIAM J. Control Optim.*, 2008.

[269] S. P. Meyn and L. J. Brown. Model reference adaptive control of time varying and stochastic systems. *IEEE Trans. Automat. Control*, 38:1738–1753, 1993.

[270] S. P. Meyn and P. E. Caines. A new approach to stochastic adaptive control. *IEEE Trans. Automat. Control*, AC-32:220–226, 1987.

[271] S. P. Meyn and P. E. Caines. Stochastic controllability and stochastic Lyapunov functions with applications to adaptive and nonlinear systems. In *Stochastic Differential Systems. Proc. 4th Bad Honnef Conference*, pages 235–257. Springer-Verlag, Berlin, 1989.

[272] S. P. Meyn and P. E. Caines. Asymptotic behavior of stochastic systems processing Markovian realizations. *SIAM J. Control Optim.*, 29:535–561, 1991.

[273] S. P. Meyn and D. G. Down. Stability of generalized Jackson networks. *Ann. Appl. Probab.*, 4:124–148, 1994.

[274] S. P. Meyn and L. Guo. Stability, convergence, and performance of an adaptive control algorithm applied to a randomly varying system. *IEEE Trans. Automat. Control*, AC-37:535–540, 1992.

[275] S. P. Meyn and L. Guo. Geometric ergodicity of a doubly stochastic time series model. *J. Time Ser. Analysis*, 14(1):93–108, 1993.

[276] S. P. Meyn, G. Hagen, G. Mathew, and A. Banasuk. On complex spectra and metastability of Markov models. In *Proc. of the 47th Conf. on Dec. and Control*, 2008. More information at https://css.paperplaza.net/conferences/scripts/abstract.pl?ConfID=32&Number=1318.

[277] S. P. Meyn and R. L. Tweedie. Stability of Markovian processes I: discrete time chains. *Adv. Appl. Probab.*, 24:542–574, 1992.

[278] S. P. Meyn and R. L. Tweedie. Generalized resolvents and Harris recurrence of Markov processes. *Contemporary Mathematics*, 149:227–250, 1993.

[279] S. P. Meyn and R. L. Tweedie. Stability of Markovian processes II: continuous time processes and sampled chains. *Adv. Appl. Probab.*, 25:487–517, 1993.

[280] S. P. Meyn and R. L. Tweedie. Stability of Markovian processes III: Foster–Lyapunov criteria for continuous time processes. *Adv. Appl. Probab.*, 25:518–548, 1993.

[281] S. P. Meyn and R. L. Tweedie. The Doeblin decomposition. *Contemporary Mathematics*, 149:211–225, 1993.

[282] S. P. Meyn and R. L. Tweedie. Computable bounds for convergence rates of Markov chains. *Ann. Appl. Probab.*, 4:981–1011, 1994.

[283] S. P. Meyn and R. L. Tweedie. State-dependent criteria for convergence of Markov chains. *Ann. Appl. Probab.*, 4:149–168, 1994.

[284] H. D. Miller. Geometric ergodicity in a class of denumerable Markov chains. *Z. Wahrscheinlichkeitstheorie und Verw. Geb.*, 4:354–373, 1966.

[285] S. Mittnik. Nonlinear time series analysis with generalized autoregressions: a state space approach. Working paper WP-91-06, State University of New York at Stony Brook, Stony Brook, N.Y., 1991.

[286] A. Mokkadem. *Criteres de melange pour des processus stationnaires. Estimation sous des hypotheses de melange. Entropie de processus lineaires.* PhD thesis, Université Paris Sud, Centre d'Orsay, 1987.

[287] P. A. P. Moran. The statistical analysis of the Canadian lynx cycle I: structure and prediction. *Aust. J. Zool.*, 1:163–173, 1953.

[288] P. A. P. Moran. *The Theory of Storage*. Methuen, London, 1959.

[289] M. D. Moustafa. Input-output Markov processes. *Proc. Koninkl. Ned. Akad. Wetensch.*, A60:112–118, 1957.

[290] P. Mykland, L. Tierney, and B. Yu. Regeneration in Markov chain samplers. *J. Amer. Statist. Assoc.*, 90(429), 1995.

[291] E. Nelson. The adjoint Markoff process. *Duke Math. J.*, 25:671–690, 1958.

[292] M. F. Neuts. Two Markov chains arising from examples of queues with state dependent service times. *Sankhyā Ser. A*, 297:259–264, 1967.

[293] M. F. Neuts. Markov chains with applications in queueing theory, which have a matrix-geometric invariant probability vector. *Adv. Appl. Probab.*, 10:185–212, 1978.

[294] Marcel F. Neuts. *Matrix-Geometric Solutions in Stochastic Models: An Algorithmic Approach*. Dover Publications, New York, 1994. Corrected reprint of the 1981 original.

[295] J. Neveu. Potentiel Markovien récurrent des chaînes de Harris. *Ann. Inst. Fourier, Grenoble*, 22:7–130, 1972.

[296] J. Neveu. *Discrete-Parameter Martingales*. North-Holland, Amsterdam, 1975.

[297] P. Ney and E. Nummelin. Markov additive processes I: eigenvalue properties and limit theorems. *Ann. Probab.*, 15(2):561–592, 1987.

[298] P. Ney and E. Nummelin. Markov additive processes II: large deviations. *Ann. Probab.*, 15(2):593–609, 1987.

[299] D. F. Nicholls and B. G. Quinn. *Random Coefficient Autoregressive Models: An Introduction*. Springer-Verlag, New York, 1982.

[300] S. Niemi and E. Nummelin. Central limit theorems for Markov random walks. *Commentationes Physico-Mathematicae*, 54, 1982.

[301] E. Nummelin. A splitting technique for Harris recurrent chains. *Z. Wahrscheinlichkeitstheorie und Verw. Geb.*, 43:309–318, 1978.

[302] E. Nummelin. Uniform and ratio limit theorems for Markov renewal and semi-regenerative processes on a general state space. *Ann. Inst. Henri Poincaré Ser. B*, 14:119–143, 1978.

[303] E. Nummelin. *General Irreducible Markov Chains and Nonnegative Operators*. Cambridge University Press, Cambridge, 1984.

[304] E. Nummelin. On the Poisson equation in the potential theory of a single kernel. *Math. Scand.*, 68:59–82, 1991.

[305] E. Nummelin and P. Tuominen. Geometric ergodicity of Harris recurrent Markov chains with applications to renewal theory. *Stoch. Proc. Applns.*, 12:187–202, 1982.

[306] E. Nummelin and P. Tuominen. The rate of convergence in Orey's theorem for Harris recurrent Markov chains with applications to renewal theory. *Stoch. Proc. Applns.*, 15:295–311, 1983.

[307] E. Nummelin and R. L. Tweedie. Geometric ergodicity and R-positivity for general Markov chains. *Ann. Probab.*, 6:404–420, 1978.

[308] S. Orey. Recurrent Markov chains. *Pacific J. Math.*, 9:805–827, 1959.

[309] S. Orey. *Limit Theorems for Markov Chain Transition Probabilities*. Van Nostrand Reinhold, London, 1971.

[310] D. Ornstein. Random walks I. *Trans. Amer. Math. Soc.*, 138:1–43, 1969.

[311] D. Ornstein. Random walks II. *Trans. Amer. Math. Soc.*, 138:45–60, 1969.

[312] A. Pakes and D. Pollard. Simulation and the asymptotics of optimization estimators. *Econometrica*, 57:1027–1057, 1989.

[313] A. G. Pakes. Some conditions for ergodicity and recurrence of Markov chains. *Operations Res.*, 17:1048–1061, 1969.

[314] K. R. Parthasarathy. *Probability Measures on Metric Spaces*. Academic Press, New York, 1967.

[315] J. Petruccelli and S. W. Woolford. A threshold AR(1) model. *J. Appl. Probab.*, 21:270–286, 1984.

[316] R. G. Phillips and P. V. Kokotovic. A singular pertubation approach to modeling and control of Markov chains. *IEEE Trans. Automat. Control*, 26(5), 1981.

[317] J. W. Pitman. Uniform rates of convergence for Markov chain transition probabilities. *Z. Wahrscheinlichkeitstheorie und Verw. Geb.*, 29:193–227, 1974.

[318] D. B. Pollard and R. L. Tweedie. *R*-theory for Markov chains on a topological state space I. *J. London Math. Society*, 10:389–400, 1975.

[319] D. B. Pollard and R. L. Tweedie. *R*-theory for Markov chains on a topological state space II. *Z. Wahrscheinlichkeitstheorie und Verw. Geb.*, 34:269–278, 1976.

[320] N. Popov. Conditions for geometric ergodicity of countable Markov chains. *Soviet Math. Dokl.*, 18:676–679, 1977.

[321] M. Pourahmadi. On stationarity of the solution of a doubly stochastic model. *J. Time Ser. Anal.*, 7:123–131, 1986.

[322] N. U. Prabhu. *Queues and Inventories*. John Wiley & Sons, New York, 1965.

[323] S. Rai, P. Glynn, and J.E. Glynn. Recurrence classification for a family of non-linear storage models. Submitted for publication., 2007.

[324] R. Rayadurgam, S. P. Meyn, and L. Brown. Bayesian adaptive control of time varying systems. In *Proc. of the 31st Conf. on Dec. and Control*, Tucson, Ariz., 1992.

[325] S. I. Resnick. *Extreme Values, Regular Variation and Point Processes*. Springer-Verlag, New York, 1987.

[326] D. Revuz. *Markov Chains*. North-Holland, Amsterdam, second edition, 1984.

[327] C. P. Robert and G. Casella. *Monte Carlo Statistical Methods*. Springer-Verlag, New York, second edition, 2004.

[328] G. O. Roberts and N. Polson. A note on the geometric convergence of the Gibbs sampler. *J. Roy. Statist. Soc. Ser. B*, 56:377–384, 1994.

[329] G. O. Roberts and J. S. Rosenthal. Geometric ergodicity and hybrid markov chains. *Electronic Comm. Probab.*, 2:13–25, 1997.

[330] G. O. Roberts and J. S. Rosenthal. Convergence of the slice sampler. *J. Roy. Statist. Soc. Ser. B*, 61:643–660, 1999.

[331] G. O. Roberts and J. S. Rosenthal. The polar slice sampler. *Stoch. Models*, 18:257–280, 2002.

[332] G. O. Roberts and J. S. Rosenthal. General state space Markov chains and MCMC algorithms. *Probab. Surv.*, 1:20–71 (electronic), 2004.

[333] G. O. Roberts and A. F. M. Smith. Simple conditions for the convergence of the Gibbs sampler and Hastings–Metropolis algorithms. *Stoch. Proc. Applns.*, 49(2):207–216, 1994.

[334] G. O. Roberts and R. L. Tweedie. Exponential convergence of Langevin distributions and their discrete approximations. *Bernoulli*, 2(4):341–363, 1996.

[335] G. O. Roberts and R. L. Tweedie. Geometric convergence and Central Limit Theorems for multidimensional Hastings and Metropolis algorithms. *Biometrika*, 83:95–100, 1996.

[336] Z. Rosberg. A note on the ergodicity of Markov chains. *J. Appl. Probab.*, 18:112–121, 1981.

[337] M. Rosenblatt. Equicontinuous Markov operators. *Teor. Verojatnost. i Primenen*, 9:205–222, 1964.

[338] M. Rosenblatt. Invariant and subinvariant measures of transition probability functions acting on continuous functions. *Z. Wahrscheinlichkeitstheorie und Verw. Geb.*, 25:209–221, 1973.

[339] M. Rosenblatt. Recurrent points and transition functions acting on continuous functions. *Z. Wahrscheinlichkeitstheorie und Verw. Geb.*, 30:173–183, 1974.

[340] W. A. Rosenkrantz. Ergodicity conditions for two-dimensional Markov chains on the positive quadrant. *Prob. Theory and Related Fields*, 83:309–319, 1989.

[341] J. S. Rosenthal. *Rates of Convergence for Gibbs Sampler and Other Markov Chains*. PhD thesis, Harvard University, 1992.

[342] J. S. Rosenthal. Correction: "Minorization conditions and convergence rates for Markov chain Monte Carlo". *J. Amer. Statist. Assoc.*, 90(431):1136, 1995.

[343] J. S. Rosenthal. Minorization conditions and convergence rates for Markov chain Monte Carlo. *J. Amer. Statist. Assoc.*, 90(430):558–566, 1995.

[344] J. S. Rosenthal. Quantitative convergence rates of Markov chains: a simple account. *Electron. Comm. Probab.*, 7:123–128 (electronic), 2002.

[345] W. Rudin. *Real and Complex Analysis*. McGraw-Hill, New York, 2nd edition, 1974.

[346] S. H. Saperstone. *Semidynamical Systems in Infinite Dimensional Spaces*. Springer-Verlag, New York, 1981.

[347] P. J. Schweitzer. Perturbation theory and finite Markov chains. *J. Appl. Prob.*, 5:401–403, 1968.

[348] E. Seneta. *Non-Negative Matrices and Markov Chains*. Springer, New York, second edition, 1981.

[349] L. I. Sennott, P. A. Humblet, and R. L. Tweedie. Mean drifts and the non-ergodicity of Markov chains. *Operations Res.*, 31:783–789, 1983.

[350] J. G. Shanthikumar and D. D. Yao. Second-order properties of the throughput of a closed queueing network. *Math. Oper. Res.*, 13:524–533, 1988.

[351] T. Shardlow and A. M. Stuart. A perturbation theory for ergodic Markov chains and application to numerical approximations. *SIAM J. Numer. Anal.*, 37(4):1120–1137, 2000.

[352] M. Sharpe. *General Theory of Markov Processes*. Academic Press, New York, 1988.

[353] Z. Šidák. Classification of Markov chains with a general state space. In *Trans. 4th Prague Conf. Inf. Theory Stat. Dec. Functions, Random Proc*, pages 547–571. Academia, Prague, 1967.

[354] K. Sigman. The stability of open queueing networks. *Stoch. Proc. Applns.*, 35:11–25, 1990.

[355] G. F. Simmons. *Introduction to Topology and Modern Analysis*. McGraw Hill, New York, 1963.

[356] R. Sine. Convergence theorems for weakly almost periodic Markov operators. *Israel J. Math.*, 19:246–255, 1974.

[357] R. Sine. On local uniform mean convergence for Markov operators. *Pacific J. Math.*, 60:247–252, 1975.

[358] R. Sine. Sample path convergence of stable Markov processes II. *Indiana University Math. J.*, 25:23–43, 1976.

[359] A. F. M. Smith and A. E. Gelfand. Bayesian statistics without tears: a sampling-resampling perspective. *Amer. Statist.*, 46:84–88, 1992.

[360] A. F. M. Smith and G. O. Roberts. Bayesian computation via the Gibbs sampler and related Markov chain Monte Carlo methods (with discussion). *J. Roy. Statist. Soc. Ser. B*, 55:3–23, 1993.

[361] W. L. Smith. Asymptotic renewal theorems. *Proc. Roy. Soc. Edinburgh (A)*, 64:9–48, 1954.

[362] W. L. Smith. Regenerative stochastic processes. *Proc. Roy. Soc. London (A)*, 232:6–31, 1955.

[363] W. L. Smith. Remarks on the paper "Regenerative stochastic processes". *Proc. Roy. Soc. London (A)*, 256:296–301, 1960.

[364] J. Snyders. Stationary probability distributions for linear time-invariant systems. *SIAM J. Control Optim.*, 15:428–437, 1977.

[365] V. Solo. Stochastic adaptive control and martingale limit theory. *IEEE Trans. Automat. Control*, 35:66–70, 1990.

[366] F. M. Spieksma. *Geometrically Ergodic Markov Chains and the Optimal Control of Queues*. PhD thesis, University of Leiden, 1991.

[367] F. M. Spieksma. Spectral conditions and bounds for the rate of convergence of countable Markov chains. Technical report, University of Leiden, 1993.

[368] F. M. Spieksma and R. L. Tweedie. Strengthening ergodicity to geometric ergodicity for Markov chains. *Stochastic Models*, 10:45–75, 1994.

[369] F. Spitzer. *Principles of Random Walk*. Van Nostrand, Princeton, N.J., 1964.

[370] D. Steinsaltz. Locally contractive iterated function systems. *Ann. Probab.*, 27(4):1952–1979, 1999.

[371] Ö. Stenflo. Uniqueness of invariant measures for place-dependent random iterations of functions. In *Fractals in Multimedia (Minneapolis, MN, 2001)*, volume 132 of *IMA Vol. Math. Appl.*, pages 13–32. Springer, New York, 2002.

[372] L. Stettner. On the existence and uniqueness of invariant measures for continuous time Markov processes. Technical report LCDS 86-18, Brown University, Providence, R.I., 1986.

[373] A. L. Stolyar. On the stability of multiclass queueing networks: a relaxed sufficient condition via limiting fluid processes. *Markov Process. Related Fields*, 1(4):491–512, 1995.

[374] C. R. Stone. On absolutely continuous components and renewal theory. *Ann. Math. Statist.*, 37:271–275, 1966.

[375] O. Stramer and R. L. Tweedie. Langevin-type models I: diffusions with given stationary distributions and their discretizations. *Methodol. Comput. Appl. Probab.*, 1(3):283–306, 1999.

[376] O. Stramer and R. L. Tweedie. Langevin-type models II: self-targeting candidates for MCMC algorithms. *Methodol. Comput. Appl. Probab.*, 1(3):307–328, 1999.

[377] D. W. Stroock and S. R. Varadhan. On degenerate elliptic-parabolic operators of second order and their associated diffusions. *Comm. Pure Appl. Math.*, 25:651–713, 1972.

[378] D. W. Stroock and S. R. Varadhan. On the support of diffusion processes with applications to the strong maximum principle. In L. M. Le Cam, J. Neyman, and E. L. Scott, editors, *Proceedings of the 6th Berkeley Symposium on Mathematical Statistics and Probability*, pages 333–368. University of California Press, Berkeley, 1972.

[379] R. S. Sutton and A. G. Barto. *Reinforcement Learning: An Introduction.* MIT Press, Cambridge, Mass., 1998.

[380] W. Szpankowski. Some sufficient conditions for non-ergodicity of Markov chains. *J. Appl. Probab.*, 22:138–147, 1985.

[381] M. A. Tanner and W. H. Wong. The calculation of posterior distributions by data augmentation. *J. Amer. Statist. Assoc.*, 82:528–540, 1987.

[382] L. Tassiulas. Adaptive back-pressure congestion control based on local information. *IEEE Trans. Automat. Control*, 40(2):236–250, 1995.

[383] L. Tassiulas and A. Ephremides. Stability properties of constrained queueing systems and scheduling policies for maximum throughput in multihop radio networks. *IEEE Trans. Automat. Control*, 37(12):1936–1948, 1992.

[384] J. L. Teugels. An example on geometric ergodicity of a finite Markov chain. *J. Appl. Probab.*, 9:466–469, 1972.

[385] Luke Tierney. Markov chains for exploring posterior distributions. *Ann. Statist.*, 22(4):1701–1762, 1994. With discussion and a rejoinder by the author.

[386] D. Tjøstheim. Non-linear time series and Markov chains. *Adv. Appl. Probab.*, 22:587–611, 1990.

[387] H. Tong. A note on a Markov bilinear stochastic process in discrete time. *J. Time Ser. Anal.*, 2:279–284, 1981.

[388] H. Tong. *Non-linear Time Series: A Dynamical System Approach.* Oxford University Press, Oxford, 1990.

[389] J. N. Tsitsiklis and B. Van Roy. An analysis of temporal-difference learning with function approximation. *IEEE Trans. Automat. Control*, 42(5):674–690, 1997.

[390] P. Tuominen. Notes on 1-recurrent Markov chains. *Z. Wahrscheinlichkeitstheorie und Verw. Geb.*, 36:111–118, 1976.

[391] P. Tuominen and R. L. Tweedie. Markov chains with continuous components. *Proc. London Math. Soc. (3)*, 38:89–114, 1979.

[392] P. Tuominen and R. L. Tweedie. The recurrence structure of general Markov processes. *Proc. London Math. Soc. (3)*, 39:554–576, 1979.

[393] P. Tuominen and R. L. Tweedie. Subgeometric rates of convergence of f-ergodic Markov chains. *Adv. Appl. Probab.*, 26:775–798, 1994.

[394] R. L. Tweedie. R-theory for Markov chains on a general state space I: solidarity properties and R-recurrent chains. *Ann. Probab.*, 2:840–864, 1974.

[395] R. L. Tweedie. R-theory for Markov chains on a general state space II: R-subinvariant measures for R-transient chains. *Ann. Probab.*, 2:865–878, 1974.

[396] R. L. Tweedie. Relations between ergodicity and mean drift for Markov chains. *Austral. J. Statist.*, 17:96–102, 1975.

[397] R. L. Tweedie. Sufficient conditions for ergodicity and recurrence of Markov chains on a general state space. *Stoch. Proc. Applns.*, 3:385–403, 1975.

[398] R. L. Tweedie. Criteria for classifying general Markov chains. *Adv. Appl. Probab.*, 8:737–771, 1976.

[399] R. L. Tweedie. Operator geometric stationary distributions for Markov chains with applications to queueing models. *Adv. Appl. Probab.*, 14:368–391, 1981.

[400] R. L. Tweedie. Criteria for rates of convergence of Markov chains with application to queueing and storage theory. In J. F. C. Kingman and G. E. H. Reuter, editors, *Probability, Statistics and Analysis*. Cambridge University Press, Cambridge, 1983.

[401] R. L. Tweedie. The existence of moments for stationary Markov chains. *J. Appl. Probab.*, 20:191–196, 1983.

[402] R. L. Tweedie. Invariant measures for Markov chains with no irreducibility assumptions. *J. Appl. Probab.*, 25A:275–285, 1988.

[403] D. Vere-Jones. Geometric ergodicity in denumerable Markov chains. *Quart. J. Math. Oxford (2nd Ser.)*, 13:7–28, 1962.

[404] D. Vere-Jones. A rate of convergence problem in the theory of queues. *Theory Probab. Appl.*, 9:96–103, 1964.

[405] D. Vere-Jones. Ergodic properties of nonnegative matrices. I. *Pacific J. Math.*, 22:361–386, 1967.

[406] D. Vere-Jones. Ergodic properties of nonnegative matrices II. *Pacific J. Math.*, 26:601–620, 1968.

[407] L. Wu and N. Yao. Large deviation principles for Markov processes via Φ-Sobolev inequalities. *Electron. Commun. Probab.*, 13:10–23, 2008.

[408] Liming Wu. Essential spectral radius for Markov semigroups I: discrete time case. *Prob. Theory Related Fields*, 128(2):255–321, 2004.

[409] L.M. Wu. Large deviations for Markov processes under superboundedness. *C. R. Acad. Sci Paris Série I*, 324:777–782, 1995.

[410] B. E. Ydstie. Bifurcations and complex dynamics in adaptive control systems. In *Proc. of the 25th Conf. on Dec. and Control*, Athens, Greece, 1986.

[411] M. Yor and B. Bru. Comments on the life and mathematical legacy of Wolfgang Doeblin. *Finance and Stochastics*, 6(1):3–47, 2002.

[412] K. Yosida and S. Kakutani. Operator-theoretical treatment of Markov's process and mean ergodic theorem. *Ann. Math.*, 42:188–228, 1941.

Index

General index

K_a-chain, 116
L^1 space, 555
L^∞ space, 555
P^n-definition of positivity, 534, 536
P^n-definition of recurrence, 533
P^n-properties, 532
σ-field, 553
    Generated, 553
    Generated by a r.v., 555
σ-finite measure, 553
τ-classification of chains, 535
τ-properties, 532
d-Cycle, 112

Absolute continuity, 75
Absorbing set, 84
    Maximal a. s., 204
Accessible atom, 96
Accessible set, 86
Adaptive control, 34
Adaptive control model, 35, 416
    V-uniform, 417
    Irreducibility, 162
    Performance, 417
    Tight, 307
Age process, 39
Aperiodic, 112, 114
    A. state, 475
    A. Ergodic Theorem, 313
    Strongly a., 112
    Topological a. state, 464
ARMA, 24, 25
Ascoli's Theorem, 557
Asymptotic variance, 422
Atom, 96
    f-Kendall a., 368
    Ergodic a., 319

    Geometrically ergodic a., 365, 368
    Kendall a., 368
Autoregressive model, 389
    ARMA, 25
    Dependent parameter (RCA), 32
    Order (k, ℓ), 25
    Order k, 24
    Random coefficient (RCA), 415

Backward recurrence time, 39, 54
    δ-skeleton, 69
    Process, 69
Balayage operator, 308
Bilinear model, 27, 389
    f-Regularity and ergodicity, 353
    Dependent parameter, 32
    Geometrically ergodic, 389, 414
    Irreducible T-chain, 152
    Multidimensional, 414
Blackwell's Renewal Theorem, 354
Borel σ-field, 553, 557
Bounded in probability, 142, 305
    On average, 288
    T-chain, 472
Brownian motion, 562

Causal controls, 34
Central Limit Theorem, 422
    Asymptotic variance, 422
    Functional, 443, 447, 563
    Martingale CLT, 563
    Random walk, 454
Chapman–Kolmogorov equations, 61, 62
    Generalized, 116
Closed sets, 556
Closure of sets, 556
Coercive function, 213, 559
Coercive sequence, 493

Symbols

$A \rightsquigarrow B$, Uniformly accessibility, 86

$A_+(x)$, States reachable from x by $\mathrm{CM}(F)$, 148

$A_+^k(x)$, States reachable from x at time k by $\mathrm{CM}(F)$, 147

$C_{x_0}^k$, Generalized controllability matrix, 153

$*$, Convolution operator, 68

$1 \otimes \pi$, Outer-product kernel, 392

$\rightsquigarrow^a B$, B is uniformly accessible using a from A, 116

$A \otimes B$, 415

A^0, Points from which A is inaccessible, 86

C^∞, Functions whose derivatives of arbitrary order exist, 26

$C_V(r) := \{x : V(x) \leq r\}$, Sublevel set of V, 188

C_n, Controllability matrix, 90

Co, Complex plane, 90

F_k, Output maps for the linear control model, 27

I_B, Indicator kernel of B, 66

I_g, Multiplication kernel using g, 126

K_{a_ε}, Resolvent kernel, 62

$L(x, A) := \mathsf{P}_x(\tau_A < \infty)$, 64

$L(x, h) = U_h(x, h)$, 296

$M_n(g)$, Martingale derived from g, 446

$N(t)$, Number of customers in queue at time t, 40

$N(t)$, Number of customers in queue immediately before nth arrival, 41

N_n^*, Number of customers in queue immediately after nth service time is completed, 43

O_w, Control set, 149, 153

O_w, Supports input in $\mathrm{CM}(F)$ and $\mathrm{NSS}(F)$ models, 29

$P(x, A)$, n-step transition probability, 61

$P(x, A)$, One-step transition probability, 52, 61

$P^n(x, A)$, n-step transition probability, 53

$P_h(x, A)$, Kernel for "process on h", 295

$Q(x, A)$, Probability that chain enters A i.o. from x, 199

$Q(x, h) = \mathsf{P}_x\{\Psi_k \in A_h \text{ i.o.}\}$, 296

$R(x)$, Conditional emptying time for dam model, 45

R_n, Residual service time immediately after customer arrival, 71

$S_n(g)$, Partial sum of $g(\Phi_k)$, 421

$T(x, A)$, Continuous component, 124

$T_{ab} = \min\{j : Z_a(j) = Z_b(j) = 1\}$, 321

U_h, Resolvent kernel, 293

$V^+(n)$, Forward recurrence time chain, 39

$V^+(t) := \inf(Z_n - t : Z_n > t, \ n \geq 1)$ Forward recurrence time process, 69

$V_\delta^+(n) = V^+(n\delta)$, Forward recurrence time δ-skeleton, 69

$V^-(t) := \inf(t - Z_n : Z_n \leq t, \ n \geq 1)$, Backward recurrence time process, 69

$V_\delta^-(n) = V^-(n\delta)$, Backward recurrence time δ-skeleton, 69

V_C, Minimal solution to (V2), 267

$\Delta = P - I$, Drift operator, 172

Γ, Distribution of a disturbance variable, 22

$\Lambda_i^*(x, A) = P(i, x; 0, A)$, Ladder chain transition probability, 70

$\Lambda_{i-j+1}(x, A) = P(i, x; j, A)$, Ladder chain transition probability, 70

$\Omega = \mathsf{X}^\infty$, Sequence space, 49

$\Omega_+(C)$, Omega limit set for $\mathrm{NSS}(F)$, 154

Φ_n, Markov chain value at time n, 3

Σ_μ, σ-field of P_μ-invariant events, 423

$\|\mu\| := \sup_{A \in \mathcal{B}(\mathsf{X})} \mu(A) - \inf_{A \in \mathcal{B}(\mathsf{X})} \mu(A)$, Total variation norm, 315

$\delta_x(A) = P^0(x, A)$, Dirac measure, 61

γ_g^2, Asymptotic variance in the CLT, 422

λ^*, Split measure on bcx, 99

$\leftrightarrow$, Communicates with, 78

$\bar{p}(M)$, Upper tail of renewal sequence, 465

$\mathbb{I}_B$, Indicator function of B, 62

$\mathbb{R}$, Real line, 553

$\mathbb{R}^n$, n-dimensional Euclidean space, 6

$\mathbb{Z}_+$, non-negative integers, 3

μ^{Leb} Lebesgue measure, 88

μ^{Leb}, Lebesgue measure on $\mathbb{R}$, 553

$\longrightarrow^w \mu_\infty$, Weak convergence of μ_k to μ_∞, 139

∇^Φ, Sensitivity process, 163

$\bar{A}$, Points from which A is accessible, 86

$\bar{A}(m)$, 86

π, Invariant measure, 229

ψ, Maximal irreducibility measure, 83

Printed in the United States
By Bookmasters